The Evolution of Organ Systems

The Evolution of Organ Systems

A. Schmidt-Rhaesa

OXFORD
UNIVERSITY PRESS

OXFORD
UNIVERSITY PRESS

Great Clarendon Street, Oxford OX2 6DP

Oxford University Press is a department of the University of Oxford.
It furthers the University's objective of excellence in research, scholarship,
and education by publishing worldwide in

Oxford New York

Auckland Cape Town Dar es Salaam Hong Kong Karachi
Kuala Lumpur Madrid Melbourne Mexico City Nairobi
New Delhi Shanghai Taipei Toronto

With offices in

Argentina Austria Brazil Chile Czech Republic France Greece
Guatemala Hungary Italy Japan Poland Portugal Singapore
South Korea Switzerland Thailand Turkey Ukraine Vietnam

Oxford is a registered trade mark of Oxford University Press
in the UK and in certain other countries

Published in the United States
by Oxford University Press Inc., New York

© A. Schmidt-Rhaesa 2007

The moral rights of the author have been asserted
Database right Oxford University Press (maker)

First published 2007

All rights reserved. No part of this publication may be reproduced,
stored in a retrieval system, or transmitted, in any form or by any means,
without the prior permission in writing of Oxford University Press,
or as expressly permitted by law, or under terms agreed with the appropriate
reprographics rights organization. Enquiries concerning reproduction
outside the scope of the above should be sent to the Rights Department,
Oxford University Press, at the address above

You must not circulate this book in any other binding or cover
and you must impose the same condition on any acquirer

British Library Cataloguing in Publication Data

Data available

Library of Congress Cataloging in Publication Data

Data available

Typeset by Newgen Imaging Systems (P) Ltd., Chennai, India
Printed in Great Britain
on acid-free paper by
Antony Rowe, Chippenham

ISBN 978–0–19–856668–7   978–0–19–856669–4 (Pbk.)

10 9 8 7 6 5 4 3 2 1

For my family
Daniela, Jelka, Jorrit,
Finn Jaro and Tjorven Finja

Contents

Preface	ix
1. Introduction	1
2. The phylogenetic frame	3
3. General body organization	34
4. Epidermls	54
5. Musculature	74
6. Nervous system	95
7. Sensory organs	118
8. Body cavities	148
9. Excretory systems	169
10. Circulatory systems	191
11. Respiratory systems	202
12. Intestinal systems	218
13. Reproductive organs	240
14. Gametes (Spermatozoa)	262
15. Final conclusions	293
References	294
Index	365

Preface

The roots of this book go back to my student time, when I was joining the vivid and productive group of Peter Ax. We studied comparative morphology in a variety of organisms with two aims: to reconstruct phylogenetic relationships and to understand the diversity of animal morphology, in particular with its historical, evolutionary background. It was especially Thomas Bartolomaeus' work on body cavities and excretory organs that influenced me much and his (unfortunately largely unpublished) habilitation thesis from 1993 can be regarded as a kind of spiritual ancestor of this book. Research like that of Thomas showed that the fascination in systematics does not stop after reconstructing a hypothesis on animal relationships, but that it continues with a discussion of what the consequences of this hypothesis are for the evolution of organisms and their organ systems.

Therefore my first thanks go to those who have intimately accompanied or supervised my scientific career, some as teachers, some as colleagues and friends. These are Peter Ax, Ulrich Ehlers, Thomas Bartolomaeus, Jim Garey, and Klaus Reinhold. Matthias Schaefer belongs to this group, although never a direct teacher, he still taught me a lot. The large group of colleagues that joined me over the years have had an important impact, although I mention here only one dear friend, Wilko Ahlrichs.

One person, who has published considerable contributions about the evolution of organ systems, especially in the latest edition of his book *Invertebrate Zoology*, is Ed Ruppert. Although knowing him personally only from a brief 'water-to-the-belly' (in my case) or 'water-to-the-trousers' (in his case) meeting inside Tampa Bay, I express my deepest respect for his contribution and authority. Other authorities, which have impressed me with their dedication to zoology are Vicky and John Pearse as well as Reinhardt Møbjerg Kristensen and Claus Nielsen. While writing on this book, hearing of Reinhard Rieger's death was shocking. His impact on comparative morphology and evolutionary biology was fundamental.

Some colleagues were so kind as to read chapters from this book and helped me with valuable corrections, additions, and suggestions. These are Wilko Ahlrichs (Oldenburg), Thomas Bartolomaeus (Berlin), Thorsten Burmester (Hamburg), Marco Ferraguti (Milano), and Ronald Jenner (Bath). Many thanks to all of you.

Other colleagues helped me by supplying images for this book or gave me valuable information, these are (in alphabetical order): Wilko Ahlrichs, Thomas Bartolomaeus, Daniela Candia Carnevali, Marco Ferraguti, Peter Grobe, Alexander Gruhl, Harald Hausen, Iben Heiner, Holger Herlyn, Alexander Kieneke, Reinhardt Møbjerg Kristensen, Juliane Kulessa, Carsten Lüter, Georg Mayer, Günter Purschke, Björn Quast, Klaus Reinhardt, Birgen Holger Rothe, Thomas Stach, Andrea Tapp, Andreas Unger. Many thanks also go to Renate Feist, who as technical assistant made some of the photos used in this book.

Then I honestly thank my students who joined me over the years. In offering them the possibility to dive into comparative morphology, I am not sure if I did them a favour. The decision for comparative morphology is everything other than a road to paradise. It requires an extensive amount of enthusiasm which I found admirably in most of my students. From all of them I want to thank Birgen Holger Rothe, in particular, for his unselfish expertise and colleagueship.

Finally, my thanks go to my family. The time of writing this book fell when my four children were all less than ten years old. Anyone who has children knows what I want to say with this. But still, my family stood behind this book and supported me and it. Thank you for that.

Trying to dive into the diversity and evolution of organ systems soon makes it clear that many different aspects, many different 'fields' of biology have to be united. Systematics and evolutionary biology are integrative fields, more than ever. One might sometimes get the feeling that currently some fields are considered to be more important than the others. I hope to show that this is not the case, that the picture is most fascinating as a whole, as an integration of morphology, genomics, paleontology, developmental biology, and more.

CHAPTER 1

Introduction

Diversity and dynamics are the two main characteristics in biology. Diversity is everywhere. Together with us, the human species, several million other species of organisms share this planet. Diversity is also found in the way these organisms are composed, from the rough body organization to its smallest, molecular components. Any such diversity is the product of a dynamic historical process; it has evolved through time over the past 3.5 billion years or more. Animal organisms (Metazoa) occurred comparatively late, probably during the period between 1,000 and 600 million years ago. Despite their diversity, all animals have several basic requirements. They have to gather and digest food, get rid of excretes, receive and process information, and so on. The animal body is made up of parts that deal with these requirements and these parts are generally called organs or organ systems.

Organs are usually defined as structures composed of more than one kind of tissue and that have a particular function within the body (Valentine 2004). Organ systems are composed of more than one organ. Organs and organ systems can be regarded as modules, from which the animal body is built (Tyler 2003). Although such definitions are straightforward and clear, they never cover the whole subject. One organ may have more than one function or its function may change during evolution (for example, the integument can absorb nutrients in some animals and therefore perform the function of the intestine). Organs usually evolve through a series, passing a cellular and then tissue organization. It is extremely exciting to search for the ancient roots of organs and organ systems that were there before the organ itself appeared. Consequently, holding on to very strict definitions would deprive us from the entire picture and I therefore use organ systems here in a very loose sense.

The inclusion of spermatozoa into a book on organ systems is, of course, beyond the loosest definition of organ systems. However, as spermatozoa evolve (in a fascinating way) and have been used in phylogenetic considerations, the temptation to include them was simply too great.

How can we make statements about the evolution of organ systems? We need solid background knowledge about the evolutionary relationships of animals, i.e. their phylogeny. If we have a tree (and 'tree' is meant here as the short form of 'hypothesis of phylogenetic relationships'), we can map the characters from organ systems onto this tree and then try to 'read' its history. I regard this as an enormously important thing. Systematists often focus mainly on the construction of trees. It is very important that this is done in the most careful and objective way possible. However, I am convinced that our imagination loves the stories and images behind the trees. A tree becomes alive when we can tell what it means, what has happened. Naming consequences of particular phylogenetic relationships for character evolution also makes it easier to evaluate conflicting hypotheses or to spot inconsistencies.

Now one might say that there is the danger of circular reasoning. We use the characters from organ systems to construct trees and then we use the trees to tell about the evolution of organ systems. But this could of course only be justified when we use only one set of characters to build trees. In fact, we use more than one set and we trust that bringing together data from several sources brings us closer to the 'real' picture. Systematics has developed remarkably during the past decades. Based on the ideas of Willi Hennig in the 1950s, sophisticated methods for phylogeny reconstruction are now available. In particular the advanced possibilities to compare DNA sequences have created an enormous amount of phylogenetic analyses. Advances in sequencing are constantly raising the number of available genomes and multi-gene approaches are adding new sets of analyses.

The evolution of some organ systems has already received great attention. This accounts in particular for the nervous system, photoreceptors, body cavities, excretory systems, and gametes, which have been used in some broad evolutionary approaches. However, in most cases the evolution of these organ systems was not derived from a tree, but instead the organ systems were arranged in a putatively logical series that was assumed to reflect their evolutionary order. For example, body cavities were divided into two types (coelom and pseudocoelom) and arranged in an evolutionary series starting from the acoelomate (lack of any body cavity) to a pseudocoelomate and finally to the coelomate condition. Such a scenario is not favoured any more, because mapping body cavity characters on a tree results in different conclusions.

Because the tree is so important for thoughts on the evolution of organ systems, I will start with a summary of phylogenetic relationships, from which the tree is derived that is used throughout the book. It will become clear that this tree has some weak branches which might require a special attention when describing the organ systems.

In the chapters on organ systems, I intend to present the diversity as well as the dynamics. In several cases, characters will be summarized in tables. The information is, of course, vast and requires that some priorities have to be set. As it is the aim to present basic lines of metazoan evolution, large monophyletic taxa (more or less equivalent to what is often called a 'phylum') are the units for considerations. As some of these taxa are very small (sometimes only one species) and others are extremely large (in species number), smaller taxa will be represented in disproportionate detail compared to larger taxa. This is particularly the case with the arthropods. A book on the evolution of arthropod organ systems could easily have the same volume as this book and still represent only a general overview.

CHAPTER 2

The phylogenetic frame

The amount of data available for phylogenetic analyses, as well as the number of tools for their analysis, increases constantly. While twenty years ago the vast majority of characters used for phylogenetic analyses came from comparative morphology, the advances in molecular biology, such as the PCR technique and automated sequencing of DNA, have made sequence data and genomic data available for analyses. Cytochemical, mainly immunocytochemical, advances opened new doors for developmental biology and comparative morphology. The possibilities of a targeted search for genes and observation of their expression patterns led to a renaissance of the evolutionary aspect in developmental biology. Finally, palaeontology has made considerable progress in finding, preparing, and interpreting fossils, including attempts to integrate fossils into phylogenetic analyses (e.g. Budd & Jensen 2000).

There are diverse analytical methods to reconstruct the phylogenetic tree (the hypothesis of phylogenetic relationships). Morphological characters are usually analysed using parsimony methods (but also likelihood-based methods have been applied, e.g. Glenner et al. 2004), for molecular characters additional methods such as distance and likelihood are used. There are attempts to combine morphological and molecular analyses (total evidence and other methods). This is supplemented by numerous further tools that help in finding confidence in tree nodes, test the suitability of data sets, and more. The different data sources and different analytical tools have led to a wide variety of phylogenetic hypotheses. Such hypotheses are sometimes congruent, but incongruence is a common phenomenon.

When a book like this tries to derive conclusions from character distributions on a tree, it must be made clear which tree is used. Some publications (e.g. Pennisi 2003, Halanych 2004) imply that there was an old tree which is now replaced by a new tree. Such an impression blurs the fact that some relationships are well supported while others are unstable. In fact, it appears appropriate not to present one singular tree, but to identify problematic regions and include these as alternatives into the discussion of the evolution of organ systems. I am convinced, nevertheless, that there is a solid backbone, along which problematic regions can be identified (Schmidt-Rhaesa 2003). To explain the tree used for this book, I will run through metazoan phylogeny in this chapter.

Besides numerous (more or less detailed) analyses dealing with certain taxa or groups of taxa, several papers have tried to summarize metazoan phylogeny, for example, Nielsen et al. (1996) and Sørensen et al. (2000) for morphological characters; Winnepenninckx et al. (1998a), Glenner et al. (2004), and Halanych (2004) for molecular characters; and Zrzavý et al. (1998), Giribet et al. (2000), and Peterson and Eernisse (2001) for combined analyses. Several zoology textbooks have started to include methods of phylogenetic systematics and results of systematic research, among these books I want to name especially Ax (1996, 2000, 2003), Nielsen (2001), Westheide and Rieger (2004, 2007), Brusca & Brusca (2003), Cracraft and Donoghue (2004), and Ruppert et al. (2004).

Although there is a diversity of analytical tools, all rest on principles for phylogenetic analysis formulated by Willi Hennig (e.g. Hennig 1950,

1966). This is not the place to present or evaluate these different methods. Only a few general remarks should be made. No method claims (or should claim) to be superior to all other methods (e.g. Jenner 2004a). The experience from the past two decades has shown that any method has its strong and its weak sides. There is no defined pathway to follow when different methods yield incongruent results. Such incongruent results are regular phenomena in phylogenetic analyses and they motivate closer investigation of the methods applied. Some of the problems currently recognized are as follows:

- **Taxon sampling**. The choice of species included into a phylogenetic analysis (Lecointre et al. 1993) as well as the number of species of one certain taxon, may have impact on the result of the analysis (see Hillis et al. 2003, Rosenberg & Kumar 2001, Rokas & Carroll 2005 for recent views and references). This impact may differ from taxon to taxon, but can be assumed to be especially important when members of a taxon are very diverse.
- **Ground pattern reconstruction**. In morphological analyses of broad phylogenetic relationships, one may integrate characters from concrete species or ground patterns, that is, hypothetical reconstructions of characters in an ancestor of a taxon (Jenner 2001, 2004b). I am convinced that the reconstruction of ground patterns is helpful in our understanding of animal evolution, because it helps to realize ancestors as real species that lived at some point in time. Especially given the aim of this book, it is inevitable that we use reconstructed ancestors or ancestral character states.
- **Secondary loss of characters**. In morphological analyses it is often difficult to distinguish between primary absence and secondary loss of a character (Jenner 2004c). This is of particular importance in morphologically simple organisms, in parasites, and when taxa underwent a rapid radiation, combined with a considerable change of characters.
- **Regaining of characters**. Although there is a rule (Dollo's rule) claiming that evolutionary pathways cannot be reversed, there are examples where characters appear to be lost and then regained. Two examples are the absence of wings in basal stick insects (Phasmida), while derived stick insects have wings that appear to be homologous to those from other pterygote insects (Whiting et al. 2003). And the probable regaining of polyps in parasitic narcomedusae (Collins 2002). In both cases it appears unrealistic to assume a convergent *de novo* evolution of wings or polyps. In contrast, it appears more likely that changes in the regulative network underlying morphogenesis have led to a silencing of certain developmental processes and a later regaining of this function. Such changes in the regulative network need not be complete absence of genes or gene products, but may be even slight temporal or spatial shifts in expression patterns (Davidson 2001). Recently, the loss and gaining of wingspots and the molecular pattern underlying these evolutionary changes has been investigated in *Drosophila* species by Prud'Homme et al. (2006).
- **Character coding**. Choosing data for a matrix to run a phylogenetic analysis of morphological characters is the most crucial and probably one of the hardest steps. Problems in morphological analyses, compared to molecular analyses are, for example, the question of whether and how to weight characters as well as problems with an absence/presence coding in matrices (see Jenner 2002). Other problems are that the data included in a matrix are rarely critically analysed and available data matrices are often recycled in other analyses (Jenner & Schram, 1999, Jenner 2001, see also Jenner 2004b for an excellent case study).
- **Molecular versus morphological data**. While coding of molecular data is comparatively easy ('only' four nucleotides are present), coding of morphological characters is much more difficult. The selective pressure on the function of several morphological structures has led to the convergent evolution of similar structures, as is evidenced in several examples. However, such problems are not absent in molecular analyses and while convergence in morphological structures can often be recognized by detailed

reinvestigations, a mutation from A to T and back to A, or an independant mutation to the same nucleotide at the same site in two species, is hard to recognize. Even the convergent evolution of proteins (and therefore the amino acid sequence and the DNA sequence) appears to be possible (examples are the convergent evolution of lysozymes [Steward & Wilson 1987, Kornegay et al. 1994] and of antifreeze glycoproteins [Chen et al. 1997]). Sometimes, the problems with the analysis of morphological characters lead to the suggestion to abandon such characters from phylogenetic analyses (e.g. Scotland et al. 2003), but this simply cannot be a scientific strategy. The only strategy is to try to combine molecular and morphological data or the results of both analyses in the best possible way (see Baker & Gatesy 2002 for a recent review and literature).

- **Resolution**. All genes used for phylogenetic analyses have a certain range in which they show good resolution (see, for example, Giribet 2002). This range may not be generally valid for all taxa in question, because evolutionary rates differ. The 18S rDNA gene is the most abundantly used gene for phylogenetic analyses and has been criticized numerous times. For example, Philippe et al. (1994) assume this gene to have a resolution of not more than 40 Mio a (which is certainly an underestimation). However, although several alternative candidates have been suggested (e.g. Friedlaender et al. 1992, 1994) and now several other genes are used (such as elongation factor 1α, myosin-heavy-chain and others), no gene has turned out to be really superior to the 18S rDNA gene. Loss of resolution can also be the reflection of a rapid radiation, during which no significant differences could manifest in gene sequences (Rokas et al. 2005).
- **Long-branch problems**. The rate with which DNA sequences change over time (i.e. evolve) differs, not only among species, but also between different regions of the DNA and even within one gene. Great differences in evolutionary rates between two species lead to differences in comparatively many nucleotides, and the alignment of such sequences becomes more and more unreliable. Moreover, such differences in evolutionary rates lead to problems in the phylogenetic analysis. This has been long known in theory (Felsenstein 1978) and because differences in evolutionary rates are illustrated in distance methods as long branches, problems caused by them are generally known as 'long-branch problems' (see reviews by Anderson & Swofford 2004, and Bergsten 2005). Typical artifacts caused by long-branch taxa are the clustering with other long-branch taxa (see, as one out of many examples, the clustering of three nematodes, a gnathostomulid, and two chaetognaths, in Littlewood et al. 1998) and the basal position of such taxa or clusters. Long-branch problems have led to some erroneous interpretations in the past (i.e. a basal position of nematodes within Bilateria as long as only *Caenorhabditis elegans* and few parasitic species were investigated), but the awareness of long-branch problems has increased now.
- **Misidentification and contamination problems**. There is rarely a chance to test whether the sequences deposited, i.e. in genbank, were determined properly. The use of a wrong sequence may be the source of errors in the analysis. There are examples for such misidentifications: Cohen et al. (1998) noted that the published phoronid sequence U12648 is a brachiopod/phoronid chimaera and Jondelius et al. (2001) note that the sequence of *Nemertinoides elongatus* (Nemertodermatida; U70083) is likely to originate from a rhabditophoran flatworm and not from a nemertodermatid. Contamination can also be a problem, because sequences may originate from food, epibionts or endosymbionts. A famous example is *Xenoturbella bocki*, which has been placed among bivalves on the basis of sequence similarities (Norén & Jondelius, 1997) that later turned out to be intestinal contents of *Xenoturbella* (Bourlat et al. 2003).
- **Multigene analyses**. With the availability of more and more DNA sequences through the sequencing of whole genomes or EST (expressed sequence tags) approaches, the number of genes available for phylogenetic analyses rises constantly. This nurtures hope that the analysis of

many genes together will produce a more correct tree than the analysis of single or few genes (e.g. Rokas et al. 2003). This hope is certainly substantial, but should not be used uncritically. The lessons learned from single genes, e.g. about varying evolutionary rates, should be transferred to multigene analyses. For example, when we know that the 'model' nematode *Caenorhabditis elegans* has a fast evolving 18S rDNA gene that is not suited for phylogenetic analyses, how sure can we be that this does not also account for other *C. elegans* genes? Because the genome of *C. elegans* was among the first to be available, it is included in several analyses (e.g. Wang et al. 1999, Hausdorf 2000, Philip et al. 2005), even to explicitly resolve the position of nematodes in the phylogenetic tree (Blair et al. 2002). Other publications, however, come to the conclusion that many *C. elegans* genes (two thirds of the genes analysed by Mushegian et al. 1998) are fast evolving and support the probably wrong tree (Mushegian et al. 1998, see also Copley et al. 2004). A simply statistical approach, assuming that the majority of genes support the correct tree, is in this case likely to be misleading. We know almost nothing about the reasons for strongly differing evolutionary rates and whether this phenomenon affects only single genes or whole genomes. One further problem of multigene analysis is to make sure that orthologous genes and not paralogs are compared with each other.

- **Incomplete fossil record**. Fossils supply very valuable data for animal evolution. However, the existence of a fossil record depends on taphonomic conditions, i.e. tissue composition, ecology, environment, and early fossilization conditions. These conditions change regionally and through time, so that fossilization, especially exceptionally good fossilization, represents more a regional and temporal snapshot than a broad and continuous representation of extinct species diversity (Butterfield 2003).

Two comments on **terminology** should be made. First, I use the term 'synapomorphy' in the sense as defined by Ax (1987). A synapomorphy is present only in two sister-taxa and is the inherited autapomorphy from their common ancestor. Identifying a character as synapomorphy is therefore essential for the recognition of sister-taxa. Hennig (1966) defined the term 'synapomorphy' not so precisely and it is often used in a broader sense. For example, hairs are commonly called a synapomorphy of mammals, but in the sense explained above, they are a synapomorphy of the taxa Monotremata and Theria. I regard the restriction in Ax's definition as helpful and will use 'synapomorphy' in this sense. I will also completely abandon the 'Linnean' hierarchical levels in this book, as there is no justification for their application in phylogenetics. In theory, it would be logical to include the genus in this rejection of hierarchical categories. As a genus is integral part of the binominal species name, its abandonment would create a restructuring and renaming of species which I do not regard as helpful. Therefore, the compromise of keeping the binomen and the genus (as a monophyletic taxon) is acceptable.

Monophyly of Metazoa

The monophyly of metazoans is rarely doubted. There are a number of characters that occur only among metazoans and therefore evolved in their common ancestor. Such characters are (see e.g. Ax 1996, Nielsen 2001): multicellularity, an ontogeny including at least a blastula stage and radial cleavage pattern, a particular type of spermatozoon (the 'primitive' spermatozoon, see Chapter 14), an oogenesis during which one oocyte and three polar bodies are formed, an extracellular matrix (ECM), focal adhaerens junctions, and septate junctions (see Chapter 4). Almost all molecular analyses also favour the monophyletic Metazoa. A probable polyphyly of Metazoa was a singular outcome of the first broad molecular analysis by Field et al. (1988) but this was a result of an immature methodology as well as low information content (Wägele & Rödding 1998).

The sister-group of Metazoa is generally believed to be found among unicellular protozoan organisms. The candidate which is discussed most abundantly is the taxon Choanoflagellata. The so-called collar complex of choanoflagellates strongly resembles the choanocytes of sponges (Nielsen 2001). It is composed of a circle of microvilli containing actin and a central cilium. This cilium bears lateral extensions, the so-called vanes. Such vanes are also present in sponges, but their homology has been doubted because of structural differences (Ax 1996, Rieger & Weyrer 1998). Vanes are described from three species of choanoflagellates (*Codosiga botrytis*, *Salpingoeca frequentissima*, and *Monosiga* sp. (Hibberd 1975), very few extensions may be present (in *Marsupiomonas pelliculata*, Jones et al. 1994) and in some species vanes appear to be absent (e.g. *Proterospongia choanojuncta*, Leadbeater 1983; and *Desmarella moniliformis*, Karpov & Leadbeater 1998). Therefore it should be made clear whether vanes were present in the choanoflagellate ancestor. Further arguments for Choanoflagellata as the sister-group of Metazoa might be the presence of a receptor tyrosine kinase (King and Carrol, 2001) and of signalling proteins, which were previously only known from multicellular Metazoa (King et al. 2003, King 2004). An alternative view is that choanoflagellates are secondarily unicellular metazoans, derived from multicellular sponges (Maldonado 2004). This hypothesis is explained below in the context of poriferan phylogeny. Analyses using 18S rDNA (Wainwright et al. 1993), heat shock protein 70 (Snell et al. 2001), and the whole mitochondrial genome (Lang et al. 2002) support a sister-group relationship between Choanoflagellata and Metazoa. Other unicellular candidates with proposed close relationships to metazoans are the scarcely known taxa Mesomycetozoea, Corallochytrea, Cristidiscoidea, and Ministeriida (Mendoza et al. 2002, Cavalier-Smith & Chao 2003).

Diploblastic (non-bilaterian) animals

The basal branching patterns among metazoans lead to four groups of animals: Porifera (sponges; Fig. 2.1), *Trichoplax adhaerens* (Fig. 2.2), Cnidaria (Fig. 2.3), and Ctenophora (comb jellies; Fig. 2.4). Sponges (about 5,500 species) live as filter feeders in marine and freshwater habitats. *Trichoplax adhaerens* is a small, flat

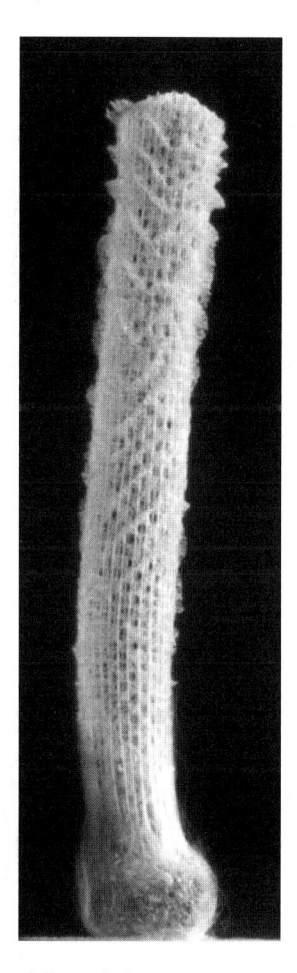

Fig. 2.1. Silicate skeleton of a hexactinellid sponge, *Euplectella aspergillum*.

organism, roughly resembling a giant, multicellular amoeba. It lives in benthic marine environments (see Syed & Schierwater 2002a for a recent reiew). A second species, *Treptoplax reptans* (Monticelli 1897) is probably not existent (Grell & Ruthmann 1991), but as there are significant genetic differences in populations of *T. adhaerens*, more than one species may hide behind this name (Voigt et al. 2004, Tomassetti et al. 2005). The roughly 8,500 species of cnidarians include corals, sea anemones, jellyfish, and the freshwater hydra. They occur in all aquatic environments. Comb jellies (about 80 species) are exclusively marine, most species being pelagic and few benthic.

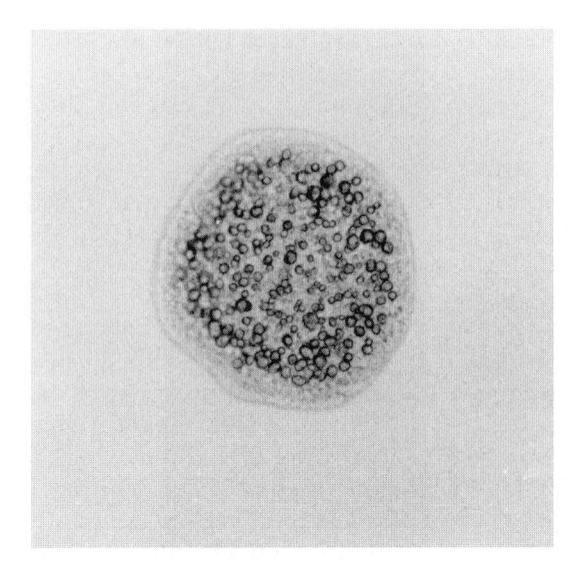

Fig. 2.2. *Trichoplax adhaerens.*

Fig. 2.4. *Mnemiopsis leidyi*, a comb jelly (Ctenophora).

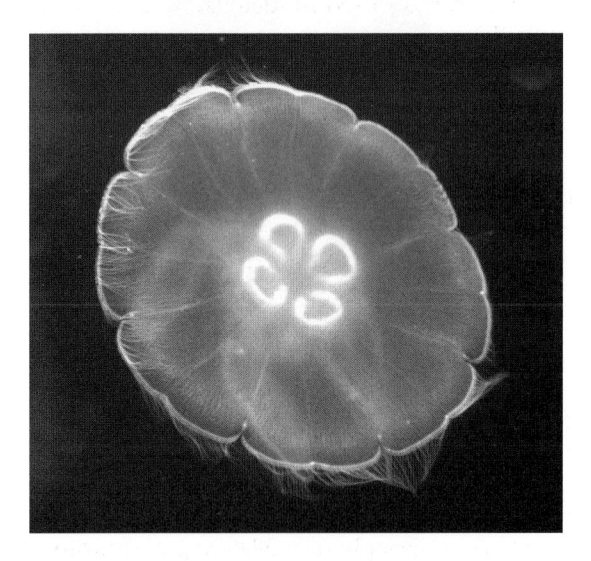

Fig. 2.3. The medusa from the scyphozoan cnidarian, *Aurelia aurita*.

There are two major problems with diploblastic metazoans. First, the monophyly of sponges is currently questioned by molecular data and second, branching patterns among the four diploblastic taxa vary considerably between morphological and molecular methods.

Until recently, Porifera were mostly considered as being monophyletic (see e.g. Böger 1988, Ax 1996, Reitner & Mehl 1996, Nielsen 2001). In several molecular analyses, however, sponges appear to be polyphyletic, with calcareous sponges (Calcarea) being more closely related to the remaining metazoans than siliceous sponges (Demospongia and Hexactinellida; Demospongia may themselves be paraphyletic, see Borchellini et al. 2004). On a closer look, the paraphyly of sponges is not convincing. The statistical values supporting a taxon (Calcarea + 'non-sponge metazoans') are rarely high (Lafay et al. 1992, Adams et al. 1999, Borchellini et al. 2000, 2001), an exception is a comparison of protein kinase C by Kruse et al. (1998). Other molecular analyses state that the problem of monophyly versus polyphyly of sponges cannot be resolved (at least with available sequences) (Rodrigo et al. 1994, Medina et al. 2001, Manuel et al. 2003) or even indicate monophyly of Porifera (Manuel et al. 2003). Additionally, a certain type of receptor tyrosine kinase appears to be unique to sponges and could support their monophyly (Skorokhod et al. 1999).

The morphological characters named as autapomorphies of a monophyletic taxon Porifera are almost all subject to discussion. Interpretation of the biphasic life cycle (with planktonic larvae and sessile adults) as an autapomorphy (Ax 1996, Reitner & Mehl 1996) contrasts the evaluation of biphasic life cycles as ancestral (see Jägersten 1972, Rieger 1994a). The organization of choanocytes in chambers, forming a water current system with inhalant and exhalant pores may be an autapomorphy of Porifera (Ax 1996, Reitner & Mehl 1996, Nielsen 2001), but the evaluation of choanocytes themselves as an autapomorphy (Ax 1996, Reitner & Mehl 1996) depends on the question of whether they are homologous to the choanoflagellate collar complex. The general pattern of a cilium surrounded by a collar of microvilli is widespread among metazoans in epithelial and receptor cells (e.g. the collar receptor or nematocytes; see Chapter 7). The lateral vanes may be homologous or not between sponges and choanoflagellates. A position of Choanoflagellata as the sister-group of Metazoa requires that the metazoan ancestor had choanocyte-like cells and that these were lost in the sister-group of sponges, that is, all remaining metazoans (few sponges have also lost their choanocytes, for example, the carnivorous cladorhizids, see Vacelet & Boury-Esnault 1995, Vacelet 2006). An alternative view pronounces similarities between sponge larvae (in which choanocytes are absent) and other metazoans, in particular regarding the presence of an outer epithelium with a basal extracellular matrix (ECM) and cilia with striated rootlets (Maldonado 2004). Striated ciliary rootlets were originally only found in larvae of calcareous sponges (Woollacott & Pinto 1995) and led Eernisse & Peterson (2004) to assume that this character supports a sister-group relationship of Calcarea to the non-sponge metazoans. Since striated ciliary rootlets are also present in larvae of (at least some) demosponges (Boury-Esnault et al. 2003), they appear to be a common metazoan character. They may be lacking in hexactinellid sponges, but only a few data are available (Boury-Esnault et al. 1999). If an organization comparable to a sponge larva is assumed for the metazoan ancestor, the adult organization of sponges would be derived. Then, however, the resemblances to choanoflagellates become problematic. They could either be convergent features or, as suggested by Maldonado (2004), choanoflagellates could be secondarily unicellular organisms derived from sponges, that is, belonging into the taxon Porifera and therefore also to Metazoa. I see not much support for this last view, because choanoflagellates, when they are included in analyses, never fall within metazoans (see references above). Additionally, their mitochondrial genome is more complex than that of metazoans (Burger et al. 2003) and a stepwise transition from unicellular over choanoflagellate and demosponge mitochondrial genomes to those from other metazoans appears plausible (Lavrov et al. 2005). As there are choanoflagellates with a life cycle including a motile, colonial phase and a sedentary, unicellular phase (Leadbeater 1983) and because there are aflagellate stages (the ciliary microtubules appear to be needed for cell division) (Leadbeater 1983; 1994), there is much space to develop scenarios for the transition to multicellular metazoans. Choanoflagellates do already carry a number of 'prerequisites' in the form of adhesion and signalling molecules (or at least of molecules that are in metazoans used as such) (King et al. 2003, King 2004).

Many morphological analyses correspond in hypothesizing a sequential emergence of the non-bilaterial (diploblastic) taxa with Porifera branching off first, followed by *Trichoplax adhaerens*, Cnidaria, and Ctenophora (see Fig. 2.5; Ax 1996, Nielsen et al. 1996, Nielsen 2001, Sørensen et al. 2000, Brusca & Brusca 2003). This is supported by the general tissue organization, the occurrence of cell–cell contacts, and other characters. Two types of epithelia, one having a dominantly nutritive and the other a dominantly protective function occur in Epitheliozoa (*Trichoplax* + Eumetazoa) (but see Syed & Schierwater 2002b for a different interpretation of *Trichoplax* morphology). In Eumetazoa (Cnidaria + Ctenophora + Bilateria), the nutritive

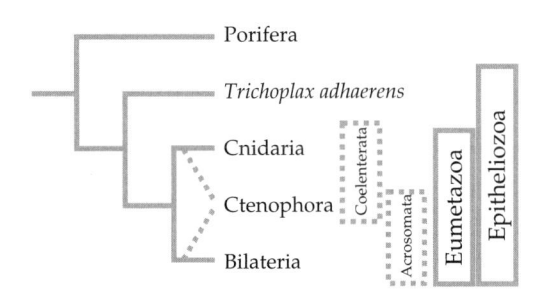

Fig. 2.5. Basal metazoan relationships, see text for explanation.

tissue (endoderm) is entirely inside the animal, connected to the outside by the blastopore. Nerve and muscle cells (in the form of epitheliomuscular cells) occur in Eumetazoa, true myocytes in Acrosomata (Ctenophora + Bilateria). Septate junctions and spot desmosomes are cell–cell contacts present in all Metazoa, belt desmosomes occur in Epitheliozoa and gap junctions in Eumetazoa (Ax 1996). This 'morphological view' is also concluded from combined analyses of morphological and molecular characters (Zrzavý et al. 1998, Petersen & Eernisse 2001). Recently, Scholtz (2004) has summarized arguments against a sister-group relationship of Ctenophora and Bilateria and in favour of a relationship between Cnidaria and Ctenophora (corresponding to the 'old' name Coelenterata) with corresponding cleavage patterns (especially the unilateral separation of blastomeres, the early determination of the oral–aboral axis, and the position of the blastopore) as the main argument. According to this hypothesis, the true myocytes and the fusion of many acrosomal vesicles to form only one are either convergent or plesiomorphic characters.

This morphological scenario (including Coelenterata, as suggested by Scholtz 2004) is not supported by molecular analyses, but neither is any particular alternative hypothesis favoured. Some analyses (e.g. Lafay et al. 1992, Philippe et al. 1994, Winnepenninckx et al. 1995, Kobayashi et al. 1996, Borchiellini et al. 1998) imply a monophyly of diploblastic animals, although this is always only weakly supported by statistical values and may be the product of long-branch-attraction (Garey & Schmidt-Rhaesa 1998). Ctenophores are often found associated with all or some sponges (e.g. Wainwright et al. 1993, Winnepenninckx et al. 1995, 1998a, Kim et al. 1999, Podar et al. 2001). Collins (1998) summarizes that molecular analyses do not favour ctenophores as a sister-group of Bilateria, but *Trichoplax* or Cnidaria may be better candidates. Cnidaria and not Ctenophora are also favoured as sister-group of Bilateria in an analysis by Wallberg et al. (2004). Eernisse and Peterson (2004) argue against the Ctenophora–Bilateria sister-group relationship because in ctenophores, fewer Hox genes were found than in cnidarians (Martindale et al. 2002), which would make cnidarians a better candidate for the sister-group of Bilateria. This variety of differing hypotheses is difficult to interpret. Often, statistical support for clusters of diploblastic taxa is not very convincing (see also Medina et al. 2001) which leads to the dominance of the morphological hypothesis in combined analyses. Additionally, no characters can be found to support any of the alternative relationships, that is, an autapomorphy for all diploblastic animals or a synapomorphy of Porifera and Ctenophora. Therefore, the phylogenetic hypothesis shown in Fig. 2.5 is taken as the basis for the following chapters, but some caution is appropriate, because Ctenophora as the sister-group of Bilateria are not supported by molecular data.

The monophyly of Cnidaria and Ctenophora is not questioned and is well supported by molecular characters and molecular data (see, e.g., Schuchert 1993, Ax 1996, Nielsen 2001, Podar et al. 2001, Collins 2002; Scholtz 2004, Collins et al. 2006). Recent analyses agree in the position of Anthozoa as the basal taxon within Cnidaria, while all other cnidarans (Tesserazoa, sometimes named Medusozoa) have a medusa stage, a stiff cnidocil, and linear mitochondrial DNA. There are problems with the placement of Cubozoa, which may be the sister-group of Scyphozoa (Schuchert 1993, Collins et al. 2006) or the sister-group of Stauromedusa within Scyphozoa (Podar et al. 2001, Marques & Collins 2004).

'Mesozoa'

The two taxa Orthonectida and Rhombozoa (the well investigated Dicyemida and the poorly known Heterocyemida) include organisms of relatively simple organization. This has led to early views of an intermediate position between unicellular organisms and multicellular animals (therefore *Meso*zoa, see e.g. Hyman 1940), but although their placement appears to be clearly within Metazoa, the exact phylogenetic position has never been found. Several authors suspected 'Mesozoa' to be a polyphyletic taxon, but because no place in the tree could be found for the two subtaxa Orthonectida and Rhombozoa, the name 'Mesozoa' persisted. Molecular investigations of 18S rDNA sequences have not been helpful, because all genes sequenced so far turned out to be fast evolving (Katayama et al. 1995, Hanelt et al. 1996, Pawlowski et al. 1996) and results are therefore unreliable. Ultrastructural investigations have produced a number of informative characters such as the presence of cilia with rootlets and ECM in both groups (Furuya et al. 1997 for Dicyemida, Kozloff 1969 [Fig. 16] and Slyusarev 1994 for Orthonectida), belt desmosomes and cuticle in Orthonectida (Slyusarev 1994, 2000) as well as septate junctions, gap junctions, and a glycocalyx in Dicyemida (Revel 1988, Furuya et al. 1997, Czaker & Janssen 1998). The absence of a basal lamina, but the intracellular presence of its characteristic components (fibronectin, laminin, collagen type IV) in dicyemids (Czaker 2000) may argue for a close relationship to Bilateria. An inclusion into Spiralia ('Lophotrochozoa') on the basis of the possession of a putative spiralian Hox-gene, *Lox-5* (Kobayashi et al. 1999) may be doubtful (Telford 2000).

Myxozoa

Myxozoans are parasitic organisms consisting of only few cells. Hosts harbour the feeding stage (trophozoite) while distribution is performed by spores (Lom 1990). Spores contain one or several polar capsules that are reminiscent of cnidarian nematocysts and have been the reason for a suspected relationship between Myxozoa and Cnidaria from time to time (see references in Siddall & Whiting 1999). After some molecular analyses (Smothers et al. 1994, Schlegel et al. 1996) found myxozoan sequences to appear among metazoan ones, the homology between polar capsules and nematocysts was newly discussed (Siddall et al. 1995; Lom & Dykova 1997, Kent et al. 2001, Zrzavý 2001) and now myxozoans are commonly regarded as derived cnidarians, probably related to Narcomedusae (but see Zrzavý & Hypša 2003). On the basis of a homology between polar capsules and nematocysts, this appears to be plausible, but several

molecular analyses fail to place myxozoans reliably due to their fast evolving genes, and Anderson et al. (1998) even conclude from the HOX gene content of myxozoans that they belong into Bilateria.

The mysterious wormlike *Buddenbrockia plumatellae*, which was described and assumed to be a mesozoan by Schröder (1910) was reinvestigated ultrastructurally (Okamura et al. 2002) and molecularly (Monteiro et al. 2002) and found to belong to the Myxozoa, among other reasons due to the possession of polar capsules (see also Okamura & Canning 2003, Zrzavý & Hypša 2003, Tops et al. 2005 for reviews).

Xenoturbella

This simply organized marine taxon includes the two species *X. bocki* (Fig. 2.6) and *X. westbladi* and is another example of a 'problematic' group that has experienced considerable journeys through the metazoan tree. The simple brain (Pedersen & Pedersen 1988, Raikova et al. 2000a), the simple sac-shaped intestine as well as the absence of excretory organs (protonephridia) make *Xenoturbella* a candidate for a basal position, for example as the sister-group of Bilateria (Ehlers & Sopott-Ehlers 1997, Raikova et al. 2000a). The extensive rootlet system of epidermal cilia and a corresponding mode of resorption of epidermal cells may argue for a close relationship between

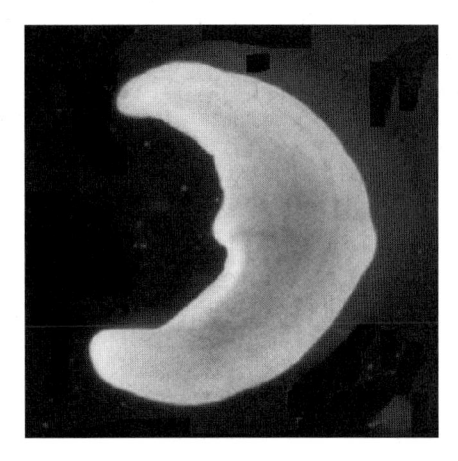

Fig. 2.6. *Xenoturbella bocki*. Photo by courtesy of Thomas Stach, Berlin.

Xenoturbella and acoelomorphs (Rohde et al. 1998, Lundin 1998, 2001). Israelsson's (1999) observations of spiral cleavage and trochophore-like larvae in *X. westbladi* would argue for an inclusion into Trochozoa (see also Israelsson & Budd 2005). The close relationship to certain bivalves advocated by 18S rDNA sequence analysis (Norén & Jondelius 1997) has turned out to be produced by an accidental sequencing of DNA from intestinal contents of *Xenoturbella* (Bourlat et al. 2003). Recent analyses place *Xenoturbella* among deuterostomes, closely related to Echinodermata and Hemichordata (Bourlat et al. 2003, 2006, see also Stach et al. 2005). I wonder whether the journey of *Xenoturbella* has ended here, but as a minimal conclusion, it appears that most analyses favour a basal position within Bilateria, although it is not known if this is basal to all Bilateria or in a basal position within Deuterostomia.

Bilateria

The Bilateria appears to us to be a very charismatic taxon, which is, for example, expressed by the fact that special names were invented for the bilaterian ancestor: it is sometimes called the protostome-deuterostome ancestor (PDA, see e.g. Erwin & Davidson 2002) or the Urbilateria (De Robertis & Sasai 1996). The reason for this might be that the bilaterian ancestor is the last common ancestor of model organisms such as *Drosophila*, *Caenorhabditis elegans* and vertebrate models such as the mouse. Especially in genetics and molecular developmental biology it is convenient to compare data from these model organisms and postulate these, when they appear in protostomes and deuterostomes, for the bilaterian ancestor. Through such procedure, it has been postulated that the bilaterian ancestor was quite complex with a segmented organization (Patel et al. 1989, Kimmel 1996, De Robertis 1997, Balavoine & Adoutte 2003), eyes (Quiring et al. 1994), appendages (Panganiban et al. 1997), and a circulatory system with heart (Scott 1994) (see also Minelli 2003a). This view contrasts with thoughts that the bilaterian ancestor was comparatively simple and microscopically small, which were influenced by the analysis of basal bilaterian taxa (e.g. Ax 1996). Davidson (2001) and Erwin and Davidson (2002) pointed out that one has to be very careful with conclusions about the ancestry of certain structures through the comparison of expression patterns. Their distinction between genes or proteins functioning in the control of differentiation genes and a morphogenetic program is rarely made. This, however, means that the presence of control genes in a tissue can be very old, while a specific morphogenetic program is comparatively new and may even evolve convergently on the basis of the control gene battery already present in a certain tissue. Erwin and Davidson (2002) conclude that uncontroversial characters of the bilaterian ancestor were 'only' an anteroposterior axis, the presence of mesodermal tissue, a gut with mouth and anus, and a central nervous system.

The monophyly of Bilateria is hardly ever questioned and appears to be well supported by almost all molecular and morphological analyses. Reconstructions of the bilaterian ancestor are, however, hotly debated. It may have been microscopically small (Ax 1996, Baguñá et al. 2001, Valentine 2004), macroscopic and polyp-like (Sauer & Kullmann 2005), large, complex, and colonial (Dewel 2000), or a combination of a small larva and a large adult (Rieger 1994a). Even when Bilateria appear to be a well supported monophylum, there are few characters that are unambiguous. The bilateral symmetry is also present in non-bilaterian taxa such as some cnidarian polyps and ctenophores (see Chapter 3). Bilateral symmetry is probably linked to directed movement (Martindale et al. 2002). The presence of a third germ layer, the mesoderm, is also controversially discussed, as individualized muscle cells occur in ctenophores and can even be present in some cnidarians (see Chapter 3). Other probable autapomorphies of Bilateria, such as the existence of protonephridia, of sheetlike layers with outer circular and inner longitudinal musculature ('Hautmuskelschlauch'), and of a brain (Ax 1996) depend on the position of acoelomorphs and the tentaculate taxa Phoronida, Bryozoa, and Brachiopoda. It appears quite clear that the bilaterian ancestor had a set of HOX genes (the HOX-cluster) patterning an anteroposterior axis of the body (Peterson & Davidson 2000, Balavoine et al. 2002), but the comparability of the HOX-cluster with the HOX genes known from non-bilaterian animals (Schierwater et al. 2002) and their exact composition is not completely clear.

There is no congruence about the basal branching patterns among Bilateria. First, there is the possibility that bilaterally symmetrical taxa such as *Xenoturbella* (see above) and Acoelomorpha (see below) are the sister-group to all remaining bilaterian taxa, for which (in the last case) the name Nephrozoa is proposed (Jondelius et al. 2002). Then there is debate about the position of the tentaculate taxa Phoronida, Bryozoa, and Brachiopoda (see below). When these are considered as closely related to Deuterostomia, the name Radialia is applied to their common taxon (Ax 1996). The sister-group of Radialia/Deuterostomia is Protostomia (Gastroneuralia), for which the monophyly is indicated in several molecular analyses, although not always strongly supported. From the morphological side, it is difficult to name characters supporting Protostomia, because most such characters are influenced by the phylogenetic position of the tentaculate taxa. Nielsen's (2001) list of potential autapomorphies is not, in my eyes, convincing. A through gut (with mouth and anus) conflicts with the interpretation of the sac-shaped gut of flatworms as plesiomorphic; multiciliate cells conflict with hypotheses assuming that monociliate taxa are basal (see Spiralia, below); and the presence of a trochophore larva is not unambiguous, because it may have evolved later (see Chapter 12). The most convincing character is the existence of a central nervous system (CNS) with brain and longitudinal nerve cords, but this is dependant on the position of phoronids and brachiopods, who have at least brainlike concentrations of nerve cells in the anterior part of the body, and on the question whether a CNS was absent in the deuterostome ancestor (see Chapter 6).

Who belongs in Protostomia? There are corresponding views that at least the two taxa Cycloneuralia and Spiralia do (I prefer to use Spiralia in contrast to Lophotrochozoa, see below under 'the tentaculate taxa'). The close relationship of Cycloneuralia to gastrotrichs (together named Nemathelminthes) or to arthropods (together named Ecdysozoa) is under current debate (see below).

Cycloneuralia/Nemathelminthes/Aschelminthes/Gnathifera

A group of animals variously named Aschelminthes, Nemathelminthes, or Pseudocoelomata has often been considered as being polyphyletic, although, with the exception of Lorenzen (1985), clear hypotheses about phylogenetic relationships appeared only after 1995 (Winnepenninckx et al. 1995, Ehlers et al. 1996, Nielsen 1995, Wallace et al. 1996). 'Aschelminthes' now appear to fall into two monophyletic taxa, Cycloneuralia and Gnathifera, while the position of Gastrotricha is disputed.

Cycloneuralia consist of the small (in terms of species numbers) taxa Nematomorpha (Fig. 2.11), Priapulida, Kinorhyncha (Fig. 2.8), and Loricifera (Fig. 2.9), as well as the ecologically successful and species rich Nematoda (Fig. 2.10). The monophyly of these taxa is well supported (see, e.g. Ax 2003) and phylogenetic analyses of relationships within some of these groups have been made: Priapulida (Lemburg 1999), Nematomorpha (Bleidorn et al. 2002) and Nematoda (summarized in DeLey &

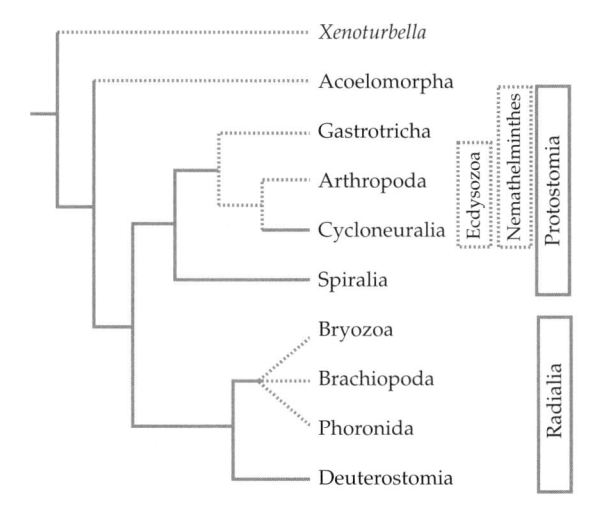

Fig. 2.7. Basal bilaterian relationships. Recently discussed relationships are represented with dotted lines.

14 THE EVOLUTION OF ORGAN SYSTEMS

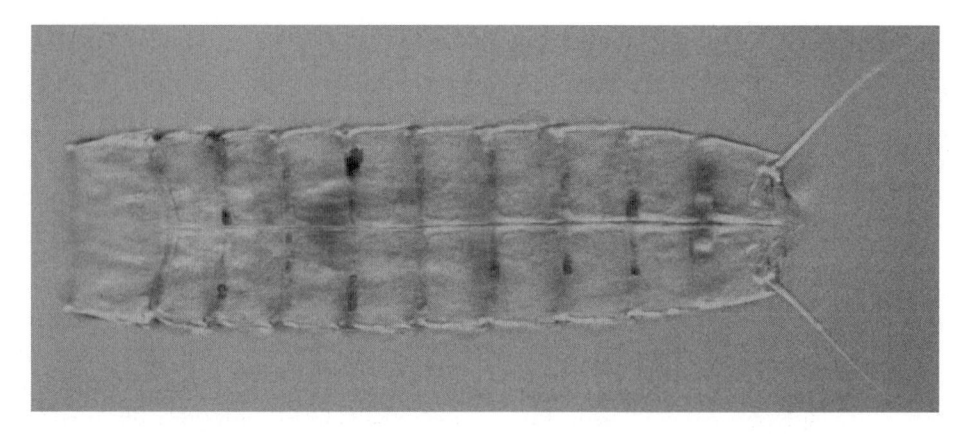

Fig. 2.8. Anterior end of the kinorhynch *Pycnophyes kielensis*, the introvert is withdrawn.

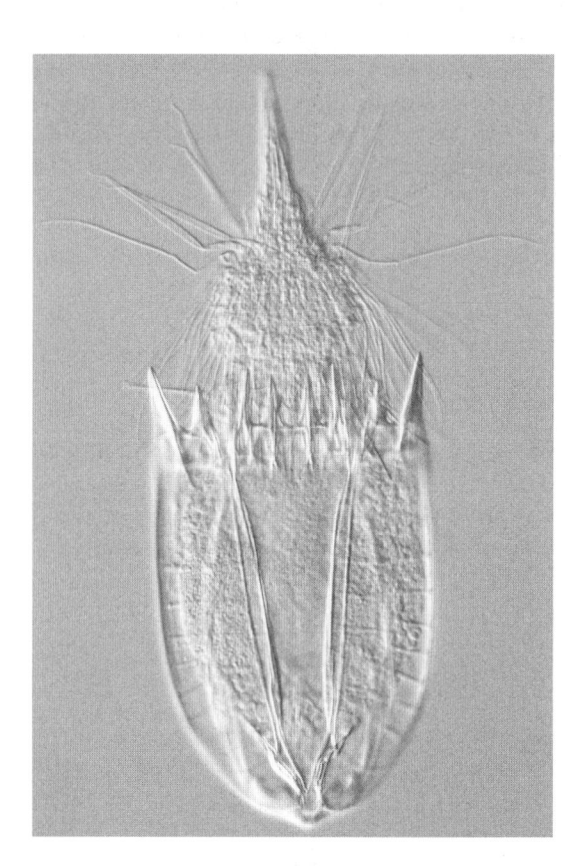

Fig. 2.9. *Nanaloricus* sp., an undescribed representative of Loricifera. Photo by courtesy of Iben Heiner, Copenhagen.

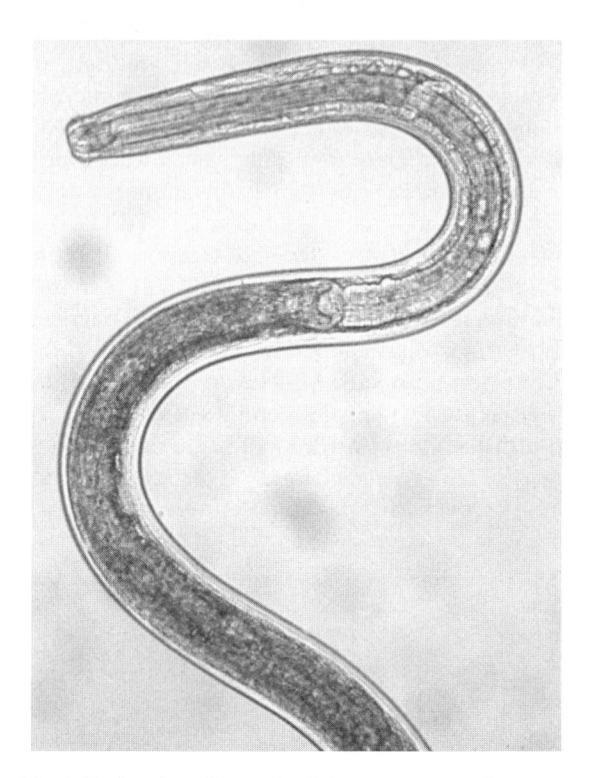

Fig. 2.10. Anterior end from a free-living, marine nematode.

Blaxter 2002, Holterman et al. 2006). Even while being small in species number, there is considerable diversity within these taxa, especially in priapulids which range in size from the meiobenthic *Meiopriapulus*-species (2–3 mm) to the macroscopic *Halicryptus higginsi* (up to 39 cm, Shirley & Storch 1999). The lately discovered Loricifera (Kristensen 1983) recently 'exploded' in species number (e.g. Gad 2004a, b, 2005a, b, Kristensen & Gad 2004, Heiner 2004, Heiner & Kristensen

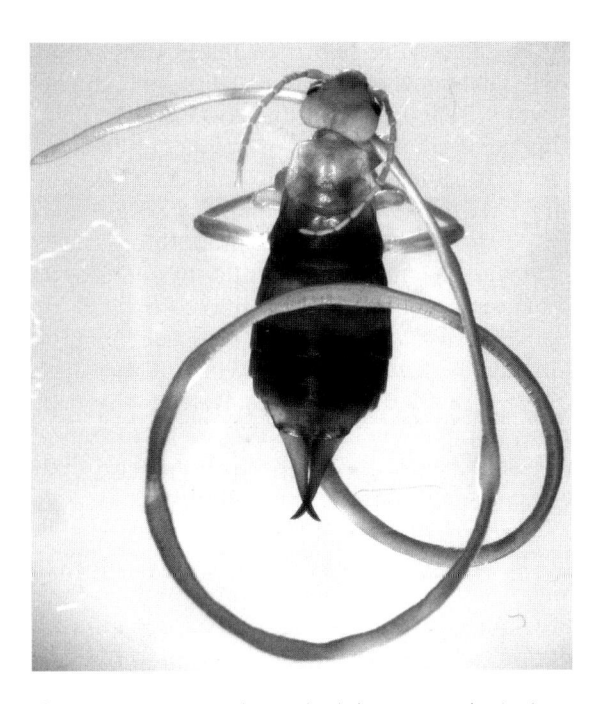

Fig. 2.11. A nematomorph, *Parachordodes gemmatus*, leaving its host (*Chelidura acanthopygia*, Dermaptera).

2005), revealing a fascinating diversity of life cycles. In nematodes, knowledge is dominated by parasitic species of economical or medical importance as well as by the free-living model organism *Caenorhabditis elegans*, whereas other free-living species, for example from the marine interstitial system, are still scarcely known. It has turned out, however, that some characters which are commonly found in textbooks as typical for nematodes are derived within this taxon and were almost certainly absent in the common ancestor of nematodes. These are, for example, a large primary body cavity (Ehlers 1994a), eutely (Malakhov 1998), and a strictly determinate cleavage (Voronov & Panchin 1998; Voronov et al. 1998; Goldstein 2001).

Most analyses correspond in regarding Nematoda and Nematomorpha as sister-taxa as well as Priapulida, Kinorhyncha, and Loricifera as belonging to a monophyletic taxon, Scalidophora (Ehlers et al. 1996, Wallace et al. 1996, Nielsen 2001, Ax 2003; Fig. 2.12), but the exact branching pattern among priapulids, kinorhynchs, and lori-

ciferans is debated (summarized by Lemburg 1999). Nematomorpha are advocated by some authors as closely related to Scalidophora (Malakhov 1980, Adrianov & Malakhov 1995, Malakhov & Adrianov 1995), but a sister-group relationship of Nematomorpha and Nematoda is more parsimonious (Schmidt-Rhaesa 1997a). Hypotheses also differ in the names applied, for example Nielsen (1995) applies the name 'Cycloneuralia' to a group including Gastrotricha, Nematoda, Nematomorpha, Priapulida, Kinorhyncha, and Loricifera, whereas 'Cycloneuralia' is adopted here according to Ahlrichs (1995) for the taxa exclusive of Gastrotricha. The fossil evidence for cycloneuralian taxa is, with the exception of priapulids (Conway Morris 1977, Wills 1998) very scarce, but fossils such as palaeoscolecids (Müller & Hinz-Schallreuter 1993, Hou & Bergström 1994, Conway Morris 1997) and *Markuelia* (Bengtson & Zhao 1997, Dong et al. 2004, 2005) are good candidates for fossil cycloncuralians (Budd 2004).

Gastrotrichs are microscopically small marine or freshwater animals (Fig. 2.13). Some of their characters, such as a muscular sucking pharynx with triradiate lumen, a terminal mouth opening, and a cuticle composed of two layers, are comparable to Cycloneuralia which makes gastrotrichs a candidate for the sister-group of Cycloneuralia (Sørensen et al. 2000, Nielsen 2001, Schmidt-Rhaesa 2002). Additionally, there are similarities in the cleavage of gastrotrichs and nematodes (Teuchert 1968, Malakhov 1994). Molecular data, in contrast, include gastrotrichs

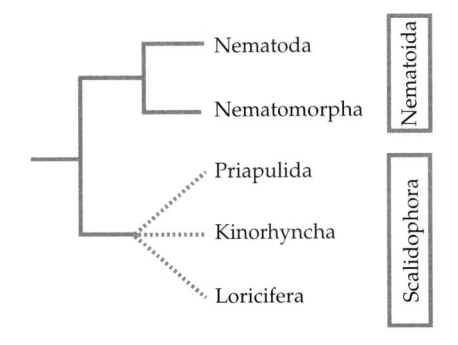

Fig. 2.12. Relationships among the Cycloneuralia.

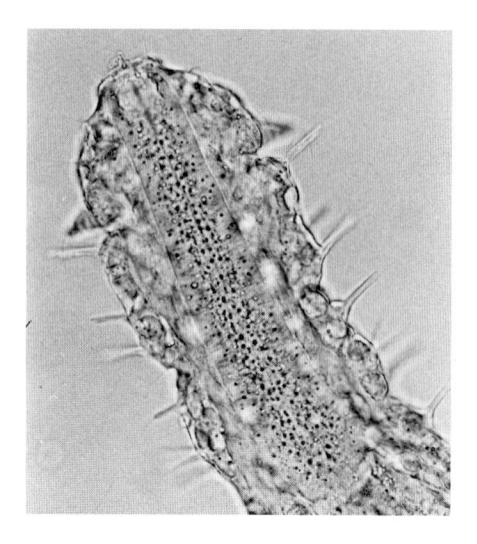

Fig. 2.13. Anterior end of the macrodasyid gastrotrich *Turbanella cornuta*.

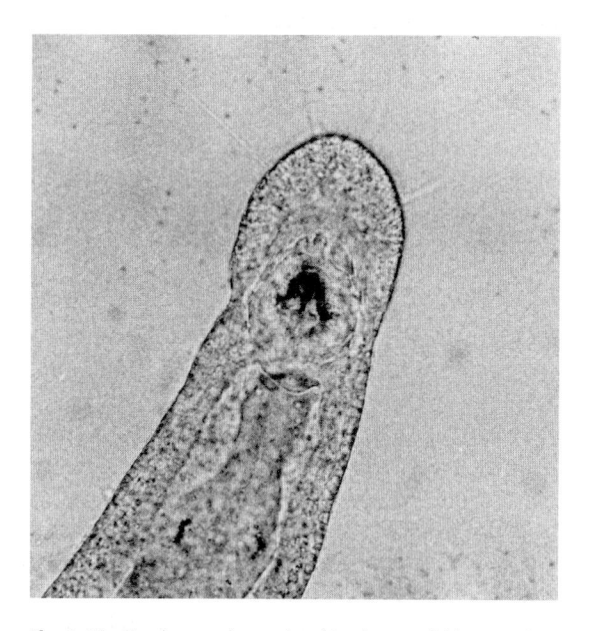

Fig. 2.14. *Gnathostomula paradoxa* (Gnathostomulida).

into the Spiralia, often in close association to Platyhelminthes or Gnathostomulida, and they are therefore often included into the 'Platyzoa' (e.g. Cavalier-Smith 1998, Halanych 2004). On a closer view, such a grouping almost never receives significant bootstrap support (Winnepenninckx et al. 1995, Littlewood et al. 1998, Peterson & Eernisse 2001, Giribet et al. 2004, Glenner et al. 2004) and gastrotrichs are not always monophyletic (Giribet et al. 2004, Glenner et al. 2004). The morphological characters that could be named to support a relationship between gastrotrichs and gnathostomulids (e.g. Balsamo 1992) are plesiomorphies. This is made more complicated, because hypotheses on the internal relationships within gastrotrichs do not correspond (Wirz et al. 1999, Hochberg & Litvaitis 2000, Manylov et al. 2004, Todaro et al. 2003, Zrzavý 2003). Better resolved phylogenetic relationships would make the reconstruction of ancestral characters for gastrotrichs and their comparison to other taxa more substantial.

The two remaining taxa that were earlier included into Aschelminthes, are Rotifera and Acanthocephala (Fig. 2.15). Although their life styles differ considerably (rotifers are predominantly free-living and acanthocephalans endoparasitic), their close relationship was hardly ever doubted, because they both share a unique character, a syncytial epidermis with an internal (intrasyncytial) layer that is functionally analogous to a cuticle (Clément & Wurdak 1991, Dunagan & Miller 1991). This leads to the name Syndermata for the common taxon. On closer examination, rotifers in the traditional sense (subtaxa Bdelloida, Monogononta, and Seisonidea) appear to be para- or polyphyletic. Acanthocephalan DNA sequences often cluster with those of a rotifer subtaxon, Bdelloida (Garey et al. 1996a, see also Garey et al. 1998, Melone et al. 1998), but these results may be influenced by problems with the sequence of the monogonont rotifer used in these analyses, *Brachionus plicatilis* (Mark Welch 2000). Two morphological characters, fibre bundles in the epidermis and dense bodies in the spermatozoa, support a sister-group relationship between Seisonidea and Acanthocephala (Ahlrichs 1997, 1998, Ferragti & Melone 1999). This is supported by one analysis of molecular analyses including a sequence from *Seison* (Herlyn et al. 2003), but other analyses including Seisonidea have different results. Mark Welch (2005) finds some support for acanthocephalans being the sister-taxon to Eurotatoria (Bdelloida + Monogononta), with Seisonidea in

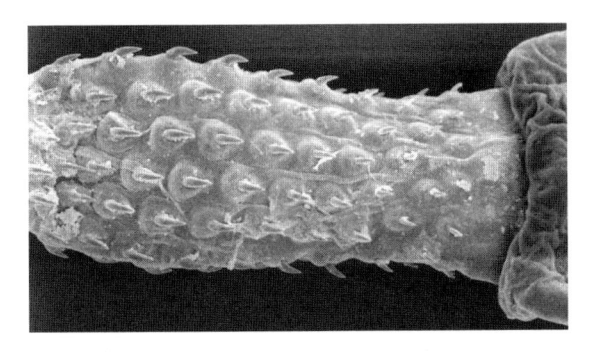

Fig. 2.15. Scanning electron micrograph of the proboscis of *Moniliformis moniliformis* (Acanthocephala).

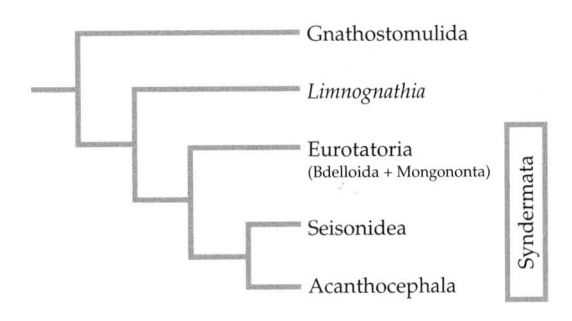

Fig. 2.16. Relationships among the Gnathifera.

a more basal position, while Sørensen and Giribet (2006) find evidence for a monophyletic group (Acanthocephala + Seisonidea + Bdelloida), named Hemirotifera. Regardless of which hypothesis will prove to be correct, a taxon Rotifera in the classical sense (Bdelloida + Monogononta + Seisonidea) is not monophyletic. This has consequences for nomenclature. When Acanthocephala and Seisonidea are sister-taxa, the name Pararotatoria has been proposed for their common taxon (Zrzavý 2001) and bdelloids and monogononts are generally named Eurotatoria. When acanthocephalans are closely related to bdelloids, the name Rotifera should be extended to include Acanthocephala (Mark Welch 2000, Sørensen & Giribet 2006).

Similarities in the jaw ultrastructure of rotifers and gnathostomulids (Fig. 2.14) have led Ahlrichs (1995) and Rieger and Tyler (1995) to hypothesize a closer relationship between Syndermata and Gnathostomulida in a common taxon Gnathifera. This has been adopted since then, although molecular analyses are not able to resolve this relationship due to the fast evolving genes of gnathostomulids (see Littlewood et al. 1998). An elegant confirmation of this hypothesis has been the discovery of *Limnognathia maerski* (Micrognathozoa) in a cold spring on Greenland by Kristensen and Funch (2000). *Limnognathia* has a jaw apparatus comparable in ultrastructure to rotifers and gnathostomulids, it has an internal layer like Syndermata, but in a cellular epidermis, and it therefore fits perfectly well into Gnathifera,

as sister-group to Syndermata (Ahlrichs 1997, Kristensen & Funch 2000, Sørensen et al. 2000, Sørensen 2003, Funch et al. 2005; Fig. 2.16). Molecular investigations have so far been not conclusive to support a clear phylogenetic hypothesis (Giribet et al. 2004).

The position of the whole group of Gnathifera is quite problematic. The cleavage of rotifers and acanthocephalans has been described as a derived spiral cleavage (Gilbert 1989) and the singular observation of a cleavage in gnathostomulids has shown it to be spiral (Riedl 1969). This places Gnathifera among spiralians, but their sister-group is unknown. Ax (1989, 1996) has proposed a sister-group relationship between Platyhelminthes and Gnathostomulida (together Plathelminthomorpha), but the characters named as supporting this relationship (hermaphroditism, internal fertilization with filiform spermatozoa) are not convincing, because these are broadly distributed characters (see also Jenner 2004b). It appears appropriate to treat Gnathifera as one branch in an unresolved spiralian polytomy, but alternative scenarios such as Gnathifera being the first branch within Spiralia may also be considered (see below, Spiralia).

Ecdysozoa or Articulata?

One of the major disputes in phylogenetic systematics is the quest for the sister-group of arthropods. The name Arthropoda is used here to include the taxa Tardigrada, Onychophora, and Euarthropoda (chelicerates, crustaceans, myriapods, and insects). The traditional view that arthropods are related to annelids (together named Articulata), based on a common mode of segmentation, has been challenged by the failure to find any molecular support for this relationship and by the hypothesis that all moulting animals (Arthropoda and Cycloneuralia, together named Ecdysozoa) share a common ancestor (Aguinaldo et al. 1997). This relationship is supported by numerous molecular analyses including several genes (e.g. Winnepenninckx et al. 1995, Aguinaldo et al. 1997, McHugh 1997, De Rosa et al. 1999, Giribet & Wheeler 1999, Garey 2001, Regier & Shultz 2001, Balavoine et al. 2002, Mallatt & Winchell 2002; Giribet 2003). Such data supported cladistic results from Eernisse et al. (1992) which already proposed that arthropods are not the sister-group of annelids, although they did not propose a close relationship of all moulting animals. The failure of some multi-gene approaches to support Ecdysozoa (as well as Articulata) (Wang et al. 1999, Hausdorf 2000, Blair et al. 2002, Wolf et al. 2004, Philip et al. 2005) may be due to problems with abundant fast evolving genes in those species for which genomic data are available (Mushegian et al. 1998), or to gene loss in nematodes (Copley et al. 2004). The exclusion of fast evolving genes or a careful attention to differing substitution rates supports Ecdysozoa (Dopazo & Dopazo 2005, Philippe et al. 2005, Philippe & Telford 2006). There has also been criticism concerning the information content of the molecular data (Wägele et al. 1999, Wägele & Misof 2001), but this does not explain the complete lack of molecular support for Articulata.

Some morphological characters appear to support the monophyly of Ecdysozoa, such as the presence of moults, a hormonal control of moults by ecdysteroid hormones, a cuticle with a comparable substructure, and occurrence of chitin in one particular layer (Schmidt-Rhaesa et al. 1998), but all these characters have been discussed because chitin, cuticles, and ecdysteroid hormones are all widely distributed (Wägele et al. 1999, Wägele & Misof 2001). It is, however, not the simple presence, but a defined substructure (of the cuticle), location (of chitin), and function (of ecdysteroid hormones) that is taken as support for Ecdysozoa. Even moulting appears to be not unique for ecdysozoans, it has been reported from leeches (Sauber et al. 1983; Berchtold et al. 1985), and dorvilleid polychaetes shed (moult?) their cuticular jaws (Paxton 2005). Moulting was proposed to be even more widespread in annelids (Pilato et al. 2005), but as it appears that in such cases not the entire cuticle, but only a superficial layer, is lost, the term 'shedding' appears to be more appropriate (Schmidt-Rhaesa 2006a). One further argument for Ecdysozoa is a specific immunoreactivity of the nervous system to horseradish peroxidase (Haase et al. 2001). An argument by Manuel et al. (2000) concerning a specific structure of β-thymosin has been shown to be not restricted to ecdysozoan taxa (Telford 2004).

The significance of segmentation as a synapomorphy of Annelida and Arthropoda has been found to be convincing due to segment substructure (Scholtz 2002) and expression patterns of *engrailed* (Prud'Homme et al. 2003), even though not all authors agree (Shankland 2003). Assuming segmentation to be a plesiomorphic feature does not really solve the problem, because 'ancient' segmentation (if existent) was unlikely to be as it is present in annelids and arthropods. Therefore the question remains: can a complex pattern such as segmentation, evolve convergently? Of course, the question also has to be: can such constant molecular support as for Ecdysozoa be artifactual? Although many textbooks have now adopted Ecdysozoa, I find a decision between the two alternative hypotheses premature, because there is (at least for my feeling) too strong support for *both* of them (see also Jenner & Scholtz 2005 for a review of the difficulties in discussing the molecular characters). Nielsen's (2003) suggestion, that Ecdysozoa are the sister-taxon to Annelida is a constructive attempt, but does not really solve the problem, because such a relationship is still not supported by the molecular data and morphological data are also still in conflict.

Arthropods

Arthropoda include the Euarthropoda as well as the taxa Onychophora (Fig. 2.18) and Tardigrada (Fig. 2.17). The application of names is not uniform and 'Arthropoda' is sometimes restricted to what is here called 'Euarthropoda' or replaced by 'Panarthropoda'. It is hardly doubted that Arthropoda and Euarthropoda both are monophyletic taxa, this is the result of almost all recent analyses (see, e.g. Wills et al. 1998, Wheeler 1997, Giribet et al. 2000, Nielsen 2001, Brusca & Brusca 2003, Eernisse & Peterson 2004, Glenner et al. 2004 for arthropod monophyly; Garey et al. 1996b, Giribet et al. 1996, Moon & Kim 1996 for a tardigrade-euarthropod relationship; Weygoldt 1986, Turbeville et al. 1991, Wheeler et al. 1993, Nielsen 1997, 2001, Ax 2000, Regier & Shultz 2001, Richter & Wirkner 2004 for euarthropod monophyly).

While onychophorans have always been linked with arthropods (sometimes even with insects and myriapods alone under the name Uniramia), tardigrades have puzzled systematists for their combination of arthropod and cycloneuralian ('aschelminth') characters. This refers, for example, to their minute size, the absence of body cavities (the report of a coelom by Marcus [1928] could not be confirmed in subsequent investigations by Eibye-Jacobsen 1996 and Hejnol & Schnabel 2005), their terminal mouth opening, and their muscular sucking pharynx with a triradiate lumen. Schmidt-Rhaesa (2001) showed that these characters could be explained as plesiomorphies when the Ecdysozoa-concept is valid. If not, such characters must be regarded as convergently evolved and then tardigrades are apparently miniaturized. When all three taxa (Tardigrada, Onychophora, Euarthropoda) are included in one analysis, a sister-group relationship between Tardigrada and Euarthropoda is favoured most often (Wills et al. 1995, 1998, Budd 1996, Nielsen et al. 1996, Wheeler 1997, Zrzavý et al. 1998, Edgecombe et al. 2000, Nielsen 2001) while few analyses favour a sister-group relationship between Tardigrada and Onychophora (Waggoner 1996, Giribet et al. 2000). While there are no morphological characters supporting the latter scheme, a sclerotization of dorsal and ventral cuticular regions, the disintegration of the muscular tube ('Hautmuskelschlauch'), and the mysterious staining of the golgi complex with bismuth (Locke & Huie 1977) support a tardigrade-euarthropod relationship (Schmidt-Rhaesa 2001). A sister-group relationship of onychophorans and euarthropods is supported, for example in Peterson and Eernisse (2001). The ventral position of the mouth, for which a shift from terminal to ventral has been elegantly demonstrated for onychophorans by Eriksson and Budd (2000), is probably a convergent character, because several fossil taxa which are likely to be intermediate between onychophorans and euarthropods have a terminal mouth opening. Many

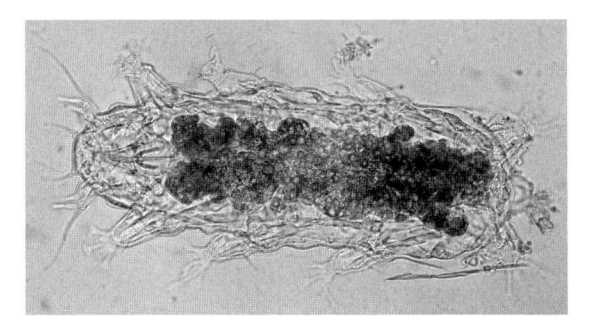

Fig. 2.17. The marine tardigrade *Batillipes mirus*.

Fig. 2.18. Undetermined onychophoran from Costa Rica (from a culture of Georg Mayer, Berlin).

such fossil taxa, including 'weird wonders' like *Hallucigenia*, *Opabinia*, and *Anomalocaris* have been integrated into a phylogenetic system of arthropods and allow the formulation of hypotheses acording to arthropod evolution (Chen et al. 1994, 1995, Hou & Bergström 1995, Budd 1996, 1997, 1998, 1999, 2001a, 2002, Dewel & Dewel 1997, Bergström & Hou 2001). These fossils do not, however, help to decide whether tardigrades or onychophorans are the sister-group of euarthropods.

The relationships within euarthropods are not clearly resolved. One particular problem is the systematic position of pycnogonids (sea spiders; Fig. 2.19), which are often assumed to belong to chelicerates due to an assumed homology of their 'chelifers' with cheliceres (e.g. Snodgrass 1938, Weygoldt 1986, Ax 2000). Cheliceres have turned out to belong to the post-ocular ('antennal') segment on the basis of *Hox*-gene expression patterns (Telford & Thomas 1998, Damen 2002). The innervation of pycnogonid chelifers, in contrast, is from the ocular segment (Maxmen et al. 2005) which makes the homology of chelifers and cheliceres unlikely and argues for pycnogonids as the sister-taxon of all other euarthropods (which is also supported by some combined molecular

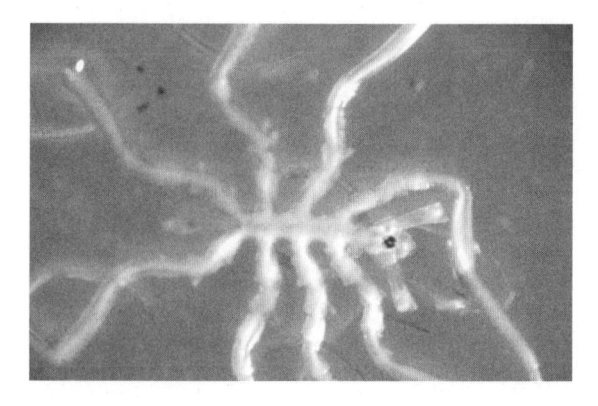

Fig. 2.19. An undetermined pycnogonid species.

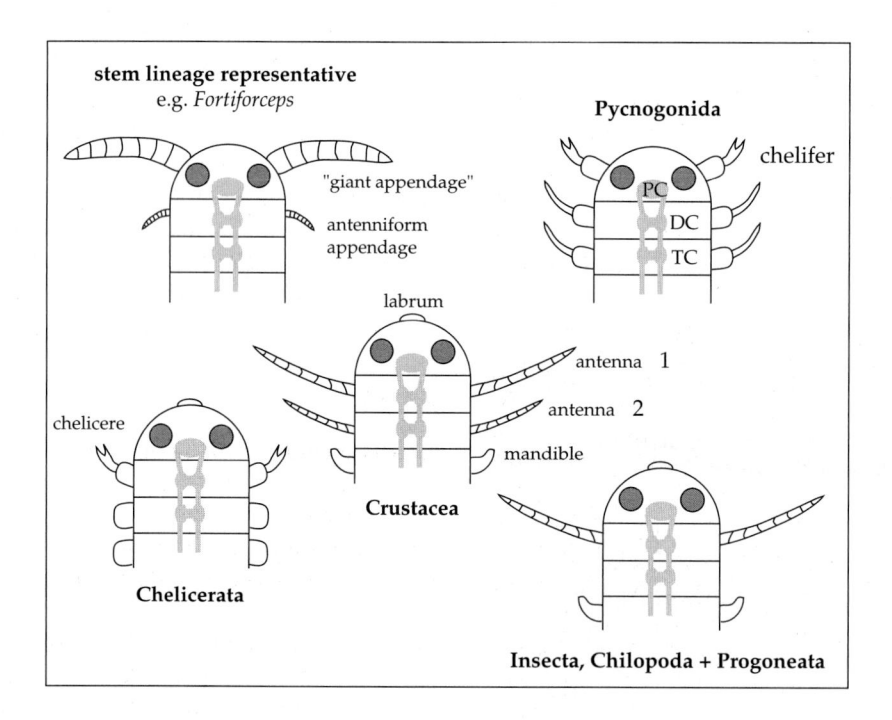

Fig. 2.20. Schematic head segmentation and insertion of appendages in fossil and extant arthropod taxa, based on the interpretations of Budd (2002) and Maxmen et al. (2005). See text for explanations (PC = protocerebrum, DC = deutocerebrum, TC = tritocerebrum). From Schmidt-Rhaesa (2006b), modified.

and morphological analyses, for example, Edgecombe et al. 2000, Giribet et al. 2001; Fig. 2.20). The results of Maxmen et al. (2005) support the hypothesis that the giant appendage present in some fossil arthropods belonged to the ocular segment, and was reduced, probably to form the labrum, during arthropod evolution (Budd 2002; Fig. 2.20). A problem with this hypothesis is, however, that a pair of appendages anterior of the chelifers has been found in a presumed Upper Cambrian pycnogonid larva (Waloszek & Dunlop 2002; see also Dunlop & Arango 2004 for a review). Comparable to pycnogonids, the position of the fossil trilobites is also debated (see Scholtz & Edgecombe 2005 for review).

All chelicerates (Fig. 2.21; exclusive of pycnogonids) are well supported as a monophylum. The monophyly of Crustacea is sometimes doubted (e.g. Moura & Christoffersen 1996), but Lauterbach (1983), Wägele (1993), Waloszek (1999) and Ax (2000) name some potential autapomorphies, and analyses by Schram and Hof (1998), Edgecombe et al. (2000) and Giribet et al. (2001) support crustacean monophyly. Myriapods are often assumed to be paraphyletic (Kraus & Kraus 1994, Wheeler 1997, Edgecombe et al. 2000, Kraus 2001), but some analyses still support their monophyly (Giribet et al. 2001). Morphological characters appear to clearly support the monophyly of insects, but this is not supported in all molecular analyses (Giribet & Ribera 2000, Nardi et al. 2003).

Several molecular analyses find a close relationship between chelicerates and myriapods and challenge the monophyly of Tracheata (rDNA: Friedrich & Tautz 1995, Min et al. 1998; mtDNA: Hwang et al. 2001; elongation factor-2: Regier & Shultz 2001; *Hox*-genes: Cook et al. 2001; mitochondrial genes: Hassanin 2006; see Stollewerk & Chipman 2006 for possible synapomorphies from neurogenesis), but there are no morphological characters supporting this relationship and other analyses favour monophyletic Mandibulata, that is, a relationship between myriapods, crustaceans and insects (Weygoldt 1986, Wheeler et al. 1993; Wägele 1993, Zrzavý et al. 1997, Edgecombe et al. 2000, Giribet et al. 2001, Kraus 2001, Richter 2002, Harzsch et al. 2005). There is growing evidence that insects and crustaceans may be closely related to each other, based on molecular (e.g. Friedrich & Tautz 1995, Regier & Shultz 1997, 2001, Cook et al. 2001, Mallatt et al. 2004, Mallatt & Giribet 2006) and combined analyses (Giribet et al. 2001), the structure of photoreceptors and characters from neurogenesis and neuroarchitecture (Paulus 2000, Harzsch & Walossek 2001, Sinakevitch et al. 2003, Harzsch et al. 2005; see also Dohle 2001, Richter 2002, Richter & Wirkner 2004). It remains unresolved, whether insects are the sister-group of all crustaceans or whether they are closely related to a subtaxon of crustaceans, in particular Malacostraca (and probably including Remipedia) (Fanenbruck et al. 2004, Harzsch et al. 2005).

Spiralia

Spiralia includes all animals with a spiral cleavage (there are exceptions, but these are almost certainly derived). Molecular analyses usually support Spiralia as a monophyletic taxon, but relationships among spiralian subtaxa become a harder topic. We can recognize four taxa which are likely to be monophyletic: Platyhelminthes (probably exclusive of Acoelomorpha, see below), Gnathifera, Nemertini, and Trochozoa. A variety of different groupings among these (and together with other taxa) has been proposed. They were excellently reviewed by Jenner (2004b)

Fig. 2.21. The tick *Ixodes ricinus* (Acari) as a representative of Euarthropoda.

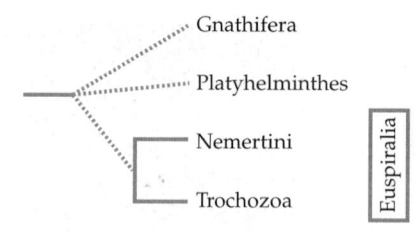

Fig. 2.22. Relationships among the Spiralia.

with the result that the relationships among these four taxa are not convincingly resolved. To name just one group that appears in several publications, I want to name Platyzoa, which includes Platyhelminthes, Gnathifera, and Gastrotricha (e.g. Cavalier-Smith 1998, Giribet et al. 2000). This is an extended version of the concept that Platyhelminthes and Gnathostomulida are sister-taxa (as Plathelminthomorpha, see Ax 1985, 1996). The morphological basis for such a relationship is weak (see discussion in Haszprunar 1996a and Jenner 2004b).

There is some support for the theory that Nemertini (Fig. 2.23) could be closely related to Trochozoa, together named Euspiralia. I prefer Euspiralia over the sometimes used Eutrochozoa, which was introduced for a taxon including annelids, molluscs, sipunculids, echiurids, and pogonophorans (Ghiselin 1988; in this sense it is synonymous to Trochozoa used here) and later extended to include Nemertini and other groups (Eernisse 1997). The characters supporting Euspiralia named by Ax (1996; one-way-gut with anus, multiciliary) are not unequivocal, because they appear in several other taxa as well, but a good alternative character has been named recently. While Nemertini do not have a typical trochophore, it has been found that large surface cells in the completely ciliated larva of *Carinoma* (Palaeonemertini, Nemertini) correspond in position and cell lineage to prototroch cells in the trochophore (Maslakova et al. 2004a, b). When these cells are homologous, a larva with some cells denominated as prospective prototroch cells would be a synapomorphy of Nemertini and Trochozoa.

One further scenario could focus on monociliarity (as it appears in gnathostomulids, Lammert

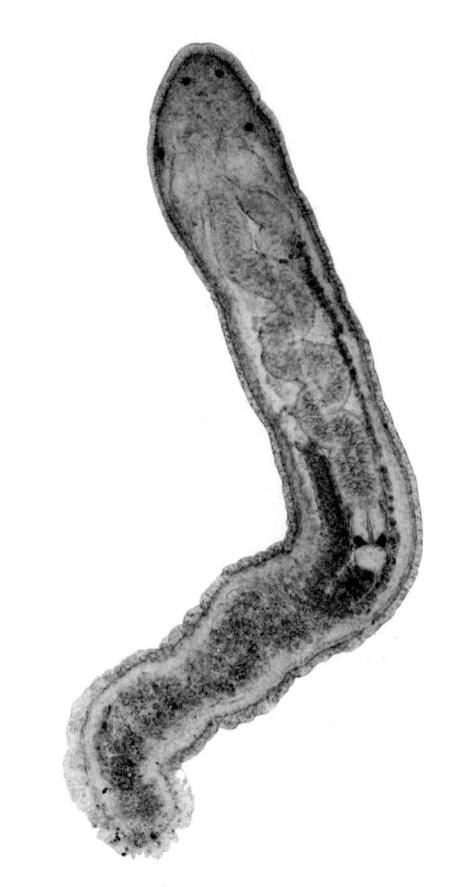

Fig. 2.23. *Prostomatella arenicola*, a small nemertean species (Nemertini).

1989) as an assumed ancestral feature. This could unite Platyhelminthes and Euspiralia by their presence of multiciliary epidermal cells. The significance of the character 'multiciliarity' is, however, quite weak, because it also occurs among deuterostomes and within Annelida. Therfore, I regard it as appropriate to present relationships among Spiralia as a trichotomy of Platyhelminthes, Gnathifera, and Euspiralia (Fig. 2.22).

Acoels and other flatworms

Flatworms, a taxon commonly named Platyhelminthes or Plathelminthes, consist of three subgroups which are each likely to be monophyletic: Acoelomorpha with the marine subtaxa Acoela and Nemertodermatida, Catenulida, a small group of freshwater, rarely

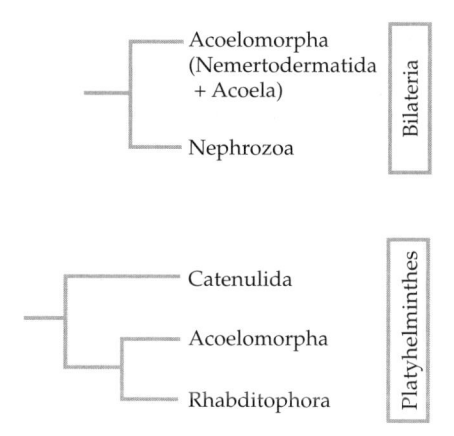

Fig. 2.24. Alternative positions of acoelomorph flatworms, either as a basal bilaterian taxon (top) or as a member of Platyhelminthes (bottom). See text for explanations.

Fig. 2.25. *Schistosoma mansoni* (Digenea), a gonochoristic parasitic representative of the Platyhelminthes.

marine, flatworms, and Rhabditophora with all remaining groups, including several free-living groups and all parasitic taxa (Fig. 2.25, see Littlewood & Bray 2001 for recent summaries). For a long time, Platyhelminthes were considered to be monophyletic (Fig. 2.24), although their potential autapomorphies are not unequivocal. Ax (1996) names as autapomorphies the multiciliarity of epidermal cells, the loss of an accessory centriole in epidermal cells, the multiciliarity of gastrodermal cells, and the existence of two cilia with 16 surrounding microvilli in the terminal cell of the protonephridium. All those characters are not exclusive for flatworms but can also be found in other groups (Smith et al. 1986, Haszprunar 1996a). The protonephridial character applies only for catenulids and rhabditophorans, because acoelomorphs do not have protonephridia. The interpretation of these characters as autapomorphies depends much on a sister-group relationship between Platyhelminthes and Gnathostomulida, as is assumed by Ax (1996), but as this relationship is not strongly supported (see above under 'Spiralia'), the characters potentially supporting Platyhelminthes are becoming weaker. Two further autapomorphies could be a certain type of stem cells (spherical neoblasts; Rieger & Ladurner 2001). Another, ambiguous character could be the presence of a biciliate sperm cell. Such sperm cells are present in acoels and (many) rhabditophorans. Ehlers (1992a) assumed that biflagellate spermatozoa evolved convergently, because monociliate sperm cells are present in Nemertodermatida, and one rudimentary cilium occurs during spermiogenesis of the (in the mature stage aflagellate) sperm cells of catenulids (see Chapter 14).

Acoels are strange worms, lacking an epithelial intestine, protonephridia, and ECM, but they appear to be closely related to the small taxon Nemertodermatida (about 10 species, see Lundin & Sterrer 2001) on the basis of a network of epidermal ciliary rootlets with corresponding branching patterns, modified cilia in which only five of the nine peripheral microtubule duplets reach the ciliary tip ('shelfed tip'), and the absence of protonephridia (Tyler & Rieger 1977, Ehlers 1985a, Smith & Tyler 1985, Ax 1996, Lundin 1997). The absence of a 'real' intestine (with epithelium) has been one of the arguments for early views considering acoels as basal bilaterians (e.g. Steinböck 1963a). This character, however, appears to be derived within acoels, because Nemertodermatida have an epithelialized gut and the basal acoel genus *Paratomella* (Ehlers 1992, Hooge et al. 2002, Petrov et al. 2004) has something like a bilayered epithelium that can be considered as a transitory stage between an epithelial and the 'typical' acoel gut (Smith & Tyler 1985, Ax 1996). Molecular analyses support the hypothesis of polyphyletic Platyhelminthes, because acoel sequences group not with other flatworm sequences, but basal to bilaterians (e.g. Ruiz-Trillo et al. 1999, Littlewood et al. 2001, Baguñà & Riutort 2004a,b; Fig. 2.24). It is a problem that the acoel sequences are generally very fast evolving and even the shortest ones are still quite long (see Ruiz-Trillo et al. 1999). This basal position appears to be valid for Acoelomorpha, that is, including Nemertodermatida, but most molecular analyses do not support the monophyly of Acoelomorpha (Ruiz-Trillo et al. 1999, 2002, Jondelius et al. 2001, 2002). There are some further data that appear to support a basal position of acoels. Their brain is different from the rhabditophoran brain and could represent a 'primitive' condition (see Chapter 6; Reuter et al. 2001a, Raikova et al. 1998, 2000b). The nervous system of nemertodermatids is also very simple (Raikova et al. 2004). In acoels, only three HOX genes have been found and this small number could reflect their basal position (Cook et al. 2004). Acoels use a different genetic code for two amino acids compared to Rhabditophora (AAA for asparagine instead of Lysin, AUA for Isoleucin instead of Methionin) (Telford et al. 2000) and they lack *let-7*, a small temporary RNA that is present in all other bilaterians (Pasquinelli et al. 2003). The only molecular analysis that shows acoels to be nested within other flatworms is from the elongation-factor-1α gene (Berney et al. 2000). Littlewood et al. (2001) reinvestigated this analysis and found it to be not unequivocal to interpret. An insertion in the EF1α-gene is present in the acoel sequence investigated by Berney et al. (2000) and a triclad sequence, but not in that of a polyclad. Littlewood et al. (2001) showed that this insertion is not present in all acoel sequences, but that other flatworms and outgroup taxa such as annelids have (slightly differing) insertions in the same location and that this insertion could well be plesiomorphic.

To conclude, I am not completely convinced about the basal position of Acoela and Nemertodermatida, because this may be due to long-branch problems. However, the morphological evidence for monophyletic Platyhelminthes is not convincing either. I will discuss Acoelomorpha and Platyhelminthes separately below, but use the name Platyhelminthes to include only Catenulida and Rhabditophora. The retention of the name exclusive of Acoelomorpha appears to be valid, because Catenulida are, according to Ehlers (1985a) and Ax (1996) the basal taxon, regardless of the inclusion or exclusion of Acoelomorpha.

Trochozoa

There are several taxa that possess a 'typical' trochophore larva (see Chapter 16 for a characterization of this): Mollusca, Kamptozoa (Entoprocta), Annelida, Sipunculida, Echiurida,

Fig. 2.26. *Mopalia* sp., a chiton (Polyplacophora) as representative of molluscs.

Pogonophora, and Myzostomida. The name Pogonophora (Siboglinidae) is used here including the taxa Frenulata, Monilifera (*Sclerolinum*), *Osedax* and Vestimentifera (Obturata) (Black et al. 1997, Halanych et al. 1998, 2001, Rouse et al. 2004, Rousset et al. 2004, Glover et al. 2005, Halanych 2005, Southward et al. 2005) Assuming that a trochophore (at least with the characteristics of the 'typical' trochophore) evolved only once, the taxon including all previous groups is named Trochozoa. One probable subtaxon of Trochozoa, whose representatives lack a trochophore, is Arthropoda, the potential sister-group of Annelida. The phylogenetic position of arthropods is discussed above (Ecdysozoa or Articulata?). In general, molecular data (especially the 18S rDNA gene) provide no good resolution among Trochozoa, probably due to rapid radiation during which not enough genetic differences were established.

From a morphological standpoint, there is no serious doubt that molluscs (Fig. 2.27) are monophyletic, but in several molecular analyses, molluscs appear as polyphyletic, although this is generally not supported by good bootstrap values. The monophyly is supported by the presence of a radula and other characters (Ax 2000, Nielsen 2001). There are some open questions concerning the phylogenetic relationships within molluscs (see, e.g. von Salvini-Plawen 1980a, Lauterbach 1984, Winnepenninckx et al. 1994, 1996, von Salvini-Plawen & Steiner 1996, Steiner 2004). The basal branching patterns among molluscs are of particular importance for hypotheses on the molluscan ancestor (e.g. Haszprunar 1992, 1996b). There is, for example, the question whether the wormlike taxa Solenogastres and Caudofoveata are sister-taxa or not (Haszprunar 2000), or whether they are basal molluscs or derived from polyplacophores (von Salvini-Plawen 1985, 2003, Scheltema 1993, 1996, Ivanov 1996, Haszprunar 2000, Scheltema & Ivanov 2002). Recently, Giribet et al. (2006) have proposed that Polyplacophora and Monoplacophora could be closely related. Annelida (Fig. 2.29) also seem to be monophyletic, but it is quite difficult to reconstruct phylogenetic relationships within the group and therefore make reliable hypotheses about ancestral features (Rouse & Fauchald 1997, 1998, Westheide 1997, Westheide et al. 1999, Purschke 2002a, Bleidorn et al. 2003, Struck et al. 2002, 2006, Jördens et al. 2004, Colgan et al. 2006). This means, for example, that we can not decide whether the ancestral annelid looked more like a polychaete (e.g. Westheide 1997) or like an oligochaete (e.g. Clark 1964). The monophyly of the smaller trochozoan taxa is undisputed.

There is evidence that the three taxa Echiurida, Pogonophora, and Myzostomida belong in the Annelida (see summary by Halanych et al. 2002).

Fig. 2.27. Mussels (*Mytilus edulis*) and a gastropod (*Nucella lapillus*) inhabiting rocky shores.

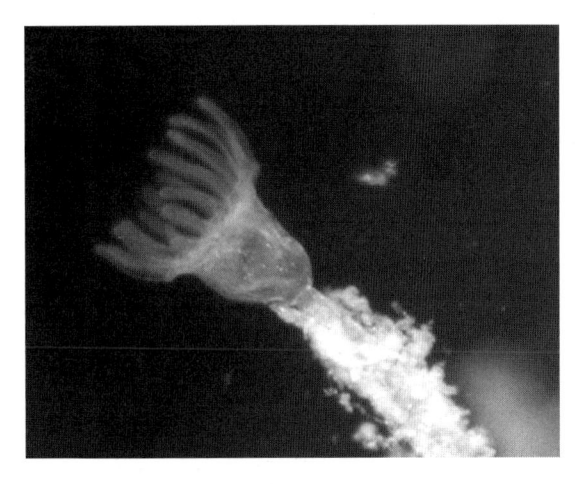

Fig. 2.28. *Pedicellina cernua* (Kamptozoa). Photo by courtesy of Andreas Unger, Bielefeld.

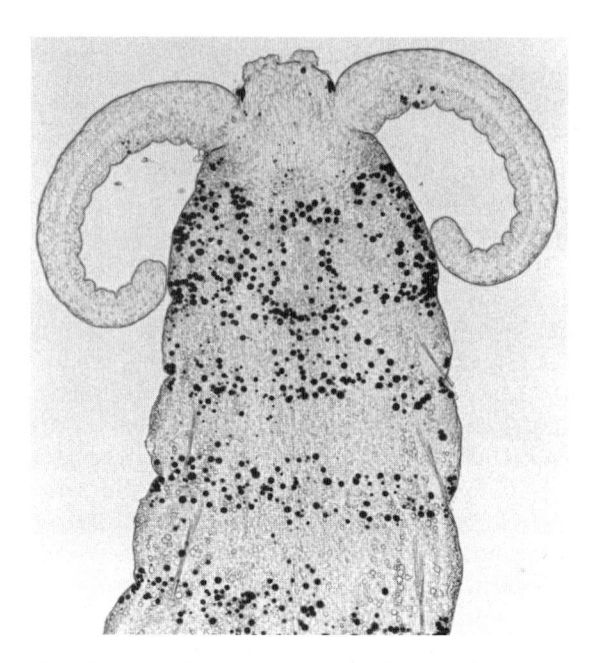

Fig. 2.29. *Protodrilus* sp., a meiofaunal polychaete annelid.

For echiurans and pogonophorans there is molecular evidence (McHugh 1997, Rouse & Fauchald 1997, Bleidorn et al. 2003, Jördens et al. 2004). Pogonophora possess annelid chaetae that even allow us to hypothesize a sister-group relationship to Sabellida (Bartolomaeus 1997). Echiurans also have chaetae (Orrhage 1971) and they have, at least during development of *Bonellia viridis*, a segmented nervous system (Hessling & Westheide 2002). In this case, the form of the chaetae does not allow an association with a particular annelid subtaxon, so the possibility that chaetae evolved in a common ancestor of echiurans and annelids is not ruled out. The segmented nervous system implies that echiurans are derived from a segmented ancestor, but it is also possible that the common ancestor of echiurans and annelids was segmented only in the nervous system and that a complete segmentation evolved only in annelids. However, the molecular evidence from the 18S rDNA gene and the elongation factor 1α gene make it more likely that echiurans are derived annelids. Myzostomes also have a segmented nervous system (Müller & Westheide 2000), as well as external segmentation and 'annelid' chaetae on parapodia (Jägersten 1936). They also have a trochophore larva (Jägersten 1939, Eeckhaut & Jangoux 1993, Eeckhaut et al. 2003, Eeckhaut & Lanterbecq 2005). Molecular data, however, do not support this relationship to annelids, but favour a relationship to Platyhelminthes (Eeckhaut et al. 2000) or, combined with morphological characters, to Cycliophora and Syndermata (Zrzavý et al. 2001). The latter hypothesis was critically reanalysed by Jenner (2003), indicating that not all available characters have been integrated in the analysis and that the proposed relationship is very sensitive to the addition of further characters.

Kamptozoa (Fig. 2.28) and Sipunculida (Fig. 2.30) are more difficult to place. Both have been considered as the sister-group of molluscs. Molluscs and kamptozoans agree in the presence of a dorsal cuticle containing chitin, a ventral creeping foot, and a lacunar system of primary body cavity interstices (Haszprunar 1996b, 2000, Ax 2000). An alternative hypothesis is that molluscs and sipunculids are sister-taxa, based on the presence of a molluscan cross during development and similarities between molluscs and a creeping larva present in some sipunculids (Scheltema 1993, 1996). Neither hypothesis is without problems. Chitin is present in colonial kamptozoans (*Pedicellina*, *Urnatella*; see Jeuniaux 1982a, b), but absent in solitary forms (*Loxosomella atkinsae*; Andreas Unger, Bielefeld, unpublished results). Therefore it is questionable whether the kamptozoan ancestor had chitin in the dorsal cuticle or whether chitin evolved within this group, convergently to molluscs. The second hypothesis relies on the assumption that the molluscan cross is a structure of phylogenetic significance (which is convincingly doubted by Jenner 2003), and on the homology of the molluscan radula sac with a protrusible pharyngeal structure in the pelagosphaera of sipunculids, as well as of the sipunculan larval lip gland with the molluscan pedal gland. There is also evidence from the structure of the mitochondrial genome that sipunculids are closely related to annelids (Boore & Staton 2002) or are even a sub-taxon of annelids, closely related to orbiniids (Bleidorn et al. 2006).

Fig. 2.30. *Sipunculus nudus* (Sipunculida).

```
┌─ Mollusca
├─ Entoprocta
├─ Sipunculida
└─ Annelida
   (including Echiurida,
   Myzostomida, Pogonophora)
```

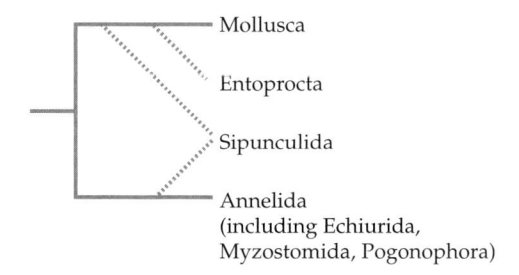

Fig. 2.31. Summary of discussed relationships among Trochozoa.

In conclusion, I regard it as justified to assume that Trochozoa are monophyletic, and that Echiurida, Pogonophora, and Myzostomida are subtaxa of Annelida, but the relationship of sipunculids and kamptozoans should be treated with some care (Fig. 2.31).

The mysterious *Symbion* (Cycliophora)

Symbion pandora was described in 1995 (Funch & Kristensen 1995), a second species, *S. americanus* was described by Obst et al. in 2006 (see also Obst et al. 2005). Despite a superficial resemblance to rotifers, the fine structure of *Symbion* does not support such a relationship (Funch & Kristensen 1995, 1997). The sequence of the 18S rDNA gene, however, supports the inclusion of *Symbion* into the Gnathifera, closely associated with eurotifers and acanthocephalans (Winnepenninckx et al. 1998b). This is supported by similarities of the musculature in the chordoid larva with some eurotifers (Wanninger 2005). Other analyses assume that the chordoid larva of *Symbion* is a modified trochophore (Funch 1996) or support a relationship to Kamptozoa (Entoprocta) based on the presence of internal budding and mushroom-shaped extensions from the basal lamina (Funch & Kristensen 1995, Sørensen et al. 2000).

The tentaculate taxa: Phoronida, Bryozoa and Brachiopoda

The phylogenetic position of three tentaculate taxa, Bryozoa (Ectoprocta; Fig. 2.32), Phoronida (Fig. 2.33), and Brachiopoda (Fig. 2.34) is, besides the Articulata-Ecdysozoa problem, another unresolved issue that spans wide regions of the metazoan tree (see Lüter & Bartolomaeus 1997, Lüter 2004 for reviews). Although the three taxa often appear as closely related, a monophyletic taxon Tentaculata (or Lophophorata) does not exist, because all characters shared are plesiomorphies and no autapomorphy can be identified. Most morphology-based considerations present the tentaculate taxa as an unresolved, probably paraphyletic, assemblage in close association with deuterostomes. Nielsen (2001, 2002a) assumes that brachiopods and phoronids are closely related, but that bryozoans are closely related to kamptozoans (entoprocts) within Spiralia. Brachiopods may not be monophyletic, because in some analyses phoronids are more closely related to inarticulate brachiopods than are articulated brachiopods (Cohen et al. 1998, Cohen 2000, Cohen & Weydmann 2005).

Based on morphological characters, Bryozoa, Phoronida, and Brachiopoda belong in close association with deuterostomes: the common taxon is named Radialia. Probable autapomorphies of Radialia are the tentacle apparatus (also called lophophore), a bipartite coelom, and a blood vascular system. A horseshoe-shaped tentacle apparatus, internally supported by a coelomic cavity, is present in most phoronids, phylactolaematan bryozoans, brachiopods and, among deuterostomes, in pterobranchs. The only difference is the position of the mouth, which is surrounded by the tentacles in the tentaculate taxa, but not in pterobranchs. The tentacles carry cilia which collect food particles in the so-called upstream-collecting system, as is likely for pterobranchs, whereas protostomes collect particles in a downstream-collecting system. The trimeric coelom that is clearly present in echinoderms, enteropneusts, and pterobranchs was long thought to be present in the tentaculate taxa, too. This supported the hypothesis of an ancestry of a coelomic trimery ('archimery', see Chapter 9). More recent investigations, however, find only two coelomic cavities in phoronids (Bartolomaeus 2001, Gruhl et al. 2005), brachiopods (Lüter 2000a), and bryozoans (Ines Wegener, Bielefeld, unpublished results). This does not necessarily contradict a relationship to deuterostomes, because the two coelomic cavities have comparable positions. The posterior one (metacoel) is the main trunk coelom, the anterior one (mesocoel) branches into the tentacles, and the anteriormost one (protocoel) could be an autapomorphy of Deuterostomia. Furthermore, there might be similarities in the mode of coelom formation, at least in brachiopods. The coelom originates from the endoderm by enterocoely in deuterostomes. In brachiopods, the coelom originates from cells that have detached from the gut (Lüter 2000a). In bryozoans and phoronids, the origin of the coelom is much less understood. Finally, phoronids, brachiopods, and deuterostomes have a blood vascular system (as a primary body cavity, see Chapter 10), which could also support their close relationship to deuterostomes (for a derivation of the funiculus in bryozoans from the blood vascular system, see Carle & Ruppert 1983).

Nielsen (summarized in 2001, 2002a) sees some spiralian traits in bryozoans and argues for a sister-group relationship between bryozoans and kamptozoans (entoprocts). The cleavage of bryozoans (based on data from *Bugula flabellata* in Corrêa 1948) is radial, but the ciliated cells of the corona, a ciliary band of the cyphonautes larva, have a lineage comparable to prototroch cells in molluscs and annelids (Nielsen 2001). This might hint at a homology between prototroch and corona and would then be a spiralian trait in the cleavage of bryozoans. The tentacles of bryozoans display, with their mechanical filter (at least in gymnolaemates and stenolaemates) a unique food capture structure (Nielsen 1987, 2002b, Nielsen & Riisgård 1998). This is in principle also present in phylactolaemates,

Fig. 2.32. *Flustrellidra hispida*, a ctenostome bryozoan. Photo by courtesy of Georg Mayer, Berlin.

the assumed basal group within bryozoans, although the laterofrontal cilia are not as long and stiff as in gymnolaemates and stenolaemates (Nielsen 2001, Riisgård et al. 2004). The principle of food capture is, however, an upstream-collecting system (as in deuterostomes). Other possible synapomorphies between bryozoans and kamptozoans are the presence of myoepithelial cells in the apical organ, correspondences in the metamorphosis, and the structure of larval eyes (Nielsen 2001).

Different molecular approaches do not support a relationship to Deuterostomia and have favoured a position within Protostomia (Halanych et al. 1995, Mackey et al. 1996, Cohen et al. 1998, de Rosa et al. 1999, Saito et al. 2000, de Rosa 2001, Mallatt & Winchell 2002, Passamaneck & Halanych 2004). The mitochondrial gene order of the brachiopod *Terbratulina retusa* resembles that of the polyplacophoran mollusc *Katherina tunicata*, but within brachiopods (as well as in molluscs), there is diversity in the gene order, which makes it uncertain whether the states in *Terebratulina* and *Katherina* are ancestral (Noguchi et al. 2000, Helfenbein et al. 2001). There is a clustering of brachiopods and phoronids together with annelids and molluscs, but a convincingly clear hypothesis of relationship is still lacking due to general resolution problems in this region of the tree. Bryozoans mostly fall outside this annelid/mollusc/phoronid/brachiopod clade, but this position again cannot be regarded as convincing. Molecular data therefore favour a position of tentaculate taxa within Protostomia, but it is, in my view, too early to name groups ('Lophotrochozoa') based on this position. Two further findings make the 'protostome-hypothesis' an interesting alternative to the 'deuterostome-hypothesis'. First, brachiopods have chaetae that are formed after the same principle as annelid chaetae (Lüter 2000b), although the secretion of extracellular material by microvilli is not an uncommon phenomenon (for example the jaw apparatus in ganthiferans is secreted in a comparable way). Second, some fossils may bridge the gap of brachiopods to molluscs.

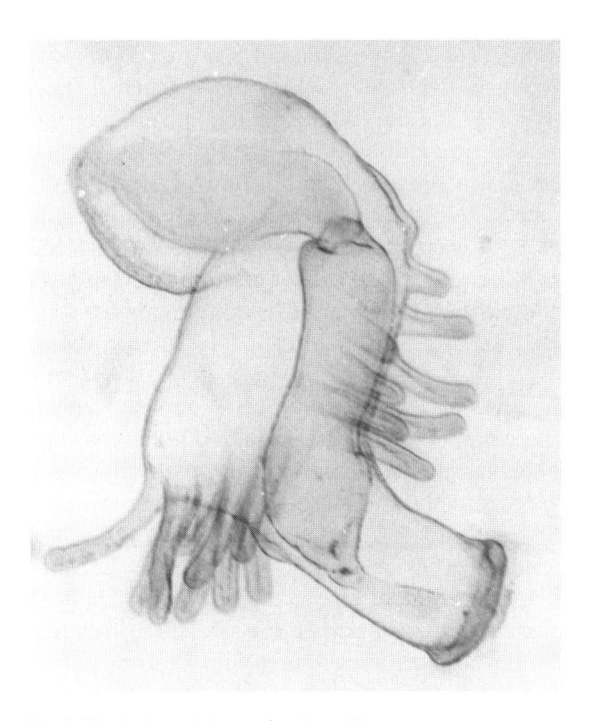

Fig. 2.33. Actinotroch larva of a phoronid.

Fig. 2.34. *Glottidia pyramidata*, a lingulid brachiopod.

Halkieriids (Conway Morris & Peel 1995) are interpreted as closely related to annelids or molluscs. It has been proposed that the two shells of a brachiopod could originate from an approaching and then flipping of the two isolated halkieriid shells (Conway Morris 1998). This corresponds to observations that both shells of the brachiopod *Crania anomala* are secreted on the dorsal side, which is obscured because the body folds before shell secretion (Nielsen 1991, see also Cohen et al. 2003). Additionally, fossils such as *Micrina* (Williams & Holmer 2002) and *Mickwitzia* (Skovestedt & Holmer 2003) bridge the gap between halkieriids and brachiopods because they have two shells (probably closer together than in halkieriids), but their shells correspond in fine details with fossil brachiopod shells.

In summary, it appears difficult to draw a conclusion. The morphology of at least phoronids and brachiopods, but also of bryozoans, appears to favour a close association with deuterostomes. The inclusion of protostomes is in contrast supported by different molecular results, few morphological characters, and probable fossil support. However, in the following, it appears best to consider both hypotheses when the organ systems of tentaculate taxa are considered.

Chaetognatha

Chaetognaths (Fig. 2.35) have been, and are still, a seriously problematic taxon for phylogenetic studies, but this appears to be slowly changing. Chaetognaths experienced long journeys through the metazoan tree, but end up abundantly close to deuterostomes, because the chaetognath mouth originates by deuterostomy (e.g. Brusca & Brusca 2003; but see Zrzavý et al. 1998 and Nielsen 2001 for alternatives). Molecular analyses have not proven to be helpful, because the chaetognath genes investigated so far are fast evolving (Telford & Holland 1993, 1997, Halanych 1996a) and regularly group with other fast evolving gene sequences (see, e.g. Littlewood et al. 1998, Peterson & Eernisse 2001). Now, there is some indication that the cleavage may show traces of spiralian cleavage (Shimotori & Goto 2001). Investigations

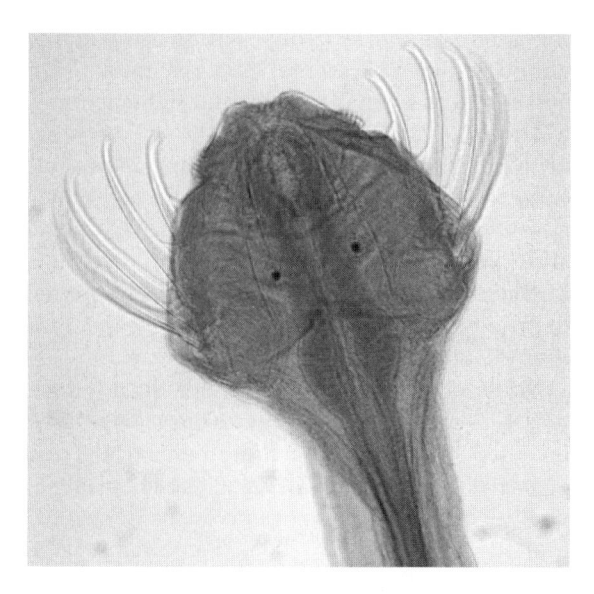

Fig. 2.35. Anterior end of *Sagitta setosa* (Chaetognatha).

of mitochondrial genes led to the conclusion that chaetognaths are either the sister-group (Helfenbein et al. 2004) or part of the protosomes (Papillon et al. 2004). This is supported by studies of multiple genes (Matus et al. 2006, Marlétaz et al. 2006). The presence of a mosaic *Hox* gene, finally, may be interpreted as a conserved stage before the splitting of the gene into a median and a posterior *Hox* gene and, because both such genes are present in bilaterians, chaetognaths may be the sister-taxon of Bilateria (Papillon et al. 2003). This last interpretation, particularly, would have considerable consequences for character evolution of, for example, coelomic cavities.

Deuterostomes

There are four taxa belonging to Deuterostomia, for which the monophyly is not really doubted, both from the morphological and the molecular perspective: Echinodermata (Fig. 2.36), Tunicata (Urochordata; Fig. 2.37), Acrania (Cephalochordata; Fig. 2.38), and Craniota (Vertebrata, *sensu* Cameron et al. 2000). Two further groups, Enteropneusta and Pterobranchia, may be not monophyletic. In the analyses of Cameron et al. (2000) and Peterson and Eernisse

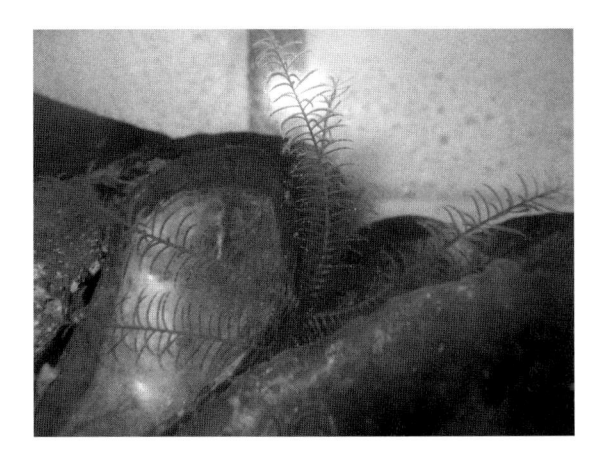

Fig. 2.36. *Antedon* sp., a hemisessile crinoid (Echinodermata).

Fig. 2.37. The colonial tunicate *Botryllus schlosseri*.

(2001) enteropneusts fall into two taxa, Harrimaniidae and Ptychoderidae, with pterobranchs being more closely related to Harrimaniidae. Pterobranchia (Cephalodiscida and *Rhabdopleura*), is regarded as paraphyletic by some authors (see below). There are sometimes problems with the internal phylogeny within these groups, for example, the question of how derived the Appendicularia are within tunicates (Stach & Turbeville 2002), or the question of whether Myxinoida and Petromyzontida are sister-taxa or not. This effects, of course, the ground pattern reconstruction, but it would go too far here to discuss internal relationships in detail.

There are, as has been shown by the excellent review by Dohle (2004), three main problems with deuterostome phylogeny: first, what are the relationships between pterobranchs and enteropneusts? second, what is the position of these two taxa among deuterostomes? and third, what is the sister-taxon of Craniota? Only the monophyly of Chordata appears to be well supported.

Pterobranchia and Enteropneusta are often summarized under the name Hemichordata, but this has been challenged by Ax (2003) who not only regards Hemichordata as being paraphyletic, but also Pterobranchia. According to him, Enteropneusta are a sister-group of Chordata (together named Cyrtotreta), the Cephalodiscida (*Cephalodiscus* + *Atubaria*) are a sister-group of Cyrtotreta (together named Pharyngotremata), and *Rhabdopleura* is, due to its lack of gill slits, a sister-group of Pharyngotremata (together named Stomochordata). The characters supporting these relationships are not unequivocal, characters of Stomochordata refer exclusively to 'hemichordate' taxa, and the presence of gill slits supporting Pharyngotremata may be a plesiomorphy because it was probably present in fossil taxa (Homalozoa). Among Homalozoa, gill slits are most evident in the cornute *Cothurnocystis* (Jefferies 1968). There is an ongoing debate about whether homalozoans are fossil echinoderms (in particular based on the presence of the stereom, a three-dimensional structure of the endoskeleton, see e.g. Philip 1979), or fossil chordates (as 'calcichordates', based predominantly on the gill slits, see e.g. Jefferies 1975, 1981, Cripps 1991, Gee 1996). The possibility must be considered that gill slits are a plesiomorphic character even of deuterostomes, and this would make their absence in *Rhabdopleura* derived.

In contrast, there are characters supporting a monophyly of Pterobranchia: the fine structure of the oral shield, a pigmented stripe separating two regions on this shield, and the stalk (Dohle 2004; although the stalk could also be an autapomorphy of Hemichordata, because a similar structure is present in young enteropneusts,

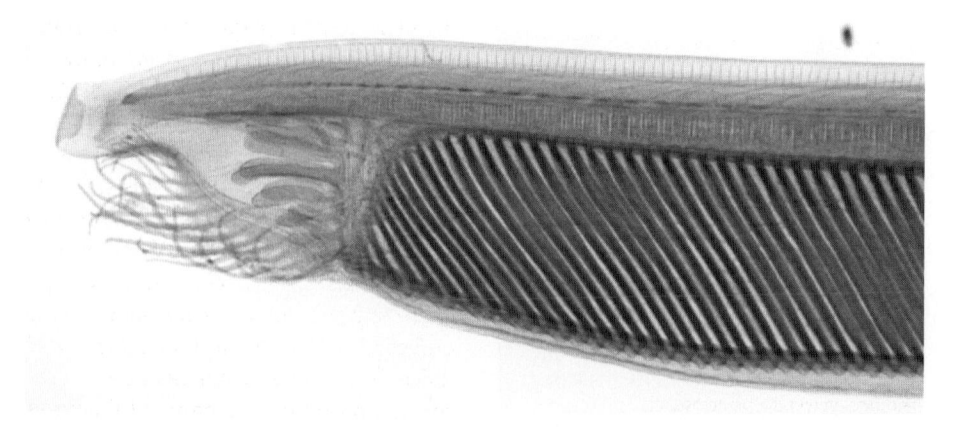

Fig. 2.38. Anterior end of *Branchiostoma lanceolatum* (Acrania). Photo by courtesy of Alexander Gruhl, Berlin.

see Burdon-Jones 1952). The stomochord with the glomerulus-pericard-complex is a potential synapomorphy of Pterobranchia and Enteropneusta (Dohle 2004), supporting monophyly of Hemichordata. There is no evidence for a homology between stomochord and chorda. The stomochord may function as a supportive structure to create pressure in the blood for filtration into the protocoel (Wilke 1972a; see Chapter 8), or as a glandular organ (Mayer & Bartolomaeus 2003). The 'lophoenteropneusts', which were long known only from deep sea photographs (Lemche et al. 1976) and which were thought to unite characters of pterobranchs (a tentaculated anterior end) and enteropneusts (a wormlike body), have turned out to be deep-sea enteropneusts without a trace of tentacles (Holland et al. 2005).

Hemichordates have often been regarded as closely related to chordates, but molecular analyses (e.g. Bromham & Degnan 1999, Cameron et al. 2000, Winchell et al. 2002 and others) as well as *Hox* gene comparisons (Peterson 2004) have found constant support for a sister-group relationship between Echinodermata and Hemichordata, together named Ambulacralia (Fig. 2.39). There are three characters that support this relationship: the larva, three pairs of coelomic cavities, and the axial organ/glomerulus-pericard-complex. Echinoderm larvae, although quite diverse at first sight, can all be derived from a hypothetical larva, the dipleurula. This closely resembles the tornaria larva of enteropneusts (but see Nezlin 2000 for different patterns in the distribution of biogenic amines and acetylcholinesterase activity). Three paired coelomic cavities are present only in these two taxa, although this trimery (also called 'archimery') has often been regarded as plesiomorphic (see Chapter 9). Finally, there are similarities in the axial complex of echinoderms (Ruppert & Balser 1986) and the stomochord-glomerulus-pericard-complex of enteropneusts (Wilke 1972b, Balser & Ruppert 1990) and pterobranchs (Dilly et al. 1986, Mayer & Bartolomaeus 2003). These similarities are regarding an excretion of 'blood' through podocytes into the protocoel in combination with a contractile coelomic organ (the pulsatile vesicle/pericardium).

Finally, the relationships of chordate taxa are disputed. Generally, Acrania and Craniota are regarded as sister-taxa (Fig. 2.39), due to the presence of a chorda as long as the body, a closed circulatory system, as well as segmental musculature and coelom. The common taxon they form has a confusing variety of proposed names: Vertebrata (Ax 2003), Chordata (Cameron et al. 2000), Notochordata (Nielsen 2001), Holochordata (von Salvini-Plawen 1989), and Myomerata (Dohle 2004). The names Vertebrata and Chordata

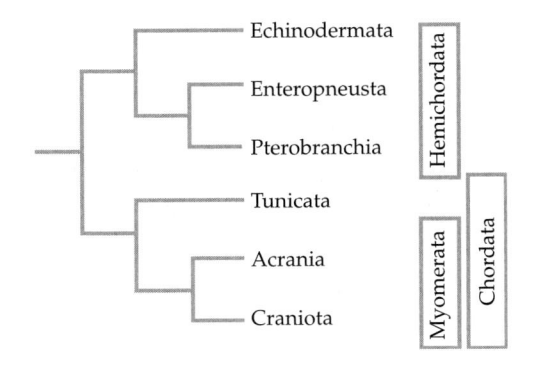

Fig. 2.39. Relationships among the Deuterostomia.

appear inappropriate because spines evolved only within Craniota, and the name Chordata is commonly applied to a taxon including Tunicata. To avoid confusion with several names referring to the chorda, I sympathize with Dohle's 'Myomerata' and will use this name here. While some analyses support Myomerata (Cameron et al. 2000, Nielsen 2001, Winchell et al. 2002, Ax 2003, Dohle 2004), some genome analyses find more support for tunicates as the sister-taxon of craniotes (Blair & Hedges 2005, Philippe et al. 2005, Delsuc et al. 2006, Schubert et al. 2006) and the common taxon is named Olfactores.

CHAPTER 3

General body organization

In this chapter, no organs or organ systems are discussed, but some thoughts are presented as to the evolution of the general body organization. This is because the general organization has an important influence on the organ systems and vice versa.

Complexity

It is a common impression that the complexity of organisms increases during evolution. Making such statements often leads to the distinction of 'simple' versus 'complex' taxa or 'lower', even 'minor', versus 'higher' taxa. I regard such distinctions as problematic, mainly because they imply a teleological thinking and order (at least as an impression) animals in a hierarchical system of different levels of organization. When we compare a recent sponge (as a representative of a presumably low level of complexity) with humans, we should not forget that both lineages leading towards the sponge and the human passed the same evolutionary time since their separation from a common ancestor. Both are equally well adapted to their specific environment.

Complexity itself is a complex topic. There can be different levels of complexity, which evolve in a coordinated manner. McShea (2002) suggested that the organization of parts into a higher-level entity leads to a decrease of complexity in these parts (the 'complexity drain'), because some of the functional demands for the parts are transferred to the higher level.

A diffuse statement that complexity grows during evolution, at least in a broad and general sense, is certainly not wrong. However, it is hard to measure morphological complexity unambiguously. There are attempts to measure complexity by the estimation of cell type diversity (see, for example, Vickaryous & Hall 2006), but it is not easy to homologize cell types over all metazoans. As an example, Valentine (2004) lists 5 cell types for sponges (Porifera) (Valentine 2004, Tab. 2.3), while e.g. Bergquist (1978) lists 14 cell types for sponges (pinacocytes can even be subdivided into 3 types, which would lead to 16 cell types). Additionally, complexity in animals may also decrease during evolution. In particular in parasitic species there is a tendency for the reduction of organs, with rhizocephalan cirripedes representing an extreme example. Finally, in several cases the male individuals are highly reduced and have hardly more than the ability to produce sperm (for example, the echiuran *Bonellia viridis*, monogonont eurotifers, and parasitic gastropods).

One should distinguish between genetic and phenotypic complexity and there is evidence that these two factors are not directly correlated with each other. Instead, there is evidence that phenotypic complexity increases to a stronger extent than genetic complexity, suggesting factors that selected for 'economy' in genomic size (called 'genetic parsimony', see Stebbing 2006). With genomic sequencing and EST-approaches, measurements of gene numbers have become more reliable. Such approaches showed that cnidarians (more precisely, the anthozoans *Nematostella vectensis* and *Acropora millepora*) have more genes than the nematode *Caenorhabditis elegans* and the insect *Drosophila* (Technau et al. 2005). There are in particular several genes in anthozoans that were previously known from only craniotes, but which are lacking in the protostome taxa. This means that the initial genetic diversity was large and was condensed during evolution. Additionally, the term 'complexity' can also be applied to the genes themselves. Craniotes and annelids (the polychaete *Platynereis dumerilii*) have genes rich in introns while those from *Caenorhabditis* or *Drosophila* have distinctly fewer introns (Raible et al. 2005). This possibly indicates that the common bilaterian ancestor also had complex, intron-rich genes, and that in nematodes and insects a loss of introns occurred. Intron-rich genes may play an important role during evolution, because they allow alternative splicing, which may lead to diverse protein products from one gene (Graveley 2001, 2005). This probably means that phenotypic complexity is more likely to be correlated with gene regulatory interactions (i.e. a genetic combinatorial complexity) than with the gene number or genomic size.

Size

The size range of multicellular animals ranges from less than one tenth of a millimetre to many metres. A biological rule, Cope's rule, states that there is an evolutionary trend towards larger body size (see Benton 2002). Such a trend is documented in the fossil record of several craniotes and insects (e.g. Kingsolver & Pfennig 2004, van Valkenburgh et al. 2004), but the reversal trend has also been documented, for example for mammals on islands (Lomolino 1985). In some environments, body size can be divided into certain classes into which the majority of taxa fit. In the benthic marine environment, microfauna (mainly unicellular organisms, almost no metazoan animals), meiofauna (animals capable of moving in the interstitial system between sand grains), macrofauna (size range of a few centimetres), and megafauna (very large animals such as fishes and birds) can be distinguished (Reise 1985). The above classification concentrates on environmental factors, but body size is also strongly connected to factors such as food and others (see La Barbera 1986). There are many taxa that contain large as well as small species, so that it is certain that body size

changed numerous times during evolution. Instead of reviewing here the size for all animal taxa, I want to highlight a few topics concerned with size: first, is the fossil record biased towards large animals? Second, where do we have indications for size decrease by heterochrony? Third, why are many parasites large compared to their free-living relatives? Fourth, was the metazoan ancestor large or small? And fifth, was the bilaterian ancestor large or small?

The fossil record is composed to a large extent by macroscopic animals. Small animals are particularly known as phosphatized fossils that can be extracted, for example, from carbonic rocks by acid treatment. Large numbers of several species of small fossils have been found, for example, in the Cambrian 'Orsten' fauna (Müller & Walossek 1991, Walossek & Müller 1997). Recently, the finding of eggs and embryos has received attention (e.g. Zhang & Pratt 1994, Bengtson & Zhao 1997, Xiao et al. 1998, Xiao & Knoll 2000, Dong et al. 2004, 2005, Donoghue et al. 2006), these range from the late Precambrian (Neoproterozoic) into the early Ordovician. Donoghue et al. (2006) investigated why there is only a comparatively small time window in which such fossils have been found, and why the majority of the Cambrian fossils seem to belong to *Markuelia*, which is suggested to be a representative of the Scalidophora (Dong et al. 2005). It appears that on the one hand *Markuelia* had a very early onset of cuticle formation, which made it better suited for fossilization than other, soft-bodied animals and that, on the other hand, the environmental conditions for an optimal fossilization were available only during a short time period. This is a general phenomenon, because several circumstances such as tissue composition, sediment type, sediment chemistry, and others have to come together to allow fossilization, especially in soft-bodied animals (Butterfield 2003). Therefore, there indeed is a bias, and the majority of small animals are likely to be lost for our observation today. Such a 'cryptic evolution' is also suspected for the Precambrian. The more or less rapid onset of macroscopic fossils with a broad taxonomic diversity can hardly be explained by the origination of all metazoans at the Precambrian/Cambrian border (apart from the fact that there are at least some Precambrian metazoan fossils). Molecular clock calculations vary between 600 and 1,200 million years for the divergence of protostomes and deuterostomes (i.e. existence of the bilaterian ancestor) (Levinton 2001, Benton & Ayala 2003, Peterson & Butterfield 2005, Peterson et al. 2005). Although such calculations have to be treated with care (see, e.g., Bromham 2003), these data still suggest that an unknown number of soft-bodied and/or small species were present, but were not fossilized.

The occurrence of microscopic and macroscopic species within one taxon requires that one of these states is derived from the other. When the ancestor was small, it is likely that evolutionary size increase to macroscopic forms was a more or less gradual process. When the ancestor was macroscopic, microscopic forms can evolve in two ways, either by a gradual decrease of size (called miniaturization, see Hanken & Wake 1993) or by the gaining of sexual maturity in larval/juvenile stages (progenesis) (Fig. 3.1; see also Gould 1977). Evidence for such phenomenona is always given by phylogenetic analyses and a comparison with closely related species. Several examples can be found among polychaetes. The Dinophilidae, for example, are a group of small polychaetes that are likely to have evolved from dorvilleid polychaetes by a gradual series of reductions, starting with (to name some taxa) *Ophryotrocha*, over *Apodotrocha*, to *Dinophilus* and *Trilobodrilus* (Westheide 1985a, Eibye-Jacobsen & Kristensen 1994; Fig. 3.2.). Because there is a gradual reduction of parapodia and segment number, as well as a retaining of larval ciliary bands, progenesis seems to be responsible for the size reduction in these cases. In nerillids, however, a more gradual decrease in size without retaining of larval features appears to be present, which therefore represents a miniaturization (Worsaae & Kristensen 2005). The evolution of macroscopic taxa from microscopic ones has been made possible in the case of *Microphthalmus hamosus*. It is likely that in *Microphthalmus* a reduction in size towards

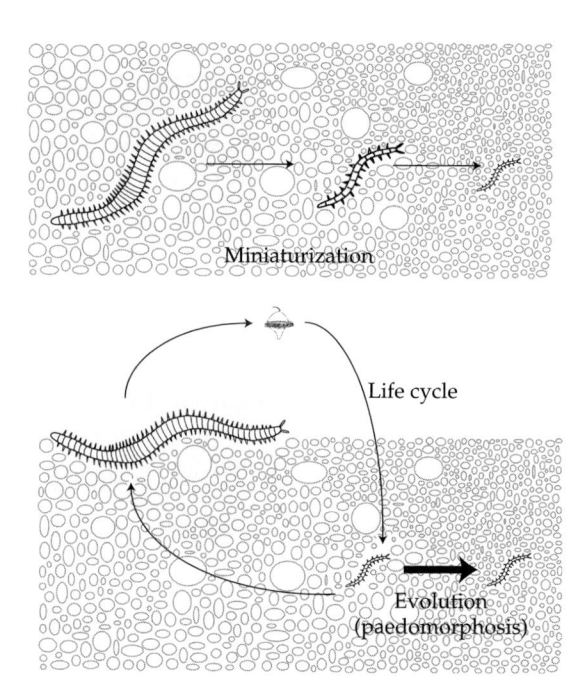

Fig. 3.1. Evolution of small body size from large-bodied ancestors by miniaturization (top) or progenesis (below). Miniaturization is the gradual evolutionary decrease of body size. In progenesis, a small developmental stage from a life cycle becomes sexually mature to form the adult of a new species during evolution. Drawing based on figures in Worsaae and Kristensen (2005).

microscopic species first occurred, during which chaetae were lost. In *M. hamosus*, an increase in size and segment number occurred, but the derivation from microscopic forms is indicated by the absence of chaetae (Westheide 1982).

There are other examples in which microscopic taxa likely evolved by progenesis or miniaturization: the interstitial cnidarian *Halamohydra*, the Bathynellacea among crustaceans, interstitial molluscs, and interstitial ascidians (Schminke 1973, 1981). Westheide (1987) assumes that progenesis is a common principle for the evolution of meiobenthic taxa, but for many taxa, any indication for this hypothesis (i.e. closely related, macroscopic species) is lacking (Gastrotricha, Nematoda, Kinorhyncha, Loricifera, Platyhelminthes, Gnathostomulida, Eurotatoria). Therefore it should be assumed that these taxa are primarily microscopic in size, i.e. that their ancestors were small. The impact of the phylogenetic position

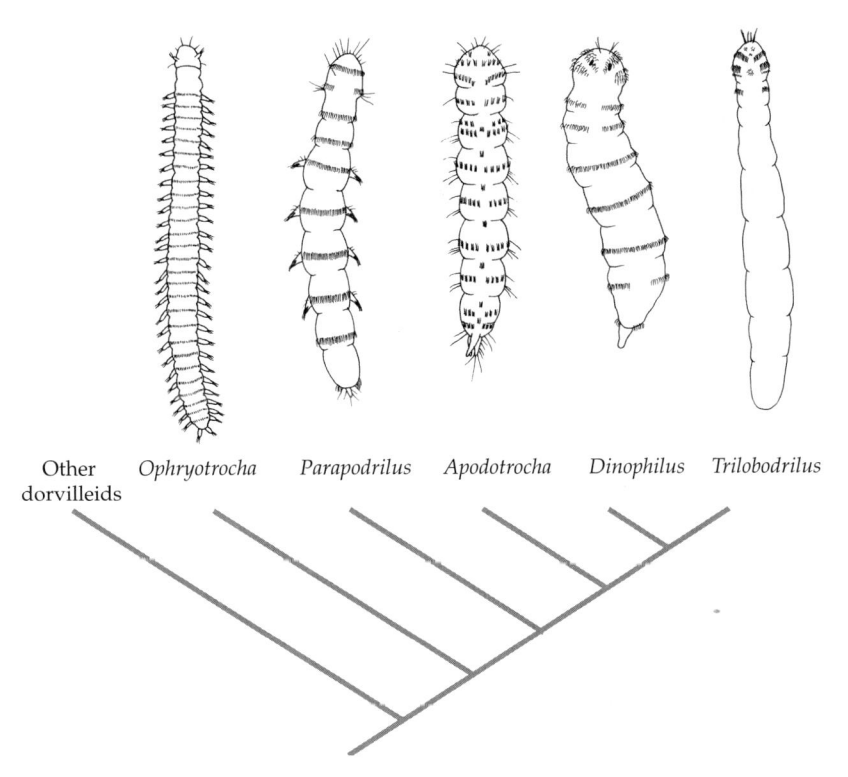

Fig. 3.2. Phylogenetic relationships of dinophilid species within Dorvilleida, showing reduction of parapodia and segment number. After Westheide (1985a), with figures after Dohle (1967), Westheide (1965, 1990), and Westheide & Riser (1983).

of a taxon for the interpretation of its body size being plesiomorphic or derived has been exemplified for tardigrades by Schmidt-Rhaesa (2001). When arthropods are considered as closely related to annelids, large body size appears to be a plesiomorphic condition and the microscopic size of tardigrades is likely to be a derived phenomenon. When arthropods are, however, related to cycloneuralians (see Chapter 2 for a discussion of these relationships), microscopic size in tardigrades can also be interpreted as a plesiomorphic character.

Many parasites are quite large, especially in comparison with closely related, free-living species. Examples are parasitic flatworms (Neodermata) versus free-living forms, acanthocephalans versus eurotifers, and parasitic nematodes versus free-living ones. In general, this evolutionary size increase in correlation with the parasitic mode of life appears to be due to an increased production of gametes. The habitat of parasites is temporarily limited and therefore the transmission from one host to the next is a central goal in parasite evolution. This has led to complicated life cycles and is always connected to the production of a high number of offspring, from which, nevertheless, only a small fraction is able to complete the life cycle. For example, it is estimated that a single female of the parasitic nematode *Ascaris lumbricoides* sheds at least 200,000 eggs per day, resulting in a lifetime production of more than 70 million eggs (Bush et al. 2001; Fig. 3.3.). Large quantities of gametes require large bodies and this is the main reason for the large body size of many parasites.

The reconstruction of the body size in two crucial stages of metazoan evolution, that is, in the metazoan and in the bilaterian ancestor, is very difficult. Because hypotheses on early metazoan evolution started from unicellular organisms, which likely aggregated into oligocellular complexes, a small size of this hypothetical ancestor

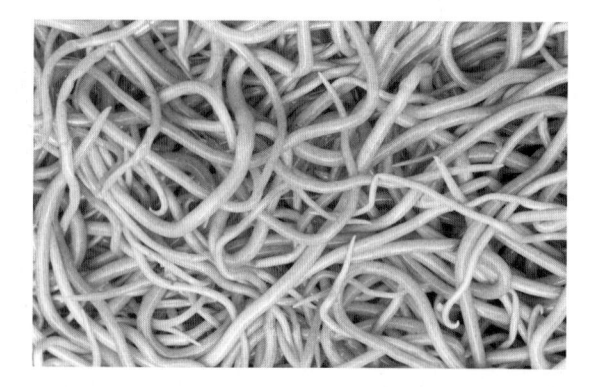

Fig. 3.3. *Ascaris suum*, a large parasitic nematode. The individual animals measure 15–25 centimetres.

can be assumed. As all basal metazoan taxa include macroscopic individuals (*Trichoplax* is somewhat intermediate), a size increase early in metazoan evolution has to be assumed. However, basal metazoans do not only have large individuals, they also have microscopically small larvae in their life cycle (together named the biphasic life cycle), which opens different possibilities for the evolution of size. The biphasic life cycle could be homologous in sponges and cnidarians (Rieger 1994a, b), indicating a single evolutionary event. When the biphasic life cycle evolved independently, the question is whether larvae or adults are comparable between sponges and cnidarians (Maldonado 2004). For the bilaterian ancestor, things are much more complex (see Rieger et al. 1991a for a brief overview). Several authors reconstruct the body size as microscopically small, mainly because many basal bilaterian taxa are small and directly developing (e.g. Ax 1996, Baguñá et al. 2001, Valentine 2004). Dewel (2000) proposes that the bilaterian ancestor was a large, colonial organism, roughly resembling pennatulaceans. Rieger (1994a, b) suggests that the biphasic life cycle is not only homologous between sponges and cnidarians, but also with those bilaterian taxa where it occurs. A conclusion is hard to draw. Among protostomes, many basal taxa (Gastrotricha, Nematoda, Kinorhyncha, Loricifera, Platyhelminthes, Gnathostomulida) have a small body size (at least in their basal representatives), suggesting

that small body size is ancestral. Deuterostomes have a biphasic life cycle (in echinoderms, enteropneusts, and most tunicates). This leaves all speculations open about which of these was the ancestral condition (see also Budd & Jensen 2000).

Symmetry

There seem to be three basic symmetrical properties (including here absence of symmetry) present in metazoan animals which appear subsequently during evolution. Adult sponges (Porifera) do not show any symmetry. Most cnidarians have a radial symmetry and all Bilateria have, as their name says, a bilateral symmetry (Fig. 3.4.). But this, of course, is only the rough picture. Within Bilateria, symmetry can be altered. The best known examples are echinoderms, which as adults have pentaradial symmetry (Fig. 3.4.). Only as larvae is the bilateral symmetry evident. Some species, such as the parasitic crustacean *Sacculina carcini* (Cirripedia), develop in the postlarval stage into an amorphous mass of tissue without any symmetry. Some parts of organisms can become radially symmetrical, such as the introvert in Scalidophora (see, e.g. Adrianov & Malakhov 2001a, b for priapulids). On the other hand, bilateral symmetry is not restricted to the taxon Bilateria. The symmetry of ctenophores is termed biradial symmetry and this includes the bilateral plane of symmetry. Among cnidarians, anthozoans (coral polyps and relatives) have bilateral symmetry (Fig. 3.4.), as well as the planula, the cnidarian larva. This makes it possible that bilateral symmetry evolved earlier than in the bilaterian ancestor. This was already assumed by Jägersten (1955, see also 1959) in his 'bilaterogastraea-hypothesis', in which a hypothetical blastula-like stage very soon turned into a benthic, bilaterally symmetrical animal (Fig. 3.5.). One argument for this scenario, used by Jägersten (1955) and by Remane (1963), is that coelomic cavities developed from gastric pockets in anthozoan polyps (see Chapter 8). This, however, requires that the bilaterian anteroposterior axis developed from an axis perpendicular to the

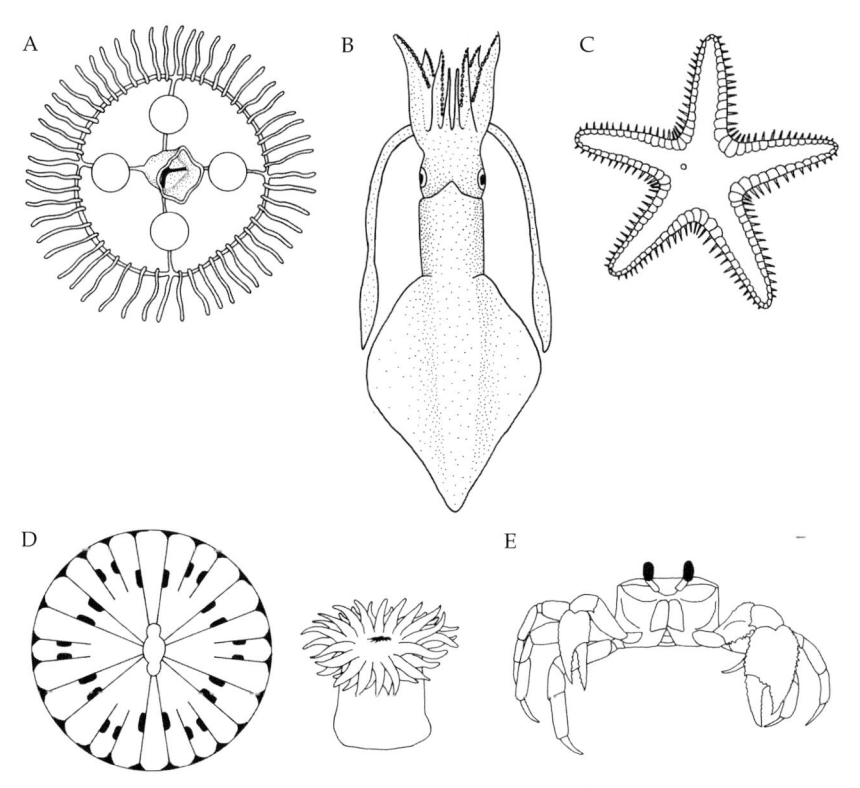

Fig. 3.4. Symmetrical properties in metazoans: A. Radial symmetry (here in the hydromedusa of *Laomedea geniculata*; after Westheide & Rieger 2007). B. Bilateral symmetry (here in the squid *Loligo forbesii*, after Hayward & Ryland 1995). C. Pentaradial symmetry (here in the asteroid *Astropecten* sp.; after Storch & Welsch 1997). D. Bilateral patterns in the internal anatomy of anthozoans; shown is the distribution of septae in schematic cross section through an actinarian polyp (after Tardent 1978). E. In many cases, certain asymmetrical patterns are superimposed on the bilateral symmetry, for example in the different claw size of the ghost crab *Ocypode quadrata* (Crustacea, Decapoda) (after photo in fieldguide).

oral-aboral axis of the polyp (or, simply spoken, that the bilaterian ancestor evolved from something like an anthozoan polyp starting to crawl sideways). This does not appear very likely, because the bilaterian anteroposterior and the cnidarian oral–aboral axis appear to be homologous (see below) and because it is not likely that a coelom was present in the bilaterian ancestor (see Chapter 8). There is some indication that symmetry depends on the lifestyle of the animal, with sessile lifestyles supporting radial symmetry and mobile lifestyles supporting bilateral symmetry (e.g. Remane 1958, Martindale et al. 2002). As a consequence, bilateral symmetry appears to be no strong character for phylogenetic implications and should be supplemented by further characters when using it as an important autapomorphy of Bilateria (e.g. with the Hox gene content, see below).

The anteroposterior axis

The anteroposterior axis (AP axis) is essential to create bilateral symmetry. It distinguishes an anterior and a posterior end and a region in between. In bilaterian animals, a certain group of genes is of fundamental importance for the establishment of the AP axis, the Hox genes (McGinnis & Krumlauf 1992). These belong to a group of genes including a characteristic nucleotide sequence, the homeobox, which codes for a protein homeodomain that acts as a DNA binding site. Homeobox genes are transcription factors and therefore important during

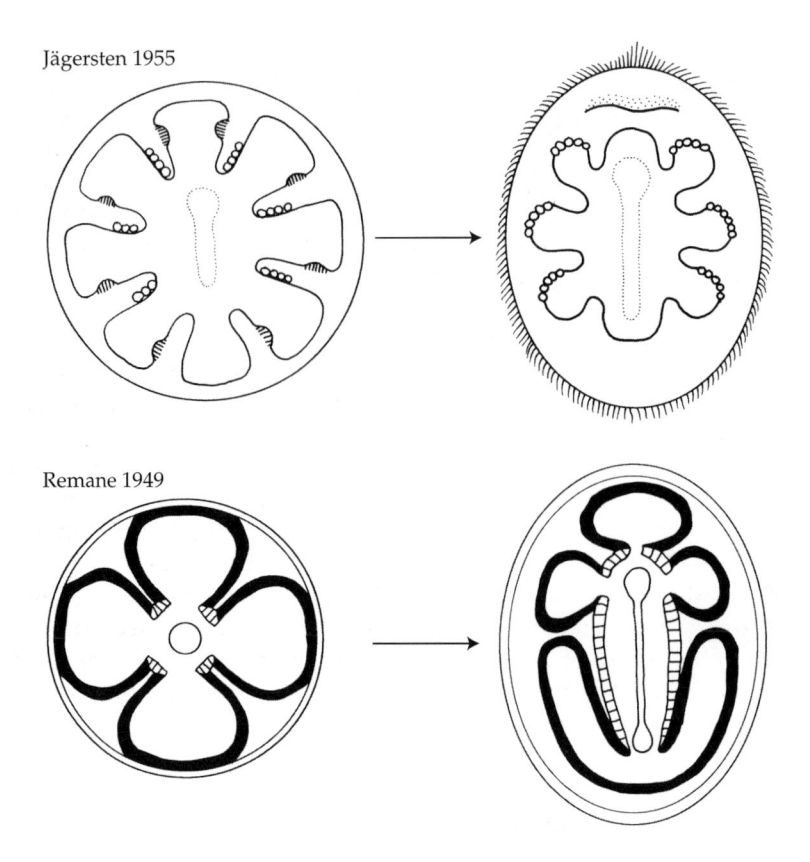

Fig. 3.5. Derivation of bilaterians from anthozoan polyps, according to drawings after Jägersten (1955, the bilaterogastraea-hypothesis) and Remane (1950). To the left are cross sections through anthozoan polyps, to the right are sagittal sections through hypothetical bilaterian ancestors.

the embryonic development of the body organization. Homeobox genes are very old and present in all eukaryotes (Bharathan et al. 1997), they can also be involved in body patterning, such as for example the gene *wariai* in *Dictyostelium* (Han & Firtel 1998). Characteristic for a number of Hox genes is that several genes are arranged in a certain order, the Hox cluster, and are expressed in the same order (colinearity of gene order and expression). Hox genes are divided into three groups (anterior, central, and posterior) and have now been found in many bilaterian taxa. This makes it likely that the ancestral Hox cluster of the bilaterian ancestor contained at least seven Hox genes (Balavoine et al. 2002). Four peculiarities should be mentioned. First, it was puzzling that nematodes seemed to have few Hox genes compared to other bilaterians. This turned out not to be true for all nematode species, but only for *Caenorhabditis elegans* (Aboobaker & Blaxter 2003a, b). Craniotes have four Hox clusters, which likely originated by two duplications of the whole Hox cluster in the craniote ancestor or even later, because in *Branchiostoma* (Acrania), there is only one Hox cluster (Garcia-Fernàndez & Holland 1994, Holland et al. 1994, Holland & Garcia-Fernàndez 1996). With the transition from bilateral to pentaradial symmetry in echinoderms, Hox genes appear to lose their conserved function in AP axis patterning and become more variable (Lowe & Wray 1997). Finally, Cook et al. (2004) and Baguñà & Riutort (2004b) investigated the Hox gene content of acoel species and found only three or four genes. Compared to the assumed content of at least seven Hox genes in the ancestral bilaterian Hox cluster, the low gene content in acoels supports their basal position as a potential sister-group of the remaining Bilateria (see Chapter 2).

Besides the Hox cluster, further Hox genes have been found. Because these also seem to be arranged in a colinearly expressed cluster, this cluster has been named the ParaHox cluster (Brooke et al. 1998) and is a paralogue of the Hox cluster. But how old are Hox and ParaHox clusters? In cnidarians, a number of Hox genes homologous to anterior and posterior Hox genes and ParaHox genes in bilaterians have been found in several cnidarian species (see Garcia-Fernàndez 2005 for a summary, but see Schierwater & DeSalle 2001 and Schierwater et al. 2002 for differing conclusions). In both ctenophores (*Cteno-Hox1*, Finnerty et al. 1996) and *Trichoplax adhaerens* (*Trox-2*, Schierwater & Kuhn 1998, Monteiro et al. 2006), so far only one Hox gene has been identified. Sponges possess several homeobox genes (Manuel & Le Parco 2000), but probably none of them is a Hox gene (one potential candidate, reported by Degnan et al. 1995, has a high sequence similarity to a tunicate Hox gene and was suspected by Holland 2001 to be a contamination). On the basis of these findings, it has been hypothesized that there was a hypothetical ancient proto-Hox gene, which created by duplication a proto-Hox cluster which, by duplication of the whole cluster, gave rise to the Hox and ParaHox clusters, originally consisting of anterior and posterior Hox genes. The central class of Hox genes evolved in the bilaterian ancestor, and within Bilateria a diversification of the clusters occurred (Finnerty & Martindale 1999, Gauchat et al. 2000, Ferrier & Holland 2001, Holland 2001, Garcia-Fernàndez 2005; Fig. 3.6.). While the first two steps are hypothetical, the last step likely occurred in a common ancestor of Cnidaria and Bilateria. The poor Hox gene content of ctenophores would support

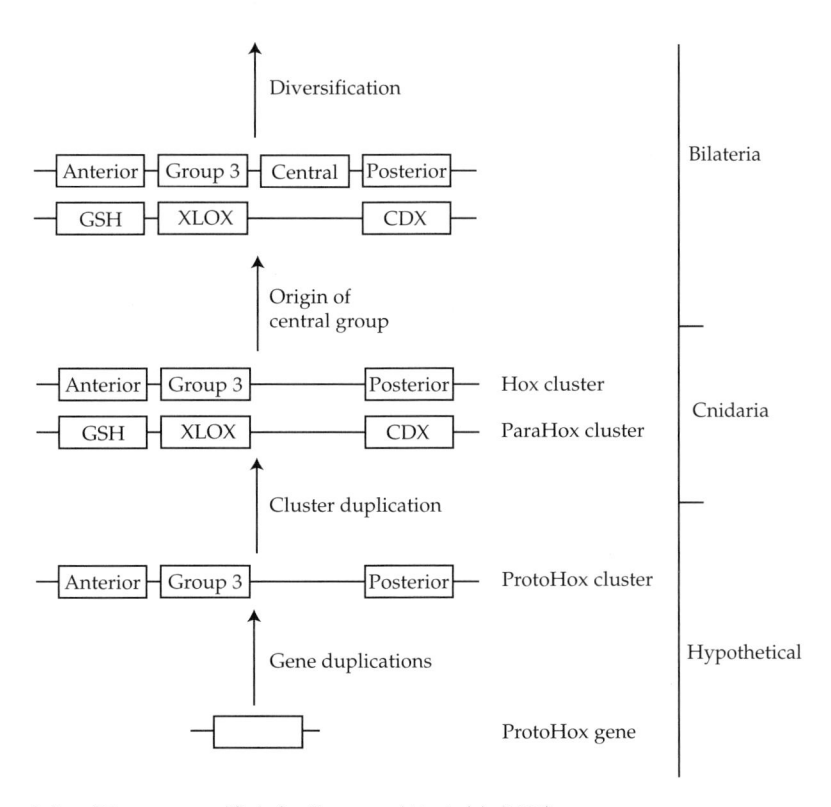

Fig. 3.6. Probable evolution of Hox genes, modified after Finnerty and Martindale (1999).

hypotheses that cnidarians and not ctenophores are closely related to bilaterians (see discussion in Chapter 2).

The cases of bilateral symmetrical patterns in non-bilaterian animals (see symmetry) make it interesting to know whether Hox genes also pattern axes in these cases. In *Nematostella vectensis* (Anthozoa, Cnidaria), Finnerty (2003) and Finnerty et al. (2004) observed localized expression patterns of five Hox genes. Three of the expressed Hox genes in *Nematostella* are homologous to anterior Hox genes of bilaterians and two to posterior genes. The anterior homologue *anthox6* is expressed in the pharyngeal endoderm and the posterior homologue *anthox1* in the aboral tip. The other three are expressed more or less in parallel along the polyp trunk. These expression patterns suggest that the oral–aboral axis of anthozoan polyps is homologous to the AP axis of bilaterians (Martindale 2005). The additional asymmetrical expression of *decapentaplegic* (*dpp*) in the pharyngeal region of *Nematostella* may indicate that there also is a precursor of the dorsoventral axis, because *dpp* is an important gene in dorsoventral patterning (see below).

The dorsoventral axis

Besides the anteroposterior axis, bilaterians have a second axis: the dorsoventral axis. Ventral is usually defined as the side of the organism that faces the substrate (or, in burrowing or swimming larvae, the comparable orientation in relation to the sediment surface). The terms ventral and dorsal become difficult to apply when during the lifecycle of an animal metamorphic shifts make it difficult to homologize larval and adult sides of the body. This happens, for example, in echinoderms where the adult buds from one region of the larva. Even if adult echinoderms have a side facing the substrate, this did not originate from the larval ventral tissue and therefore the terms 'oral' and 'aboral' are better suited in the case of adult echinoderms.

It appears that there is a common pattern of dorsoventral axis formation, comparable to the Hox-genes in the anteroposterior axis (De Robertis & Sasai 1996, Gerhart 2000). This involves, in the embryonic ectoderm of *Drosophila*, the genes *decapentaplegic* (*dpp*) and *short gastrulation* (*sog*), in craniotes their homologues *bone morphogenetic protein-4* (*BMP-4*) and *chordin* (*chd*). However, the expression of these genes is inverted in comparison between *Drosophila* and craniotes, with dorsal *dpp* and ventral *sog* expression in *Drosophila*, but dorsal *chd* and ventral *BMP-4* expression in craniotes (Arendt & Nübler-Jung 1994, Jones & Smith 1995). This suggests an inversion of the dorsoventral axis in comparison of these two taxa (see Nübler-Jung & Arendt 1994). The question now is, when exactly did such an inversion happen and how is it performed?

In comparing entropneusts with chordates, especially with *Branchiostoma* (Acrania), it appears that the dorsal epibranchial ridge in the pharynx is homologous to the ventral endostyl in the pharynx of *Branchiostoma* and, vice versa, the ventral hypobranchial ridge in enteropneusts is homologous to the dorsal epibranchial groove in *Branchiostoma* (Nübler-Jung & Arendt 1999, Ruppert et al. 1999). This implies a homology of the ventral nerve cord of hemichordates with the dorsal neural tube of chordates, rejecting ideas that a short hollow part of the nervous system, the dorsal collar cord, is homologous to the neural tube due to its hollow character (see also Ruppert 2005; Chapter 6). Lowe et al. (2006) point out that the development of the (diffuse rather than centralized) nervous system in enteropneusts is not affected by different levels of *BMP*-expression and they conclude that the deuterostome ancestor did not use the molecule patterning of the dorsoventral axis (i.e. the *BMP-chordin* gradient) to also pattern the nervous system. The homology of other structures, such as the heart and the stomochord/chorda dorsalis between enteropneusts and chordates is controversial. Due to similarities between the heart (better heart-glomerulus-complex) of hemichordates, and the axial hemal vessel of echinoderms (see Chapter 10); and due to structural, positional, and functional differences between the stomochord of hemichordates and the chorda dorsalis of chordates (Ruppert et al. 1999),

I agree in doubting the homology of these structures.

Why and how such an inversion occurred is quite speculative, but the clues to understanding it are likely to be found in the embryonic and larval development. Arendt and Nübler-Jung (1997) note that the sequence of prospective tissues along the animal–vegetal axis corresponds in polychaete and amphibian zygotes. A shift of the later dorsoventral axis from the early animal–vegetal axis is probably related with the mode of gastrulation. In one model, the shifting blastopore pushes the prospective neuro-ectoderm into different positons that later become ventral or dorsal (see Arendt & Nübler-Jung 1997). Another model, preferred by Arendt & Nübler-Jung (1997), starts from regarding amphistomy as the basal mode of blastopore fate. Amphistomy means that the blastopore becomes slitlike and the margins fuse in the middle, resulting in a separated anterior and posterior opening, which later becomes the mouth and anus. This mode is realized in some taxa (certain polychaetes, onychophorans), but in many protostomes the blastopore becomes only the mouth while the anus breaks through secondarily (which would be explained by a unilateral closure of the slitlike blastopore in the amphistomy model). In deuterostomes, a similar blastopore closure leads to the anus developing from the blastopore while the mouth breaks through secondarily. Because the blastopore is located in the neural region, that is, in tissue becoming nerve tissue, the mouth (and mostly anus) in protostomes, and the anus in deuterostomes, are on the neural side, whereas the mouth in deuterostomes breaks through in non-neural tissue. If the animal tends to project the mouth towards the substrate, it has, simply spoken, to turn the whole body over. The problem with the correllation of gastrulation and dorsoventral inversion is, however, that the inversion was expected to happen in the stem lineage of deuterostomes, but, as explained above, it took place much later, in the stem lineage of chordates.

Van den Biggelaar et al. (2002) propose another solution to the problem. The blastopore originates in protostomes and deuterostomes in the same location within the embryo, but the subsequent development differs. In deuterostomes, the embryonic anteroposterior axis is extended, pushing the blastopore to the posterior end of the embryo. In protostomes the initial anteroposterior axis does develop more irregularly, including the migration of originally ventral cells to the dorsal and lateral side. This embryonic migration of cells from ventral to dorsal (and from dorsal to ventral) can explain the different expression patterns of dorsoventral-specific genes.

Lacalli (1996) and Ruppert et al. (1999) point out that basal deuterostomes do not show a marked preference for a certain axis. Larvae of echinoderms and hemichordates often spiral as they swim (Lacalli 1996) and the larval axial properties are confused in adult echinoderms and ascidians. Adult acranians, finally, have an inverted position in the sediment, with their ventral side (as corresponding to the larva) showing up. These are not satisfying answers, but they vaguely support the idea that an axial inversion appears to have happened in the chordate stem lineage (see Ruppert 2005 for coding this character as a potential chordate autapomorphy).

Germ layers

The tissues forming an animal body can be divided into three groups according to their embryonic source, and are named the germ layers: ectoderm, endoderm, and mesoderm. While sponges are problematic in this respect, cnidarians and ctenophores possess only two of these germ layers, ectoderm and endoderm (Fig. 3.7.), and are therefore often collectively named the diploblastic animals. The mesoderm is present in Bilateria, therefore they are sometimes called triploblasts. Ectoderm and endoderm are usually organized as epithelial layers, while mesoderm can be epithelial or a compact, three-dimensional tissue. This has fundamental implications for the organization of diplo- and triploblastic animals, with diploblastic taxa having an essentially epithelial organization and triploblastic taxa having a more solid, voluminous tissue organization. Body size is not affected by the number of

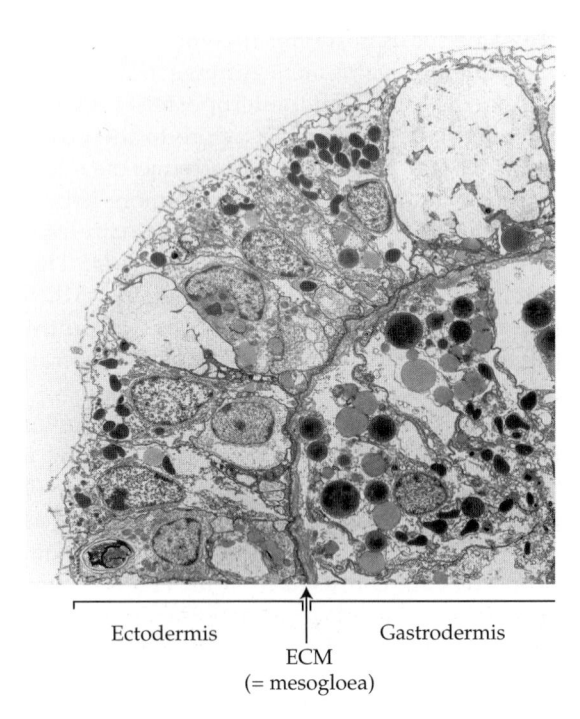

Fig. 3.7. Diploblastic organization of the planula larva of *Hydractinia echinata* (Hydrozoa, Cnidaria); transmission electron micrograph.

germ layers present, as cnidarians and ctenophores can grow to a large size. A compact organization, however, creates several transport problems (nutrients, oxygen, excretes) in the bilaterian body, for which special organs had to develop.

Germ layers develop during embryogenesis, and the distinction into different germ layers is useful to determine the source of a tissue during development. I will focus here briefly on the development of germ layers and then discuss the 'roots' of mesoderm in diploblastic animals.

The first cleavages in many zygotes lead to a hollow sphere, where a cavity, the blastocoel, is surrounded by one layer of cells, the blastula. Therefore, the external epithelium, the ectoderm, is the first to form. In some sponges (Fell 1997), cnidarians (Martin 1997), and nemerteans (Henry & Martindale 1997), the sphere is filled with cells and therefore contains no blastocoel. This is probably derived from the hollow blastula (the coeloblastula), either by the ectodermal cells becoming so large that they touch each other basally, or by an invasion of cells from the epidermal layer to the interior. The next process is the formation of the intestinal tract. In many cases, the coeloblastula invaginates in one position (the blastopore) and the invaginated epithelium becomes the endoderm, surrounding the primordial intestine, the archenteron. This process is called gastrulation. However, such an invagination gastrula is not always realized and the archenteron can be formed in a number of different ways. For example, in cnidarians, cells can migrate into the blastocoel either in one particular region (uniplar ingression) or in many places (multipolar ingression), followed by a re-arrangement of immigrated cells into the endodermal epithelium (Fig. 3.8.). The ectodermal epithelium can also subdivide and give off the endodermal epithelium by delamination. When a sterroblastula is present, internal cells can re-arrange into the endodermal epithelium, as in the case of cell ingression (see Tardent 1978 and Martin 1997 for summary; Fig. 3.8.). The same diversity of processes is also present in sponges (see Maldonado 2004 for review), but here no archenteron is formed. The exception are demosponges from the taxon Halisarcida, in which an invagination is formed (in addition to multipolar ingression), which nevertheless appears to be present only for a certain time (Maldonado 2004). Among others, von Salvini-Plawen and Splechtna (1979) regard the division into an outer epithelial layer and internal cells as basically comparable between sponges and cnidarians (but see Ereskovsky & Dondua 2006 for a different view) and therefore argue that the development of diploblasty and the development of the archenteron are two different processes, which only in cases of the invagination gastrula are coincident (see also Jenner 2006).

Mesoderm can also originate in a variety of modes. The terms ectomesoderm and endomesoderm describe the sources of mesoderm from either of the two other germ layers. Mesoderm can originate in an epithelial mode by the separation of spherical cavities from the (undifferentiated) endoderm (enterocoely) or by the ingression of cells between ecto- and endoderm

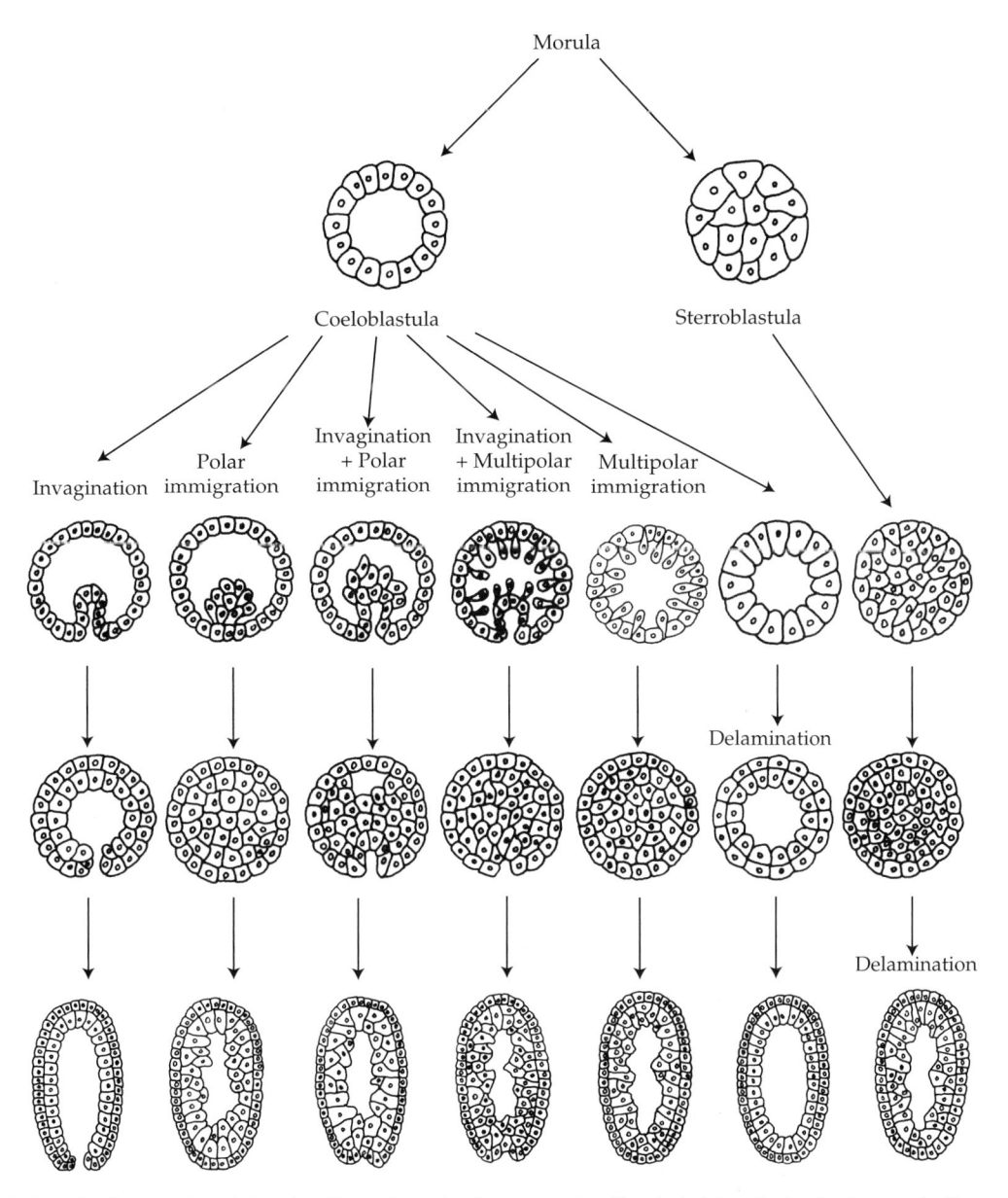

Fig. 3.8. Diversity of patterns in cnidarians, by which a bilayered embryo can develop. These include invagination, immigration of cells, combinations of these two, or delamination (after Tardent 1978).

(the mesenchymal mode, see Technau & Scholz 2003). The distinction between ecto- and endomesoderm rests on the assumption that endomesoderm contributes to the majority of mesoderm and is therefore the 'true' mesoderm (Hyman 1951). Such views underestimate the contributions of ectomesoderm to the formation of organs (see Ruppert 1991a for examples) and it appears questionable whether the different sources have any importance for the further development of the mesoderm. Ruppert (1991a) therefore argues against an ontogenetic definition of the mesoderm, but in favour of a topological definition, calling anything mesoderm that is positioned between ectoderm and endoderm. Within Bilateria, mesoderm development is partly conserved in spiralians, where the 4d blastomere, called the mesentoblast, gives rise to the entire endomesoderm, with differences only in its temporal development (van den Biggelaar et al. 1997, Boyer & Henry 1998). Ectomesoderm plays a role in larvae as well as in adults, and the dual origin of mesoderm (as ecto- and endomesoderm) appears to be an ancient pattern of the spiralian ancestor (Boyer et al. 1996, Boyer & Henry 1998).

Ruppert (1991a) claims that all metazoan animals, that is, including the 'diploblastic' ones, are in fact triploblastic. In poriferans, *Trichoplax* and cnidarians, single cells such as amoebocytes, archaeocytes, muscle cells, or others can be found below the pinacoderm (in sponges), between upper and lower epithelium (in *Trichoplax*) or between the epidermis and gastrodermis (in Cnidaria) (see also Hyman 1940). In comb jellies (Ctenophora), syncytial muscle fibres traverse the ECM between epidermis and gastrodermis (Hernandez-Nicaise & Amsellem 1980, Hernandez-Nicaise 1991). This indicates that another aspect of the term mesoderm is not properly defined: can single cells already be regarded as mesoderm? The name meso*derm* already suggests that it should be restricted to tissues, that is, to complexes of cells. Additionally, one particular structure in cnidarians has been regarded as mesodermal tissue: the entocodon (e.g. Hyman 1940) or Glockenkern (e.g. Tardent 1978) of hydrozoan medusae. In several species of Hydrozoa, medusae bud off directly from either the polyp or, exceptionally, from the medusa (for example, in the anthomedusa *Rathkea octopunctata*, budding is possible from the polyp and from the manubrium of the medusa; Fig. 3.9.). During this process, ectodermal material, the entocodon, is separated from the epidermis and lies between the epidermis and gastrodermis

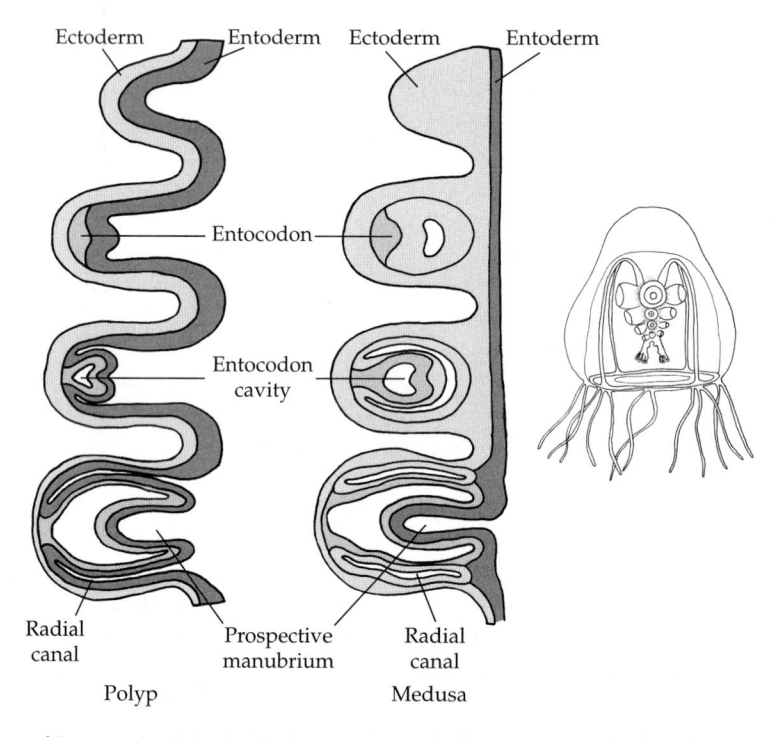

Fig. 3.9. The development of the entocodon during budding in the hydrozoan *Rathkea octopunctata* (whole medusa to the right), either through budding in the polyp (left) or in the medusa (centre). Ectoderm, entoderm, and their derivates are shaded in different grey-scales. Note that radial canals can develop from different sources. Modified after Tardent (1978) and Werner (1984).

(see, e.g., Glätzer 1971, Boelsterli 1977). This entocodon develops a cavity by schizocoely (the development of an epithelial layer around a cavity from a compact mass of cells; see Chapter 8). Eventually, the entocodon cavity opens and the epithelium forms the subumbrellar ectoderm of the new medusa. In several hydroids, medusae remain attached to the polyp and show different degrees of reduction. This includes the persistence of the entocodon cavity (i.e. it does not open to the exterior) and in such cases it can be used for fertilization and for brood protection of developing embryos (for example, *Tubularia crocea*, Tardent 1978). Interestingly, the development of radial canals differs in *Rathkea octopunctata*, depending on whether the medusa buds off from a polyp or from a medusa (Tardent 1978). In the first case, the radial canals grow from an extension of the polyp gastrovascular cavity towards the bud and are therefore lined by endodermal epithelium (Fig. 3.9.). In the second case, radial canals develop independantly as blindly ending structures from ectodermal cell material and then gain contact to the prospective manubrium, which again is an outgrowth of the 'mother medusa' gastrovascular system. This is one example of how the same structure can be formed in different ways, utilizing tissues of different origin.

The entocodon has stimulated speculations on the origin of the mesoderm (Boero et al. 1998, Spring et al. 2002, Seipel & Schmid 2005) and poses the question whether entocodon and mesoderm are homologous structures. I suspect the answer is no. The entocodon is restricted to hydrozoans (and may be an autapomorphy, see Collins 2002) and it is only a transient stage towards an epithelial organization of the 'daughter medusa'. While hydrozoans were once candidates for a basal position within Cnidaria, it now seems much more likely that anthozoans occupy this position, due to their absence of medusae, the (plesiomorphic) circular mtDNA and the 'simple' cnidocysts without a stiff cnidocil (Schuchert 1993, Bridge et al. 1995). Therefore, the entocodon should be regarded as analogous and not homologous to the mesoderm (see also Collins et al. 2006 for a similar conclusion).

Another approach to characterizing mesoderm is through searching for genes known to be involved in mesoderm formation in bilaterian animals; and several such genes have been found to be expressed in cnidarians (Technau 2001, Technau & Scholz 2003, Martindale et al. 2004). It remains, however, questionable whether these genes are really 'mesoderm markers' that indicate homologous tissues. *Brachyury*, for example, is expressed during mesoderm formation in bilaterians and in cnidarians in the blastopore region, that is, where the ectoderm and endoderm meet and from where, at least in some taxa, mesodermal cells invade the blastocoel (Technau 2001). Technau (2001) suggested that the primary function of *brachyury* is the specification of the blastopore, probably by regulation of very basic cellular features like cell–cell interaction, cell adhesion, or cell motility (Technau & Scholz 2003). There are more genes involved in such basic cellular movements. In *Nematostella vectensis*, there is an ingression of cells at the blastopore during invagination, somewhat resembling the ingression of mesodermal cells in bilaterian taxa. These cells do not detach completely and re-arrange into epithelial tissue soon after (Kraus & Technau 2006). The transcription factors *Snail* and *Forkhead* are involved in this process (Fritzenwanker et al. 2004). This movement of cells from an epithelial context into a subepithelial position has been called the epithelial–mesenchymal transition (EMT) and is crucial for bilaterian development (Savagner 2005). It can be argued that the basic cellular mechanisms for mesoderm formations, but not mesoderm itself, were already present in cnidarians.

Segmentation

Repetitive patterns of organs or whole body regions are common phenomena in metazoan animals. In the most advanced form, as in annelids and arthropods, these repetitive patterns are realized in all organs, except for the intestine, and such an organization is generally called 'segmentation'. It is, however, very difficult to define or characterize segmentation in relation to other repetitive patterns, and it is difficult to decide whether segmentation occurred once or more than one time during evolution.

First of all, the terminology is not used in a standard manner. If one looks in older publications, the terms 'metamery' or 'merism' are used and the body parts resulting from a metameric organization are then referred to as 'segments' (e.g. Bateson 1894). To give a few examples of the confusing terminology, Hyman (1951) distinguished between 'true segmentation' and 'superficial segmentation', Clark (1980) between 'metameric' and 'pseudometameric' structures. Willmer (1990) uses 'segmentation' dogmatically for arthropods, annelids, and vertebrates, 'metamerism' for cestodes, and 'serial repetition' for the remaining cases. Now, 'segmentation' is either broadly used, in the extreme even covering the repetitive subdivisions of the scyphozoan polyp during strobilation (Kroiher et al. 2000), or it is restricted to only very complex repetitive patterns (e.g. Scholtz 2002). These examples show how difficult it is to define segmentation and to draw boundaries between different patterns. Budd (2001b) has pronounced that even very advanced segmental patterns could not occur suddenly during evolution, but must be derived from more inconspicuous patterns. Consequently, it can be expected that the transition between simple and advanced segmental patterns is not well defined. Let us now take a brief look at repetetive patterns in animals other than annelids, arthropods and chordates.

Repetitive patterns are very widespread. Some examples are external annulations (with repetitive attachment points for musculature) in some nematodes (Francis & Waterston 1991, Hardin & Lockwood 2004), repetitively arranged ovarian pouches in nematomorphs (Lanzavecchia et al. 1995), and external annulations as well as repetitive structure of the intestine in the nemertean *Annulonemertes minusculus* (Berg 1985). Flatworms, especially the free-living seriates and the parasitic forms, exhibit several examples of repetitive structures in the nervous system, excretory organs, intestinal diverticula, gonads, or copulatory organs (Lang 1882, Ax 1958, Clark 1980; Fig. 3.10.). Within cestodes, Olson et al. (2001) showed that the metameric organization of eucestodes (which is composed in particular of an external annulation and repetitive reproductive organs) evolves from a non-metameric organization as is present in Caryophyllidea, via a multiplication of reproductive organs as in Spathebothriidea, and a final separation of these repeated compartments into proglottids in Eucestoda. Von Salvini-Plawen (1969), in explaining repetitive structures in molluscs, especially in *Neopilina* and polyplacophores, showed that it is likely that the repetitive organization of one organ system has influence on other systems as well. He pointed out that in flatworms repetitive intestinal diverticula spatially restrict the distribution of dorsoventral muscles, which become fewer than in species without diverticula. Within molluscs, a comparable arrangement of dorsoventral muscles is found in Solenogastres and probably is the starting point for a further reduction in number in polyplacophores (16 pairs) and *Neopilina* (8 pairs). Especially in *Neopilina*, further organ systems such as gills, excretory organs, and gonads are repetitively arranged, but, although this was first taken as an indication of a common segmented ancestor of annelids and molluscs (Lemche & Wingstrand 1959, Götting 1980), it is more plausible to assume that repetitive patterns evolved within molluscs (Von Salvini-Plawen 1969, Clark 1980, Haszprunar 1996b, Friedrich et al. 2002, Wanninger & Haszprunar 2002). A very far-reaching repetitive body organization is present in kinorhynchs (Fig. 3.11.), as is expressed in the denomination of the body units as 'segments' by several authors (while others prefer the term 'zonite'). Here, integument, musculature, nervous system, and glands are repetitively arranged into 11 clearly separated body units (see Schmidt-Rhaesa & Rothe 2006).

Clark (1980) states that there is a general tendency in long and narrow animals to replicate organs and arrange them, for simple geometrical reasons, in a serial pattern. Replication in one organ may cause other organs to follow.

In annelids and arthropods, repetition is far-reaching and ideally includes an external annulation, a pair of coelomic sacs, a pair of metanephridia, a segmentally arranged musculature, a pair of ganglia, and one pair of

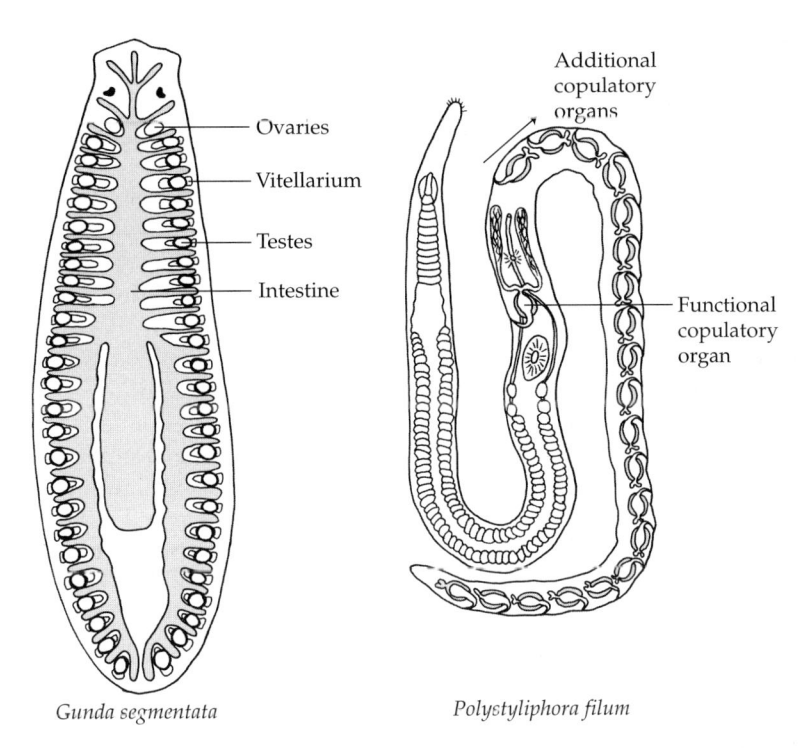

Fig. 3.10. Two examples for repetitive patterns in flatworms. In *Gunda segmentata* (Tricladida), vitellaria, testes, and intestinal diverticula are serially arranged (after Lang 1882), in *Polystyliphora filum* (Proseriata), the copulatory organs are serially arranged (after Ax 1958).

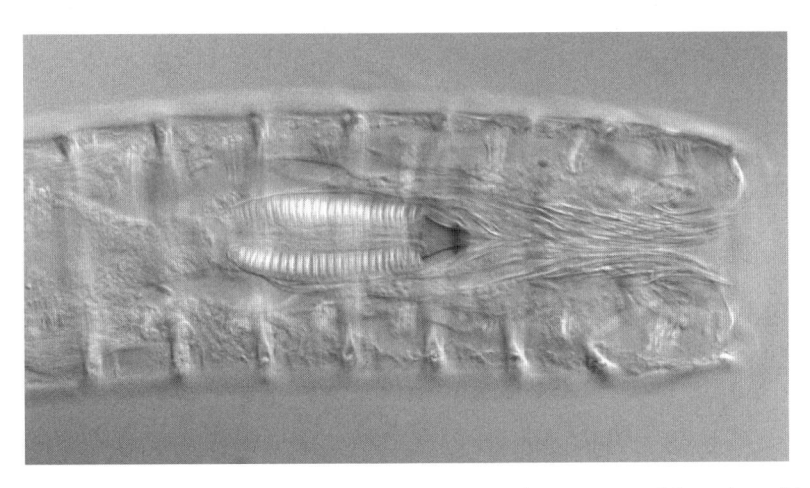

Fig. 3.11. The anterior eight trunk segments of the kinorhynch *Pycnophyes kielensis*. The introvert is withdrawn, but scalids and pharynx are in focus. Photo by Birgen H. Rothe & A. Schmidt-Rhaesa, Bielefeld.

appendages per segment (Scholtz 2002). It must be noted that such a complete list of segmentally arranged structures is only rarely realized. There are, for example, among annelids, several polychaetes in which the coelomic sacs are fused to varying degrees (e.g. Arenicolidae, Tomopteridae), the ventral nerve cords lack segmental ganglia (e.g. in *Enchytraeus crypticus*, Hessling & Westheide 1999), or appendages are lacking in several species. The difficulties in annelid systematics (see Chapter 2) make it hard to determine whether the ancestral annelid had the complete series of segmentally arranged structures. Among arthropods, only onychophorans show most segmental characteristics, although there are several outer annulations per segment and the commissures in the nervous system do not reflect segmental patterns (Schürmann 1995). Segments are formed in a posterior growth zone in both arthropods and annelids. During development, several developmental regulatory genes are expressed and are assumed to be involved in segment formation. These are in particular the segment polarity genes *engrailed* and *wingless* and pair-rule genes such as *even-skipped*, *hairy* and others. Especially in arthropods, the expression patterns appear to be very conserved (Davis & Patel 1999, Seaver 2003), while they are more variable in annelids. *Engrailed*, for example, is expressed in the ectoderm of arthropods, but in ectoderm and mesoderm in annelids (e.g. Lans et al. 1993 for *Helobdella triserialis*). Seaver (2003) summarized *engrailed* expression in annelids and concluded from expression in the nervous system and ablation experiments that *engrailed* probably has no role in segment formation in annelids. However, as these findings come mostly from clitellates, and as expression patterns in the polychaete *Platynereis dumerilii* appear to be comparable to arthropods (Prud' Homme et al. 2003), one may argue that annelids originally formed segments in the 'arthropod way', but altered this mode during annelid evolution.

Repetitive patterns are also present in Myomerata (Acrania + Craniota), especially in the musculature, coelom, and the nervous system. *Engrailed* is also expressed in craniotes (e.g. Patel et al. 1989).

On the basis of *engrailed* expression, it has been proposed that segmentation evolved in the bilaterian ancestor and was conserved in annelids, arthropods, and Myomerata (Kimmel 1996, De Robertis 1997, Balavoine & Adoutte 2003). This would mean that repetitive patterns in other taxa are more or less reductions from a quite complex segmentation, which is in contrast to, for example, explaining the occurrence of repetitive patterns in Platyhelminthes and Mollusca as evolved within these taxa. Furthermore, as the three segmented taxa occupy more or less derived positons in the tree, it appears questionable why a complex segmentation evolved (in the bilaterian ancestor) then got lost in several basal lineages and was conserved only in derived lineages. But how likely is the alternative explanation of a multiple evolution of repetitive and segmental patterns? I think it is quite likely.

The anteroposterior body axis is a characteristic of the bilateral body organization (see above). An anteroposterior axis means that during development an anterior and a posterior end must be defined and the tissue in between must also be regionalized. This may be done by repetition of information. It has been shown in models (Meinhardt 1986) and in practice (e.g. Carroll et al. 2005) that molecules can create iterated expression patterns, for example, through an expression series of gap genes, pair-rule genes, and segment polarity genes. I believe that such a regionalization of the AP axis is very important for the organization of bilateral animals. This may be reflected, for example, by the repetitive occurrence of embryonic motorneurons in nematodes (Walthall 1995) or in the repetitive occurrence of founder cells for the muscular grid in flatworms (Ladurner & Rieger 2000). If we regard basic molecular equipment for a repetitive regionalization of the AP axis as ancestral, this could be the basis from which further repetitive and more complex segmental patterns evolved several times in parallel. As an example, genes like *engrailed* are also present in non-segmented animals (e.g. in echinoderm mesoderm and nervous system, Byrne et al. 2005 and in the shell glands of molluscs, Moshel et al. 1998, Wanninger & Haszprunar 2001), showing that important segmentation genes are also present (with a different function) in non-segmented animals.

On the basis of these explanations, I regard a complex segmentation of the bilaterian ancestor as unlikely, and assume that repetitive patterns evolved several times in parallel (Fig. 3.12.)

```
┌─────────────────────┐   ┌─────────────────────┐
│ Multiple evolution of│   │ Multiple evolution of│
│  metameric patterns │   │  complex segmental  │
│                     │   │      patterns       │
└─────────────────────┘   └─────────────────────┘
     ↑   ↑   ↑   Within Bilateria  ↑   ↑   ↑
┌───────────────────────────────────────────────┐
│ Regulatory genes for creation and regionalization│
│         of the anteroposterior axis            │
└───────────────────────────────────────────────┘
              Bilaterian ancestor
```

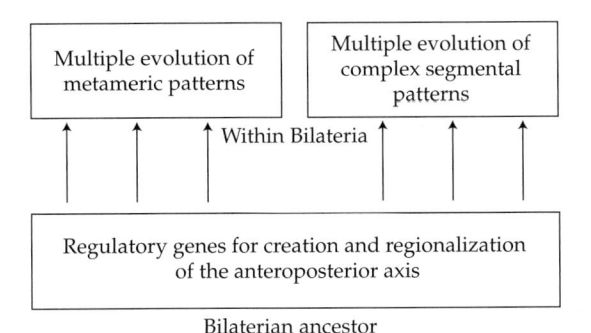

Fig. 3.12. Hypothesis explaining a possible convergent origin of 'complex' segmental patterns; see text for explanation.

(see Minelli 2003a for a corresponding conclusion). Such a statement does not solve the Articulata-Ecdysozoa conflict (see Chapter 2), but it may offer a starting point to explain how complex segmental patterns could evolve convergently.

Skeletons

The structures most commonly regarded as skeleton are the internal bones or the external cuticle of euarthropods, but these are only two examples of skeletal structures. From a general perspective, the most important structure for skeletal functions is the extracellular matrix (ECM). Within this matrix, crystals such as calcium carbonate, or hardening structures such as chitin or tanned proteins can be integrated to make the whole structure hard. Hardening of the external ECM, the cuticle, occurs to a varying degree in cycloneuralian taxa and in arthropods. The cuticular plates of kinorhynchs, the loriciferan lorica, and euarthropod tergites and sternites are prominent examples. Such skeletons are called exoskeletons. When hardening processes occur in the internal ECM, endoskeletons result. Endoskeletons with calcified components are present in craniotes and echinoderms (though very close to the surface). In non-calcified cases, the ECM includes rigid components and is usually combined with an internal pressure of the cells surrounded by the ECM. Such internal pressure is performed, equivalently to plant cells, by large vacuoles in cells. With ionic differences between vacuole and external medium, the pressure of the fluid within vacuoles can be raised. This gives the cell a certain stiffness, which can be optimized by surrounding the vacuolated cells with a rigid ECM. Vacuolated tissue is widespread, but often occurs only in few species of a larger taxon. Examples for vacuolated epithelia are some acoels (e.g. *Convoluta pulchra*, Ladurner & Rieger 2000; *Haplogonaria* sp., Ax 1966), gastrotrichs (*Macrodasys*, Teuchert 1978), flatworms (catenulids, Ehlers 1985b; *Cystiplex axi*, Ax 1966), polychaetes (*Hesionides*, Ax 1966), and the meiobenthic gastropod *Philinoglossa* (Bartolomaeus 1993a). Vacuolated tissue can also be present in the endoderm of several flatworms (Ax 1966), in pharyngeal cells of gymnolaemate bryozoans (Mukai et al. 1997), in larvae ('chordoid larva') and adults of *Symbion pandora* (Cycliophora, Funch 1996) and in tunicates (Burighel & Cloney 1997). When the ECM around a vacuolated cell is very prominent, cartilage (in craniotes and in the chorda of chordates) results. Cole & Hall (2004) list some criteria after which a tissue should be named cartilage. Examples for cartilage in invertebrates (or at least tissue coming very close to the definition) are the odontophore in gastropods such as *Busycon*, the branchial cartilage in *Limulus* (Chelicerata), cranial cartilage in cephalopods, tentacular cartilage in some polychaetes, and a lophophore cartilage in the brachiopod *Terebratulina* (Hall 2005).

Besides such intracellular vacuoles, fluid-filled body cavities perform skeletal functions. They function as cushions, which are not compressible, but deform in response to muscular pressure. Together with the muscular system, such cavities form important systems called the hydroskeleton. Apart from skeletal functions, muscular deformation of the cushion—and the resulting changes in shape—have important functions in locomotion (burrowing of the earthworm, eversion of introverts) or other movements such as the extension of tentacles (for example, in bryozoans). The hydroskeleton, in particular the segmentally arranged hydroskeleton, plays an important role in several evolutionary scenarios.

An animal does not always have the same type of skeleton during its entire life. During development, the type of skeleton can change. For example, during moulting in euarthropods, the new cuticular exoskeleton is not immediately stiff, but requires a certain amount of time, during which the animals usually are very vulnerable. Taylor & Kier (2003) have shown that in the decapod crustacean *Callinectes sapidus* the body shape (in particular of the chelipeds) is maintained during moulting by an increased internal fluid pressure, that is, by a hydroskeleton. Once the exoskeleton is hard, the internal fluid pressure decreases and the exoskeleton takes over the skeletal function.

Locomotory appendages

Locomotion in metazoan animals is very diverse and ranges from ciliary movement, undulating movements, peristalsis, eversion of body parts acting as anchors, to leech-like movements. Locomotory appendages occur in four groups of animals: echinoderms, craniotes, arthropods, and annelids. In echinoderms, the numerous tiny feet (which are lacking in crinoids) are extensions of the water-vascular system and moved by the fluid stream within this system. Some deep-sea holothurians may additionally develop foot-like extensions that allow a walking movement. In craniotes, two pairs of appendages are formed, which develop into fins, wings, or extremities. Both types of extremities evolved within their taxa (echinoderms and craniotes) and the basal representatives are lacking such locomotory appendages. Arthropods have either simple cone-shaped appendages such as in onychophorans, tardigrades, and fossil 'lobopods', or 'segmented' appendages as in euarthropods. It appears likely that the simple 'lobopod' with terminal claws is the plesiomorphic structure. In annelids, several polychaetes have appendages (named parapodia) that are often composed of two rami and include internal (acicula) and external chaetae. Such parapodia are reduced to varying degrees and are also completely lacking in several taxa (meiobenthic species as well as macroscopic ones, for example, the lugworm *Arenicola marina*). The question of whether parapodia belong in the ground pattern of annelids depends much on the phylogeny of the group, which is still not settled (see Chapter 2). Clark (1964) assumes from functional reasons that burrowing with the aid of coelom and musculature was ancestral and that parapodia evolved within the group. On the other hand, Westheide (1997) assumes that parapodia were present in the annelid ancestor and that they played an essential role in the development of segmentation. In the context of a close relationship between annelids and arthropods, it has been suspected that their common ancestor may have had appendages, from which both lobopodia/arthropodia as well as parapodia evolved (Lauterbach 1978). From the structure of both types of appendages, I see no reason for such comparisons. Lobopodia are ventrolateral extensions with claws, parapodia are lateral extensions with chaetae; therefore neither the position nor the structural composition agree with each other. The only common character of arthropod appendages and annelid parapodia is the expression of the gene *distal-less* (Panganiban et al. 1997). Such expression, however, is not restricted to annelids and arthropods, but also occurs in craniote appendages, echinoderm ambulacral feet, and the ampulla of tunicate larvae (a structure developed for attachment prior to metamorphosis). It has been proposed that all such appendages are homologous (Panganiban et al. 1997), but it appears more plausible that the *distal-less* is a regulatory gene that can be ubiquitously recruited when outgrowths with a proximo-distal axis are formed (see also Minelli 2003b).

Parasitism

It should be mentioned here briefly that a parasitic lifestyle can alter the general shape of the body considerably. As has been noted above (under 'size'), reproductive success is extremely important for parasites and, therefore, the reproductive system is often developed extraordinarily well. In contrast, there are several cases, in

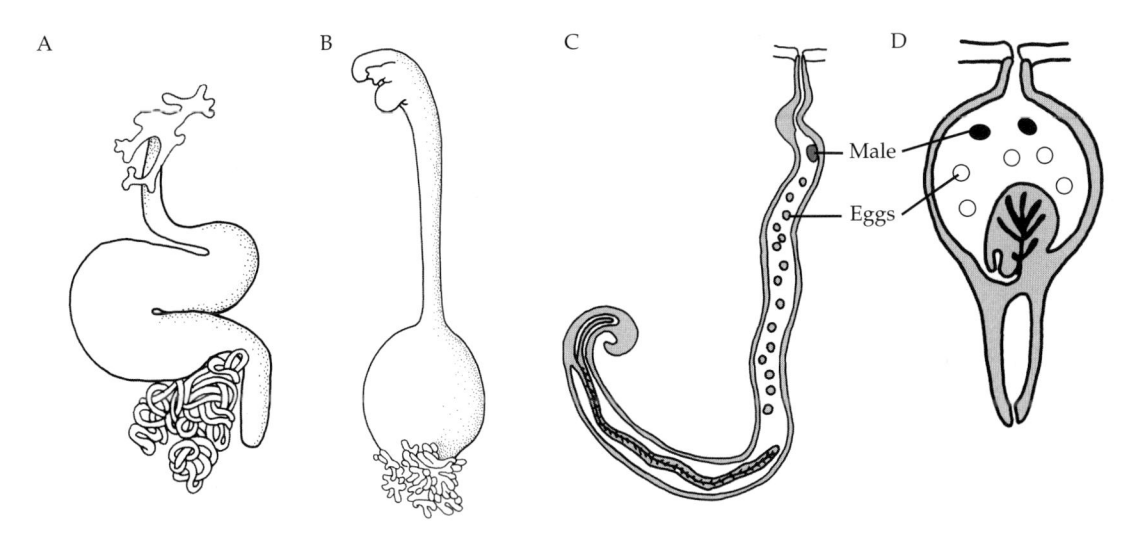

Fig. 3.13. Examples of strongly modified body organization in parasites. A. Female *Lernaeocera branchialis* (Copepoda). B. Female *Sphyrion lumpi* (Copepoda). C. Female and dwarf male of *Entocolax* sp. (Gastropoda). D. Female and dwarf male of *Thyonicola* sp. (Gastropoda) (after Lützen 1968, Westheide & Rieger 2007).

which the intestinal system is weakly developed (nematomorphs) or completely reduced (cestodes, acanthocephalans). In these cases, the parasites absorb nutrients over the body surface. Some parasitic taxa change their general shape of the body so much that it can be difficult to assign them to certain taxa: rhizochephalan cruscaceans, parasitic copepods (Crustacea; Fig. 3.13.), pentastomids (related to branchiuran crustaceans), and parasitic snails (e.g. Entoconchidae, see Lützen 1968; Fig. 3.13.). The parasitic taxa Myxozoa, Myzostomida, Dicyemida, and Orthonectida are still problematic concerning their phylogenetic position.

CHAPTER 4
Epidermis

The epidermis is the tissue that mediates between the external environment and the internal of the body. It always has the organization of a sheet of cells. In most cases, this sheet is a monolayer of cells, but several layers of cells can also be present. Epidermal cells are, as all epithelial cells, polarly organized and they are are able to secrete extracellular material both towards the inside of the body (basally) as well as towards the outside (apically). Secretions towards the outside are called glycocalyx and cuticle, while the term extracellular matrix (ECM) is restricted in the following to any kind of extracellular material basal of the epidermal cells (Fig. 4.1.). This ECM continues into the body (see Chapter 8), but in this chapter only the layer of ECM basal of the epidermal cells is treated.

The epidermis has many different functions. One is to keep the organism together and give it something like a distal border. Another function is to control the flow of substances, both from the outside inwards and from the inside outwards. The epidermis often has an important role in protection. Nutrients and oxygen can enter the body via the epidermis, excretes and CO_2 leave the body through it. The epidermis may play a role in osmoregulation, thermoregulation, and is an important location for sensory structures. Finally, epidermal structures may function as antagonists of the muscle system.

There is reason to believe that something like an epidermis was the first organ system to be found in metazoan animals. There are different models of how multicellular animals evolved from unicellular ones (see e.g. von Salvini-Plawen 1978, Rieger 1994b, 2003 and Rieger & Weyrer 1998 for reviews), but the most likely one is that cells remain attached to each other after cell division and in this way formed multicellular, probably spherical, compartments. Two things are important for this association of cells: a basal ECM on which cells attach and cell-cell junctions that keep cells attached to each other. Such spherical forms give the possibility of a separate internal compartment from the external surrounding, in which conditions can be controlled and maintained. This, in turn, is essential for further development.

Tyler (2003), in his review on epithelia, stressed the importance of epithelia for the whole body organization. Because epithelia are the first differentiations of cells during embryogenesis, epithelia have even been regarded as a 'default stage' of organization, which is, for example, important in the discussion of cancer (Frisch 1997). Tyler (2003) develops a model of four steps in epithelial evolution, based mainly on the development of the mammalian embryo

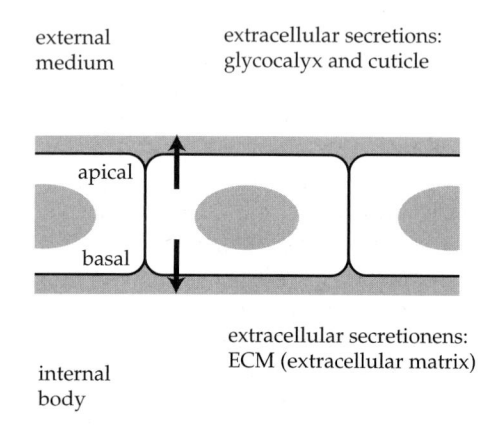

Fig. 4.1. Epithelial cells can secrete extracellular material to their apical and basal sides.

(Fleming & Johnson 1988): first, the development of cell polarity, second, the evolution of integrins that link the cells to ECM, third, the aggregation of cells into sheets using cadherins and other cell adhesion molecules, and fourth, the localization of cadherins and adhesion proteins in cell–cell junctions.

Cell–cell junctions

In general, there are three types of cell-cell junctions that all have a different function in the epidermal cell (see Fig. 4.2.). 'Anchoring junctions' function in stability and are also responsible for movements of epidermal cells during development. 'Occluding junctions' are diffusion barriers, and 'gap junctions' allow communication between cells.

Anchoring junctions are composed of transmembrane proteins and intracellular anchor proteins, to which elements of the cytoskeleton, either actin or intermediate filaments, attach. They can be present between cells or between a cell and extracellular matrix. When actin filaments are present, a cell–cell contact is called an 'adhaerens junction' and a cell–ECM contact a 'focal adhesion'. When intermediate filaments are present, the cell–cell contacts are called 'desmosomes' and the cell–ECM contacts 'hemidesmosomes' (terminology after Alberts et al. 2002). The transmembrane and anchor proteins cause a thickening in the cell membrane which can usually be well recognized in electron microscopical images. The distinction between actin and intermediate filaments, however, is not always made and the terminology explained above is not consistently applied in the literature.

In sponges (Porifera), no intermediate filaments have been found, but actin is a common cytoskeletal element (Pavans de Ceccatty 1986). Spot-like cell–cell junctions are present which represent, according to the definition above, adhaerens junctions (focal adhaerens junctions). They appear, for example, between exopinacocytes (Simpson 1984). In *Trichoplax adhaerens*, both intermediate filaments and actin are reported by Behrendt & Ruthmann (1986). Anchoring junctions are present and are connected to bundles of microfilaments (Ruthmann et al. 1986), but the nature of these filaments is not further stated. Because these junctions completely surround each epidermal cell on its apical side, they are probably comparable to (and were termed by Ruthmann et al. 1986) 'belt desmosomes', which occur in representatives of Eumetazoa (Fig. 4.3.). Intermediate filaments (and, of course, actin) occur, with the exception of arthropods, in all representatives of Eumetazoa (Bartnik & Weber 1989, Erber et al. 1998) and therefore, in these taxa both desmosomes and adhaerens junctions are present.

Occluding junctions appear in two forms, as 'septate junctions' in invertebrates (Fig. 4.4.) and as 'tight junctions' in craniotes. In tight junctions, transmembrane proteins of the protein family of claudins, especially occludin, directly interact with those from the neighbouring cell and therefore create a diffusion barrier both between cells and within the cell membrane. Therefore the flow of substances between cells is effectively regulated and the 'fluid' cell membrane is divided into a proximal and a distal compartment. Septate junctions in invertebrates are recognizable electron-microscopically by their ladder-like appearance. They include the

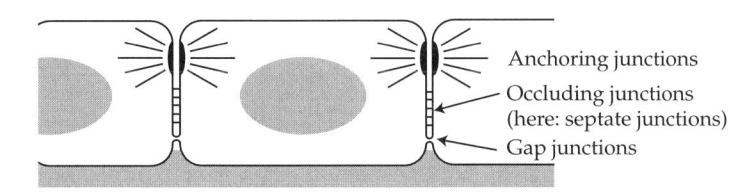

Fig. 4.2. The three types of cell–cell junctions. Anchoring junctions are often in an apical position in the epidermis, followed by occluding junctions.

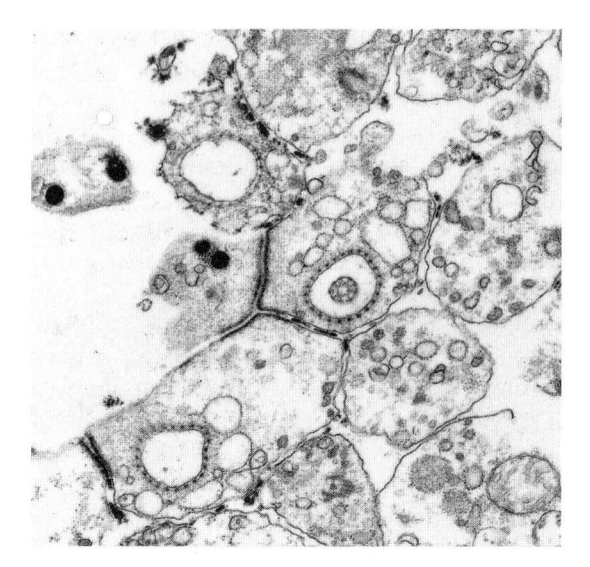

Fig. 4.3. Belt desmosomes in *Trichoplax adhaerens*. This section shows part of the belt desmosomes as bold lines running around three cells, the remaining part is out of the section plane.

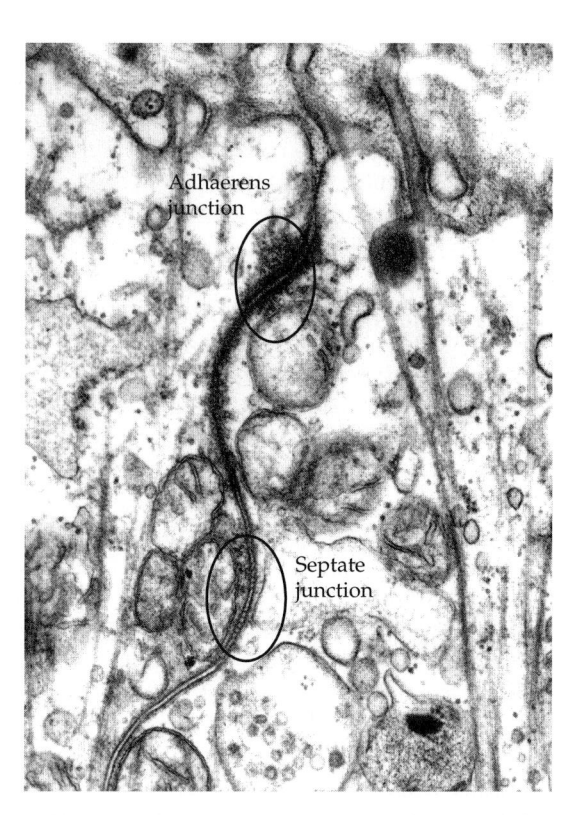

Fig. 4.4. Septate junctions in the epidermis of *Euclymene oerstedii* (Polychaeta, Annelida). Photo by courtesy of Andrea Tapp, Bielefeld.

transmembrane protein Neurexin IV and are linked in the cell to proteins such as Scribble, Lethal giant larvae, and Discs large (Knust & Bossinger 2002). Green & Bergquist (1982) suggested that tight junctions evolved from septate junctions by changes in a 'spacing factor' between cell membranes. Indeed, components of craniote tight junction are also present in non-craniote animals. Claudins have been found in protostomes (*Drosophila* and *Caenorhabditis*), probably localized in septate junctions (Asano et al. 2003, Behr et al. 2003). The cnidarian *Hydra* expresses a protein, HZO-1, which is homologous to a mammalian tight junction component, ZO-1 (Fei et al. 2000). On the other hand, only one particular craniote structure, the so-called paranodal glial-axonal junction at the nodes of Ranvier in myelinated nerve cells (Wiley & Ellisman 1980), is proposed as a homologue to septate junctions, because it expresses Caspr, a homologue to the *Drosophila* Neurexin-IV (Tepass et al. 2001).

Septate junctions occur (with the exception of craniotes) in all metazoans (Nielsen 2001) and tight junctions are an autapomorphy of craniotes. In the epidermis, anchoring junctions are often in a distal position, close to the cell surface and occluding junctions are also distal, but below anchoring junctions. In sponges, septate junctions are present, but their position is not in the epidermis. Instead, they are found between choanocytes and between spicule secreting cells (Ledger 1975, Nielsen 2001). In sponge larvae, septate junctions have not been found (Maldonado 2004).

Gap junctions connect neighbouring cells by, simply speaking, canals between both cells. These canals are composed of complexes of the transmembrane protein connexin, six of which form a connexon (see, e.g., Unwin 1987, Kumar & Gilula 1996). The six subunits of a connexon can shift their position relative to each other and so open or close a central lumen. Connexons from neighbouring cells are in register and their lumen therefore continues from one to the next cell. Gap junctions have only been found in Eumetazoa, for example in the taxon Cnidaria (Wood 1985), but not in *Trichoplax adhaerens* or in sponges (Nielsen 2001).

Therefore, different types of cell–cell contacts appear sequentially during early metazoan evolution, as far as phylogenetic relationships as presented in Fig. 2.5. are correct. The metazoan ancestor had focal adhaerens junctions and septate junctions. In the epitheliozoan ancestor, belt desmosomes evolved and in the eumetazoan ancestor, gap junctions were present. In the vertebrate ancestor, septate junctions were replaced by tight junctions (Fig. 4.25.).

Extracellular matrix (ECM)

The subepidermal extracellular matrix, to which the term ECM is restricted in the following, is compopsed of polysaccharides and proteins. Chains of polysacharids, the glycosaminoglycans, form a gel-like substance, into which proteins are embedded. There are a number of proteins, of which collagen, fibronectin, and laminin have important functions. ECM proteins are modular in structure and composed of different domains, some of which are shared by different proteins. Duplication and domain shuffling probably played an important role in the evolution of ECM proteins (Engel et al. 1994). Collagens are a large family of proteins, most of which are fibrous. They are composed of triplets of collagen α-chains, of which 25 types are known. Although this allows a vast number of different collagens, only about 20 types are known. For example, in vertebrates, types I, II, III, V, and XI are dominant (see Alberts et al. 2002). Collagens are connected to cellular transmembrane proteins, mostly integrins, by the protein fibronectin.

Collagens can be fibrillar or non-fibrillar (see, e.g., Mayne 1984). Sponges (Porifera) possess a special form of collagen named spongin (Garrone 1998) that contains short collagen chains (Exposito et al. 1991). The genes coding for these collagen chains show similarities to non-fibrillar collagens of other metazoans, for example collagen type IV (see below) and collagens from the nematode cuticle (Garrone 1998). Sponges also possess fibrillar collagen (Garrone 1978, Mackie & Singla 1983). Garrone (1998) suggested that collagen probably evolved as a non-fibrillar, membrane-bound molecule, but that fibrillar collagen types occurred very soon.

In several animals, distinct layers can be differentiated within the ECM, especially a very dense layer just below the epidermal cells. For this layer, the names basal lamina, basement membrane, and lamina densa are applied (see, e.g., Laurie & Leblond 1985, Pedersen 1991). I regard the term basement membrane as unfortunate, because confusion with the cellular membrane should be avoided. In the following, the term basal lamina will be used. Within this basal lamina two layers, an electron lucent lamina lucida and a darker lamina densa, can sometimes be recognized. In comparison with the remaining ECM (sometimes called lamina fibroreticularis), the basal lamina contains special proteins: a unique type of collagen (type IV) and laminin. A basal lamina is certainly present in Bilateria and was evaluated as an autapomorphy of this taxon (Ax 1996), but components of the ECM are also found in non-bilaterian animals. From the hydrozoans *Hydra vulgaris* and *Podocoryne carnea* (Cnidaria, Hydrozoa), immunocytochemical investigations reveal the presence of collagen type IV and laminin (Sarras et al. 1991, Schmid et al. 1991; Fig. 4.5.). Genes similar to genes coding for collagen type IV are also present in sponges (Boute et al. 1996, Aouacheria et al. 2006).

What is the function of the ECM? Because epidermal (and other) cells are tightly attached to the ECM by hemidesmosomes and focal adhesions, it serves as the substrate on which epithelial sheets can be organized. Cells can also migrate into the ECM and can follow special pathways within (see Chapter 3). Within the body, ECM has further functions (see Chapter 9).

ECM is found, with two exceptions, in all metazoan animals. The exceptions are *Trichoplax adhaerens* and acoels. In acoels, however, small amounts of material between cellular junctions connecting epidermal and muscle cells are

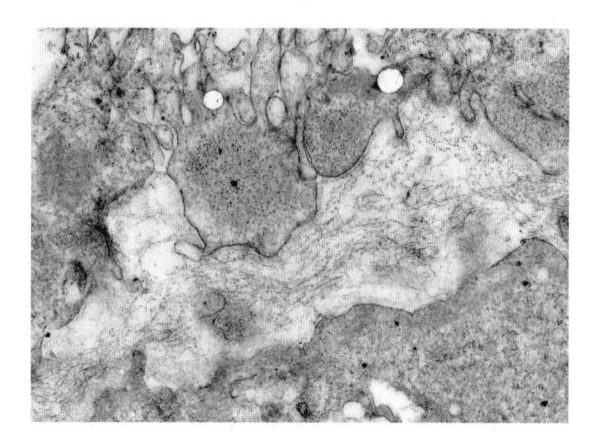

Fig. 4.5. ECM with fibrous collagen between epidermal and gastrodermal cells in *Hydra* sp. (Hydrozoa, Cnidaria).

interpreted as remnants of ECM by Tyler & Rieger (1999). Because acoels possess an extensive network of ciliary rootlets, it is assumed that this network functionally replaces the basal ECM, which was therefore reduced (Ehlers 1992a). An ECM is present in sponges, including collagen (Garrone 1998), fibronectin (Labat-Robert et al. 1981, Akiyama & Johnson 1983), and integrin (Brower et al. 1997, Pancer et al. 1997). Therefore, it must be assumed that the absence of ECM in *Trichoplax adhaerens*, and the absence or scarcity in Acoela, are derived.

Metazoan animals are not the only multicellular organisms. Besides higher plants and fungi there are filamentous Cyanobacteria, Myxobacteria with multicellular fruiting bodies, green algae such as members of the taxon Volvocales, and other examples (Bonner 1998, Kaiser 2001). An extracellular matrix also exists in most of these cases, for example as the cell wall in plants or as a basal ECM in *Volvox* (see, e.g., Kirk 1998, Kirk & Nishii 2001 for *Volvox carteri*). The structure of the ECM is, however, fundamentally different. Therefore, the structure of the ECM in multicellular animals is unique and can be regarded as an autapomorphy of Metazoa. The single components of the ECM, however, can be traced beyond metazoans. Genes with a strong similarity to collagen coding genes were found in extracellular material covering fungal appendages (Celerin et al. 1996) and probably in choanoflagellates (Karpova et al. 2003). A fibronectin-like molecule has been found in the unicellular apicomplexan *Eimeria* (del Cacho et al. 1997), as well as an integrine-like molecule (with strong immunoreactivity against anti-chicken integrin; Lopez-Bernad et al. 1996).

Glycocalyx and cuticle

Epidermal cells can secrete a variety of extracellular material to the outside. Such material can detach from the epidermis as mucous or tubes or it can remain in close contact with the epidermis. Only the latter structures will be covered here. It is, as will be explained below, likely that first epithelium-bound extracellular structures were glycoproteins linked to polysaccharides, and that during evolution this structure served as a matrix for the deposition of other molecules such as collagen and chitin which led to diverse types of cuticles. To the first structure, the name 'glycocalyx' (surface coat, fuzzy coat) is generally applied, although it was originally extended to cover a wider variety of extracellular structures (Bennett 1963). As will be explained below, there can be transitions between a glycocalyx and a 'real' cuticle which makes the application of the term 'cuticle' problematic. Some authors call any extracellular structure on the apical side of epidermal cells a cuticle, including the glycocalyx. As the glycocylx seems to be a plesiomorphy for metazoans (see below), I prefer to distinguish such structures from the more complex cuticles and therefore apply the term cuticle only to structures which are more solid in appearance and composed of additional molecules compared to a glycocalyx, even if there are cases in which this distinction is not unambiguous.

A glycocalyx is a layer of proteins and sugars on the outside of epidermal cells. Proteins are proteoglycans or glycoproteins which are integrated into the membrane, or are external but attached to transmembrane proteins. These proteins are linked to polysaccarid chains. A glycocalyx can be observed as a filamentous network on the epidermal cells by transmission electron microscopy (Figs. 4.6., 4.7.). However, the standard fixation with glutaraldehyde often destroys the glycocalyx (Rieger 1984), so that failure to observe a glycocalyx is no proof for the absence of it. A glycocalyx seems to be a common structure of eukaryotic cells (Alberts et al. 2002). It has been described for most metazoans, including sponges (Bagby 1970, Harrison & de Vos 1991) and cnidarians (Thomas & Edwards 1991). The glycocalyx is the only outer extracellular structure in the taxa Platyhelminthes (Rieger et al. 1991b), Gnathostomulida (Rieger & Mainitz 1977), *Limnognathia maerski* (Kristensen & Funch 2000), Nemertini (Turbeville 1991), Phoronida (Fig. 4.6.; Herrmann 1997), the echinoderm taxa Crinoida, Astroida and Echinoida (Holland & Nealson 1978, Holland 1984, Cavey & Märkel 1994, Chia & Koss 1994, Heinzeller & Welsch 1994), Enteropneusta, Pterobranchia (Benito & Pardos 1997), and Acrania (Ruppert 1997). When the cuticle does not cover the whole animal, the glycocalyx is usually present in the remaining parts of the epidermis, such as on the tentacles of bryozoans (Mukai et al. 1997) and kamptozoans (Nielsen & Jespersen 1997) or on the 'inner' epithelium of brachiopods (Williams 1984, 1997). Sometimes larval forms have a glycocalyx when adults develop a cuticle, for example, larvae of bryozoans (Hughes & Woollacott 1978). Glycocalyx and cuticle are not mutually exclusive structures and often the glycocalyx remains present on the outside of the cuticle (Fig. 4.7.; see, e.g., Westheide & Rieger 1978 for polychaetes). The only cases in which neither a glycocalyx nor a cuticle have been reported are *Trichoplax adhaerens* and ctenophores (Rieger 1984).

According to Rieger (1984), a glycocalyx is in most cases secreted by epidermal microvilli and can be found as a loose network between microvilli. As sponges do not have abundant microvilli in their pinacocytes, the glycocalyx is secreted from the whole surface of the cells. The decision about which of these modes is ancestral, depends on the question of whether the metazoan ancestor had epidermal microvilli or not.

As the glycocalyx is the plesiomorphic structure, it appears plausible to assume that it served as a matrix for the deposition of further extracellular molecules, forming a cuticle. This allowed the external extracellular material to play an important role, especially in protection and as a structural element. A cuticle often covers the whole body, but can also only be present in specific body parts.

There are a few reports of non-bilaterian species in which cuticle-like structures are found. In the sponge *Aplysina cavernicola* (Demospongia) and some others there is a cuticle-like thickening (Simpson 1984). Some anthozoans (Cnidaria) are capable of extracellular secretions comparable to cuticles. Representatives of the Zoantharia have an extracellular covering of the body which might represent a cuticle. Representatives of the

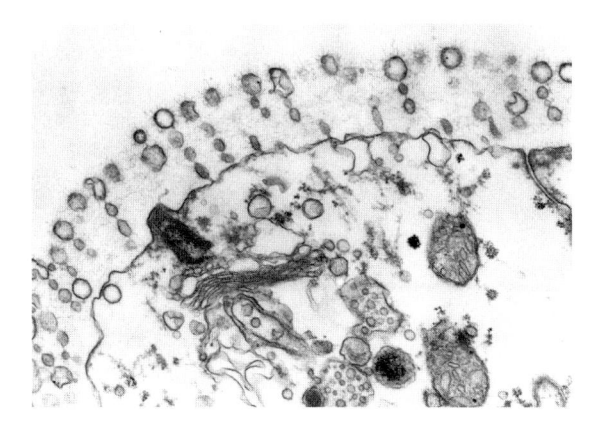

Fig. 4.6. Epidermal microvilli secrete a glycocalyx in *Phoronis ovalis* (Phoronida). Photo by courtesy of Alexander Gruhl, Berlin.

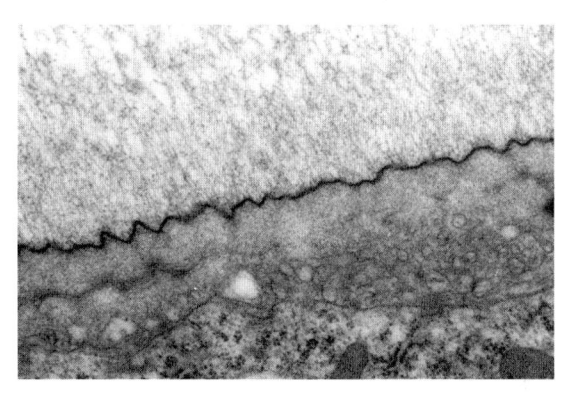

Fig. 4.7. Larval cuticle with apical glycocalyx in parasitic stages of *Paragordius varius* (Nematomorpha).

taxa Antipatharia and Gorgonacea (sea fans) secrete extracellular skeletons that are about 90% composed by a collagene-like molecule named gorgonin (Goldberg 1976, Fautin & Mariscal 1991). The anthipatharian skeleton also contains chitin (Goldberg 1978). The famous potential of corals to secrete calcium carbonate is not into a preformed cuticular matrix (as for example in molluscs). These cuticles or cuticle-like structures in sponges and cnidarians seem to have evolved within the taxa and to have been absent from their ancestor. Among bilaterians, cuticles appear in several taxa and shall be reviewed briefly.

A cuticle is present in all taxa belonging to the Cycloneuralia. In nematodes, cuticles exhibit a broad variety of structures and layers (see, e.g. Wright 1991) and some attempts have been made to interpret this diversity in a phylogenetic context (e.g. Inglis 1964, Maggenti 1979, Malakhov 1994). Taking into account problems of phylogenetic relationships within nematodes, it remains very difficult to make statements about the evolution of the nematode cuticle and the probable ancestral state. An almost constant feature of the nematode cuticle is a trilaminate epicuticle, i.e. a layer composed of three very thin sublayers, two of which are electron-dense and the medium one electron-lucent (Wright 1991). Of interest for the comparison with other cycloneuralian taxa is the structure of the pharyngeal cuticle of *Oesophagostomum* species (Strongyloidea). Below the epicuticle is a homogeneous layer which is in its basal part finely striated and contains chitin (Neuhaus et al. 1997a). Apart from the eggshells, from which chitin is also known, this is the only location of chitin in the nematode cuticle. Main components of the nematode cuticle are non-filamentous collagens cross-linked by disulphides and cuticulins, which are structural proteins with dityrosine residues (Brivio et al. 2000).

Horsehair worms (Nematomorpha) have two different types of cuticle, a larval cuticle is replaced by an adult cuticle around the time of emergence from the host (Schmidt-Rhaesa 2005). The larval cuticle is composed of three layers, a fine fibrillar inner layer, a homogeneous layer, and an outer thin darkly-stained layer (Schmidt-Rhaesa 1996a). Chitin has been detected in the basal layer of the marine nematomorph *Nectonema munidae* (Neuhaus et al. 1996). The adult cuticle consists of crossed layers made up of large fibres which are probably not collageneous, but consist of some other protein (Fig. 4.8.; Protasoni et al. 2003). There is evidence for cross-linking by dityrosine compounds (Brivio et al. 2000). In kinorhynchs, there is a finely granular and chitinous layer and a trilaminate epicuticle (Neuhaus et al. 1996, Neuhaus & Higgins 2002). An additional fine fibrillar layer can be present in some species. In priapulids and loriciferans, there is a basal fine fibrillar layer containing chitin, followed by a homogeneous layer and a trilaminate epicuticle (Neuhaus et al. 1996, 1997b, Lemburg 1998).

It seems that a cuticle with three layers (basal finely striated—including chitin, median homogeneous, and apical trilaminate) is common in most taxa within Cycloneuralia (Nematoda, Nematomorpha, Priapulida, Kinorhyncha, Loricifera) and can be assumed as an ancestral character. In nematodes, cuticles diversified considerably and chitin became restricted to the pharyngeal cuticle (whether this is valid for all, most, or few species is unknown). In nematomorphs, the cuticle is replaced in adults by a new type of cuticle, and in kinorhynchs chitin is restricted to a granular layer.

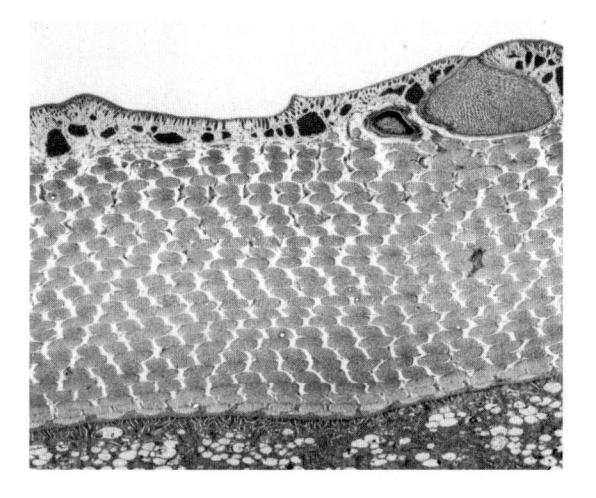

Fig. 4.8. Adult cuticle with crossed layers of large fibres in free-living stages of *Paragordius varius* (Nematomorpha).

A comparison of the cuticular structure of cycloneuralians with arthropods in the context of the Ecdysozoa-hypothesis has shown that both taxa are quite comparable and that the cuticular structure might support a monophyletic taxon Ecdysozoa (Schmidt-Rhaesa et al. 1998). Trilaminate epicuticles are abundant in arthropods (e.g. Bacetti & Rosati 1971, Neville 1975, Gnatzy & Romer 1984, Wright & Luke 1989). Below this epicuticle, a homogeneous and a basal fibrillar layer with chitin are often present (e.g. Hackman 1984, Neville 1984, Greven & Peters 1986, Wright & Luke 1989). The terminology of layers, however, varies considerably.

The cuticular structure of gastrotrichs, the possible sister-taxon of Cycloneuralia or, if accepted, of Ecdysozoa, shows some similarities. It is composed of two layers, an inner, proteinaceous one and an outer one which consists of few to multiple copies of the trilaminate layer (Rieger & Rieger 1977, Ruppert 1991b). Epidermal microvilli are rare (Ruppert 1991b), but cilia are present on the ventral side. They traverse the cuticle and are completely surrounded by it (Fig. 4.9.; Rieger & Rieger 1977, Ruppert 1991b). This is very rare among metazoans (see below). The cuticle in gastrotrichs, which is devoid of chitin, may be homologous to the epicuticle and to the (usually) chitin-free homogeneous layer (see also Schmidt-Rhaesa 2002).

A cuticle is found in all trochozoan taxa. In molluscs, a cuticle is present in the basal taxa Solenogastres, Caudofoveata (Scheltema et al. 1994), and Polyplacophora (Eernisse & Reynolds 1994). This cuticle covers the dorsal part of the body and includes calcareous spicules. The calcareous shell that covers the dorsal side of the remaining molluscs (taxon Conchifera) is secreted into a previously formed organic matrix, the periostracum (Watabe 1984, 1988). The periostracum includes glycoproteins (e.g. conchin). Such a periostracum is, of course, also present on the shell plates of polyplacophorans (Eernisse & Reynolds 1994). It may be assumed, that a dorsal cuticle was present in the ancestor of molluscs and became specialized as a periostracum accompanying shell evolution. The remaining parts of the molluscan body do not possess a cuticle, but a glycocalyx is present between microvilli of the epidermal cells (Bubel 1984). Chitin is present in the cuticle of basal molluscs (e.g. von Salvini-Plawen & Nopp 1974) as well as in the periostracum of other molluscs (Jeuniaux 1982a).

The body is covered with a cuticle in colonial kamptozoans, with the exception of the atrium (the region between tentacle bases containing mouth and anus) and the ciliated parts of the tentacles. It consists of an electron dense outer layer with 'caps' formed by transcuticular microvilli and a basal, fibrillar layer (Nielsen & Jespersen 1997). Chitin is present in the cuticle, at least in stiffer parts of the stalk (Jeuniaux 1982a, b, Nielsen & Jespersen 1997). Chitin seems to be absent in solitary kamptozoans (*Loxosomella atkinsae*; Andreas Unger, Bielefeld, unpublished results), making it questionable whether chitin in the cuticle is a ground pattern character of kamptozoans.

The cuticle in the taxa Annelida (including Pogonophora), Sipunculida, and Echiurida is of a comparable structure. In general, there is an inner region containing non-striated collagen fibres and an outer, homogeneous layer. The whole cuticle is traversed by epidermal microvilli (Fig. 4.10.) which produce epicuticular projections, the 'caps' already described for kamptozoans (Fig. 4.11.; see e.g. Sylvia Richards 1984, Gardiner 1992, Gardiner & Jones 1992, Jamieson 1992, Hausen 2005a for Annelida;

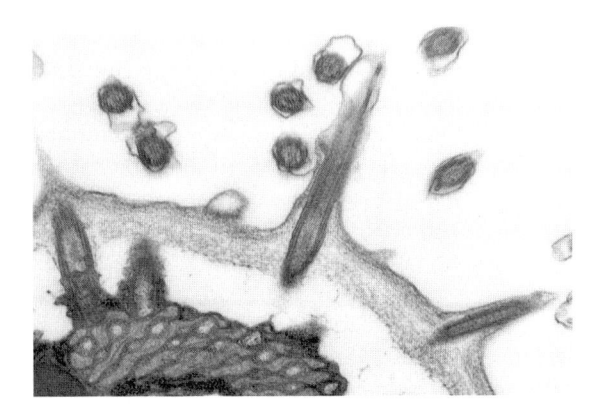

Fig. 4.9. The outer layer of the cuticle surrounds the ventral cilia in gastrotrichs (species shown here is *Chaetonotus maximus*). Photo by courtesy of Alexander Kinecke, Oldenburg.

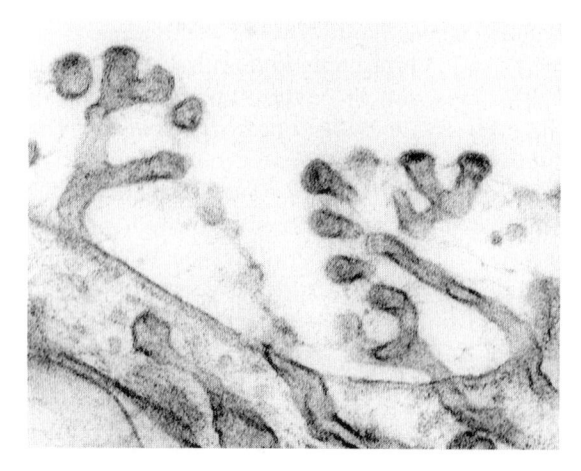

Fig. 4.10. Epidermal microvilli traverse the cuticle and branch apically in *Protodrilus* sp. (Polychaeta, Annelida).

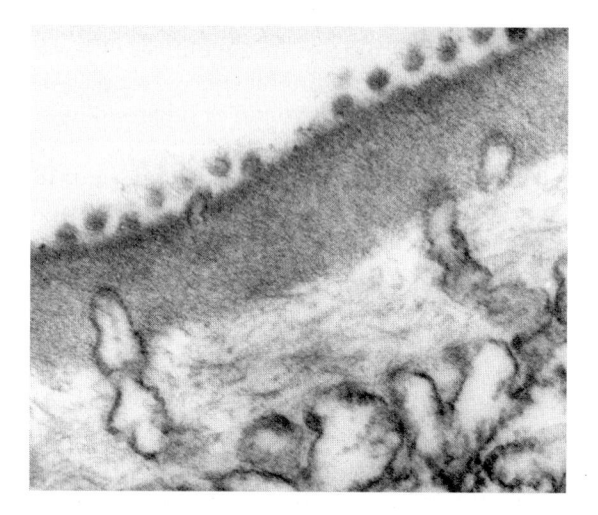

Fig. 4.11. Epidermal microvilli traverse the two-layered cuticle and form apical caps on the cuticular surface in *Poecilochaetus serpens* (Polychaeta, Annelida).

Storch 1984, Pilger 1993 for Echiurida; Rice 1993 for Sipunculida). Microvilli often branch within the cuticle. The cuticle can be very thin or absent in some regions of the body, for example where epidermal cilia are present. In large 'oligochaete' species, several layers of large fibres can be present (Rieger 1984, Gustavsson 2001). The cuticle does not contain chitin, this is only present in special epidermal structures, the chaetae (see below).

Bryozoans and brachiopods secrete in some parts of the body an organic cuticle, the periostracum, that contains mucopolysaccarids and chitin (Williams 1984). This serves as the matrix for calcium carbonate deposition in cyclostome and cheilostome bryozoans as well as in articulate and some inarticulate brachiopods. Lingulid brachiopods (such as *Lingula, Glottidia, Discinisca*) possess a shell composed of chitin, calcium phosphate, and collagen (Williams et al. 1994, Williams 1997).

Within Echinodermata, crinoids, asteroids, and echinoids have epidermal microvilli, among which a glycocalyx is present. In ophiuroids and holothurians, there is a more complex structure (Byrne 1994, Smiley 1994), which might be named a cuticle, although nothing is known about the molecular components. In ophiuroids, fibrillar elements can be recognized and cap-like structures above the microvillar tips are present (Byrne 1994). Given the phylogenetic relationships within echinoderms, it appears likely that a glycocalyx between epidermal microvilli is the ancestral condition and that cuticle-like structures evolved in parallel in ophiuroids and holothurians.

Tunicates are characterized by their 'mantle', the tunic, which connects the animal to the substratum. As the tunic remains in close contact to the epidermis, it is valid to call it a cuticle, at least according to the definition above. There are, however, some unique features. The tunic is composed of a dense outer and a fibrous inner layer that contains fibres of a polysaccharide named tunicin which is related to cellulose (Burighel & Cloney 1997). A cellulose-like structural element is unique among animals. Furthermore, cells are present in the tunic, again a unique feature for the external extracellular matrix (Turon et al. 2005).

Finally, chaetognathes have a homogeneous cuticle in some parts of the head, probably as a protection from captured prey. Additionally, fibrous parts (in *Spadella*) or an additional, electron-lucent region can be present (Shinn 1997).

The brief review above reveals a very diverse picture of cuticular structures in metazoans. It seems convincing to assume that a glycocalyx is a plesiomorphy of metazoans. As assumed by Rieger (1984), this glycocalyx served as a matrix for the further deposition of molecules. Cuticles evolved independently several times. Mapping the presence of cuticles onto the phylogenetic tree results in the following scenario (Fig. 4.25.): cuticle-like structures evolved among non-bilaterian animals only within anthozoan cnidarians and probably within sponges. The bilaterian ancestor was devoid of a cuticle and contained only a glycocalyx between epidermal microvilli. Within protostomes, a cuticle evolved in the ancestor of Nemathelminthes, first as a bilayered cuticle without chitin and then, in the ancestor of Cycloneuralia, including a third, basal layer with chitin. If Ecdysozoa is regarded as a valid taxon, this last character is a synapomorphy of Arthropoda and Cycloneuralia. A cuticle evolved again in the ancestor of Trochozoa, as a collageneous cuticle traversed by microvilli. Chitin in the dorsal cuticle appears to have evolved in the ancestor of Lacunifera (Kamptozoa + Mollusca). The phylogenetic relationships of phoronids, bryozoans, and brachiopods are not clear and the cuticular structure of bryozoans and brachiopods gives no hint to solve this problem. If these three taxa are related to the Deuterostomia, the cuticle in bryozoans and brachiopods would be an independant acquisition. If they belong into the Trochozoa, the absence of a cuticle in phoronids would have to be interpreted as a secondary loss. The ancestor of deuterostomes did not possess a cuticle, but only a glycocalyx. Within echinoderms, cuticle-like structures evolved twice, in ophiuroids and holothurians. A cuticle (as the unique tunic) evolved again independently in the tunicate ancestor.

The problem with chitin

Chitin is a ubiquitous polysaccharide (N-acetylglucosamine) that is not restricted to metazoan animals. In fungi, chitin is an important element of the cell wall (see, e.g., Aronson 1965, Cabib et al. 1996). Therefore, it seems plausible that the basic metabolic components for chitin synthesis were present in the metazoan ancestor.

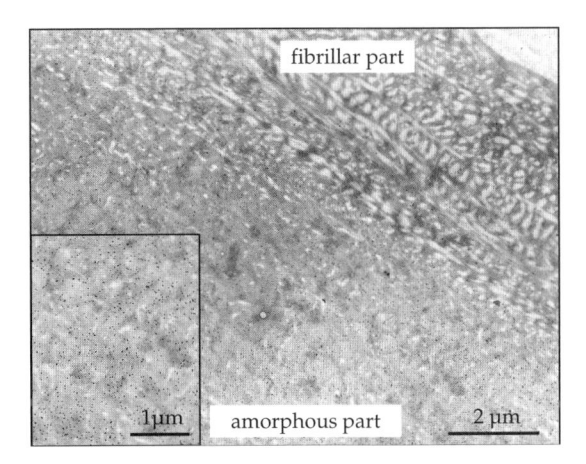

Fig. 4.12. Chitin labelling with gold-labelled wheat germ agglutinin in *Urnatella gracilis* (Kamptozoa). Black gold particles indicate chitin in the basal amorphous layer, but not in the apical layer. Insert shows magnification of amorphous layer. Photo by courtesy of Andreas Unger, Bielefeld.

Chitin is, as a cuticular component or as a secretion product (i.e. in a tube), broadly distributed (see above, additional literature Jeuniaux 1975, 1982a). Chitin can be detected in several ways. There is a chitosan reaction, enzymatic digestion with chitinase, X-ray diffraction of chitin crystals, and a reaction with gold-labelled wheat germ agglutinin (Fig. 4.12.; Peters & Latka 1986). Only the last method can detect the localization of chitin within a cuticle, because it is applied to and observed on ultra-thin sections. Only the X-ray diffraction can differentiate between different types of chitin.

Chitin has been used to support phylogenetic hypotheses (i.e. the monophyletic Ecdysozoa and the sister-group relationship between Kamptozoa and Mollusca) but this has also been heavily criticized. Therefore, the question is whether chitin can be used as a phylogenetical marker or not. In my mind the answer is yes, but with restrictions. First, it cannot be the simple presence of chitin, but a specific localization and probably also a special type of chitin, that potentially bears phylogenetic information. Second, phylogenetic hypotheses should not rest on a character concerning chitin alone, but when other characters favour a certain relationship, chitin can support this hypothesis. What does this mean for the problems mentioned above?

Although the potential of chitin synthesis is obviously widespread, this does not mean that chitin is a component of any cuticular structure. For example, within Trochozoa, chitin is present either in the cuticle (in Kamptozoa and Mollusca) or absent in the cuticle itself and present only in chaetae of Echiurida and Annelida. In such a case it is justified to hypothesize the secretion of chitin into the cuticle as a potential synapomorphy of Kamptozoa and Mollusca, and the integration of chitin into the chaetae as a potential synapomorphy of Echiurida and Annelida.

Concerning the Ecdysozoa problem, two aspects have to be mentioned. As has been discussed above, one can argue that cuticles of arthropods and cycloneuralians share a common substructure. Chitin is almost always restricted to the inner, finely fibrous layer. Additionally, the type of chitin (as far as this has been shown) is α-chitin (Lotmar & Picken 1950, Shapeero 1962, Neville 1975). The only known exception is the pentastomid *Raillietiella gowrii*, where β-chitin occurs (Karuppaswamy 1977). On the other hand, chitin in trochozoans is β-chitin, as has been shown in molluscs and in annelid chaetae (Lotmar & Picken 1950, Okafor 1965). The only exception here is that α-chitin occurs in the operculum of the tube of the polychaete *Pomatoceros triqueter* (Bubel et al. 1983). These differences in localization and type of chitin bear, in my mind, valid phylogenetic information, but phylogenetic hypotheses should not rest on this character alone.

Moulting

The potential taxon Ecdysozoa has received its name from the fact that all its representatives moult. Moulting is well known in euarthropods and also occurs in onychophorans (Storch & Ruhberg 1993) and tardigrades (Dewel et al. 1993). Nematodes constantly moult four times, nematomorphs moult once (Schmidt-Rhaesa 2005), kinorhynchs moult, as far as has been documented, six times (Neuhaus & Higgins 2002), priapulids and loriciferans moult several times (Kristensen 1991, Storch 1991). Additionally, there is evidence that both taxa, Arthropoda and Cycloneuralia, use similar hormonal pathways for their moulting. Apart from the report of ecdysteroid hormones which have been found in several nematodes (see literature in Schmidt-Rhaesa et al. 1998), arthropod hormones can be used to induce a premature moulting in nematodes (Dennis 1976, Warbrick et al. 1993). It must, however, be stressed that the knowledge of the physiological control in cycloneuralians is only fragmentary and restricted to nematodes.

Ecdysteroid hormones are widespread and appear in different functional contexts, for example in gamete maturation. There are some annelids which also show moulting. The leech *Hirudo medicinalis* moults its cuticle and this moulting is induced by ecdysteroid hormones (Sauber et al. 1983, Berchtold et al. 1985). Eunicid polychaetes have cuticular jaws which they moult (Paxton 2005). However, these seem to be derived phenomena within Annelida. Other cases such as the shedding of symbiotic bacteria together with parts of the cuticle in *Tubificoides benedeii* ('Oligochaeta') (Giere et al. 1988) and partial loss of cuticular layers (Pilato et al. 2005) cannot be considered as moulting, but are better termed shedding. It is obvious that the widespread ecdysteroid hormones can be utilized in the context of moulting, but this occurred at least two times in metazoan animals. Moulting alone is not a sufficient character to support the monophyly of Ecdysozoa, but has to be discussed in combination with further evidence.

Special cuticular structures

The potential to form a cuticle makes it possible to produce, besides the cuticle covering the epidermis, further cuticular structures. A cuticle can be soft and flexible, but it can serve as a matrix, into which hardening or sclerotizing elements can be integrated. These elements may be, among others, chitin and even iron. The cuticle can form all kinds of extensions such as bristles, spines, hooks, scales, and so on. Receptors, especially, are often located in special cuticular structures (see Chapter 7).

As the ectodermal epidermis usually forms the anterior and posterior parts of the intestinal system, as well as the distal part of reproductive systems, cuticular structures such as jaws, teeth, and stylets are often formed in these regions. It should, however, be noted that not all such structures are really cuticular. Hooks on the anterior end of acanthocephalans, for example, are formed from the subepidermal ECM (Herlyn & Ehlers 2001), and copulatory hard structures in the male reproductive system of many flatworms are either intracellular or also formed by the subepidermal ECM (Ehlers 1985a).

One cuticular structure must be presented in detail: the chaetae of annelids (Fig. 4.13.), echiurans, and brachiopods. In annelids, chaetae are very diverse, but are always formed according to the same principle. A specialized epidermal cell, the chaetoblast, secretes cuticular material from the surface of numerous microvilli. As secretion of cuticular material proceeds, the forming chata is pushed out of the epidermis. Microvilli retreat and secrete further material on the basis of the chaeta (reviewed by Bartolomaeus 1997a, Hausen 2005b). By special arrangements of the microvilli, as well as by positional shifts of the chaetoblast, special structures such as small teeth and a curvation of the whole structure can be produced (Fig. 4.13; Bartolomaeus 1997a). Chaetae are the only cuticular structures in annelids that possess chitin (β-chitin, see Lotmar & Picken 1950). Chaetae similar to those in annelids occur in echiurans (Orrhage 1971, Storch 1984). This is also compatible with a sister-group relationship between Annelida and Echiurida as with an inclusion of Echiurida into the Annelida. Of particular interest is that brachiopods have chaetae which are formed in a similar way as in annelids. Chaetae are present in most brachiopod species as larval or adult chaetae (Lüter 2000b, 2001). The similar structure of chaetae in the taxa Annelida/Echiurida and Brachiopoda could support the hypothesis of a closer relationship. The phylogenetic position of brachiopods is controversial, but as current hypotheses alter between an inclusion into the Radialia or a relationship to molluscs, but not to annelids (see Chapter 2), it seems that chaetae evolved convergently.

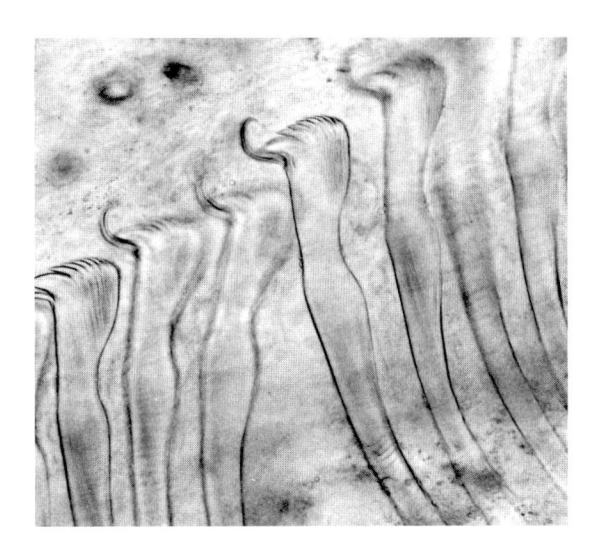

Fig. 4.13. Hooked chaetae of an unidentified polychaete.

The Epidermal Cell

Epidermal cells can express a variety of different forms, such as supporting cells, gland cells, or sensory cells. In most metazoans, the epidermis forms a monolayer of cells (Fig. 4.14). Exceptions are chaetognaths and vertebrates, which have a multilayered, so-called 'stratified' epidermis. In monolayers, the thickness can vary considerably. Abundant gland cells often increase the thickness, such as in enteropneusts. In clitellate annelids, the epidermis in the gland-rich region of the clitellum is thicker than in the remaining body. In contrast, the epidermis can be very thin, especially when a cuticle is present. Nematodes shift their epidermal nuclei into lateral epidermal cords, so that the remaining parts can be quite thin. In nematomorphs, the epidermis is relatively thick in early developmental stages during which nutrients are absorbed from the host through the epidermis. Later stages form a more or less impermeable cuticle and the epidermis decreases in size (Schmidt-Rhaesa 2005). In this way, mediated by abundant intermediate filaments in the thin epidermal cells, the longitudinal musculature and the cuticle can form an effective functional union.

Epidermal cells are rarely the arrangements of cubic cells as they often occur in schematic

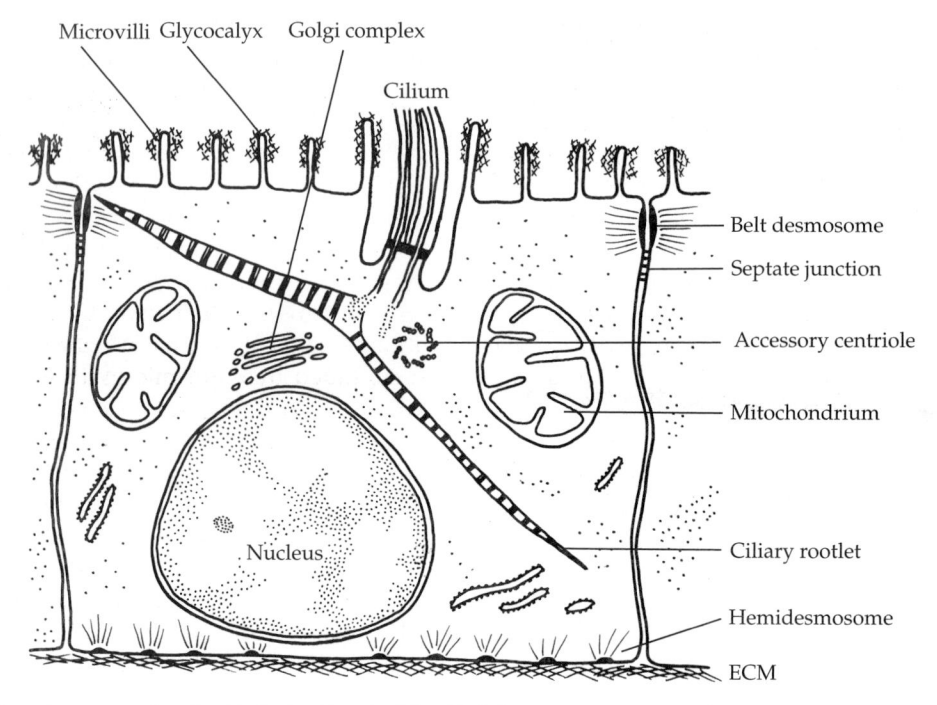

Fig. 4.14. Schematic representation of an idealized epidermal cell (modified after Rieger & Mainitz 1977).

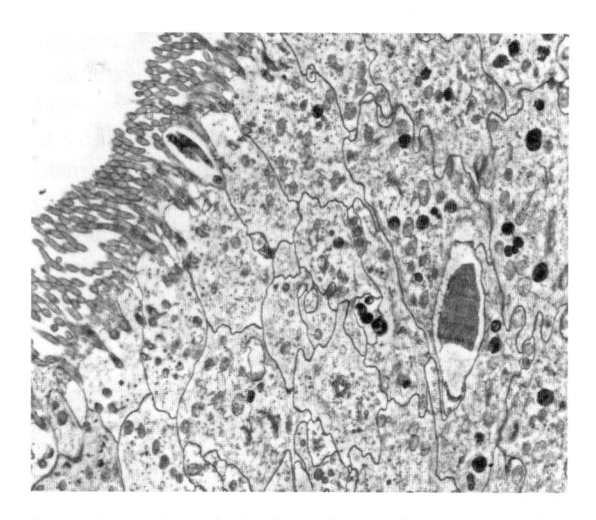

Fig. 4.15. Pseudostratified cuticle with heavily interdigitating cells in *Lineus ruber* (Nemertini).

representations. Instead, epidermal cells are often more or less interdigitating. This may make it hard to recognize the monolayered structure. The epidermis can even appear stratified, but still each epidermal cell has a connection to the basal ECM and to the apical surface. Such epithelia are called pseudostratified. Examples are the nemertean epidermis (Fig. 4.15.; Turbeville 1991) or the epidermis in the proboscis region of entropneusts (Benito & Pardos 1997). Pseudostratified epithelia are, for example, important in models concerning the evolution of individualized muscle cells (see Chapter 5).

Epidermal microvilli and cilia are two important structures of the apical side. Both are extensions of the cell membrane. Microvilli usually occur to increase the surface area, such as to ingest nutrients or to increase the amount of membrane-bound photopigments in photoreceptors, but are also important in secretion of a glycocalyx and cuticle. Actin filaments are present in microvilli. Cilia always contain microtubules in a special, but variable distribution pattern. The most common type is the $9 \times 2 + 2$ pattern which means that there are nine peripheral duplets of microtubules and two single central ones (Figs. 4.16., 4.17.). With this formula (periphery + centre), any deviating pattern can be expressed. Cilia with this pattern occur in all

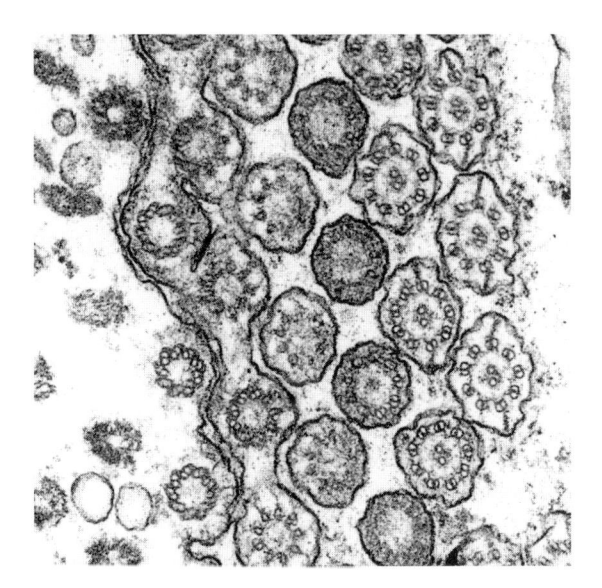

Fig. 4.16. Section through the attachment of comb cilia to epidermal cells in *Beroe* sp. (Ctenophora). The section shows cilia above the epidermal surface with a 9 × 2+2 pattern of microtubuli (to the right) and the transition to the intraepidermal basal body with a 9 × 3+0 pattern (towards the left).

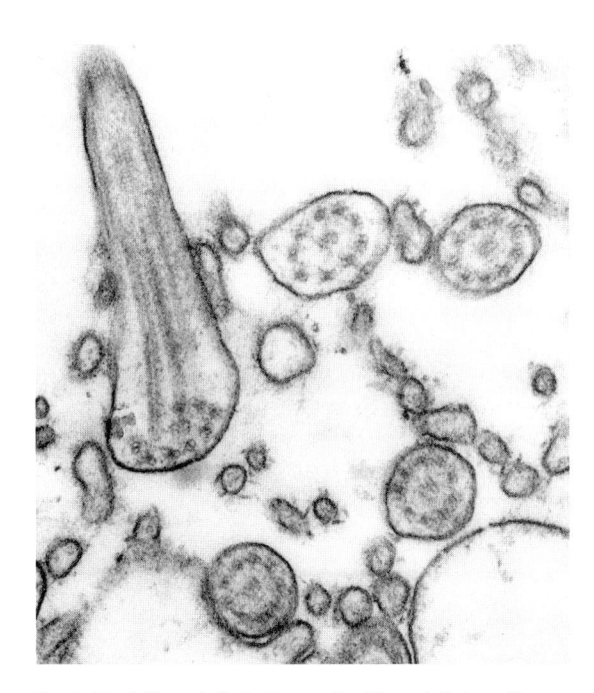

Fig. 4.17. Epidermal cilia in *Lineus ruber* (Nemertini). One cilium is sectioned in a bend, so that some microtubuli are sectioned perpendicular and others in longitudinal direction.

eukaryotes (Nielsen 1987). There are further fine structural details such as dynein arms (see, e.g., Sanderson 1984) and more, but these seem not to vary much within metazoans. Both microvilli and cilia are capable of movement. In microvilli, this is performed by actin alone while in cilia the microtubules are responsible for movement. Microtubules can slide along each other in a way roughly comparable to the interaction between actin and myosin (Sanderson 1984). In cells with only one cilium (monociliate), the cilium is often surrounded by eight microvilli (e.g. Rieger & Mainitz 1977). This association between one cilium and eight microvilli is especially evident in modified epidermal cells, such as terminal cells of protonephridia (see Chapter 8), or in sensory cells such as collar receptors (see Chapter 7).

The function of cilia is to create a current in a fluid. Ducts, be it from the excretory or the reproductive system, are often ciliated. Cilia are important for feeding in many groups. In filter feeders, they often create a water current through which particles approach the sites where they can be captured. Captured particles are often transported along ciliated regions towards the mouth opening. For example, the echiurid *Bonellia viridis* 'sweeps' the sediment surface with its proboscis, and organic matter is transported by ciliary bands along the long proboscis to the mouth opening. Mussels also transport pellets of particles filtered on their gills towards the mouth. There are many more examples. Smaller animals use cilia for locomotion. The majority of microscopically small larvae are ciliated, either completely, or restricted to one or more ciliary bands. Several microscopically small animals also move with their cilia, for example gastrotrichs and free-living flatworms. Spermatozoa often move by ciliary action, but there are also aciliary types (see Chapter 14). Finally, receptor cells often contain cilia as an important part of the receptive system (see Chapter 7).

Ciliated epidermal cells and a cuticle are (more or less) mutually exclusive structures, but a few exceptions are known (see below). In the taxa Cycloneuralia and Arthropoda, all epidermal cilia are reduced with the exception of cilia in receptor cells. This reduction is one potential autapomorphy of the taxon Ecdysozoa (Schmidt-Rhaesa et al. 1998). All other taxa in which a cuticle occurs (for example, in molluscs and annelids), have ciliated regions in which no cuticle occurs. The exceptions, in which an external cuticle and cilia occur together, are gastrotrichs, lobatocerebrids, and orthonectids. In gastrotrichs, the cilia are surrounded by cuticle (or at least by the epicuticle) (Fig. 4.9.; Ruppert 1991b). Lobatocerebrids, which are assumed to belong to annelids, possess a fibrillar cuticle which is traversed by cilia from multiciliate cells (Rieger 1981), and in orthonectids, cilia pass through a thin cuticle (Slyusarev 2000). Cuticle and traversing cilia have also been found in the anterior, ectodermal part of the intestinal system in some cases: in the pharyngeal cuticle of some polychaetes (*Hesionides*, *Microphthalmus*, Westheide & Rieger 1978; *Glycera tridactyla*, Purschke 2005), in the esophagus of kamptozoans (Nielsen & Jespersen 1997), and in the pharynx of some acoels (Solenofilomorphidae, Todt & Tyler 2006).

In several taxa, some cilia are in close contact and beat together as compound cilia (Nielsen 1987; see Fig 4.18.). This occurs in some anthozoan larvae, in phoronid larvae, in all trochozoan taxa, and in enteropneust larvae (Nielsen 1987). Compound cilia can either combine cilia from several monociliate cells, or from one or several multiciliate cells (Nielsen 1987). In chaetonotoid gastrotrichs, several ventral cilia can be 'bound' together by the covering epicuticle and form cirri (Ruppert 1991b). In ctenophores, cilia within the comb rows are connected to form plates (Fig. 4.19.). Within the cilia, special structures called 'compartmenting lamella' occur on opposing sides (Fig. 4.20.). These structures of neighbouring cells are in register and are connected by an extracellular cement-like substance (Hernandez-Nicaise 1991). Beroid ctenophores possess additionally enormous ciliary structures called macrocilia on the inner side of their lips. These are composed of numerous $9 \times 2 + 2$ arrangements of microtubule which are altogether surrounded by the

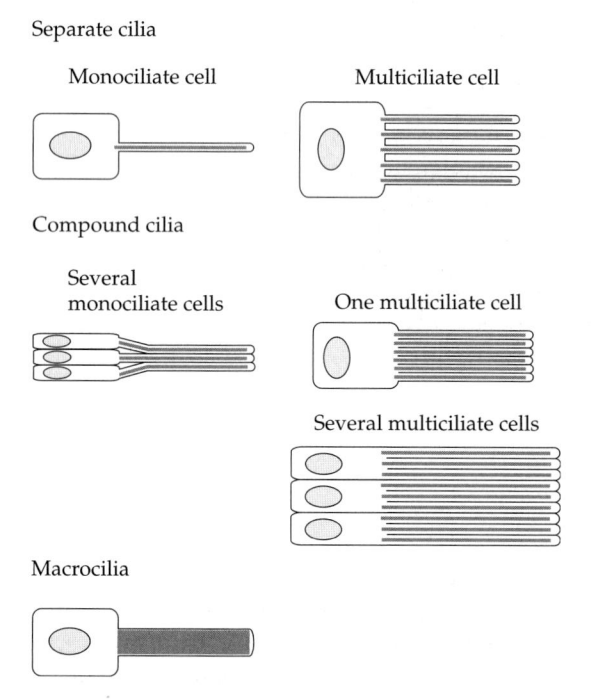

Fig. 4.18. Overview on different types of ciliation pattern (modified after Nielsen 1987).

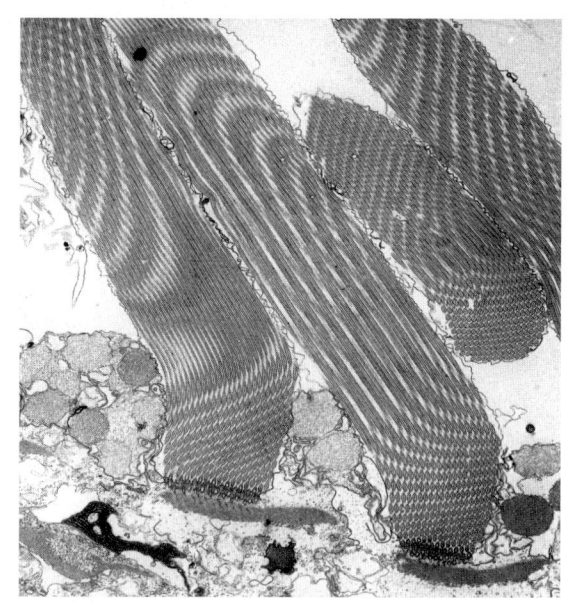

Fig. 4.19. Comb rows of ctenophores are composed of numerous cilia attached to each other (*Beroe* sp.).

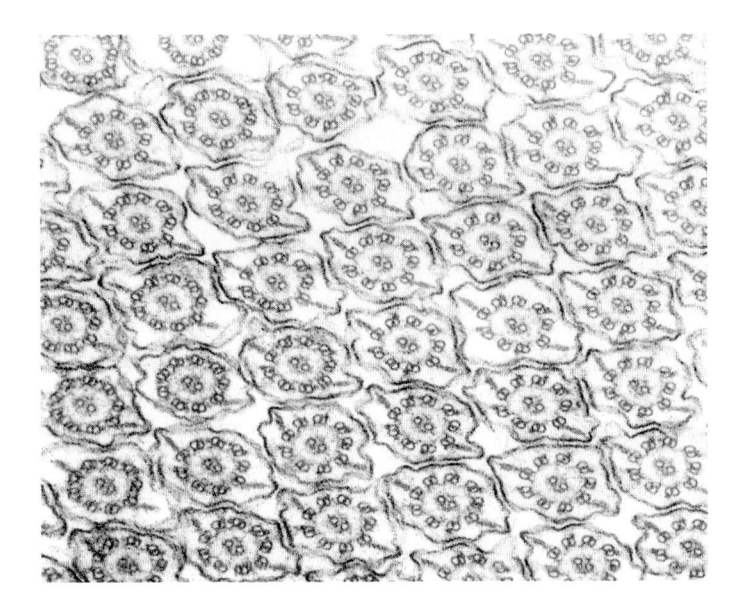

Fig. 4.20. Cilia in comb rows of ctenophores show lateral extensions that are responsible for the adhesion to neighbouring cilia (*Deroe* sp.).

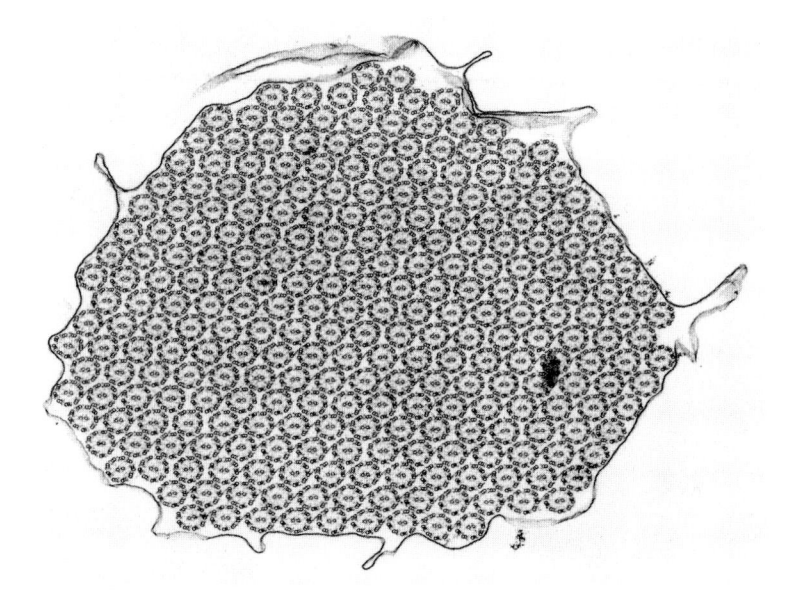

Fig. 4.21. In macrocilia of beroid ctenophores, several 9 × 2 + 2 arrangement patterns of microtubuli are surrounded by one common cell membrane.

cellular membrane (Fig. 4.21.; e.g. Hernandez-Nicaise 1991, Tamm & Tamm 1993).

In three taxa, not all microtubules reach the tip of the cilium. In acoelomorph flatworms (Acoela + Nemertodermatida) microtubule dupletts 4-7 terminate before the tip and create a visible 'step' (Tyler 1979, Rieger et al. 1991, Ax 1996). This is also the case in *Xenoturbella bocki* (Franzén & Afzelius 1987) and in enteropneusts (Benito & Pardos 1997), although here the numbering of microtubules ending before the tip is not mentioned. Because all three taxa were considered to be not closely related, it was assumed to be an independently evolved character in each taxon. However, some recent hypotheses on the relationships of acoelomorphs and *Xenoturbella* might change this (see Chapter 2). Acoelomorpha are discussed as the sister-taxon of Bilateria (Ruiz-Trillo et al. 2002) as is *Xenoturbella* (Ehlers & Sopott-Ehlers 1997). Alternatively, comparisons of the 18S rDNA gene suggest a closer relationship between *Xenoturbella* and enteropneusts (together with pterobranchs and echinoderms) (Bourlat et al. 2003).

The distribution pattern of microtubules changes towards the base of the cilium. The central microtubules terminate and triplets appear in the periphery ($9 \times 3 + 0$ pattern). Such an arrangement is characteristic for centrioles. Centrioles have some capacity in organizing microtubules. An accessory centriole, perpendicular to the ciliary basal structure, is often present. Apart from forming the basal structure of cilia, centrioles are essential for cell division. The absence of an accessory centriole is often connected to the inability of epidermal cells to divide, for example in flatworms (Ehlers 1985a) and in nematodes. In nematodes, however, this may not account for the whole group. Divisions of epidermal cells occur in enoplid nematodes (Malakhov 1998) and (modified) centrioles have been found in *Capillaria hepatica* (Wright 1976).

A further characteristic of cilia is the presence of striated rootlets that start from the ciliary basal structure (Figs. 4.22. 4.23.). Such rootlets are present in all representatives of the Epitheliozoa. Among sponges, striated rootlets are absent in adults, but occur in larvae of Calcarea, in some, but not all larvae of Demospongia (e.g. Woollacott & Pinto 1995, Boury-Esnault et al. 2003, Maldonado 2004). They are absent in the few hexactinellid larvae

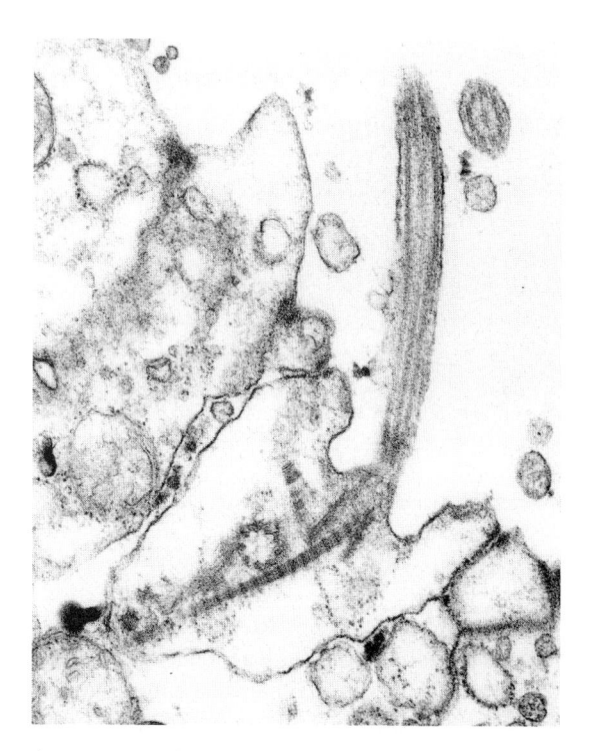

Fig. 4.22. Monociliated epidermal cell of *Trichoplax adhaerens*, showing the striated ciliary rootlet and accessory centriole.

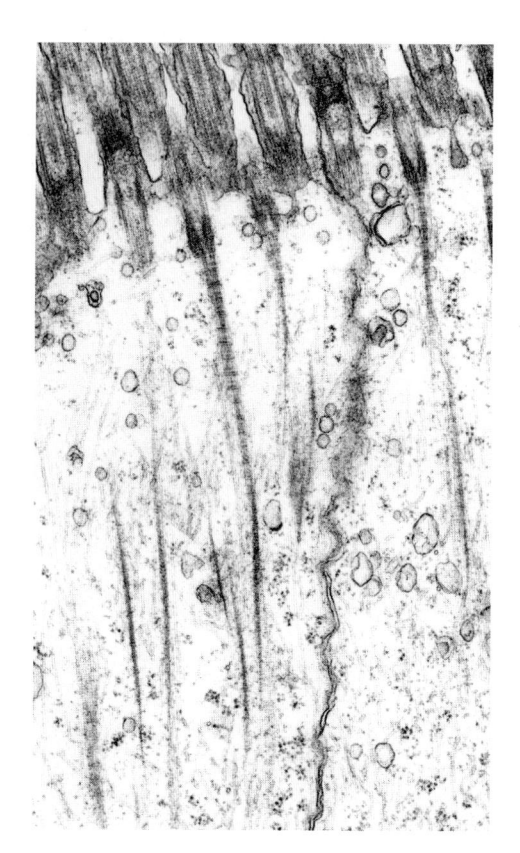

Fig. 4.23. Deep striated ciliary rootlets in the epidermis below comb rows of *Beroe* sp. (Ctenophora).

investigated so far. Therefore, striated ciliary rootlets can be assumed as a character present in the metazoan ancestor. Some choanoflagellates possess rootlets with a striation pattern (Karpov & Leadbeater 1998), but according to Maldonado (2004), this is not homologous to the metazoan striation.

In representatives of the flatworm taxon Acoelomorpha (Nemertodermatida + Acoela), the ciliary rootlets develop a complex system (Hendelberg & Hedlund 1974, Ax 1996). There is a strong anterior and a weaker posterior rootlet from which two fibrous bundles originate. In acoels, two additional branches originate from the anterior rootlet. The rootlet structures contact each other and form a complex network that probably has a structural function.

Choanocytes in sponges carry special structures, the vanes. These are extracellular lateral appendages on two sides of the choanocyte cilium (Mehl & Reiswig 1991, Weissenfels 1992). Their function is obviously to create a stronger water current. Comparable structures also exist on the cilium of choanoflagellates (Hibberd 1975), although only one species has been examined in detail. Are these vanes homologous structures in poriferans and choanoflagellates? Nielsen (2001) assumes that choanocytes, including vanes, are indeed homologous, while Ax (1996) assumes a parallel evolution, indicated for example by a different fine structure of vanes. Finally, extracellular appendages on the cilium are also present in fungi (the 'tinsel-type' cilium; Kole 1965).

Number of cilia per cell

Epidermal cells can contain one or more cilia and it is generally assumed that monociliarity is the plesiomorphic condition. The reason is that sponges, *Trichoplax adhaerens*, and cnidarians have monociliate epidermal cells (Nielsen 1987, 2001). This is not exactly correct, because a few exceptions exist. A few hydropolyps and anthozoans such as *Aglantha digitale* have more than one cilium per cell (Nielsen 1987), and in the larva of the demosponge *Sigmadocia caerulea* biciliary cells exist (Maldonado 2004). Ctenophores have multiciliate cells (Nielsen 1987). Within Bilateria, there are several taxa, which are either exclusively monociliate (such as Gnathostomulida, Phoronida, Brachiopoda, Echinodermata) or in which larvae have monociliate epidermal cells but adults change to multiciliarity (Bryozoa, Enteropneusta). Pterobranchia generally have monociliate cells, but biciliarity can occur (Mayer & Bartolomaeus 2003). In gastrotrichs, monociliarity seems to be the plesiomorphic condition, while multiciliate cells in some taxa are derived (Rieger 1976). More difficult to evaluate is the occurrence of monociliate cells in larvae of the polychaetes *Owenia fusiformis* and *Magelona mirabilis* as well as on tentacles of adult *Owenia fusiformis* (Bartolomaeus 1995). All other representatives of the taxon Trochozoa have multiciliate cells, even in the trochophore larva (Nielsen 1987). Therefore, multiciliate epidermal cells seem to be the plesiomorphic condition within Trochozoa, and monociliarity in the two species named above therefore must be a derived condition, – a secondary monociliarity (Bartolomaeus 1995).

Do sponges have a 'real' epidermis?

When the epidermis is presented here as an organ system, this contradicts often read notions that sponges do not have organs or tissues comparable to other metazoans. Indeed, pinacocytes of adult sponges lack several characteristics which are present in the epidermis of epitheliozoan metazoans. But the organization of the external layer of the sponge larvae, especially, appears to be comparable to Epitheliozoa in several respects. Cells are monociliate, rest on a basal matrix and have adhaerens junctions (Leys & Degnan 2002, Boury-Esnault et al. 2003). Therefore, outer epithelia in all metazoan animals are homologous, and it is not wrong to call the larval epithelium of sponges as well as the pinacoderm of adults, which is derived from this epithelium, an epidermis.

Syncytial epidermis

In most animals the epidermis retains a cellular condition, but in some taxa cell borders can disintegrate and give way to a multinucleate syncytium. Epidermal syncytia occur in several parasitic taxa such as acanthocephalans, parasitic flatworms (Neodermata), and parasitic nematodes. This might suggest an advantage of a syncytium for the parasitic lifestyle, either for protection or for nutrient uptake. However, the functional advantage is unclear and the syncytial epidermis is not restricted to parasitic taxa. *Seison* and rotifers are included with acanthocephalans in the taxon Syndermata due to the possession of a syncytial epidermis. In this case, a syncytial epidermis might be of advantage for an internal hard structure, which is equivalent to a cuticle, but located within the syncytium (Fig. 4.24.; Ahlrichs 1997). However, the internal layer is not restricted to the Syndermata, but also occurs in its sister-taxon, *Limnognathia maerski*, where it is present in a cellular epidermis (Kristensen & Funch 2000).

Parasitic flatworms of the taxon Neodermata possess a unique epidermis called the neodermis which replaces the ciliated epidermis. This replacement can be traced in the digenean life cycle. When the first larval stage, the miracidium, enters the intermediate host, the ciliated epidermal cells are shed. Subepidermal stem cells have then already formed a syncytial 'new' epidermis, the neodermis, which persists during the following larval stages and in the adults. This syncytial neodermis is an autapomorphy of the Neodermata, but some free-living flatworms also show a tendency towards a syncytial epidermis. Ehlers (1985a) documents this, for example, by rudimentary membranes in the apical regions of the epidermal cells in the kalyptorhynch *Marirhynchus longaseta* (Ehlers 1985a, figure 10) and by the almost completely syncytial epidermis of *Provortex tubiferus* ('Dalyellioida') (Ehlers 1985a, figure 11).

Finally, syncytial tissues are present in hexactinellid sponges, but the reasons for this organization are unknown.

Conclusions

The ancestor of all metazoans likely had an epidermis with a basal extracellular matrix (ECM), an apical extracellular glycocalyx, and one cilium with a striated rootlet per cell. The ECM is an autapomorphy of Metazoa, but its components (such as collagen, fibronectin, and laminin) seem to be inherited, because they can be found also in non-metazoan taxa. In the bilaterian ancestor, a dense layer evolved within the ECM, the basal lamina (lamina densa). The glycocalyx is also a plesiomorphy of metazoans. It served as a matrix for the secretion of further molecules, thereby forming a cuticle, several times convergently. The most abundant cuticle molecules are collagen and chitin, but the physiological pathways for their production must have been present already in the metazoan ancestor. Cell–cell junctions appeared in a sequence of focal

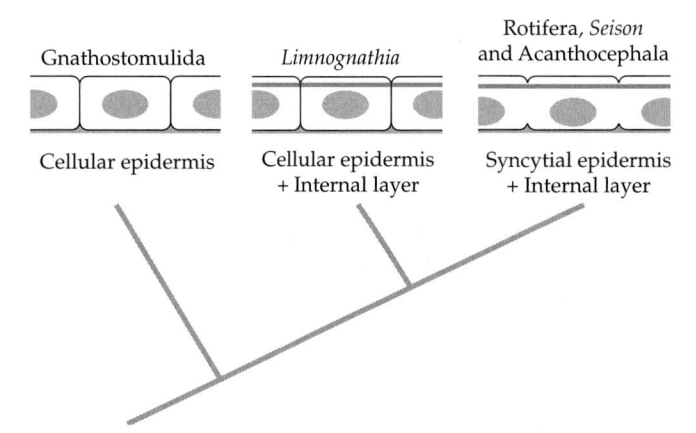

Fig. 4.24. Evolution of the epidermis within Gnathifera.

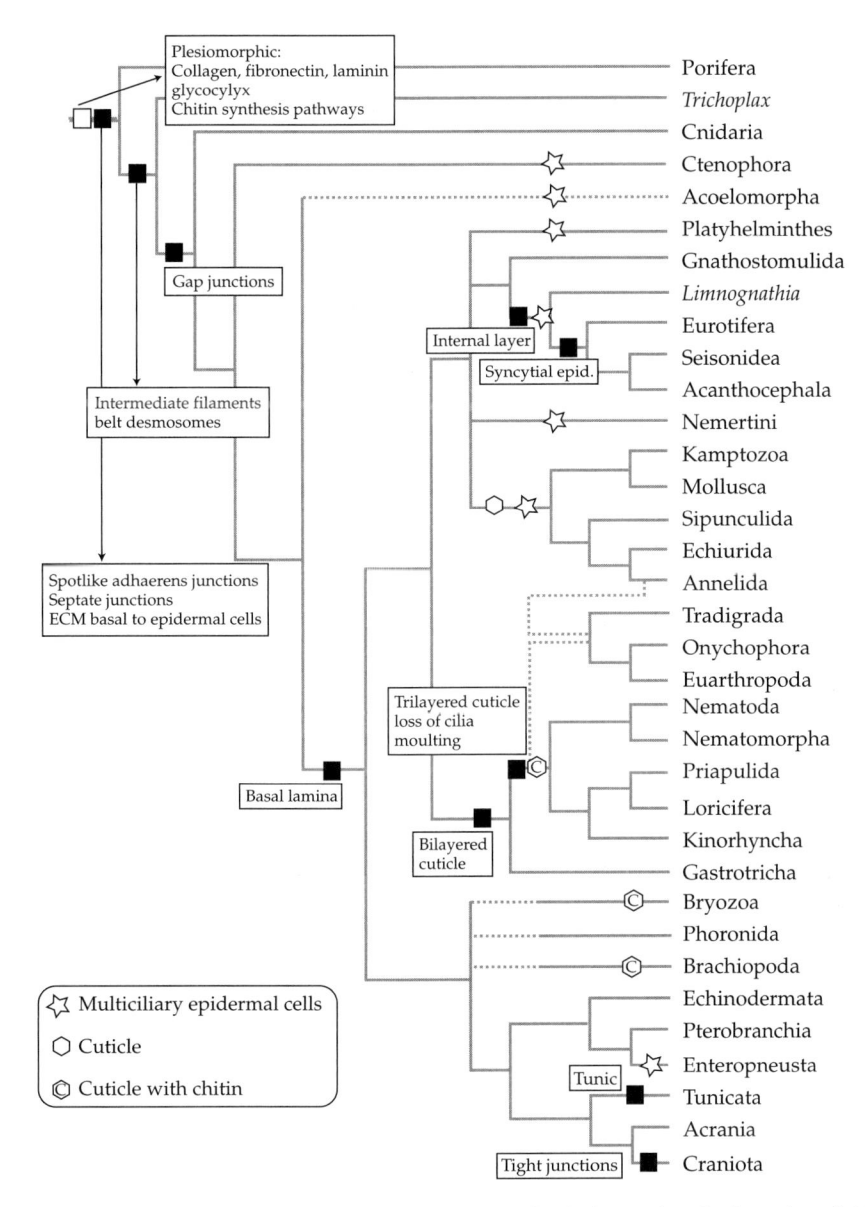

Fig. 4.25. Summary of the evolution of epidermal characters and the distribution of multiciliary epidermal cells, cuticle, and chitinous cuticle. See text for explanation.

adhaerens junctions and septate junctions in the metazoan ancestor, as belt desmosomes in the epitheliozoan ancestor, and as gap junctions in the eumetazoan ancestor. Cells were originally monociliate, but shifted to multicellularity several times. The reversal, a shift from multiciliate to monociliate cells, probably happened within polychaetes.

CHAPTER 5

Musculature

Muscles are organs capable of contraction and are therefore involved in motion, not only in locomotion, but also in motion of internal parts of the body such as the intestine or the heart. The molecular basis for muscle contraction is the association of two proteins, actin and myosin. Actin—at least the muscular actin—is a filamentous protein, while myosin has a filamentous 'tail' and one or two 'heads'. The myosin head includes binding sites for actin. Under dephosphorilation of ATP, the myosin heads change conformation, leading to a shift in position. This is accompanied by a temporary detachment from actin and results in a new binding of the myosin head on the actin filament adjacent to the former position (see Alberts et al. 2002 for a more detailed description). This movement of myosin along an actin filament (well known as the sliding filament model, going back to Huxley & Hanson 1954) is the basis for muscular contraction. The molecular and cytological roots for the evolution of muscles as recognizable organs can be traced back deep in the tree of life.

Actin and myosin are 'old' molecules

The basis for an understanding of the evolution of musculature is the cytoskeleton. Actin is, together with tubulin, a basic component of the cytoskeleton of eukaryotes. Both actin and tubulin are dynamic filamentous proteins, capable of elongation through polymerization and shortening. This makes them ideally suited to form an intracellular network and function in cellular stability as well as in cellular motion, for example during fission. While it was long thought that cytoskeletons are absent in prokaryotes, they have now been found to be present, at least in long and slender bacteria. These have helicoidally arranged filamentous proteins named MreB and Mbl, both of which are closely related to actin and can be regarded as their predecessors (Jones et al. 2001, van den Ent et al. 2001). Another molecule with a cytoskeletal function in prokaryotes, FtsZ, is related to tubulin (Nogales et al. 1998). It has been hypothesized that the stability caused by a cytoskeleton might have been an advantage during the engulfment of other cells (Hartman & Fedorov 2002) which led to a fusion of both cells (endosymbiosis, see Margulis 1981).

In eukaryotes, actin and tubulin are associated with motor-proteins, which can move along the filamentous pathways of the cytoskeleton. Motor-proteins associated with actin are from the myosin family, those associated with tubulin are dyneins and kinesins. Myosins underwent a vast diversification during evolution. While, for example, five myosin genes are known from the yeast *Saccharomyces cerevisiae*, the slime mold *Dictyostelium discoideum* possesses 12, the vascular plant *Arabidopsis thaliana* 17, and humans 40 genes (Berg et al. 2001, see also Thompson & Langford 2002). Motor-proteins seem to be absent in prokaryotes. The primary function of the actin–myosin system may be intracellular transport of organelles and other compartments. This function is retained by all eukaryotes (see, e.g., Sheetz & Spudich 1983 for the alga *Nitella* and Adams & Pollard 1986 for the protozoan *Acanthamoeba*). The intracellular function of actin and myosin, therefore, was present in the metazoan ancestor, but within metazoans, this system soon evolved into an intercellular system. It is interesting to note that, although there are two systems, both composed of a filamentous pathway and an associated mobile component, only the actin–myosin system has succeeded to form intercellular systems, while the tubulin–dynein–kinesin system remained intracellular.

An intercellular functioning of actin and myosin is, however, not unique to animals. It has been shown that in the multicellular alga *Volvox*, actin and myosin are involved in the inversion which occurs during development. New spherical *Volvox* colonies (this term is probably most appropriate) originate by a local invagination of the mother sphere, leading to small spherical colonies inside the mother sphere (Kirk 1998). Due to this mode of development, locomotory cilia are directed to the inside of the sphere, but before the daughter spheres completely detach from the mother sphere, an inversion takes place to expose the 'right' side (the one with cilia) to the outside. Actin and myosin are involved in this inversion (Nishii & Ogihara 1999).

From myocytes to epitheliomuscle cells to fibre muscle cells—muscle evolution in diploblastic animals

It seems clear that the metazoan ancestor inherited from its unicellular descendants an actin cytoskeleton and motor-proteins of the myosin superfamily. Within metazoans, these two molecules were arranged into effective contractile units, the muscles. The basic trends for muscle evolution are already expressed in the diploblastic taxa.

In sponges, both actin and myosin have been shown to be present (Harrison & De Vos 1991, Kanzawa et al. 1995). It has been long known that actin is enriched in a certain cell type named myocytes (see Bagby 1966 and references therein). Myocytes occur, for example, around the osculum of *Tedania ignis*, where they form something like a sphincter (Bagby 1966). Myocytes can also be distributed elsewhere and those in the pinacoderm and osculum of freshwater sponges such as *Ephydatia muelleri* appear to be responsible for coordinated contractions associated with expelling wastes. Here, it was shown that actin filaments form tracts traversing the cell and leading to intercellular junctions (Sally Leys, University of Alberta, Edmonton, personal communication). Therefore, adjacent myocytes can form functional units.

In *Trichoplax adhaerens*, actin appears to be present especially in the fibre cells between the dorsal and the ventral epithelium (Thiemann & Ruthmann 1989, Grell & Ruthmann 1991), but it is not clear whether these cells can be regarded as homologous to the sponge myocytes. In cnidarians, muscular elements are an important part of the body organization. Tentacle movement, contraction of the body in polyps, and swimming in medusae are performed by muscular action. Due to the epithelial organization of cnidarians, muscle-cells are not isolated, specialized cells, but are an integrative part of the epithelia, the so-called 'epithelio-muscle-cells' (EMC). This means that epithelial cells, in the epidermis as well as in the gastrodermis, concentrate actin and myosin filaments in their basal part (Fig. 5.1.). These basal parts often bulge into the basal extracellular matrix (ECM). These basal parts can form, together with those from adjacent EMCs, muscle strands or sheets (Fig. 5.1.). Some of the most extreme examples are the 'retractors' in the septa of anthozoan polyps. Even when these are visible under low magnification and appear as muscle strands (Fig. 5.2.), they are entirely composed of EMCs. In polyps of hydrozoans, cubozoans, and scyphozoans, the muscular processes of EMCs form distinctive muscle fibres. In hydrozoans, these fibres are oriented parallel to the oral–aboral axis of the polyp, i.e. longitudinally, while they are circularly oriented in the gastrodermis. In (at least some) scyphopolyps and cubopolyps some myofilament containing cells have lost their connection to the epidermis (Werner et al. 1976, Chia et al. 1984, Werner 1984, Lesh-Laurie & Suchy 1991). Such cells are called here 'fibre muscle cells' ('myocytes' according to Lesh-Laurie & Suchy 1991). Fibre muscles are the only muscular cells in ctenophores, no EMCs have been found in this group.

There are two models for the transition from EMCs to fibre muscle cells (Fig. 5.3.). From the mode, how the basal part of many cnidarian

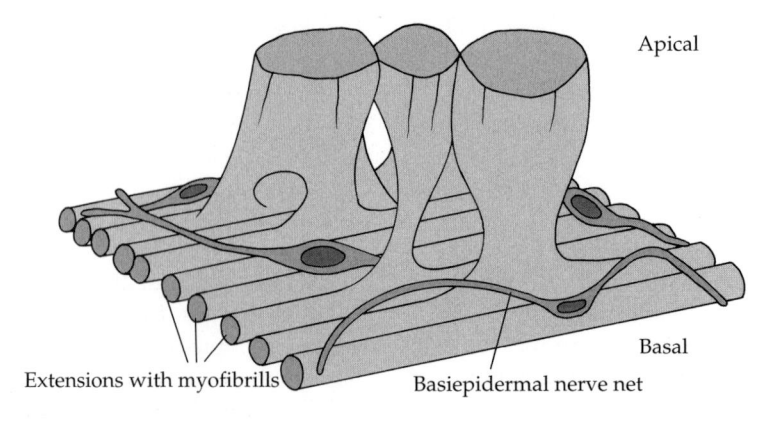

Fig. 5.1. Schematic representation of cnidarian epitheliomuscle cells (EMCs) with long and highly ordered basal myofilamentous parts. After Mackie & Pasano (1968).

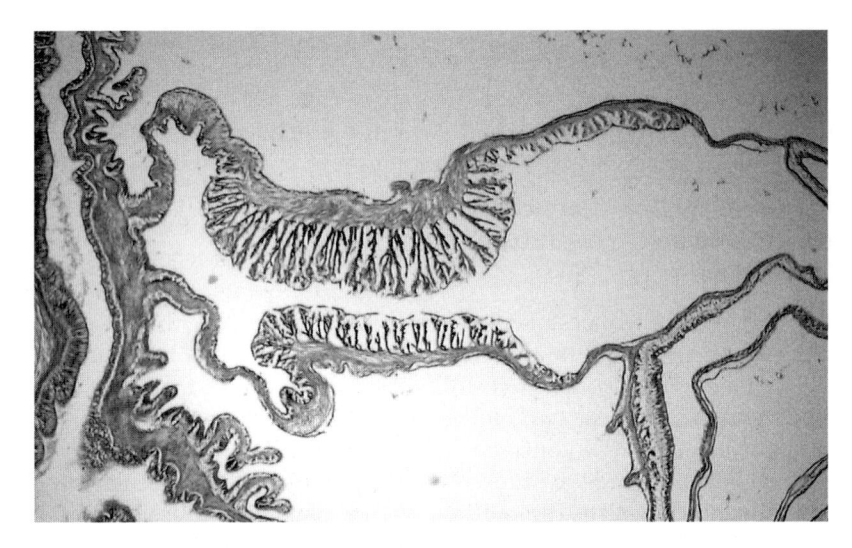

Fig. 5.2. Cross section through the trunk of a solitary sea anemone, *Methridium senile* (Anthozoa), showing septa with bulging portions that contain voluminous epithelio-muscle cells acting as retractors.

EMCs bulges into the ECM, it may be assumed that EMCs invade the ECM and then become detached apically. This is supported by observations, how muscular processes bulge into the ECM during contraction of tentacles in *Chrysaora quinquecirrha* (Perkins et al. 1971). The other model includes a pseudostratified stage, in which some epithelial cells have a voluminous basal myofilamentous part, but still reach the apical side by small extensions. Equally, epithelial cells with no myofilaments are voluminous in the apical part, but still contact the basal ECM. By secreting an ECM between the myofilamentous and the epithelial parts, both become separated as individual layers. This model is realized in the evolution of coelom epithelia (see Chapter 8; Rieger & Lombardi 1987), but for the origin of fibre muscle cells in diploblastic animals, an individual ingression of EMCs into the ECM appears to be more likely.

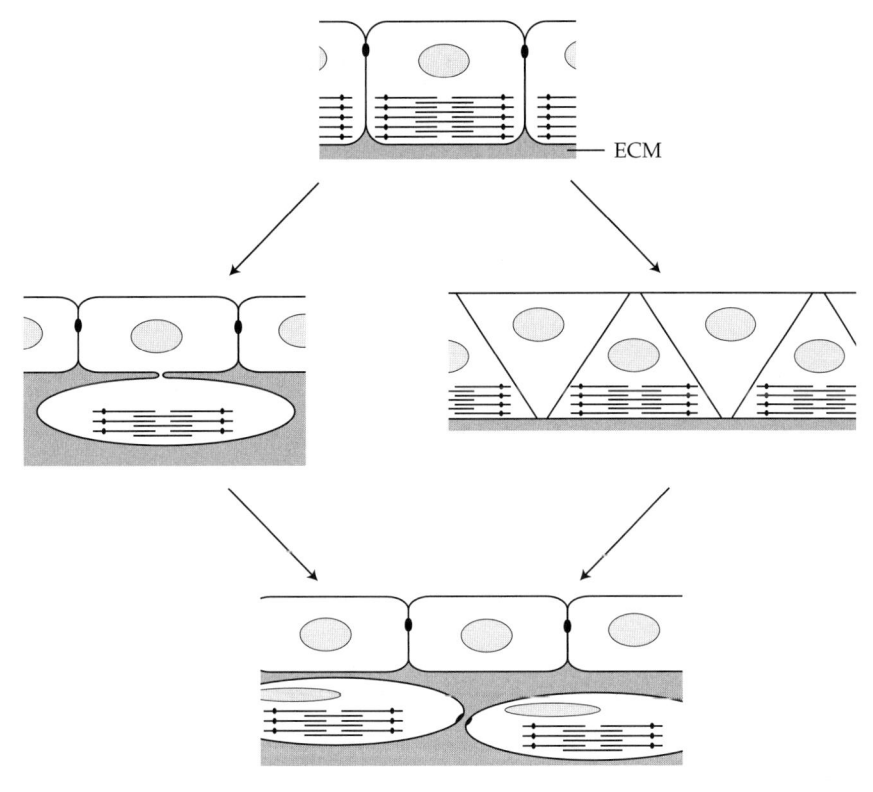

Fig. 5.3. Two models for the evolution of fibre muscle cells. To the left, the basal part invades the ECM and subsequently separates from the apical part. To the right, the epithelium becomes pseudostratified and then cells are separated by ECM into apical epithelial and basal muscle cells.

Myofilament arrangement patterns

The transition from an intracellular system where myosin 'travels' along actin pathways to myocytes and other types of muscle cells, that is, systems capable of effective contraction, was made by a specific, linear arrangement between actin and myosin. Actin is bound to structures with a stabilizing function, the Z-elements. A bipolar myosin molecule moves between two actin molecules (see Fig. 5.12.). This arrangement, including a Z-element – actin filament – myosin filament – actin filament – Z-element is called 'sarcomere' (this is a minimal definition of a sarcomere, often the term sarcomere is applied only when several individual actin–myosin systems are in register and are as such clearly recognizable in electron miscrographs and even with light microscopy). By the movement of myosin relative to the two actin filaments the whole sarcomere shortens, which leads to a contraction of the whole cell. A concatenation of sarcomeres and several muscle cells leads to an increase of power and allows an organism to contract for a significant distance and deal with stronger forces, e.g. in locomotion of large-bodied animals where ciliary movement is not effective.

Sarcomeres can be arranged in different ways that characterize three main types of muscles (Fig. 5.4.). In 'smooth muscles', the sarcomeres are irregularly distributed, that is, Z-elements are small dotlike structures (sometimes named Z-dots or dense plagues) and not aligned with other Z-elements. When Z-elements are aligned in a row perpendicular to the longitudinal axis of the cell, the muscle is 'cross-striated'. Z-elements can also be arranged obliquely to the longitudinal axis of the cell, this type is called 'oblique striation'. Because Z-elements are often arranged helicoidally around a cytoplasmatic center of the cell, obliquely striated muscles are sometimes also called helicoidal muscles.

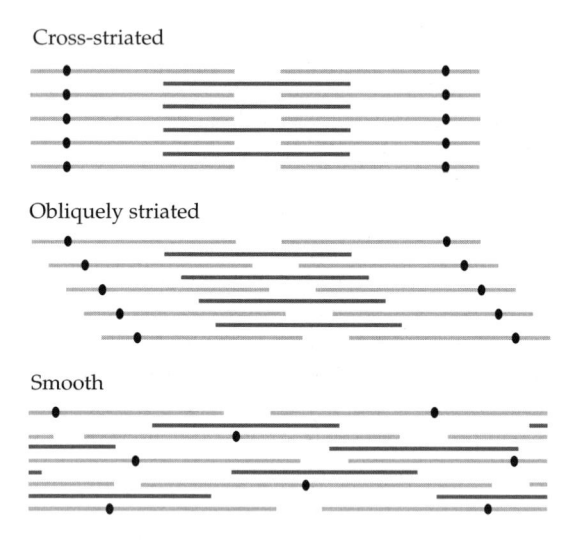

Fig. 5.4. Arrangement of sarcomeres into cross-striated, obliquely striated, and smooth musculature.

In cross-striated muscles of (at least adult) Euarthropoda, Chaetognatha, and Myomerata, the aligned Z-elements are tightly interconnected to form Z-lines or, more appropriately, Z-discs. In other muscles with an apparent cross-striation (e.g. in gnathostomulids and kinorhynchs), singular Z-bodies are aligned in a line which is never so regular as is the Z-disc. This seems to be also the case for some insect larvae (Candia Carnevali & Valvassori 1981). In obliquely striated muscles, Z-bodies, or slightly elongate Z-rods, are staggered with respect to their neighbouring Z-elements, creating the obliquely striated appearance.

In obliquely striated muscle cells, myofilaments are often located in the periphery, surrounding a central cytoplasmatic core. Three main types of muscle cells can be distinguished (two of which have already been mentioned by Schneider in 1860). In circomyarian cells, the filaments cover the entire periphery and therefore completely surround the cytoplasmatic core (Fig. 5.5., 5.6.b). In platymyarian muscle cells the myofilaments are concentrated on one (usually the apical) side of the cell while nucleus and cytoplasm are located on the other side

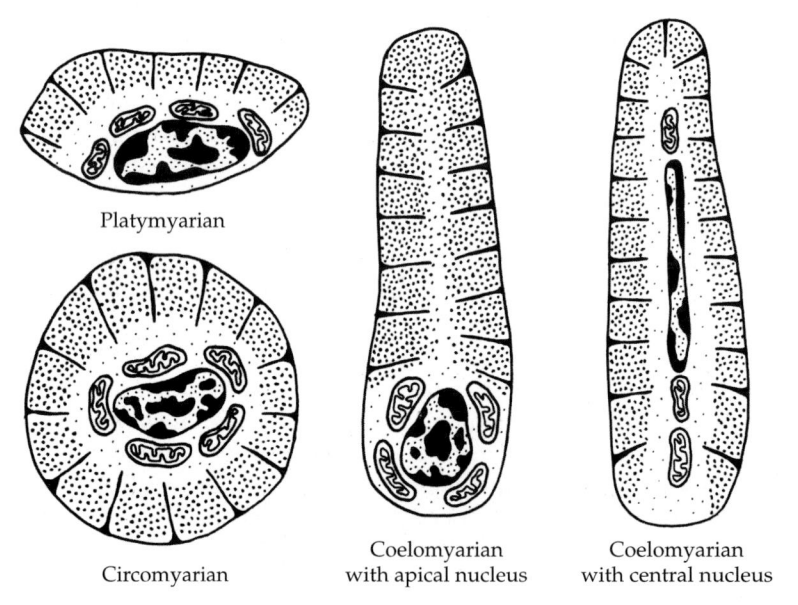

Fig. 5.5. Myofilament arrangement patterns in obliquely striated muscle cells as platymyarian, circomyarian, coelomyarian with apical cytoplasmatic core, and coelomyarian with central cytoplasmatic core. Modified after Lanzavecchia et al. (1988).

(Fig. 5.5.). Coelomyarian cells are elongated along a proximo–distal axis (in relation to a cross section of the animal) and myofilaments occur in a U-shaped pattern along the lateral and the distal side of the cell (Fig. 5.5., 5.6.a). Their cytoplasm is located proximally and in a narrow core between myofilaments. This core can also be reduced, so that myofilaments from both sides of the cell attach to each other (Fig. 5.5.). A border between these two sides, indicating that such muscle cells are derived from coelomyarian cells, is usually visible (e.g. Hope 1969 for the nematode *Deontostoma californicum*).

The distribution of such muscle types has been intensively studied in nematodes and in annelids. The body wall of nematodes contains only longitudinally arranged muscle cells. The type of muscle cell present appears to depend on the size of the animal. Small species often have few (as few as four) platymyarian muscle cells while large species often have a high number (up to 600 in *Ascaris*) coelomyarian muscle cells (see, e.g., Chitwood & Chitwood 1950). Such relation to body size is repeated during development. In *Ascaris*, fewer platymyarian cells are present in juveniles, they multiply and differentiate to coelomyarian cells during development (Thurst 1968, Stretton 1976). The same has been shown for nematomorphs, which as juveniles have few platymyarian cells and as adults many coelomyarian longitudinal muscle cells (Schmidt-Rhaesa 2005). Circomyarian muscles are also (but rarely) present in nematodes (e.g. in *Leptosomatum acephalatum*, see Timm 1953) and nematomorphs (e.g. in *Nectonema munidae*, see Schmidt-Rhaesa 1998; Fig. 5.6.b). In annelids, all three types of muscle cells, and derivates from them, occur (see Lanzavecchia et al. 1988, Gardiner 1992). In oligochaetes and polychaetes there are often numerous longitudinal muscle cells in the body wall and they resemble at first sight nematode coelomyarian cells (see Orrhage 1962). Annelid muscle cells often differ from nematode cells in lacking a central cytoplasmatic core and they are often regarded as derived from circomyarian cells (Lanzavecchia et al. 1988, Gardiner 1992). In hirudineans, clear circomyarian muscles are present in the body wall (Lanzavecchia 1977, Fernández et al. 1992). It appears, therefore, that the distribution of filaments within the obliquely striated cell (as well as the number of cells) depends on body size.

There are some special cases of striation patterns worth noting. Myofilaments may be arranged with different distributions within one cell, this is known from some annelids, and a phoronid (see Lanzavecchia et al. 1988). In chaetognaths, the massive longitudinal trunk muscles are composed of cross-striated muscle cells, but among them is a special type named secondary muscle (about one per cent of trunk wall muscles; Duvert 1991, Casanova & Duvert 2002). In these, 'typically' composed sarcomeres alternate with sarcomeres containing only one type of filament (therefore the secondary muscle is also called heterosarcomeric muscle).

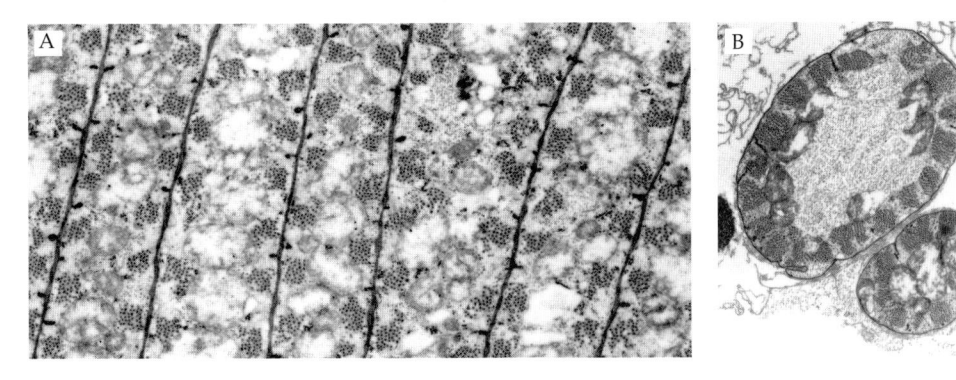

Fig. 5.6. Coelomyarian (A) and circomyarian (B) muscles in the nematomorph *Nectonema munidae*. From Schmidt-Rhaesa (Invertebrate Biology 117, 1998), with kind permission of Blackwell Publishing, Oxford.

Immunocytochemical investigations have shown that actin and the regulatory protein tropomyosin are found in these atypical sarcomeres, but not myosin and paramyosin (Royuela et al. 2003). Another special pattern was found in some of the longitudinal muscles in the third segment of the kinorhynch *Pycnophyes kielensis*. Here, the sacromere length decreases from anterior to posterior from 7.5 µm to 2.5 µm within the same muscle (Schmidt-Rhaesa & Rothe 2006).

Functions of striation patterns

Why are there different striation patterns in muscle cells? The three types differ in their measurable capacities concerning the degree of contraction, the force developed during contraction, and the speed of contraction, although such a statement is a generalization and the parameters mentioned above depend on more factors than the striation pattern alone. In general, cross-striated muscle cells have shorter filaments compared to smooth and obliquely striated muscles. Shorter filaments reach the limits of their contraction quickly, therefore they are very fast but they can develop force only in a limited range. The minimal length of a sarcomere is determined by the complete overlapping of actin and myosin filaments, while the maximal length is defined by the minimal overlap of both filaments, that is, the sum of actin and myosin filament lengths. Tension can, in cross-striated muscles, only be developed in the range between maximal and minimal length, and therefore only in a comparatively small, well defined range. Such a limitation of range is provided in animals with a skeleton, either the internal skeletons working as lever systems in craniotes or external skeletons in euarthropods.

There are, however, exceptions from such a strict range of concentration. There are cases in which a contraction beyond the minimal sarcomere length is possible in cross-striated muscles with Z-discs. The Z-discs can be perforated and allow myosin filaments to pass through during contraction. This phenomenon is called 'supercontraction' and has been shown in euarthropods, for example soft-bodied insect larvae (Osborne 1967, Beinbrech 1998), chaetognaths (Casanova & Duvert 2002), and in tongue muscles in chameleons (Herrel et al. 2002).

There are more factors than just sarcomere length that determine the contraction properties of a muscle. Candia Carnevali and Saita (1976) compared ventrolateral and dorsal muscles of the pill woodlouse *Armadillidium vulgare* (Isopoda, Crustacea). As the name says, *Armadillidium* can roll up to a completely round ball. This is done by the contraction of ventrolateral muscles, which should therefore be able to develop more tension compared to the dorsal muscles which unroll the animal. Ventrolateral muscles have longer myosin filaments than dorsal ones (7 µm long, 200Å thick compared to 4 µm/150Å, respectively), they have a higher density of actin filaments around each myosin filament (actin to myosin filament ratio from 6:1 to 7:1 compared to 4:1), and the density of myosin filaments is higher (610 filaments per μm^2 compared to 510 filaments per μm^2). In another paper, Candia Carnevali & Valvassori (1982) investigated the diplopod *Glomeris marginata*, the 'pill millipede', which is able to roll up like *Armadillidium vulgare*. Here, different properties of muscles are performed by the potential of ventrolateral muscles to supercontract.

In soft-bodied animals, cross-striated muscles (with aligned Z-elements) appear to occur in correllation with fast movement in a defined, predictable range. The (ten-armed) squid *Loligo vulgaris*, for example, has obliquely striated muscles in its eight arms, but cross-striated muscles in the two arms that are adopted for prey capture and are capable of rapid elongation (Kier & Curtin 2002). Among flatworms, which generally have smooth muscles, cross-striated muscles occur in the the tail of cercaria of several digenean species (Lumsden & Foor 1968, Rees 1971, Chapman 1973, Nuttman 1974). Cercaria are the last larval stage, actively moving from the mollusc intermediate host to the vertebrate final host. This is performed by a regular beating of the tail, which appears to be a typical case for cross-striated muscles.

Obliquely striated muscles show much greater variability in contraction. Hirudinean muscles, for example, can be stretched about five times their contracted length and still develop tension (Lanzavecchia et al. 1988). It seems that the dotlike shape of Z-elements (Z-bodies) and their staggered arrangement are responsible for this phenomenon. It is likely that myosin can move beyond the Z-body during contraction and therefore contract beyond the minimal sarcomere length as defined above. During stretching, a slight shearing might allow contact to adjacent actin filaments (for the shearing partners hypothesis, see Lanzavecchia 1981, Lanzavecchia et al. 1988) and therefore lead to the development of tension during stretching above the maximal sarcomere length as defined above. In contrast to cross-striated muscles which are (not exclusively, but typically) found in animals with some kind of solid skeleton, obliquely striated muscles are often associated with hydroskeletons (Lanzavecchia 1977, 1981).

T-tubules and sarcoplasmatic reticulum

We know from the fine structure of craniote cross-striated muscle cells that two kinds of membraneous systems can be present in muscle cells: the t-tubules and the sarcoplasmatic reticulum (SR). T-tubules are infoldings of the cell membrane, and the SR is a derivative of the golgi complex and therefore an intracellular membraneous system. The membrane of the t-tubules carries action potentials deep into the muscle fibres. They are closely associated with the SR that surrounds each muscle fibre. The contacts between T-tubules and SR are marked by large proteins and are called dyads or triads (referring to the contact of a t-tubule and one SR, or one t-tubule and two SR cisterna, respectively). The stimulation of certain proteins in the t-tubule membrane activates large calcium channels located in the SR membrane which then release calcium ions to the cytoplasm of the muscle cell. This is the signal for sarcomeres to contract. The SR is therefore a reservoir for calcium ions. In fact, a SR has been found in almost all muscles (see Table 5.1.), although t-tubules are not always present.

Membraneous systems are very weakly developed in muscle cells of diploblastic taxa. In cnidarian epitheliomuscle cells, both t-tubules and SR may be absent, but Smith (1966) notes that some small vesicles close to the mitochondria may represent the SR (see also Keough & Summers 1976, Singla 1978a). In ctenophore myocytes, Hernandez-Nicaise (1991) describes a large network of the endoplasmatic reticulum, which probably corresponds to the SR. In many taxa, especially from basal bilaterians (*Xenoturbella*, Acoelomorpha, platyhelminths, gnathostomulids, nemerteans; see Table 5.1.), the SR is only relatively weak. One may speculate that this must have an influence on the amount of released calcium ions and therefore on contraction speed and duration. Hill et al. (1978) approached this question in investigating the longitudinal body wall muscles in the holothurian *Isostichopus badionotus*, which has only scarce SR and no t-tubules. Incubation in a calcium-free solution leads to an inability to contract after only ten hours. This suggests that there must be other storage systems for calcium and the contraction capacities of muscles are not directly evident from the amount of SR. However, in euarthropods and craniotes, with their well developed t-tubules and SR in combination with cross-striated muscles, there appears to be a correllation between the contraction speed and the amount of membraneous systems present (Smith 1966). Craniote fast muscles, for example, have well developed SR and t-tubules, while slow muscles have no t-tubules and only weak SR (Smith 1966). Josephson and Young (1987) found that the duration of twitches in cicadas is corellated to the amount of t-tubules and SR, as well as to myofibril size. There are also cases in which the SR is extremely abundant, although the function is not clear. In the leg muscles of wasps the SR forms numerous sheets around the muscle fibre (Weaving & Cullen 1978).

Evolution of striation patterns

It is tempting to try and draw conclusions on the evolution of striation patterns, but this has to be done with extreme caution. First, each type of striation pattern appears to be widespread in the phylogenetic tree. Second, each striation type has functional advantages and appears to occur (often, but not always) in response to such functional needs. It should be mentioned briefly, that it is not always straightforward to recognize the exact type of striation and, therefore, misinterpretations are possible. Rosenbluth (1972) has already indicated that the oblique striation pattern is sometimes hard to distinguish from smooth muscles. This also accounts for the recent staining methods of actin with phalloidin labelled with a fluorescent dye. While the cross-striated pattern is clearly visible (see, e.g., Müller & Schmidt-Rhaesa 2003), obliquely striated muscles appear to be smooth. Obliquely striated muscles may also be confused with cross-striated muscles when sectioned in a certain plane. In the XY-plane (according to Rosenbluth 1972) Z-bodies are aligned in line (see Fig. 5.7.).

Let us take a brief view on the distribution of striation patterns among metazoans (for summary see Table 5.1.).

In cnidarian epitheliomuscle cells, generally no ordered arrangement pattern can be recognized so that these cells are considered as smooth (see, e.g., West 1978, Thomas & Edwards 1991). There are, however, several exceptions. Hydromedusae often contain a cross-striated arrangement of fibrils in their subumbrella (Fig. 5.8.; e.g. Chapman et al. 1962, Keough & Summers 1976, Germer & Hündgen 1978, Singla 1978b, Lesh-Laurie & Suchy 1991). Regions of cross-striation have also been found in the tentacle tips of cubopolyps (Chapman 1978, Werner 1984) and different muscles of the scyphopolyp of *Aurelia aurita* (Chia et al. 1984), but in both examples the striated pattern fades into a smooth pattern. In ctenophores, smooth muscles predominate, but cross-striated muscles have been found in the prehensile tenilla on the tentacles of *Euplokamis* sp. (Mackie et al. 1988). These tentilla are capable of very fast coiling and uncoiling and function in food capture. This distribution of muscle cell types in non-bilaterian animals makes it likely that an unordered arrangement of sarcomeres (smooth muscles) is the plesiomorphic condition, but that sarcomeres can 'easily' be aligned to a cross-striated pattern when functional needs (i.e. regularly repeated movement in a defined range) are present.

In almost any bilaterian taxon, more than one type of striation is realized, but often one particular type is predominant. In gastrotrichs, the situation is already complex. Most species appear to have obliquely striated muscles (see Ruppert 1991b, Hochberg & Litvaitis 2001a), but several species (from the genera *Dactylopodola, Xenodasys, Neodasys, Draculiciteria, Musellifer*) have cross-striated muscles (Fig. 5.9.; Ruppert 1991b, Hochberg & Litvaitis 2001a, Hochberg

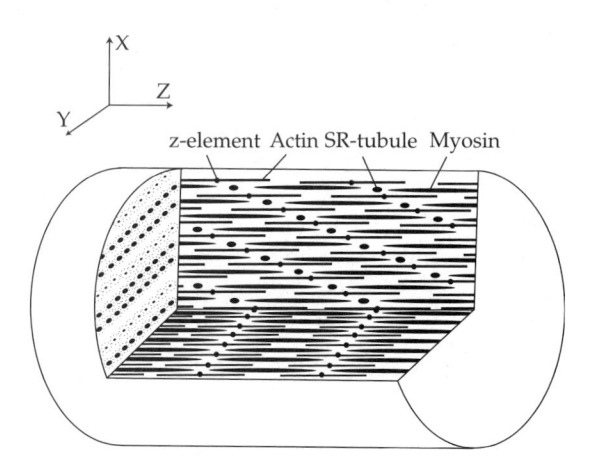

Fig. 5.7. Arrangement of filaments from an obliquely striated muscle fibre in different sections. After Rosenbluth (1972).

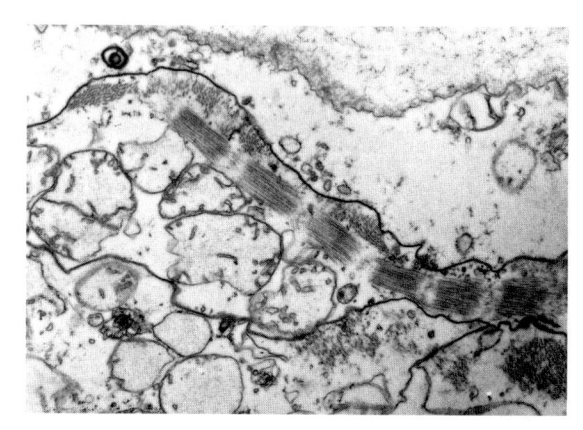

Fig. 5.8. Cross-striated epitheliomuscle cell in the bell margin of the medusa from *Obelia* sp. (Cnidaria, Hydrozoa).

2005). *Lepidodasys* sp. has smooth muscles (Ruppert 1991b). In several cases, the type of striation can be correllated with the locomotory behaviour, especially with escape reactions (Ruppert 1991b). These are rapid, but predictable repeated ventral flexions in *Dactylopodola* and *Xenodasys*, resulting almost in jumping. *Turbanella* (with obliquely striated muscles) escapes by crawling backwards with stretching the posterior part of the body to attach the posterior adhesive tubules. *Lepidodasys*, in contrast, is generally slow. These behaviours correspond well with the characteristic properties of the striation patterns (see above). Obliquely striated and cross-striated muscles occur in both gastrotrich subtaxa, Macrodasyida, and Chaetonotida, so that an ancestral condition can not be reconstructed and it appears that the type of striation is influenced more by function than by phylogeny.

Such cases, as exemplified here for gastrotrichs, occur in many taxa. Flatworms and acoelomorphs have smooth muscles (Rieger et al. 1991b), but, as mentioned above, the tail of digenean cercaria, which performs a regular beating, contains cross-striated muscles. Cross-striated muscles have also been found in the acoel *Convoluta pulchra* (Tyler & Rieger 1999). All three types of muscle cells have been found in eurotifers, molluscs, annelids, bryozoans, brachiopods, and tunicates (see Table. 5.1.). In fact, the restriction to only one striation pattern appears to be the exception. This is quite remarkable in euarthropods. Here, the plasticity in striation patterns was obviously lost and all muscles are cross-striated (Figs. 5.10, 5.11.). On second view, there is, however, some flexibility within this system. Soft-bodied larvae, for which obliquely striated muscles could be predicted based on the characteristics outlined above, have cross-striated muscles, but with independant Z-elements which create some more flexibility

Fig. 5.10. Cross-striated muscle in longitudinal section from an expanded abdominal extensor muscle in the lobster, *Homarus gammarus*. Photo by courtesy of Andreas Unger, Bielefeld.

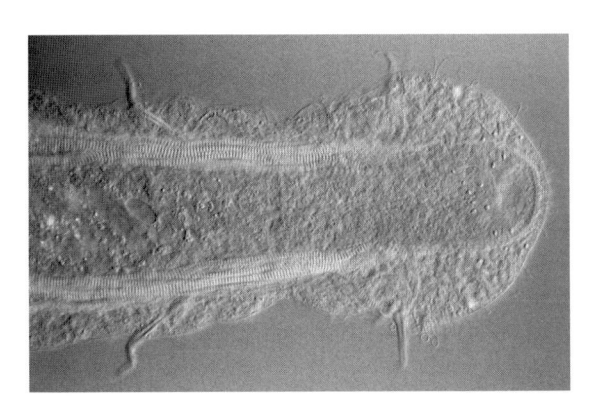

Fig. 5.9. Cross-striated longitudinal muscles in the anterior end of *Dactylopodola baltica* (Gastrotricha). Photo by Birgen H. Rothe & A. Schmidt-Rhaesa, Hamburg.

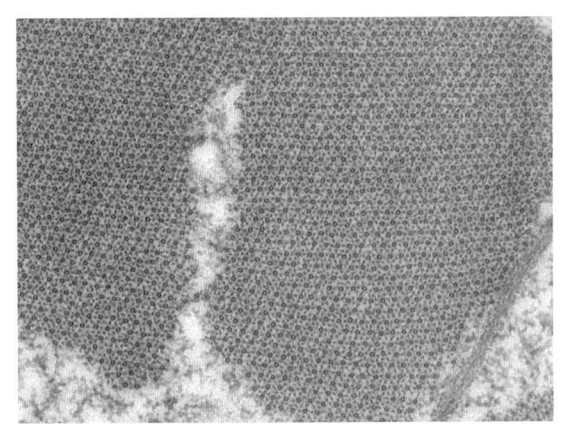

Fig. 5.11. Cross section through the crusher muscle from the lobster (*Homarus gammarus*) showing arrangement of thick (myosin) and thin (actin) filaments. Photo by courtesy of Andreas Unger, Bielefeld.

than a Z-disc (Candia Carnevali & Valvassori 1981). Additionally, the potential of several cross striated muscles for supercontraction (Beinbrech 1998) overcomes the limitations of cross-striation.

The functional constraints for the occurrence of certain types of striation pattern become additionally evident when two types of striation occur in one and the same muscle (bundle). A well known example is the adductor muscle of lamellibranch bivalves, which has a smooth and a cross-striated portion (see Morse & Zardus 1997 and references therein). This is also the case in adductor muscles of brachiopods where articulate brachiopods have smooth and cross-striated portions, while lingulids have a combination of smooth an obliquely striated muscles (James 1997). In several muscles of the bryozoan *Asajirella gelatinosa*, for example in the tentacle, only the basis is cross-striated while the apical portion is smooth (Mukai et al. 1997).

Some muscles have been described as 'intermediate' or 'showing transition' to another type. In Platyhelminthes and Nemertini, for example, indications of oblique striation have been observed among the dominant smooth musculature (Turbeville & Ruppert 1985, Rieger et al. 1991b). The musculature of tardigrades has been considered as intermediate between oblique and cross-striated (Dewel et al. 1993, Walz 1974, 1975a). According to Schmidt-Rhaesa & Kulessa (2007), only a few muscles show cross-striation, while the majority do not. In onychophorans, few sarcomeres are aligned, and such clusters of aligned sarcomeres are distributed irregularly with respect to each other (Camatini et al. 1979, Lanzavecchia & Camatini 1979). This pattern may be considered as a kind of precursor of cross-striation as present in euarthropods.

Finally, the plasticity of striation types can be inmagined from transdifferentiation experiments. Under certain conditions, isolated cross-striated muscles from the hydormedusa *Podocoryne carnea* differentiate into different cell types, including smooth muscles (Schmid & Alder 1984, 1986).

The question remains whether a certain direction in the evolution of striation patterns can be identified. There are some taxa with predominant smooth musculature, in which either obliquely striated or cross-striated muscles occur in likely derived cases (Cnidaria, Platyhelminthes, Nemertini, Mollusca, Bryozoa, Brachiopoda, Pterobranchia). There are also cases in which a derivation of cross-striated from obliquely striated muscles appears likely (in priapulids and probably also in annelids) and where oblique striation is derived from cross-striation (nematomorphs). The first case accounts, for example, for the cross-striated longitudinal muscle cells in the larva of *Tubiluchus corallicola* (Higgins & Storch 1989), while obliquely striated muscles are predominant in other priapulids (Storch 1991). In the second case, nematomorph larvae have cross-striated muscles (Müller et al. 2004), but in adults these are obliquely striated (Lanzavecchia et al. 1979). In conclusion, it appears that, as long as Z-elements are dotlike Z-bodies, they are extremely variable and can form, according to functional requirements, different patterns of arrangement. Only when Z-elements become tightly interconnected to form Z-discs, does this plasticity appear to become lost.

Molecular components other than actin and myosin

The cross-striated skeletal muscle cells of vertebrates are quite well known, better than most muscles in invertebrates. Apart from actin and myosin, several other protein components shape the contractile system, some of which still remain uncharacterized. For example, the Z-elements, arranged as a planar disc, contain f-actin, α-actinin, CapZ, zeugmatin, amorphin and other proteins (Fig. 5.12.; Vigoreaux 1994). The Z-discs hold actin filaments in place, interconnect adjacent filaments, and prevent their depolymerization. The other, free ends of the actin filament are stabilized by tropomodulin. Nebulin determines the length of the actin filaments (Fowler et al. 2006). The myosin filaments are also anchored in the Z-disc by an elastic protein called titin.

The molecular components of obliquely striated and smooth muscles are not as well

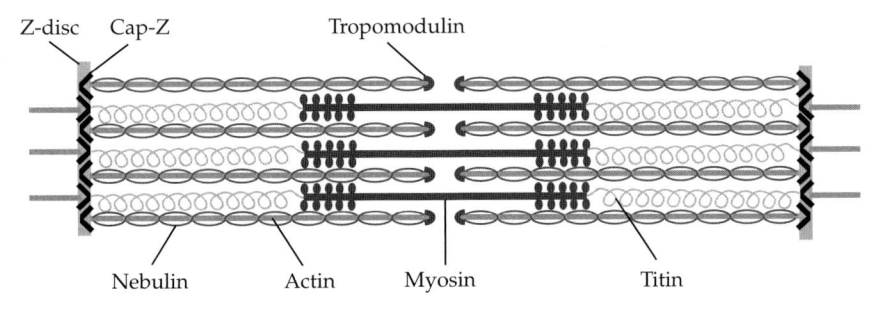

Fig. 5.12. Schematic representation of several molecules in a craniote sarcomer. Modified after Alberts et al. (2002).

studied as in vertebrates and so the distribution of components is not completely clear. It appears as if some components of the Z-elements are restricted to invertebrates (kettin and others, see Vigoreaux 1994) while others are restricted to vertebrates (zeugmatin, amorphin, and others, see Vigoreaux 1994). Nebulin, titin, and α-actinin appear to be widespread (see, e.g., Royuela et al. 2000a for their occurrence in mites). To date, too few details appear to be known to make fundamental statements about the distribution of muscular proteins other than actin and myosin (see Hooper & Thuma 2005 for review), and phylogenetic analyses based on muscular proteins (Oota & Saitou 1999) are based on data from very few taxa. Royuela et al. (2000b) tried to identify proteins as markers of a certain muscle type. They investigated different muscle types for the occurrence of troponin, which was known from cross-striated muscle cells in craniotes, and of caldesmon and calponin which were known from smooth muscles of craniotes. In invertebrates, however, obliquely striated muscles in the annelid *Eisenia foetida* contained troponin, but the obliquely striated buccal muscle of the mollusc *Helix aspersa* contained caldesmon and calmodulin.

One molecule that is widespread among invertebrates is paramyosin (Paniagua et al. 1996). It forms the core of the the myosin filaments and therefore obviously stabilizes it so that myosin filaments can be longer than without paramyosin. This means that tension can be maintained over a broad contraction range, but contraction is comparably slow due to the length of the filament. Paramyosin may already be present in cnidarians (Perkins et al. 1971, Chapman 1978) and was detected in flatworms (Ishii & Sano 1980), annelids (Camatini et al. 1976), molluscs (Pante 1994), nematodes (Epstein et al. 1985), brachiopods (Eshleman et al. 1982), nematomorphs (Swanson 1971), euarthropods (Bullard et al. 1973, de Villafranca & Haines 1974), echinoderms (Byrne 1994, Heinzeller & Welsch 1994), and in the chorda of *Branchiostoma lanceolatum* (Guthrie & Banks 1970), which is a derivative of musculature.

Antagonistic systems

It is important to remember that muscles are capable of developing force only in one direction. Relaxation does not automatically lead to a restoration of the original shape of the muscle. Therefore, muscles always function as antagonistic systems. In addition to this, muscles need an antagonist onto which muscular force can be transferred. Such antagonists can be hard elements of the body such as hard skeletons (in chordates and echinoderms) or the cuticle (in cycloneuralians, arthropods, and others). When hard elements are lacking, muscular action is transferred to fluid-filled compartments. In a sense, this is the case in compactly organized animals, where fluid is in the cells, and in the small intercellular interstices, but in many animals larger fluid-filled cavities are present (see Chapter 8). An enclosed mass of fluid is an ideal antagonist for muscular action, because fluid can hardly be compressed and therefore has to expand into another direction. A well known example for such interaction between a fluid-filled compartment and the musculature is realized in the locomotion of earthworms. Here, the whole system is segmentally arranged into units, each composed of paired coelomic sacs, circular and longitudinal musculature (Fig. 5.13.). Contraction of the longitudinal musculature will cause the coelomic sacs to shorten, but extend into the radial direction, i.e. the segment becomes short but thick. Contraction of circular musculature will make the segment long but thin. In this way, waves of contraction can be produced that aid in burrowing.

Also very important is the attachment of musculature. As muscle cells (if they are not epithelio-muscle cells) are completely surrounded by ECM, their primary attachment is to ECM. They may then attach to bones or mesodermal skeletal elements, but when they attach to the cuticle, there is always the epidermis that separates muscles from cuticle. In these cases, the epidermis is usually very thin and a dense array of intermediate filaments crosses the cell from basal to apical, making force transduction from the musculature to the cuticle very effective (see, e.g. Lai-Fook & Beaton 1998 for insects, Wright 1991 for nematodes and Müller et al. 2004 for nematomorphs).

The arrangement of body wall muscles in bilaterian animals

The muscular systems of most organisms are complex, but there is one widespread feature which is probably plesiomorphic and from which several muscular patterns can be derived. This is a continuous layer of musculature under the subepidermal ECM which is composed of outer circular and inner longitudinal muscles (Fig. 5.13, 5.14.). This whole complex has been named with the german word *Hautmuskelschlauch*, for which I will use the abbreviation HMS in the following. An HMS is abundantly distributed in animals without hard skeletons, such as platyhelminths, priapulids, onychophorans, nemerteans (Fig. 5.14.), sipunculids, echiurids, annelids (Fig. 5.13.), phoronids, and holothurians. Because it is assumed that the bilaterian ancestor was wormlike and had no hard skeleton, it appears quite likely that it also had an HMS.

Fig. 5.13. Transmission electron micrograph of the arrangement of the HMS in an undetermined 'oligochaete' clitellate (Annelida).

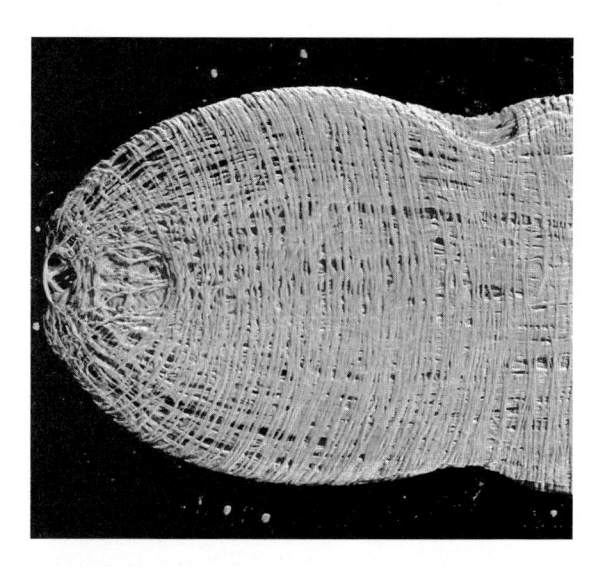

Fig. 5.14. The HMS in the anterior end of *Ototyphlonemertes* sp. (Nemertini), visualized by phalloidin staining and confocal laser scanning microscopy. Photo by Birgen H. Rothe & A. Schmidt-Rhaesa, Bielefeld.

The embryonic origin and development of the HMS has been studied in only very few taxa, but this may yield some information about its evolutionary origin. A study by Ladurner & Rieger (2000) is very informative on the muscle development in an acoel, *Convoluta pulchra*. Staining of actin with phalloidin showed that in early embryonic stages actin is present only intracellularly in the region of the adhaerens junctions. The first muscles are detected after half of the time between egg-laying and hatching has passed. These first muscles are isolated fibres, oriented circularly around the body and arranged at regular distances from each other. Such muscle fibres fuse to form four to five bands of circular musculature. Longitudinal muscle fibres appear briefly after circular fibres in the same mode. The result is a gridlike pattern of circular and longitudinal musculature. This grid is then, in a second phase of development, used as a template, along which further muscle fibres occur. These fibres line up in parallel to the primary grid and may use this as a guide. These 'secondary' fibres eventually become more abundant and form a complex, but regular, grid of circular and longitudinal muscles. The same accounts for muscle development in the sipunculid *Phascolion strombus* (Wanninger et al. 2005a). Here, circular muscles occur synchronously and not successively during development. These results are interesting for two reasons. First, it may be speculated that the regular occurrence of relatively few fibres at equal distances is correlated to the anteroposterior patterning of a bilaterian, elongate animal. Regionalization is very important for animals with an anteroposterior axis, and mechanisms comparable to the pair-rule genes (see, e.g., Carroll et al. 2005) could be responsible for creating the regularly iterated starting points of the muscular grid (see also Chapter 3). Second, Ladurner & Rieger (2000) point out that the growing of muscle fibres along a grid of primary fibres (founder cells) appears to be a widespread pattern, as has been shown especially macrostomid flatworms (Reiter et al. 1996) and in leeches (Jellies & Kristan 1988, 1991).

It is sometimes speculated that the HMS may be derived from the sheets of muscles formed by EMCs in cnidarian polyps, because these form one sheet of longitudinal and one sheet of circular musculature. Because the arrangement of muscular sheets is reversed in the HMS compared to cnidarian polyps (epidermal longitudinal and gastrodermal circular musculature), this derivation appears not to be very plausible.

The grid-like nature of the HMS is very evident in *Xenoturbella* (Ehlers & Sopott-Ehlers 1997, Raikova et al. 2000a) and acoelomorphs. In acoels, there is some variation of the musculature. While only few species have only a 'simple' muscular grid, longitudinal fibres may bend posterior of the ventral mouth opening or additional diagonal muscles may be present (Hooge 2001).

In gastrotrichs, the principal arrangement of outer circular and inner longitudinal musculature is still present, but muscles are not arranged into sheets, but rather into discrete bundles (Ruppert 1991b). There is considerable variation of the body wall musculature (see Hochberg & Litvaitis 2001a-c, 2003, Hochberg 2005, Leasi et al. 2006). Among Cycloneuralia, circular musculature is completely reduced in nematodes and in nematomorphs. It may be speculated that this is due to a comparatively strong cuticle that does not allow deformations of the body wall. The sole presence of longitudinal musculature results in an undulating movement. In nematodes, the musculature is interrupted by four epidermal cords (dorsal, ventral, and lateral, in some species, there can even be more cords; Chitwood & Chitwood 1950). In *Nectonema*, the marine nematomorph genus, a dorsal and a ventral epidermal cord are present (see cross sections in Schmidt-Rhaesa 1999) which is, as these cords are also present in nematodes, the plesiomorphic condition. In the freshwater nematomorph taxon Gordiida, the dorsal epidermal cord is reduced and the ventral epidermal cord containing the ventral nerve cord is shifted (during development, Schmidt-Rhaesa 2005) to a submuscular position, thus creating a completely circular sheet of longitudinal musculature, hardly interrupted by a thin ventral lamella leading to the nerve cord. In priapulids, the HMS is well developed (see e.g. Rothe et al. 2006 for *Tubiluchus troglodytes*) and at least in the macroscopic species, it aids burrowing by peristaltic contraction waves. The main burrowing organ is, however, the introvert, the anterior part of the body which can be completely retracted into the body and everted again. Retraction is due to special retractor muscles, but eversion is a process of pumping body fluids into the anterior body region. This is performed by a contraction of the HMS muscles. This principle, evagination of the anterior body part (the introvert) by fluid pressure, and invagination by special retractor muscles, has evolved convergently in different taxa (Acanthocephala, Sipunculida). It is also present in the close relatives of priapulids, in kinorhynchs, and loriciferans and was therefore probably present in their common ancestor. Kinorhynchs and loriciferans possess comparatively hard cuticular exoskeletons and have therefore reduced the HMS to isolated bundles in loriciferans (Kristensen 1991) or to isolated longitudinal bundles in kinorhynchs (Müller & Schmidt-Rhaesa 2003, Rothe & Schmidt-Rhaesa 2004). Circular musculature is absent in kinorhynchs, but there is dorsoventral musculature, which is responsible for the creation of body fluid pressure to evaginate the introvert. It may be speculated whether the dorsoventral musculature is derived from circular musculature by a partial detachment from the body wall, but the few developmental data of the musculature (Schmidt-Rhaesa & Rothe 2006) do not help in this question.

The dissociation of the HMS into isolated muscles, in combination with the evolution of a hard exoskeleton, can also be observed within arthropods. Onychophorans with their flexible cuticle have a well developed HMS. Additionally, they develop special muscles for the extremities. In the microscopic tardigrades, only longitudinal muscles and extremity muscles are present (Fig. 5.15.; Dewel et al. 1993, Schmidt-Rhaesa & Kulessa 2007), which attach at the margin of each segment and are therefore responsible for a differential movement of segments in relation to each other. In euarthropods, such segmental longitudinal musculature is also present in some isolated bundles, but the additional musculature undergoes a string diversification and specialization.

Among Spiralia, the HMS is again conserved in a number of taxa. In catenulid flatworms, it is a comparatively simple grid (Hooge 2001), while the musculature in the remaining flatworms (Rhabditophora) becomes more complex and often includes an additional layer of diagonal musculature (Hooge 2001). Among Gnathifera, the form of the HMS shows considerable variation. In gnathostomulids, outer circular and inner longitudinal muscles are present, but arranged in isolated bundles (Lammert 1991, Müller & Sterrer 2004). In *Limnognathia maerski*, circular musculature appears to be reduced, while longitudinal and dorsoventral musculature are present as isolated bundles (Kristensen & Funch 2000). In rotifers and *Seison*, which have

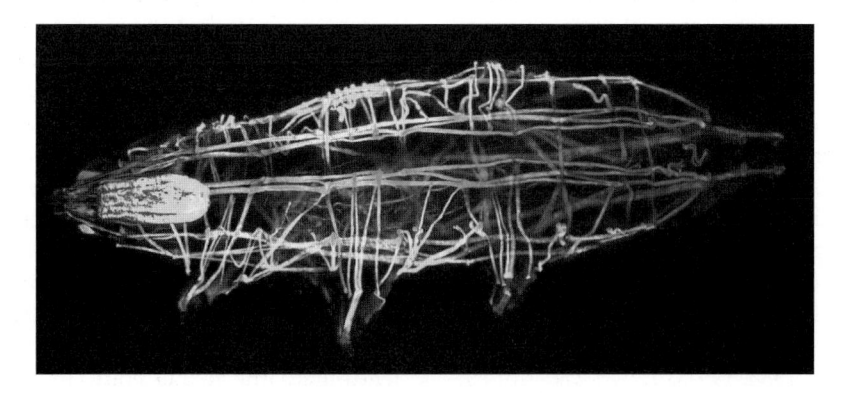

Fig. 5.15. The musculature of the tardigrade *Milnesium tardigradum* consists of isolated muscles running in longitudinal direction and into the appendages. Phalloidin staining and confocal laser scanning microscopy. Photo by Juliane Kulessa & A. Schmidt-Rhaesa, Bielefeld.

a comparatively hard skeleton, only a few muscles are associated with the body wall, the majority of muscles run as isolated bundles through the body cavity and act, for example, as retractors of the corona (Clément & Wurdak 1991). Acanthocephalans again have a clear HMS, its main function may be the eversion of the proboscis in the mode explained above (for priapulids). The muscles show some special features, they have been assumed to be hollow and even connected to the epidermal lacunar system (Miller & Dunagan 1977, Dunagan & Miller 1991), but this could be a misinterpretation, because investigations of *Macracanthorhynchus hirudinaceus* (Archiacanthocephala), *Acanthocephalus anguillae* (Palaeacanthocephala), and *Paratenuisentis ambiguus* (Eoacanthocephala) have shown muscles to be circomyarian and therefore contain a cytoplasmatic core (Herlyn et al. 2001, Holger Herlyn, Mainz, pers. commun.).

In the flexible and often macroscopic nemerteans, the muscular HMS is extremely important. It is well developed and often contains additional muscle layers (Turbeville 1991). In palaeonemerteans, even something like a 'double HMS' is present. Here, both circular and longitudinal muscles are present within the epidermis, in addition to the 'normal' subepidermal muscles (Turbeville & Ruppert 1983). In kamptozoans, subepidermal musculature is restricted to longitudinal muscles that form a continuous sheet in the stalk, but radiate into isolated bundles in the calyx (Wanninger 2004). Within molluscs, an HMS is present in the basal, vermiform taxa Caudofoveata and Solenogastres (Scheltema et al. 1994), with an additional layer of obliquely oriented fibres between the outer circular and the inner longitudinal musculature. With the evolution of the shell, the muscular system undergoes strong modifications. Longitudinal musculature is found as an 'enrollment muscle' in the lateroventral foot region. This may be a derivative from the HMS, and already some solenogastres show an individualization of paired ventral longitudinal muscles (von Salvini-Plawen 1972). This enrollment muscle is reduced in Ganglioneura (gastropods, cephalopods, bivalves, and scaphopods). Molluscs possess additional dorsoventral muscles. These are present in a variable number (depending on the body length) in Solenogastres, but more or less reduced in Caudofoveata (von Salvini-Plawen 1972). With the evolution of eight shell plates, as expressed in polyplacophorans, the number of dorsoventral muscles is reduced to sixteen pairs (two pairs per plate), monoplacophorans have eight pairs, and Ganglioneura one pair. This can be seen as a series of reduction (von Salvini-Plawen 1969), from numerous to one pair. Such a reduction series does not favour the interpretation of the serial arrangement of dorsoventral muscles, as present in polyplacophorans and monoplacophorans, as an indication for a segmented body organization. This is also supported by the observation that the serial dorsoventral muscles in polyplacophorans are derived from numerous muscles during development (Wanninger & Haszprunar 2002).

Echiurids and sipunculids have a well developed HMS, but within each taxon some variation exists, such as a tendency to dissolve the continuous muscular layer into single bundles or the presence of additional muscle layers. In the echiurid genus *Ikeda*, the sequence of layers is inverted, with an outer longitudinal and an inner circular layer (and an innermost oblique layer; Pilger 1993). In annelids, a grid-like HMS is present only in a few, probably mainly microscopic species (see Fig. 18 in Hooge 2001). In clitellates, the longitudinal musculature forms a complete sheet, but in polychaetes, longitudinal muscles are arranged in distinct bands (Purschke & Müller 2007). In a number of polychaete taxa, circular musculature is reduced (see Tzetlin et al. 2002, Filippova et al. 2005, Purschke & Müller 2007) and this has even been assumed to be the ancestral condition for annelids (Tzetlin & Filippova 2005).

In the tentaculate taxa and in deuterostomes, we do not find an HMS in its simple form of sheets of circular and longitudinal muscles. However, both elements are present in the vermiform taxa, that is, in phoronids, holothuroids (Echinodermata), and enteropneusts, so that we

90   THE EVOLUTION OF ORGAN SYSTEMS

can assume that this is also derived from an HMS of the bilaterian ancestor. In these taxa, we find a thin outer layer of circular musculature and isolated bundles of inner longitudinal muscles. In taxa which live in tubes (Pterobranchia), casts (Bryozoa) or have a hard skeleton (Brachiopoda, Echinodermata exclusive holothurians, Tunicata), the HMS is dissolved and the musculature is quite specialized. With the evolution of the chorda, we find a specialization of the muscles. Serially arranged packets of muscles, the myotomes, are arranged on both sides of the

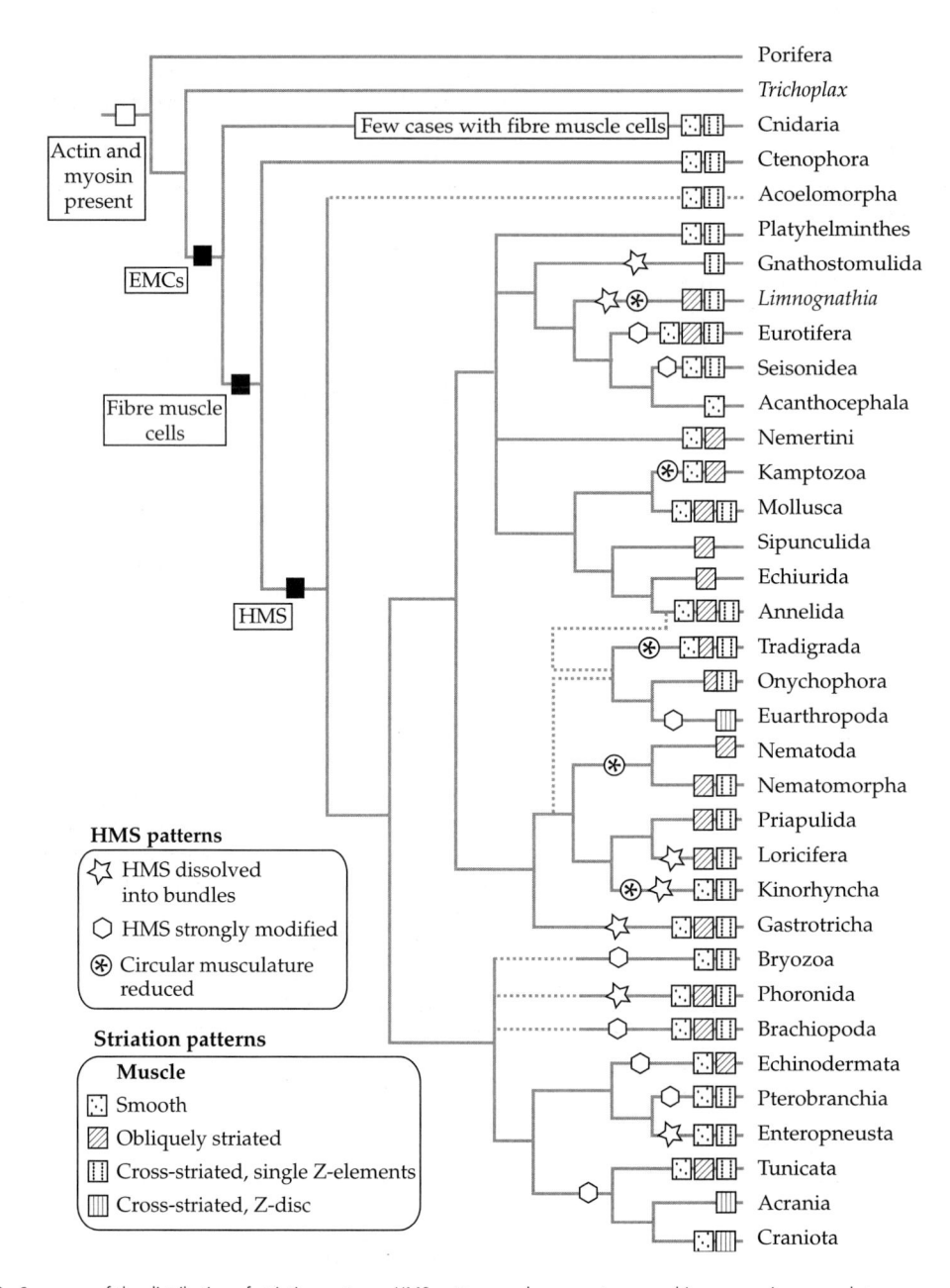

Fig. 5.16. Summary of the distribution of striation patterns, HMS pattern, and some autapomorphies concerning musculature.

Table 5.1 Occurrence of striation patterns, t-tubuli (tt), and sarcoplasmatic reticulum (SR). In cases where more than one striation pattern is present, the dominant pattern (when evident) is highlighted in bold.

	Striation pattern	t-Tubules (tt) and SR	Reference
Cnidaria	**Smooth** Cross in many hydromedusae, polyp of *Aurelia* (Chia et al. 1984), tentacle tips of cubopolyp (Werner 1984)	Both absent (SR may be present as few vesicles; Smith 1966)	Fautin & Mariscal 1991, Lesh-Laurie & Suchy 1991, Thomas & Edwards 1991
Ctenophora	**Smooth** Cross in *Euplokamis* tentilla, (Mackie et al. 1988)	tt absent SR present[1]	Hernandez-Nicaise & Amsellem 1980, Hernandez-Nicaise 1991
Xenoturbella	Smooth	tt absent SR present but weak	Ehlers & Sopott-Ehlers 1997
Acoelomorpha	**Smooth** Cross muscles close to the brain in *Convoluta pulchra* (Tyler & Rieger 1999)	Both absent (see figures in Ehlers 1994b, Tyler & Rieger 1999, Gschwentner et al. 2003)	Rieger et al. 1991b, Tyler & Rieger 1999
Gastrotricha	Cross Oblique Smooth	tt in association with cross striated muscles when these are present SR present	Ruppert 1991b
Nematoda	Oblique	tt absent in small species, present in large ones. SR weak in small, abundant in large species	Rosenbluth 1972, Lanzavecchia 1981, Wright 1991

Nematomorpha	Oblique
	  in adults
	Cross
	  in larvae (Müller et al. 2004)
		tt absent,
		SR present but weakly developed
			Lanzavecchia et al. 1977, Schmidt-Rhaesa 1998

Priapulida	**Oblique** Cross in longitudinal muscles of *Tubiluchus* larva (Higgins & Storch 1989)	Both present	Storch 1991, Candia Carnevali & Ferraguti 1979
Kinorhyncha	**Cross** Smooth few small muscles	Both present	Kristensen & Higgins 1991, Müller & Schmidt-Rhaesa 2003, Rothe & Schmidt-Rhaesa 2004
Loricifera	Cross Oblique	tt probably absent, SR absent in some muscles (trunk muscles) but present in others (mouth tube retractors) (Kristensen & Heiner, Copenhagen, pers. comm.)	Kristensen 1991

Platyhelminthes	**Smooth**
	  sometimes an indication of oblique striation (Rieger et al. 1991b)
	Cross
	  in tail of cercaria (Nuttman 1974 and others)
		tt absent
		SR present, but weakly developed
			Rieger et al. 1991b

Gnathostomulida	Cross	tt absent SR weak in Bursovaginoida, lacking in Filospermoida	Lammert 1991

Table 5.1 (*Contd*)

	Striation pattern	t-Tubules (tt) and SR	Reference
Limnognathia	Oblique Cross muscles in pharyngeal apparatus	both absent	Kristensen & Funch 2000

Eurotifera	Smooth
	    foot retractors of *Philodina*, ventral
	    circular muscle in *Asplanchna* and
	    others
	Cross
	    corona retractors in *Trichocerca*,
	    retractors in several species
	Oblique
	    longitudinal retractors in *Philodina*,
	    Brachionus, *Rhinoglena* and others
		tt probably absent
		SR present
			Clément & Amsellem 1989,
			    Clément & Wurdak 1991

Seisonidea	Cross
	    in mastax muscles
	Smooth
	    in retractor, other data unknown
		Both probably absent	Ahlrichs 1995

Acanthocephala	Smooth	?[2]	Crompton & Lee 1965

Symbion	Oblique
	Cross
	    in male reproductive system
		tt probably absent, SR well
		    developed in cross-striated
		    cells of male reproductive system,
		    in obliquely striated muscles
		    probably absent
			Funch & Kristensen 1997

Nemertini	**Smooth**
	Oblique
	    indications of oblique striation in
	    some palaeonemertean species
		tt absent
		SR present, but weak and in
		    periphery
			Turbeville & Ruppert
			    1985, Turbeville
			    1991

Mollusca	**Smooth**
	Oblique
	    dominant type in cephalopods, in other
	    groups rare, e.g. in pericardial glands of
	    scallops
	Cross
	    part of adductors in lamellibranch
	    bivalves, rarely in gastropods,
	    bivalve and cephalopod heart
		tt often absent, but present at
		    least in some prosobranch
		    gastropods SR present, usually
		    in periphery
			Watts et al. 1981, Nicaise & Amsellem
			    1983, Eernisse & Reynolds 1994,
			    Voltzow 1994, Budelmann et al.
			    1997, Morse & Zardus 1997

Kamptozoa	Oblique
	    in stalk muscles of *Barentsia gracilis*
	Smooth
	    in star-cell complex of colonial species
		tt absent
		SR present as vesicles under
		    sarcolemma
			Reger 1969
			Emschermann 1969

Sipunculida	Oblique	Both present	de Equileor & Valvassori 1977, Rice 1993
Echiurida	Oblique	tt probably absent, SR present	Pilger 1993

Annelida	**Oblique**
	    as 'double oblique striation' in
	    polychaetes
	Cross
	Smooth
		both present	Lanzavecchia et al. 1988,
			    Gardiner 1992, Jamieson 1992

Table 5.1 (*Contd*)

	Striation pattern	t-Tubules (tt) and SR	Reference
Tardigrada	'intermediate' between smooth and oblique Cross in Arthrotardigrada (Kristensen 1978), few muscles in Eutardigrada (Schmidt-Rhaesa & Kulessa 2007)	tt absent SR present, in periphery	Dewel et al. 1993, Walz 1974, 1975, Schmidt-Rhaesa & Kulessa 2007
Onychophora	'intermediate' between oblique and cross	Both present	Camatini et al. 1979, Lanzavecchia & Camatini 1979
Euarthropoda	Cross (with Z-discs)	Both present	Beinbrech 1998, Fahrenbach 1999

Bryozoa	Smooth
	  retractors, vestibular sphincter in *Asajirella*, tentacle muscles in *Cryptosula* and *Asajirella*, buccal muscles in *Crisia*
	Cross
	  vestibular sphincter in *Electra* and *Membranipora*, tentacle muscles in *Flustrellidra* and cyclostomes, buccal muscles in *Cryptosula*
		tt probably absent, SR present (from figures in Mukai et al. 1997)	Mukai et al. 1997

Brachiopoda	Smooth Cross Oblique brachiopods possess adductor muscles which have different striation patterns in different branches: cross and smooth in articulate brachiopods, oblique and smooth in lingulids	tt absent SR present, but weakly developed	Reed & Cloney 1977, James 1997

Phoronida	Smooth
	  entire musculature in *Phoronis ovalis*
	Oblique
	  longitudinal muscles of larva (actinotocha)
	Cross
	  some hood muscles of larva
		Both probably absent (A. Gruhl, Berlin, pers. comm.)	A. Gruhl, Berlin (pers. comm.), see also Turbeville & Ruppert 1985
			Santagata 2002
			
			Santagata 2002

Chaetognatha	**Cross** (with Z-discs) Smooth few muscles in lateral field cells	Invaginations of the cell membrane lead to large flattened structures, these are probably homologous to t-tubuli SR present	Duvert & Savineau 1986, Shinn 1997

Echinodermata	**Smooth**
	  most muscles, e.g. coelothel muscles in crinoids, arm muscles in ophiuroids, longitudinal body wall muscles in holothuroids
	Oblique
	  brachial muscles in crinoids
		tt probably absent
		SR present, but generally weakly developed
			Hill et al. 1978, Byrne 1994, Heinzeller & Welsch 1994, Smiley 1994
			
			Candia Carnevali et al. 1986, Heinzeller & Welsch 1994

Enteropneusta	Smooth Cross in pericardium of *Saccoglossus kovalevskii*	Probably both absent (from figures in Benito & Pardos 1997)	Castellani & Saita 1974, Balser & Ruppert 1990, Benito & Pardos 1997

Table 5.1 (*Contd*)

	Striation pattern	t-Tubules (tt) and SR	Reference

Pterobranchia	Smooth
	Cross
	  few fibrils in tentacles and mesocoelic
	  pore of *Cephalodiscus*
		?	Benito & Pardos 1997

Tunicata	Smooth
	  in somatic muscles of adult ascidians
	Cross
	  in heart of adults, tadpole larva of
	  ascidians, locomotory muscles of salps
	Oblique
	  in muscular bands of doliolids
	  (Thaliacea) (Bone & Ryan 1974)
		tt absent
		SR described in some cases (e.g.
		  ascidian heart), but absent in
		  e.g. ascidian tadpole larva
			Smith 1966, Kalk 1970, Oliphant &
			  Cloney 1972, Bone & Ryan 1973,
			  Nunzi et al. 1979, Burighel &
			  Cloney 1997

Acrania	Cross (with Z-discs)	Both absent	Ruppert 1997
Craniota	Cross (with Z-discs) Smooth	Both present	Different sources

[1] Hernandez-Nicaise (1991) describes a large network of the endoplasmatic reticulum, this probably corresponds to the SR.

[2] Fine structural descriptions of the acanthocephalans musculature are very rare. Crompton & Lee (1965, p.360) note that the 'sarcolemma is seen to be a folded structure' in *Polymorphus minutus*, which may indicate at weakly developed t-tubules. T-tubules and SR are probably absent from *Paratenuisentis ambiguus* and *Macracanthorhynchus hirudinaceus* (figures in Herlyn et al. 2001, Herlyn 2002).

chorda, and together this system produces undulating left–right movements. This system is found in the tail of the mobile tadpole larva of ascidians (and in adult appendicularians), as well as in acranians and in basal craniotes. Acranians can be viewed as a combination of both life stages of tunicates: they live hemisessile as filter feeders, but at the same time are capable of sudden and strong muscular escape reactions with the chorda/muscle system. Within craniotes, this system is still important in most fishlike groups, but becomes modified with the strong pronounciation of extremities in tetrapods.

Conclusions

The acto–myosin system, which is basic for muscular action, evolved before the metazoan ancestor. Within metazoans, muscles evolve from the cellular to a tissue level. First, myofibrils are present in epithelial cells (epitheliomuscle cells) but then they invade the ECM and become fibre muscle cells. The cytological patterns of the musculature, in particular the arrangement patterns of Z-elements, actin and myosin, are very flexible. An unordered pattern of sarcomeres (smooth muscle) is probably the plesiomorphic condition and we have to assume multiple transitions to different striation patterns. The basic arrangement of muscle cells appears to be a grid of outer circular and inner longitudinal musculature, the *Hautmuskelschlauch* (HMS). This HMS is conserved in almost all wormlike animals (although muscles often form discrete bundles instead of the plesiomorphic sheets), while in animals with skeletons the HMS dissolves into more specialized muscle systems.

CHAPTER 6

Nervous system

Nervous systems conduct information in a directed way through the body. This is done by electrical and/or chemical signals and by cells specialized for these functions. The conduction of information does not depend exclusively on the nervous system, as electrical impulses can also be distributed throughout the body, but then in a more or less untargeted way. Additionally, chemical substances are also used as signalling tools within other systems, for example, as hormones in the circulatory system. Nervous systems may be regarded as specializations for fast and targeted information conduction.

The evolution of the nervous system has attracted several researchers and has resulted in a considerable literature, from which Hanström (1928), Bullock & Horridge (1965), Lentz (1968), and Rehkämper (1968) should be mentioned as broad approaches.

Pre-metazoan nervous components

A nerve cell is characterized by more or less long extensions, the dendrites and axons, but their intracellular components are not basically different from other cells. Transmembrane ion canals and a vesicle transport system are important for nerve cells. These are of course very general cytological features. Among the molecules used in nerve cells, especially as chemical transmitters to bridge the gap between different nerve cells, several are also found outside the metazoa (Mackie 1990; see below). One characteristic for nerve cells is the electrical conduction of information by action potentials 'travelling' along the dendrites and axons. However, as reviewed by Leys et al. (1999), electrical signalling as a general phenomenon is not restricted to metazoan animals, but also occurs in plants. To give an example, the feeding of caterpillars on tomato leaves creates action potentials travelling to neighbouring leaves, where they trigger the synthesis of toxic substances (Wildon et al. 1992, Rhodes et al. 1996).

A number of genes are presumed to be specific for the nervous system of humans, but orthologues from a considerable fraction of these genes have been found to be very ancient, predating the fungi/metazoan split (Noda et al. 2006). The exact function of these genes and probable functional changes, however, are not known. The presence of components from the eumetazoan nervous system in non-metazoan organisms is not surprising, and in fact only means that many of its components were present for a long time and only their assembly in particular cells makes a cell a nerve cell.

Sponges: signalling without a nervous system

In sponges, no tissue or cells with any resemblance to nervous systems in other animals can be found. Nevertheless, sponges are able to conduct information through their body. Some sponges contract regularly (Parker 1910, Nickel 2004) or are able to stop the feeding current in a coordinated manner (Mackie 1979). The reason for such coordinated actions are electric signals (Lawn et al. 1981), which have been particularly shown in hexactinellids, where the spreading throughout the sponge body is facilitated by the syncytial organization (Leys & Mackie 1997, Leys et al. 1999, Leys 2003).

Additionally, some molecular components typical for nervous systems can already be found in sponges. Weyrer et al. (1999) found immunoreactivity against serotonin in the demosponge *Tedania ignis*. Müller & Müller (1999) found crystallin and a metabotropic glutamate receptor in the demosponge *Geodia cydonium*, two molecules that are known from the sensory system of other metazoans. The function of these molecules in sponges, however, is unknown.

The 'simple' nerve net in cnidarians and ctenophores

As any textbook tells, the 'first' nerves recognizable at a cellular level are found in cnidarians. This means that nerve cells evolved in the eumetazoan ancestor. One very important aspect is that cells are able to transfer information from one cell to another. This is done in synapses, which can transfer the signal either directly (in electrical synapses) or via chemical transmitters (chemical synapses). Both electrical and chemical synapses occur in cnidarians (Fautin & Mariscal 1991, Lesh-Laurie & Suchy 1991, Thomas & Edwards 1991). In cnidarians, many chemical synapses are bidirectional, meaning that synaptical vesicles including transmitter substances are present on both sides of the synapse and impulses can be conducted in both directions (Horridge & Mackay 1962, Anderson & Schwab 1981, Anderson 1985). Such synapses are often called symmetrical synapses, but asymmetrical synapses, allowing signal conduction in only one direction, also occur, for example in the planula larva of the anthozoan *Anthopleura elegantissima* (Chia & Koss 1979). Asymmetrical synapses require the development of a differentiated pre- and postsynaptical membrane. The presynaptical membrane is specialized for transmitter release and 're-capture', while the postsynaptic membrane is specialized for signal reception and modulation. Asymmetrical synapses are the only synapses in bilaterians.

The general architecture of the nervous system in cnidarians as well as in ctenophores is that of multipolar nerve cells forming a net-like structure (Fig. 6.1.). The nerve cells are located

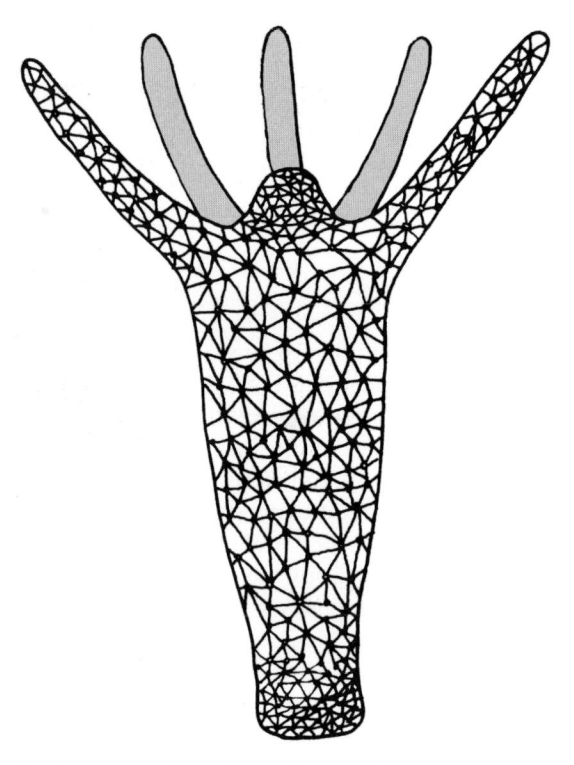

Fig. 6.1. Nerve net (plexus) in Cnidaria. In the figured polyp, different densities of the plexus can be recognized (after Ruppert et al. 2004).

basiepithelially, that is, they meander among the basal portions of both epidermal and gastrodermal cells (Fig. 5.1., 6.2.). The nerve cells are connected to neurosensory cells that are an integral part of the epidermis. What is called here simply a 'nerve cell' is sometimes termed a 'ganglion cell' (e.g. Hyman 1940), but as 'ganglion' is used here with a particular meaning (see below), such terminology is inappropriate. The neurons in cnidarians work in the same way as those in other animals, and the basal physiological processes are similar (Anderson & Schwab 1983). In ctenophores, the organization of the nervous system as a basiepithelial net is principally the same as in cnidarians (Hernandez-Nicaise 1991).

However, although a simple network is likely to be the ancient condition, there is a trend in cnidarians and ctenophores in which parts of the net are pronounced in some way, creating diversity within the net. In some cases it has been

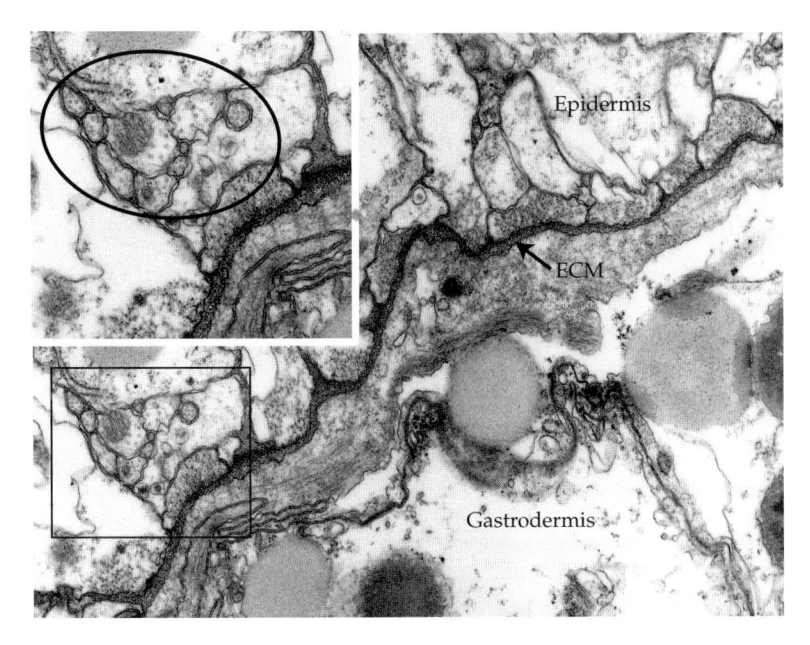

Fig. 6.2. Transmission electron micrograph from the epidermis and ECM (mesogloea) of the planula larva of *Hydractinia echinata* (Cnidaria, Hydrozoa), showing basiepidermal nerves (see circle in magnified subset).

shown that such pronounced neurons play an important role in specific processes, for example as giant axons for coordinated swimming (see below). Some examples for neuron diversity are:

- The nerve net in the epidermis and the gastrodermis usually shows a different neuron density (Thomas & Edwards 1991).
- In polyps, there is often a higher density of neurons around the mouth (Hyman 1940, Grimmelikhuijzen & Westfall 1995).
- In the hydrozoan species *Polyorchis penicillatus* (Singla 1978b, Spencer 1979) and *Aglantha digitale* (Singla 1978a, Weber et al. 1982), giant neurons are present and connected to muscles in the umbrella. In both species, the giant neurons control swimming and have therefore sometimes been called swimming motor neurons (Donaldson et al. 1980, Roberts & Mackie 1980, Spencer 1981).
- Two giant axons are present in the siphonophore *Nanomia bijuga* (Mackie 1973, 1978; Fig. 6.3.).
- In the tissue next to the rhopalia (umbrellar sensory complexes) of the scyphozoan medusa *Cyanea capillata*, there is a giant fibre nerve net (Anderson & Schwab 1981).
- In medusae of both Scyphozoa and Cubozoa, there are marginal 'ganglia', better called marginal centres, which have a pacemaker function and therefore divide the nerve net into a slow (diffuse) and fast part (Anderson & Schwab 1982, Thomas & Edwards 1991).
- In polyps and medusae of cubozoans, there are well differentiated nerve rings (a double gastrodermal/epidermal ring) (Werner et al. 1976, Chapman 1978).
- There is regionalization in the nervous system by a different content in neuropeptides (Grimmelikhuijzen & Westfall 1995).
- In ctenophores, there are concentrations along the comb rows, a special labial nerve in *Beroe*, and tentacular nerves in cydippids *Pleurobrachia* and *Hormiphora* (Hernandez-Nicaise 1973, 1991; Fig. 6.4.). All such nerves remain in their architecture within the shape of the nerve net, but are recognizable as thickened pathways.

The many patterns of nerve net diversification in cnidarians and ctenophores are not all

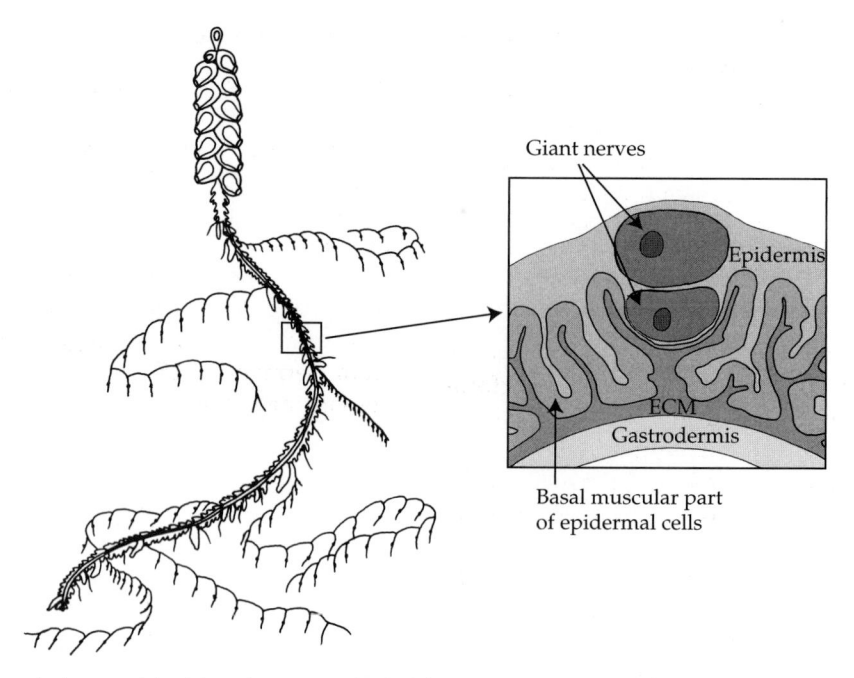

Fig. 6.3. Giant nerves in the stem of the siphonophore *Naomia* (to the left is *N. cara*, after Werner 1984). Cross section to the right after Mackie (1978). Only the two giant nerve cells are shown at cellular level.

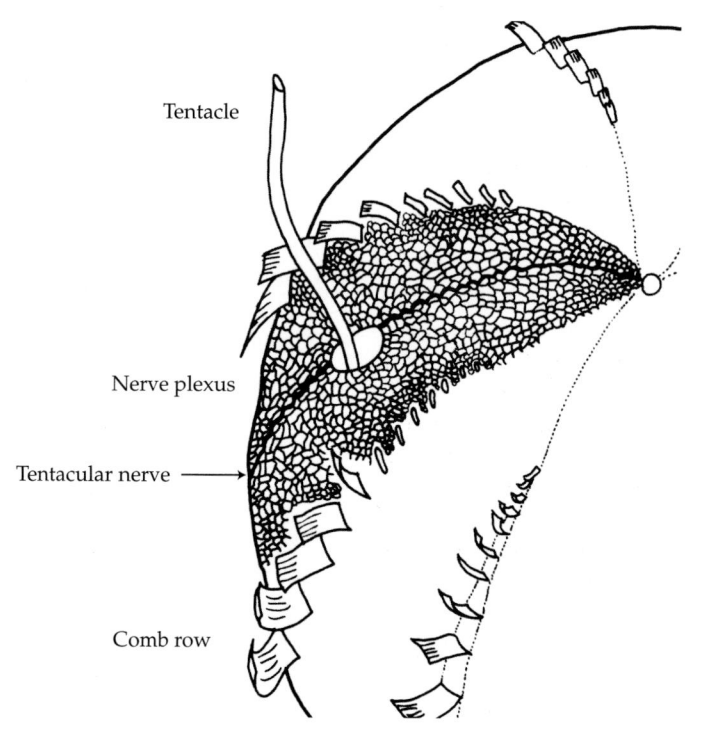

Fig. 6.4. Nerve net in a cydippid ctenophore, shown for one segment between two comb rows. In the median, a tentacular nerve is pronounced within the 'usual' pattern of the nerve net. Modified after Hernandez-Nicaise (1973).

comparable with each other and therefore do not reflect an evolutionary sequence, but rather a broad trend towards a diversification of the nerve net and to the pronounciation of particular nerves among the net. This trend was also present in the bilaterian ancestor and led to a much stronger diversification of nervous systems.

Bilateria: brains and cords

The reconstruction of the nervous system of the bilaterian ancestor is not at all easy to perform. At first sight it appears that there is a logical connection between bilateral symmetry, directed locomotion, a concentration of sensory cells at the anterior end, the accumulation of neurons into a brain, and the presence of longitudinal nerves leading into the body. But this may not be so. Therefore, we will first take a look at its subtaxa, before trying to make up our minds about the nervous system of the bilaterian ancestor.

Before going further, a few terminological issues should be mentioned. I call any accumulation of neurons in the anterior end a 'brain'. 'Cords' are strong strands composed of numerous neurons (or at least their axons). When neuronal material is accumulated into visible concentrations, this is called a 'ganglion' (in this sense, the brain is also a ganglion, the cerebral ganglion). The somata of nerve cells can be distributed along the nerve cords (*Markstrang*) or can be concentrated in ganglia. One pair or few pairs of ganglia can co-occur with the *Markstrang*-nervous system, but only when the concentration of somata into ganglia is the basic principle of the entire nervous system is a 'ganglionated nervous system' present. There can be different 'subsets' of the nervous system. Nerves that are positioned either between epidermal cells or basal to them, but still somehow in connection with the body wall, are called the 'peripheral nervous system' (PNS). Nervous elements that are positioned deeper in the body are called the 'central nervous system' (CNS), and those associated with the intestinal system are called the 'stomatogastric nervous system'. Only PNS and CNS will be discussed in the following.

The orthogon of spiralians

In 1924, Reisinger introduced an important term 'orthogon' into the literature. It is important, because it was assumed (e.g. Hanström 1928, Reisinger 1972) that such an orthogon represents an ancestral organization of the nervous system. An orthogon is a 'system of several... longitudinal nerve cords... with connecting, pseudometamerically arranged circular commissures' (Reisinger 1972, translated from German by ASR; see Fig. 6.5.). The most convincing orthogons, or derivations of it, are

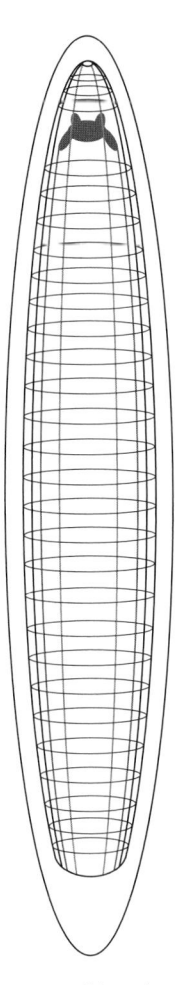

Fig. 6.5. Generalized structure of the orthogon with a regular system of longitudial and circular nerves. Modified after a figure from Reisinger (1972) of the platyhelminth *Minona trigonopora* with four pairs of longitudinal nerve cords.

found among the Spiralia and should be briefly reviewed here, before comparing it with the nervous systems in other taxa.

Acoelomorpha
Acoelomorphs have a very variable nervous system and show more or less orthogonal patterns. Several acoels have four pairs of longitudinal nerves, something like a brain in the anterior end, and a number of commissures between the longitudinal nerves that form a more or less regular pattern (Bullock & Horridge 1965 and Reisinger 1972 present a figure of *Convoluta* sp. from Delage 1886). Species such as *Anaperus* sp., however, show strong modifications in having only longitudinal nerves (Westblad 1949). Investigations of the nervous system with antibodies against neural components could not show 'picture-book' orthogons (see, e.g., statement in Reuter et al. 2001a), but some orthogonal patterns are present. First of all, there are only a limited number of investigations, but these already reveal a variation within both Acoela and Nemertodermatida. Nemertodermatids have, at least in *Nemertoderma* species, a ringlike concentration of nerves with anti-serotonin immunoreactivity in the brain (Raikova et al. 2000b, Reuter et al. 2001b). In *Meara stichopi*, only one pair of cells with anti-serotonin immunoreactivity is present (Fig. 6.6.; Raikova et al. 2000b, Reuter et al. 2001b). The remaining nervous system is intra- or basiepidermal (Ehlers 1985a, Raikova et al. 2004) and in some cases also submuscular, such as paired, broad ventrolateral accumulations of fine nerve fibres in *Meara stichopi* (Fig. 6.6.; Raikova et al. 2000b, Reuter et al. 2001b). There are circular (*Meara*) and a high number of longitudinal (*Nemertoderma*) elements, but the orthogon is not very evident here, except that it is probably indicated by regularly branching peripheral nerves in *Meara* (Raikova et al. 2000b). Acoels have a brain, this is either a collection of nerves more strongly condensed than the remaining ones, but still with individually identifiable paths (such as in *Faerlea glomerata* or *Childia groenlandica*; Fig. 6.6.)

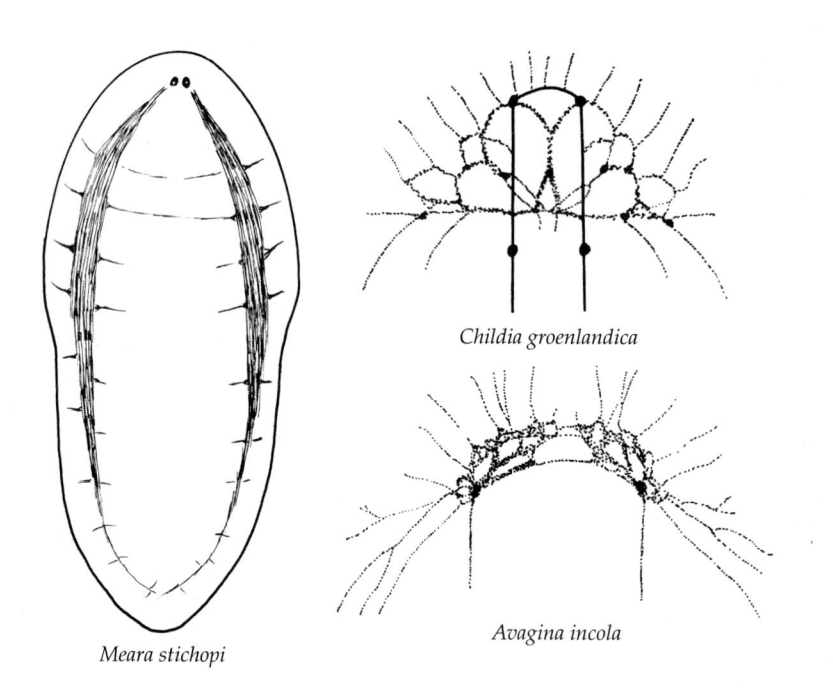

Fig. 6.6. Patterns of the nervous system in acoelomorphs. In *Meara stichopi* (Nemertodermatida), a pair of serotinergic cells is present, from which a broad bundle of fibres originates (after Raikova et al. 2000b). In *Childia groenlandica* and *Avagina incola* (Acoela), variable degrees of fibre concentration are present in the brain region (after Raikova et al. 1998 and 2001a).

or a further condensing with nerves forming a bridge-like structure as in *Avagina incola* (Fig. 6.6.; Raikova et al. 1998, Reuter et al. 2001a, b). The arrangement of nerves in the anterior end, in particular in *Faerlea glomerata*, resembles an orthogon, but in the remaining peripheral nervous system, as far as has been observed, these patterns are much less evident (Reuter et al. 2001a).

Platyhelminthes
Flatworms clearly have an orthogon (e.g. Fig. 6.7.), but this can be quite variable (Rieger et al. 1991b, Reuter & Gustafsson 1995, Halton & Gustafsson 1996). Up to three plexuses can be present: one basiepidermal, one subepidermal, and one submuscular. The brain and orthogon are in the submuscular position. The number of longitudinal nerve cords and the number of connecting commissures is very variable already within basal taxa such as Catenulida (Ehlers 1985a), making it difficult to choose a particular number as the probable ancestral condition. Reuter et al. (1998) distinguish two different types of longitudinal nerves, main and minor ones. The brain is bilobed. Because its development starts before the development of the longitudinal nerves, Reisinger (1972) assumes that it is not a specialized part of the orthogon, but an independant structure. The brain probably has an anteroposterior regionalization, because of the differential expression of three homeobox genes (Umesono et al. 1999).

Gnathifera
In the gnathiferan taxa, a bilaterally organized or unpaired dorsal brain and two ventrolateral nerve cords are present, additionally there can be concentrations in the form of ganglia (Lammert 1991, Clément & Wurdak 1991, Dunagan & Miller 1991, Müller & Sterrer 2004, Hochberg 2006; see Table 6.1.). According to figures, the longitudinal nerve cords of gnathostomulids, and the brain in bursovaginoid gnathostomulids, are basiepidermal (Kristensen & Nørrevang 1977, Lammert 1991), while the brain of filospermoid gnathostomulids and the complete nervous system of eurotifers, seisonids, and acanthocephalans is subepidermal. In *Limnognathia maerski*, the position of the nervous system is unclear, because the ECM is not clearly visible (Kristensen & Funch 2000). In gnathostomulids, the longitudinal nerve cords are connected only in the posterior region of the body (Lammert 1991, Müller & Sterrer 2004). These patterns can be derived from an orthogon without problems, but as there is no representative of gnathiferans with a more obvious orthogon, such derivation is speculative and depends on the phylogenetic position.

Nemertini
In nemerteans, there is a bilaterally symmetrical brain with paired dorsal and ventral lobes, which are both connected to the corresponding part on

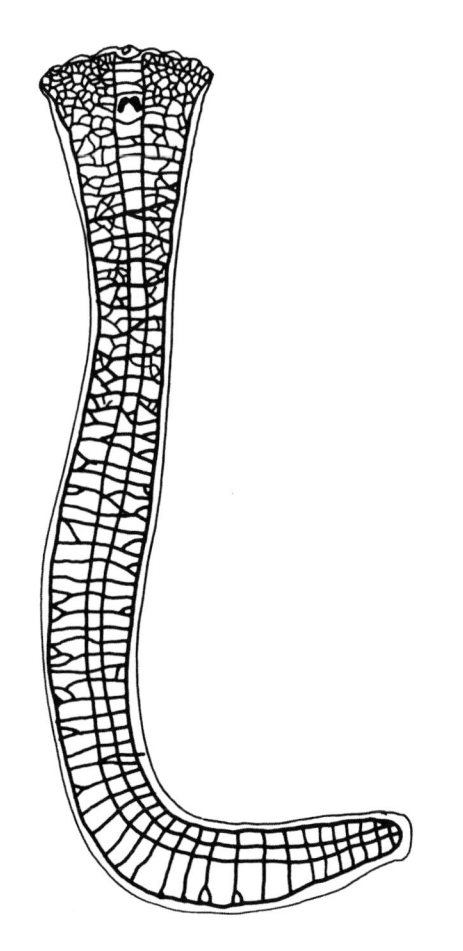

Fig. 6.7. Orthogonal pattern in the metacercaria of the digenean platyhelminth *Apharyngostrigea cornu*. After Reisinger (1972).

the other side and therefore surround the foregut (Turbeville 1991, Senz & Tröstl 1997). From the brain originates at least one pair of lateroventral longitudinal nerve cords, in some species there can be further longitudinal nerve cords. There also is a basiepidermal nerve plexus, which is connected to the longitudinal nerve cords. It is not clear whether the longitudinal nerve cords are connected by commissures. According to Reisinger (1972), regular circular commissures are present, additionally to paired subdorsal and unpaired median nerves, in Drepanophorida and in pelagic nemerteans, while Turbeville (1991) mentions only one posterior commissure.

Kamptozoa
The nervous system of Kamptozoa is composed of a paired cerebral ganglion, and further small ganglia at the base of the tentacles and close to lateral sensory organs (Nielsen & Jespersen 1997). There are several smaller nerves, one pair of which is probably homologous to the ventral or ventrolateral nerve cords in other spiralians (Fuchs et al. 2006).

Mollusca
The basal taxa Solenogastres, Polyplacophora, and Monoplacophora have a bilaterally symmetrical brain, from which four longitudinal cords arise (Eernisse & Reynolds 1994, Scheltema et al. 1994, Haszprunar & Schaefer 1997). These ventral and lateral cords are connected by a number of commissures, making it likely that this structure is derived from an orthogon (Reisinger 1972; Fig. 6.8.). In all other molluscs, the Ganglioneura, a strong concentration of the nervous system into ganglia and fewer cords takes place.

Annelida
The nervous system is strongly structured by the segmental body organization, but it can be derived from an orthogon. The dorsal brain is bilaterally organized, there is a yet unresolved debate whether it is segmentally organized or not segmented (Orrhage & Müller 2005). The brain connects to the ventral nerve cord with connectives around the foregut. There are two ventral or ventrolateral main nerve cords, with nerve cell somata concentrated in segmental pairs of ganglia. The ganglia on both sides in each segment are connected by a variable number of commissures (Müller 2006). In addition, there is a peripheral nervous system with an intraepidermal or subepidermal position (Orrhage & Müller 2005). The number of peripheral longitudinal nerves varies from none to seventeen. In the latter case, *Saccocirrus papillocerus*, the longitudinal and circular nerves form a perfect

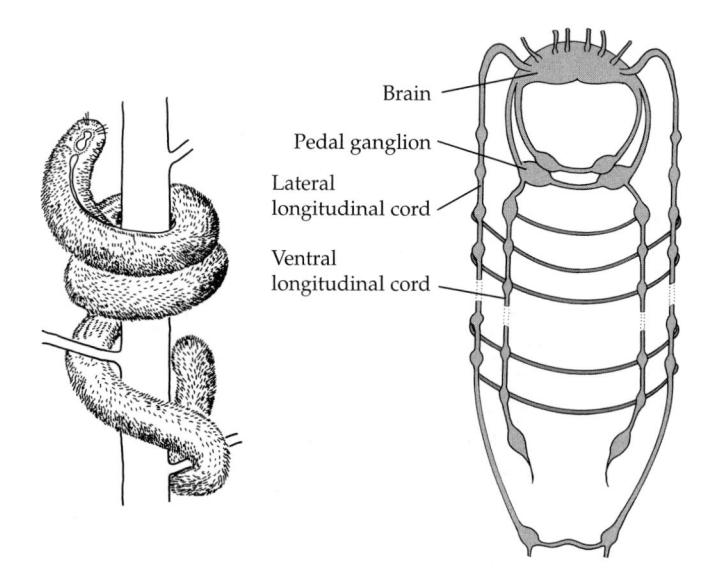

Fig. 6.8. Basal molluscan nervous systems show orthogonal patterns, exemplified by the nervous system of Solenogastres (whole animal is *Rhopalomenia aglaopheniae*, nervous system is from *Genitoconia atriolonga*; both figures after von Salvini-Plawen 1971).

grid, identical with the orthogon (Orrhage & Müller 2005; Fig. 6.9.). Some trochophores, especially from species in the polychaete genera *Polygordius* and *Lopadorhynchus*, have an orthogonal organization of the nervous system (Hanström 1928). Some annelids, such as the enchytraeid *Enchytraeus crypticus* (Clitellata) reveal, at least during development, no ganglionic character of the ventral nerve cord (Hessling & Westheide 1999).

Echiurida
In echiurids a dorsal brain is absent. The ventral nerve cord is swollen to form a subesophageal ganglion in the anteriormost region, from there circumesophageal connectives originate, but they do not connect to a dorsal ganglion (Hessling & Westheide 2002). The nervous system is metamerically organized in larvae of *Echiurus echiurus*, in adults of *E. abyssalis* (Baltzer 1931), and in larval and adult males of *Bonellia viridis* (Hessling & Westheide 2002). In *E. abyssalis*, complete rings of nerves originate at regular distances from the ventral nerve cord and lead to sensory papillae, which are arranged in rings (Baltzer 1931; Fig. 6.10.). The metameric patterns have been taken as an indication that echiurids are derived from segmental ancestors (which can be the case when they are derived annelids or when they are the sister-group of annelids), but it can also be explained as a remnant of an ancestral orthogonal organization of the nervous system.

Sipunculida
There is no trace of an orthogon in sipunculids. The ventral longitudinal cord is unpaired. Numerous nerves branch off, but they show no regular pattern. There is a subepidermal plexus of nerves, but no elements, neither longitudinal nor circular, are particularly pronounced (Rice 1993).

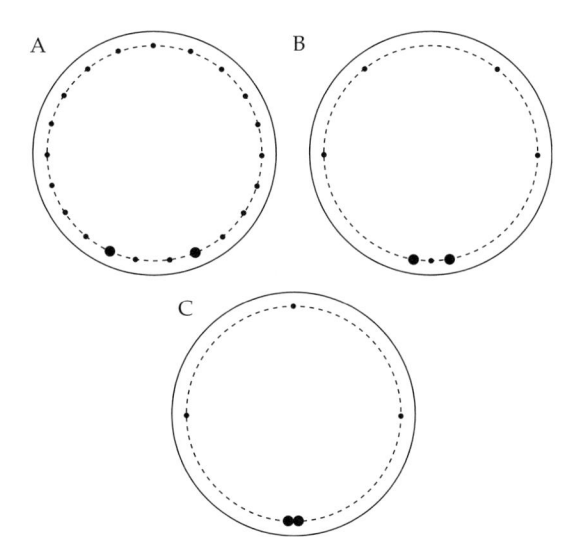

Fig. 6.9. Diversity in the number and location of longitudinal nerve cords in three polychaete species: *Saccocirrus papillocerus* (A), *Scoloplos armiger* (B) and *Nerilla antennata* (C). After Orrhage & Müller (2005).

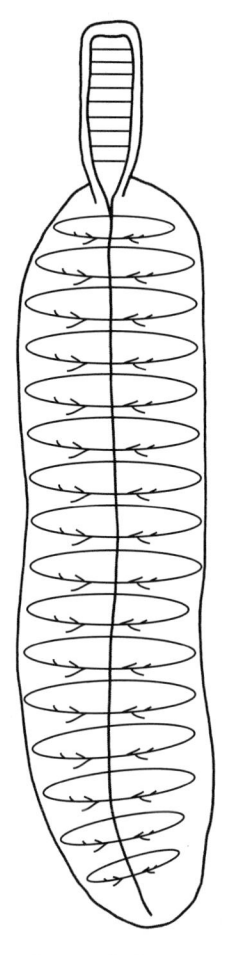

Fig. 6.10. Orthogonal patterns in the nervous system of *Echiurus* sp. (Echiurida). After Hanström (1928).

In conclusion, the derivation of spiralian nervous systems from an orthogon appears to be plausible and an orthogon appears to be an ancestral feature. It is, however, clear that there are numerous and various trends to alter or reduce the orthogonal pattern. Ancestral orthogonal patterns are rarely realized. Platyhelminthes, which often serve as a model for the orthogon, pose one major problem. When the orthogon is derived from the basiepidermal nervous system of the eumetazoan ancestor, it is difficult to explain why the orthogon in platyhelminths is submuscular while the basiepidermal plexus is still present (at least in some species). Probably, parts of the plexus have separated here and sunk inwards, but this means that the orthogon in platyhelminths shows derived features in this respect. On the other hand, polychaetes, such as *Saccocirrus papillocerus*, show a much more typical orthogon.

Do gastrotrichs, cycloneuralians, and arthropods have an orthogon?

When the spiralian ancestor had an orthogon, can we trace this character further back? If the protostome ancestor had an orthogon, at least some representatives of the Gastrotricha, Cycloneuralia, and Arthropoda should show some traces of it.

The evidence for an orthogon is not as clear as in spiralians. The patterns with closest resemblance to the orthogon are found in segmentally arranged nervous systems, that is, in arthropods and in kinorhynchs. In both groups, there are segmental pairs of ganglia, which are connected within the segment by commissures and between segments by connectives. In arthropods, there are nerves leading from the ganglia to the periphery, but there are no additional longitudinal nerve cords and the arrangement of nerves in the periphery does not show orthogonal patterns. In kinorhynchs, however, Kristensen & Higgins (1991) speak of an orthogonal arrangement of the nervous system, because there are, additionally to the paired ventral nerve cords, six further longitudinal nerves (middorsal, lateral, and lateroventral pairs). All longitudinal nerves probably have ganglia, and the midventral and middorsal ones are connected by commissures (Kristensen & Higgins 1991, for *Echinoderes aquilonius*; unpublished observations using immunocytochemical staining in *Pycnophyes kielensis* by Rothe & Schmidt-Rhaesa 2004). In animals with a segmental body organization (and kinorhynchs are regarded here to represent at least advanced segmental patterns) it is hard to determine whether 'rope-ladder-like' ventral nerve cords with commissures and connectives are remnants of an orthogonal pattern, or just the result of the superimposition of segmental patterns on organ formation. Kinorhynchs, with their eight pairs of longitudinal nerves and two sets of commissures, come more close to a derived orthogonal pattern.

In nematodes, priapulids, and loriciferans, the nervous system is intraepidermal and contains a paired (loriciferans) or unpaired (nematodes, priapulids) ventral nerve cord as well as additional longitudinal nerves: these are 13 in priapulids, 10 in loriciferans, and 5 or more in nematodes (van der Land & Nørrevang 1985, Kristensen 1991, Storch 1991, Wright 1991). In the nematode *Caenorhabditis elegans*, the longitudinal nerves are connected by commissures (White et al. 1986; Fig. 6.11.), and although this pattern is not regular and left and right side of the body differ in the number of commissures (the right side has several more), this arrangement somewhat resembles an orthogon. Regularly arranged circularly oriented nerves have not been described in these taxa. In nematomorphs, only representatives of the marine genus *Nectonema* have an intraepidemal (large) ventral and a (small) dorsal nerve cord, the terrestrial Gordiida only have a subepidermal ventral nerve cord (Schmidt-Rhaesa 1996b), which is translocated from an intraepidermal to the subepidermal position during development (Schmidt-Rhaesa 2005). A peripheral nerve plexus is present (Schmidt-Rhaesa 1996b), but it does not contain further pronounced nerves. The system of intraepidermal nerve cords in nematodes, priapulids and loriciferans, especially, is congruent with a derivation from an orthogon, but it seems that circular elements are generally reduced. The brain of cycloneuralians is of additional phylogenetic importance. While the spiralian brain is always located dorsally and has a central neuropil with peripheral somata (see Table 6.1.), the cycloneuralian brain surrounds the anterior part of the intestine in a ring-like manner (this is the name-giving character) and is divided in longitudinal direction into three parts with anterior somata, intermediate neuropil, and posterior somata (White et al. 1986 for Nematoda; Nebelsick 1993 and Neuhaus 1994 for Kinorhyncha; Rehkämper et al. 1989 for Priapulida; Kristensen 1991 for Loricifera). Spiralians also have a circumintestinal structure of the anterior nervous system, but in contrast to cycloneuralians the brain is dorsal, with 'only' smaller connectives leading to the ventral nerve cord. An exception to the cycloneuralian pattern are nematomorphs, which have a subesophageal brain with only a very thin supraesophageal commissure (Schmidt-Rhaesa 1996b), but this condition is derived in comparison to all potential relatives.

Gastrotrichs, with their questionable phylogenetic position (see Chapter 2), show hardly any resemblance to an orthogonal arrangement of nerves. There are two lateroventral nerve cords, but circular elements or further longitudinal nerves are either short, scattered, or absent (Ruppert 1991b, Hochberg & Litvaitis 2003, Rothe & Schmidt-Rhaesa, unpublished observations using immunocytochemical staining). The brain is not ring-like, but may be regarded as a semicircle, where a dorsal commissure connects lateral nerve cell somata and from which the longitudinal cords originate at the lateral endpoints (Fig. 6.12.; Rothe & Schmidt-Rhaesa, unpublished observations using immunocytochemical staining). Two ultrastructural investigations differ concerning the distribution of somata and neuropil. While Teuchert (1977) describes in *Turbanella cornuta* an anteroposterior sequence (somata – neuropil – somata) comparable to cycloneuralians, Wiedermann (1995) described a more even distribution of somata dorsal and ventrolateral of the neuropil in *Cephalodasys maximus*. In our own, unpublished investigations

(B.H. Rothe & A. Schmidt-Rhaesa), we find the brain consisting mainly of a commissure dorsal of the pharynx, with a variable number of lateral somata. The lateroventral longitudinal cords originate in the region of the somata. There is additionally a fine subpharyngeal commissure.

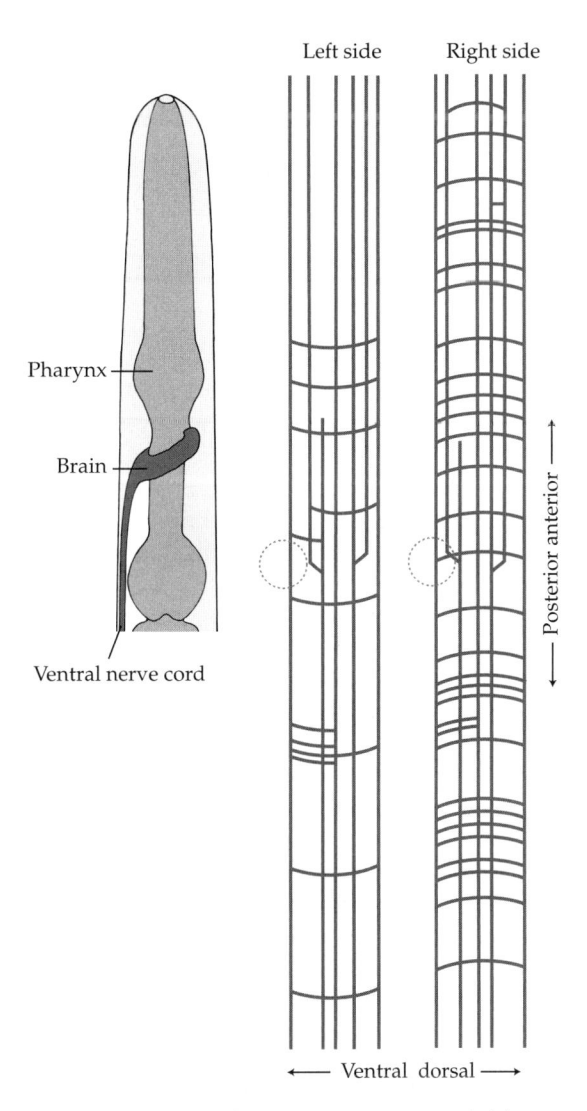

Fig. 6.11. Architecture of the nervous system in *Caenorhabditis elegans* (Nematoda), showing circumpharyngeal brain and ventral nerve cord (left) and the unsymmetrical arrangement of longitudinal and semicircular nerves (right). The ventro-median opening of the female reproductive system is in the encircled area. Modified after White et al. (1986).

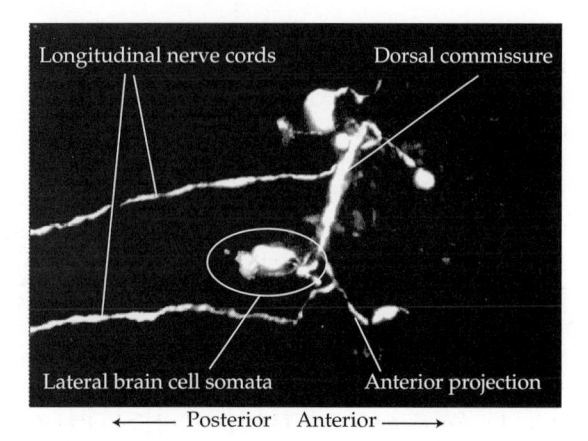

Fig. 6.12. Brain structure in *Turbanella cornuta* (Gastrotricha). Nerve cell somata are located laterally, connected by a dorsal commissure. Immunocytochemical staining of serotonin and confocal laser scanning microscopy. Photo by Birgen H. Rothe & A. Schmidt-Rhaesa, Bielefeld.

In arthropods, the nervous system has been such an important character for phylogenetic considerations that the marginal view presented here is only a small appetizer. For deeper insights into nervous system evolution, in particular in euarthropods, Hanström (1928) as an older reference, as well as Kutsch & Breidbach (1994), and Harzsch (2006) are referred to as summaries and literature references.

The brain of euarthropods is composed of dorsally located ganglia from three segments (proto-, deuto- and tritocerebrum). In contrast to older opinions, this tripartition is also present in chelicerates (Damen et al. 1998, Telford & Thomas 1998). Deuto- and tritocerebrum innervate head appendages such as antennae and cheliceres. It has been recently shown that the cheliferes, chelicere-like appendages in pycnogonids, are innervated by the protocerebrum (Maxmen et al. 2005; Fig. 2.20.). This fits into hypotheses that the pair of giant appendages of fossil stem lineage euarthropods (such as *Anomalocaris* and *Fortiforceps*) have to be regarded as a pre-antennal appendage, likely also innervated by the protocerebrum (Budd 2002; Fig. 2.20.).

In onychophorans and tardigrades, the segmental character of the brain is much disputed. In adult onychophorans, the brain shows no traces of segmentation (Schürmann 1987), but Pflugfelder (1948) observed that during embryonic development three consecutive ventral organs form ganglia which then fuse to form the brain. Eriksson and Budd (2000) were also unable to identify segmental patterns in the onychophoran brain and even consider it as derived from a circumintestinal brain. However, recently Strausfeld et al. (2006a, b) stated that the brain of onychophorans shows some similarities to the chelicerate brain (although such characters may be plesiomorphies rather than synapomorphies). In tardigrades, a segmental character of the dorsal brain is not immediately evident, but Dewel and Dewel (1996) tried to homologize sub-regions of the brain with ganglia from up to four segments. Zantke et al. (2007), in contrast, could not observe a tripartition in the tardigrade brain based on immunocytochemical staining with antibodies against acatylated α-tubulin. The ventral nerve cords of onychophorans differ from those in euarthropods in having a large number of comissuers (9–10 per segment, Schürmann 1995) and in lacking ganglia. Instead, the somata are distributed along the nerve cords, which therefore represent a *Markstrang* (Schürmann 1995, Georg Mayer, Berlin, pers. comm.). In tardigrades, four ganglia (corresponding to the four pairs of appendages) are present, but commissures are lacking (Zantke et al. 2007). The conditions in onychophorans and tardigrades, particularly with the current sketchy knowledge, make it hard to reconstruct the structure of the nervous system in the arthropod ancestor. This, however, is important for comparisons with the annelid nervous system in a comparison of the articulate/ecdysozoan hypothesis (see Chapter 2). It cannot be excluded that the arthropod ancestor had a nervous system that did not represent a 'textbook rope-ladder-like system'.

Nervous systems in the tentaculate taxa

The nervous system of bryozoans, phoronids, and brachiopods is not very complex, but more or less comparable among the three taxa. There always is an unpaired ganglion, from which several nerves originate (see Table 6.1.), these can be organized into a fine nerve ring at the tentacle base or can lead to an additional, smaller ganglion. An intraepidermal plexus has been described for at least some parts of the body in phoronids and brachiopods (Herrmann 1997, James 1997). It remains unclear, whether this organization reflects a primary 'simple' condition or whether it is derived from other patterns, in particular due to the sessile or hemisessile lifestyle.

Nervous systems in deuterostomes

Deuterostome nervous systems, especially those of echinoderms and hemichordates, include several characters regarded as ancestral, and it is even sometimes stated that the echinoderm nervous system does more closely resemble a cnidarian nervous system than, for example, that of protostomes or chordates.

Echinodermata

The nervous system in Echinodermata lacks concentrations such as the brain or other ganglia. There are two subsystems, the ectoneural system develops in the epidermis, while the hyponeural system develops in the coelom epithelia (Cobb 1995). Both systems together form nerve cords, in particular circumoral and radial nerves. The nerve cords are composed of adjacent regions of epidermis and coelom epithelium, in which there is a strong concentration of intraepithelial nerves. Therefore, the nervous system can be regarded as entirely intraepithelial, despite the fact that recognizable nerves are present. Within the nerve cords, the ectoneural and the hyponeural system remain spatially separated and are connected only by few nerves (see, e.g. Märkel & Röser 1991, Mashanov et al. 2006; Fig. 6.13.). An intraepidermal nerve plexus is additionally present, at least in echinoids (Cavey & Märkel 1994) and astroids (Chia & Koss 1994). In ophiuroids, echinoids, and holothurians the circumoral and radial nerves originate through tubular infoldings of the epithelia, somewhat comparable to the neurulation (the process by which the dorsal nerve cord of chordates develops). This process was observed long ago (von Ubisch 1913) and results in the tubelike

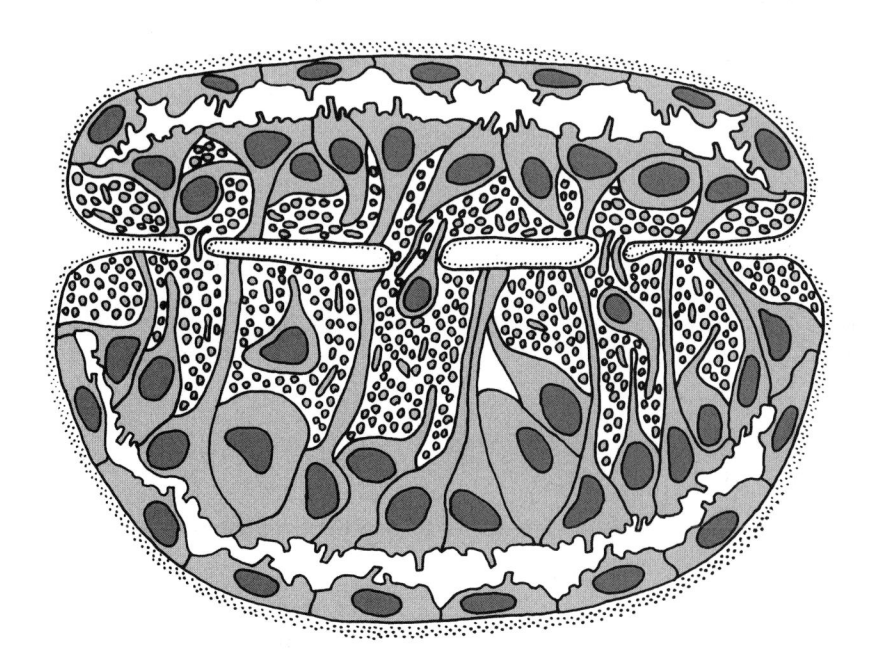

Fig. 6.13. Combination of ectoneural (lower part) and hyponeural (upper part) systems into one radial nerve in holothurians (Echinodermata). Both systems are arranged epithelially around a lumen. After Mashanov et al. (2006).

character of both ectoneural and hyponeural systems (see, e.g., Mashanov et al. 2006 for holothurians; Fig. 6.13.). Crinoids also have intraepidermal nerve cells in the epidermis as well as in the coelom epithelia (Heinzeller & Welsch 1994), but while their ectoneural system remains intraepidermal, two other systems are located in the ECM. One system (which is probably homologous to the hyponeural nervous system) forms an oral ring and radial nerves, while the other is aborally located and is composed of a large cup-shaped region and radial nerves. It is called the entoneural nervous system and acts as an important motor-system in crinoids.

Hemichordata
In hemichordates, the nervous system is dominated by an intraepidermal plexus, within which some elements are pronounced. These are a dorsal and a ventral longitudinal nerve cord in the trunk of enteropneusts as well as a dorsal proboscis nerve (Fig. 6.14.). Of particular interest is the dorsal region of the collar, in which a hollow, subepidermal collar cord is present (Bullock 1940, Silén 1950, Dilly et al. 1970, Cameron & Mackie 1996). In Ptychoderidae, this cord is hollow and opens to the exterior by pores (Benito & Pardos 1997). A brain-like concentration is absent. In Pterobranchia, there are concentrations of the plexus in the oral shield and tentacles as well as a ganglion-like concentration at the tentacle base (Dilly 1975).

Chordata
Chordates are characterized by the common presence of a hollow dorsal nerve cord, the neural tube. In Tunicata, this is present only in the tadpole larvae (see Nicol & Meinertzhagen 1988 for development), young salps (Lacalli & Holland 1998), and in adult appendicularians. In the anterior end, there is a cerebral vesicle and, in front of it, a sensory vesicle containing a statocyst and (when present) a photoreceptor. In adult ascidians, the nervous system is highly modified, probably due to their sessile life-style. There is a brain, from which several nerves originate and lead into the body (Burighel & Cloney 1997, Mackie & Burighel 2005). The brain is accompanied by a neural gland, from which a so-called dorsal strand extends (Fig. 6.15.). Around this dorsal strand is a net of nerves in different stages of neurogenesis, this is called the dorsal strand plexus (Mackie 1995). In Acrania, the neural tube is present along the whole length of the animal. It is hollow and opens through a small pore, the Kölliker's pit, to the outside. In

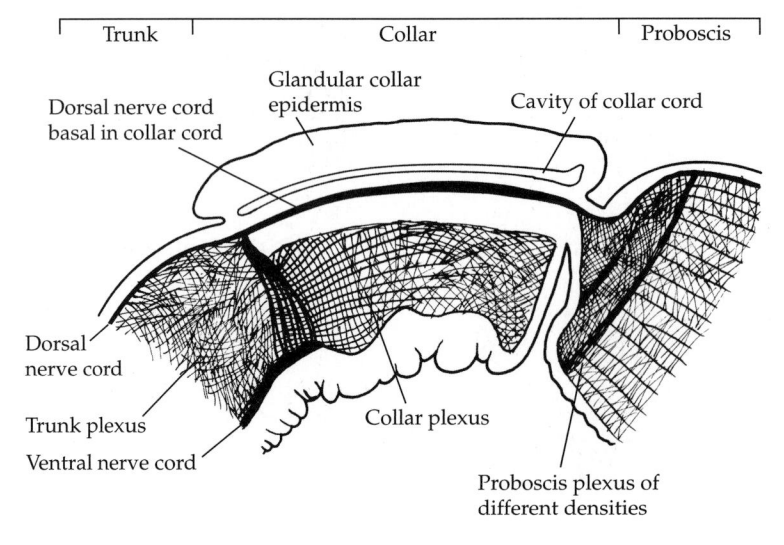

Fig. 6.14. Schematic representation of the nervous system in the anterior end of *Saccoglossus cambrensis* (Enteropneusta). The collar cord lies in the basal part of the hollow collar cord. After Knight-Jones (1952).

the anterior region, a slight swelling is called the brain or, more carefully, a cerebral vesicle. Metameric dorsolateral nerves leave the neural tube, leading to the myotomes and other organs. In the neural tube, the nerve cells are epithelially organized and most of them have cilia directed into the tube cavity (Ruppert 1997). In Craniota, the neural tube is closed and a considerable brain with five regions develops in the anterior end (see Striedter 2005 for evolution of the brain within craniotes).

It appears that the deuterostome ancestor has taken over from diploblastic ancestors an epithelial organization of the nervous system, with some condensations. A brain was probably not present, because it is absent in echinoderms, enteropneusts, and acranians. According to this view, the ganglion in pterobranchs is a derived feature. However, it may still be argued that a brain was lost due to the sessile or hemisessile mode of life in most basal deuterostomes. The hollow neural tube is an autapomorphy of chordates. Somewhat problematic is the collar cord of enteropneusts. In the animal, it is in a dorsal position (when the orientation in the sediment and the position of the mouth opening are considered). It has therefore been assumed that it could be homologous to the chordate neural tube. However, the possibility of a dorsoventral inversion speaks against such a homologization (see Chapter 3) as well as the fact that the collar cord does not appear to have any coordinating function (Cameron & Mackie 1996). Although the neural tubes in larval ascidians and acranians have no or only slight swellings, expression patterns of several genes indicate that the anterior part of the neural tube is already regionalized in three consecutive parts (Wada et al. 1998, Meinertzhagen & Okamura 2001, Nieuwenhuys 2002, Mazet & Shimeld 2002, Shimeld & Holland 2005). As Lowe et al. (2003) have shown, comparable gene expression patterns are also present in the enteropneust nervous system, so that the anteroposterior patterning of the nervous system can be an even older character. It is interesting that some nerves in deuterostomes develop by tubiform invaginations: the neural tube of chordates, the collar cord of enteropneusts, and also radial nerves in echinoderms. It is hotly debated whether this phenomenon can be regarded as homologous (Haag 2005, 2006) or not (Ruppert 2005, Nielsen 2006).

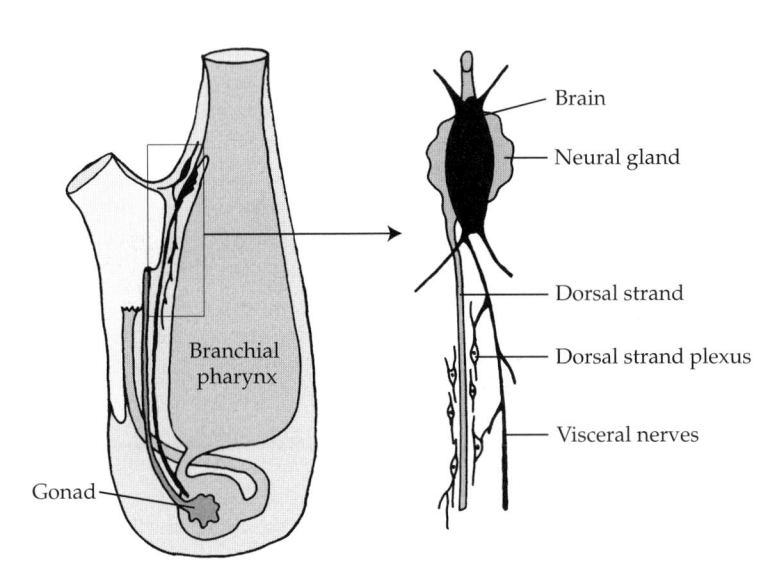

Fig. 6.15. Overview (left) and detail (right) of the nervous system in adult ascidians (Tunicata). The nervous system is shown in black. After Mackie (1995).

Nervous system evolution in Bilateria: some scenarios

As has been noted above, it appears plausible to assume that the bilateral ancestor acquired its bilateral symmetry by combining a directed locomotion with a clear anteroposterior differentiation in body organization (see Chapter 3). If this was so, it can be assumed that there is a strong selective pressure towards an advanced structural patterning of the nervous system as well, and we expect a centralisation of nerves in the anterior end (a brain) as well as a differentiation of more strongly and weakly pronounced neuronal paths. This is also the conclusion from comparing features of brain and nerve cord patterning during development in protostomes and deutrostomes (Arendt & Nübler-Jung 1996, Reichert & Simeone 2001, Ghysen 2003, see also Lacalli 1994 and Nielsen 1999). From taking a look at the phylogenetic tree, however, this expectation depends strongly on the phylogenetic hypothesis that is favoured.

1. Acoelomorpha are basal bilaterians
When acoelomorph's are the sister-group of all remaining Bilateria, and not members of the Platyhelminthes, it appears that a brain with longitudinal nerve cords is indeed a bilaterian autapomorphy. Acoels have been taken, for example, by Reisinger (1972), as strong supporters of the orthogon-hypothesis, but as has been shown above, their nervous system is very diverse. It contains, however, a dorsal concentration of fibres and somata (the brain). Raikova et al. (1998) named the brain a 'commissural brain'. Therefore, the ancestral condition of the brain may have been a concentration of dorsolateral nerve cells which are connected by a more or less concentrated commissure composed of axons. The gastrotrich brain resembles such a brain in some respects, it appears to be composed of a strong dorsal commissure and dorsolateral nerve cells (see above). Assuming that gastrotrichs are closely related to cycloneuralians, the evolution of the cycloneuralian circumintestinal nerve ring could be imagined by an extension of the semicircular commissure (as present in gastrotrichs) to the ventral side and its subsequent ventral closure. This would bring the longitudinal nerve cords from a lateral position (as in gastrotrichs) to a ventral position (as in cycloneuralians). In spiralians (and, if Ecdysozoa are valid, in parallel in arthropods) the brain evolves in a different direction, forming a more compact mass of axons, around which the nerve cell somata form almost a sheath.

In this scenario, we have to assume that in the radialian/deuterostome branch of the tree the nervous system was first reduced and later flourished again (in craniotes). Regardless of the position of the tentaculate taxa (see Chapter 2), the absence of a brain and longitudinal nerves in echinoderms and hemichordates is likely to be a reduction, probably connected with the sessile or hemisessile mode of life.

The presence of only intraepidermal nerves in *Xenoturbella* (Pedersen & Pedersen 1988) with the absence of any concentration (at least their components with anti-serotonin and anti-FMRFamide immunoreactivity, see Raikova et al. 2000a) are in agreement, as well with a position as the sister-group to Acoelomorpha + remaining Bilateria as with a close relationship to basal deuterostomes, in particular echinoderms and hemichordates (see Chapter 2).

2. Acoelomorpha are flatworms and the tentaculate taxa are protostomes
If acoelomorphs are not basal bilaterians, but belong within the Platyhelminthes, the interpretation of nervous system evolution depends much on the phylogenetic position of the tentaculate taxa. When, as suggested by molecular analyses (see Chapter 2), bryozoans, phoronids, and brachiopods are spiralians, then we face the situation that in one bilaterian branch, the Protostomia, a brain and longitudinal nerve cords are present, while in the other, the Deuterostomia, such characters are absent in the basal taxa, in particular in echinoderms and hemichordates. This makes it at least possible that brain and longitudinal ventral nerve cords are not a bilaterian feature, but evolved in the protostome ancestor. The bilaterian as well as the deuterostome ancestor would, according to this

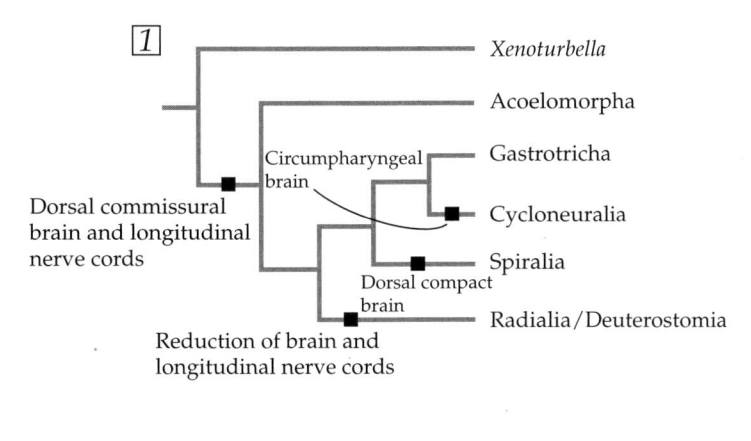

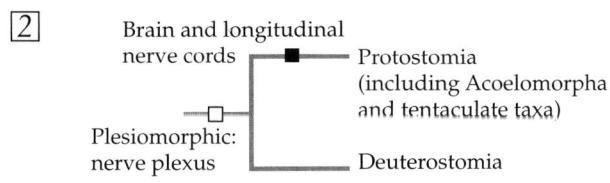

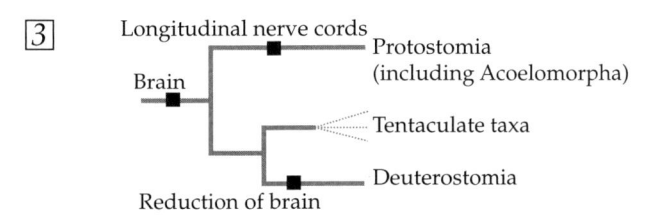

Fig. 6.16. Three senarios for nervous system evolution in Bilateria. See text for explanation.

scenario, have taken over the basiepidermal nerve plexus. Only within deuterostomes did a dorsal longitudinal nerve cord and an anterior concentration of the nervous system evolve.

3. Acoelomorpha are flatworms and the tentaculate taxa are related to Deuterostomia

When bryozoans, phoronids, and brachiopods are related to Deuterostomia, i.e. when a taxon Radialia exists, the evolution of at least a brain in the bilaterian ancestor appears to be more likely. Despite the fact that the tentaculate taxa are sessil or hemisessil, they have a concentration of the nervous system which may be homologous to a brain. If this is so, then the brain would be a bilaterian autapomorphy, but longitudinal cords evolved later and independantly.

Short remarks on nervous system development

When larvae are present, the architecture of the nervous system usually differs between larvae and adults. During metamorphosis, there is often a replacement of major parts of the larval nervous system by adult components. This can be performed in a more or less gradual reorganization as in hydrozoan planulae (Martin 2000), or in a 'catastrophic metamorphosis' as, for example, in bryozoans (Wanninger et al. 2005b). A common character of diverse larvae is the possession of an apical organ, being composed of a ciliated tuft at the apical tip of the larva and an underlying concentration of nerve cells (see Nielsen 2001). Such apical organs occur in at least some of the larvae of cnidarians (Martin & Koss 2002), polyclads (Ruppert 1978a), trochozoans, phoronids, bryozoans, brachiopods, echinoderms, and hemichordates (Nielsen 2001). Usually, several nerves originate from the apical organ and run to different parts of the larval body. Serotinergic nerves have been found to be particularly associated with ciliary bands (see, e.g., Hay-Schmidt 1990a-c, 1992, Beer et al. 2001, Santagata & Zimmer 2002, Voronezhskaya et al. 2002, Cisternas & Byrne 2003, Nezlin & Yushin 2004, Wanninger et al. 2005b). A probable phylogenetic difference is that in protostomes, the larval apical organ or adjacent nervous tissue are usually integrated into the adult system, while in deuterostomes and 'tentaculate' taxa this is not the case (Nielsen 2001). However, one has to be careful with phylogenetic conclusions from this argument. The homology of larvae in different taxa is not certain (see Chapter 12) and, when larvae are not homologous, apical organs must have evolved several times. Most 'tentaculate' taxa and basal deuterostomes have either no, or no significant, brain into which the apical organ can be integrated. Furthermore, even during a 'catastrophic' metamorphosis, part of the larval nervous tissue can be integrated into the juvenile nervous system, as has been shown by Manni et al. (1999) for ascidians and by Santagata (2002) for phoronids.

Unusual innervation patterns

It is well known that in nematodes nerves do not lead to the muscles, but instead muscle processes lead towards the nerve cords to form synapses there (e.g. Bird & Bird 1991). This is an unusual pattern of innervation, but it is not restricted to nematodes. It has also been reported to be present in nemerteans (Turbeville & Ruppert 1985, Turbeville 1991), as very short muscular processes in gastrotrichs (Teuchert 1977 for *Turbanella cornuta*), in echinoderms (Cobb & Laverack 1967 for *Echinus* sp.), and in Acrania (Flood 1966, 1968 for *Branchiostoma lanceolatum*).

The evolution of neurotransmitters

Animals use a wide variety of different neurotransmitters in chemical synapses. The main reason for the existence of such diversity may be that this allows a better distinction between different incoming information in a target neuron (Squire et al. 2003). Additionally, for each transmitter there are different receptors in the post-synaptic membrane (see, e.g. Ribeiro et al. 2005). This leads to differentiated responses to the same transmitter, thereby multiplying the diversity of incoming signals. One major requirement for a substance to be used as a neurotransmitter is that it is easy to synthesize. Therefore, neurotransmitters are either small molecules such as peptides, or they are larger molecules but derived from substances available from 'usual' metabolic pathways in the cell. These can be amino acids (e.g. γ-aminobutyric acid = GABA), choline esters (e.g. acetylcholine), or biogenic amines which are derived from tryptophan or tyrosine metabolism. The most well known biogenic amines are catecholamines (dopamine, octopamine, adrenaline, noradrenaline), serotonin (5-hydrotryptamine) and histamine. The presence or absence of specific neurotransmitters has been checked for a number of taxa and some phylogenetic conclusions have been drawn from it. Most transmitters, however, are very broadly distributed and are obviously very old molecules (Walker et al. 1996).

Because neurotransmitters, in particular the biogenic amines, are so closely related to abundant metabolic substances, it is not surprising that they have been found in unicellular organisms and in sponges, which have no synapses. This accounts for catecholamines (dopamine, adrenaline, and noradrenaline) and serotonin (Turlejski 1996, Weyrer et al. 1999, Coppi et al. 2002, Eichinger et al. 2002). In cnidarians, neuropeptides are abundant and it has been stated that the nervous system is essentially peptidergic, that is, using mainly neuropeptides as transmitters (Grimmelikhuijzen 1983, Grimmelikhuijzen et al. 1991, Grimmelikhuijzen & Westfall 1995, Shaw 1996). However, several biogenic amines have also been shown to

Table 6.1 Summary of nervous system organization in Bilateria. Largely ignored here is the stomatogastric nervous system and several 'smaller' nerves, e.g. those leading from the brain to anterior sensory organs. LNC = longitudinal nerve cord

	General structure	Brain	Plexus	Longitudinal nerves	References
Xenoturbella	Intraepidermal plexus	Absent	Intraepidermal plexus	Absent	Pedersen & Pedersen 1988, Raikova et al. 2000a
Acoelomorpha	Brain and variable number of LNCs and commissures	Dorsal brain with differing structure (see text)	Most components are basiepidermal, but not forming a complete plexus	Variable number of (mostly) intraepidermal LNCs with some commissures	Raikova et al. 2000b, Reuter et al. 2001a,b
Gastrotricha	Dorsal brain, 1 pair of ventrolateral LNCs	Dorsal commissure with lateral somata	Absent	1 Pair of intraepidermal, ventrolateral LNCs	Teuchert 1977, Ruppert 1991b, Wiedermann 1995
Nematoda	Ring-shaped brain, unpaired ventral LNC, additional ganglia in posterior end	Ring-shaped, anterior and posterior somata and median neuropil	Some intraepidermal nerves are present, but not organized as a plexus	Intraepidermal unpaired ventral LNC, additional finer longitudinal nerves	White et al. 1986, Bird & Bird 1991, Wright 1991
Nematomorpha	Brain, unpaired dorsal (only *Nectonema*) and ventral LNCs	Subesophageal brain with small dorsal commissure	Present	Intraepidermal unpaired dorsal (only *Nectonema*) and ventral LNCs	Schmidt-Rhaesa 1996
Priapulida	Ring-shaped brain, unpaired ventral LNC, caudal ganglion	Ring-shaped, anterior and posterior somata and median neuropil	Not described	Intraepidermal unpaired ventral LNC, additional finer longitudinal nerves	Rehkämper et al. 1989, Storch 1991
Kinorhyncha	Ring-shaped brain, paired ventral LNCs with metameric ganglia	Ring-shaped, anterior and posterior somata and median neuropil	Not described	Intraepidermal paired ventral LNC with metameric ganglia, additional finer longitudinal nerves	Kristensen & Higgins 1991
Loricifera	Ring-shaped brain, paired ventral LNCs, additional ganglia including caudal ganglion	Ring-shaped, anterior and posterior somata and median neuropil	Not described	Intraepidermal paired ventral LNC, additional finer longitudinal nerves	Kristensen 1991
Platyhelminthes	Dorsal brain and orthogon	Bilobed, subepidermal, central neuropil	Up to 3 plexuses with intra- and subepidermal positions	Variable number	Rieger et al. 1991, Reuter & Gustafsson 1995
Gnathostomulida	Dorsal brain, 1 pair of LNCs, buccal ganglion	Unpaired, basi- (Bursovaginoidea) or subepidermal (Filospermoidea), central neuropil	Not described	1 Pair or up to 3 pairs (in *Rastrognathia*), basiepidermal	Lammert 1991, Müller & Sterrer 2004
Limnognathia maerski	Dorsal brain, 1 pair of LNCs, thoracal and caudal ganglia	Slightly bilobed, position unknown, central neuropil	Not described	? Pair, probably subepidermal	Kristensen & Funch 2000

Table 6.1 (Contd)

	General structure	Brain	Plexus	Longitudinal nerves	References
Eurotifera	Dorsal brain, ventral mastax ganglion, 1 pair of LNCs, additional caudal ganglion	Bilobed, subepidermal, central neuropil	Not described	1 Pair, subepidermal	Clément & Wurdak 1991
Seisonida	Dorsal brain, ventral subesophagal ganglion, 1 pair of LNCs	Unpaired, subepidermal, central neuropil	Not described	1 Pair, subepidermal	Ricci et al. 1993, Ahlrichs 1995
Acanthocephala	Dorsal brain, 1 pair of LNCs, in males additional genital and bursal ganglia	unpaired, subepidermal, central neuropil	Not described	1 Pair, subepidermal	Dunagan & Miller 1991
Cycliophora (free-living stages)	Brain, paired ventral LNCs and additional nerves	Bilobed dorsal brain with central neuropil	Not described	1 Pair of ventral LNCs	Funch & Kristensen 1997, Wanninger 2005
Nemertini	Dorsal brain, at least 1 pair of LNCs, peripheral plexus	Bilobed, circumintestinal, subepidermal, central neuropil	Basiepidermal plexus	1 Pair, subepidermal, sometimes additional nerves, e.g. unpaired dorsal and ventral	Turbeville 1991
Kamptozoa	Brain, probably 1 pair of ventral LNCs, additional ganglia e.g. at tentacle base	Bilobed, probably subepidermal	Several small nerves, but probably no plexus	Probably 1 pair of nerves is homologous to the ventral LNC in other taxa	Nielsen & Jespersen 1997, Fuchs et al. 2006
Mollusca	Basal condition: brain and 2 pairs of LNCs, connected by commissures (= orthogonal organization), additional ganglia can be present. Derived is a stronger concentration of the nervous system.	Bilobed, subepidermal, central neuropil	Epithelial nerves originate from LNC, but a complete plexus is not described	2 Pairs, subepidermal	Eernisse & Reynolds 1994, Scheltema et al. 1994, Haszprunar & Schaefer 1997
Sipunculida	Dorsal brain, unpaired ventral LNC, from which peripheral nerves originate	Bilobed, subepidermal, central neuropil	Subepidermal plexus	1 LNC, unpaired, subspidermal	Rice 1993
Echiurida	1 Pair of ventral LNCs with subesophageal ganglion and numerous nerves originating from LNCs	Absent	Not described, peripheral nerves (probably subepidermal) are present but may not form a plexus	1 Pair, probably subepidermal	Pilger 1993

Taxon	Brain	Peripheral nervous system	Longitudinal nerve cords (LNCs)	References
Annelida	Dorsal brain, 1 pair of main ventral LNCs and variable number of additional LNCs. Ganglionated nervous system with segmental pairs of ganglia and commissures.	Bilobed, intra- (some polychaetes) or subepidermal (remaining taxa) position, central neuropil	Present, but to different extent.	1 Main pair, intra- (some polychaetes) or subepidermal (remaining taxa) position, in polychaetes variable number of additional intraepidermal LNCs, orthogonal organization	Golding 1992, Jamieson 1992, Fernández et al. 1992, Orrhage & Müller 2005, Müller 2006
Tardigrada	Brain, circumintestinal connectives, paired LNCs with metameric ganglia (without commissures)	Dorsal brain with central neuropil	Several peripheral nerves, but not organized as plexus	1 Pair, with segmental ganglia	Dewel et al. 1993, Dewel & Dewel 1996, Zantke et al. 2007
Onychophora	Brain, circumintestinal connectives, paired ganglionated LNCs with metameric commissures	Dorsal brain with central neuropil	Several peripheral nerves, but not organized as plexus	1 Pair, no ganglia, 9–10 commissures per segment	Schürmann 1987, 1995
Euarthropoda	Brain, circumintestinal connectives, paired ganglionated LNCs with segmental commissures	Dorsal three-segmented brain with central neuropil	Several peripheral nerves, but not organized as plexus	1 Pair, with segmental ganglia and commissure	E.g. Babu 1985, Fahrenbach 1999, Farley 1999, Felgenhauer 1999, Sandeman 1982
Chaetognatha	Brain and five further ganglia in head region, large ventral ganglion with numerous radiating nerves	Dorsal, unpaired brain (cerebral ganglion) with central neuropil	Intraepidermal plexus	No 'true' LNCs, but longitudinal connectives from brain to ventral ganglion and caudal nerves originating at its posterior end	Shinn 1997
Bryozoa	Brain, visceral ganglion, several individual nerves	Unpaired cerebral ganglion	Not described	Absent	Mukai et al. 1997
Phoronida	Brain, nerve ring at tentacular base, peripheral plexus and several identifiable nerves are present	Unpaired cerebral ganglion	Basi- to subepidermal plexus	Absent	Herrmann 1997
Brachiopoda	Only brain (inarticulates) or additional supraentric ganglion (articulates), several individual nerves and plexuses (e.g. in mantle) are present	Unpaired subentric ganglion	Basiepidermal plexus at least in some regions of the mantle	Absent	James 1997
Echinodermata	Circumoral and radial nerves	Absent	Basiepidermal plexus present in some taxa	Absent	Cobb 1995, Mashanov et al. 2006

Table 6.1 (*Contd*)

	General structure	Brain	Plexus	Longitudinal nerves	References
Enteropneusta	Plexus with condensations in form of longitudinal nerves, tubelike collar cord	Absent	Intraepidermal plexus	Present as concentrations within plexus	Bullock 1940, Silén 1950, Benito & Pardos 1997
Pterobranchia	Plexus with condensations in form of nerves, collar ganglion	Brain as collar ganglion	Intraepidermal plexus	Absent	Dilly 1975, Cameron & Mackie 1996, Benito & Pardos 1997
Tunicata	Brain with several nerves and 'dorsal strand plexus'. Larvae: cerebral vesicle with dorsal neural tube	Unpaired brain with central neuropil	Plexus only along the dorsal strand	Absent	Koyama & Kusunoki 1993, Mackie 1995, Mackie & Burighel 2005
Acrania	Dorsal neural tube with anterior 'brain', metameric dorsolateral nerves	Hollow brain as thickened region of the neural tube	Absent	Absent	Ruppert 1997
Craniota	Dorsal neural tube with anterior brain, metameric dorsolateral nerves, numerous further nerves	Hollow brain with 5 regions as anterior region of neural tube	Peripheral nerves present, but not organized as plexus	Absent	Westheide & Rieger 2004, Striedter 2005

be present, such as dopamine, adrenaline, norephidrine, and serotonin (Hay-Schmidt 2000, Westfall et al. 2000, Leitz 2001, Anctil et al. 2002, Anctil & Bouchard 2004), and these may play locally important roles. The presence of GABA is also likely in cnidarians (Leitz 2001).

Two transmitters that probably evolved later are acetylcholine and octopamine. Acetylcholine is abundant in bilaterias but is likely to be absent in diploblastic animals, and it has therefore been evaluated as an autapomorphy of Bilateria (Nielsen 2001). There are some reports of acetylcholine or acetylcholinesterase in cnidarians (see, e.g., Lentz 1968), but these findings are not unambiguous and several targeted investigations were unable to detect these molecules (see Martin & Spencer 1983 for a summary). Octopamine is another broadly distributed transmitter in bilaterians, which has been shown for a number of taxa (Pflüger & Stevenson 2005) and may also be an autapomorphy of Bilateria. Octopamine is present in craniotes only in traces and is here functionally replaced by noradrenaline (Pflüger & Stevenson 2005).

Conclusions

Nervous systems evolved in the ancestor of Eumatazoa. Several components of nervous systems, as well as forms of signal transduction, can be found in sponges and non-metazoan organisms. Within Eumetazoa, some general trends can be observed, such as from:

- diffuse nets to defined nerves;
- a system without coordinating centres (brain) to systems with a brain;
- a peripheral to a centralized system (although a peripheral system usually persists in addition to a central system);
- an undirected (with bidirectional synapses) to a directed system (with unidirectional synapses);
- a regular grid (orthogon) to more specialized innervation patterns (in protostomes).

These are called trends, because several of these processes probably occurred more than one time in parallel and to different degrees. The reconstruction of nervous system patterns, especially concerning the ancestor of Bilateria, depends strongly on some recently open questions of phylogenetic relationships and can therefore (at this point of time) not be solved satisfactorily. In general, however, the nervous system has considerable information content for phylogenetic studies.

Transmitter substances are problematic for phylogenetic conclusions because they are molecules derived from abundant and usual metabolic pathways. They also have a broad distribution among animals, which makes it hard to find out where they evolved. It may be more promising to concentrate on receptors for transmitter substances, because these are more numerous than the transmitters themselves and probably yield more phylogenetic resolution.

CHAPTER 7

Sensory organs

Sensory organs are those organs or structures that receive information (stimuli) and translate this information into a signal recognizable to the nervous system. The received information can be very variable, with light, mechanical stimuli, taste, odour, and sound being the most familiar ones to us as humans, but many more are possible (e.g. electric fields, temperature, pressure etc.).

Despite the fact that there are many different stimuli and many specialized sensory structures, the basic principles of signal transduction are surprisingly similar (see, e.g., Fain 2003). The recipients for a stimulus are membraneous proteins (so-called sensory receptor molecules) that change their conformation in response to the stimulus (Fig. 7.1.). The effective answer is the opening of transmembrane ion channels that change the intra- and extracellular ion concentrations and thereby create action potentials. The sensory receptor proteins can themselves be ion channels and therefore directly receive information and transform it, this is done, for example, in mechanoreceptors. In many cases signal transduction is more complicated, with further effector molecules and second messengers being incorporated into a signal transduction cascade before the information reaches the ion channels. The exceptions to this generalized mechanism are electroreceptors, which do not need a signal translation.

Because the sensory receptor molecules and the ion channels are membrane proteins, receptor cells typically contain extensions of the cell membrane in the form of microvilli or, more abundantly, cilia. Ciliary receptors are involved in almost all kinds of reception. It is often possible to suspect a sensory function of a structure, for example when ciliated cells have a connection to the nervous system. This is the state of knowledge for a large number of receptor structures, and physiological experiments that verify their sensory function are often lacking. On the other hand, particular responses of animals to stimuli (e.g. light or touch) can often be observed but the according receptors cannot be found.

A common distinction of sensory cells is between primary (sensory cells send axons to a nerve cell) and secondary cells (direct contact between sensory and nerve cell soma) as well as of free endings of nerve cells which can also be sensory. As cnidarians have primary and secondary sensory cells, it cannot be decided whether one of the two types evolved earlier

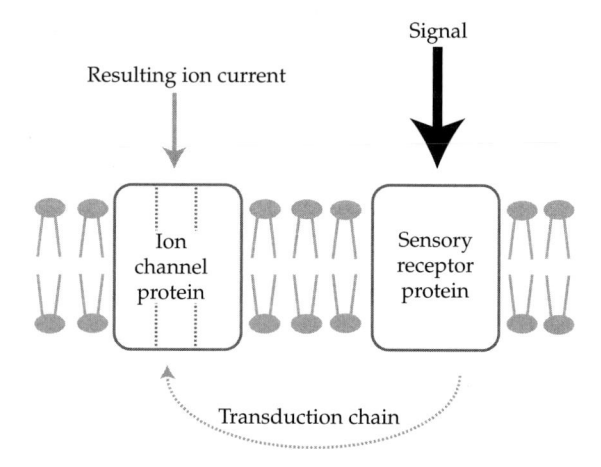

Fig. 7.1. General principle of the translation of different signals. Signals modify the transmembraneous sensory receptor protein, which activates a transduction chain that finally opens ion channels. The sensory receptor protein may also be itself an ion channel.

than the other one, and a distinction between primary and secondary sensory cell is neglected in the following. Sensory cells can form structures or even organs of rising complexity, in these cases further cells, such as support cells or gland cells, can be involved.

As a detailed analysis of each sensory structure would exceed the available space of this chapter, I present the diversity of sensory structures in a summary table only, discuss ciliary sensory cells and statocysts briefly, and then focus on photoreceptors, because these have always been an important topic in evolutionary biology.

Overview on the basic principles in sensory structures

Before taking a look at the receptive structures themselves, it is helpful to briefly inform about the different kinds of reception. The following information is based on standard physiological textbooks, in particular on Smith (2000) and Fain (2003).

- **Mechanoreception.** The common principle is the mechanical change in conformation of transmembraneous proteins. These also have an extra- and an intracellular domain, which are linked to further structural components. These can be the cytoskeleton (internal) and the extracellular matrix (external). In the case of the touch receptor of the nematode *Caenorhabditis elegans*, the external domain is linked to the cuticle and the internal domain to the intracellular microtubuli (Chalfie & Sulston 1981, Tavernarakis & Driscoll 1997). The shear between extra- and intracellular components, caused by even slight movements, is thereby effectively transduced to the transcuticular protein. The protein acts itself as an ion channel and therefore directly produces ion currents necessary for the emergence of action potentials. Besides direct mechanical stimuli such as touch, stretching etc., mechanoreceptors are also important for the detection of medium motion (air, water current, blood). They are also basal components of balance organs (statocysts) and of acoustic organs. Mechanoreceptors in this wide sense are found on the body surface as well as inside the body.
- **Chemoreception.** In chemoreceptive cells, a membraneous receptor protein is sensitive to particular molecules. Contact with such molecules leads to a change in conformation, which in turn starts a transduction cascade in which a second messenger finally opens ion channels (Fig. 7.1.). Chemoreception is extremely important when reproduction or coordinated behaviour are initiated by pheromones or other substances. It is also important in any kind of olfaction and taste, or in enteroreceptors that control the functional state of internal organs. An example for chemoreception is the human nose, in which about a thousand different receptors, each encoded by its own gene, are present (Firestein 2001).
- **Photoreception.** To be photosensitive, a membraneous protein requires a prosthetic group, the chromophore. This protein-chromophore complex makes conformational changes in response to light and and starts a transduction cascade which finally opens ion channels (see below for more details).
- **Thermoreception.** The entire process of thermoreception is not fully resolved, but there is evidence that thermosensitive membraneous proteins are involved, which finally open ion channels in a way comparable to other receptive structures.
- **Electroreception.** Electrosensitive cells do not need mediation by membraneous proteins. Nevertheless, a structural organization is required in which currents can flow through the electrosensitive cells. They cause hyper- or depolarization which directly influences synapses to nerve cells (electrosensitive cells are secondary receptor cells).
- **Magnetoreception.** Very few details are known from this kind of sense, but it appears that cells containing magnetite (Fe_3O_4) are essential for the reception of magnetic fields.

With the exception of thermo- and electroreception, all senses can already be detected from organisms other than metazoans. For example, magnetoreception and chemoreception were studied in bacteria, and mechanoreception in the ciliate *Paramecium*. This means that the molecular pathways for sensory structures are often very old (see also Vinnikov 1982). The main difference is that in unicellular non-metazoans the signal transduction directly aims at the target structure within the same cell, while in metazoans (or, better, eumetazoans, because only these have sensory cells) the signal transduction aims (with the exception of electroreceptors) at ion channels, which are the central interface between the sensory and nervous system.

Summary of sensory structures in metazoans

The following summary of sensory structures concentrates on the morphologically recognizable structures and neglects, to a large extent, speculations on the function of these structures. It is briefly noted whether statocysts and photoreceptors are present in species of the listed taxa, even if they are present only in few species or particular life stages like larvae. A more detailed discussion follows below.

As can be seen from Table 7.1., ciliated sensory cells are present in sensory structures of all eumetazoan taxa. In many cases, individual receptor cells are present as integrated parts of the epidermis, and it can be assumed that a specialization of particular epidermal cells to contain receptor molecules and transduce this information to the nervous system was the first step in the evolution of sensory cells. The function of the first sensory cell cannot be reconstructed, but it seems likely that at least mechano- and chemoreception

Table 7.1 Summary of sensory structures. Statocysts (st) and photoreceptors (pr) are only indicated as present or absent (cases where only few representatives have the structure are also coded as 'present'), for further information see text.

		st	pr
Cnidaria	**Ciliary epidermal receptors** as collar receptors with cilia and surrounding regular or modified microvilli (e.g. Tardent & Schmid 1972, Vandermeulen 1974, Golz & Thurm 1994, Westfall et al. 1998), collar receptors are also present in the gastrodermis (Raikova 1995). Assumed function is mechano- and chemosensory (Fautin & Mariscal 1991, Thomas & Edwards 1991).	x	x
	Nematocytes include a sensory apparatus, which is comparable to the epithelial collar receptors, especially in anthozoans (Schmidt & Moraw 1982, Fautin & Mariscal 1991), while the ciliary apparatus (cnidocil) in Tesserazoa (Medusozoa) is modified (e.g. Westfall 1970, Holstein & Hausmann 1988, Lesh-Laurie & Suchy 1991, Thomas & Edwards 1991). Modifications are loss of ciliary basal apparatus and accessory centriole as well as modification of circumciliary microvilli (Ax 1996). In some cases further modifications such as the multiplication of ciliary microtubuli occur (Cormier & Hesslinger 1980 for *Physalia physalis*).		
	Internal ciliary receptor: A probably internal (i.e. basiepidermal) receptor with a cilium and surrounding collar of microvilli is described from *Ceriantheopsis americanus* (Peteya 1973).		
Ctenophora	**Ciliary epidermal receptors**, cilia with complex basal structure but without surrounding microvilli (Mackie et al. 1988, Hernandez-Nicaise 1991, Tamm & Tamm 1991). Complex **apical organ** at aboral pole including statocyst, sensory and nervous cells within a 'ciliary dome'. Also present are lamellate bodies with an assumed photosensitive function and epithelial papillae with a probable function in pressure reception (Tamm 1982, Hernandez-Nicaise 1984, 1991).	x	?
Xenoturbella	Ciliary epidermal receptors are probably present (briefly mentioned in (Pedersen & Pedersen 1986), but are not descibed in detail.	x	–
Acoelomorpha	**Ciliary epidermal receptors**, including collar receptors (Bedini et al. 1973, Smith & Tyler 1986, Rieger et al. 1991b, Todt & Tyler 2007).	x	?
Gastrotricha	**Ciliary epidermal receptors** as collar receptors, microvilli remain under cuticle, epicuticle covers cilia (Ruppert 1991b). Cells can be individual or grouped (in sensory bristles). They occur in the anterior end or in adhesive tubules.	–	x
	Cephalic chemoreceptor composed of sensory cells and microvilli in a capsule, some microvilli pass through the cuticle (Teuchert 1976a, Gagne 1980).		

Table 7.1 (contd)

		st	pr
Nematoda	**Cuticular ciliary receptors** as either bristles or setae (in particular the cephalic sensillae; Fig. 7.2) or as pores (amphids, phasmids in Secernentea, further body pores) (Fig. 7.3; summaries in Coomans 1979, Wright 1980, 1991, Jones 2002). Receptors can be mechano- and chemoreceptors (Wright 1983) and some can also perform secretory (glandular) functions (e.g. McLaren 1976, Nebelsick et al. 1992, Bauer-Nebelsick et al. 1995). **Free endings of nerve cells** (Jones 2002). **Internal ciliary receptors:** stretch receptors (Lorenzen 1978, Hope & Gardiner 1982), touch-cell receptors (Chalfie & Sulston 1981), internal ciliary receptors (Wright 1980).	–	x
Nematomorpha	Transcuticular epidermal projections in some terrestrial species (*Paragordius varius*, *Chordodes nobilii*) probably have a sensory function (Schmidt-Rhaesa 2004, Schmidt-Rhaesa & Gerke 2006). In the marine *Nectonema*, four giant cells in the anterior end are likely to be sensory in function (Schmidt-Rhaesa 1996a).	–	–
Priapulida	**Cuticular ciliary receptors** as flosculi and scalids. Flosculi are cuticle covered collar receptors with a flower-shaped external occurrence (Moritz & Storch 1971, Lemburg 1995a). The scalids on the introvert contain different types of receptors (including one type with a coiled cilium) (Lemburg 1995b). In the anterior body region, some further cuticular/ciliary sensory structures can be found such as a receptor complex containing a so-called pin-cilium receptor (Lemburg 1995a).	–	–
Kinorhyncha	**Cuticular ciliary receptors** as flosculi, scalids, sensory setae, and sensory spots (Kristensen & Higgins 1991). Flosculi and sensory spots are cuticle-covered collar receptors, in the case of sensory spots within the cuticle and with an apical pore.	–	x
Loricifera	**Cuticular ciliary receptors** as flosculi, scalids and larval setae (Kristensen 1991).	–	–
Platyhelminthes	**Ciliary epidermal receptors**, including collar receptors, receptors are mono- or multiciliary, sometimes concentrated in special areas in head region (e.g. MacRae 1967, Brooker 1972, Lyons 1972, Storch & Abraham 1972, Ehlers & Ehlers 1977, Rieger 1981, Sopott-Ehlers 1984, Rieger et al. 1991b, Rohde & Watson 1995).	x	x
Gnathostomulida	**Ciliary epidermal receptors** with collar of 8 microvilli (Lammert 1991), additional occurrence in intestinal epithelium. A diversity of types is known (including branched cilia or cilia in a perforated epidermal pit (Lammert 1986). Cells can be individual or aggregated, e.g. in sensory bristles (Lammert 1991). Spiral ciliary receptor in the anterior end with one long coiled cilium and unknown function (Lammert 1984).	–	–
Limnognathia maerski	**Ciliary epidermal receptors**, some with stiff cilia. Probably also present in intestinal epithelium (Kristensen & Funch 2000).	–	?
Eurotifera	**Ciliary epidermal receptors** among wheel organ, in dorsal 'antenna', on foot, in few pores, and in males as a penis receptor (Clément & Wurdak 1991). **Free nerve endings** (Clément & Wurdak 1991).	–	x
Seisonida	**Ciliary epidermal receptors** among wheel organ, additional ciliary receptors in intestinal system. **Internal ciliary receptors** in head and trunk. **Unciliated receptor** in and lateral of dorsal 'antenna'. (all data from Ahlrichs 1995).	–	–
Acanthocephala	**Unciliated receptors** with free nerve endings in an apical and lateral sensory organ (Dunagan & Miller 1983, Herlyn et al. 2001) and in the male reproductive organs (Dunagan & Miller 1991).	–	–
Cycliophora	**Ciliary epidermal receptors** are present only in free-living stages. In the dorsal ciliated pits of the chordoid larva ciliary receptors are concentrated together. Feeding stages are devoid of peripheral sensory structures, only in the ciliated mouth ring collar receptors have been found (Funch & Kristensen 1997).	–	–
Nemertini	**Ciliary epidermal receptors**, either individualized or concentrated in slits or groves (Storch & Moritz 1971, Turbeville 1991). **Ciliary cerebral organs** with canal opening to the exterior and nervous connection to the brain; chemosensory and neuroendocrine function is assumed (Ling 1969, 1970, Ferraris 1985).	x	x

Table 7.1 (contd)

		st	pr
Kamptozoa	**Ciliary epidermal receptors** are mentioned, but not further described by Nielsen & Jespersen (1997).	–	–
Mollusca	**Ciliary epidermal receptors** are present in several different forms; **Solenogastres:** frontal setae of compound cilia, dorsofrontal pit, atrial sense organ and pedal pit (Haszprunar 1986, Scheltema et al. 1994); **Caudofoveata:** frontal cilia (Scheltema et al. 1994); **Polyplacophora:** cilia are found within the complex girdle sensory organ (Fischer et al. 1980); **Gastropoda:** chemo- and mechanoreceptive cells in epidermis, especially on the tentacles (Voltzow 1994); **Bivalvia:** sensory cells are present in certain body regions such as mantle edge, siphon, or tentacles (Morse & Zardus 1994); **Scaphopoda:** sensory cells on posterior mantle (Reynolds 1992); **Cephalopoda:** sucker mechanoreceptors, in the crista/cupula-system within coleoid statocysts, in the coleoid lateral line analog, in neck proprioreceptors and in several kinds of chemoreceptors (Budelmann et al. 1997).	x	x
	Osphradia are chemoreceptive sense organs including ciliated sensory cells, they are well developed in prosobranch gastropods (e.g. Haszprunar 1985a,b, Voltzow 1994), but also present in aplacophorans (Haszprunar 1987a; not all authors agree to homology with osphradia), chitons, bivalves, (Haszprunar 1987b) and the cephalopod *Nautilus* (Haszprunar 1987a).		
	Non-ciliary sensory structures are the pedal commissural sac in Solenogastres (unknown function; Haszprunar 1986), free nerve endings and muscle proprioreceptors (see, e.g., Budelmann et al. 1997).		
Sipunculida	**Ciliary epidermal receptors** are abundantly present, they are often located in papillae or pits (Storch 1984, Rice 1993). Epidermal organs are composed of ciliary sensory and glandular cells (Rice 1993), nuchal organs and cerebral organs are also complex organs composed of ciliated and other cells (Purschke et al. 1997).	–	x
Echiurida	The presence of epidermal receptors is suspected, but not documented ultrastructurally (compare Pilger 1993), except for monociliary epidermal cells with a proposed sensory function in *Maxmuelleria lankesteri* (McKenzie & Hughes 1999).	–	–
Annelida	**Ciliary epidermal receptors** are abundantly present as isolate receptor cells, in clusters (e.g. on papillae) or together with other cell types in more complex sense organs. Ciliated sensory cells can penetrate the cuticle or remain below the cuticle, they can be mono- or multiciliary. In several, but not all cases, they are organized as collar receptors with a circumciliary ring of microvilli.	x	x
	Sense organs with ciliary receptor cells are nuchal organs, dorsal and lateral organs. For summaries and references see Fernández et al. (1992), Jamieson (1992), Verger-Bocquet (1992), Purschke (2005) and Purschke & Hausen (2007).		
Tardigrada	**Cuticular ciliary receptors** are present in some body regions, especially in the anterior end (e.g. in heterotardigrades in the form of cirri) (Walz 1975b, 1978, Kristensen 1981).	–	x
Onychophora	**Cuticular ciliary receptors** are present in different types on the entire surface, with higher concentrations on the 'antennae' (Storch & Ruhberg 1977, 1993, Bittner et al. 1998).	–	x
Euarthropoda	**Cuticular ciliary receptors** are abundant and very diverse. They occur as bristle-like mechano- and/or chemical receptors (perforated cuticle in the case of chemoreceptors), as flat campaniform sensilla, as chordotonal organs, as stretch or acustic receptors, as slit sense organs in arachnids, or as thermo-/hygrosensitive receptors (Fig. 7.3, 7.4; e.g. Haupt 1979, Ache 1982, Bush & Laverack 1982, Barth & Blickhan 1984, Steinbrecht 1984, 1998, Kerkut & Gilbert 1985, Barth 1986, 2002, Govind 1992, Keil 1998). The cilia involved in sensory structures have a 9x2+0 pattern of microtubuli, this is regarded as an autapomorphy of Euarthropoda (Paulus in Westheide & Rieger 2007).	x	x
Chaetognatha	**Ciliary epidermal receptors** are present in different forms, among them the ciliary fence receptors composed of numerous collar receptors and the corona ciliata on the dorsal surface of the anterior end. The retrocerebral organ is assumed to be sensory in function and may be composed of strongly branched cilia (summary in Shinn 1997).	–	x
Phoronida	**Ciliary epidermal receptors** are present as collar receptors in the actinotroch larva (Hay-Schmidt 1989) and on the tentacles of adults (Pardos et al. 1991, 1993), sensory cells are also present in the apical organ of the actinotroch larva (Santagata 2002).	–	–

Table 7.1 (contd)

		st	pr
Brachiopoda	**Ciliary epidermal receptors** are present as collar receptors on the median tentacle of the larva of *Lingula anatina* (Lüter 1996). In other species and adults, the presence of ciliary sensory structures is likely (James 1997), but not convincingly documented. The (acellular) chaetae are assumed to participate in reception, but the transduction of stimuli to the nervous system is unknown (James 1997).	x	x
Bryozoa	The presence of ciliary epidermal receptor cells is likely, especially on the tentacles (Ruppert et al. 2004), but has not been sufficiently documented.	–	–
Echinodermata	**Ciliary epidermal receptors** are present in all taxa as individual cells or in groups as pads or papillae (see Pentreath & Cobb 1982, Byrne 1994, Cavey & Märkel 1994, Chia & Koss 1994, Heinzeller & Welsch 1994, Smiley 1994 for summaries). At least in several cases ciliary sensory cells are organized as collar receptors (Holland 1984).	x	x
Enteropneusta	Ciliated epidermal receptors occur in the integument (Knight-Jones 1952), but have not been documented very well. Ciliated sensory cells are also present in the apical organ of the tornaria larva (Dautov & Nezlin 1992).	–	x
Pterobranchia	Receptors are not well documented, but from figure 63 in Benito & Pardos (1997) it is likely that collar receptors occur on the tentacles of *Rhabdopleura* sp.	–	–
Tunicata	**Ciliary epidermal receptors** are present individually or in groups (Bone & Mackie 1982, Burighel & Cloney 1997), cupular sense organs are specialized ciliary receptors in which sensory cilia project into a gelatinous cone, the cupula (Bone & Ryan 1978).	x	x
Acrania	**Ciliary epidermal receptors** are present as individual monociliary cells in the epidermis and in the atrium, ciliary pits, and glandular/sensory organs as well as the mechano- or pressure-sensitive corpuscles of de Quatrefage also contain ciliary sensory cells (see Ruppert 1997 for review and further literature).	–	x
Craniota	The several integumental receptors mostly include free nerve endings. Ciliary receptors are present as cupular sense organs in the lateral line of fish-like craniotes, in the olfactory, taste, and vestibular/acoustic system (see, e.g., Kardong 2005).	x	x

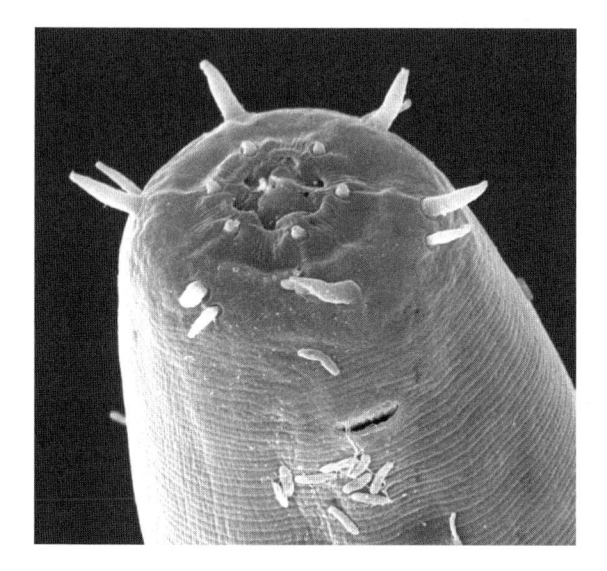

Fig. 7.2. Anterior end of the freshwater nematode *Tobrilus gracilis* with three rings of head sensilla. The slit-like pore present behind the head sensilla leads to an other type of sense organ, the amphid.

were present very early on (for photoreception see discussion below). Structurally, it appears likely that the first sensory cell was organized as a 'collar receptor', that is, with a cilium and a collar of microvilli (Fig. 7.5.). The basal number of microvilli is probably eight. This type of receptor cell is very abundant and is present in several basal taxa. There is variation in this cilium/microvilli-system, in particular microvilli can be 'typical' microvilli, or become long and stiff (and are then often confusingly called stere-o*cilia*). Todt and Tyler (2007) assume that structural differences in collar receptors are so large that they cannot be assumed to be homologous. Whether or not this is appropriate, collar receptors display a characteristic cilium surrounded by a ring of microvilli, a common cellular architecture present also in sponge choanocytes as well as in other cell types of bilaterian animals—such as protonephridial terminal cells.

124 THE EVOLUTION OF ORGAN SYSTEMS

Starting from individual ciliary receptor cells, there are numerous parallel trends to combine sensory cells with other sensory cells, or further cells in more complex sensory structures. Statocysts and photoreceptors appear to be among the first complex sensory structures to evolve (see below). When a (more or less solid) cuticle is present, peripheral receptors have to be integrated into the cuticle. Usually, cilia either project into cuticular

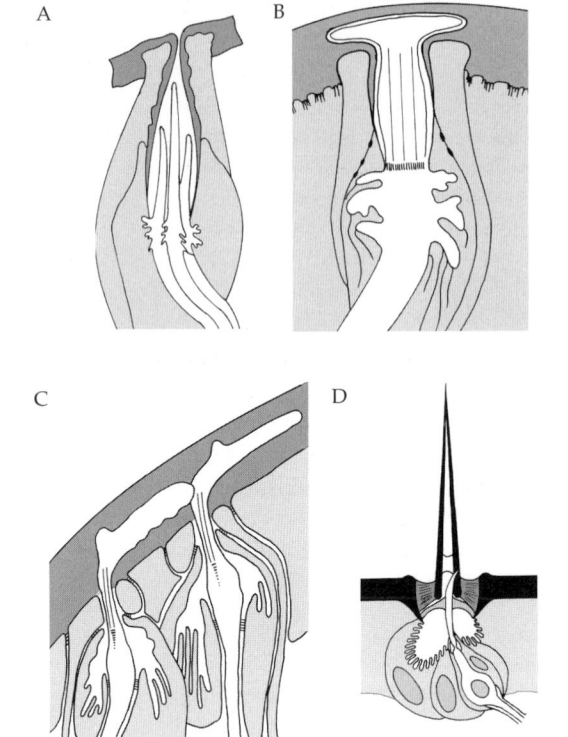

Fig. 7.3. Four different kinds of receptors associated with the cuticle in nematodes (A–C) and euarthropods (D). A. Sensory cilia project into a cavity connected to the exterior by a pore (amphid of *Dipetalonema viteae*, after McLaren 1976). B. Modified sensory cilium projects into the cuticle (head sensilla of *D. viteae*, after McLaren 1976). C. Sensory cilia project into and bend within the cuticle (head sensilla of *Caenorhabditis elegans*, after Wright 1980). D. Sensory cilium terminates in cuticle of mobile bristle (after Keil 1998).

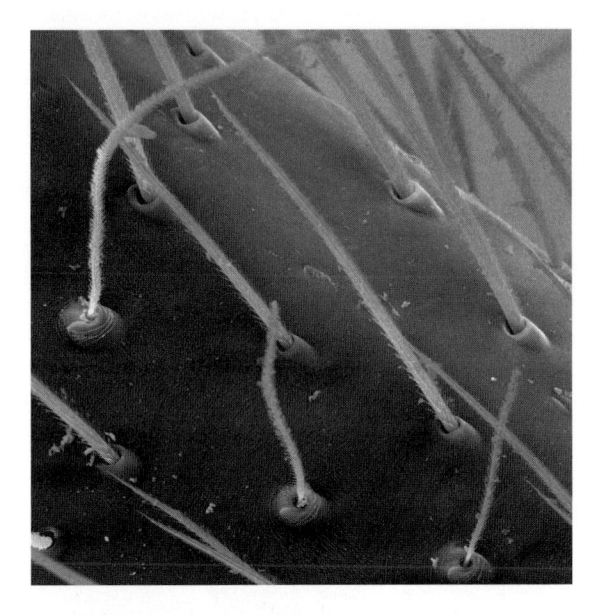

Fig. 7.4. Mechanoreceptors in spiders. Each cuticular bristle originates in a pit in which it is mobile.

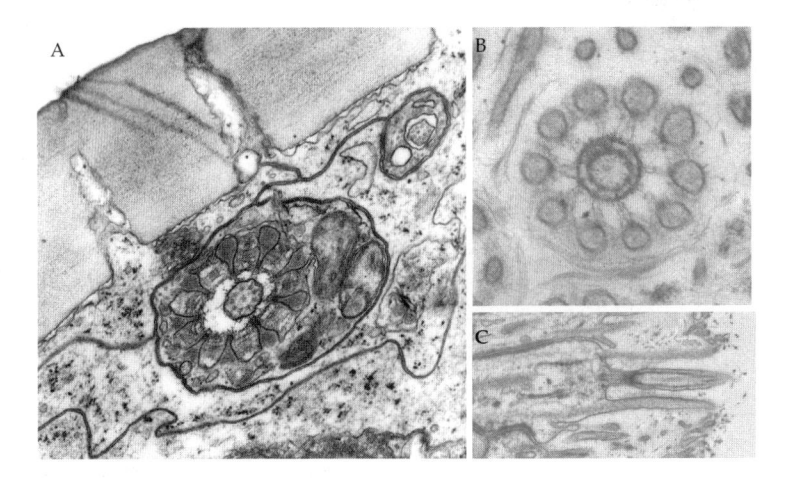

Fig. 7.5. Collar receptor

extensions of various forms or they project into intra- or subcuticular holes which receive information through pores from the external environment (Fig. 7.3.). It has sometimes been noted that cuticular ciliary receptors in cycloneuralians (especially in nematodes) and arthropods resemble each other (e.g. Kristensen 1981), but these resemblances appear to be structural consequences of the possession of a cuticle and therefore themselves do not necessarily hint to a common ancestor.

Static sense organs

Static sense organs provide information on the orientation of the body, in most cases in reference to gravity. In a few cases of aquatic animals, the opposite of gravity, the ascending of a gas bubble, is used for positional information. Gravity can be perceived in some ways (e.g. by special cuticular hairs in insects, see Horn 1985), but most animals have a special type of organ, the statocyst, that functions according to a common principle. Statocysts are usually ciliary receptors, in which the cilia of receptor cells are stimulated by either anorganic material or by a fluid of comparably high viscosity. The anorganic material can be present within a cellular vacuole (then the cell is called a 'lithocyte') or it is an extracellular aggregation in a chamber surrounded by cells (in this case it is called a 'statolith'). In the latter case, either several smaller, or one larger, statolith can be present, smaller statoliths are often called statoconia. The organ containing a lithocyte or a statolith is called a 'statocyst'. The movements of the statolith according to gravity selectively stimulate particular regions of receptor cells and thereby send positional information to the nervous system (Figs. 7.6., 7.7.).

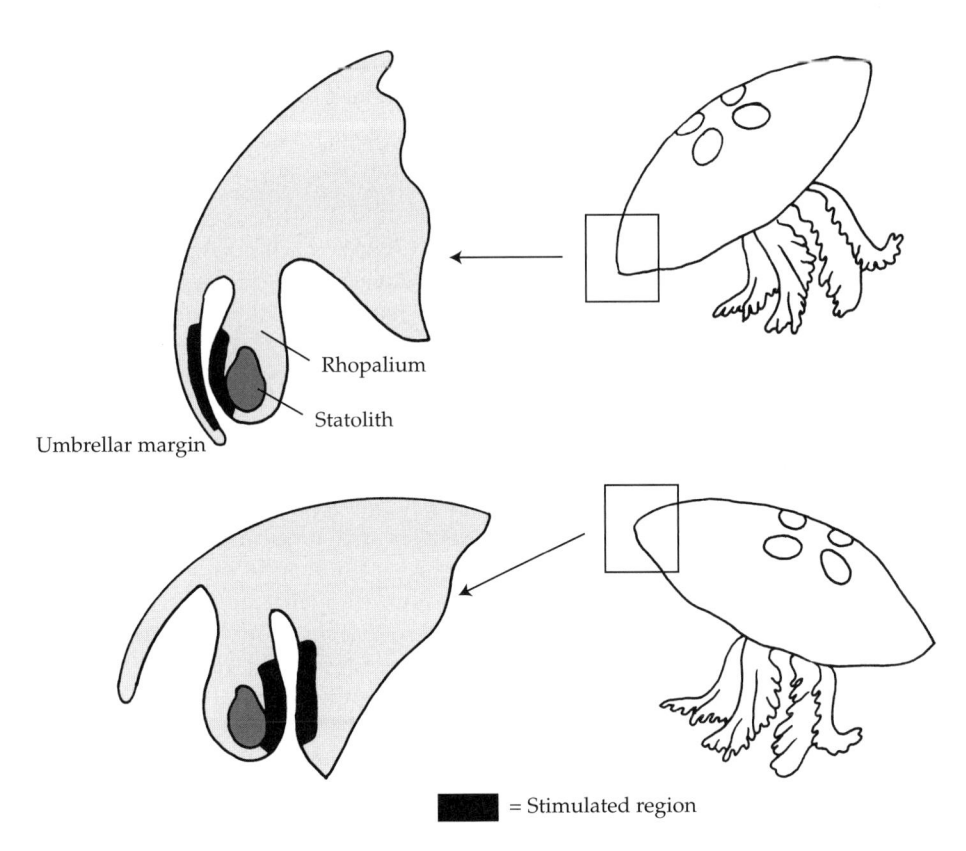

Fig. 7.6. Principle of statocysts, demonstrated in the club-shaped rhopalia of Aurelia aurita (Cnidaria, Scyphozoa). The 'heavy' statolith responds to gravity and stimulates different regions (black) according to the position of the medusa.

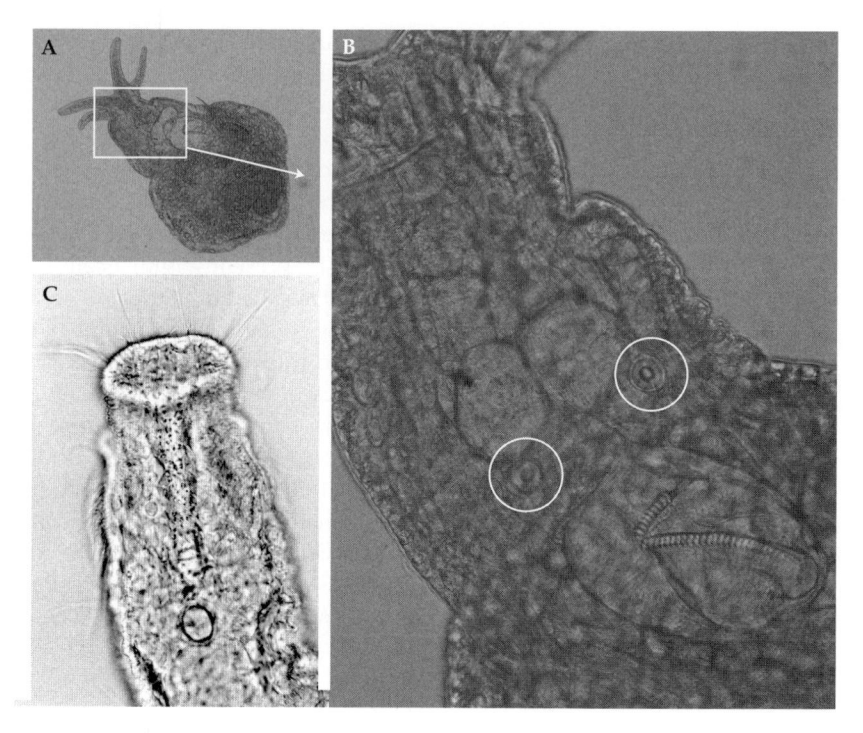

Fig. 7.7. A,B. Paired statocysts in an undetermined meiobenthic gastropod, situated between brain and pharynx with radula. C. Unpaired statocyst in the anterior end of *Parotoplana* sp. (Proseriata, Platyhelminthes).

Cnidaria

Statocysts are present in many medusae, but are absent in the polyp stage. Exceptions are the (not further investigated) statocysts reported from some 'aberrant' benthic hydrozoans such as *Tetraplatia volitans* or *Halammohydra* species (Swedmark & Teissier 1966, Tardent 1978), which are probably modified medusae, and a statocyst lacking ciliary structures in the hydropolyp *Corymorpha palma* (Campbell 1972). All statocysts contain intracellular statoliths, but their structure differs considerably. In some trachy- and narcomedusae, the statocyst is in a club-like extension on the outer bell surface. Both the club and the adjacent epithelium can carry receptor cells that are stimulated by the swinging of the club (Horridge 1969, Singla 1975). In some trachymedusae, such as *Rhopalonema velatum*, the club swings inside a closed chamber (Fig. 7.8.). In some Leptomedusae there is no swinging club, but a large, statolith-carrying cell next to one (*Eirene viridula*) or several receptor cells (Fig. 7.8.; Singla 1975, Germer & Hündgen 1978). The chamber is either open or closed. Although Horridge (1969) arranges the different types of hydrozoan statocysts in a putative evolutionary order, they differ so much in structure, origin (ecto- or endoderm) and position that it is more likely that they evolved several times in parallel. In Rhopaliophora (Scyphozoa + Cubozoa), statocysts are present in the complex rhopalia (which also include photoreceptors) on the bell margin. Rhopalia are club-shaped and hang in pits, so that they touch the umbrellar epithelium when swinging (Fig. 7.8.). The terminal region of each rhopalium contains several lithocytes, the sensory cells are present in the rhopalian epithelium and in the umbrellar epithelium of the pit (Horridge 1969 for *Nausithoë punctatum*, Pollmanns & Hündgen 1981 for *Aurelia aurita* and Laska-Mehnert 1985 for the cubozoan *Tripedalia cystophora*). The statolith is in most

SENSORY ORGANS 127

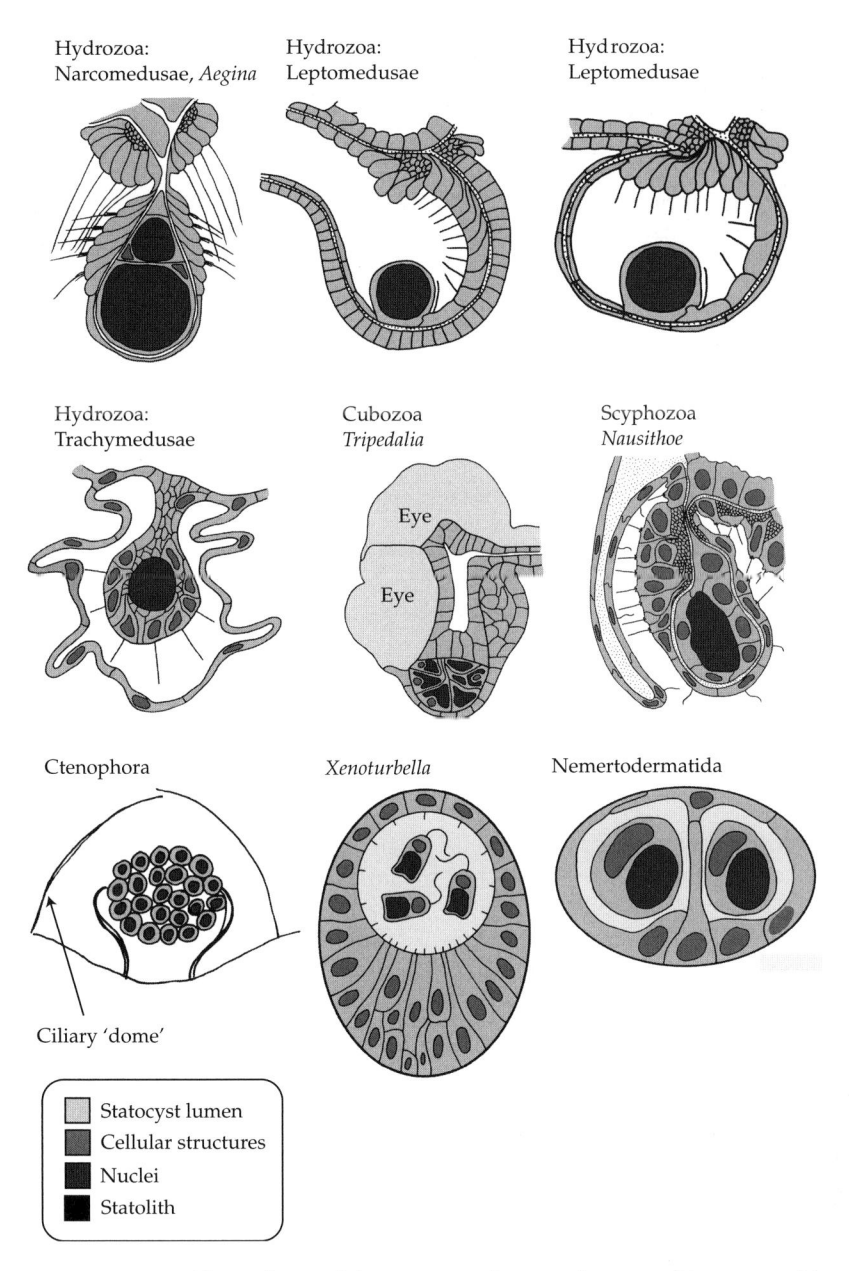

Fig. 7.8. Overview on statocyst structure. Colour coding: very light grey: statocyst lumen, medium grey: cellular structures (lithocyte cytoplasm, peripheral and nerve cells), darker grey: nuclei (omitted in some figures), black: statoliths. Sensory cilia are indicated, when present. Figures modified after Singla 1975 (*Aegina*, Leptomedusae), Horridge 1969 (Trachymedusae, *Nausithoë*), Laska-Mehnert 1985 (*Tripedalia*), Tamm 1982 (Ctenophora) and Ehlers 1991 (*Xenoturbella*, Nemertodermatida).

cases composed of calcium sulfate (Chapman 1985) or calcium sulfate hemihydrate (Tiemann et al. 2006), which is, at least to a large extent, taken up from sea water during development (Spangenberg 1968). In fewer cases, the statolith can be composed of other minerals such as magnesium and/or calcium phosphate (Singla 1975, Chapman 1985).

Ctenophora
The apical organ includes a static sense organ. There is a group of more than hundred lithocytes lying within in a chamber ('dome') formed by non-motile cilia (Fig. 7.8.; Tamm 1982). These lithocytes rest on four groups of balancer cilia, which have sensory properties (Tamm 1982).

Xenoturbella
A statocyst is present in the epidermis of the anterior end. It is composed of a capsule surrounding many ciliated cells around a central cavity. In this cavity, several ciliated lithocytes are present (Fig. 7.8.). Nervous processes penetrate the capsule (Ehlers 1991). It can be assumed that the capsule cells are sensory in function, but it is unique as well as puzzling why the lithocytes are ciliated (Ehlers 1991 therefore avoids calling these cells lithocytes).

Acoelomorpha
In all acoelomorphs the statocyst is intracerebral in position. Excepting *Paratomella* (Ehlers 1992a), it is enclosed by an extracellular capsule. There is a variable (Nemertodermatida) or low (two in Acoela) number of unciliated peripheral cells surrounding a central lumen, in which one (Acoela) or two (Nemertodermatida) lithocytes are present (Figs. 7.8., 7.9.; Ferrero 1973, Ehlers 1991, Ferrero & Bedini 1991). The nemertodermatid statocyst is divided by extensions of the peripheral cells into two chambers, producing a characteristic bipartite statocyst (Fig. 7.8.). In acoels, the capsule is perforated by nervous projections, these were not found in nemertodermatids (Ehlers 1991). The statocysts show characteristic features, but as cilia (and contact to the nervous system in nemertodermatids) are lacking, the exact transduction of the signal is unknown.

Platyhelminthes
Statocysts are present in catenulids, proseriates (Fig. 7.7.), and *Lurus*, a genus of Dallyellioida (Rhabdocoela). In catenulids, the capsulated statocyst is postcerebral in position and contains two to four peripheral cells surrounding a lumen with one to six statoliths (Fig. 7.9.; Ehlers 1985c, 1991). The extracellular capsule is perforated by nervous projections. In those proseriates, where a statocyst is present it is anterior of the brain. An extracellular capsule surrounds several peripheral cells as well as accessory cells. In the central lumen is one large lithocyte (Fig. 7.9.). The capsule is perforated by nerve projections (Ferrero et al. 1985, Ehlers 1991). In both taxa, the putatively receptive peripheral cells are not ciliated, comparable to acoelomorphs. In *Lurus*, the statocyst appears to be composed of a large central lithocyte with three statoliths, surrounded by peripheral and nerve cells (Rohde et al. 1993). The peripheral cells have cilia, which run parallel to the outer membrane of the lithocyte. Additionally, what appears to be a second lithocyte with three smaller statoliths is present adjacent to the large lithocyte. The statocysts in platyhelmiths are so different in position and structure that Ehlers (1991) assumes that they evolved convergently.

Nemertini
Statocysts occur only in species of the interstitial genus *Ototyphlonemertes*. One pair of statocysts is present posterior of the brain. It is composed of an extracellular capsule, several peripheral cells, and one central chamber cell tightly enclosing three (exceptionally two to six) statoliths (Fig. 7.9.; Brüggemann & Ehlers 1981). Nerve projections pierce the capsule and contact the chamber cell. There are no ciliated cells involved in the statocyst. It is assumed that the chamber cell is the receptive part of the statocyst.

Mollusca
Statocytes are present in all conchiferan taxa (Figs. 7.7., 7.9.). In most cases they are positioned close to the pedal ganglia. Peripheral cells surround a fluid-filled lumen containing numerous small or one large statolith (Fig. 7.9.). The lumen can be connected to the exterior by a ciliated duct. The peripheral cells can differ in composition, but they always contain the ciliated receptive cells. The statocysts of monoplacophores and scaphopods are not sufficiently investigated (see Haszprunar & Schaefer 1997 and Shimek & Steiner 1997), for summaries on bivalve and

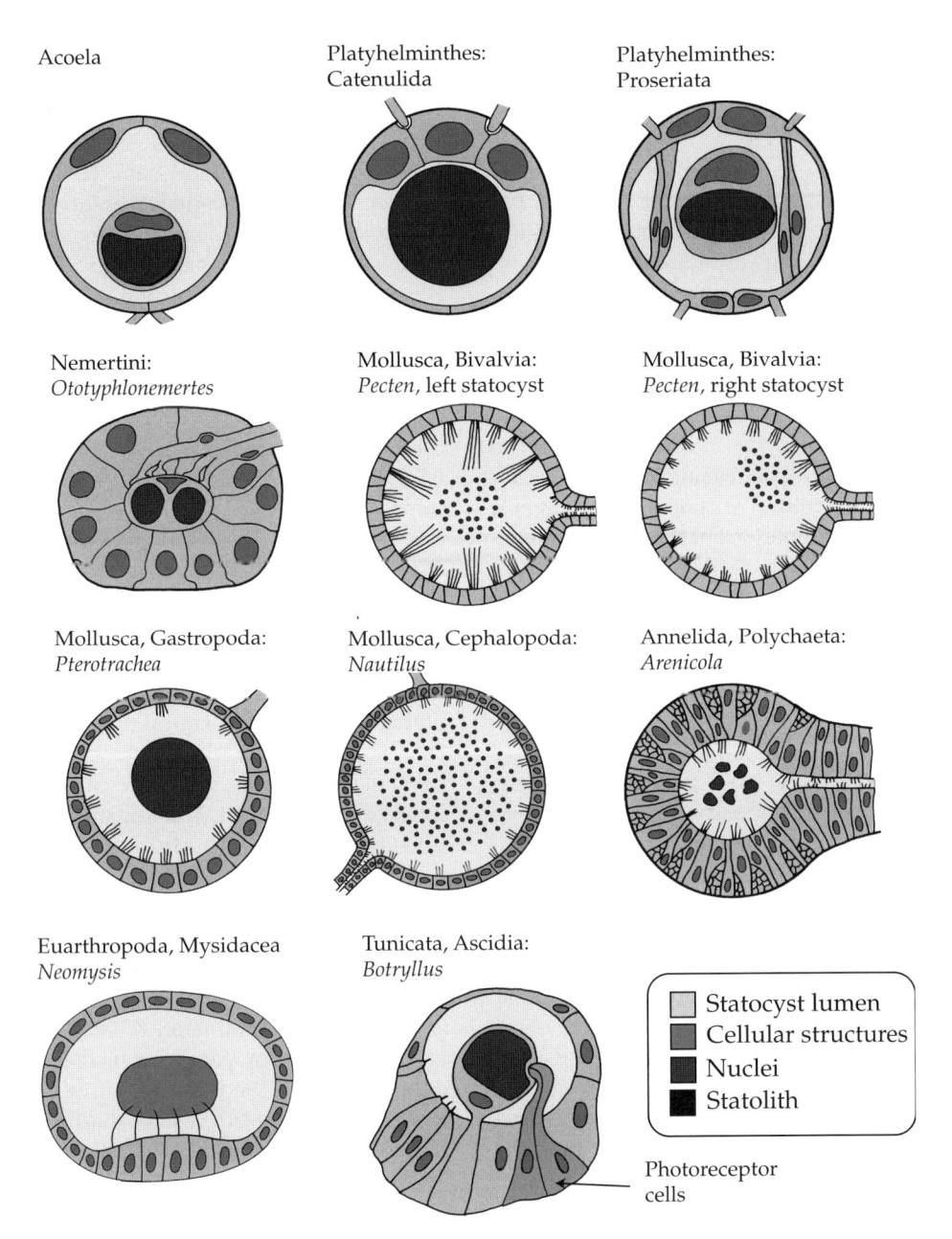

Fig. 7.9. Overview on statocyst structure. Colour coding as in 7–8. Figures modified after Ehlers 1991 (Acoela, Catenulida, Proseriata), Brüggemann and Ehlers 1981 (*Ototyphlonemertes*), Barber 1968 (all Mollusca), Storch and Schlötzer-Schrehardt 1988 (*Arenicola*), Espeel 1985 (*Neomysis*), and Sorrentino et al. 2000 (*Botryllus*).

gastropod statocysts see Barber (1968), Voltzow (1994) and Morse & Zardus (1997). In cephalopods, the statocyst of *Nautilus* corresponds to the described conchiferan pattern (Fig. 7.9.; Barber 1968, Budelman et al. 1997, Neumeister & Budelmann 1997), but within coleoid cephalopods, statocysts become very complex organs, incorporating within the crista/cupula-system an angular acceleration receptor (Young 1989, Budelmann 1994, Budelmann et al. 1997).

Annelida
Statocysts are present in comparably few species of polychaetes. They occur in, for example, larvae of *Lanice conchilega* (Heimler 1983) and in adults of burrowing or tube-living species (Storch & Schlötzer-Schrehardt 1988, Verger-Bocquet 1992, Purschke 2005). In the ultrastructurally investigated *Arenicola marina* statocyst (see Storch & Schlötzer-Schrehardt 1988), peripheral cells (supportive, glandular, and sensory) surround a central cavity, which is connected to the exterior by a ciliated canal. In the centre are several small statoliths. In clitellates, a statocyst has been described only from the marine enchytraeid *Grania americana* (Locke 2000). It is composed of four chambers with a statolith in each chamber. All cilia involved show an unusual pattern of eight peripheral and one central duplett of microtubuli.

Euarthropoda
Statocysts occur in the basal segment of the first antenna of several malacostracan crustaceans: in Mysidacea, astacuran and brachyuran Decapoda, and some isopod species (Sekiguchi & Terazawa 1997). Probably the organ of Bellonci in amphipods also functions as a statocyst (Steele 1984). Statocysts have different positions in the body (uropods in Mysidacea, dorsal cephalon in Amphipoda, dorsal telson in Isopoda, basal antennal segment in Decapoda). Some are open to the external environment (astacuran decapods, e.g. Sekiguchi & Terazawa 1997; isopods have a slit-like opening, see Rose & Stokes 1981), the others are closed. The statolith can be a mineral concretion (see, e.g., Wittmann et al. 1993 for Mysidacea), but astacuran decapods use sand grains as statoliths (Sekiguchi & Terazawa 1997). Sensory structures are cuticular ciliary mechanoreceptors, these are usually positioned in cushions or pits. In mysidaceans, the tips of the sensory bristles enter the statolith (Fig. 7.9.; Espeel 1985), in for example, isopods, they are in contact with the statolith but do not enter (Rose & Stokes 1981).

Brachiopoda
Statocysts occur in larvae and probably adults of inarticulate brachiopods (see James 1997) and probably in the articulate *Terebratulina transversa* (Cavey & Wilkens 1982). Lüter (1997) described statocysts from pelagic juveniles and sessile subadults of *Lingula anatina*. They are composed of a layer of peripheral monociliary cells surrounding a central lumen. Statoliths could not be oberved in the lumen, but may have been dissolved during preparation.

Echinodermat
Statocysts occur in some species of holothurians (see Smiley 1994), but ultrastructural descriptions are lacking.

Tunicata
Lithocytes are present in ascidian larvae, appendicularians, and doliolids. Fine structural details are only known from ascidians and appendicularians. In ascidian larvae the sensory vesicle, which is a part of the neural tube, contains gravity receptive and photoreceptive structures (in some cases, only the statocyst is present, see Svane & Young 1991). The lithocyte is integrated into the epithelium of the sensory vesicle (Fig. 7.9.). It is a large bulging cell with a vacuole containing a melanin droplet (Eakin & Kuda 1971, Bone & Mackie 1982, Torrence 1986, Sorrentino et al. 2000). Ciliated sensory cells are present next to the statocyte (Sorrentino et al. 2000). In the colonial species *Botryllus schlosseri*, gravity and photoreceptor fuse to form one organ, in which the photoreceptive cells appear to use the statolith as a shading pigment (Fig. 7.9.; Grave & Riley 1935, Sorrentino et al. 2000). The statocyte in the appendicularian *Oikopleura dioica* is almost similar to the one in ascidian larvae, but in contrast to ascidian larvae the lithocyte contains material with a substructure of concentric rings (Holmberg 1984).

Craniota
The static and auditory system are combined into a complex sensory organ in the inner ear (see, e.g., Kardong 2005 for details). The static part of the system (the vestibular system)

includes structures for rotational (one to three canals starting from the utricle) and for linear (the saccule) movements. Receptors are hair cells—they have a tuft of microvilli with one cilium in the periphery of this tuft. The hair cells are either surrounded by a gelatinous cupula, which is deformed by movements of the endolymph (the liquid within the inner ear), or they carry statoliths (which are here called otoliths). The statoacustic organ develops from particular neurogenic placodes (the acoustico-lateralis placode). Certain domains in the ectoderm and the presence of hair cells in tunicates make it probable that such regions (but not the statocysts in the sensory vesicle) are homologous to the acoustico-lateralis placode (Manni et al. 2004, Mazet et al. 2005). This shows that the static structures in craniotes are not homologous to the tunicate statocysts and evolved their static properties independantly.

Is there evidence that statocysts evolved only once or several times in parallel? The fine structure is so diverse that even within certain taxa such as cnidarians or platyhelminths it can be suspected that statocysts evolved several times in parallel. Even if Horridge's (1969) scenario of hydromedusan statocyst diversity does not represent evolution, it might provide an idea of how more or less complex types of statocysts can develop from simpler structures. It appears plausible that the secretion of 'heavy' material, and the arrangement of mechanoreceptive cells in its vicinity, occurred several times in parallel.

Photoreceptors

Photoreception is performed by the same general principles that account for most kinds of receptor structures (see above). Photosensitive intramembraneous proteins make a conformational change when stimulated by light, this signal is transduced intracellularly to other components. Taking all living organisms into account, there are several different molecules that are photosensitive, most of which are used in photosynthesis (see, e.g., Hellingwerf et al. 1996 for procaryotes). The proteins used for vision in metazoans are opsins, they function in combination with a prosthetic group (also called chromophore) called retinal. Together they form 'rhodopsin'.

The presence of rhodopsin can be traced back a long way. Rhodopsin has been found in up to four different forms in archaebacteria and eubacteria, where it is called bacteriorhodopsin. In the archaebacterium *Halobacterium salinarum*, two rhodopsins act as proton- or as chloride-pumps and the two other rhodopsins function in phototaxis by starting a short trandsuction cascade that effects the motility of the flagellum (Spudich et al. 2000). Rhodopsins are also found in fungi, where they have the same two functions: as a proton-pump (Waschuk et al. 2005, Brown & Jung 2006) and in phototaxis of 'zoospores' (Saranak & Foster 1997). In some algae (especially in *Chlamydomonas*, but also in e.g. *Volvox*), rhodopsin is used for phototaxis (Ebnet et al. 1999, Hegemann et al. 2001, Sineshchekov et al. 2002, 2005).

The rhodopsins in archaebacteria and metazoan animals show some differences, for example, in the amino acid sequence (Smith 2000, Terakita 2005), but they also share some characteristics, for example, there are always seven transmembrane α-helices and the chromophore is in each case bound covalently in a particular position, the binding pocket (Spudich et al. 2000). Archaebacteria use all-*trans* retinal as a chromophore, this changes on light stimulation to 13-*cis* retinal. Metazoans use another stereoisomer of retinal, 11-*cis* retinal, which changes to all-*trans* retinal. Another difference is the signal transduction. In metazoans, a G-protein is the first molecule in the transduction cascade, while G-proteins are not present in archaebacteria. During the transduction cascade, the signal can be varied. For example, in arthropods, light stimulation of rhodopsin leads to the opening of cation channels, while in craniotes such channels are closed (Fain 2003). The protein family of opsins is very large, with more than 1,000 different opsins having been described so far (Terakita 2005). They fall into different groups ('subfamilies') with different functional properties, especially concerning the associated G-protein (Terakita 2005).

In many cases, photoreception is performed by specific organs or structures often called eyes or ocelli. However, many metazoans lacking defined eyes are nevertheless capable of showing a response to light. Specimens in several taxa (Cnidaria, Mollusca, Crustacea, Echinodermata, and probably many others) have been shown to have a general 'dermal' photosensitivity (Millot 1969, Yoshida 1979, Taddei-Ferretti & Musio 2000, Musio et al. 2001, Martin 2002). It is assumed that nerves are photosensitive, which means that such nerves likely contain photopigments in their membrane (Millot 1969, Yoshida 1979). A response to light is also present in sponge larvae, which have no photoreceptive structures (Leys & Degnan 2001, Leys et al. 2002), although the region of photoreception can be narrowed down to a posterior ring of monociliated cells in the larva of the demosponge *Reneira* sp. (Leys & Degnan 2001).

Specialized structures for photoreception are not unique to animals. There are several cases (most well known are euglenids), in which unicellular organisms form recognizable 'eyespots' including a pigment shading of the receptive structures (see, e.g., Walne & Arnott 1967, Dodge 1991, Hausmann et al. 2003). Within metazoans, structurally recognizable photoreceptors occur only in eumetazoans.

Summary of eumetazoan photoreceptors

From the overwhelming information on eumetazoan photoreceptors, the following summary will try particularly to distill information about the position of photoreceptors, the arrangement of cells (receptor, pigment, supportive cells; Fig. 7.10.), and the type of receptive structure (microvilli, cilia, or both). The terminology for visual sense organs is not used consistently, with, for example, 'photoreceptor' meaning either the entire organ (e.g. Wolken 1971) or just the receptive cell (e.g. Eakin & Hermans 1988). The term 'eye' may be restricted to those structures including pigment (Arendt & Wittbrodt 2001). I use 'photoreceptor' here as synonymous to termini eye and ocellus for the entire organ that

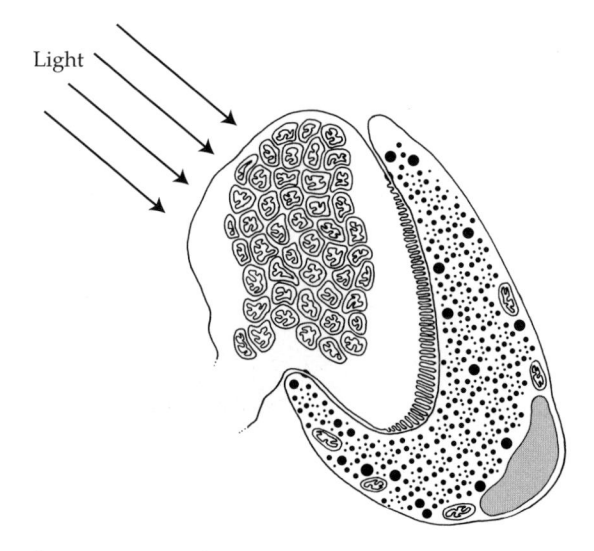

Fig. 7.10. Function of pigment-cup photoreceptors: light stimulates microvilli (as in this case) or cilia of the receptor cell, which is surrounded by one or more pigment cells. In the case shown here, the photoreceptor of the larva from the polychaete *Spirorbis spirorbis*, numerous mitochondria in the receptor cell likely act as a kind of lens. After Bartolomaeus (1992b).

performs visual reception. The single elements are called receptor, pigment, and supportive cells.

An important distinction is between 'inverse' and 'everse' eyes (Fig. 7.11.). In everse eyes, the receptor cells are integrated into a layer of pigment cells. Their receptive structures project into the cup formed by the pigment cells, whereas their connection to the nervous system is 'behind' the pigment cup. The path of the incoming light, its translation, and the direction of the nerves, is therefore more or less linear. In inverse eyes, the pigment cells alone form a cup and the receptor cells enter the cup from its margin. This means that the path of information is first with the light into the cup, but the nerves first run a bit towards the light and only then turn and run in the direction of the body centre.

The general types of photoreceptors can be arranged into an order of increasing complexity and 'visual potential'. Simple photoreceptors are thereafter flat spots which can only detect differences in light intensity. More advanced are cup-shaped photoreceptors, which can additionally detect the direction of light (see Land 1991 for a

Everse: *Helix* (Gastropoda)   Inverse: *Lineus* (Nemertini)

➤ Path of light
➢ Path of transformed signal

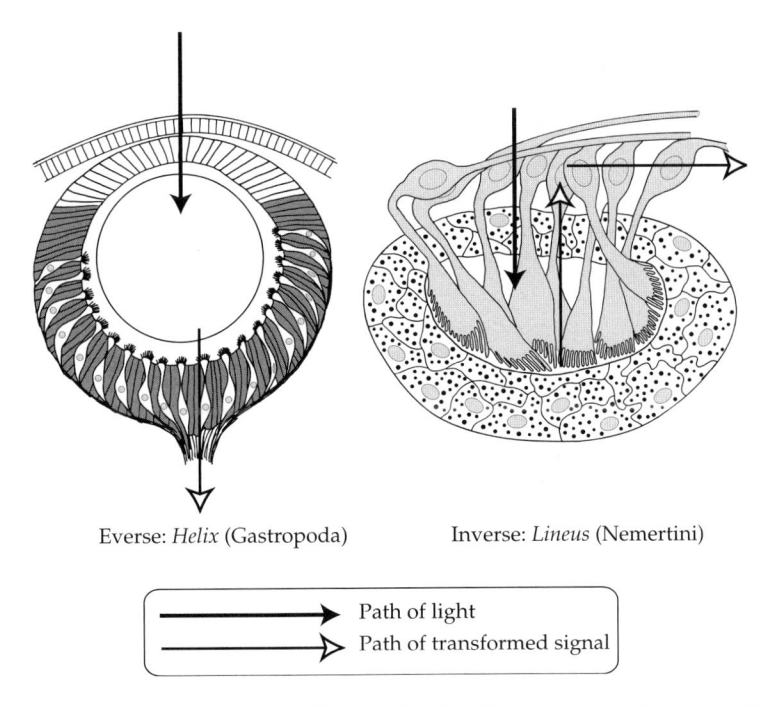

Fig. 7.11. Comparison of an inverse and an everse eye. They differ in the direction of incoming and transformed signal. These are more or less in line in everse photoreceptors, while they are partly opposing in inverse photoreceptors. Figures modified from Land (1984) and Vernet (1970).

more detailed account). When the cup becomes deeper and the tissue closes except for a small pore, a camera lucida eye is formed which can project an image onto the sheet of pigment cells. The presence of a lens is a further optimization of this system (see, e.g., Goldsmith 1990, Wolken 1995, Land & Nilsson 2004, Warrant & Nilsson 2006 for further details). It has to be tested whether such a 'logical' and physical order of photoreceptors does really mirror evolution. The type of the receptor cell has been of considerable importance for evolutionary considerations, because some cells use cilia as carrier structures of the photosensitive proteins while others use microvilli.

In the taxa not mentioned here, no photoreceptors have been found, although several of them show a general sensitivity to light.

Cnidaria

Photoreceptors are present only in medusae. There is only one known exception; this is the polyp of the stauromedusa *Stylocoronella riedli*, which has simple photoreceptors on its tentacles (Blumer et al. 1995). Medusae have a diversity of photoreceptor types, ranging from flat pigmented spots (Fig. 7.12a.), to cup-shaped ones, to those with lenses. The receptive structures are always cilia, which are in several cases apically branched (Singla 1974, Martin 2002). Hydrozoan medusae have photoreceptors either on their tentacles or on the bell margin, the number is sometimes high (>100). Photoreceptors are flat sheets, cups, or cups with lenses (Singla 1974, Martin 2002). Scypozoan and cubozoan medusae have photoreceptors in their rhopalia on the bell margin. These complex sensory organs (see static sense organs) often contain different types of eyes, including flat and cup-shaped ones and those with lenses. Cnidarian photoreceptors are usually composed of receptor and pigment cells, which both occur together in one layer. In only a few cases, such as in the lens-eyes of the cubozoan *Carybdea marsupialis*, do pigment granules occur in the proximal parts of the receptor cells and not in separate cells (Martin 2002). In some

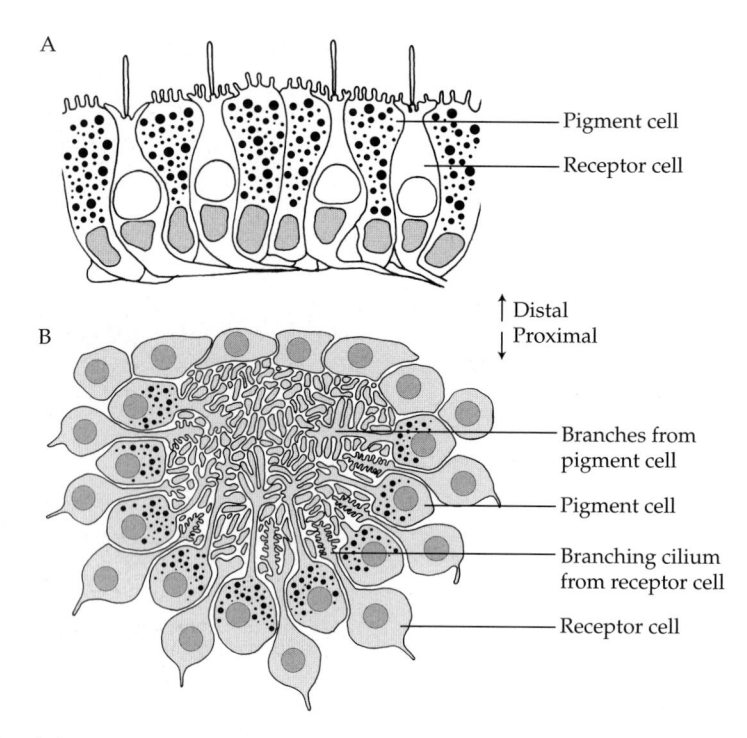

Fig. 7.12. Two examples of photoreceptors in Cnidaria. A. Flat photoreceptor as sheet of alternating receptor and pigment cells in *Leuckartiara octona* (Hydrozoa). B. Complex everse photoreceptor in *Bougainvillia* sp. (Hydrozoa). Both figures after Singla (1974).

cases, pigment cells are endodermal cells while receptor cells are, as in all other cases, ectodermal (e.g. in the small eye of *Aurelia aurita*, see Yamasu & Yoshida 1973, Pollmanns & Hündgen 1981). In these cases the photoreceptors are inverse. Everse photoreceptors are more abundant, but inverse eyes occur in both Scypohozoa (e.g. *Aurelia aurita*) and Hydrozoa (e.g. *Tiaropsis multicirrata*, Singla 1974). The sensory cilia with their branches often fill the interior of the cup-shaped receptors, but in some cases (e.g. *Bougainvillia principis*, Singla 1974; Fig. 7.12b.) the pigment cells also send extensions into the cup lumen. In cubozoan eyes, lenses and, in the case of *Carybdea marsupialis*, a cornea, are present (Yamasu & Yoshida 1976, Martin 2002). The lens can project an almost perfect image on the retina, although this is not positioned in the exact focal layer (Nilsson et al. 2005). The cornea is a thin sheet of epithelial cells completely covering the eye, the lens is cellular and contains crystalline proteins (Martin 2002). Receptor cells are arranged here, without additional pigment cells, in a dense layer (retina).

Ctenophora

There is response to light, but photoreceptors have not been unequivocally identified. The lamellate bodies on the base of the apical organ are often suspected as being photoreceptive. These are cells with some cilia pointing into an intracellular cavity (Hernandez-Nicaise 1991). The ciliary membrane has numerous extensions, leading to the lamellate bodies.

Acoelomorpha

Probable photoreceptors have been described from only one acoel species, *Convoluta convoluta* (Popova & Mamkaev 1985, as cited in Rieger et al. 1991b). A photoreceptive function is assumed, because there is a cup-shaped pigment cell, into which a few neurons project. These have, however, neither microvilli nor cilia. Some further cells may have a photoreceptive function, but these cells have no resemblance to 'usual' photoreceptive cells and lack both microvilli and cilia (Ehlers 1985a).

Gastrotricha

Photoreceptors are present in several species, but may not be recognized easily, because some are not (or only very weakly) pigmented. When present, there is one pigment cell forming a cup into which the processes of a few receptor cells project (Liesenjohann et al. 2006). The receptive structure in all photoreceptors described so far appears to be a highly branched cilium (Fig. 7.13.; Teuchert 1976a, Ruppert 1991b, Hochberg & Litvaitis 2003, Liesenjohann et al. 2006).

Nematoda

Several species of mainly marine, but also freshwater, nematodes from the taxa Enoplida and Chromadorida have photoreceptors. These are paired structures with nervous connection to the ventral nerve cord present in the anterior body region. The photoreceptor appears to be always composed of two cells, a cup-shaped pigment cell and a receptor cell inside the cup. The receptive structure is described in all cases as stacks of lamellae (see Siddiqui & Viglierchio 1970, Croll et al. 1975, Van de Velde & Coomans 1988). In *Diplolaimella* sp. (Monhysterida), a lens of amorphous material is present (Van de Velde & Coomans 1988). Several further species possess no particular photoreceptors but pigment-shaded photosensitive neurons in the amphids (Jones 2002).

Nematomorpha

No convincing description of a photoreceptor is known from nematomorphs. A questionable report of an eye in the freshwater species *Paragordius varius* (Montgomery 1903) has not been confirmed. The giant cells in the anterior end of the marine *Nectonema* species contain regions with several microvilli (Schmidt-Rhaesa 1996a), but a photosensitive function of these regions is only hypothetical.

Kinorhyncha

Many kinorhynchs are photosensitive (Kristensen & Higgins 1991) and the receptive

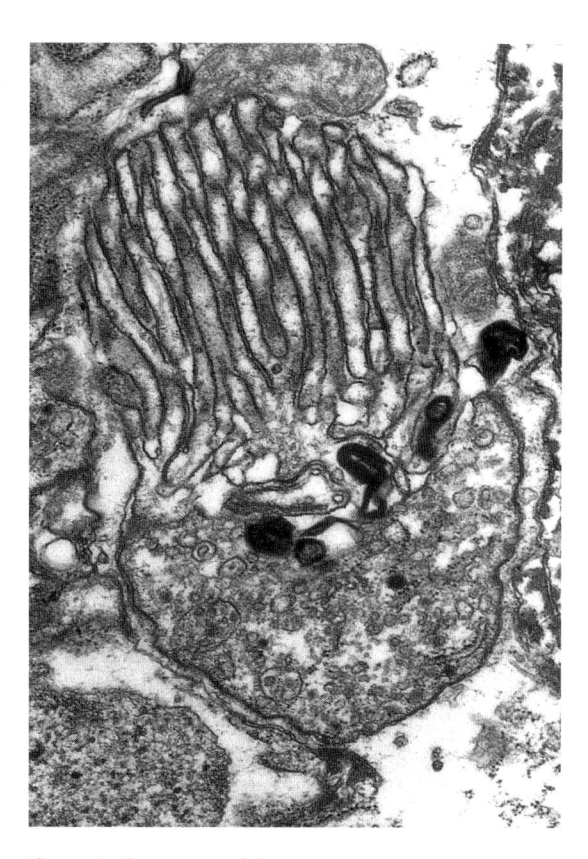

Fig. 7.13. Photoreceptor of the gastrotrich *Dactylopodola baltica*, showing basal cytoplasmatic part of the receptor cell (lower part) and microvilli-like extensions (upper part) as the receptive structures. These are, however, not microvilli, but branches from a modified cilium.

structures are probably cephalic sensory organs. In *Pycnophyes* species, the cephalic sense organ is composed of one receptor and one enveloping cell (Neuhaus 1997). The presumably receptive structure is a single cilium that is partly swollen and has several branches (Neuhaus 1997). Pigment is not present in *Pycnophyes* species, but obviously in other kinorhynch species, and in some species a local thickening of the cuticle may function as a lens (Kristensen & Higgins 1991).

Platyhelminthes

Photoreceptors are widespread among flatworms. They are found either in association with the brain (cerebral eyes) or in other places in the anterior end. Most photoreceptors lie below the epidermal ECM or even below the subepidermal musculature. They are almost always pigment-cup ocelli and rarely simple flat pigment spots (Rieger et al. 1991b). The majority of the cup-shaped receptors are inverse, with an uninterrupted cup of pigment cells into which receptor cells run through the apical opening. Some species have lenses (e.g. the oncomiracidium of the monogenean *Entobdella solae*, Kearn & Baker 1973) and others have, instead of a cup of pigment cells, a cup made of a cell with refractive platelets (in the monogenean *Polystoma integerrimum* and related species, Fournier & Combes 1978). It is notable that the putative basal platyhelminth taxon, Catenulida, has no photoreceptors, at least none comparable to other taxa. There are single nerve cells which contain a large, light-refracting mitochondrion (Ruppert & Schreiner 1980); these probably have a photoreceptive function. The receptive structures in flatworms are microvilli as well as cilia. Microvilli are more abundant, occurring in species of the taxa Macrostomida, Polycladida, Prolecithophora, Seriata, in free-living rhabdocoels, larval monogeneans, and digeneans (Pike & Wink 1986, Fried & Haseeb 1991, Rieger et al. 1991b). Ciliary or ciliary/microvillar photoreceptors have been found in some macrostomids (Palmberg et al. 1980), polyclad larvae, (Eakin & Brandenburger 1981a, b, Lanfranchi et al. 1981) and proseriates (Sopott-Ehlers 1982). The polyclad larvae possess three eyes. Two of them migrate through the basiepidermal ECM towards the brain and become cerebral eyes (Lanfranchi & Bedini 1986). While Lanfranchi et al. (1981) found both cerebral eyes of *Thysanozoon brocchii* and *Stylochus mediterraneus* larvae to have cilia, Eakin & Brandenburger (1981a,b) found that the left cerebral eye of the *Pseudoceros canadensis* larva has microvilli and cilia whereas the right eye has only microvilli.

Limnognathia
Photoreceptors may be present in the anterior end, but are insufficiently described (see Kristensen & Funch 2000).

Eutotifera
Photoreceptors are present in several species. In monogononts, they can be located in close association to the wheel organ or, as cerebral eyes, integrated into the brain (summaries in Clément & Wurdak 1984, 1991). Both types can be present within the same species, e.g. in *Asplanchna brightwelli*. The wheel organ photoreceptors are composed of a cup-shaped pigment cell and a receptor cell with either microvilli (e.g. *Filina longiseta*, *Rhinoglena frontalis*) or modified cilia (e.g. *Asplanchna brightwelli*, *Trichocerca rattus*). A lens composed of a lipid droplet is present in some species (e.g. *Filina longiseta*). The cerebral eyes are also composed of a cup-shaped pigment cell, but here the pigment cell is distal of the receptor cell and its process to the brain. The receptive structures are stacks of membranes originating from a neuron. In bdelloids, cerebral eyes are present in *Philodina roseola*, but here the receptive structures are modified cilia.

Nemertini
Many nemertean species have one, two or more pairs of eyes (Fig. 7.14a.). They are composed of a cup of pigment and receptor cells. In *Lineus ruber*, the pigment cells form a layer completely surrounding the receptive part of the receptor cells, but they are thicker on one side and therefore still form a functional cup (Fig. 7.11.; Vernet 1970). The receptor cells enter through the thinner, distal part of the pigment layer and the eye can therefore be regarded as inverse. The receptive structures appear to be microvilli, but cilia may also contribute to reception. In *Lineus ruber*, Vernet (1970) described ciliary rootlets in the slender part of the receptive cells and a branching into numerous microvilli distal of this part. This could mean that the branches are not microvilli, but a modified branching cilium. Storch & Moritz (1971) showed in the same species that the receptive cells have a broad apical surface, from which numerous microvilli originate, and that one or few cilia are present among them. Whether or not these cilia contain photopigment is not known. In addition, Vernet (1974) found putative cerebral photoreceptors in two *Lineus* species, these consist of several branching cilia.

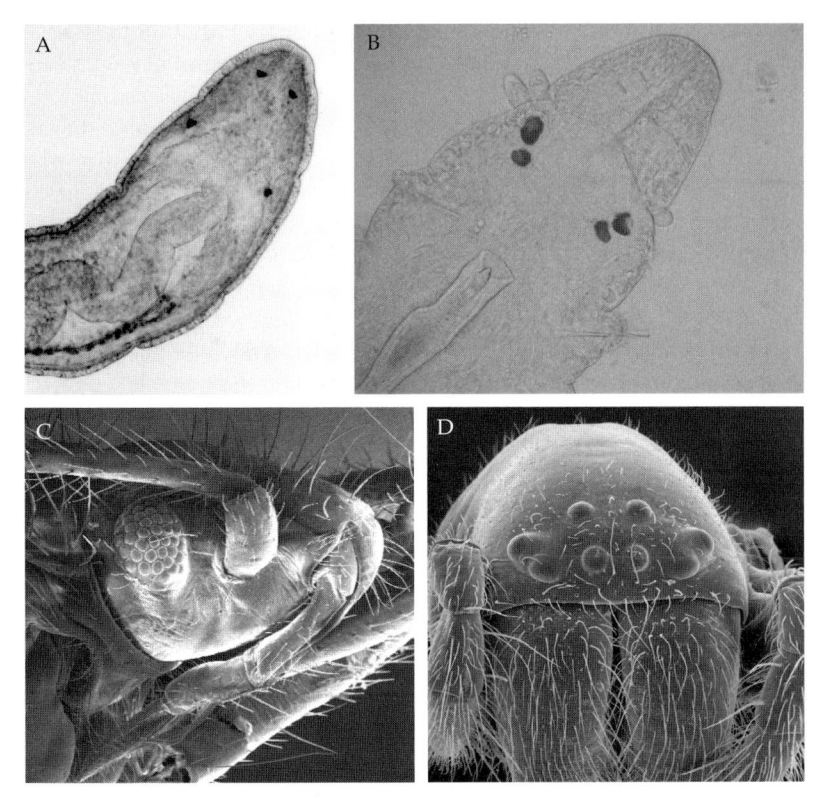

Fig. 7.14. Some examples for photoreceptors. A. The meiofaunal *Pseudostomella arenicola* (Nemertini) has two pairs of eyes. B. The polychaete *Exogone* sp. also has two pairs of eyes. C. Lateral complex eyes in the bug *Afrocimex constrictus* (Heteroptera, Insecta) (animal collected by Klaus Reinhardt, Sheffield). D. Eight single eyes in the spider *Amaurobius* sp. (Araneae).

Kamptozoa
Photoreceptors are only present in larvae of kamptozoans. There is only one description from *Loxosomella harmeri* (Woollacott & Eakin 1973). The paired photoreceptors are positioned close to the frontal organ. Each is composed of one receptor cell, one cup-shaped pigment cell and a lens cell. The processes of the receptor cell to the nervous system are not described. The receptive structures are numerous cilia.

Mollusca
Photoreceptors have not been found in aplacophorans, monoplacophorans, or scaphopods, but are present in great diversity in the remaining groups (see, e.g., Hamilton 1991, Messenger 1991). Photoreceptors of gastropods and cephalopods are in most cases in the head region, while those of chitons and bivalves are in different locations. Photoreceptors range from being composed of few cells to more than 20 million cells in *Octopus* (Land 1984). In chitons, photoreceptors are present in larvae and adults. Larval photoreceptors (which are still present for a while after metamorphosis, Eernisse & Reynolds 1994) are composed of a cup of pigment cells with few interspersed receptor cells, whose microvilli fill the cup (Rosen et al. 1979, Fischer 1980, Bartolomaeus 1992a). Few cilia are also present in the receptor cells. In adults, photoreceptors are present in the aesthetes, which pierce the shell plates and are probably complex sensory and secretory structures. There are, among other cell types, central pigment cells and peripheral receptor cells (Fischer & Renner 1978, Baxter et al. 1987, 1990). Receptor cells have

many microvilli and sometimes a few cilia (Fischer 1978). In *Acanthochiton fascicularis*, the receptor cells differ according to their location on the shell plate: in the central region they have microvilli, but in the lateral region they have cilia (Fischer 1979). In gastropods, photoreceptors are present in the veliger larva and in adults of many species. Larval photoreceptors have been described from only a few species, they contain a pigment cup composed of one or several pigment cells and a lens. In the veliger of *Aporrhais* sp. and *Bittium reticulatum* the receptor cells have one (*Aporrhais*) or two (*Bittium*) cilia with a folded ciliary membrane (Blumer 1994), while the veliger of *Lacuna divaricata* has two receptor cells, one with many microvilli and one cilium and another with several cilia and short microvilli (Bartolomaeus 1992a). Adult gastropods have either 'simple' pigment cups as in patellids (Marshall & Hodgson 1990), almost closed pigment cups (pinhole eyes) in *Haliotis* (Tonosaki 1967), but most abundantly everse pigment-cup eyes with lenses (Fig. 7.11.; Land 1984). Heteropod gastropods have very derived photoreceptors (see Dilly 1969, Land 1984). Lenses may be soft and hardly refractive (as e.g. in *Helix pomatia*) or hard and spherical (as e.g. in *Littorina* and *Lymnaea*). Gastropod eyes have microvilli as receptive structures, but cilia are often present in the receptor cells (von Salvini-Plawen & Mayr 1977, Land 1984). While cilia are often single and appear to be more or less rudimentary, several cilia are present (together with microvilli) in photoreceptors of *Viviparus viviparus* (Zhukov et al. 2006). Entirely ciliary photoreceptors have been found, in addition to rhabdomeric eyes in the head (Katagiri et al. 1995), in the 'pulmonate sea slug' *Onchidium verruculatum* (Yanase & Sakamoto 1965). Cephalopods have either pinhole eyes (*Nautilus*) or complex eyes with cornea, pupil, iris, and lens (Coleoida), which resemble craniote eyes in many details, but have a different development (see Budelmann et al. 1997 for summary and references). The receptive structures are exclusively microvilli, in coleoids these are highly ordered into rhabdomeres. Rhabodomeres of four cells together form rhabdomes and because the retina is composed of numerous such rhabdomes, the eye is sometimes called a 'compound eye with one common lens'. Some bivalves have eyes, but due to the reduction of a head they are in different positions on the body. Cerebral eyes are present in *Mytilus edulis*, they are everse pigment-cups, the receptor cells have microvilli and a single cilium (Rosen et al. 1978). Besides these, there are photoreceptors on the siphonal tentacles of *Cerastoderma* (*Cardium*) *edule* and eyes in the mantle margin in *Pecten* and arkshells. Arkshells have two different kinds of eyes along the mantle edge, one being compound eyes with a ciliary receptive structure and one a 'simple' pigment cups with microvilli as receptive structures (Nilsson 1994). *Cerastoderma* photoreceptors are everse pigment-cups with cilia as receptive structures (Barber & Wright 1969). The eyes of *Pecten* are very complex, having a cornea, a cellular lens, a two-layered retina, a reflecting layer (called argenta), and a pigment layer (Barber et al. 1967). The two layers of the retina contain different receptor cells, in the distal layer these are cilia and few microvilli directed towards the light and in the proximal layer these are microvilli and few cilia directed away from the light. The photopigments of both layers have a similar spectral absorption, but the rhodopsines as well as the transduction cascade are different (Kojima et al. 1997, summary in Fain 2003). However, the functional background of such an arrangement is not yet clear.

Sipunculida

Photoreceptors are present on the introvert of several sipunculid species; they range from simple invaginations of the epidermis to cerebral eyes (Rice 1993). Only cerebral eyes have been investigated ultrastructurally by Hermans & Eakin (1969, 1975). As the cerebral eyes lie at the end of a cuticle-lined canal (the ocular tube) opening to the external, it may be suspected that sipunculid eyes display a series of different grades of sinking into the body. The photoreceptors of *Phascolosoma agassizii* are composed of a cup of pigment cells, with receptor cells being interspersed among them (Hermans & Eakin 1969). The receptor cells project many microvilli as well as one cilium into the proximal cavity of the ocular tube. Among them is also a 'plug' of cuticular material, whether this serves as a lens is not known.

Annelida
There is a diversity of photoreceptors in annelids, especially in polychaetes. If not otherwise stated, information is based on reviews by Eakin & Hermans (1988), Verger-Bocquet (1992), Purschke (2005) and Purschke et al. (2006). In polychaetes, the most abundant photoreceptors are located close to or within the brain and are therefore called cerebral eyes. They sometimes originate in the trochophore larva, but adult cerebral eyes may either be persistent from the larva or new formations (e.g. in *Platynereis dumerilii*, Rhode 1992). The larval cerebral eyes are located directly underneath the epidermis and are not separated from it by ECM (Bartolomaeus 1992b), therefore it might be assumed that those cerebral eyes that persist in adults have an epidermal origin and then sink into the brain region. Larval and adult cerebral eyes can be composed from minimally two cells (one pigment and one receptor cell; Fig. 7.10.) up to many cells. In all cases, the pigment cell or cells form a cup into which the receptive structures project. Purschke (2005) noticed a correlation, in which eyes composed of few cells are inverse while large multicellular eyes are everse (with Flabelligeridae being an exception). In addition to pigment and receptor cells, lenses (cellular or acellular) and a cornea can be present. There are several cases in which unpigmented cerebral eyes are present, but these are likely derived from pigmented eyes. The receptive structures in cerebral eyes are microvilli, but a more or less rudimentary cilium is often present. Besides the cerebral eyes, further photoreceptors can be present in polychaetes, either as unpigmented ciliary eyes in the anterior region of the prostomium or in the brain; as branchial eyes on the tentacles of several sabellids and serpulids; as segmental eyes on trunk segments in some opheliids, eunicids, syllids, and sabellids; or as pygidial eyes in some sabellids. While the microvilli-bearing cerebral eyes are regarded as homologous, the ciliary cerebral eyes use different photopigments and transduction chains (e.g. in *Platynereis dumerilii*, Arendt et al. 2004, see below). In most cases, receptor cell and supportive cell together form an extracellular cavity into which the receptive cilia project. In few cases among polychaetes, the cilia project into an intracellular cavity of the receptor cell which develops from an invagination of the cell surface. This structure is called a phaosome (Fig. 7.15.). Phaosomes are the only known receptive structures in clitellates (Jamieson 1992, Fernández et al. 1992). Here, several eyes can be found in close association with the epidermis. In oligochaete species, pigment is often lacking and present only in Naididae (Purschke 2003), but in leeches, pigment cells form a cup, in which several phaosomous receptor cells are present (Hansen 1962, Röhlich & Török 1964, Clark 1967). The receptive structures are microvilli, but a pair of centrioles (indicating a reduced cilium) can be present (Clark 1967). In *Eisenia foetida*, the intracellular cavity is not closed, but connected to the exterior by a canal (Jamieson 1992). This probably represents an early stage in the development of phaosomes, where the internal cavity is not sealed off from the exterior. Purschke (2003) develops a hypothetical evolutionary sequence, which derives phaosomes as well as cerebral photoreceptors from epidermal cells (Fig. 7.15.).

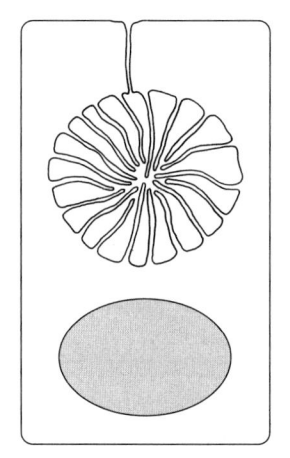

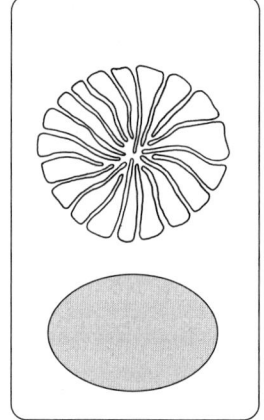

Fig. 7.15. Schematic structure of phaosomes, to the left with canal connecting to the outer cell membrane, to the right with reduced canal.

In pogonophorans, phaosomes are also present, they have been found in the epidermis of *Siboglinum fiordicum* (with only microvilli, Nørrevang 1974) and *Oligobrachia gracilis* (with microvilli and one cilium, Southward 1984).

Tardigrada
Many eutardigrades and some heterotardigrades have eyes, but ultrastructural data are only present from the marine eutardigrade *Halobiotus crispae* (Kristensen 1982) and the terrestrial eutardigrade *Milnesium tardigradum* (Dewel et al. 1993). Both are located close to the brain and consist of a cup-shaped pigment cell and few receptor cells, in which cilia and/or microvilli are present.

Onychophora
There is a pair of eyes at the base of the 'antennae'. It is an everse eye with pigment cells forming a cup and receptor cells being interspersed among them. From the receptor cells, thin processes from which numerous microvilli originate run into the internal of the cup (Eakin & Westfall 1965, Mayer 2006a). Interestingly, there are basal structures of one or two cilia in the basal region of these processes. Whether this means that the processes could be modified cilia is not clear, usually the onychophoran photoreceptors are regarded as rhabdomeric. The pigment cells also have microvilli and a cilium.

Euarthropoda
There are two types of eyes in euarthropods, the median eyes (several other names can be applied) and the lateral compound eyes (Fig. 7.14c.). Three or four median eyes are present in many pycnogonids, chelicerates, crustaceans, and insects (Gruner et al. 1993, Elofsson 2006). They are more or less pigment-cup ocelli. Pigment cells and/or other light-reflecting cells (the tapetum) form a cup, in which receptor cells are located. In pycnogonids, the nuclei of the receptor cells are distal of the receptive structures, but the axons pass through the pigment layer and the eye therefore is everse (Fig. 7.16a.; King 1973, Heß et al. 1996). The receptive structures are microvilli; they form a 'brush' at the longitudinal side of the slender cells. When the brush borders of several receptor cells face each other, the resulting central microvillous 'rod' is called a rhabdome. The median eyes can, especially in insects, form lenses, either by a local thickening in the cuticle or within special corneagenous cells. The compound eyes are composed of several to many single ocelli which resemble the median eyes. Compound eyes are found in trilobites (e.g. Fortey 2000, Clarkson et al. 2006), crustaceans (e.g. Shaw & Stowe 1982), insects (e.g. Caveney 1998), among chelicerates only in xiphosurans, and among chilopods only in Notostigmophora. In the latter case, it has been discussed whether the compound eyes in e.g. *Scutigera* are homologous to other compound eyes or are 'pseudocompound' (see Müller et al. 2003 for review). In the simplest and probably basal type, which is found in Xiphosura (Fahrenbach 1969, Battelle 2006; see Harzsch et al. 2006 for development), there are between four and twenty receptor cells and numerous pigment-bearing and cornea-secreting cells. The number of these cells is reduced during euarthropod evolution to eight receptor cells and two corneageneous cells in mandibulates. Another structure, the crystalline cone, is present in mandibulates. The exact structure of compound eyes including their nervous projections into the brain is very informative for phylogenetic relationships within euarthropods and has, for example, led to the Tetraconata-hypothesis (see, e.g., Paulus 2000, Dohle 2001, Harzsch & Waloszek 2001, Richter 2002, Bitsch & Bitsch 2005, Regier et al. 2005, Harzsch 2006). For reconstructions of the ancestral arthropod eye, it is very important whether pycnogonids are the sister-taxon of all other euarthropods or a taxon within (see Chapter 2). Because pycnogonids have only median eyes, compound eyes may not belong to the euarthropod ground pattern when pycnogonids are the basal euarthropod taxon. Usually, compound eyes as well as median eyes are assumed to be an autapomorphy of euarthropods (e.g. Ax 2000, Nielsen 2001).

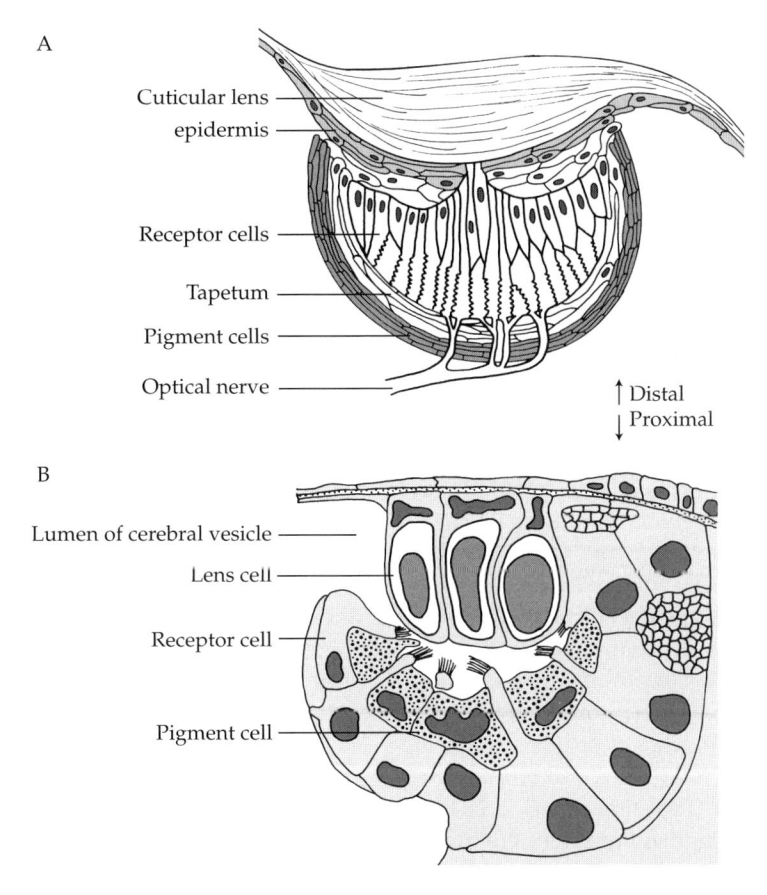

Fig. 7.16. A. Photoreceptor in pycnogonids (Euarthropoda), after King (1973). B. Photoreceptor structure in those ascidian larvae (Tunicata) where statocyst and photoreceptor are separated. After Barnes (1971).

Chaetognatha
Most chaetognaths have eyes on the dorsal side of the head; these can either be 'direct', with the receptive structures being oriented towards the outside (present in the genus *Eukrohnia*) or 'indirect' with the receptive structures being directed in various directions towards a central pigment cell (present in sagittids and spadellids). The direct eyes may or may not contain pigment. In the indirect eyes, several receptive cells project their lamellate receptive structures into indentations of the pigment cell (Eakin & Westfall 1964, Goto & Yoshida 1984, Shinn 1997). The receptive structure is a highly modified cilium. Originating from a ciliary basal body, the cell widens and contains proximally a so-called conical body and distally numerous parallel lamellae. In the direct eyes, the receptive structure is also a modified cilium with a proximal conical body and apical microvilli-like elaborations of the cell membrane (which are nevertheless derivations from a cilium). Apically, a lens covers the receptive cell (Ducret 1978, Shinn 1997).

Bryozoa
Some bryozoan larvae have pigmented photoreceptors, these have been described for several species (Woollacott & Zimmer 1972, Hughes & Woollacott 1978, 1980). The paired photoreceptors are at the margin of the apical plate. Some are flat, others are in epidermal pits. The receptive structure is made up by a tuft of unbranched cilia. The

receptor cells themselves contain pigment in some species (Hughes & Woollacott 1980), while in *Bugula neritina* (Woollacott & Zimmer 1972) and *Scrupocellaria bertholetti* (Hughes & Woollacott 1978) pigment is present in additional cells.

Echinodermata

Many echinoderms are photosensitive, but photoreceptors have been described only from asteroids and from one holothurian species (Yoshida et al. 1984). Although echinoids have no recognizable photoreceptors, six opsin genes have been found in the genome of *Strongylocentrotus purpuratus* (Raible et al. 2006). Asteroids have eyes at the tips of their arms. Many (>100) photoreceptors are located together in one 'optic cushion'. Each single photoreceptor is an everted pigment-cup and the receptor cells have one cilium and numerous microvilli (Vaupel-von Harnack 1963, Eakin & Brandenburger 1979, Penn & Alexander 1980). In the holothurian *Opheodesoma spectabilis*, the photoreceptors are located in pigmented patches close to the tentacular nerves. Receptor cells have one cilium and numerous microvilli. They are surrounded by supporting cells, but pigment granules are found in the underlying nerves (Yamamoto & Yoshida 1978). In ophiuroids, a potential photoreceptor has been identified in *Ophiocoma wendtii* (Hendler & Byrne 1987, Aizenberg et al. 2001). The receptive structures appear to be free endings of nerve cells, which receive light through special structures in the skeleton of the dorsal arm plate.

Enteropneusta

Photoreceptors are only known from the tornaria larva and have been described from *Ptychodera flava* (Brandenburger et al. 1973). They are everse pigment-cup photoreceptors and the receptor cells have both a cilium and microvilli.

Tunicata

Photoreceptors are present in larval ascidians (but are probably lacking in appendicularians, see Holmberg 1984) in close association to the statocyst within the cerebral vesicle. In most species, statocyst and photoreceptor are spatially separated in the cerebreal vesicle (e.g. *Amaroucium constellatum*, Barnes 1971; *Ciona intestinalis* and *Distaplia occidentalis*, Eakin & Kuda 1971). In these species, there is a cup-shaped arrangement of pigment cells, with several underlying receptor cells projecting through the pigment cells into the cup lumen (Fig. 7.16b.). The receptive structures are several cilia. Additionally, there are three lens cells covering the cup opening. In *Botryllus schlosseri* and related species, there is an association between photoreceptor and statocyst (Fig. 7.9.; Grave & Riley 1935, Sorrentino et al. 2000), which can probably be interpreted in the way that the photoreceptive cells use the cup-shaped statolith of the statocyst for shading. In some species of the genus *Styela*, one pigment cell without photoreceptive structures is in the position, where in other species the photoreceptor is located (Fig. 7.17.; Ohtsuki 1990). It appears likely that the presence of separated photoreceptor and statocyst is the ancestral condition (and an autapomorphy of Ascidia), whereas in

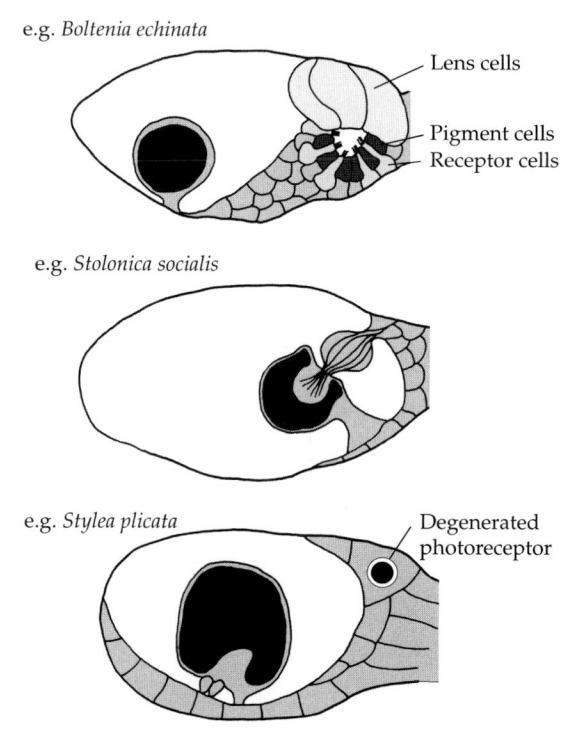

Fig. 7.17. Three different patterns of the statocyst/photoreceptor complex in the cerebral vesicle of ascidian larvae. After Ohtsuki (1990).

the Styelidae this pattern was either conserved, altered as in *Botryllus*, or reduced (Fig. 7.17.; see Sorrentino et al. 2000 for a summary of species and their sensory organs). In adults, photoreceptors are known from *Ciona intestinalis*, where they are located around the siphons (Dilly & Wolken 1973). The receptor cells, which have microvilli and one cilium, are located in the epidermis and are supplemented by a subepidermal cup of pigment cells. Salps (Salpida, Thaliacea) have dorsal photoreceptors, associated with a neural ganglion, but not integrated into the nervous system as in ascidian larvae (Gorman et al. 1971). The receptive structures here are microvilli.

Acrania

Several photoreceptive structures are present in the neural tube of acranians (summarized in Ruppert 1997). An unpaired frontal ocellus is located at the anterior end of the neural tube, close to the neuropore. There are some pigment

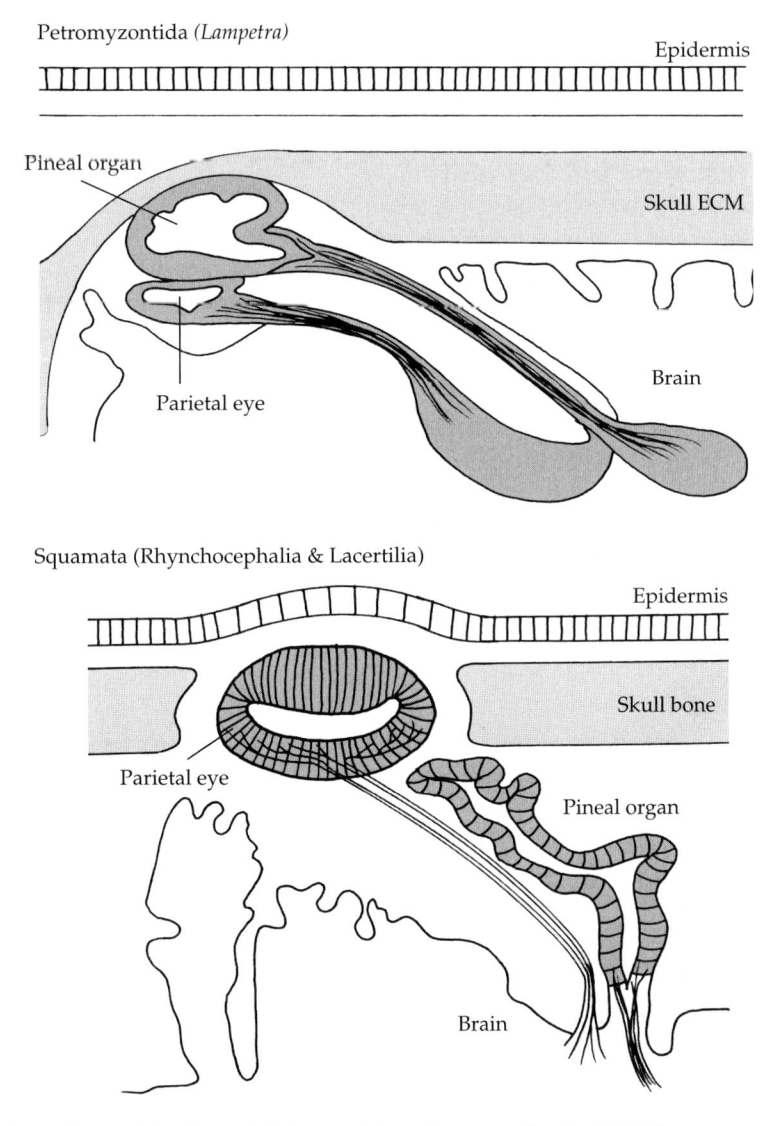

Fig. 7.18. Structure of parietal eye and pineal organ in Petromyzontida and Squamata. After Starck (1982).

cells and in the close neighbourhood a few sensory cells with cilia. In the anterior dorsal wall of the neural tube are photoreceptive Joseph cells which are ciliated, but also contain many microvilli. Along the neural tube are many (up to 1,500 per side) photoreceptors ('Hesse ocelli'), each composed of one cup-shaped pigment cell and one receptor cell with one cilium and many microvilli. Additionally, so-called lamellar cells in the anterior neural tube may also be photosensitive.

Craniota

Three types of eyes occur in craniotes, the paired lateral eyes and two unpaired eyes (Fig. 7.18.). The anterior of the unpaired eyes, the parietal eye, is present in Petromyzontidae and several sauropsids, and is particularly well developed in the tuatara (*Sphenodon*, Rhynchocephalia). It can also be found in some fossil craniotes and traces can be found embryologically in several other craniotes. The parietal eye is an epithelial-bound vesicle positioned under the dorsal epidermis of the head. The distal cells of the vesicle may function as a lens, the proximal ones contain the receptor cells and pigment cells (also called supportive or glia cells) (Eakin 1973, Jenison & Nolte 1979). The eye is everse. Receptive structures are branched cilia, the branches form regular disc-like stacks. The posterior of the unpaired eyes, the pineal eye (pineal gland), is present in most craniotes, but only in some it has a photoreceptive function (several fishes, anurans, and squamates, see, e.g., Klein 2004, Ung & Molteno 2004), but in others it evolved into an endocrine gland (secreting the hormone melatonin). In many cases, these functions are combined, for example the pinealocytes in the pineal gland of birds are photosensitive (Natesan et al. 2002) and may be involved in the regulation of biological clocks. The lateral eyes are complex eyes with cornea, pupil, iris, and lens (Fig. 7.19.), very similar in structure to the eye of coleoid cephalopods,

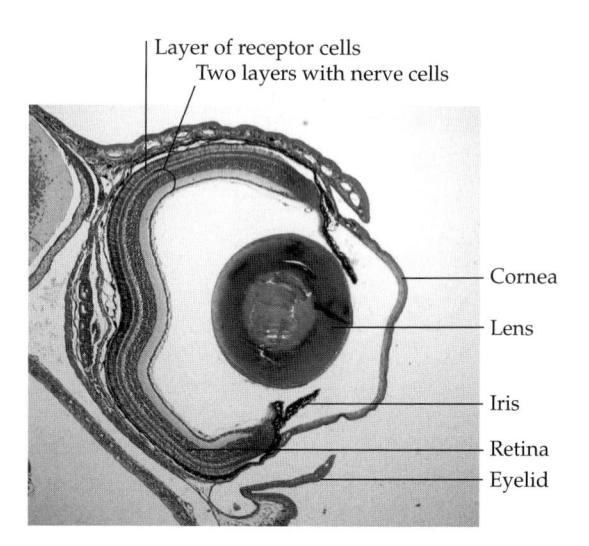

Fig. 7.19. Cross section through the lateral eye of *Rana temporaria* (Anura, Amphibia).

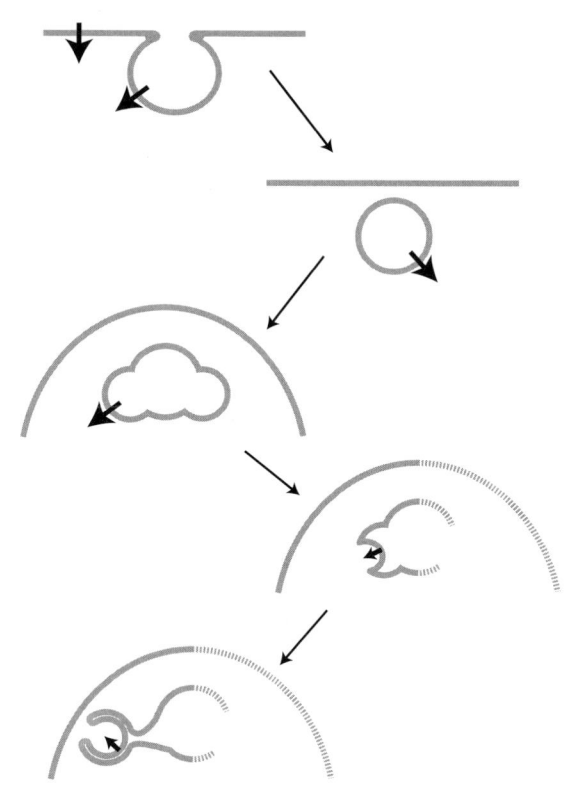

Fig. 7.20. Schematic development of the eyes in craniotes, explaining the presence of inverse eyes. The small, bold arrows indicate the direction of signal transduction from sensory to nerve cells. After neurulation, sensory cells are internal in the tube and nerve cells peripheral. The eye develops from lateral extensions of the brain, which invaginate to form the eye-cup. The signal transduction in the retina is consequently from the 'lower' receptor cells to the nerve cells. Numerous axons are bundled and pass through the entire retina as the optical nerve (not shown here).

but differing in being inverse instead of everse. This is explained by the mode of eye development, starting from neurulation, during which the cell polarity is reversed (Fig. 7.20.). Several kinds of receptor cells are present, adapted to light intensity (rods) or colour vision (cones), but all have modified cilia as receptive structures (see, e.g., Walls 1963, Polyak 1968).

Table 7.2 Summary on photoreceptor characters in eumetazoan taxa. For explanations and references see text. The distinction between inverse and everse eyes is only applied here when more than one pigment cell is present. Receptive structure: ci = only cilia, mv = only microvilli (rhabdomeric), mv/ci = many microvilli with single or rudimentary cilia, ci/mv = several cilia and few microvilli. A '+' indicates that two structures are present within the same animal.

	Shape	Inverse or everse	Receptive structure
Cnidaria (Hydrozoa, Scyphozoa, Cubozoa)	Flat or pigment-cup, sometimes with lens	Inverse or (rarely) everse	ci
Ctenophora	Phaosomous-like	–	ci
Gastrotricha	Pigment-cup	–	ci
Nematoda	Pigment-cup	Inverse	ci
Kinorhyncha	2 cells with intercellular lumen, structure unknown in pigmented eyes	?	ci
Platyhelminthes	Flat or pigment-cup, sometimes with lens	Inverse	mv, ci or mv/ci
Eurotifera			
Wheel organ eyes	Pigment-cup, sometimes with lens, probably also phaosome-like	Inverse	mv, ci
Cerebral eyes	Pigment cup distal of receptor cells	–	ci or other
Nemertini	Pigment-cup	Inverse	mv/ci
Kamptozoa, larva	Pigment-cup	–	ci
Mollusca			
Polyplacophora, larva	Pigment-cup	Everse	mv/ci
Polyplacophora, adult	In aesthetes, as lateral components	–	mi, ci or mv/ci
Gastropoda, larvae	Pigment-cup with lens	Everse	ci, ci + mv or mv/ci + ci/mv
Gastropoda, adult	Flat, pigment-cup or pinhole-eye, often with lens	Everse	mv, mv/ci
Cephalopoda	Pinhole-eye (Nautilus) or complex lens eye (Coleoidea)	Everse	mv
Bivalvia, *Mytilus*	Pigment-cup	Everse	mv/ci
Bivalvia, *Cerastoderma*	Pigment-cup	Everse	ci
Bivalvia, *Arca, Barbatia*	Pigment-cup (2 kinds)	Everse	either mv or ci
Bivalvia, *Pecten*	Pigment-cup with 2 retinas and lens	Inverse	ci/mv + mv/ci
Sipunculida	Pigment-cup	Everse	mv/ci
Annelida			
Polychaeta, rhabdomeric cerebral eye	Pigment-cup, sometimes with lens	Inverse or everse	mv, mv/ci
Polychaeta, ciliary Cerebral eye	receptor cell with support cell, but no pigment cell	Inverse (often in small eyes) or everse (often in large eyes)	ci
Clitellata	Phaosome	–	mv
Tardigrada	Pigment-cup	–	mv/ci
Onychophora	Pigment-cup	Everse	mv/ci
Euarthropoda			
Median eyes	Pigment-cup, sometimes with lens	Everse	mv
Compound eyes	Composed of several ocelli, each ocellus with pigment cells, receptor cells and lens	Everse	mv

Table 7.2 (*Contd*)

	Shape	Inverse or everse	Receptive structure
Chaetognatha	Central pigment or pigment absent	–	ci
Bryozoa, larva	Flat or pigment-cup	Everse	ci
Echinodermata			
Asteroida	Pigment-cup	Everse	mv/ci
Holothuroida (*Opheodesoma*)	Pigment in nerves underlying receptor cells	–	mv/ci
Enteropneusta, larva	Pigment-cup	Everse	mv/ci
Tunicata	Pigment-cup with lens (or derived types)	Everse	ci
Acrania, frontal ocellus	Receptor cells with neighbouring pigment cells	–	ci
Joseph cells	Only receptor cells in neural tube wall	–	ci/mv
Hesse ocelli	Pigment-cup composed of 1 pigment cell and 1 receptor cell	–	mv/ci
Craniota, median eyes	Pigment-cup	Everse	ci
Lateral eyes	Complex lense eye	Inverse	ci

Evolutionary comparison of photoreceptors

As can be derived from the summary above, photoreceptors of eumetazoan animals are very diverse (Table 7.2.). They can be found in different positions (in the epidermis, associated with the brain, or in other positions), they can be flat, cup-shaped, or spherical; with a pinhole, with a lens, and sometimes even with an iris; they can have pigments or not; receptive structures can be ciliary, rhabdomeric (with microvilli), or combined; and they can be inverse or everse. Eakin (1963, 1965, 1968, 1979, 1982) has put much emphasis on the type of receptive structure, which he regarded as almost diagnostic, with protostomes having rhabdomeric photoreceptors while cnidarians and deuterostomes have ciliary photoreceptors. As there are a number of derivations (i.e. protostomes with ciliary and deuterostomes with rhabdomeric photoreceptors), some authors have questioned Eakin's evolutionary conclusions. For example, Vanfleteren & Coomans (1976) and Vanfleteren (1982) regarded these differences to be more quantitative than qualitative and proposed that cilia are generally important for the development of outgrowths on the cellular surface. Therefore, the rhabdomeric type of photoreceptor is regarded as derived from a ciliary type, with species displaying more or less complete reductions of the cilia. In fact, pure rhabdomeric photoreceptors, that is, those including only microvilli and no trace of cilia, are the exception rather than the rule (Table 7.2.).

While both Eakin and Vanfleteren/Coomans assumed a common origin of photoreceptors with subsequent evolutionary change, von Salvini-Plawen & Mayr (1977; see also von Salvini-Plawen 1982) regarded photoreceptors as so diverse and so scattered in distribution that they might have evolved several times in parallel, between 40 and 65 times.

Apart from the structural knowledge of photoreceptors, the biochemical properties have become much clearer since then. The discovery that the mouse *small eye* gene and the *Drosophila eyeless* gene are homologous (Quiring et al. 1994) and that *eyeless* can induce a so-called ectopic formation of eyes in untypical locations such as on the legs and antennae of *Drosophila* (Halder et al. 1995) led to the discovery of homologous genes (*Pax6* genes) in many animals (see, e.g., Gehring & Ikeo 1999, Arendt & Wittbrodt 2001, Gehring 2001, 2004, 2005 for summaries). It therefore became likely that at least some factors are common to all kinds of eyes.

A multiple *de novo* evolution of eyes or the homology of all eyes are not the only alternatives and it is quite clear that the real picture must be somewhere in between. Arendt & Wittbrodt (2001), Arendt (2003) and Arendt et al. (2004) have presented a more differentiated picture of photoreceptor evolution. They distinguished not only between ciliary and rhabdomeric eyes, but also between cerebral and not cerebral (usually epidermal) eyes and between inverse and everse eyes. With some exceptions, inverse cerebral eyes are present in protostome larvae, they are either transferred to the adult or are replaced during development by everse eyes. Such eyes are, again with some exceptions, rhabdomeric (Arendt & Wittbrodt 2001). Chordate cerebral eyes are everse and ciliary, but ciliary cerebral eyes are also present in polychaetes and nemerteans (Arendt et al. 2004). A closer investigation of the visual pigment and the transduction chain has shown that the ciliary cerebral eye in *Platynereis dumerilii* (Polychaeta) shows more similarity with craniote ciliary receptor cells than to the *Platynereis* rhabdomeric eyes (Arendt et al. 2004). Furthermore, the visual pigment and transduction chain typical for rhabdomeric eyes are present in craniote eyes, namely in the so-called retinal ganglion cells. This suggests that rhabdomeric and ciliary eyes have their typical forms of rhodopsin and of the transduction chain components. It further suggests that both rhodopsins and the respective transduction cascades were present in the common bilaterian ancestor. The bilaterian ancestor likely had even more opsins than these two kinds of rhodopsins, indicating several previous gene duplications (Terakita 2005).

Taken together, a fascinating picture starts to emerge. It appears that the molecular inventory of photoreceptors not only evolved early in metazoan evolution, but had already diversified into a microvillous and a ciliary 'kit' early on. Such kits must have been present in the ancestors of each higher taxon, but they are not expressed in numerous cases (those, in which no photoreceptors can be found). The kits may enable a cell to turn into a receptor cell by expressing the photoreceptive protein in the membrane and the transduction chain in the cytoplasm. Whether pigment cells or lens-forming cells have a comparable common molecular background is largely unknown, but taking the many different forms into account, I would doubt this (or assume it only in smaller scales, i.e. for single taxa or small groups of taxa. See for example the presence of βγ-crystallin, a craniote lens protein in the statolith of the tunicate *Ciona intestinalis*, Shimeld et al. 2005). Therefore, photoreceptors appear to be composed of a mosaic of homologous and not homologous components.

Nilsson & Pelger (1994) addressed the question of how fast changes in eye construction can be performed during evolution. In a model they estimated the number of generations in which a simple photosensitive patch of epithelium can turn into a complex lens eye of the craniote type. They result in less than 40,000 generations, which equals a period of less than half a million years (assuming a generation time of one year). In geological terms, this is a very brief period.

One may finally ask why, among the variety of eye types, similar types appear to occur in parallel. This accounts for pigment-cup eyes, for pinhole eyes and, as the most well known and convincing example, for the complex lens eyes of coleoid cephalopods and craniotes. Conway Morris (2003) has given a convincing answer to this. Photoreceptors function in a variety of shapes, but this variety is restricted by physical properties. Therefore, the pathway in which an eye can improve its capability is fixed. Several taxa may go down such paths in parallel but each will stop at similar, yet analogous stations. Therefore, in those cases where complex lens eyes with an iris and so on are able to evolve, they must resemble each other, because other designs make no sense in terms of physical properties.

CHAPTER 8

Body cavities

In several animals, the body is composed of cells that are in close contact to each other, separated only by ECM and forming what we call a 'compact organization'. In many animals, however, we find some kind of cavity, either as small interstices between cells, as tubelike systems, as spacious cavities, or as serially repeated units. Such cavities are involved in a variety of functions, ranging from skeletal (as a hydroskeleton), to transport, excretion, and reproduction.

Body cavities are perhaps the organ system, for which evolutionary patterns have been discussed most vividly and most contradictorily. Body cavities have been central in several concepts of phylogenetic relationships, starting with Haeckel (1874) and Hertwig and Hertwig (1882). Before discussing the most important of these hypotheses, the terminology used in this chapter should be made clear.

Body cavities—types and terms

Structurally, two types of body cavities can be distinguished: they are called 'primary' and 'secondary' body cavities. They differ in the way the central lumen is bordered. In primary body cavities, an extracellular matrix (ECM) borders the entire cavity, whereas in secondary body cavities, this is a cellular layer (an epithelium) which itself rests on an ECM (Fig. 8.1.). It has turned out that such a division is a good starting point for the discussion of body cavities. Since transmission electron microscopy has been used, the distinction of the two types of body cavities is in most cases easy, even in cases where the ECM (lining the primary body cavity), or the epithelium (lining the secondary body cavity), are very thin.

The secondary body cavity is commonly named the 'coelom', a term that will also be used here because it is unequivocal. The primary body cavity is sometimes called a 'pseudocoel'. This term is used in concepts of a classification of animals into acoelomate (no body cavity), pseudocoelomate, and coelomate (see below) animals, and is then often regarded as characteristic for animals that are united as pseudocoelomates (rotifers, nematodes, and others). This neglects the fact that primary body cavities are very broadly distributed and occur in many 'coelomate' animals (e.g. as the circulatory system) or during embryogenesis (as blastocoel). I will therefore use the more descriptive term of primary body cavity instead of 'pseudocoel'.

Body cavities are located in mesodermal tissue and are therefore present only in bilaterian animals. The only exception here is the blastocoel, which develops as a hollow space in the spherical blastula. During further development it may persist or disappear. What makes the matter a bit complicated is that all three possibilities (absence, primary, and secondary cavity) can be present during the life cycle of one and the same animal, as an ontogenetical succession or together side by side in a particular stage. In special cases, primary and secondary body cavities can even fuse to form what is called a mixocoel, or an acoelomate condition can be derived from a coelomate condition.

As mentioned above, body cavities can have a variety of functions. One of the most important

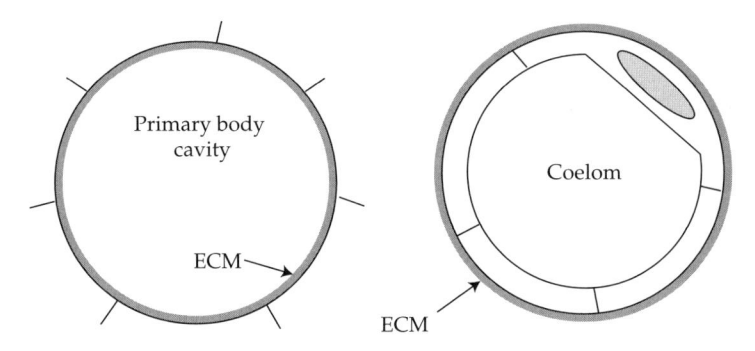

Fig. 8.1. Distinction between primary and secondary body cavities on the basis of the position of extracellular matrix (ECM) or an epithelium.

is the transport of substances by fluid. This leads us to the borders of the definition of body cavities. Fluids can also float in a compact body where clearly detectable body cavities are absent. The fibrous layer of the ECM, the lamina fibroreticularis, is a network of fibres that create numerous tiny interstices. Water and small molecules, but not large molecules or cells can pass through this part of the ECM. This kind of selective permeability is, by the way, an essential prerequisite for a filter function of ECM in excretory organs (see Chapter 9). However, such 'intra-ECM' interstices are not regarded as body cavities here. Only when clearly visible clefts are recognizable is the term body cavity used. This definition becomes problematic when gel filled cavities are present as, for example, in larvae of enteropneusts, echinoderms (Strathmann 1989), and phoronids (Bartolomaeus 2001). It is not completely clear whether such gels should be regarded as equivalent to fluid or as part of an ECM. They may constitute extremely extended ECMs, in which fibres of the lamina fibroreticularis are pushed far apart from each other.

The gonads are a special problem. Quite often, they are epithelially-lined cavities into which mature gametes are released. Following this definition, it is not wrong to call these cavities a coelom. A hypothesis, in which the coelom originates from gonads (the 'gonocoel hypothesis') will be discussed below. As gonads are in most animals well separated from other body cavities and have a single function in the reproductive system, I will exclude them from the discussion of body cavities.

Absence of body cavities—the acoelomate condition

When all mesodermal cells in the body attach on each side to other cells, this organization is called acoelomate or, better, compact. Of course cells do not attach directly with their membranes, but via ECM. As a simplified rule (which does not hold in all cases), body cavities occur in large bodied animals, but are absent in small animals. Therefore, when large animals have microscopically small larvae, these are often acoelomate and develop a body cavity during their development. There are also small animals with body cavities (rotifers, meiobenthic priapulids) and large-bodied animals without body cavities (polyclads and parasitic species among flatworms). As has been mentioned above, absence of body cavities does not mean absence of fluid currents through the body. This is performed through the fibrous part of the ECM.

Primary body cavities

As has been mentioned above, primary body cavities are lined by ECM. This can best be explained by their development from cracks and fissures within the ECM. If we imagine an acoelomate body organization and increase during development the pressure of the interstitial fluid, the most likely structure to tear is the

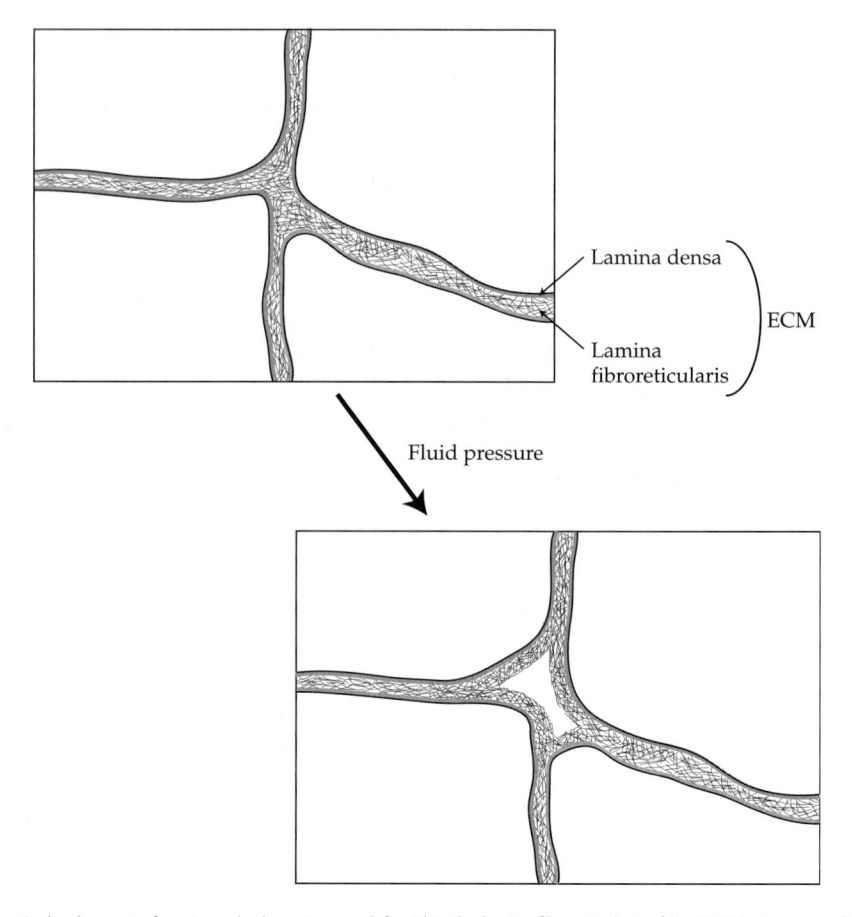

Fig. 8.2. Ontogenetic development of a primary body cavity as a cleft within the lamina fibroreticularis of the ECM by increase of fluid pressure.

lamina fibroreticularis of the ECM (Fig. 8.2.). In this sense, the primary body cavity can be regarded as a cleft *within the ECM*, created by the very simple mechanism of fluid pressure. This appears to be a comparatively simple process, which makes it questionable whether primary body cavities can be regarded as homologous.

In the development of many animals, a hollow blastula stage is formed during early embryogenesis. Such a hollow blastula is even regarded as an autapomorphy of Metazoa (Ax 1996) and is present already in some sponges (see summary by Maldonado 2004). The cavity inside the blastula is called the blastocoel and is a primary body cavity. The blastocoel can persist during further development to be continuous with the primary body cavity in adult animals, which has led Hyman (1951, p.22) to define a pseudocoel as a persistent blastocoel. In many cases, however, the blastocoel vanishes during development and adult primary body cavities are formed independently. Therefore, the definition of a primary body cavity as a persisting blastocoel is too restrictive (see also Ruppert 1991a).

Primary body cavities can have the form of numerous small interstices, such as in molluscs; they can be spacious, as in priapulids and large nematodes; or they can be restricted to form vessels, as in the circulatory system of annelids. Ruppert (1991a) distinguished according to the size of the cavity lacunes (small cavities), sinuses, and the hemocoel. His smallest category, interstitia, describes the space between fibrills in the ECM, such orgaization is included here in the acoelomate condition (see above). Primary body cavities coexist with secondary body cavities in a number of taxa.

Secondary body cavities—the coelom

A coelom is always lined by an epithelium (also called the coelothel). It is therefore directly bordered by cells and not by ECM. This has to be emphasized, because it has led to confusion in the past. Nematodes, for example, were regarded by some authors (e.g. Remane 1963, Siewing 1976) to have a coelom, because the cavity is surrounded by an epithelially organized layer of longitudinal musculature. But because the muscle cells are completely surrounded by ECM, the body cavity is lined by ECM and not immediately by the cells, and it is therefore a primary body cavity and not a coelom. Whether the bordering tissue beyond the ECM is organized as an epithelium or not is not relevant for the definition of a coelom.

Coeloms originate in two ways. They can be derived from an already epithelially organized tissue, the endodermal intestine, and this process is called 'enterocoely' (Fig. 8.3.). In this case, parts of the intestine form pockets that later separate to form hollow compartments, the coelom. Or, in the other case, coelom originates from

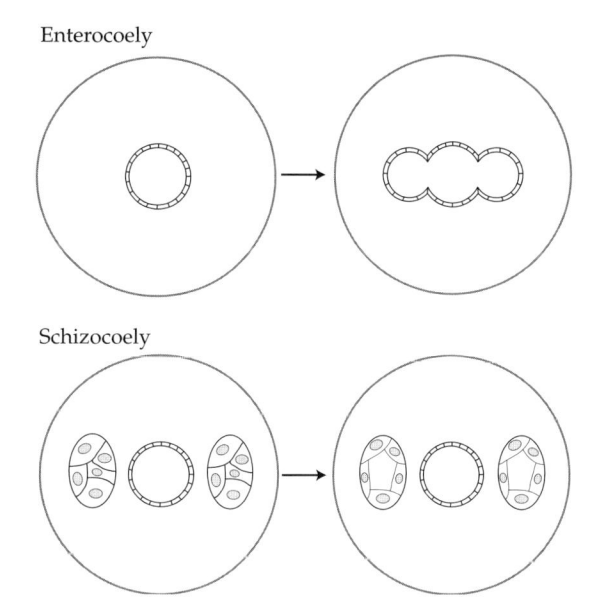

Fig. 8.3. Development of coeloms by enterocoely (origin from the intestinal epithelium) or by schizocoely (epithelial organization of a compactly preformed tissue).

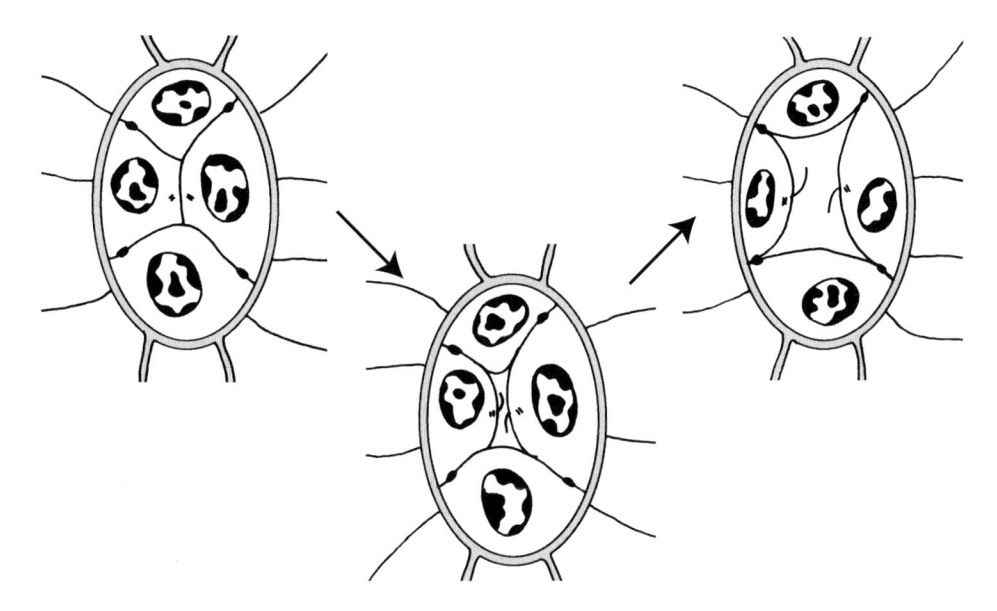

Fig. 8.4. One example for schizocoely: the development of *Magelona* sp. (Polychaeta, Annelida). The tissue strand is surrounded as a whole by ECM (grey), cell–cell contacts are in the peripheral region of the cells. Fluid influx pushes the cells apart, two cells project cilia into the lumen. After Turbeville (1986)

compact masses of mesodermal cells that transform to an epithelial organization and then surround a cavity. This process is called 'schizocoely' (Fig. 8.3.). In these cases, the single cells are not completely surrounded by ECM, but ECM only surrounds the entire cell mass. Cells within this mass rest on the ECM with their basal side, and form adhaerens junctions between each other at their apical side. When fluid pressure between cells increases, the cells become organized epithelially around a central lumen (Fig. 8.4.).

The epithelium in a coelom, the coelothel, can be variable in structure. It can be a myoepithelium, that is, a layer of myoepithelial cells (Fig. 8.5.). It can also be a layer of epithelial cells lacking myofilaments, and in this case it is called a peritoneum (Fig. 8.5.). Coelothel cells can be ciliated or lack cilia. One broadly distributed feature of coelothels is the possession of podocytes. These are specialized peritoneal cells which form fingerlike extensions that interdigitate intensively (Fig. 8.6.). Podocytes occur where fluid moves into the coelom from adjacent structures, in particular during excretory processes (see Chapter 9). The fingerlike extensions maximize the intercellular space, through which fluids can enter the coelom.

There are cases in microscopically small annelids, in which the development of a coelom is arrested at the stage where an epithelially organized, but still compact, cell mass is present (Fransen 1980, Smith et al. 1986). In such cases, the animals are acoelomate in organization, but this is clearly derived from a coelomate condition as evidenced by close relatives (which have a coelom) and by embryology. Another such special case is the coelom of acranians which forms as serially iterated cavities surrounded by a myoepithelium. During development, the myoepithelium expands until it almost obliterates the coelomic space (see Ruppert 1997).

The mixocoel

In adults of onychophorans and euarthropods, a spacious body cavity is present that has the characteristics of a primary body cavity. In addition, small coelomic compartments (sacculi) are present in association with the excretory organs.

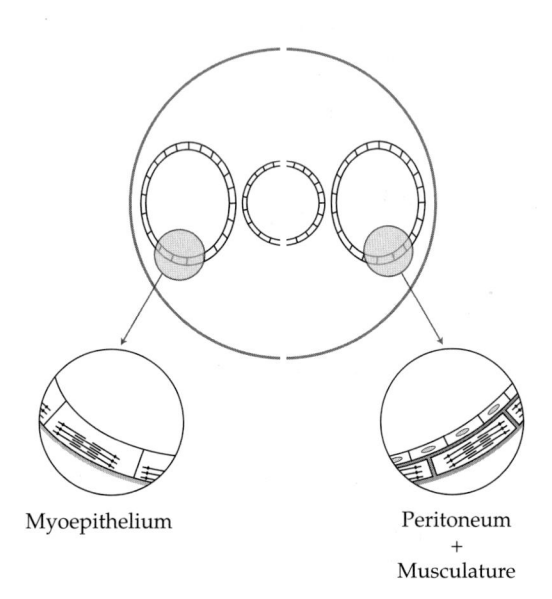

Fig. 8.5. Different structure of a myoepithelium (left) and a peritoneum with separate muscular layer (right).

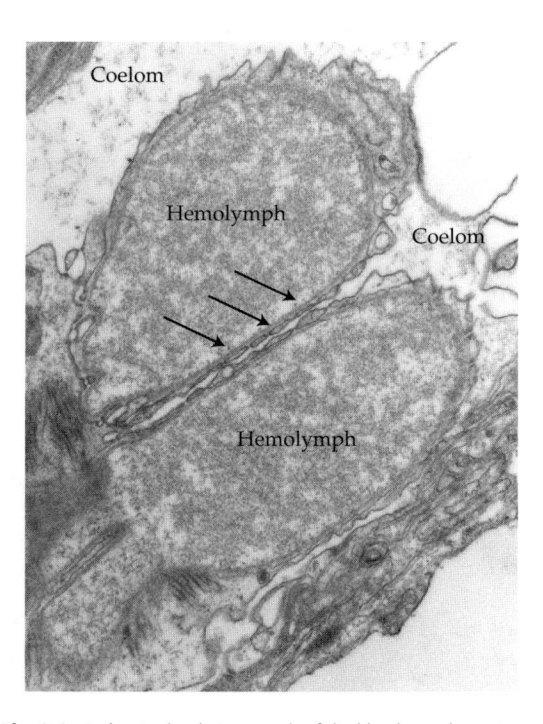

Fig. 8.6. Podocytes bordering vessels of the blood vascular system in vicinity to the coelom in the polychaete *Fabricia sabella*. Photo by courtesy of Thomas Bartolomaeus, Berlin.

Onychophorans possess a pair of such sacculi in each postoral segment, while in euarthropods, sacculi occur only in a few anterior segments. During early development, segmentally arranged pairs of coelomic cavities are present in both taxa. Histological investigations of the development have made it probable that these coelomic cavities break up and that the coelothel migrates into several positions to form structures such as heart, musculature, and others (see Anderson 1973 for summary).

Taking into account the different nature of primary and secondary body cavities, the process of a fusion of these two cavities appears to be unlikely and the possibility remains that such a fusion resulted from some kind of misinterpretation. In onychophorans, for example, Bartolomaeus & Ruhberg (1999) suspected that the embryonic coeloms may remain small during development and may then represent the sacculi in adults. Recently, however, Mayer et al. (2004) have shown that the 'old' interpretations of a fusion of primary and secondary body cavity are indeed correct. By investigating differently developed segments of an embryo of the onychophoran *Epiperipatus biolleyi*, they showed that a primary body cavity develops adjacent to the embryonic coelom and that this coelom really breaks up (Fig. 8.7.). At this stage, a cavity is present that is lined in part by ECM (the former primary body cavity) and in part by an epithelium (the former secondary body cavity). In this case, the term *mixocoel* is therefore appropriate.

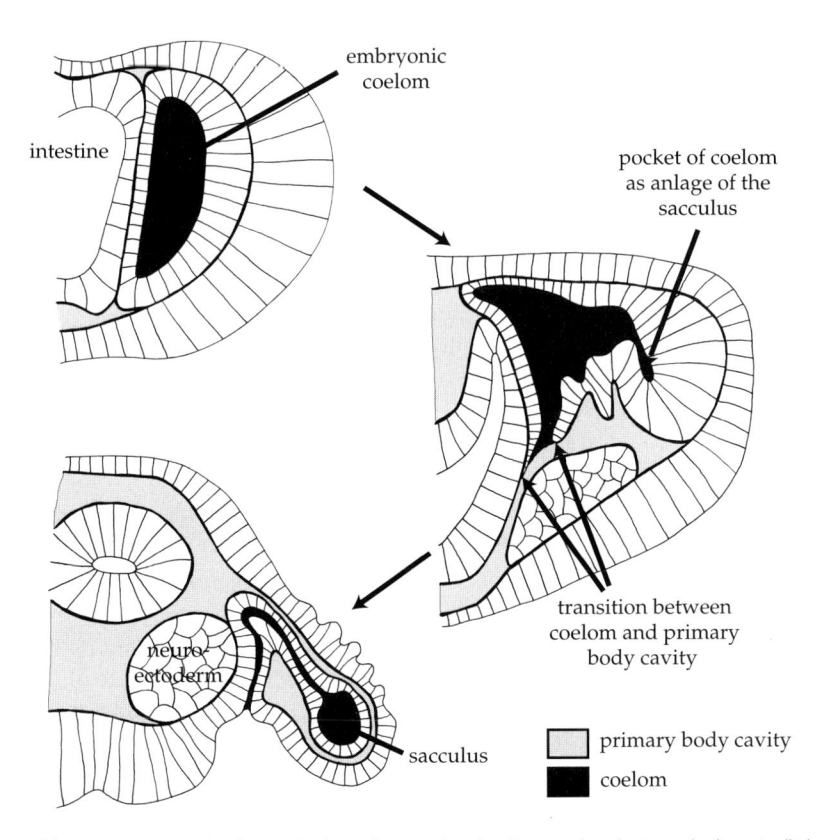

Fig. 8.7. Development of the mixocoel in onychophorans. In the embryo, coelom (medium grey) and primary body cavity (lighter grey) are both present. During development, the coelom opens ventrally. The sacculus is formed by an outgrowth of part of the coelom epithelium. ECM is symbolized by bold black lines. Drawing after figures in Mayer (2006b).

Functions of body cavities

Body cavities are a comparatively simple way to enlarge the body size, because no (in the case of primary body cavities) or comparatively little tissue (in the case of secondary body cavities) is involved in large voluminal changes. Because fluid is hardly compressable, a fluid-filled cavity can ideally perform skeletal functions to maintain a certain body shape (hydroskeleton). Furthermore, muscular action on the cavity will force the fluid to move into another direction and therefore lead to changes of the body shape. This principle is very abundantly realized in animals with body cavities, for example in the eversion of structures such as introverts. While the retraction of such eversible structures is usually performed by special retractor muscles, the eversion is performed by the pressure of fluid into the eversible structure through the contraction of muscles in the main body.

Body cavities often contain different contents. These can be molecules, cells, or even aggregates of cells. By movement of the body cavity fluid, the contents can be transported to different target tissues, and therefore body cavities can have a transport function. When the body cavities are more or less restricted in space they form vessels and represent a circulatory system (see Chapter 10). Body cavities, especially coeloms, often play an essential role in excretion. Hemolymph or blood is filtered into the coelom which serves as a reservoir and as a modifier of the so-called primary urine. In annelids, the coelom can also store gametes.

The epithelium of coeloms is often metabolically active. Welsch (1995) has summarized a number of known activities such as reabsorption, transport, osmoregulation, secretion, phagocytosis, and storage of lipids and glycogen.

Body cavities in animals

A body cavity is completely lacking in all diploblastic animals, with the exception of a primary body cavity being transitorily present as the blastocoel in several blastulae (the blastula can also be a compact sterroblastula). In basal or assumed basal bilaterian taxa (*Xenoturbella*, Acoelomorpha, and Gastrotricha), a body cavity is absent and the organization of the body is compact. Among Spiralia, Platyhelminthes and Gnathostomulida are acoelomate, even in the larger species. *Limnognathia maerski* is probably also acoelomate, some observed cavities are likely to be artefactual (Kristensen & Funch 2000).

Nematoda

For nematodes, a significant primary body cavity is often regarded as being characteristic, but it is present predominantly in large, parasitic species. Small species, such as enoplids, lack a body cavity and are therefore acoelomate (Wright 1991, Ehlers 1994a). As large species are derived within nematodes (Lorenzen 1994, Blaxter et al. 1998), the nematode ancestor might also have been acoelomate. The notion that nematodes have a coelom (Remane 1963, Bleve-Zacheo et al. 1975, Siewing 1976) is mainly due to fact that the musculature in nematodes is a monolayer and therefore has an epithelial character. The body cavity (when present) therefore appears to be surrounded by an epithelium, but ultrastructural investigations have shown that there is ECM between the cavity and the muscle cells, making the cavity a primary body cavity. Free cells (improperly named coelomocytes) are present (when a body cavity is there), and probably have a phagocytotic function (Bird & Bird 1991).

Nematomorpha

Nematomorphs are acoelomate as larvae (Zapotosky 1974, 1975, Jochmann & Schmidt-Rhaesa 2007), but develop a primary body cavity during development (Lanzavecchia et al. 1995, Schmidt-Rhaesa 2005). This cavity becomes spacious in females, but in males it is reduced during further development by massive growth of parenchyma, a tissue with a presumed storage function for nutrients taken up from the host (Schmidt-Rhaesa 2005). In some cases, this may again lead to an acoelomate organization (Fig. 8.8.), but in most cases, a small peri-intestinal cavity remains.

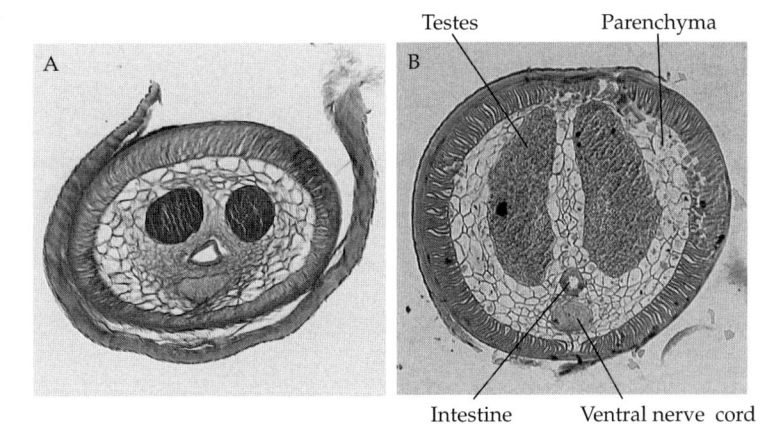

Fig. 8.8. Acoelomate condition in Nematomorpha, shown in cross sections of males from the species *Pseudochordodes bedriagae* (A) and *Gordius* sp. (B). The cuticle is partly dislocated in both sections.

Priapulida
Priapulids have a very spacious primary body cavity (Storch 1991) which is even present in the meiobenthic species such as *Meiopriapulus* (Storch et al. 1989) and *Tubiluchus* (Storch et al. 1985). Notions that priapulids have a coelom (Shapeero 1961) are again misinterpretations. Only in the neck region of *Meiopriapulus fijiensis* is a small cavity surrounding the foregut lined by an epithelium and therefore is a coelom (Storch et al. 1989). This small coelom coexists with the more spacious primary body cavity. Free cells ('coelomocytes') are abundant in priapulids, they act as erythrocytes and amoebocytes (Schreiber et al. 1991, Storch 1991).

Kinorhyncha
Reports of a spacious body cavity are based on artefacts (see Kristensen & Higgins 1991), but some interstices of the primary body cavity are present, particularly in the anterior body region (Neuhaus 1994) where they provide space to withdraw the introvert. Free cells (amoebocytes) are present in these interstices (Neuhaus 1994).

Loricifera
Some taxa have a tiny primary body cavity (such as *Nanaloricus*) while others (Pliciloricidae) have a larger one (Kristensen 1991). This is largest when the introvert is everted, and therefore likely functions as a reservoir to withdraw the introvert, as in kinorhynchs.

Tardigrada
In tardigrades, the reports of coelomic cavities go back to an observation by Marcus (1928) who stated that segmentally arranged coelomic pouches originate by enterocoely. This statement could not be confirmed in later reinvestigations on *Halobiotus crispae* (Eibye-Jacobsen 1996) and *Thulinia stephaniae* (Hejnol & Schnabel 2005), and tardigrades appear to be acoelomate throughout all stages of their development.

Onychophora and Euarthropoda
Onychophorans and euarthropods have a spacious body cavity which has, in adults, characteristics of a primary body cavity (Fig. 8.9.). Additionally, they have small segmental coelomic sacs, the sacculi (in almost all onychophoran segments, but restricted to few anterior segments in euarthropods) as part of the excretory system. From light-microscopical observations, it has been stated that embryos possess coelomic cavities that disintegrate during development and fuse with the primary body cavity to form a mixocoel (see Anderson 1973 and literature therein for onychophorans and arthropods; exemplary Weygoldt 1958 for *Gammarus pulex*, Crustacea). As explained above,

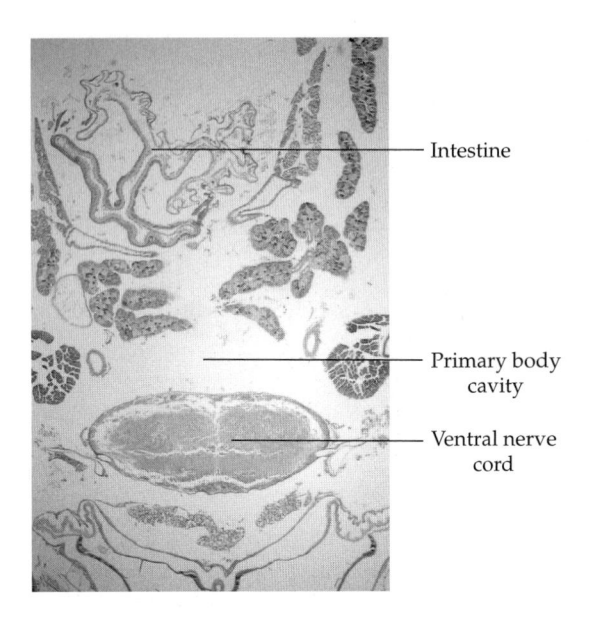

Fig. 8.9. Organs in Euarthropod are surrounded by a spacious body cavity, as shown in a cross section through the cricket *Gryllus firmus* (Insecta).

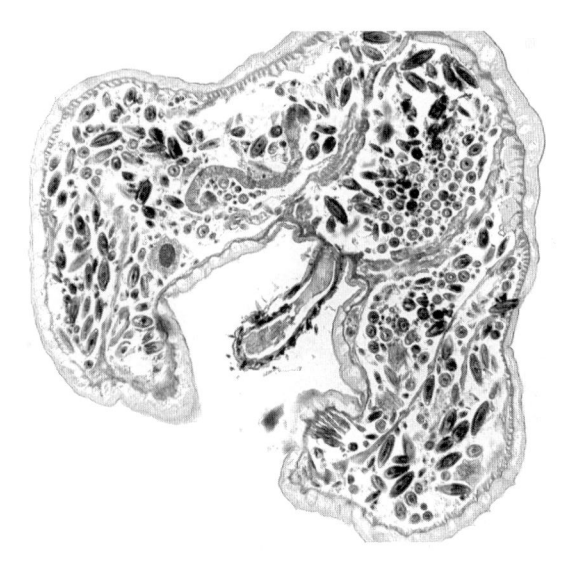

Fig. 8.10. In Acanthocephala, a large primary body cavity is present, in which oocytes and embryos of different developmental stages are present. The shown cross section is from the palaeacanthocephalan *Corynesoma* sp., the proboscis (in centre) projects laterally from the main body in this genus.

Mayer et al. (2004) showed unequivocally, by using transmission electron microscopy, that such a mixocoel indeed is formed (Fig. 8.4.). The former coelomic wall becomes integrated into several tissues. The dorsal parts form the heart and the so-called pericardial septum. The distal-lateral wall forms most of the somatic musculature (appendage musculature, longitudinal and circular musculature), the proximolateral wall forms the midgut musculature and the gonads. The ventral portions, finally, form the segmental organs (Anderson 1973). The sacculus of these segmental organs constitutes the only coelomic cavity of adults. It is formed during development from part of the remaining coelomic epithelium (Fig. 8.4.; Mayer 2006b).

Syndermata (Eurotifera, Seisonidea, and Acanthocephala)

All syndermatans possess spacious primary body cavities. In rotifers, no free cells are present in the body cavity (Clément & Wurdak 1991). In acanthocephalans, the body cavity extends between the body wall musculature and the reproductive system. In female palaeacanthocephalans, the ligament sacs which contain the gametes break up during development and therefore the body cavity contains gametes (Fig. 8.10.; Miller & Dunagan 1985, Crompton 1999). In Archi- and Eoacanthocephala the ligament sacs remain intact and the body cavity does not contain gametes.

Nemertini

Nemerteans are unique among invertebrates in possessing an epithelially lined circulatory system which is by definition a coelom. They also have a second coelomic cavity, the rhynchocoel, which serves as the space into which the proboscis can be retracted (Turbeville 1991). Muscular pressure on the fluid in the rhynchocoel everts the proboscis. Both coeloms can be in close contact when special rhynchocoel vessels exist, and in *Carinoma tremaphoros*, podocytes have been observed in this contact region (Turbeville 1991). During development, the blood vessels originate from compact cellular strands that are surrounded by ECM (Turbeville 1986). The cells of the strand are, as far as observed by Turbeville (1986), not connected by adhaerens junctions and they do not contain myofilaments. A primary body cavity appears not to be present or to be restricted to very small spaces. Bartolomaeus (1993a) detected a small primary body cavity in freshly hatched *Lineus viridis* (misinterpreted in Bartolomaeus 1985 as ECM).

Kamptozoa
The organization of kamptozoans appears to be compact, but numerous small interstices of primary body cavity, the so-called lacunae, are present (Nielsen 2002a). This is a potential synapomorphy with molluscs (see below).

Mollusca
As in kamptozoans, there are numerous lacunae of primary body cavity in molluscs (von Salvini-Plawen & Bartolomaeus 1995). This lacunar system is canalized in the dorsal region to form a heart with adjacent vessels. This heart is surrounded by a small coelomic cavity, the pericard (Fig. 8.11.; see, e.g., Bartolomaeus 1996 for *Philinoglossa helgolandica*). Bartolomaeus (1993a) compared the structure of the pericard in *Lepidochiton cinereus* (Polyplacophora), *Philinoglossa helgolandica* (Gastropoda) and *Ensis directus* (Bivalvia). It is bordered by a peritoneum in *Philinoglossa*, but by a myoepithelium in *Lepidochiton* and *Ensis*. Podocytes are present in all species. The development of the pericard confirms older results: paired anlagen fuse and form a coelomic cavity by schizocoely (Moor 1983, von Salvini-Plawen & Bartolomaeus 1995).

Sipunculida
There are two coeloms in sipunculids, a spacious trunk coelom and an additional tentacle coelom (with the exception of *Golfingia minuta*, see Bartolomaeus 1993a). Large, burrowing species also have extensions of the trunk coelom passing through the muscle layer—the dermal canals (Ruppert & Rice 1995). Within coeloms of both taxa, several types of free cells, amoebocytes and erythrocytes occur (Valembois & Bioledieu 1980, Dybas 1981, Rice 1993). Among these are the multicellular ciliated urns, which are described for

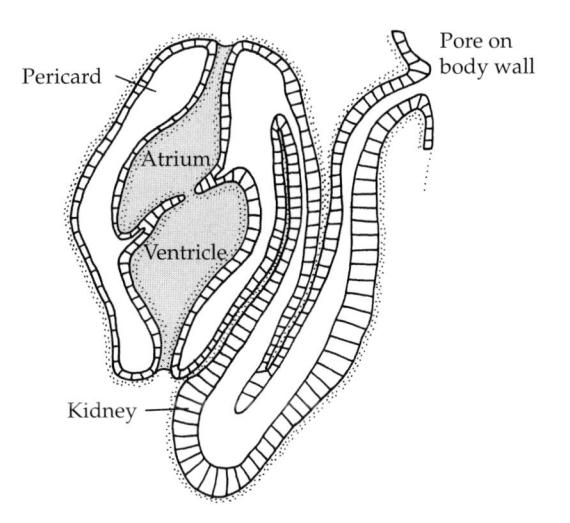

Fig. 8.11. Structure of the mollusc heart, as shown by a reconstruction from the gastropod *Philinoglossa helgolandica*. ECM is dotted and primary body cavity is shaded in grey. After Bartolomaeus (1996).

some sipunculids (free moving in the genera *Sipunculus* and *Phascolosoma*, but urns fixed to the coelothel may be further distributed). Ciliated urns appear to be phagocytotic in function and clear the coelom of foreign particles (Dybas 1976). The coelothel has different regional characteristics. It is bordered on the distal side by a peritoneum and on the proximal side by a myoepithelium. Podocytes have shown to be present (Bartolomaeus 1994).

Echiurida
Echiurans possess one spacious body cavity in the form of a coelom. It contains free moving cells of different types (Dybas 1981). According to Siewing (1985), there are also urns comparable to the ones in sipunculids. As in sipunculids, the coelothel is a peritoneum on the distal, but a myoepithelium on the proximal, side (Bartolomaeus 1994).

Annelida
In annelids, coelomic cavities are segmented and one pair of coeloms occurs (ideally) in each segment (Fig. 8.12.). In some polychaetes, coeloms from adjacent segments fuse to form larger

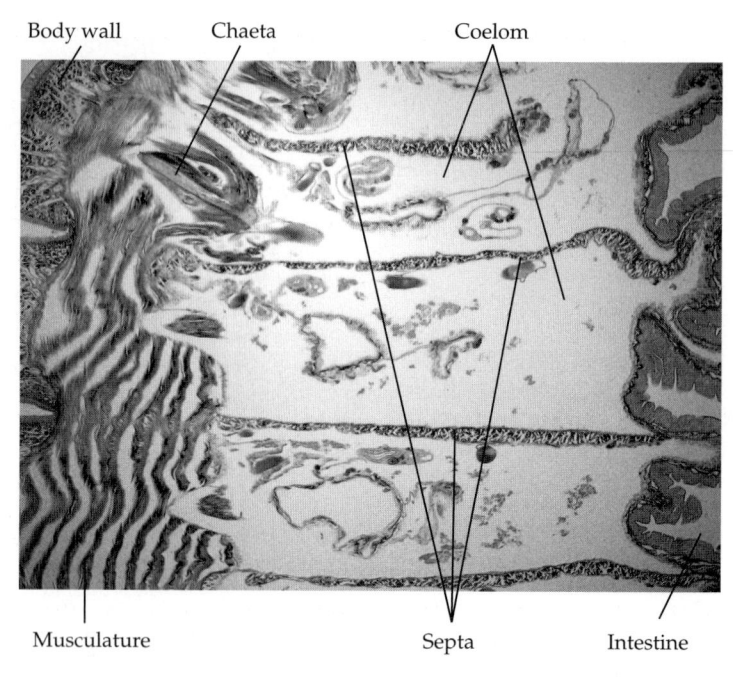

Fig. 8.12. Segmental coelomic sacs in *Lumbricus terrestris* (Clitellata, Annelida), shown in a transverse section from one body side. The content within the coeloms are parts from the metanephridial system.

coelomic compartments, for example, in the 'thorax' region of *Arenicola marina* or in the entire coelom of *Tomopteris helgolandica*. In small, meiobenthic species, the coelomic cavity can be obliterated: in these cases, the coelomic epithelium is present, but it surrounds only minute spaces or no lumen at all (Fransen 1980, Smith et al. 1986; see discussion above). This supports well the hypothesis that small annedids evolved by heterochrony, because the epithelial organization of the coelomic wall without a cavity represents an early stage in coelom development. In leeches, the coelom becomes modified in adults and forms a network of canals. The fine structure of the coelomic wall differs widely within annelids. Bartolomaeus (1993a, 1994) has compared a number of different polychaete species and concluded that a complete myoepithelium is probably the plesiomorphic condition (see also Fransen 1988). Within annelids, a peritoneum evolves in several taxa. The proximal wall of the coelom is always a myoepithelium, but the distal wall can be a myoepithelium (as, for example, in *Eulalia viridis*); a peritoneum with adjacent musculature, separated by ECM (as, e.g., in *Tomopteris helgolandica*); or a mixture of muscle and non-muscular, peritoneal cells (as, e.g., in *Diplocirrus glaucus*) (Bartolomaeus 1994). Podocytes are present in several species with a peritoneum, and myoepithelia with fenestrated regions have been shown in several species (Bartolomaeus 1993a, 1994). The development of coeloms starts from a compact strand of mesodermal tissue, which is as a whole surrounded by ECM. By the formation of apical adhaerens junctions the strand receives an epithelial organization which unfolds when fluid influx pushes the cells apart (Bartolomaeus 1993a).

Phoronida, Bryozoa, and Brachiopoda
The 'tentaculate' taxa all possess coelomic cavities. Originally it was assumed that a tripartite body organization, associated with three pairs of coelomic sacs, was present (see, e.g., Zimmer 1978 for phoronids), but this could not be confirmed in recent investigations. In the phoronid *Phoronis*

muelleri, Bartolomaeus (2001) showed that larvae (the actinotrocha) possess only one coelomic cavity while a second, the tentacle coelom, originates during metamorphosis. Adults only have these two coelomic cavities (see Bartolomaeus 2001 for a discussion of contradictory observations, Gruhl et al. 2005). In the preoral hood of the larva, the ECM is voluminous and has a gel-like consistence (Bartolomaeus 2001). Together with the musculature, this structure was probably misinterpreted as a coelomic cavity. The coelothel appears to be a myoepithelium (Pardos et al. 1991), and podocytes are present in the coelothel of phoronids (Storch & Herrmann 1978). A primary body cavity is present as the circulatory system. In brachiopods, two coelomic cavities, the trunk and the tentacle coelom, are present, but a third coelomic cavity is absent (Lüter 2000a). These coeloms originate from a single anlage composed of cells detached from the endoderm; this may represent a special case of enterocoely (Lüter 2000a). Nielsen (1991) described a different development of coelomic cavities in *Neocrania anomala*, where he observed that four pairs of coelomic sacs are formed. A primary body cavity is present in brachiopods as the hemal system (Nielsen 2001), podocytes have not been described. In bryozoans, an externally tripartite body is only present in Phylactolaemata, while other taxa lack the epistome, a small process close to the mouth. There are two coelomic cavties, the trunk and the tentacle coelom, which are both incompletely separated from each other (Mukai et al. 1997). In phylactolaemates, the epistome contains a cavity lined by an epithelium, but this opens into the tentacle coelom (Mukai et al. 1997, Nielsen 2001). Therefore, bryozoans do not possess three clearly separated coelomic cavities, but rather one cavity that is roughly partitioned into two or three compartments. Podocytes have not been described. A primary body cavity is present in the funicular system (see Chapter 10). Different types of coelomocytes are present (see Mukai et al. 1997 for summary and references).

Chaetognatha

In chaetognaths, three consecutive coelomic cavities are present in adults. During embryogenesis, first an unpaired head coelom is formed, followed by one paired trunk coelom which is later subdivided into an anterior and a posterior part. The coelothel is regionally differently organized as musculature, myoepithelial cells, or non-muscular peritoneal cells (Shinn 1997). A primary body cavity is present as a small hemal system between intestinal epithelium and coelothel, close to the hemal sinus there are podocytes in the coelothel (Shinn 1997).

Echinodermata and Hemichordata

In echinoderms and hemichordates, a tripartite organization is present, for which I will use the neutral terms proto-, meso- and metacoel, although in echinoderms further terms are used. All coelomic cavities develop by enterocoely. With the development of a pentaradial organization in echinoderms, the coelomic cavities become modified, in particular by a stronger development of the cavities on the left side of the body. A myoepithelium as the coelomic epithelium is widespread, it can be in the form of flat cells with or without cilia or as prismatic, ciliated cells (Welsch 1995). This latter type occurs especially where fluid movement by cilia is important, for example, in the stone canal of echinoderms (Balser & Ruppert 1993) or in the ciliated pits of crinoids (Grimmer & Holland 1979). Besides myoepithelia, a peritoneum can be present as well as intermediate forms. These are pseudostratified epithelia, that is, monolayers, in which all epithelial cells reach the apical (facing the coelom) as well as the basal (facing the ECM) side, but in which the main part of a cell is either basal or apical. Myoepithelial cells are generally basal, while cells free of myofibrils are apical (e.g. Cavey 2006). This can be interpreted as an intermediate step in the evolution of a peritoneum with surrounding musculature from a myoepithelium (Rieger & Lombardi 1987; see Fig. 5.3.). It is interesting to note that the coelomic cavities can develop in different ways in enteropneusts and echinoderms, that is, from three separate pairs of enterocoelic pouches, from one unpaired and one paired pouch, or from only one pair of pouches (Remane 1963). Podocytes are present in echinoderms (Welsch &

Rehkämper 1987, Balser & Ruppert 1993), enteropneusts (Balser & Ruppert 1990), and pterobranchs (Dilly et al. 1986, Mayer & Bartolomacus 2003). A primary body cavity is present in echinoderms and hemichordates in the form of the circulatory system.

Tunicata
In tunicates, the only coelomic cavity present is the pericard, which is bordered by a myoepithelium (Fig. 10.9.; Kalk 1970): a primary body cavity is present in the form of the circulatory system.

Table 8.1 Distribution of body cavities in Bilateria. Explanation and references in the text.

	Primary body cavity	**Secondary body cavity (= coelom)**
Xenoturbella	–	–
Acoelomorpha	–	–
Gastrotricha	–	–
Nematoda	(Presumed primary absence, present in large species)	–
Nematomorpha	Central cavity	–
Priapulida	Large central cavity	Small coelom around foregut, only in *Meiopriapulus fijiensis*
Kinorhyncha	Small cavity in anterior region	–
Loricifera	Small cavity in anterior region	–
Platyhelminthes	–	–
Gnathostomulida	–	–
Limnognathia	–	–
Eurotifera	Large central cavity	–
Seisonidea	Large central cavity	–
Acanthocephala	Large central cavity	–
Nemertini	Very small spaces can be present (in large species?)	Rhynchocoel and blood vessels
Kamptozoa	Lacunar system	–
Mollusca	Lacunar system	Pericard
Sipunculida	–	Tentacular and trunk coelom
Echiurida	Circulatory system	One large cavity
Annelida	Circulatory system	Segmental cavities (and derivations from this)
Tardigrada	–	–
Onychophora	Mixocoel (fusion of primary and secondary body cavity during development)	Segmental sacculi
Euarthropoda	Mixocoel	Sacculi in anterior segments
Chaetognatha	Small hemal system	Three cavities
Bryozoa	Funicular system	Two pairs of cavities
Phoronida	Circulatory system	Two pairs of cavities
Brachiopoda	Circulatory system	One subdivided cavity
Echinodermata	Circulatory system	Three pairs of cavities
Enteropneusta	Circulatory system	Three pairs of cavities
Pterobranchia	Circulatory system	Three pairs of cavities
Tunicata	Circulatory system	Pericard
Acrania	Circulatory system	Several cavities
Craniota	–	Pericard, visceral coelom, circulatory system, Bowman's capsule

Table 8.2 Comparison of characteristics of the coeloms in different taxa.

	Coelom	Coelomic epithelium	podocytes	free cells	reference
Priapulida	Small coelom around foregut in *Meiopriapulus fijiensis*	Non-muscular, with microvilli	Absent	Absent (free cells in primary body cavity, but not in coelom)	Storch et al. 1989, Storch 1991
Nemertini	Rhynchocoel and coelomate blood vessels	Non-muscular peritoneum, cilia present in *Carinoma tremaphoros* rhynchocoel	Present in *Carinoma tremaphoros*, absent in *Micrura leidyi* and *Zygeupolia rubens*	Present (in circulatory system)	Turbeville & Ruppert 1985, Turbeville 1991
Mollusca	Pericard	Myoepithelium or peritoneum	Present	Absent	Von Salvini-Plawen & Bartolomaeus 1995
Sipunculida	Trunk and tentacle coelom	Distal peritoneum, proximal myoepithelium	Present	Present	Rice 1993, Bartolomaeus 1994
Echiurida	One trunk coelom	Distal peritoneum, proximal myoepithelium	Present	Present	Bartolomaeus 1994
Annelida	Segmentally iterated pairs of coeloms (derivations: coeloms fused or reduced)	Assumed ancestral state: myoepithelium, within annelids both myoepithelium and/or peritoneum	Present	Present	Fransen 1988, Bartolomaeus 1994
Onychophora	Segmental pairs of coelomic sacs during development, segmental sacculi in adults	Non-muscular epithelium in coelomic sacs and sacculi	Present in sacculi	Absent	Storch et al. 1978, Storch & Ruhberg 1993, Bartolomaeus & Ruhberg 1999
Euarthropoda	Segmental pairs of coelomic sacs during development, segmental sacculi in few segments of adults	Non-muscular epithelium in coelomic sacs and sacculi	Present	Absent	Anderson 1973
Chaetognatha	Head coelom and 2 consecutive, paired trunk coeloms	Myoepithelium, musculature and non-muscular peritoneal cells	Present	Absent	Shinn 1997
Bryozoa	1 Paired coelom with incomplete tripartition	Mixture of myoepithelium and peritoneum in tentacles	Absent	Present	Mukai et al. 1997
Brachiopoda	1 Pair of coelomic anlage, 2 coelomic cavities in adults	Myoepithelium in tentacles, peritoneum in trunk coelom	Absent	Present	Reed & Cloney 1977, James et al. 1992, Üter 2000, Nielsen 2001
Phoronida	2 Pairs of coelomic cavities	Myoepithelia with rudimentary cilia in tentacle coelom and cilia in trunk coelom	Present	Few coelomocytes	Storch & Herrmann 1978, Herrmann 1997, Bartolomaeus 2001

Table 8.2 (contd)

	Coelom	Coelomic epithelium	Podocytes	Free cells	Reference
Echinodermata	3 Pairs of coelomic sacs	Both myoepithelium and peritoneum can be present	Present	Present	Rieger & Lombardi 1987, Balser & Ruppert 1993, Cavey 2006
Pterobranchia	3 Pairs of coelomic sacs	Myoepithelium	Present	Present	Dilly et al. 1986, Benito & Pardos 1997, Mayer & Bartolomaeus 2003
Enteropneusta	3 Pairs of coelomic sacs	Myoepithelium	Present	Present	Balser & Ruppert 1990, Benito & Pardos 1997
Tunicata	Pericard	Myoepithelium	Absent	Absent	Kalk 1970
Acrania	Several coelomic cavities	Regionally different: myoepithelium or peritoneum	Present	Present	Ruppert 1997
Craniota	Pericard, somatic coelom, blood vessels, Bowman's capsule	Myoepithelium or peritoneum	Present (in Bowman's capsule)	Present	Welsch 1995

Acrania

In acranians, segmental coelomic cavities develop through enterocoely. During development, they produce musculature and several small coelomic cavities (such as, for example, the fin box coeloms, gill bar coelom, velar coelom, and others, see Ruppert 1997). The myoepithelium of the dorsal part develops into the segmentally arranged musculature and narrows down the coleomic lumen almost completely (Ruppert 1997). In other parts, the coelothel can be organized either as a myoepithelium or as a non-muscular peritoneum (Ruppert 1997). Podocytes are present (Brandenburg & Kümmel 1961). Primary body cavity exists in the form of the circulatory system.

Craniota

In craniotes, the process of coelomogenesis is comparable to acranians, but during development two main coelomic cavities are formed: the pericardium around the heart and the somatic coelom surrounding most internal organs (Starck 1982). Additional coelomic cavities are the endothelially lined circulatory system and the bowman's capsule in the vertebrate kidney, which is richly equipped with podocytes (Junqueira et al. 2005).

Conclusions

The presence of a compact organization, primary body cavity, and coelom are distributed in a fairly scattered manner over the phylogenetic tree (Fig. 8.13.), making it very difficult to reconstruct the evolution of body cavities. In anticipation of the discussion of different hypotheses below, I will try to answer some central questions.

1. *Is the acoelomate condition ancestral or derived?*

There are two clear examples where the acoelomate condition is derived from a coelomate condition: in meiobenthic annelids and in the myotomes of acranians. There is, however, no reason to generalize this conclusion for all acoelomate taxa. When a coelothel is formed, but later occluded by reduction of the lumen, ECM remains on the basal side of the cells. In contrast, in acoelomate animals mesodermal cells are surrounded on all sides by ECM. There is no indication that such an organization can be derived from a coelomate condition. Most taxa that are acoelomate have a more or less basal position in the tree (Fig. 8.14.), making it likely that this is the plesiomorphic state.

2. *How often did primary body cavities evolve?*

Primary body cavities are not formed by peculiar cells or tissues, they can occur anywhere among mesodermal tissue by comparatively simple mechanisms. Therefore, their scattered distribution is best explained by a multiple origin. Taxa such as 'Pseudocoelomata' do not exist and even within the Cycloneuralia (which was formerly part of 'pseudocoelomates'), a primary body cavity probably evolved three times convergently: within nematodes, in nematomorphs, and in Scalidophora (Priapulida, Kinorhyncha, Loricifera).

3. *Are coeloms homologous structures?*

In contrast to primary body cavities, coeloms are formed by a certain mesodermal tissue, an epithelium that in many cases includes a characteristic substructure—the podocytes. We can approach the question about the homology of coeloms from two sides: from their phylogenetic distribution and from their structural comparison.

Coeloms are scattered across the phylogenetic tree (Fig. 8.14.). A comparison of the hypothesis that coeloms are ancestral and were reduced several times, with the hypothesis that coeloms evolved convergently several times, shows that the multiple evolution appears slightly more parsimonious than the multiple reduction (Fig. 8.13.). Taking into account that there are uncertain regions in the tree and that some taxa, such as chaetognaths, were excluded, this comparison alone is not convincing. What can be said is that, among protostomes, basal taxa appear to lack coeloms while coeloms are present in more or less derived taxa.

The question remains whether coeloms are structurally similar and are organs of a convincing complexity. I have some problems in recognizing such similarities. First, the positions and

164 THE EVOLUTION OF ORGAN SYSTEMS

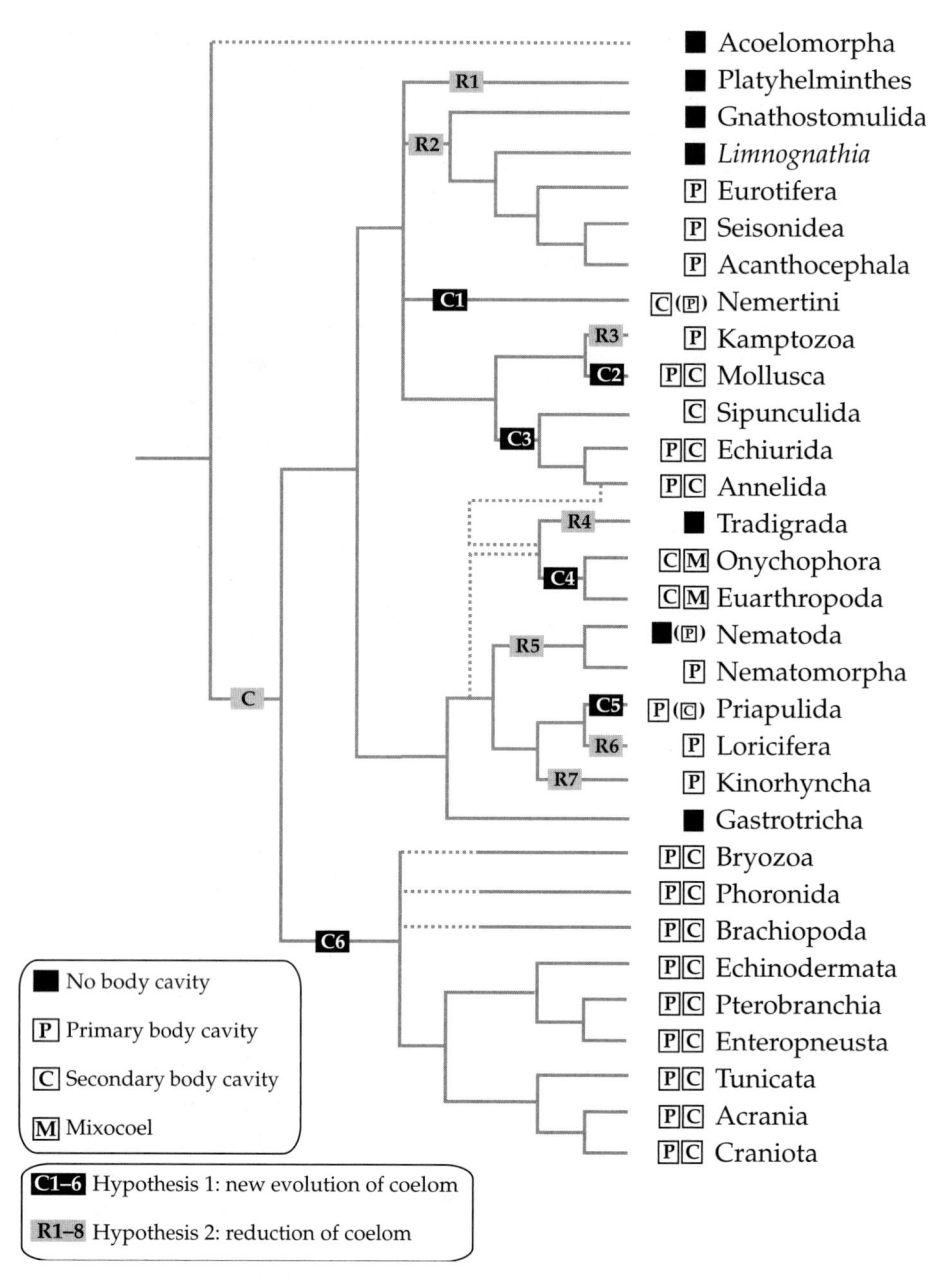

Fig. 8.13. Comparison of two different hypotheses of coelom evolution: convergent evolution of coeloms or ancestral condition of a coelom with subsequent reductions. See text for explanation.

functional contexts of coeloms are different among taxa, either as a hydroskeleton, as a pericard, as part of the excretory system, or something else. The presence of two coelomic cavities at the same time in nemerteans makes it difficult to decide whether only one of these is homologous to the single coelom in other organisms and, if so, which of the two. There are differences in the development of coeloms, they develop either by enterocoely or by schizocoely.

Even among schizocoely, Bartolomaeus (1993b) notes differences between nemerteans and molluscs on the one hand and annelids on the other hand: in nemerteans and molluscs, the coelom develops in a strand of non-muscular cells which is separated from surrounding musculature by ECM; in annelids it develops among differentiated myocytes. Finally, coeloms are not very complex structures. Their lining varies a lot, but appears to be derived in most cases from a

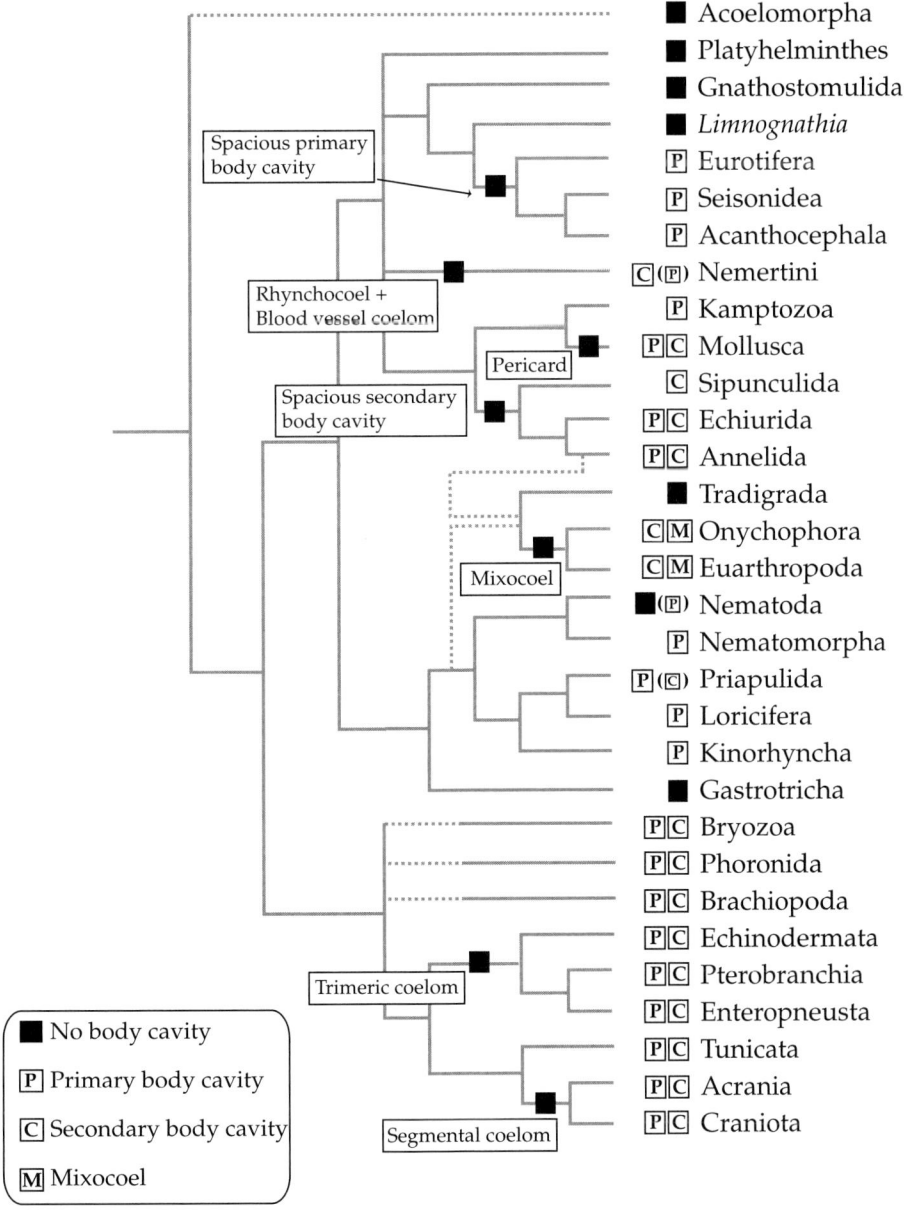

Fig. 8.14 Distribution of body cavities among bilaterian taxa. Logos in parenthesis mean: tiny primary body cavity present in at least some species of Nemertini, primary body cavity derived within Nematoda, small coelom in addition to primary body cavity only in *Meiopriapulus fijiensis* (Priapulida).

myoepithelium. Myoepithelial cells in general are plesiomorphic structures. Podocytes and metanephridia are not convincingly complex, they are quite simple structures: podocytes are interdigitating coelothel cells, a feature common in other epithelia, i.e. in the epidermis. Additionally, there is diversity in podocyte structure. While in many cases podocytes are flat, interdigitating cells, there are examples (e.g. in onychophorans, see Storch & Ruhberg 1993), where the cell soma is below the epithelium and the processes form the interdigitating structure. Such structural differences do not per se argue for an independant origin, but may supply hypotheses of an independant origin of coeloms based on other data. Metanephridia are the transition regions to ducts connecting coeloms to the outside, and in these regions the ciliation is higher compared to neighbouring regions. Moreover, a convergent evolution appears to be probable (see Chapter 9). Finally, there is a strong functional constraint on the presence of podocytes and metanephridia. Therefore, I regard it unlikely that coeloms of all bilaterian animals are comparable and evolved very early (see also Kristensen 1995, p. 41 and Valentine 2004, p. 498 for similar conclusions).

Considering all these questions, few convincing characters concerning the evolution of body cavities remain to be named (Fig. 8.14.). It appears that among Gnathifera, the spacious primary body cavity evolved in the syndermatan ancestor from an acoelomate condition. The specialization into a rhynchocoel and an epithelial circulatory system are autapomorphies of nemerteans. The pericard appears to be an autapomorphy of molluscs and a spacious coelom as a hydroskeleton evolved in a common ancestor of sipunculids, echiurids, and annelids. The mixocoel is characteristic for Onychophora + Euarthropoda. A segmental coelom appears to have evolved at least two times, in Annelida and in Myomerata. A sister-group relationship between Echinodermata and Hemichordata makes it likely that the trimeric coelom organization evolved in their common ancestor. The two coelomic cavities in phoronids and brachiopods could be homologized with the meso- and metacoel of echinoderms and hemichordates and support, in addition with a (kind of) enterocoely, a closer relationship to deuterostomes. This is more questionable for bryozoans, which seem to have only one coelom for which the development is largely uncertain.

The enterocoel hypothesis and the ancestry of the coelom

Several authors (summary in Remane 1963) derive the coelom directly from the pockets of the gastrovascular cavity in cnidarians (see Fig. 3.5.). It is hypothesized that these pockets become detached from the gut and therefore form coelomic cavities by enterocoely. The number of hypothetical coelomic compartments varies from two (Heider 1914), to four (Remane 1963), six (Jägersten 1955, 1959), and numerous (Sedgwick 1884). The assumption of six pockets corresponds to the number of coelomic cavities in echinoderms, hemicordates and the originally assumed number in the tentaculate taxa, and in the (then) incompletely known pogonophorans. This led Ulrich (1950) to name this organization archicoelomate, and assume it as an ancestral bilaterian character (see also Ulrich 1973, Siewing 1980).

The archicoel-hypothesis has been weakened by findings that the tentaculate taxa do not, on closer examination, have a trimeric organization of body cavities. Pogonophorans were only regarded trimeric as long as their annelid-like, segmented posterior end was not discovered. Therefore, the trimeric coelom is realized only in echinoderms and hemichordates and may therefore even be an autapomorphy of their common ancestor.

But even when a trimeric organization of body cavities is unlikely, several authors regard the coeloms of different taxa as homologous, and therefore a taxon Coelomata as monophyletic. This is exemplified by questions such as *Are platyhelminthes coelomates without a coelom?* (title of the paper by Balavoine 1998) or to the claimed decision between Ecdysozoa and Coelomata as alternative concepts by investigating the genome of nematodes (*Caenorhabditis elegans*), arthropods (*Drosophila*), and vertebrates (e.g. Philip et al. 2005). Apart from the question of how parsimonious it is to assume a unique evolution of a coelom, and whether the coelom is a structure of such complexity that its (multiple) convergent evolution appears unlikely (see above), any hypothesis on the ancestry of the coelom must explain how primary body cavities and the acoelomate condition are derived from the coleomate condition. Because an acoelomate condition often precedes the coelom formation during development, the acoelomate condition could be reached by heterochrony. However, there are some difficulties with this interpretation. Rieger (1986) has favoured this interpretation, based on the observation that in early mesodermal stripes of the polychaete *Polygordius* sp. the organization of cells can be epidermal as well as mesenchymal. His distinction between these two forms is made by differences in the cell polarity and in the positon of cell–cell contacts. It does not include the positon of ECM. Mesodermal stripes are surrounded as a whole by ECM, whereas in an acoelomate animal every cell is surrounded by ECM. The derivation of an acoelomate condition from a mesodermal stripe must therefore require that cells within the stripe secrete ECM, but there is no indication for this.

Considering primary body cavities, an origin by heterochrony would only be possible when a primary body cavity was the persisting blastocoel. As this is not the case for all primary body cavities, the derivation of primary body cavities from coelomate animals by heterochrony is unlikely.

There is one interesting finding that appears to support the ancestry of coeloms. Chen et al. (2004) have described a tiny animal (< 180μm), *Vernanimalcula guizhouena*, from the precambrian Doushantuo fauna (about 580–600 Mio a). If correctly interpreted (see Bengtson & Budd 2004 for doubts), single cells and epithelia are visible in this animal and it appears as if paired coelomic sacs are present on both sides of the intestine. This animal appears to be the earliest bilaterian found to date and if the structure really is a coelom, it appears likely that coeloms are quite ancestral in bilaterians. It is, however, remarkable that *Vernanimalcula* is quite unique in its combination of a small body size and comparably large coeloms.

Acoelomate – pseudocoelomate – coelomate: a phylogenetic succession?

It was by the authority of Libby Hyman that a tripartition of animals into acoelomate, pseudocoelomate, and coelomate animals became somewhat like a phylogenetic concept. Hyman adopted this tripartition from Schimkevitch (see Hyman 1940, p. 35) and explicitly regarded them as 'grades of structure' (1940, p. 35) which 'do not . . . entirely correspond to taxonomic relationships' (1951, p.23). Despite such clear statements, several authors have adopted this tripartition as an evolutionary sequence and translated it into a tree, implying an evolution from the absence of a coelom, to a pseudocoel and a coelom (e.g. Ruppert & Barnes 1994, fig. 20.3; and Pechenik 2005, fig. 2.10). Even if not strictly meant as such, Hyman's concept implied the monophyly of Pseudocoelomata and of Coelomata. Several authors have doubted the monophyly of Pseudocoelomata (Ruppert 1991a, Kristensen 1995) and according to recent hypotheses they include at least two monophyla, Cycloneuralia and Gnathifera, with different positions in the tree (see Chapter 2). The monophyly of coelomates has been discussed above and also seems to be uncertain. Furthermore, as has also been explained above, coeloms are never derived from primary body cavities, while both primary and secondary body cavities can be derived from the acoelomate condition. Therefore, the sequential evolution of body cavities appears to be unlikely.

The gonocoel hypothesis: a derivation of the coelom from gonads?

The gonocoel theory homologizes gonads and coelomic cavities and derives coeloms from gonads. This hypothesis was first formulated by Hatschek (1878) and especially elaborated by

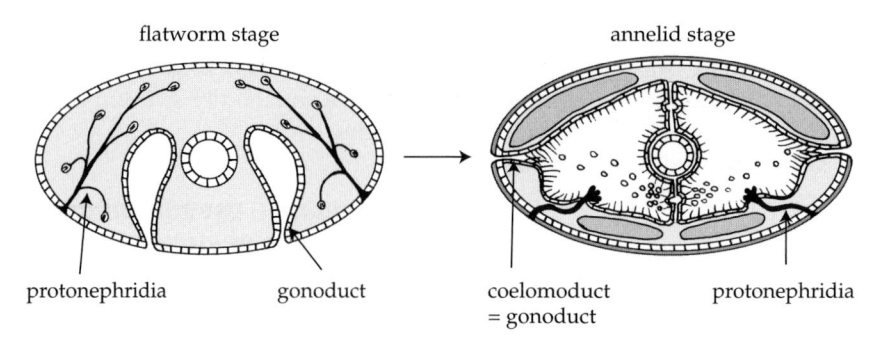

Fig. 8.15. The gonocoel theory: Transition from gonads in flatworms to coeloms in annelids. Figures after Goodrich (1845).

Goodrich (1895, summary in 1945). It rests primarily on a comparison of Platyhelminthes and Annelida and assumes that, because the coelom epithelium gives rise to the germ cells in annelids, it can be derived from a compact gonad by an extension in volume due to an accelerated growth of the epithelium (Fig. 8.15.). It is assumed the segmental organization of annelids is already preformed by the repetitive gonads in some flatworms and nemerteans. In flatworms, the excretory system (protonephridia) and the gonads have separate openings, and this is also assumed to be the ancestral condition in annelids. Within annelids, both excretory and gonocoel duct fuse. Goodrich (1945) distinguishes several such fusion products, the nephromixia.

There are two problems with the gonocoel theory. First, it neglects the fact that in several taxa gonads and coeloms co-exist, but are not fused (hemichordates, echinoderms, basal condition in molluscs). This could be explained when only the *annelid* coelom is regarded as derived from a gonad, but not for coeloms in other taxa. Second, the interpretation of ducts in annelids as 'protonephromixia', 'metanephromixia', and 'mixonephria' is necessary (for Goodrich), because the gonocoel theory is used axiomatically. Bartolomaeus (1999) critically discusses the assumptions of Goodrich and comes to a different interpretation for the evolution of ducts and excretory organs in annelids (see Chapter 9).

In conclusion, none of the previous hypotheses appears to be well supported, and it appears questionable whether the evolution of body cavities can be explained by far-reaching hypotheses including few evolutionary steps, or whether body cavities are more flexible and dynamic structures that evolved several times convergently.

CHAPTER 9

Excretory systems

Excretory organs eliminate metabolic waste products from the body. Because such products are often dissolved in fluids, excretory organs can have an important impact on the water budget and therefore on osmoregulation of the animal.

As biological membranes are semipermeable systems, water can leave or enter the body when the concentrations of soluble substances are different between the interior (the body) and exterior (the surrounding medium). Water will always tend to bring concentrations on both sides of a semipermeable membrane to the same level. This process is called osmosis. In marine animals, the osmotic differences between interior and exterior are generally minimal. It appears logical that this is because living organisms evolved in the marine environment. When organisms enter freshwater or land, the osmotic differences become more dramatic. In freshwater, the concentration of soluble substances within the body is higher than in the surrounding medium, so water will constantly enter the body and has to be transported out again. On land, strong outflow of water with dissolved excretes will lead to dessication problems and strategies to save water have to be developed.

Some of the metabolic endproducts, such as CO_2 and water, can leave the body relatively easily. Molecules containing nitrogen pose more problems. Such substances build up, for example, when amino acids are digested. Although amino acids are essential, animals can store only a limited amount and have to digest them when larger amounts are consumed. The first step is the splitting off of the NH_3^+-group. Therefore, ammonia is a primary waste product. Ammonia is toxic to cells already in moderate concentrations and its accumulation must be avoided. Because ammonia is very soluble, it diffuses easily in aqueous media. If an efficient elimination of ammonia by diffusion is not possible, it has to be transferred into other nitrogeneous substances with lower toxicity. These can be a variety of substances, among which urea, uric acid and guanin are the most abundant.

Both problems, osmoregulation and elimination of nitrogeneous substances, are combined in the same structures, the excretory organs. Excretory organs are found only among bilaterian animals. Among the few exceptions in which excretory organs are absent are acoelomorphs and *Xenoturbella*. Both of these taxa have been proposed to have a basal position within Bilateria. This would mean that excretory organs evolved in the ancestor of Bilateria, exclusive Acoelomorpha, and/or *Xenoturbella* and Jondelius et al. (2002) have proposed the name Nephrozoa for such a taxon.

It appears that the loose organization of sponges – and the epithelial organization of *Trichoplax*, cnidarians, and ctenophores – allows osmoregulation and elimination of nitrogeneous waste products in the form of ammonia on a cellular level, that is, by diffusion from the cells where such waste products build up directly into the sea water. No excretory organs or even simpler structures suggesting an excretory function are found in these taxa. The lack of excretory organs has not prevented sponges and cnidarians entering freshwater.

With the exception of *Xenoturbella* and acoelomorphs, the more compact organization of triploblastic animals creates problems because, for example, ammonia produced in the centre of the animal may concentrate to toxic levels on its way to the exterior. We find different kinds of excretory organs among bilaterian animals, but before we take a look at their distribution, we have to discuss the principles of excretory organs.

Two basic principles occur in excretory organs: 'active transport' and 'ultrafiltration' (see, e.g., Schmidt-Nielsen 1997). They are either combined, or active transport is the only method. Ultrafiltration always takes place from one compartment containing fluids into another (Fig. 9.1.). This means that body cavities play an essential role for excretory organs, but filtration can also be performed in compact bodies from the interstitial fluid within the ECM into a canal system. The filtration structure is always the extracellular matrix (ECM). Its fibrous construction serves as an ideal filter, holding back large particles such as large proteins, but allowing the passage of water, ions, urea, and other small molecules. Excretory organs working by ultrafiltration are called 'nephridia'. Active transport is the secretion or the resorption of substances by the cell (Fig. 9.1.). From an energy perspective, active transport is always more costly compared to ultrafiltration, which is driven by different pressures in the two compartments. Ultrafiltration is selective in the sense of separating large from small molecules, but it is unselective concerning the small molecules that can pass through the filter. Therefore, it has to be combined with active transport, by which at least water, salt, and other substances important for the body are reabsorbed. The filtrated fluid is called 'primary urine', after it is modified it is called 'secondary urine'. Even with the combination of ultrafiltration and active reabsorption, this kind of excretion appears to be energetically similar or superior to active transport alone, because excretory organs based on ultrafiltration are realized in diverse types and are abundantly distributed. Whether they all go back to one common ancestral design will be discussed later.

Among excretory organs working with ultrafiltration, we can distinguish two basic types,

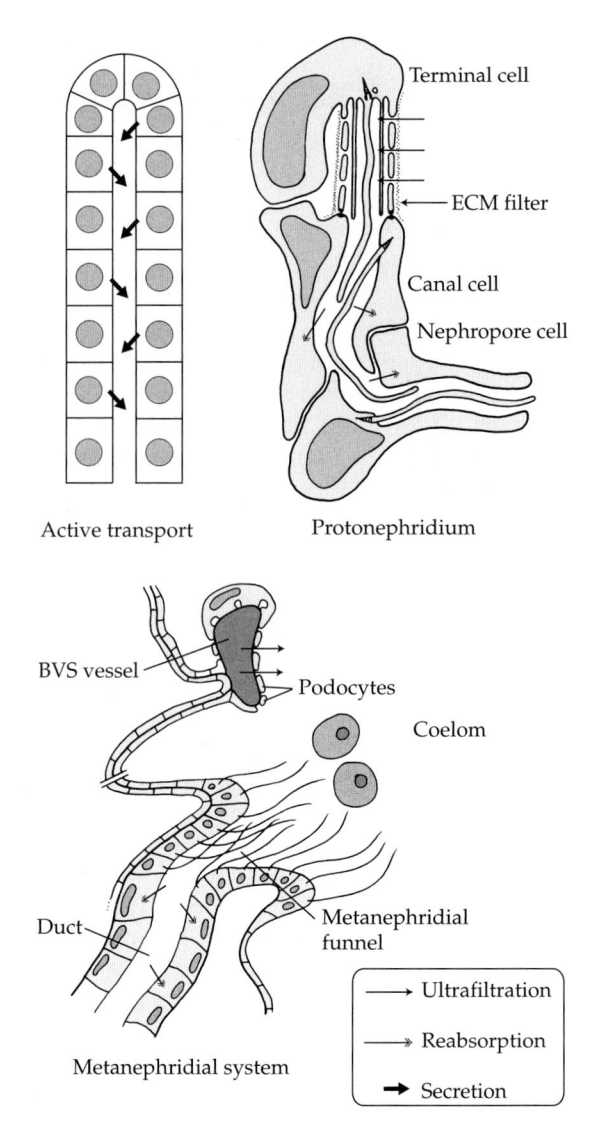

Fig. 9.1. Schematic representation of main principles in excretion. In active transport excretes are secreted by cells into a lumen of an epithelially lined tube. In a protonephridium, filtration occurs through the ECM covering slits or pores in the terminal cell (arrows). The filtrate flows through a canal system, in which reabsorption of water and ions is possible (double-headed arrows). In metanephridial systems, filtration occurs from a blood-vascular system vessel through podocytes into the coelom. Modification takes place in the canal following the metanephridial funnel. Figures partly after Bartolomaeus and Ax (1992).

'protonephridia' and 'metanephridial systems' (Fig. 9.1.). Protonephridia are typical for acoelomate animals, but can also be present in animals with a primary body cavity or even a coelom. They consist of two elements, the so-called terminal cell (sometimes called solenocyte) and a canal leading to a porus on the body surface (Fig. 9.1.; Wilson & Webster 1974, Brandenburg 1975, Bartolomaeus & Ax 1992). The terminal cell is formed by (usually) one single cell, part of which is cup-shaped and surrounds a central lumen (Figs. 9.2.). This central lumen is continued by the canal through the attaching first canal cell. The terminal cell is the site of ultrafiltration. It is completely surrounded by ECM and contains in the cup-shaped part pores or slits which are covered by ECM (Fig. 9.2.). In these regions, ECM is the only barrier between the intercellular fluid outside of the terminal cell and the canal system of the protonephridium, and it is therefore the site of ultrafiltration. There can be different structures to stabilize the terminal cell and prevent its collapse, especially in the cytoplasmatic rods separating the slits. These can be ciliary roots, microtubules, or actin filaments (Ruppert & Smith 1988, Xylander & Bartolomaeus 1995). Besides pores and slits, the site of filtration can also be the ECM between microvilli from two adjacent cells, either between terminal cells or between terminal and canal cell. These structures are called a weir (Fig. 9.2.). Filtration is driven by the beating of one or more cilia of the terminal cell (Kümmel 1975). The cilia are often surrounded by microvilli, to distinguish them from microvilli that may form the filtration structure they are called 'inner microvilli' and those carrying the filtration structure 'peripheral microvilli'. The canal with its constituting canal

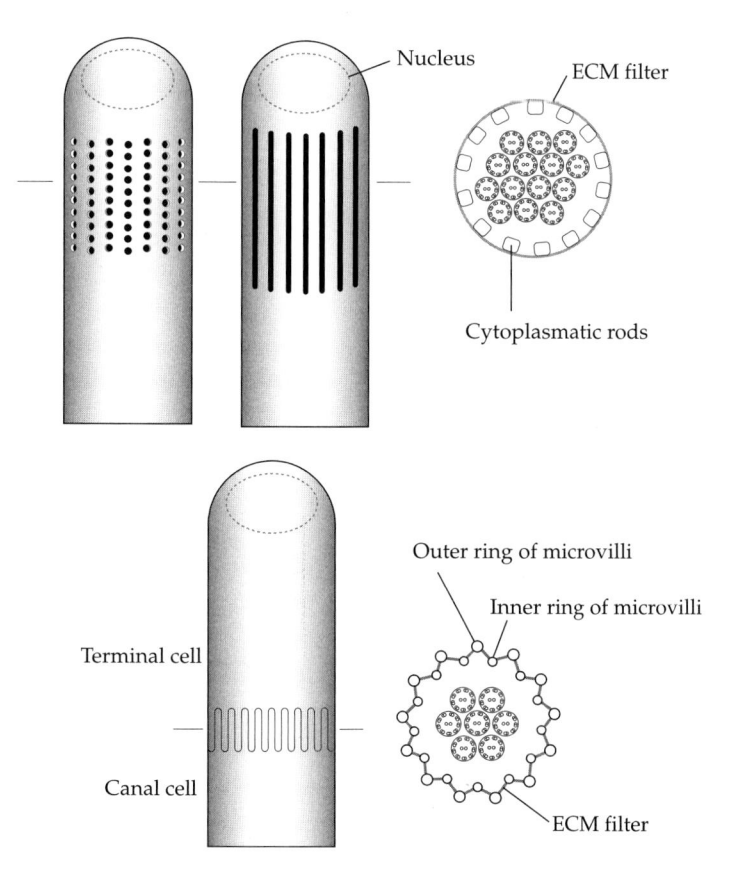

Fig. 9.2. Schematic representation of different designs in terminal cells. On top are cells with pores or slits; cross section at indicated level shows the filter composed of ECM covering the pores or slits. Below is a weir, where the filtration area is composed of microvilli from the terminal and the canal cell cross section is from the level indicated.

cells is the site of active transport by reabsorption and probably also by additional secretion.

Where protonephridia are flexible in filtering fluid from different compartments (ECM, primary body cavity, coelom) into a canal, metanephridial systems always filter from one fluid-filled cavity (in many cases the circulatory system) into the coelom (Fig. 9.1.). As in protonephridia, ECM is the only filtration structure. Filtration occurs into the coelom between specialized coelothel cells, the podocytes, which interdigitate strongly and thereby increase the volume of intercellular gaps, which are only covered by ECM and through which ultrafiltration can occur (Fig. 8.6., 9.3.). In those cases, where fluid to be filtered comes from an endothelialized system, that is, the blood vessels in vertebrates, these epithelia have pores, so that ECM remains the only barrier between blood vessel and coelom. From the coelom, excretes are led out by a canal system to pores on the body wall. In almost all cases, the transition between coelom and canal is a ciliated funnel. Such funnels may be very conspicuous, as in several annelids, but in many other taxa they are simpler. Only in euarthropods cilia are lacking in the funnel region, but it can be assumed that in this case ciliation was lost, because onychophorans have ciliated funnels. In the canal, the filtrate is modified by reabsorption of ions and water. This is another important difference to protonephridia. Whereas in protonephridia the site of filtration and modification are immediately adjacent, they are spatially separated in metanephridial systems. The metanephridial canal can also be used to release gametes stored in the coelom. There are some terminological problems with metanephridial systems. The term 'metanephridium' is usually restricted to the ciliated funnel (together with the canal) while the term 'metanephridial system' was introduced by Ruppert & Smith (1988) to include the site of ultrafiltration. Ruppert & Smith's (1988) definition includes muscular action as the driving force for ultrafiltration but does not mention ciliated funnels. I will summarize here as metanephridial systems all cases, in which filtration occurs into a coelom in the region of podocytes and in which outflow with modification of the filtrate is performed in a canal system regardless of whether a ciliated funnel is present or not.

The function of excretory organs using ultrafiltration has been shown by experiments measuring salt concentrations (Zerbst-Boroffka & Haupt 1975), accumulation of iron dextran (Smith & Ruppert 1988), or colloidal gold particles (Xylander & Bartolomaeus 1995).

With this information we can take a closer look at the distribution of excretory organs among bilaterian animals.

Excretory organs in Bilateria

Special structures with an excretory function are not present in diploblastic animals, in

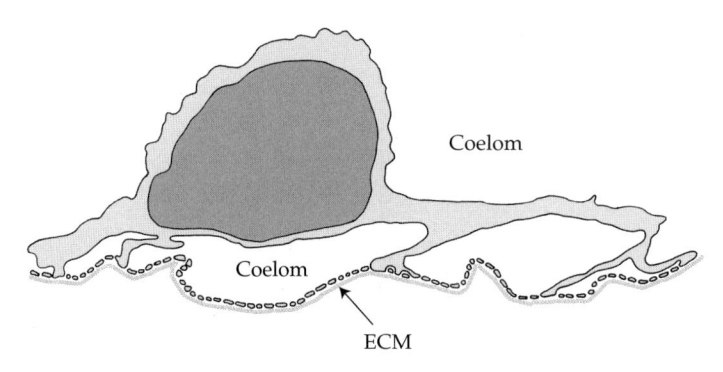

Fig. 9.3. Drawing of a podocyte from the onychophoran *Peripatopsis moseleyi*, showing numerous extensions of the podocyte soma which are covered by ECM. After Storch and Ruhberg (1993).

acoelomorphs (Acoela and Nemertodermatida), or in *Xenoturbella*. Possible exceptions are specialized cells with a probable excretory function, the dermonephridia. These have been found only in the acoel genus *Paratomella* (Ehlers 1992b). Almost all remaining bilaterian taxa have excretory organs in some form which I will review in brief in the following (for summary see Table 9.1.). From the variety of substructures, special attention will be given, in the case of protonephridia, to the terminal cell, especially the number of cilia and microvilli and the shape of the filtration structure. In metanephridial systems, special attention will be given to the location of podocytes and the structure of the transition area between coelom and canal.

In the case of protonephridia, some terms have to be explained in advance. A 'terminal cell' is an individual cell that performs filtration and attaches to the canal system. When more than one terminal cell together comprise the filtration structure, this is called a 'terminal organ'. In the

Table 9.1 Characteristics of terminal cells in protonephridia of bilaterian animals. Table includes number of cilia, number of microvilli directed into the lumen of the terminal cell, and the shape of the filtration site.

	Cilia	Microvilli	Filtration site	Selected references
Gastrotricha Macrodasyida	1	8	Pores or slits	Teuchert 1973, Neuhaus 1987, Ruppert 1991
Gastrotricha Chaetonotida, Paucitubulatina	2	16	Pores or slits	Brandenburg 1962, Kieneke et al. 2007
Loricifera	1	9	Microvilli	Kristensen 1991
Kinorhyncha	2	Many	Microvilli	Neuhaus 1988, Kristensen & Hay-Schmidt 1989, Kristensen & Higgins 1991
Priapulida, larva	1	Many	Between attaching terminal cells	Lemburg 1999
Priapulida, adult	Many	Many		Kümmel 1964, Storch & Alberti 1986, Lemburg 1999
Platyhelminthes Catenulida	2	Varying number	Slits	Ehlers 1985a, 1994c
Platyhelminthes Rhabditophora	4 or more	–	Slits or 2-cell weir	Ehlers 1985a
Gnathostomulida	1	8	Slits	Lammert 1985
Limnognathia	1	9–10	?, probably microvillar weir	Kristensen & Funch 2000
Eurotifera	2 to many	Few to about 20	Slits	Clément & Wurdak 1991, Ahlrichs 1995
Seisonidea	Many	–	Pores	Ahlrichs 1993b
Acanthocephala	Many	–	Probably pores leading into spongy system	Dunagan & Miller 1986a, b, 1991
Symbion	8–9	Many	?	Funch 1996
Nemertini	Many	Many	Meandering slits	Bartolomaeus 1985, 1988, Turbeville 1991
Kamptozoa	Many	Many	Meandering slits between adjacent terminal cells	Franke 1993
Mollusca, larva	Many	Many	Slits	Brandenburg 1966, Bartolomaeus 1989a
Echiura, dwarf male of *Bonellia*	Many	Absent	Porous system of terminal cell	Schuchert 1990
Annelida Polychaeta, larvae	1–many	Many	Slits	Smith & Ruppert 1988, Bartolomaeus & Ax 1992
Annelida Polychaeta, adults	1–many	Many	ECM between microvilli	Smith & Ruppert 1988
Annelida Myzostomida	6–9	10 per cilium	Slits	Pietsch & Westheide 1987
Phoronida, larva	1	7–9	Slits	Bartolomaeus 1989c

case of a syncytial organization, the proximal end of a protonephridium is called the 'terminal region'. When several terminal cells are close together, but still function individually, this is called a 'terminal complex'. When the filtration structures are longitudinal slits, the cytoplasmic regions between the slits are called 'cytoplasmatic rods'. When adjacent cells form microvilli which are interdigitating with microvilli from the other cell, such a structure is called a 'weir'.

Gastrotricha

Gastrotrichs have protonephridia. One pair has been found in all chaetonotid gastrotrichs investigated so far, while macrodasyids have from two to eleven serially arranged pairs (Teuchert 1967, 1973, Neuhaus 1987). In *Turbanella cornuta*, three or four terminal cells are clustered and connect to a common canal, in other species investigated, each protonephridium has only one terminal cell. Macrodasyidan terminal cells have one cilium which is surrounded by a collar of eight microvilli. The terminal cell contains pores (probably *Turbanella cornuta*, Teuchert 1973) or slits (*Dactylopodola baltica* and *Mesodasys laticaudatus*, Neuhaus 1987). In *D. baltica* and *M. laticaudatus* the whole protonephridium was shown to be composed of three cells, a terminal cell, a canal cell, and a nephridiopore cell (Neuhaus 1987). Only few chaetonotids (Paucitubulatina) have been investigated ultrastructurally, but it appears that a terminal cell with two immediately attaching rings of eight microvilli surrounding a cilium could be widespread in this taxon (Brandenburg 1962, Kieneke et al. 2007). Whether such a condition arose by a duplication of the ciliary/microvillous part of the terminal cell or by the fusion of two terminal cells is unknown.

Nematoda

In nematodes, nephridia are absent and excretion is assumed to be by diffusion over the body wall as well as by two structures: the renette cell (the ventral cell, excretory gland) which is widely distributed among nematodes, and the excretory canal system which is present only in secernentean nematodes (Chitwood & Chitwood 1950, Bird & Bird 1991, Gibbons 2002).

In secernenteans, both systems can be present in parallel. Both systems open by a pore to the external, the renette cell is usually one large cell and the excretory canal is also one large cell in the form of an 'H'. No filtration structures are present and excretion must be by active transport. Apart from excretion, further functions have been shown to be performed such as osmoregulation and secretion of diverse products (Bird & Bird 1991), therefore the term secretory-excretory system might be more appropriate. The main molecule excreted is ammonia and to a lesser content urea (Wright 1998).

Loricifera

Loriciferans possess protonephridia as excretory organs. The few available data indicate that one pair of protonephridia is present with a terminal cell bearing one cilium and nine (inner) microvilli surrounding this cilium (Kristensen 1991). The filtration structure seems to be formed by cytoplasmatic rods of the terminal cell (Kristensen 1991, Fig. 44). There is some indication that the excretory canal may join the reproductive system (forming an urogenital system) and, in *Nanaloricus mysticus*, may even lead into the rectum (forming a cloaca) (Kristensen 1991).

Kinorhyncha

In kinorhynchs and priapulids, protonephridia are peculiar because there are no singularly functioning terminal cells, but several terminal cells fuse and form terminal organs (Fig. 9.4.). Each cell in the terminal organ projects cilia and microvilli into a common lumen. In kinorhynchs, the filtration site is formed between cytoplasmatic rods or peripheral microvilli, probably originating from several terminal cells. In *Echinoderes aquilonius*, one terminal organ is composed of three terminal cells (Kristensen & Higgins 1991), in *Pycnophyes greenlandicus* of eleven cells (Kristensen & Hay-Schmidt 1989) and in *P. kielensis* of twenty-two cells (Neuhaus 1988). All terminal cells are biciliate, some microvilli are present at least in *P. kielensis* (Neuhaus 1988). The excretory canal opens with a pore on the body wall in segment eleven.

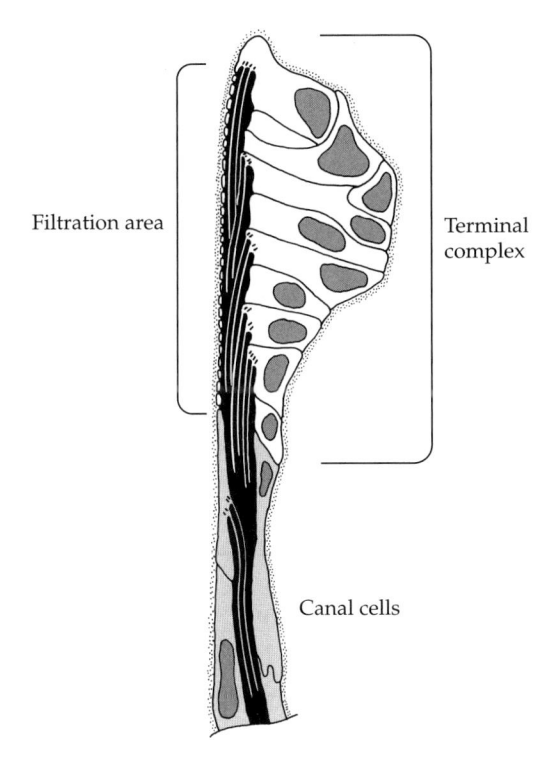

Fig. 9.4. Terminal organ in the protonephridial system of the kinorhynch *Pycnophyes kielensis*. Several terminal cells together form the filtration area. After Neuhaus (1988).

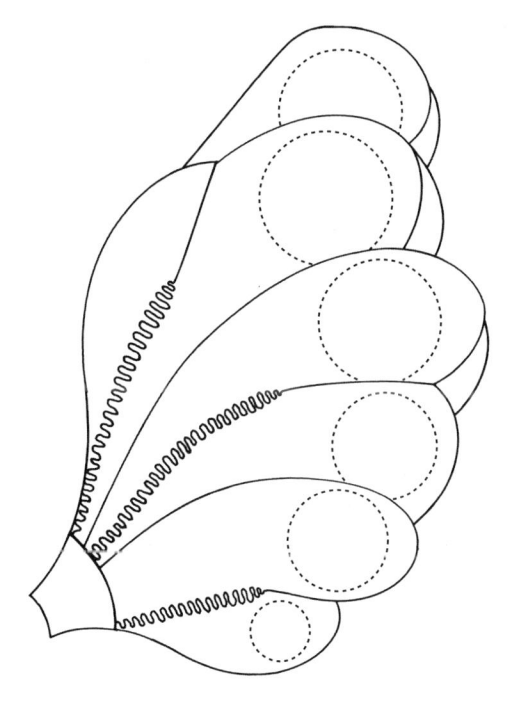

Fig. 9.5. Terminal organ in the protonephridial system of the priapulid *Halicryptus spinulosus*. The filtration area is formed in the interdigitating contact zone of adjoining terminal cells. Modified after Lemburg (1999).

Priapulida
Priapulids are exceptional in several respects. The excretory canal joins the gonoduct and therefore a urogenital system is present. The excretory canal branches multiple times and leads to the many terminal organs. This whole structure was in earlier publications called a 'solenocyte tree' (e.g. Nørrevang 1963). As in kinorhynchs, terminal organs are present, but they have a different structure. Each terminal organ is composed of two (*Tubiluchus philippinensis*, Alberti & Storch 1986) or several intensely connected terminal cells (Kümmel 1964, Storch et al. 1989, Lemburg 1999 for *Halicryptus spinulosus*, *Priapulus caudatus* and *Meiopriapulus fijiensis*). The filtration sites are formed along the attachment sites of neighbouring terminal cells in the form of undulating lines (Fig. 9.5.). In larvae, at least of *Halicryptus spinulosus* and *Priapulus caudatus*, each terminal cell is monociliated, but adults are multiciliated (Lemburg 1999). Numerous microvilli project into the lumen of the terminal cell, but not in a regular pattern. The terminal cells project into the primary body cavity.

Platyhelminthes
In flatworms, only protonephridia are present. They are variable in shape and contain important phylogenetic information (Ehlers 1985a, Rohde et al. 1995). Catenulids are special in possessing an unpaired, median protonephridium, in all other platyhelminthes there are paired protonephridia (varying from one to numerous pairs). The terminal cell of catenulids is biciliary, has either a few short or up to 21 longer microvilli, and the filtration structure are longitudinal or traverse slits (Ehlers 1985a,

1994c). Among rhabditophoran flatworms, there are no microvilli surrounding the cilia of the terminal cell. The cells are always multiciliate, with a minimum of four cilia per cell. The filtration structure is either formed as slits in the terminal cell, or microvilli from the terminal cell and the adjacent canal cell alternate to form a weir (Fig. 9.2.). In these cases, one can often distinguish an outer (from the canal cell) and an inner (from the terminal cell) ring of microvilli. The microvilli of both cells are connected to each other by adhaerens junctions. Such weirs are present in the parasitic taxa (Neodermata) and in some free-living forms, among macrostomids and in proseriates. With a rise in body size, there is usually a more complex protonephridial system with numerous terminal organs, a widely branched canal system, and numerous excretory pores (see summary in Ehlers 1985a).

Gnathostomulid
Protonephridia are present in gnathostomulids, their number ranges from one pair (*Rastrognathia macrostoma*, Kristensen & Nørrevang 1977) to five pairs (adults of *Gnathostomula paradoxa*, Lammert 1985). Ultrastructural observations are restricted to *G. paradoxa* and *Haplognathia rosea* (Graebner 1968, Lammert 1985). Both species correspond in having protonephridia composed of three cells, a monociliated terminal cell, a canal cell, and a nephropore cell. The cilium in the terminal cell is surrounded by eight microvilli, the terminal cell has slits as filtration sites.

Limnognathia
Kristensen and Funch (2000) report two pairs of protonephridia in *Limnognathia maerski*, each consisting of four terminal cells, two canal cells, and one nephridiopore cell. All cells appear to be monociliated. The terminal cell contains nine to ten stiff microvilli surrounding the cilium. The microvilli continue over the terminal cell into the canal. Between the microvilli, electron-dense extracellular material is present which might be involved in filtration. How fluids enter the lumen of the terminal cell is unknown.

Eurotifera
Eurotifers have one pair of protonephridial systems with a complex canal system and few or more (up to a hundred in certain *Asplanchna* species) terminal regions. The protonephridia project into the primary body cavity. The organization is syncytial and consists of terminal and canal syncytia, where, for example, the terminal syncytium is composed of at least three nuclei (Ahlrichs 1993a). The filtration structures in the species investigated to date are longitudinal slits, separated by cytoplasmic columns. There are always many cilia which are connected by septate junctions to form a flame bulb (Clément 1968, Ahlrichs 1995). The minimal number of cilia is two (in *Encentrum marinum*, Ahlrichs 1995), but most species have higher numbers, for example, about one hundred cilia in *Notommata copeus* (Clément 1968). Microvilli are present in the terminal syncytia, either as a few, irregularly distributed microvilli – as in *Encentrum marinum* (Ahlrichs 1995), or as a regular ring proximal of the cytoplasmic colums that form the filter – as in *Asplanchna brightwelli* (Warner 1969). Their function is probably to support the filtration structure, in species with a high number of cilia they can have a considerable diameter compared to the outer cytoplasmatic columns and the cilia.

Seisonidea
The excretory system of *Seison annulatus* has been described by Ahlrichs (1993b). There is one pair of protonephridia, each composed of three regions: a terminal syncytium, a canal syncytium, and a nephropore cell. The terminal syncytium is very long and there are two connections to the canal syncytium. The terminal syncytium contains eight terminal regions, each of which has pores as filtration sites (Fig. 9.2.), is multiciliary, and lacks microvilli.

Acanthocephala
Excretory organs are lacking in most acanthocephalans, they have been found as protonephridia only in species of the Oligacanthorhynchidae (Archiacanthocephala) and are probably restricted to this taxon. Archiacanthocephala are basal in several (Near et al. 1998, García-Varela

2000, Monks 2001), but not all (Herlyn et al. 2003), phylogenetic analyses of Acanthocephala. This may indicate that protonephridia were present in the acanthocephalan ancestor but were reduced in all taxa except for Oligacanthorhynchidae. There are two types of protonephridia. Species such as *Macracantorhynchus hirudinaeus* have a dendritic excretory system, in which the canal branches multiple times and leads to numerous multiciliate terminal cells (Dunagan & Miller 1986a). In the other, the capsular type, which is present for example in *Oligacantorhynchus taenioides* (Dunagan & Rashed 1988, Dunagan & Miller 1991), numerous terminal cells surround a bladder-like capsule. In both types, the excretory ducts lead into the gonoducts, forming a urogenital system. The terminal regions are syncytial and contain more than one nucleus. They project into the primary body cavity. The filtration sites are probably pores that lead into an extended canal system, the spongy layer (Dunagan & Miller 1986a), and from this into the lumen of the terminal region. The number of cilia per terminal syncytium is high, microvilli appear to be absent. In *Gigantorhynchus echinodiscus* there is a rudimentary excretory organ composed of one (terminal?) cell, but a duct appears to be absent (see Dunagan & Miller 1986b, 1991).

Cycliophora
Excretory organs have only been found in the pandora larva. Here, one pair of protonephridia is present, but the structure is not completely known. Funch (1996) has found one terminal and one canal cell, the terminal cell has 8–9 cilia and several microvilli. The location of the filtration structure was not observed.

Nemertini
Nemertini possess one or more pairs of protonephridia. In *Tubulanus annulatus*, there appears to be a close connection with the circulatory system, for the terminal part of the 'protonephridium' lies in close contact to the vessel wall which is fenestrated in these regions (Fig. 9.6.; Jespersen & Lützen 1987). Therefore, only ECM separates the lumen of the vessel

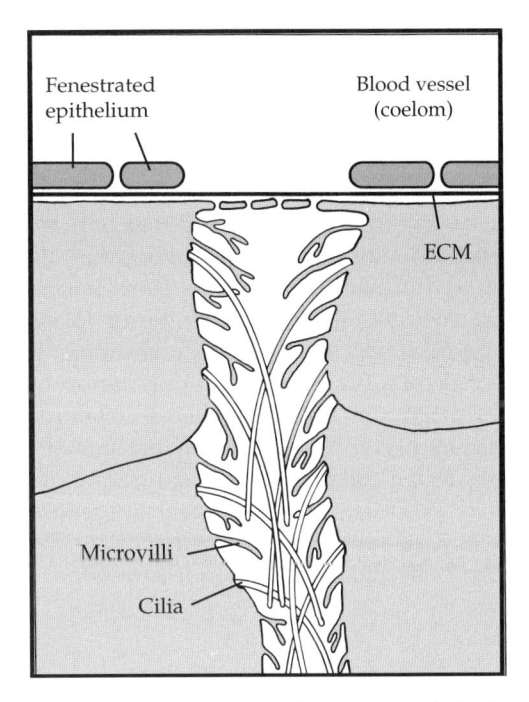

Fig. 9.6. Schematic representation of the excretory region in the nemertean *Tubulanus annulatus* between circulatory system and excretory canal. Modified after Jespersen and Lützen (1987).

and that of the terminal part from each other. Although termed a protonephridium by Jespersen & Lützen (1987), the nephridium of *Tubulanus* is unique and differs from protonephridia in other nemerteans. The terminal part of the duct is composed of several cells forming interdigitating extensions towards the blood vessel. Whether these have to be interpreted as podocytes or as meandering slits is unclear, but at least no single terminal cell as in other nemerteans is present, but a terminal part composed of several cells. The close contact between excretory organ and blood vessel is not present in all species, for example in *Prostomatella arenicola*, it could not be shown (Bartolomaeus 1988). Furthermore, the protonephridial system develops earlier and is functional without the circulatory system in *Lineus viridis* (Bartolomaeus 1985). The filtration structure of the terminal cells are meandering slits (Bartolomaeus 1985, 1988). Terminal cells are always multiciliate (Turbeville 1991, 2002), but develop from monociliary cells (at least in *Lineus viridis*, Bartolomaeus 1985). There are numerous microvilli among the cilia.

Kamptozoa
Marine species possess one pair of protonephridia, while the single freshwater species *Urnatella gracilis* has a paired system of branching canals and many terminal organs, each composed of two terminal cells (Emschermann 1965). The protonephridium of marine forms is composed of a terminal organ (consisting of two terminal cells), one canal and one nephropore cell. The filtration structure is between interdigitating (microvilli-like) cellular extensions of the two terminal cells, forming meandering clefts (Franke 1993). The terminal organs are multiciliated (e.g. about 60 cilia in *Loxosomella fauveli*, Franke 1993) and contain short microvilli.

Mollusca
Protonephridia have been found in larvae of polyplacophorans (Bartolomaeus 1989a), gastropods (Brandenburg 1966, Bartolomaeus 1989a), and bivalves (Brandenburg 1966). There appears to be one pair of protonephridia in marine species, but more than one terminal cells in most, but not all, freshwater gastropods (e.g. *Acylus fluviatilis*, *Planorbarius corneus*, *Physa* sp., and *Lymnaea stagnalis*, light microscopical investigation by Meisenheimer 1899, *Limax flavus*, Brandenburg 1966). Terminal cells are multiciliate, have irregularly distributed, short microvilli and the filtration structure are slits in the terminal cell (Brandenburg 1966, Bartolomaeus 1989a). In the opisthobranch gastropod *Aeolidia papillosa*, there is one terminal cell, two canal cells, and one nephropore cell, in the polyplacopore *Lepidochiton cinereus*, there are up to fifteen canal cells (Bartolomaeus 1989a). In adult molluscs, a metanephridial system is present. The basal process of excretion is a filtration from the primary body cavity into the coelom, the pericard (see Fig. 8.11.). The sites of filtration are podocytes. From the pericard, the filtrate is led to the body wall through pericardioducts. Such ultrafiltration has been shown experimentally, for example by Hevert (1984) for lamellibranch bivalves. Podocytes in the pericard epithelium are described from solenogastres (Reynolds et al. 1993), polyplacophorans (Økland 1980, 1981), monoplacophorans (Haszprunar & Schaefer 1997), gastropods (Potts 1975), bivalves (Morse & Zardus 1997), scaphopods (Reynolds 1990), and cephalopods (Schipp & Hevert 1981). In all members of Eumollusca (all molluscans exclusive aplacophorans), the pericardioduct is specialized for reabsorption and active secretion. This structure is named the kidney. Active secretion by the kidneys has been shown ultrastructurally and experimentally (see Andrews 1988, Morse & Reynolds 1996 for summaries). The presence of active transport in the ducts of nephridia is a common phenomenon, but in eumolluscs, it is much more strongly developed than in other taxa.

Sipunculida
Either one single (genera *Phascolion* and *Onchnesoma*) or one pair of metanephridia are present in sipunculids. The ciliated funnel leads into an often inflated, bladder-like region, from which a dead-end tubule with a heavily folded epithelium originates. The excretory pore is in the anterior, bladder-like part, close to the ciliated funnel (Rice 1993). Podocytes as sites for ultrafiltration have been found close to vessels of the circulatory system (Pilger & Rice 1987, Pinson in Rice 1993), but also on the peritoneal lining directly adjacent to the metanephridia (Rice 1993). In the first case it appears that filtration is from the primary body cavity into the coelom, in the second case, the path of filtration is not completely clear. There is evidence for modification of the coelomic fluid within the metanephridia (Pinson in Rice 1993). The metanephridia also serve for the release of gametes.

Echiurida
In the microscopic dwarf male of *Bonellia viridis*, the excretory organ is a protonephridium, which is associated with the coelom (Schuchert & Rieger 1990a). The proximal part of the protonephridium projects into the coelomic cavity. The multiciliate terminal cells (named crown cells by Schuchert & Rieger 1990a) are not covered by the coelomic epithelium, they are separated from

the coelomic lumen only by ECM. The terminal cells contain an intensive system of interstices which open to the exterior by clefts (which are covered by ECM). Fluid can enter the protonephridial lumen from the coelom either through these clefts or between the terminal cells. In both cases, it has to pass the ECM. According to Goodrich (see Goodrich 1945), the larva of *Echiurus* sp. has monociliate protonephridia, but no ultrastructural re-investigation is available. In all adult echiurans (except for the *Bonellia* male), metanephridial systems are present. Ultrafiltration is presumably performed from the circulatory system into the spacious coelom through podocytes (Bartolomaeus 1994). Ciliated funnels are present in two positions in the coelom. Few (or up to 400 pairs in *Ikeda taenioides*, Pilger 1993) funnels are specialized for the storage and shedding of gametes. A high number (usually 200–300, with a minimum of 12 per sac in *Echiurus abyssalis*, see Baltzer 1931, and up to 8,500 in female *Bonellia viridis*, as estimated by Harris & Jaccarini 1981) of ciliated funnels are present on the so-called anal sacs. These are paired evaginations of the hindgut. Experiments have shown that coelomic fluid passes through the ciliated funnels into the anal sacs (Baltzer 1931, Harris & Jaccarini 1981). It is assumed that modification takes place within the anal sacs.

Annelida
The excretory systems in annelids are extremely diverse. Generally, trochophore larvae (when present) possess one pair of protonephridia, while adults have metanephridial systems. Among polychaetes, there are several taxa, which also have protonephridia as adults.

- **Polychaeta**. Trochophore protonephridia are composed of minimally three cells (terminal, canal, and nephropore cell, as in *Magelona mirabilis*, Bartolomaeus 1995 and in *Spirorbis spirorbis*, Bartolomaeus & Ax 1992), but in most species the number of cells is higher. The terminal cells are either monociliate or multiciliate (Smith & Ruppert 1988, Smith 1992, Bartolomaeus & Ax 1992, Bartolomaeus 1993a). Microvilli usually surround the cilia, their number varies but appears to be always more than eight. The filtration structures are usually clefts in the terminal cell. An exception among larval polychaetes is the mitraria larva of *Owenia fusiformis*, which possesses monociliated podocytes lining a small coelomic body cavity that opens to the exterior by a ciliated duct (Smith et al. 1987). In adults, the larval protonephridia are often retained in the adults as so-called head kidneys (Bartolomaeus & Quast 2005). They are located in the prostomium, behind the photoreceptors. In the postlarval segments, the earliest anlage of an excretory organ is represented by a few cells surrounding a small canal (Fig. 9.7.; see Bartolomaeus & Quast 2005). Ciliogenesis starts soon after this stage. The distal end of this canal grows towards the epidermis and finally pierces it to form the nephridial pore. The proximal region can develop in different ways. In some species (e.g. *Nereis pelagica*, *N. diversicolor* [Nereidae] and *Pectinaria koreni* [Terebellida]), it fuses with the coelom epithelium and opens into the coelom as a ciliated funnel, the metanephridium (Fig. 9.7., 9.8.; Bartolomaeus 1993a). In other cases, the distal cell develops into a modified terminal cell, representing a protonephridial condition. One cilium is surrounded by a ring of long microvilli, which span ECM between them (Fig. 9.9.). Such terminal cells are present in small meiobenthic species with a reduced coelom (Brandenburg 1970, Westheide 1985b, 1986, Clausen 1986) or in macroscopic taxa with a reduced blood vascular system (e.g. Hausmann 1981 for *Anaitides mucosa*; Smith 1992). In these latter cases, they project into the coelom (Fig. 9.9.). The exception are nephtyids, which have a well developed blood vascular system, but possess, uniquely among polychaetes, extracellular hemoglobin. Terminal cells might be more effective in preventing hemoglobin loss than metanephridia. In adults of the taxa with a well developed blood vascular system, the protonephridial stage is replaced by metanephridia in adults. There are different modes of such a replacement. In some taxa (e.g. Alciopidae, Phyllodocidae, Pisionidae), ciliated funnels are temporarily formed during the reproductive period (Fig. 9.7.; see Stecher

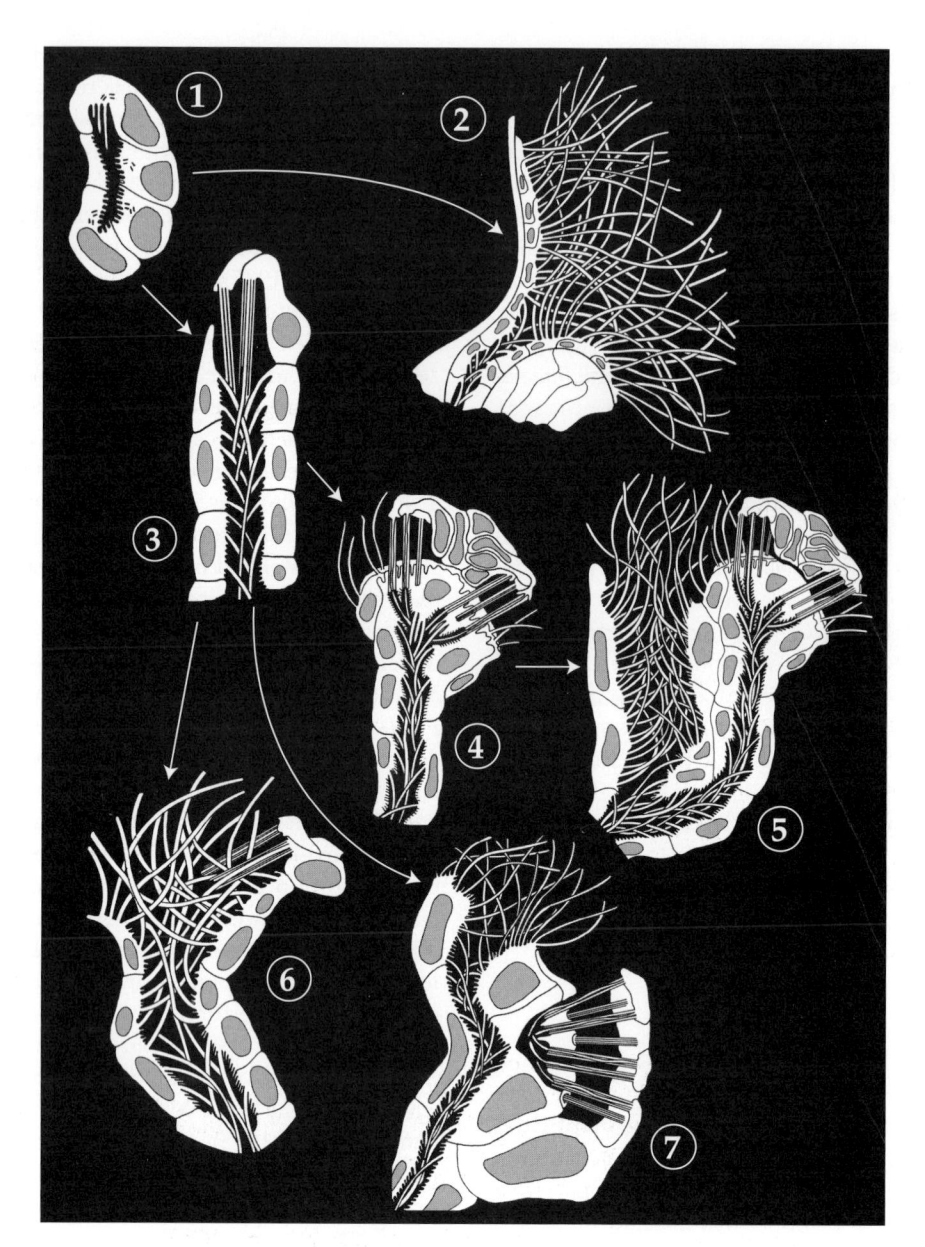

Fig. 9.7. Correspondence of protonephridial and metanephridial systems in polychaetes. In all cases, development starts with an anlage composed of few cells (1). During further development, this either directly opens to the coelom with a metanephridial funnel (2) or terminal cells develop in the proximal part (3). The filtration region of such protonephridia is within the coelom (4). These protonephridia may remain present throughout life, but during reproduction temporary metanephridial funnels open from the canal region into the coelom (5), examples are *Phyllodoce* and *Eulalia*. Protonephridia can also be replaced by the metanephridial funnel opening in the proximal region of the duct and subsequently degenerate (6), for example in *Pholoe* and *Harmothoe*. In *Tomopteris*, protonephridia direct into a small blind lumen and are probably nonfunctional, a metanephridial funnel is present close to this region (7). Compiled using figures from Bartolomaeus (1993a) and Bartolomaeus and Quast (2005).

1968 for *Pisione remota*, Olive 1975 for *Eulalia viridis*, Bartolomaeus 1989b for *Anaitides mucosa*). During this period, the excretory canal branches and one branch leads to the terminal cells while the other leads into the coelom via the ciliated funnel. In *Pholoe inornata* (Sigalionidae) and *Harmothoe sarsi* (Polynoidae), the proximal canal cells of the protonephridium proliferate and open as a ciliated funnel into the coelom while the terminal cells degenerate (Fig. 9.7.; Bartolomaeus & Ax 1992, Bartolomaeus 1993). In *Tomopteris helgolandica*, the anlagen of protonephridium and metanephridium develop simultaneously and close to each other, but the protonephridial terminal cells do not receive contact to the canal (Fig. 9.7.; Bartolomaeus 1997b). This condition appears to be present throughout life. In *Sabellaria cementarium* (Sabellariidae), the terminal cell develops into podocytes (Smith & Ruppert 1988). This appears to be an exception, because usually podocytes develop independantly from the nephridial anlagen. In summary, excretory organs develop from anlagen with few cells, which either develop directly into metanephridia, into protonephridia, or via protonephridia to metanephridia. Bartolomaeus & Quast (2005) assume that the direct development of the anlage into the metanephridium is the plesiomorphic condition, and it might be discussed whether the 'protonephridial stage' is homologous to the head kidneys or to protonephridia in other bilaterians.

- **Myzostomida**. In *Myzostoma cirriferum*, five pairs of protonephridia have been found (Pietsch & Westheide 1987). Each is composed of three multiciliary terminal cells, each of the 6–9 cilia is surrounded by 10 microvilli. The sites of filtration are clefts in the wall of the terminal cells. A canal composed of one canal cell leads to a porus.
- **Pogonophora**. Structures with an assumed excretory function are present in several

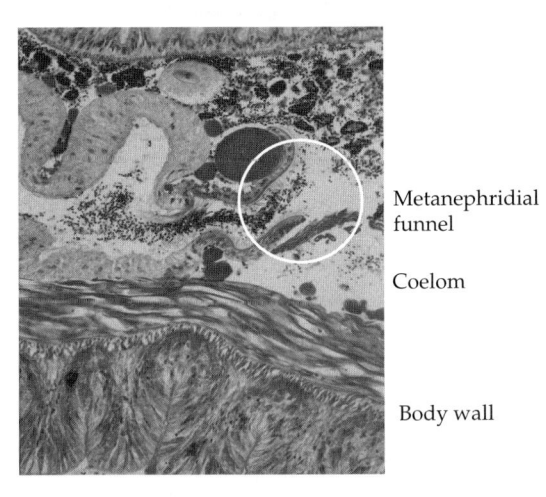

Fig. 9.8. Metanephridial funnel (in white circle) in the polychaete *Clymenura clypeata*. Coelom is to the right, the canal to the left. Photo by courtesy of Thomas Bartolomaeus, Berlin.

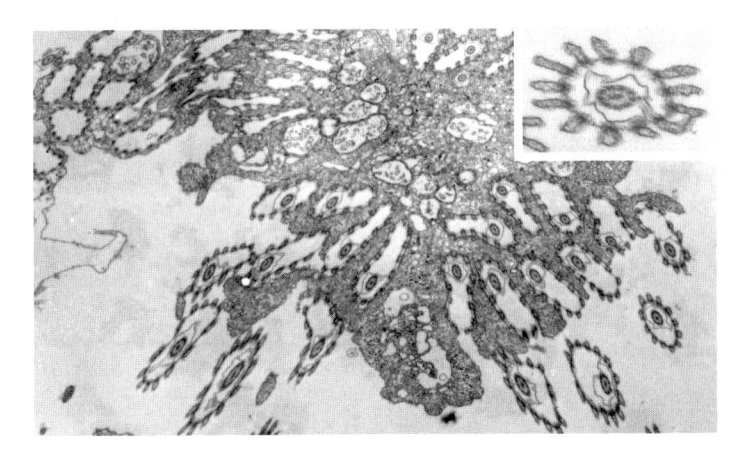

Fig. 9.9. In *Tomopteris helgolandica* and some other polychaetes, the cilium and circumciliary microvilli of the terminal cell pass through the coelom (the photo shows some free in the coelom, others are sectioned where they are surrounded by cells). Insert shows section through one terminal cell. Photo by courtesy of Thomas Bartolomaeus, Berlin.

pogonophorans. The exact structure is not completely clear (see Southward 1993). In *Siphonobranchia lauensis*, ducts lead from the medial coelom to the body surface. They are in close contact with the ventral blood vessel, but filtration structures such as podocytes have not been described. In *Oligobrachia gracilis*, there is no connection between coelom and duct, but here a structure roughly resembling a terminal cell has been found which contains many microvilli and probably cilia (Southward 1993). A similar structure is present in *Siboglinum ekmani*; here, two cilia are regularly surrounded by approximately 10 microvilli (Fig. 20B in Southward 1993). As the filtration structure is not clear, excretory organs in pogonophorans require further investigation. The excretory organs of vestimentiferans appear to be modified coelomoducts, associated with a blood vessel (Gardiner & Jones 1993), but as in pogonophorans, the pathway of excretes is incompletely known.

- **Clitellata**. Some clitellate embryos possess nephridia which have been termed protonephridia by Anderson (1973). These organs are present before the coelom and blood vascular system are formed. Quast & Bartolomaeus (2001) showed for embryos of the leech *Erpobdella octoculata* that these organs are composed of three cells and open into the blastocoel (primary body cavity) and this may also be valid for 'oligochaete' nephridia. Adult 'oligochaetes' always possess metanephridia, in most species there is one pair per segment. These metanephridia develop independantly from the larval nephridia. Podocytes at the transition between blood vascular system and coelom have been found in several taxa (Jamieson 1992, Hansen 1995). The metanephridium is either separated from a coelomoduct, which serves as gonoduct, or gametes are released (in few anterior segments) through the metanephridia. In leeches, Fernandéz et al. (1992) report a protonephridium in a six-day old embryo of *Hirudo medicinalis*. In *Erpobdella octoculata*, this organ is a blindly ending, unciliated tube (Quast & Bartolomaeus 2001). Homology to nephridia in other annelids is possible only by their position. Adult leeches possess metanephridia, but there is no contact between the ciliated funnel and canal. In Rhynchobdelliformes, which still possess a primary blood vascular system, the funnel and canal are close together but apparently unconnected (Fernandéz et al. 1992). In the other taxa, which have only the secondary blood vascular system composed of the derived coelom (see Chapter 10), the metanephridial funnel has no connection to the excretory duct. It is assumed that it acts as a phagocytotic organ and probably also propells the blood (Fernandéz et al. 1992). The production of primary urine is by secretion of the canal cells, in particular so-called canaliculus cells (Zerbst-Boroffka 1975, Zerbst-Boroffka & Haupt 1975).

Tardigrada
Species from the tardigrade subtaxon, Eutardigrada, produce excretes by active transport in the malpighian tubules which lead into the intestine at the border between mid- and hindgut. From ultrastructural investigations (Dewel & Dewel 1979, Greven 1979, Weglarska 1980), there is evidence for active secretion of excretes in the proximal part of the malpighian tubules and modification in the distal part.

Onychophora
Each segment contains one pair of metanephridial systems, there can be some variation in shape between different segments (Storch & Ruhberg 1993). The coelomic part of the system is composed of a small coelomic cavity, the sacculus, the epithelium of which contains a high number of podocytes (Seifert & Rosenberg 1976, Storch et al. 1978). It develops from part of the embryonic coelothel tissue after its breakup to form a mixocoel (see Fig. 8.7.; Mayer 2006b). A heavily ciliated funnel leads into an aciliate duct, in which several regions can be distinguished (Storch et al. 1978). A bladder-like extension is present in the distal region of the duct, close to the nephridiopore at the inner base of the legs. It is evident from this structure that primary urine forms by filtration from the mixocoel into the sacculus and is further modified during its passage down the duct.

Euarthropoda

Different forms of excretory structures are present in euarthropods. Comparable to onychophorans, a system composed of a sacculus and an excretory duct is present in chelicerates (as coxal glands, see, e.g., Alberti & Coons 1999, Coons & Alberti 1999, Farley 1999, Felgenhauer 1999), crustaceans (as antenna and maxillary nephridia, see, e.g., Riegel & Cook 1975, Walter & Wägele 1990), in chilopods and progoneates (as maxillary nephridia, see Minelli 1993 for chilopods), and in basal insects (as labial nephridia, e.g. Altner 1968, François 1998). In contrast to onychophorans, they are restricted to few anterior segments. *Limulus* is special in forming one coxal gland, which is composed of the sacculi from segments 2–5 (Fahrenbach 1999). The sacculi contain podocytes, but the funnel is not ciliated. The excretory duct can be structured into different regions. In arachnids and insects, malpighian tubules are present which are, comparable to tardigrades, blindly ending tube like extensions of the intestinal system, into which excretes are actively secreted (e.g. Eichelberg & Wessing 1975, Wessing & Eichelberg 1975, Phillips 1981, Bradley 1998). However, in arachnids, malpighian tubules are of endodermal origin but in insects they are ectodermal, though not cuticularized. In aquatic species, a considerable part of excretion (and osmoregulation) is performed through not, or hardly, cuticularized epithelia, for example the gills (see Riegel & Cook 1975 for references). With terrestrialization (particularly in web spiders and insects) further mechanisms for a positive water budget had to be developed (see Maddrell 1982 for a summary of insect osmoregulation). In spiders, Seitz (1986) and Felgenhauer (1999) list, besides malpighian tubules and coxal glands, nephrocytes, interstitial cells, guanocytes, intestinal diverticula, or the stercoral pocket as sites of excretion and osmoregulation. To save water, excretes are often barely solulable in water, such as uric acid or guanine. Guanine, for example, is stored in guanocytes in the periphery of intestinal diverticula and is externally visible by their white colour as the cross in the garden spider (*Araneus diadematus*) (Fig. 9.10.).

Fig. 9.10. The cross of the garden spider *Araneus diadematus* is made up of guanocytes.

Chaetognatha

There are no special excretory organs in chaetognaths, and excretion is assumed to take place over the body wall (Shinn 1997).

Phoronida

Protonephridia are present in larval phoronids (the actinotrocha), while adults have a metanephridial system. Protonephridia are paired and project into the blastocoel. Ultrastructural descriptions are available for *Phoronis muelleri* (Hay-Schmidt 1987, Bartolomaeus 1989c). The canal branches and leads to terminal complexes, each composed of about 25 (according to Hay-Schmidt 1987) or more than 30 (according to Bartolomaeus 1989c) individual terminal cells and 10–15 additional accessory cells (Fig. 9.11.; accessory cells reported only by Bartolomaeus 1989c). Each terminal cell is monociliated and the

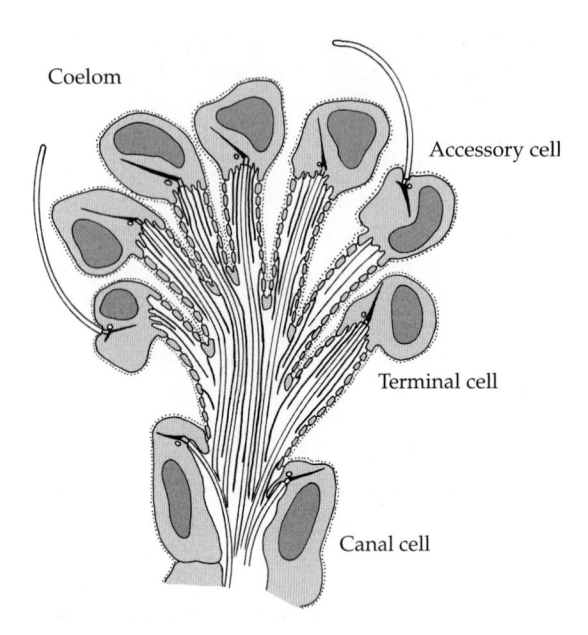

Fig. 9.11. The terminal complex in the actinotrocha (Phoronida) is composed of terminal and accessory cells. After Bartolomaeus (1989c).

cilium is surrounded by 7–9 microvilli. The filtration structures are clefts in the terminal cell. Accessory cells are almost identical to terminal cells with the exception that their single cilium is not directed towards the lumen of the cell, but into the blastocoel. Additionally, the cilium is not surrounded by microvilli. During metamorphosis, the terminal complexes and proximal branches of the canal are cast off and degenerate, only the distal duct remains and ends blindly (Bartolomaeus 1989c). Later, this duct receives contact to the metacoel and coelothel cells form the ciliated funnel (Bartolomaeus 1989c). Therefore, the metanephridium of adult phoronids is composed of an ectodermally derived duct, which is the remaining distal protonephridial duct and the mesodermally derived ciliated funnel.

Bryozoa

Excretory structures are neither known from larval nor from adult bryozoans. Mukai et al. (1997) summarize vague hints for excretory functions of some structures (such as coelomocytes), but none of these has been convincingly tested.

Brachiopoda

A metanephridial system has been found as the only excretory organ in brachiopods, even in pelagic larvae (e.g. *Lingula anatina* and *Calloria inconspicua*, Lüter 1997). There seems to be no protonephridial 'precursor' of the metanephridia, but there is evidence for a bipartite composition of metanephridia, as in phoronids. Lüter (1997) observed in larvae of *Calloria inconspicua* ducts directing towards the anlage of the coelom. This observation corresponds with a sharp morphological distinction between canal cells and cells of the ciliated funnel (Lüter 1995) and makes it likely that the canal is ectodermally derived and the funnel mesodermally. In adults, metanephridia are present as one pair, opening into the metacoel. Only the Rhynchonellidae possess two pairs of metanephridia.

Hemichordata

Ultrastructural investigations of enteropneusts (Wilke 1972b, Balser & Ruppert 1990) and pterobranchs (Dilly et al. 1986 for *Cephalodiscus gracilis*, Mayer & Bartolomaeus 2003 for *Rhabdopleura compacta*) have shown that the excretory organs of both taxa are similar in structure. The excretory complex includes the hemolymph vessels which are non-endothelial (and therefore the primary body cavity) and a coelomic cavity, the protocoel. The hemolymph passes a heart which is surrounded by a small coelomic cavity, the pericard and from there enters a sac-like extension, the glomerulus (see Fig. 10.8.). The glomerulus bulges into the protocoel and the coelothel is differentiated here into podocytes. It is assumed that the heart creates the hemolymph pressure necessary for ultrafiltration from the glomerulus into the protocoel. From here, fluid can be secreted by paired pores. Modification of the ultrafiltrate, the primary urine, is likely, but has not been shown. Ruppert & Balser (1986) showed that excretion already takes place in the larva of enteropneusts, the tornaria. Here, a coelomic cavity is coexistent with the blastocoel (as primary body cavity) and opens to the outside through a short ciliated duct and a pore, the hydropore. A small second coelomic cavity functions as a pulsatile vesicle and probably drives a fluid stream from the blastocoel through podocytes of the coelomic epithelium into the coelom. From there, a considerable outstream of fluid through the hydropore was measured by Ruppert & Balser (1986).

Echinodermata
Excretory processes and structures appear to be quite comparable to those described for hemichordates (Nielsen 2001). The larva (bipinnaria) of the asteroid *Asterias forbesi* exhibits an almost identical blastocoel/coelom/hydropore system as described above for the enteropneust tornaria (Ruppert & Balser 1986). In adults, coelomic cavities become more strongly modified than in hemichordates, but it was found that podocytes are present in the coelomic epithelium lining extensions of the axial sinus (which is homologous to the protocoel) (Welsch & Rehkämper 1987). Comparable structures have also been found in crinoids (Balser & Ruppert 1993) and holothurians (Balser et al. 1993). These extensions are surrounded by blood sinuses (as primary body cavity), so that filtration from the blood into the axial sinus appears likely. From here, excretes can leave the body through the madreporite. In holothurians, the whole body surface, as well as the respiratory trees, is assumed to contribute in excretion (Ruppert et al. 2004).

Tunicata
There are no complex excretory organs present in tunicates. Excretion was shown to be present at the cellular level. In molgulid ascidians, nitrogeneous waste is accumulated in a vesicle called the renal sac; in corellids and ascidiids, excretory vesicles form clusters near the gut (Burighel & Cloney 1997). Probably certain glands, such as the pyloric gland and the neural gland, can also have an excretory function.

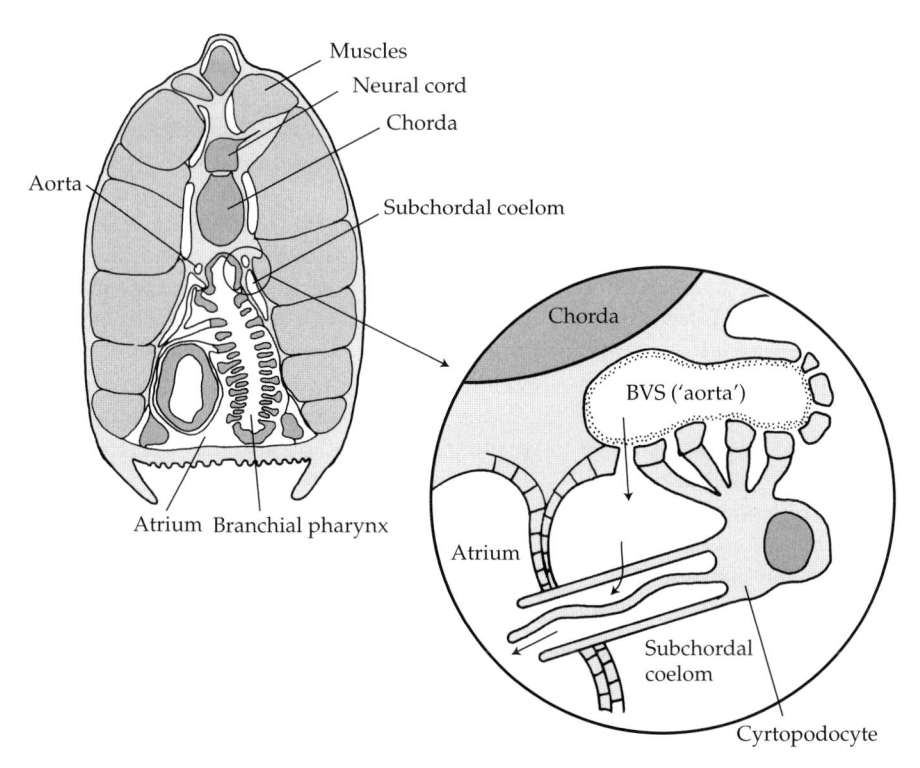

Fig. 9.12. Cyrtopodocytes in *Branchiostoma* (Acrania) are specialized coelom epithelium cells that connect the blood vascular system and the atrium through the subchordal coelom. One part is formed like podocytes, the other contains a cilium and circumciliary microvilli. Cross section after Ruppert (1997).

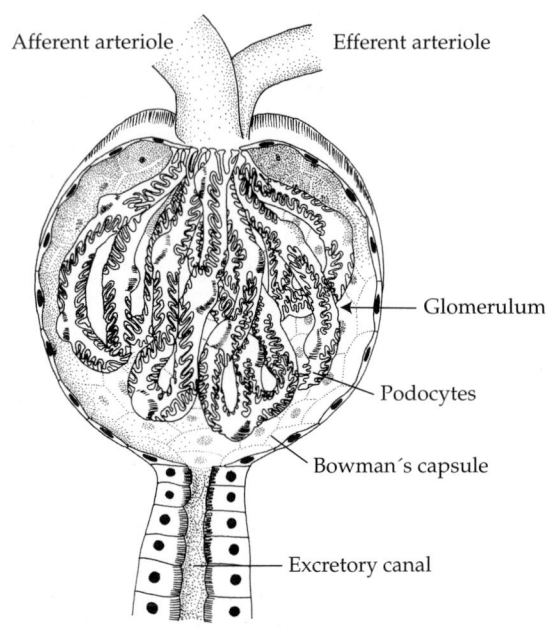

Fig. 9.13. Glomerulum in the human kidney. The afferent and efferent vessels form a capillary glomerulus bulging into a coelomic compartment, the Bowman's capsule. The vessels are densely covered by podocytes. The filtrate leaves the capsule through the excretory duct. After Junqueira et al. (2005).

Acrania

The excretory structures are present as numerous branchial nephridia and as the unpaired Hatschek's nephridium (Ruppert 1997). Both nephridia are similar in structure. A special cell, the 'cyrtopodocyte' arises as part of the coelomic epithelium (in the case of branchial nephridia the subchordalcoelom) (Fig. 9.12; Stach & Eisler 1998). Its epithelial components are podocytes that surround the glomerular plexus, a part of the blood vascular system (which is the primary body cavity). From the cell soma of the cyrtopodocyte, which bulges into the coelom, one long cilium which is surrounded by 10 microvilli emerges (Fig. 9.12.; Brandenburg & Kümmel 1961, Ruppert 1997). They traverse the coelomic cavity and project into a nephridial canal that leads into the atrium. The microvilli are surrounded by ECM (Ruppert 1997, Stach & Eisler 1998). This structure implies that blood is filtered from the blood vascular system through podocytes into the coelom, and from there through the canal created by the cyrtopodocyte microvilli into the nephridial duct. During this passage, it has to pass a second filter around the microvilli. Although the cilium/microvilli of the cyrtopodocytes broadly resemble terminal cells of protonephridia, it is more likely that cyrtopodocytes are highly specialized coelothel cells. In summary, the excretory system of acranians is a metanephridial system.

Craniota

In all craniotes, a kidney is present which is composed of subunits, the nephrons. Each nephron is a metanephridial system, in which ultrafiltration from the blood into a coelomic cavity is the mechanism of formation of primary urine. Arteries of the blood vascular system form capillaries (the glomerulus) that bulge into a small coelomic compartment, the Bowman's capsule (Fig. 9.13.). The epithelium of the blood vessels is fenestrated and Bowman's capsule contains numerous podocytes (e.g. Farquhar 1982, Kluge & Fischer 1990). Therefore, blood has to pass only the ECM from the capillaries into the coelom. Modification of the filtrate takes place in the canal leading from Bowman's capsule into larger collecting ducts. The transition between capsule and duct is, at least in cyclostomes, ciliated (Kluge & Fischer 1991). In the assumed ancestral state, the archinephros, nephrons were segmentally arranged (Ruppert 1994). Filtration took place into the segmental coelomic cavity, before the dorsal part of it separated as the Bowman's capsule (Ruppert 1994). Primitive kidneys are found among cyclostomes, but within gnathostomes a regionalization occurs and the functional kidney develops from several posterior nephrons. In males, nephrons anterior of the kidney gain contact to the reproductive system and serve to release spermatozoa. In both sexes, urogenital systems are present, that is, the ducts from the kidney join the reproductive system. There are a few cases in which no glomerulus is formed and no ultrafiltration occurs. This occurs, for example, in some antarctic fish (Eastman et al. 1979) and makes sense taking into account the abundance of glycoproteins in the blood which serve as antifreeze proteins. The size of these glycoproteins is so small that they would pass through the ECM and a complete reduction of ultrafiltration prevents their loss or the energy expensive reabsorption.

Before discussing evolutionary conclusions concerning excretory organs, three important models have to be presented. For all three models, the explanation of the diversity of excretory organs in annelids presents a central challenge.

Evolution of nephridia: the Goodrich model

In a number of publications, summarized in 1945, Goodrich presented a hypothesis on the evolution of excretory organs that is closely connected to the gonocoel theory (see Chapter 8, 13). Goodrich assumed that coeloms originated from the gonad epithelium and are, as such, present in all bilaterian taxa. The gonad/coelom always has a duct to release gametes to the outside, the gonoduct/coelomoduct. Nephridia are formed independantly from the gonoduct, either as protonephridia or, when the coelom expands in size, as metanephridia. Because Goodrich assumes such states as being ancestral, all other states have to be interpreted as derived. This is especially important in annelids. Among the broad diversity of nephridia, separated nephridioducts and gonoducts exist in capitellid polychaetes and in several oligochaetes, which is taken by Goodrich as a proof for his hypothesis. The many other patterns are interpreted as different stages of fusion between gono- and nephridioduct, leading to what Goodrich calls nephromixia. An assumed support for such a fusion are differentiated stainings of funnel and canal in histological preparations that are interpreted as indicative of a different origin of these structures.

There are several critical points in Goodrich's hypothesis. First, it depends upon the gonocoel hypothesis which appears not to have much support (see Chapters 8 and 13). According to Bartolomaeus (1993a), funnel and duct cells do not differ from each other ultrastructurally and it appears that the metanephridial funnel develops from the proximal canal cells when the canal tears open with the growing coelomic space (Bartolomaeus 1999, Bartolomaeus & Quast 2005). The differential staining observed by Goodrich may be a simple product of a different density of ciliary roots in funnel and canal cells. Finally it is questionable whether the condition of separate ducts in annelids is an ancestral pattern.

Evolution of nephridia: the Ruppert & Smith model

Ruppert & Smith (1988, see also Smith & Ruppert 1988, Smith 1992) develop a functional model in which the type of nephridium depends primarily on body size and not on phylogeny. Protonephridia function as 'cilia-mediated filtration of extracellular fluid' (Ruppert & Smith 1988, p. 252,) while metanephridial systems 'function by muscle-mediated filtration of vascular fluid into a coelomic space' (p. 252). Therefore, the type of nephridium can be predicted by the presence or absence of a circulatory system. This prediction is well fitting and good examples are those polychaetes which lack a circulatory system as adults and consequently retain their larval protonephridia (see above). Protonephridia in general are regarded as not homologous, following conclusions by Wilson & Webster (1974) and Brandenburg (1975), their occurrence is based on functional requirements and not on phylogeny. Protonephridial and metanephridial canals are identical in structure and protonephridia and metanephridia are believed to be interchangable, depending on function (e.g. body size).

The functional correlation between body size, presence of circulatory system and coelom, and the type of nephridia, has been supported by many observations. It only has to be discussed whether protonephridia and metanephridia are really interchangable according to functional requests or whether their occurrence is additionally influenced by phylogeny.

Evolution of nephridia: the Bartolomaeus & Ax model

Bartolomaeus & Ax (1992) summarize thoughts on the evolution of nephridia already outlined by Bartolomaeus (1989c) and further explained and modified by Bartolomaeus (1993a, 1999). This model stresses the evolutionary distribution of nephridia. Protonephridia are regarded as homologous throughout all bilaterian taxa in which they appear. Despite their variability, there is always a tripartition into a terminal cell (terminal organ/terminal region), a canal, and a nephropore. The filtration structure is always part of the terminal cell and covered by ECM. Bartolomaeus & Ax (1992) reconstruct the basal type of protonephridium as consisting of three cells (one terminal, one canal, and one nephroporus cell), with the terminal cell containing one cilium surrounded by eight microvilli. Of special interest is the situation in polychaetes (Fig. 9.7.). The larval protonephridium (and adult head kidney) is a 'standard' protonephridium. According to Bartlomaeus & Ax (1992), protonephridia are also retained in the segments of adults. In large species, they develop temporary funnels for reproduction connecting the protonephridial canal with the coelom (as in *Pisione*, *Eulalia*, or *Anaitides*). Within polychaetes, there is a tendency to develop the funnel earlier in development and by this process the protonephridia are suppressed. Therefore, examples such as *Tomopteris*, where a protonephridium is present but appears to be non-functional because it is not connected to a duct; or the degeneration of terminal cells after the opening of the funnel as in *Pholoe* and *Harmothoe*; are interpreted as intermediate stages. Conditions such as the presence of separated genital and nephridial ducts are interpreted as secondary derivations. This interpretation implies that metanephridia evolved within annelids and are not homologous with metanephridia in other taxa. Support for this comes from the comparison of the development of metanephridia in phoronids and annelids. In annelids, the ciliated funnel develops from duct cells, while in phoronids it is composed of coelothel cells.

Bartolomaeus (1999) and Bartolomaeus & Quast (2005) modifiy this model, stressing that all nephridial organs in annelids go back to an anlage composed of a few cells that surround a small lumen into which they project cilia. This lumen is passively opened by the force of fluid pressure from the development of the adjacent coelom, and forms a metanephridium. If this is not the case, that is, when no coelom is formed, the anlage develops into a protonephridium or at least into a stage resembling a protonephridium. Because the ancestor of annelids is thought to already have a spacious coelom, this would mean that the metanephridial system was present in the annelid ancestor. Problems in interpreting the evolution of excretory organs in annelids are complicated by the fact that the annelid phylogeny is not clearly resolved.

Conclusions

Excretory organs working only with active transport are scattered across the phylogenetic tree and are structurally hardly comparable (Fig. 9.14.). They are therefore assumed to have independant origins from each other. Nephridia, that is, excretory organs based on ultrafiltration, are much more broadly distributed. Although there is considerable variation in the structure of protonephridia (such as differences in where exactly the filtration structure is formed and where the terminal parts project into), there are also similarities (Bartolomaeus & Ax 1992, see above) which makes it likely that they go back to one common ancestor. As protonephridia are only present in protostomes and phoronids, their origin depends on the positon of phoronids. If they are a member of the Radialia, that is, closely related to deuterostomes, then protonephridia should have been present in the bilaterian ancestor (probably excluding *Xenoturbella* and Acoelomorpha). If phoronids are protostomes (see Chapter 2), then it is likely that protonephridia evolved in the protostome ancestor.

The 'simplest' organization of protonephridia is composed of three cells. Although simple is not always plesiomorphic, the presence of such protonephridia in gastrotrichs, gnathostomulids, and in the larva of *Magelona mirabilis* (Annelida) makes it likely that this is the plesiomorphic condition. If so, then the presence of one cilium and eight circumciliary microvilli also appear

EXCRETORY SYSTEMS 189

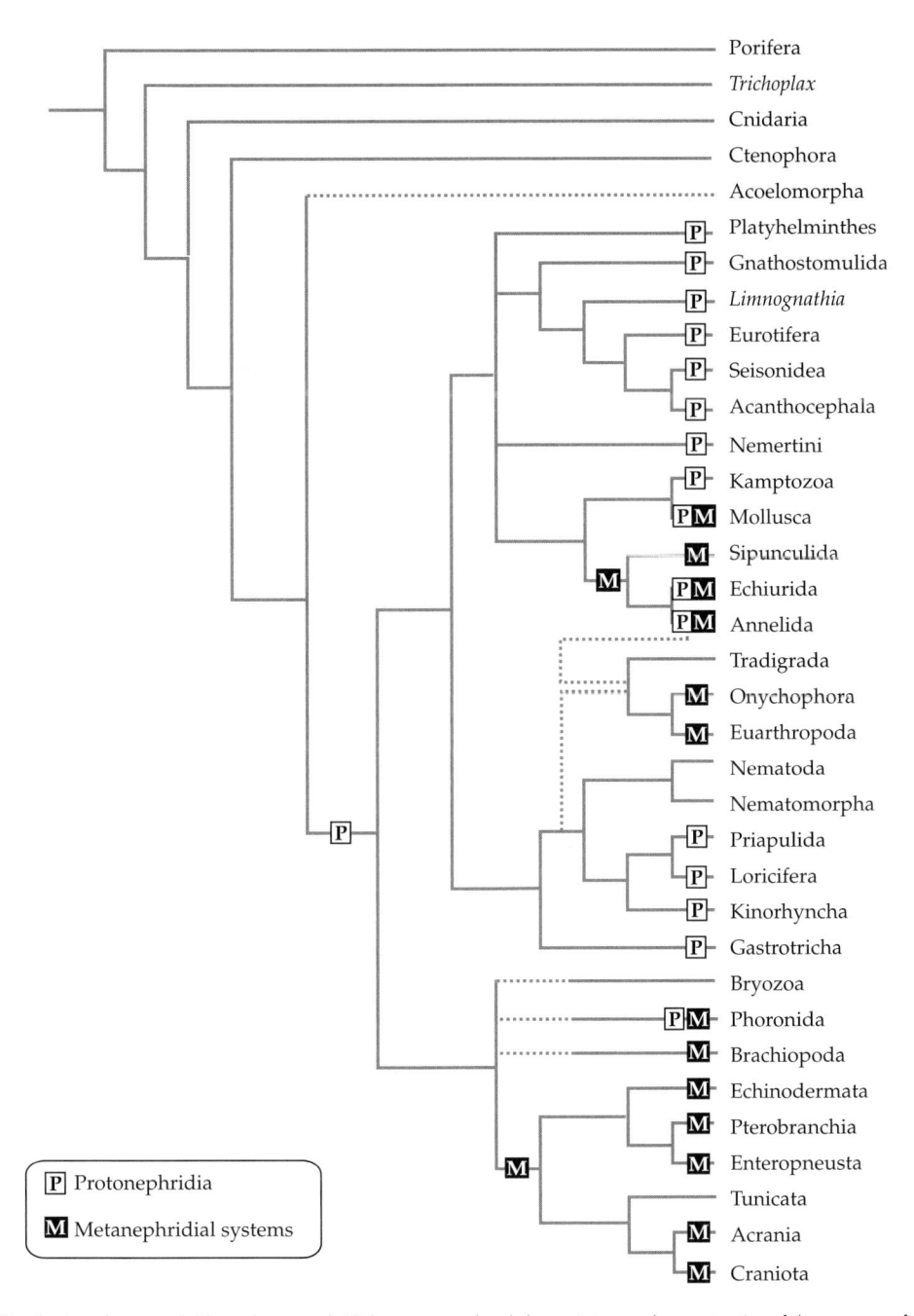

Fig. 9.14. Distribution of protonephridia and metanephridial systems on the phylogeneic tree and reconstruction of the presence of these systems in some ancestors.

to be ancestral features, as assumed by Bartolomaeus & Ax (1992). Other cells such as choanocytes of sponges or collar receptors (Chapter 7) also have such characteristics and this might indicate that terminal cells are modified epidermal cells.

Metanephridial systems have been assumed to evolve convergently by some authors. Because metanephridial systems depend on the presence of a coelom, an independant evolution of coeloms (see Chapter 8) would require a convergent evolution of metanephridial systems. The different composition and development of metanephridia in annelids and phoronids (Bartolomaeus & Ax 1992) may also indicate a convergent evolution. In Chapter 8 it was argued that podocytes, although being conspicious cells, may in fact not be extremely complex structures (so that their convergent evolution does not appear unlikely). This certainly accounts for the ciliated funnels. In many cases, this is a rather inconspicuous region, hardly discernable from peritoneal or duct cells. Therefore, their convergent evolution appears not to be unlikely.

Imagining a convergent evolution of metanephridial systems, mapping their distribution on the phylogenetic tree implies that they evolved at least three times (Fig. 9.14.): in molluscs, in the ancestor of Annelida + Sipunculida + Echiurida, and in the deuterostome ancestor. The interpretation of their occurrence in the tentaculate taxa and in arthropods (except tardigrades) depends on the exact phylogenetic position of these taxa. Because of positional and structural correspondences between metanephridial systems of echinoderms and hemichordates, Nielsen (2001) assumed that they are homologous. Whether this system was further developed in myomerates (Acrania + Craniota) is a different question.

CHAPTER 10

Circulatory systems

While in small or epithelially organized organisms the distribution of nutrients and oxygen is performed by diffusion processes, this becomes a problem with growing size and complexity. Therefore, many animals have invented some kind of circulatory system, in which fluid floats through the body and by this transports substances.

It is not easy to delimit the exact borders for the definition of a circulatory system. Limited movement of fluids is already possible in the interstices between cells, that is, between collagen fibres within the ECM, and such a fluid stream may already aid the distribution of substances. Because cavities are best suited for fluid streaming, primary and secondary body cavities are the most important components of circulatory systems. Large cavities serve to mix fluids and distribute them into different regions, but this process is undirected. Distribution is more effective when it occurs in restricted systems in the form of defined vessels (see also Ruppert & Carle 1983).

A vessel can ideally be imagined as a tube-like structure, in which fluid will flow in a certain direction. When the entire circulatory system is composed of such vessels, this system is called a 'closed circulatory system'. When vessels are only present in some part of the system, and fluid enters a reticulate system of lacunae or an open body cavity, the system is called an 'open circulatory system'.

In those cases where defined vessels are present, they represent in most cases a primary body cavity, that is, they are bordered by ECM. As such they are delimited by adjacent structures. For example, the dorsal and ventral vessels in annelids are defined by the position of the adjacent coelomic epithelia, although these epithelia do not belong to the vessel itself (Fig. 10.1.). In only four cases (Nemertini, Cephalopoda, Hirudinea, and Craniota), do vessels have their own epithelium and are therefore coelomic cavities. Circulatory systems lined by ECM are called, according to Ruppert & Carle (1983), 'blood vascular systems' (BVS), while endothelially lined systems are called here 'coelomic circulatory systems'. It appears useful to me to also apply distinctive terms to the fluid within these two types of systems. For this chapter, I will use the term 'hemolymph' for fluids within a BVS, and 'blood' only for fluids in coelomic circulatory systems. In practice, the term blood is often used when a more or less defined system of vessels is present, while in more

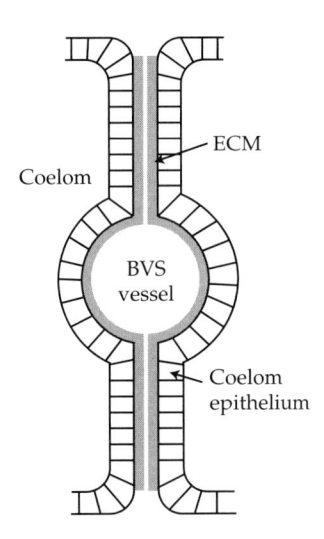

Fig. 10.1. Schematic representation of the relationship between blood vascular system (BVS) and coelom. The BVS forms as primary body cavity between ECM bordering the coelom epithelium.

open systems the fluid is named hemolymph. As the border between open and closed systems is more or less gradual, I regard it as better to use the type of body cavity for the definition of terms. It may sound illogical to call a system a *blood* vascular system when it does not contain blood, but this is based on the usage of available names.

The functions of circulatory systems are variable. The dominant function is transport of nutrients, oxygen, and excretes. In some taxa, circulatory systems are also used for hormone transport, temperature control, gamete reservoir, or even as brood space for embryos.

Circulatory systems in Metazoa

In the following, the compact organization will be neglected, although it allows a certain amount of fluid circulation within the lamina fibroreticularis of the ECM.

Cnidaria
In cnidarians, oxygen uptake and excrete release are performed directly by the epithelia but, especially in large polyps and in medusae, the distribution of nutrients is more problematic. Therefore, the gastrovascular system is often spacious, leads into the polyp tentacles, or is developed into a system of canals through which gastrovascular contents can float.

Large nematodes, Nematomorpha, Priapulida, Eurotifera, Seisonidea
In taxa with a (more or less) spacious primary body cavity, a relatively simple, undirected movement of fluid is possible. The driving force of this movement is the general body musculature and the dilution of substances by diffusion.

Acanthocephala
A peculiar lacunar system is present in the syncytial epidermis and musculature of acanthocephalans. It consits of main canals, from which smaller rami branch off, either in a regular or in an irregular pattern (Dunagan & Miller 1991). There are very few ultrastructural data, but the lacunar system appears to be neither primary nor secondary body cavity, but rather clefts in the syncytial epidermis (Holger Herlyn, Mainz, pers. commun.). According to Dunagan & Miller (1991), the lacunar system connects via pores to the outside and probably distributes nutrients absorbed through these pores in the body. Additionally, at least in some species, there is a 'rete system' between layers of the body wall muscles (Miller & Dunagan 1985, Dunagan & Miller 1991), which is said to be 'thin-walled', although ultrastructural data are not available. Movement of the fluid within the lacunar and the rete system is caused by the general body musculature. It has been proposed that the lacunar system is continuous with the interior of the hollow mucles (Miller & Dunagan 1976, 1977 for *Oligacanthorhynchus* and *Macracanthorhynchus* species), although Herlyn et al. (2001) found muscles in different species to be compact and not connected to the lacunar system (see Chapter 5).

Platyhelminthes
No circulatory system is present in flatworms, but within this taxon it can be observed how increase of body size poses problems for the distribution of substances. It can be assumed that the flat structure of the large flatworms allows the diffusion of oxygen and carbon dioxide through the integument and throughout the body. Distribution of nutrients is performed by an increasingly branched intestine, as present in triclads or digeneans and especially in polyclads (Fig. 10.2.), where intestinal diverticula reach almost any region of the body. In adult digeneans, there is a so-called lymphatic system in the parenchyma that may also have an assumed function in the distribution of substances (Fried & Haseeb 1991).

Nemertini
Nemerteans are, besides cephalopods and leeches, unique among invertebrates in possessing a coelomic circulatory system, that is, vessels lined by an epithelium (Turbeville 1991). Minimally, there are two lateral longitudinal vessels which are joined anteriorly and posteriorly, but this pattern can be complicated by the presence of further longitudinal and connecting vessels

CIRCULATORY SYSTEMS   193

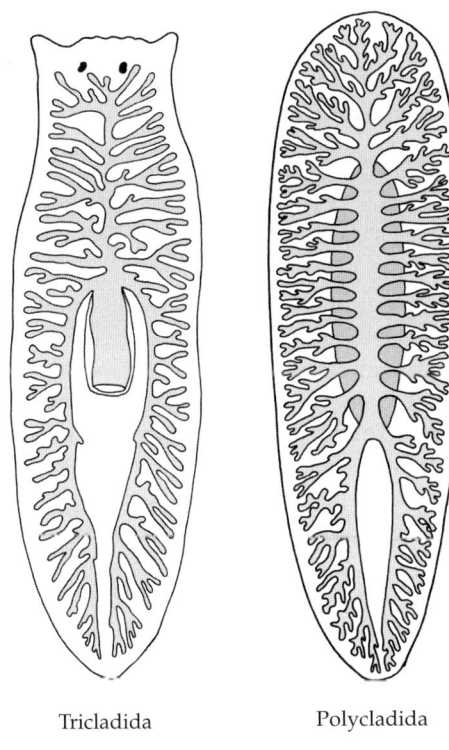

Fig. 10.2. Extension of the intestinal system by multiple branching in triclad and polyclad flatworms. Figures after Ax (1996).

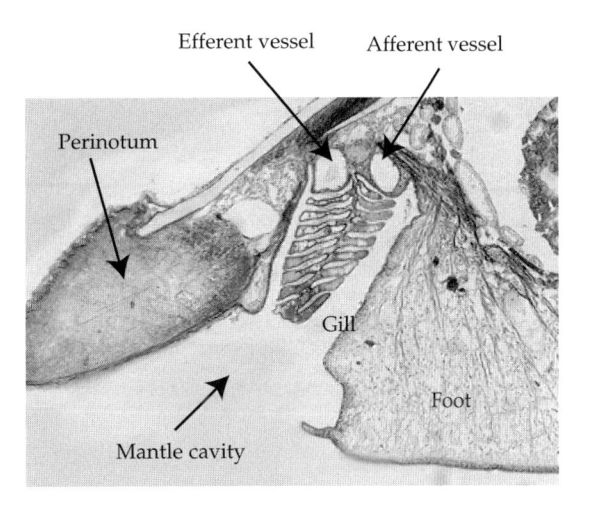

Fig. 10.3. Lateral cross section through a polyplacophore, *Lepidochitona cinereus*, showing efferent and afferent lacunar vessels above the gill.

(Turbeville 1991). The vessels originate by schizocoely (Turbeville 1986). A considerable number of possible functions has been correlated with the circulatory system in nemerteans (summarized by Turbeville 1991), including gas exchange, peptide circulation, transmission of fluid pressure changes, transport of neuroendocrine secretions (Ferraris 1985), and possibly excretion (at least in *Tubulanus*, Jespersen & Lützen 1987; see Chapter 9).

Mollusca
The circulatory system in all molluscs is a blood vascular system, which includes a heart composed of a central sinus, surrounded by a coelomic pericard. Due to myofilaments in the pericard wall, its contraction can propel hemolymph in an anterior direction. Most molluscs have an open system, where hemolymph enters sinuses and lacunae, but there is a roughly directed flow from the heart into the body, then through the gills (when present), and back to the heart. In all taxa, a dorsal vessel (the 'aorta') originates from the heart, and vessels are present as afferent and efferent vessels in the gills (Fig. 10.3.). Other vessel-like restrictions are rare. The heart is generally composed of one ventricle and two auricles, which receive hemolymph from the paired gills (see Scheltema et al. 1994 for aplacophorans, Haszprunar & Schaefer 1997 for Monoplacophora, Eernisse & Reynolds 1994 for Polyplacophora and Jones 1983 for Gastropoda and Bivalvia). When the number of gills is reduced to one, as in many gastropods, one auricle is also reduced. Cephalopods have evolved an almost closed system (Fig. 10.4.), in which arteries and veins can be distinguished. Both are connected by capillaries. This is obviously an adaptation to an increase in size and an increased metabolic rate (Wells 1983). Additionally, in cephalopods there is a trend towards endothelialized vessels. Both incomplete and complete endothelia have been recorded (Budelmann et al. 1997). There is one further case of endothelia in a mollusc circulatory system: the vessels next to ganglia in the pulmonate gastropod *Helix pomatia* have been reported to have an endothelium (Pentrath & Cottrell 1970). Both cases appear to be clearly derived from a blood vascular system, that is, from an unepithelialized state.

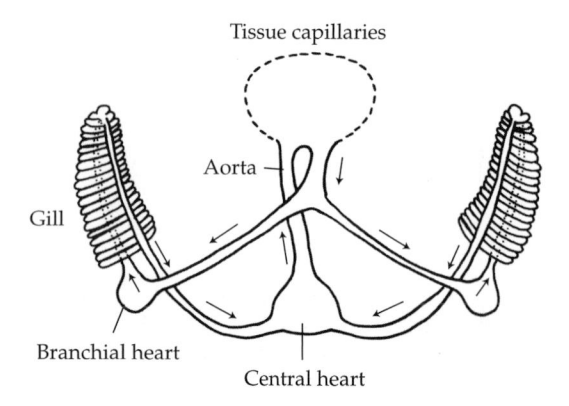

Fig. 10.4. Schematic circulatory system in cephalopods. After Ruppert et al. (2004).

Kamptozoa
Within the ECM, there is a system of small lacunae (or blood vascular system) which allows the circulation of fluid (Nielsen 2002a). This circulation is undirected and probably generated by general muscle contractions. Emschermann (1969) assumed that the 'star-cell complex', a muscular structure at the transition of calyx and stalk in all colonial kamptozoans, functions as a heart. Emschermann (1969) observed regular contractions and reconstructed a flow of fluid into the stalk. This makes sense concerning the demand of nutrients in the (growing) stalk, but Bartolomaeus (1993a) doubted the interpretation, first because it is not clear whether or how a current back from the stalk into the calyx is performed, and then because the complex is probably covered by ECM instead of allowing free through-flow of fluid. An alternative function, however, is not known.

Sipunculida
There is no restricted circulatory system in sipunculids, but the two coelomic cavities perform some of the functions from circulatory systems. Especially in the tentacular coelom, there appears to be a directional flow of fluid and there is a contractile element, the compensation sac (two sacs in some species) (Hyman 1959, Rice 1993, Ruppert & Rice 1995). The size and shape of the tentacles and the introvert (and the extension of the tentacular coelom) depend on the living habits of the species and are best developed in 'tentacle breathers' (Ruppert & Rice 1995). There is evidence for transport of oxygen between the coelomic cavities (see Chapter 11) and substances are distributed within the body by the spacious body cavity.

Echiurida
With the exception of *Urechis caupo*, which lacks a circulatory system, echiurids possess an almost closed blood vascular system (Pilger 1993). There are vessels in most parts of the body, especially in the proboscis and as a short dorsal and long ventral vessel (Amor 1973). The connection between the posterior end of the ventral vessel and the dorsal vessel is not known, but is assumed to be through small sinuses, therefore representing an 'open' fraction of the whole system. The vessels are bordered by ECM (Bosch 1975, 1984) which is followed by a myoepithelium. However, it is questionable whether this myoepithelium participates in fluid propulsion, or whether this is performed by the peristaltic contractions of the body musculature. The vessels in the extremely long proboscis of *Bonellia viridis* show a special structure in showing muscular rings (Bosch 1984). These may help to stabilize the vessels when the proboscis is extended (such extension can cover extreme lengths and ends in a helicoidal coiling of the proboscis, see Schembri & Jaccarini 1977). The hemolymph is colourless, but nevertheless respiratory pigments have been found in free 'blood cells' (see Chapter 11).

Annelida
The circulatory system of annelids is a blood vascular system. It is often considered to be a closed system, but in fact it is an *almost* closed system in the sense that the major part is organized as vessels, but in some regions, particularly close to the intestine, a system of sinuses is present. The two most important vessels are the dorsal vessel, in which hemolymph is propelled anteriorly, and the ventral vessel, in which it moves posteriorly (Fransen 1988, Gardiner 1992). These vessels are connected by ring vessels in the septa and by a sinus system around the intestine (Fig. 10.5.). Several further vessels support the appendages,

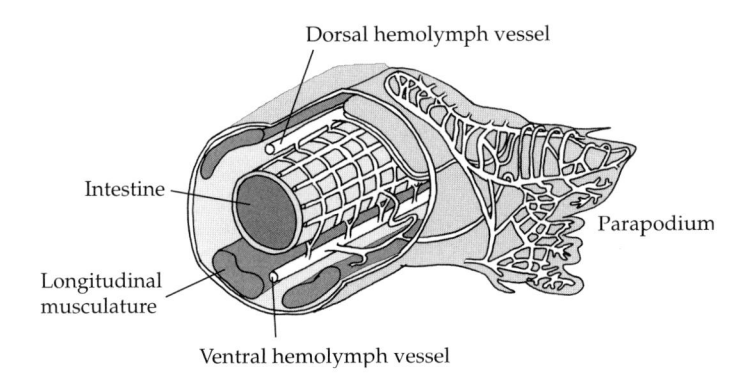

Fig. 10.5. Schematic representation of the circular system in polychaetes with well developed parapodia. After Ruppert et al. (2004).

the head and the gonads. Especially in clitellates, there is an intensive capillary system under the integument, indicating oxygen uptake over the entire integument. All vessels are formed between the epithelia of adjacent coelomic cavities. Because those epithelia are usually muscular (see Chapter 9), almost any region of a vessel is capable of some kind of peristaltic movement. In some regions, for example in the dorsal vessel of most species or in the lateral vessels of *Lumbricus*, the adjacent coelomic epithelium can be more muscular than usual and therefore serve as a pump ('heart'). The blood vascular system is lacking in small annelid species, but it is also lacking in some large polychaetes, for example, glycerids (Smith 1992). Westheide (1997) assumes that the septa evolved in annelids primarily to form blood vessels which, in turn, supplied the appendages. Leeches (Hirudinea), with the exception of *Acanthobdella*, are exceptional in possessing a coelomic circulatory system made up by the modified coelom.

Onychophora

Onychophorans possess an open circulatory system. The spacious body cavity, a mixocoel (see Chapter 8), is divided by a perforated transversal septum, the pericardial septum. Within the dorsal sinus lies the heart, a long tube with paired segmental openings, the ostia. The heart itself is composed of myoepithelial cells which are covered on both sides by ECM (Nylund et al. 1988). The heart lumen therefore represents the primary body cavity. It appears to develop from dorsal parts of the disintegrating embryonic coelomic epithelium which move further dorsal and enclose the primary body cavity (Anderson 1973). According to Pass (1991), antennal vessels arise from the anterior part of the heart, but these are the only vessel-like structures in the circulatory system of onychophorans.

Euarthropoda

The circulatory system of euarthropods resembles that of onychophorans. In general, the body cavity is separated by a pericardial septum and there is a dorsal tubular heart with ostiae. In chelicerates and tracheates, the heart is almost the only defined structure within the circulatory system, while in crustaceans several vessel-like sinuses create a roughly directed stream of hemolymph (Fig. 10.6A.). These vessels are tubular extensions from the heart. The absence of vessels in insects is related to the extensive tracheal system which extends into the entire body (Fig. 10.6B.; Wasserthal 1998). In crustaceans, the circulatory system is particularly well developed in larger species (e.g. Martin & Hose 1992), but less developed in smaller species (e.g. Hessler & Elofsson 2001 for the cephalocarid *Hutchinsoniella macracantha*). For example, in decapod crustaceans, the hemolymph leaves the heart through a cephalic and segmental arteria. These open in a system of lacunae and sinuses which finally collect hemolymph and lead it through the gills and from there back to the heart. The heart is muscular and is likely to be derived from embryonic coelomic wall tissue (see, e.g., Weygoldt 1958 for the amphipod *Gammarus pulex*). Wilkens (1999) assumes that the well developed systems have evolved from less complex systems with fewer vessels.

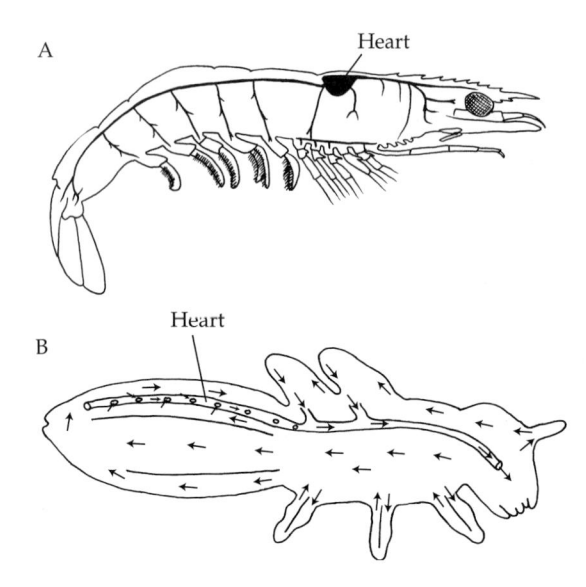

Fig. 10.6. Schematic representation of the circulatory system in malacostracan crustaceans (A) and insects (B). A. after McLaughlin (1980), B. after Westheide and Rieger (2007).

Chaetognatha
There is no elaborate circulatory system in chaetognaths, but Shinn (summarized in 1997) describes a restricted primary body cavity called the hemal system, which is present in the central part of the trunk, mainly around the intestine. The posterior part is directed towards the ovary and the anterior portion appears to fade in the anterior trunk region. There is evidence for ultrafiltration by podocytes in the coelomic epithelium (Shinn 1997) and it is hypothesized that the hemal system may receive dissolved metabolites from the intestine and distribute them by peristaltic waves of the trunk musculature anteriorly towards the head and posteriorly towards the ovary.

Bryozoa
The coelomic cavities may be used for circulation, but it is not clear whether the contribution of the coelom is 'needed', for example, for oxygen transport, or whether it plays a minor role as a transport system (see Mukai et al. 1997). The metacoel of bryozoans surrounds a tube-like primary body cavity, the funiculus, through which nutrients may be exchanged between individuals in a colony. The funiculus represents a blood vascular system (Carle & Ruppert 1983).

Phoronida
The circulatory system of phoronids is a blood vascular system, composed of some vessels and sinus networks, the latter occurring around the intestine and the gonads. There is a horseshoe-shaped vessel at the basis of the tentacular apparatus, from which one or two lateral and one median vessel run into the body, and several blindly-ending vessels into the tentacles (Fig. 10.7.; Herrmann 1997). Because phoronids live in tubes, it is assumed that the tentacles are the major site for oxygen uptake, which is stored in the hemolymph by hemoglobin (Garlick et al. 1979) and transported into the body. The lacunar network around the intestine suggests that the circulatory system may also distribute metabolites. There is no heart, but contraction waves have been observed, these are generated by the myoepithelia bordering the coeloms, between which the vessels (themselves representing the primary body cavity) are located.

Brachiopod
A circulatory system is present as a blood vascular system, it is delimited in several regions by the coelomic epithelia. Myoepithelial cells of these adjacent epithelia act as pumping organs ('hearts'), located in the stomach region. An intensive system of lacunae is present around the intestine, runs into the tentacles, and forms numerous branches within the mantle (James 1997). From this distribution it can be assumed that the circulatory system distributes nutrients from the intestine and oxygen from the tentacles, especially to the gonads in the mantle.

Hemichordata
Both enteropneusts and pterobranchs have a well defined blood vascular system as vessels and sinuses, but this system is much better developed in enteropneusts compared to pterobranchs (Benito & Pardos 1997). Both taxa correspond in possessing a dorsal and a ventral vessel, and a basiepidermal lacunar network. In the dorsal vessel, hemolymph moves anteriorly into a

CIRCULATORY SYSTEMS 197

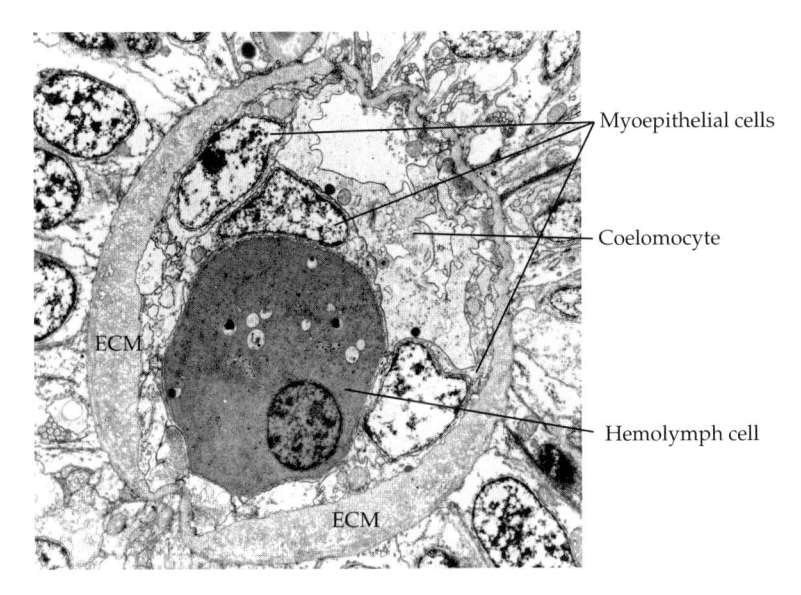

Fig. 10.7. Cross section through the tentacle of *Phoronis ovalis*, showing close association of blood vascular system and coelom. Both cavities are almost completely obliterated by floating cells (coelomocyte and hemolymph cell). Photo by courtesy of Alexander Gruhl, Berlin.

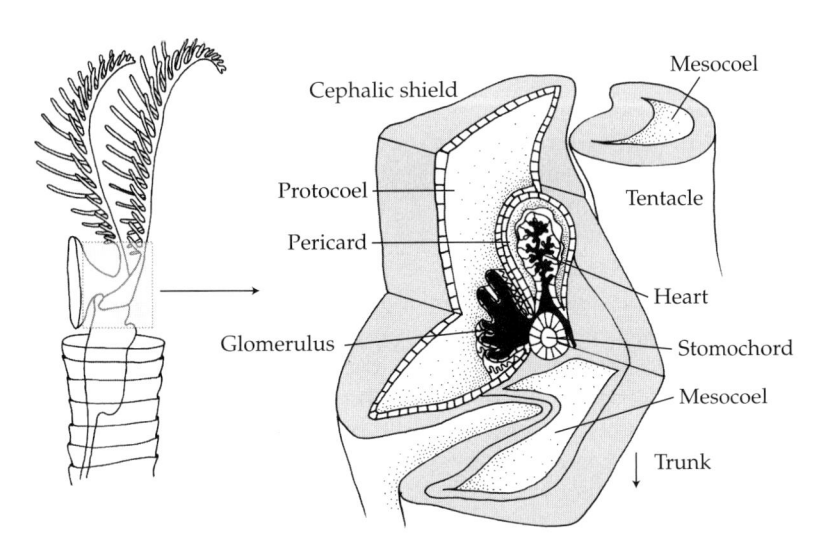

Fig. 10.8. The heart-glomerulus-complex in the pterobranch *Rhabdopleura compacta*. As indicated to the left, the right is a reconstruction from the region where tentacles, cephalic shield, and trunk meet. For explanations see text. After Mayer and Bartolomaeus (2003).

central sinus which is surrounded by a small unpaired coelomic cavity, the pericard (Fig. 10.8.). The myoepithelial cells of the pericard act as the propulsive force for the hemolymph, and therefore this structure can be called a heart (although the dorsal and ventral vessel are also surrounded by coelomic myoepithelia and may therefore contribute to propulsion). From the heart, the hemolymph enters another sinus called the glomerulus, which bulges into the protocoel (see Wilke 1972, Balser & Ruppert 1990 for enteropneusts, Dilly et al. 1986 for *Cephalodiscus* and Mayer & Bartolomaeus 2003 for *Rhabdopleura*). The presence of podocytes in the protocoelomic epithelium suggests ultrafiltration (see Chapter 9). From the glomerulus, a ventral vessel leads hemolymph posterior into the body. In entropneusts, there are additional, smaller vessels, for example, supporting the gill slits.

Echinodermata

In echinoderms, both the elaborate coelomic cavities and blood vascular system, usually named hemal system, are used as circulatory systems (see Harrison & Chia 1994). The hemal system is a primary body cavity and is delimited in most spaces by the epithelia of adjacent coeloms. In general, the hemal system is composed of a vertical sinus and the axial hemal vessel, from which three radial vessels originate (named hyponeural, gastric, and genital in asteroids). Asteroids, ophiuroids, and echinoids have a pumping organ, the heart, associated with the axial vessel. This originates as a hollow sphere, the larval pulsatile vesicle (Ruppert & Balser 1986), and later gains direct contact to the aboral part of the axial vessel. In holothuroids and crinoids, a heart is absent, but several vessels in holothuroids can contract and propel the hemolymph. The hemal system in holothuroids is more complex than in other echinoderms, in particular the taxon Aspidochirotida, which have a dense network (called rete mirabile) around the intestine and the left respiratory tree (e.g. Herreid et al. 1976 for *Stichopus moebii*), suggesting an intensive exchange of nutrients and oxygen. Although the circulatory systems (hemal system and coeloms) are well developed in echinoderms, their function is not always clear. At least oxygen transport appears to be of minor importance, because respiratory pigments are present only in ophiuroids and holothurians.

Tunicata

There is an open blood vascular system and a heart. The heart originates as a compact mass which forms a coelomic cavity by schizocoely. This hollow vesicle invaginates, creating a coelomic sheath (the pericardium) around a primary body cavity sinus, the hemocoelic sinus, or heart (Nunzi et al. 1979). The lower lips of the pericardium do not fuse, but the hemocoelic sinus is enclosed here by ECM (Fig. 10.9.; Oliphant & Cloney 1972). There are a number of different cells floating in the hemolymph which have a number of different functions (see Burighel & Cloney 1997).

Acrania

There is a well developed blood vascular system in acranians, forming an almost closed circulatory system (Ruppert 1997). There is no distinct heart, propulsion is performed by some of the vessels, and those in close connection with the coelom are usually especially contractile, because of the adjacent muscular coelom epithelium. There are large vessels, the dorsal one leading hemolymph caudally and the ventral one leading anteriorly. There are a number of connecting sinuses, especially through the gills, and a plexus around the intestine (Rähr 1979). From

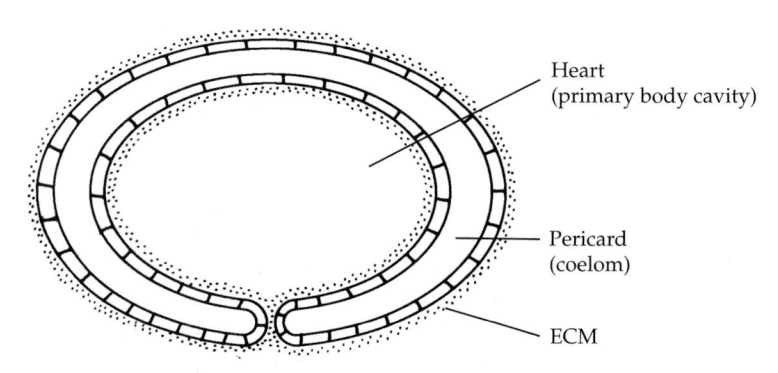

Fig. 10.9. Schematic cross section through the heart in tunicates. Modified after Oliphant and Cloney (1972).

the interstine a sinus, the vena portae, collects hemolymph and leads it to a ventral extension of the intestine, the hepatic cecum. From there, hemolymph flows back into the main circuit through the vena hepatica. The names of the vessels indicate that the whole system closely resembles the craniote circulatory system.

Craniota
The blood vessels in craniotes have an epithelium and are therefore a coelomic circulary system. The system is closed and resembles the one in acranians. Within craniotes, major modifications are connected to the transition from gill to lung respiration (Fig. 10.10.; see Starck 1982). Fish-like craniotes have a tube-like heart composed of a sinus venosus, an atrium, and a ventricle. A truncus arteriosus branches into the gills, from which vessels lead into the head and into the body, where they supply all organs and finally lead back to the sinus venosus. Amphibian larvae already have two atria. With the growth of the lung during amphibian development, the first branchial vessel leads to the lung. The undivided ventricle pumps 'mixed' blood with portions from the oxygen-poor blood from the body and the oxygen-rich blood from the lung into the body. In amniotes, there is a strong tendency towards a separation of the ventricle into two chambers, which is completed convergently in birds and mammals. This also leads to a reduction in the number of the former branchial vessels. In a simplified circulatory system of birds and mammals (Fig. 10.11.), the heart pumps oxygen-rich blood into the head and the body. In the body, blood is distributed into several organs. Only from the intestine is there is a connection to the liver, the vena portae. From all other organs, oxygen-poor blood is collected, led back to the heart and from there into the lungs.

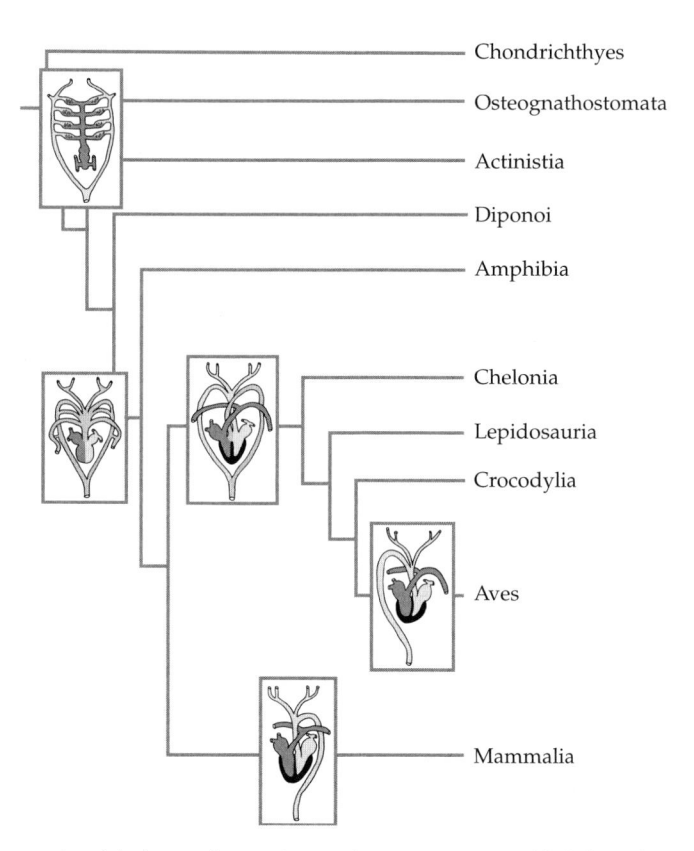

Fig. 10.10. Schematic representation of the heart and connecting vessels in craniote taxa. Modified after Wehner and Gehring (1995).

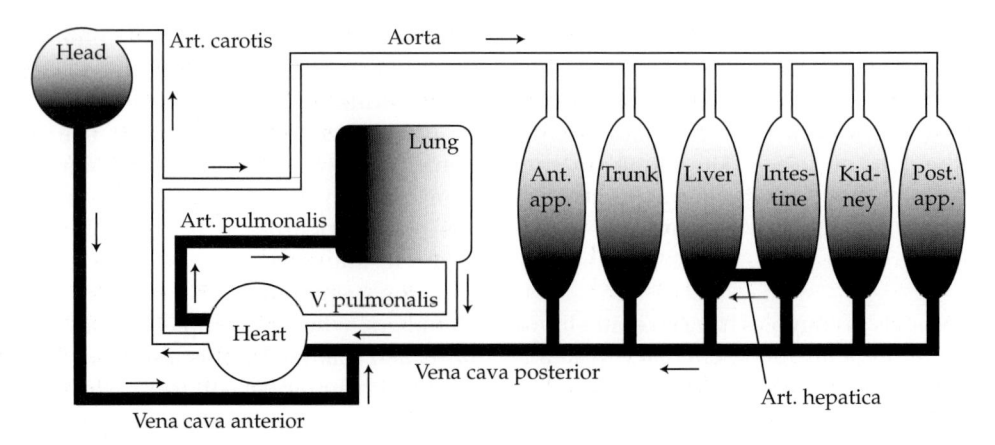

Fig. 10.11. Simple schematic representation of the circulatory system in mammals and birds. After Westheide and Rieger (2007). Oxygen-rich blood is white, oxygen-poor blood is back.

Table 10.1. Overview on circulatory systems in metazoans. BVS = blood vascular system (primary body cavity), CCS = coelomic circulatory system. 'Intermediate BVS' means that there are a number of vessels, but also lacunar systems, usually around the intestine. For references see text.

	Nature of circulatory system	Open or closed system
Acanthocephala	Epidermal and subepidermal lacunar system, fine structure is not clear	—
Nemertini	CCS	Closed
Mollusca	BVS (epithelialized vessels present in Cephalopoda and *Helix*)	Open
Kamptozoa	BVS	Open
Sipunculida	No extended circulatory system, but coeloms, especially tentacular coelom, act as circulatory systems	(Closed)
Echiurida	BVS	Almost closed
Annelida	BVS	Almost closed
Onychophora	BVS	Open
Euarthropoda	BVS	Open
Chaetognatha	BVS	Spatially restricted to hemal sinus around central intestine
Bryozoa	BVS (funiculus)	Spatially restricted to funiculus
Phoronida	BVS	Intermediate
Brachiopoda	BVS	Intermediate
Hemichordata	BVS	Intermediate
Echinodermata	BVS	Intermediate
Tunicata	BVS	Open
Acrania	BVS	Closed
Craniota	CCS	Closed

Conclusions

From the previous summary on circulatory systems, it becomes clear that there is a strong connection between body cavities and circulatory systems. Either, body cavities themselvs distribute fluids, or the epithelia of coeloms create spatially restricted spaces in the primary body cavity which then serve as a circulatory system. There may be a general tendency from open to closed systems (within Mollusca, within Trochozoa towards annelids, within Deuterostomia towards Myomerata), but this is no universal trend, as can be seen by the poorly developed circulatory system in euarthropods. However, euarthropods show how closely the extent of a circulatory system is connected to its functional needs. When, for example, the tracheal system is dominant, there is no need for an extensively developed circulatory system. The advantages of a closed system are that a smaller total amount of hemolymph/blood is needed and that a higher pressure can be applied to the fluid (Budelmann et al. 1997).

Propulsion of the fluid is by muscular action and this is in many cases performed by the myoepithelia of adjacent coeloms. This may account to 'simple' vessels, or develop into strongly muscular structures that we then call hearts. It appears that hearts evolved convergently in several taxa, especially in molluscs, hemichordates, tunicates, and craniotes. Hypotheses on the origin of the circulatory system rely on the discussion of the evolution of the coelom (see Chapter 8), but I regard it as unconvincing to assume a homology and common origin of circulatory systems.

The expression of homologous genes in *Drosophila* and craniotes (Bodmer & Venkatesh 1998, Tanaka et al. 1999) may stimulate a different view, which in the extreme leads to the hypothesis that the bilaterian ancestor had a circulatory system including a contractile vessel (De Robertis & Sasai 1996). I regard it more likely that the respective genes were present in a common ancestor, but had either different functions or regulated very basal cellular processes. Further examples are the vascular endothelial growth factors (VEGF), which are important in the development of blood vessel epithelia in craniotes. Homologous genes are expressed in the cnidarian *Podocoryne carnea* and probably shape the tube-like extension of the gastrovascular system into the tentacles (Seipel et al. 2004). Here, it also appears that the genes originally had general patterning functions, and cannot be regarded as markers for a certain structure.

CHAPTER 11

Respiratory systems

Almost all animals need oxygen for their metabolism, although there are cases of tolerance towards low oxygen contents and even anaerobic metabolic pathways in some species. Oxygen is present in the air and in the water. It can enter the body over the whole surface by diffusion. Within the body, the diffusion decreases with distance to the surface and is dependant on the composition of the tissue and the consumption of oxygen on its way through the body. Attempts have been made to calculate the diffusion distances in nematodes, because their bodies are almost cylindrical (Atkinson 1976, 1980). Following these calculations, nematodes with a diameter of 100 µm will receive sufficient oxygen throughout their body only by diffusion (Nicholas 1991). Although such calculations can only be taken as rough indications, it becomes clear that the increase of size requires further strategies to supplement the 'normal' diffusion of oxygen through the body surface. These strategies are the evolution of circulatory systems to distribute oxygen from the region of uptake throughout the animal (see Chapter 10), optimization of oxygen uptake by carriers with high (but reversible) affinity to oxygen, and by an increase of the surface area itself. Especially when circulatory systems form (at least partly) closed systems with a directed flow of hemolymph or blood, or when the body is covered by a cuticle, it appears useful to design a certain region where the hemolymph or blood is oxygenated. In these cases, special respiratory structures are formed. In general, all respiratory structures are similar in the sense that they expose as much surface as possible to the external medium and keep the tissue between the external medium and the internal circulatory system as thin as possible. In general, respiratory structures functioning in water are called 'gills' while those functioning in air are called 'lungs'. Gills are usually external structures, that is, outgrowths of the epithelium, while lungs are internal structures.

Respiratory organs and pigments in Metazoa

In a number of taxa, neither respiratory structures nor respiratory pigments have been detected. While it is quite convincing to assume that respiratory structures did not, or only to a small extent, escape the existing investigations, respiratory pigments have not been looked for in several taxa. As one example, among gnathostomulids there are reddish species (*Haplognathia rosea*, Sterrer 1969 as *Pterognathia rosea*), although respiratory pigments are unknown from this taxon. Therefore, we might expect a further distribution of pigments than is currently known. In the following account, those taxa not mentioned lack report of respiratory organs or respiratory pigments.

Gastrotricha
No particular respiratory structures are known, but Colacino and Kraus (1984) found hemoglobin in *Neodasys* sp., where it is located in mesodermal cells surrounding intestine and muscles. In another gastrotrich, *Turbanella ocellata*, no hemoglobin was found (Colacino & Kraus 1984).

Nematoda
There are no respiratory structures in nematodes and oxygen uptake appears to be possible through the cuticle. This is sufficient in small

nematodes, but several large, parasitic nematode species have adopted to low oxygen levels. Intracellular and extracellular hemoglobins, sometimes called nemoglobins (Blaxter 1993), have been shown to be present in many parasitic and a few free-living species (summary in Blaxter 1993 and Wright 1998). A comparison between the two free-living marine species *Enoplus brevis* and *E. communis* indicates a possible correspondence between environmental oxygen content and the amount of respiratory pigments: *E. brevis*, who lives in low oxygenated muddy sediments in estuarines has many more respiratory pigments than *E. communis* who lives in more oxygenated sediments (Ellenby & Smith 1966, Atkinson 1976, 1980). Most respiratory pigments are present intracellularly, but *Ascaris suum* has large quantities of extracellular hemoglobin in the primary body cavity (Wright 1998). This has an extremely strong affinity, but low dissociation rate, to oxygen. It has been shown that the extracellular hemoglobin functions enzymatically to oxidize nitric oxide (NO) and thereby keeps the body cavity free of oxygen (Minning et al. 1999), probably to keep the anaerobic pathways functional. It is interesting that *Ascaris* also contains hemoglobin in the cuticle and in the body wall. Both these hemoglobins have a lower affinity to oxygen compared to the extracellular hemoglobin, but much higher dissociation rates. This suggests that *Ascaris* has a localized uptake of oxygen, possibly to supply the musculature, but keeps the central region of the body oxygen-free. A very unique function of a hemoglobin was found in *Mermis nigrescens* (Mermithida), where it forms intracellular crystals that surround the photoreceptor and therefore have a light-shadowing function (Burr et al. 2000).

Priapulida
Priapulids from the genera *Priapulus* and *Priapulopsis* have caudal appendages which are highly branched (Fig. 11.1.). Fänge and Matthisson (1961) held *Priapulus caudatus* under low oxygen supply and observed a muscular constriction of the appendage after the water was again oxygenated. The primary body cavity continues into the caudal appendages and it is evident that the contraction of the appendage pushes oxygenated hemolymph into the body. The body cavity contains cells, some of which include haemerythrin as a respiratory pigment (Fänge 1950, Fänge & Åkesson 1951, Weber et al. 1979). Caudal appendages are lacking in other taxa, but hemerythrin is also present (at least in *Halicryptus spinulosus*, Schreiber et al. 1991). The long caudal appendage of *Tubiluchus*-species may not be of respiratory function but rather an adaptation to the meiobenthic life, as such tail structures are abundantly distributed in different meiobenthic taxa to hold onto sand grains (Ax 1966).

Platyhelminthes
There are no respiratory structures in flatworms, and Jennings (1988) noted that the surface area versus body volume ratio in flatworms should allow the oxygen supply of the whole body by simple diffusion. This is aided by the dorsoventral flattening which decreases the distances from the body surface to any internal body region. Respiratory pigment (monomeric intracellular hemoglobin) is known from several species, but with the exception of *Phaenocora* sp. (in Vernberg 1968), respiratory pigments appear to be present particularly in parasitic species (as well those belonging to the Neodermata as

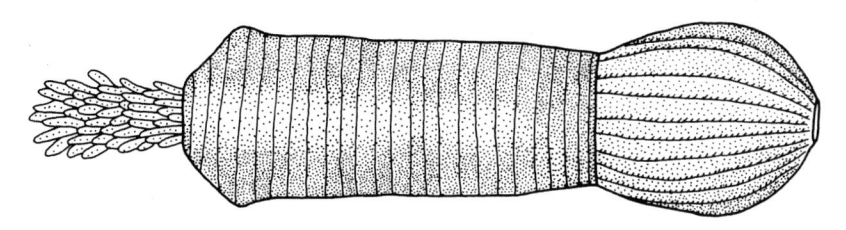

Fig. 11.1. Drawing of *Priapulus caudatus*, showing the caudal appendage with assumed respiratory function.

non-neodermatans; Lee & Smith 1965, Vernberg 1968, Phillips 1978, Tuchschmid et al. 1978, Jennings & Cannon 1985, 1987). As endoparasitic or endosymbiotic species live in oxygen-poor environments, the possession of hemoglobin may enable them to compete with the host for 'their' oxygen. As one example, three pterasteri-colid flatworms have been found to possess hemoglobin while this is lacking in their starfish hosts (Jennings & Cannon 1985).

Nemertini
Nemerteans have no respiratory organs and oxygen uptake presumably takes place over the entire surface. Considering the large size of many species and the presence of a well developed circulatory system, there are surprisingly few details known about respiratory pigments in nemerteans. A very small hemoglobin molecule is known from *Cereratulus lacteus* (Vandergon et al. 1998, Pesce et al. 2002a). According to Fänge (1969), hemoglobin can also be present in blood cells (in species of *Drepanophorus*, *Barlasia* and *Euborlasia*) as well as extracellularly in the blood plasma of *Polia sanguirubra*. Wittenberg et al. (1965) reported hemoglobin from the nervous tissue of *Amphiporus* sp.

Mollusca
The ancestral respiratory organs in molluscs are gills. Although one basal taxon, the Solenogastres, lacks gills, one pair is present in the other aplacophoran taxon, Caudofoveata (see Table 11.1. for a summary of mollusc respiratory organs). Each gill is structured like a feather with a central trunk and lateral extensions to both sides, this structure is called a bipectinate ctenidium. With the exception of neopilinids, which possess monopectinate ctenidia, and scaphopods, which possess no gills at all, all other marine representatives have bipectinate ctenidia (Table 11.1.). In polyplacophorans (chitons) and in nautiloid cephalopods, their number was raised, in some chitons up to 88 pairs. With the exception of cephalopods, ctenidia are ciliated and are therefore able to create a water current. In the region of the gills, the open circulatory system has a more regular structure, forming afferent and efferent vessels which allow an effective distribution of oxygenated hemolymph into the body (Fig. 10.3.; see Chapter 10). Within bivalves and gastropods, respiratory structures are the subject of exciting evolutionary transitions. In bivalves, only the basal protobranchs have bipectinate ctenidia, while in the remaining taxon Metabranchia, the gill trunk fused with the upper mantle cavity, while the lateral extensions became fine, long filaments. Each filament hangs down into the mantle cavity, bends up and contacts either the mantle or the visceral body in the foot region, therefore separating within the mantle cavity two chambers. This type of gill is called lammelibranch gill. By directing the water current through the gills from the lower into the upper chamber, food particles are filtered. The gills therefore have, besides respiration, nutritive function. Ciliary beating collects food particles and mucus and transports it to the so-called food groves on the lowest point of each gill filament. Within these food groves they are transported towards the mouth. In gastropods, the possession of the shell and especially the presence of gills and anus within the same mantle cavity, appears to create some problems, because feces may pollute water before respiration. Basal marine gastropods exhibit a variety of solutions to create a directed water current that first passes the gills and then the anus. This is excellently described by Ruppert et al. (2004) and includes the presence of slits (e.g. *Pleurotomaria*) or perforations (e.g. *Haliotis*, Fissurellidae) in the shell, or the evolution of secondary gills (Patellogastropods). What prevails is that one gill is reduced and the water current runs from left to right through the mantle cavity, first passing the remaining left gill, then the anus, and then leaving on the right side (Fig. 11.2.). This remaining gill is originally bipectinate, but becomes monopectinate in a number of taxa. In some cases, as in the slipper shell (*Crepidula fornicata*), the gill filters food, comparable to metabranch bivalves. In opisthobranchs, there is a strong tendency to reduce the ctenidium and to form new gill structures. With the exploration of land by pulmonates, gills became unnecessary

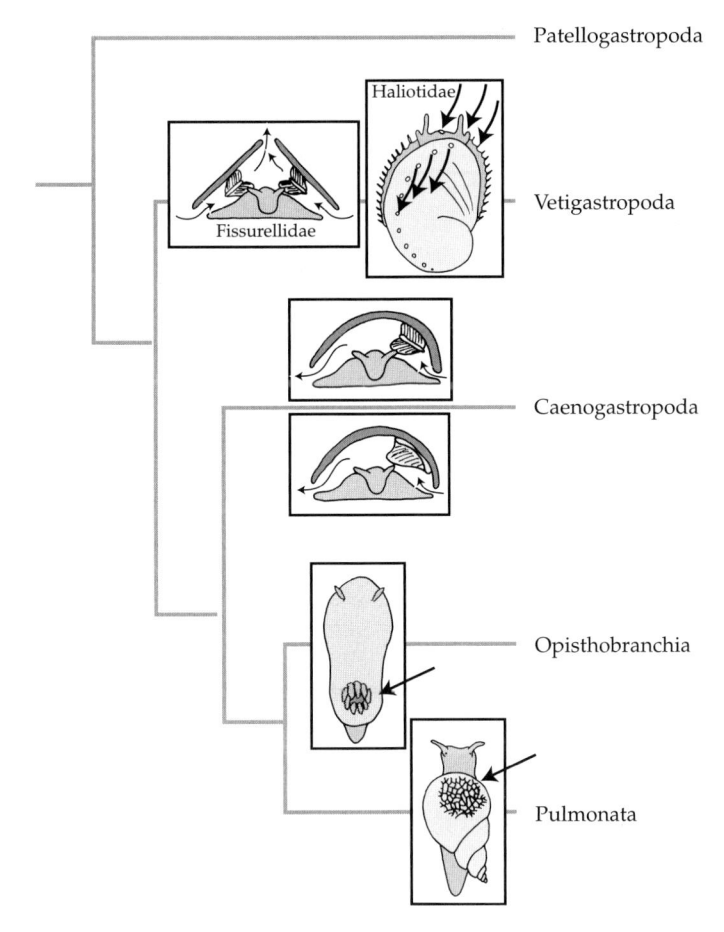

Fig. 11.2. Selected possibilities of water current (arrows) through the mantle cavity of certain gastropods. Arrow in ophistobranch drawing indicates secondary gills, arrow in pulmonate drawing indicates the 'lung'. Drawings modified after Ruppert et al. (2004).

Table 11.1 Diversity of respiratory structures within molluscs.

Solenogastres	Absence of gills
Caudofoveata	1 pair of bipectinate ctenidia
Polyplacophora	6–88 pairs of bipectinate ctenidia
Neopilinida	3–6 pairs of monopectinate ctenidia
Gastropoda: assumed groundpattern	1 pair of bipectinate ctenidia
Gastropoda: Pulmonata	Lungs
Cephalopoda: Nautiloidea	2 pairs of bipectinate ctenidia
Cephalopoda: Dibranchiata (Coleoida)	1 pair of bipectinate ctenidia
Scaphopoda	Reduction of gills
Bivalvia: Protobranchia	1 pair of bipectinate ctenidia
Bivalvia: Metabranchia	Filamentous gills

and were reduced. Increased oxygen uptake is now performed in a specialized region of the mantle cavity, which often shows some kind of surface extension and is called the lung.

Molluscs generally possess hemocyanins as respiratory pigments in the hemolymph, but they can also have hemoglobin. Hemoglobin occurs primarily as intracellular, small, mostly monomeric molecules in different tissues, for example, in the radular muscle or in nerve tissue (Wittenberg et al. 1965, Bonaventura & Bonaventura 1983), intracellularly in body cavity cells or, only rarely, as large, polymeric extracellular molecules, for example in the gastropods

Planorbarius corneus and *Helisoma trivolvis* (Terwilliger 1980, Bonaventura & Bonaventura 1983). In general it appears that hemocyanins are the main oxygen transport pigments (at least in most species) and hemoglobins remain mainly intracellular to support oxygen supply for certain tissues.

Sipunculida

Respiration can be concentrated on the introvert, or the tentacles, or it occurs over the whole body surface (Ruppert & Rice 1995). In species living in crevices, the tentacles are directed towards the open water and are assumed to be the major region for respiration (Ruppert & Rice 1995). Within the tentacle and the trunk coelom, hemerythrocytes are found which contain hemerythrin as respiratory pigment (Klippenstein 1980, Rice 1993). This is the most abundant type of coelomocytes in sipunculids. The hemerythrins in the tentacle coelom and the trunk coelom differ from each other and a third type is present in the muscles (Klippenstein 1980). Rice (1993) does not exclude that complete hemerythrocytes can be exchanged between the two coelomic cavities. Manwell (1960) compared the content of hemerythrin in the tentacle coelom (denoted as 'vascular system which supplies the tentacles', but because a circulatory system is lacking, this must be the tentacle coelom) versus the coelomic fluid in two species with a different ecology. In *Dendrostomum*, which is endobenthic but sticks its tentacles out of the sediment to take up oxygen, the affinity of the coelomic fluid to oxygen is higher than that of the hemolymph. This ensures that oxygen, taken up by the hemolymph in the tentacles, diffuses into the coelom. In *Siphonosoma*, which lives in a tube and takes up oxygen with the whole body surface, the tentacles are only needed for food uptake, and therefore the affinity for oxygen of the hemolymph and the coelomic fluid is almost similar. Therefore, it appears that according to the lifestyle, certain regions of the body play a dominant role in oxygen uptake, and that the exchange of oxygen is performed by different affinities of the hemerythrins in the two coeloms.

Echiurida

There is no specialized respiratory structure in echiurids, and oxygen is absorbed over the entire surface of the body, presumably also through the anal sacs and, in *Urechis caupo*, also through the cloaca which is regularly filled with water (Pilger 1993, Julian et al. 1996). Echiurids possess hemoglobins as respiratory pigments. These are found, according to Pilger (1993), within free cells in the circulatory system, the blood cells (hemolymph cells would be more appropriate); although Vinson & Bonaventura (1987) report three different hemoglobins from coelomocytes, received by washing the coelom of *Thalassema mellita*. In *Urechis caupo*, where a circulatory system is absent, such blood cells are found in the coelom (Pilger 1993) and the hemoglobin is a tetramer (Garey & Riggs 1984). In all other species the hemoglobins are, as far as investigated, mono- or dimers (Vinson & Bonaventura 1987 for *Thalassema mellita*). In addition, hemoglobin appears to be present in the muscle tissue (Pilger 1993).

Annelida

Annelids generally take up oxygen over their whole body surface. Polychaetes in particular possess several appendages, and often dorsal appendages on the parapodia are specialized as gills (Storch & Alberti 1978, Gardiner 1988). Such gills are predominantly found in species living in tubes or in large-bodied, burrowing species (Ruppert et al. 2004). Gills can be present along most of the body or concentrated at the anterior end as, for example, in cirratulids. Storch & Alberti (1978) have documented a variety of different internal structures of the gills. They all contain part of the circulatory system (as primary body cavity). Such hemal lacunae can be formed either between a central extension of the coelomic cavity into the tentacles and the subepidermal ECM (as in *Scalibregma inflatum*), as vessels entirely enclosed by the coelomic epithelium (as in *Malacoceros fuliginosus*), or the coelom can be absent and only a central hemal vessel present from which extensions almost invade the epidermal cells (as in *Dendronereides heteropoda*) and in this way reduce the distance between external

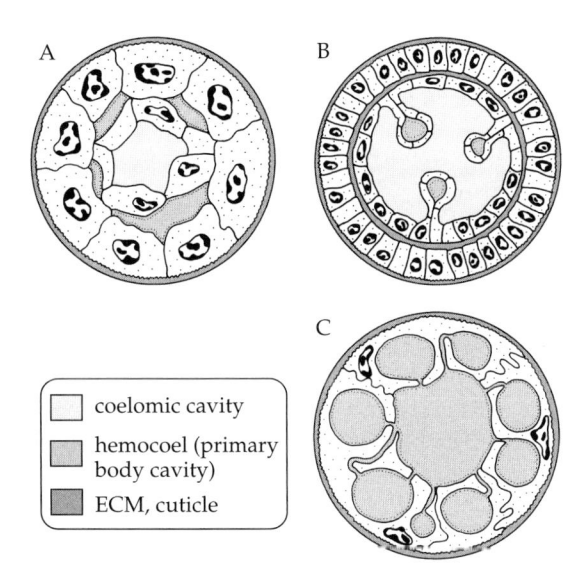

Fig. 11.3. Different structures of annelid gills as seen in cross sections. A. *Malacoceros fuliginosus*, B. *Scalibregma inflatum*, C. *Dendronereides heteropoda*. After Storch and Alberti (1978).

medium and hemal vessel to a minimum (Fig. 11.3.). The scale-worms (including, e.g., Aphroditidae and Polynoidae) possess dorsal scales. Between the dorsal body surface and the scales (elytra), a water current is created by cilia or, in *Aphrodite*, by the movement of the scales (Ruppert et al. 2004). Annelids have a wide variety of respiratory pigments. Several types of hemoglobins occur (e.g. Weber 1978, Garlick 1980, Dewilde et al. 1996, Zal et al. 1996), including smaller intracellular and large extracellular ones. Sabellidae, Flabelligeridae, and Serpulidae possess a certain type of hemoglobin which differs in having a heme with a formyl-group instead of a vinyl-group (van Holde 1997). This molecule, called chlorocruonin, is greenish in colour. Even hemerythrin has been found in *Magelona papillicornis* (Wells & Dales 1974).

Onychophora
There is a well elaborated system of tracheae in onychophorans, through which air can enter the body. The openings of the tracheae are scattered and numerous, up to 75 openings have been counted per segment (see Storch & Ruhberg 1993). Tracheae are covered by a thin cuticle and it appears that oxygen diffuses through the cuticle and the tracheal epidermis into the body. The abundance of trachaea suggests that oxygen is distributed throughout the body. The hemolymph possesses different types of hemocytes (Storch & Ruhberg 1993) and hemocyanin is known to be present in the hemolymph (Kusche et al. 2002).

Euarthropoda
Respiratory organs are present in most euarthropods and are only lacking in smaller species. The marine euarthropods (Xiphosura and Crustacea), as well as the extinct trilobites (Whittington & Almond 1987, see also Fortey 2000) have gills as appendages on the basal part (the coxa) of certain legs. Gills as extensions from the legs can be traced back even further to fossils of the euarthropod stem lineage. They are present, for example, in the Cambrian species *Kerygmachela kierkegaardi* (Budd 1993, 1999), *Parapeytoia yunnanensis* (Hou et al. 1995), and *Opabinia regalis* (Budd 1996) (Fig. 11.4.). With terrestrialization, new respiratory organs evolved. The extinct eurypterids probably possessed two kinds of respiratory organs, gills homologous to those in Xiphosura and so-called 'Kiemenplatten', which indicate that they might have been partly terrestrial (Manning & Dunlop 1995). In arachnids (the terrestrial chelicerates) respiratory organs are originally so-called book lungs, in which numerous lamellae project into a protected chamber that opens to the exterior by a slit. Probably such book lungs evolved from internalization of book gills (Scholtz & Kamenz 2006). Book lungs are present in Scorpiones (four pairs), Uropygi (one or two pairs), Amblypygi (two pairs) and basal aranean taxa (two pairs), while they are replaced by tracheae—tube-like invaginations of the body surface—in derived spiders (Neocribellata), Pseudoscorpiones, Solifugae, Opiliones, and Acari. Tracheae are also present in chilopods, progoneates, and insects, and have been name-giving for a common taxon Tracheata. However, as the monophyly of Tracheata has been recently challenged (see Chapter 2), a convergent evolution of tracheae (apart from a certain independant

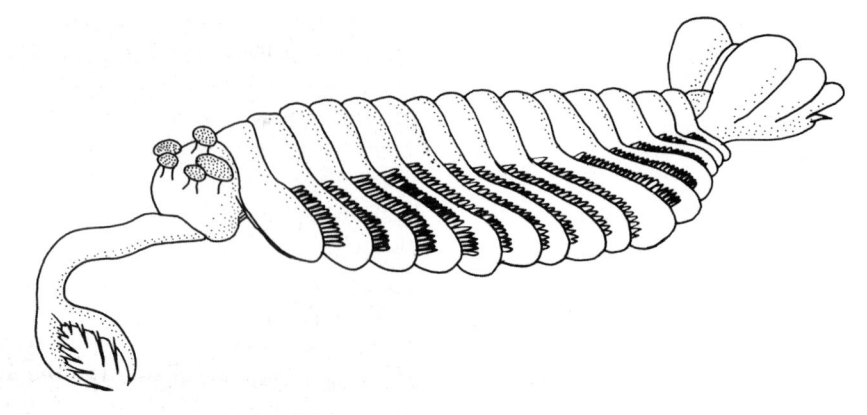

Fig. 11.4. Reconstruction of the Cambrian stem lineage arthropod *Opabinia regalis*, showing presumed gills on the lateral extensions (after Budd 1996).

evolution among arachnids) has become probable. This is supported by the finding that a structural analysis of tracheae cannot indicate arguments for their common origin (Hilken 1998).

The respiratory pigment abundantly present in euarthropods is hemocyanin (see Burmester 2002 for a summary), but in branchiopod crustaceans and some insects, hemoglobin is present (Terwilliger 1980, Ilan et al. 1981, Mangum 1983, Vinogradov 1985).

Phoronida
There are no respiratory structures in phoronids, but an intracellular hemoglobin has been shown to be present (Garlick et al. 1979).

Brachiopoda
There are no specialized respiratory organs, but the hemolymph of lingulids contains hemerythrin (Manwell 1960, Joshi & Sullivan 1973), this is present in hemolymph cells ('blood cells').

Hemichordata
Enteropneusts possess a series of gill slits in the pharyngeal region connecting the intestinal lumen with the external. These gill slits filter food particles from the water and a respiratory function is also assumed, because vessels of the blood vascular system (BVS) are present close to the gill slits (Benito & Pardos 1997). Respiratory pigments have not been clearly identified, but the BVS fluid contains particles that correspond in size with hemoglobin (Benito & Pardos 1997). In pterobranchs, only the *Cephalodiscus* species possess one pair of gill pores, while these are completely lacking in *Rhabdopleura*. It can be assumed that oxygen is taken up by the tentacular apparatus and is distributed through the coelom either by diffusion or by movement of the coelomic fluid.

Echinodermata
As most echinoderms are rather large animals with a peripheral endoskeleton, they all depend upon further respiratory structures (see Shick 1983 and Ruppert et al. 2004 for summaries). Respiration is, in all taxa, performed by the tube feet, which are thin-walled and contain extensions of the water-vascular system (see, e.g., Hajduk 1992 for ophiuroid tube feet). In crinoids, tube feet are the only respiratory structures, and this may be explained by the comparably large surface area due to the tentacles. In asteroids, there are further small outpockets of the body wall, the papulae (Fig. 11.5.). These resemble tube feet, but do not have terminal suckers and include extensions of the perivisceral coelom. Ophiuroids have special invaginations on the circumoral surface, the bursae (Pentreath 1971). Echinoids have peristomial gills, which are extensions of the body wall on the oral side of the animal. Holothurians, finally, have a range of possibilities for respiration. Because the

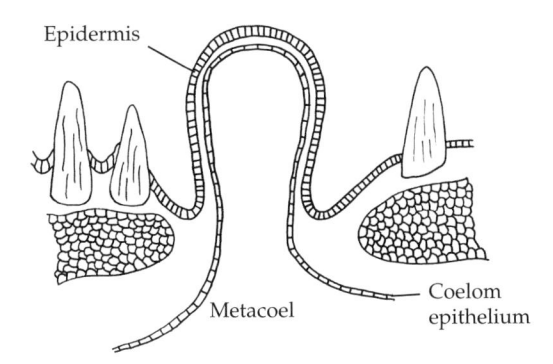

Fig. 11.5. Structure of papulae in asteroid echinoderms, showing how minimal the tissue between external and coelom is (after Ruppert et al. 2004).

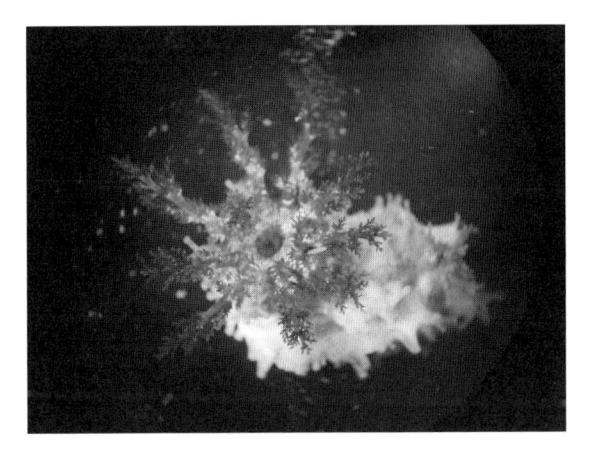

Fig. 11.6. *Cucumaria sp.*, a mediterranean holothurian (Echinodermata) with tentacles around the mouth opening, which aid in respiration. Photo by courtesy of Peter Grobe, Berlin.

endoskeleton is reduced to small, isolated plates, the whole body surface can be used in oxygen uptake and is the most important site in some, thin-walled species. Those species which have highly branched circumoral tentacles can use these for oxygen uptake (Fig. 11.6.). Important structures, however, are the internal respiratory trees (or water lungs). These branch from the cloaca, which contracts and thereby constantly pumps water into the respiratory trees (Newell & Courtney 1965, Brown & Shick 1979). The respiratory trees are only lacking in species of the taxa Elasipodia and Apodida. Most echinoderms appear to contain no respiratory pigments, hemoglobin has only been found in hemocytes of several holothurians (presumably in Dendrochirotida and Molpadida, but not in Aspidochirotida and Apoda; Manwell 1966, Terwilliger & Read 1972, Smiley 1994). Hemoglobin is also present in those ophiuroid species that lack bursae, such as *Hemipholis elongata*, *Ophiactis rubropoda*, and *O. virens* (see Byrne 1994).

Tunicata

There are no respiratory organs in tunicates. Although a circulatory system and several types of floating cells are present (see Chapter 10), this hemolymph does not carry oxygen (Baldwin et al. 1984). In the draft genome of *Ciona intestinalis*, genes were found for hemocyanin (Dehal et al. 2002) and for four different globins (Ebner et al. 2003). The *Ciona* globins cluster together and are related to insect and craniote globins (Ebner et al. 2003), but a respiratory function could not be shown (T. Burmester, Hamburg, pers. commun.).

Acrania

Although acranians possess a well developed branchial basket and an almost closed circulatory system (see Chapter 10), which intensively passes through the gills, no respiratory pigments are known and the blood is colourless (Ruppert 1997). Instead, it is assumed that oxygen is taken up over the whole surface and the wall of the atrium.

Craniota

Gill slits in the pharyngeal region of the intestine, which are also present in (at least) tunicates and acranians, are taken over to the craniote ancestor. But while the role of the gill slits in respiration is probably minor in tunicates and acranians, it becomes more pronounced in craniotes. Hagfishes (Myxini) and lampreys (Petromyzontida) have spherical extensions of the gill slits which allow respiration while the mouth is filled with food (Romer 1977). This indicates that respiration plays an important role in the gills (Fig. 11.7.). Most fish-like craniotes have well developed and branched gills for respiration. Their number varies from four to

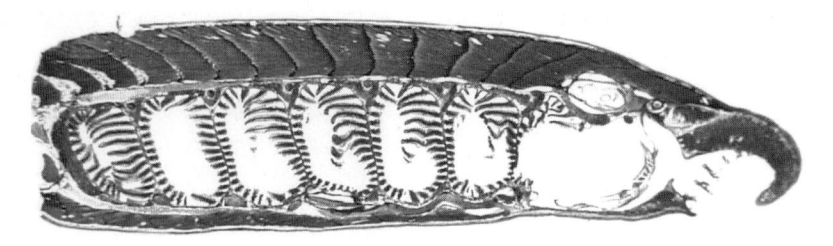

Fig. 11.7. Longitudinal section through a young cyclostome, *Lampetra fluviatilis* (Petromyzontida), showing gill chambers. Photo by Birgen H. Rothe & Andreas Schmidt-Rhaesa.

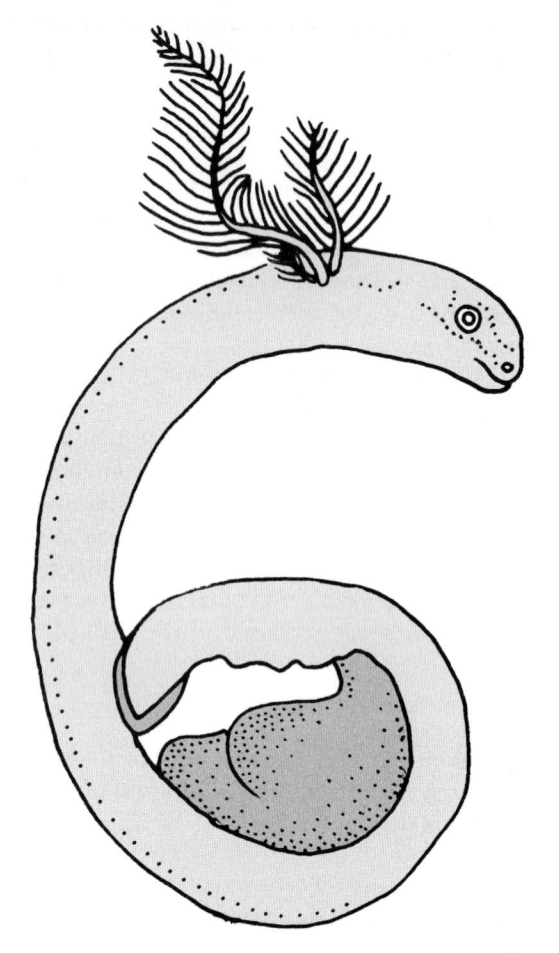

Fig. 11.8. Gymnophione larva with external gills (after Himstedt 1996).

seven, but five pairs is the most common number. Gills are also present in all larval amphibians, even of the terrestrial Gymnophiona (Fig. 11.8.; Himstedt 1996).

Terrestrial craniotes breathe air with the aid of lungs. Lungs had already developed as paired ventral pockets from the intestine in the ancestor of Osteognathostomata (Fig. 11.9.; Duncker 2004, Mickoleit 2004). They are still present in *Polypterus* (Cladistia) and in lungfishes (Dipnoi), both of which live in habitats where the oxygen content can become very low or which even dry out. In actinopterygian fishes, apart from Cladistia, the ventral intestinal pocket migrates dorsally and becomes the swim-bladder, a mainly hydrostatical organ. Starck (1982) assumes that this was performed with the invasion of open water in which the oxygen content is much more stable than in shallow shore regions. In several fish-like craniotes, however, accessory respiratory structures to breathe air were developed, for example, as pockets in various positions of the pharyngeal region, or the swim-bladder remains connected to the intestine by the ductus pneumaticus and is used secondarily as a respiratory organ (examples are *Amia calva* [Halecomorphi], *Lepisosteus* [Ginglymodi] and *Arapaima* [Teleostei]; Starck 1982, Duncker 2004, Mickoleit 2004). In terrestrial tetrapods, the lungs receive a constant increase in internal structuring, ranging from comparatively small structured lungs in amphibians to the highly branched lungs of craniotes. The most exceptional lungs were developed in birds, which with their ability to fly, depend on an increased oxygen consumption. They manage this by an increased efficiency of the lung which is, in contrast to the other craniotes, a stiff organ. Ventilation of air is performed by a system of air sacs (Fig. 11.10.). Air passes the lungs during inhalation and exhalation, and is therefore constantly in respiratory action, in contrast to the lungs of other craniotes, where oxygen uptake decreases during exhalation (Duncker 1971, 2000, 2001).

There is a diversity of hemoglobins in craniotes, including myoglobin, cytoglobin, and neuroglobin (Hankeln et al. 2005). Hemoglobin is contained intracellularly in erythrocytes, which are present in all craniotes except for ice fishes (Cannichthyes) (Ruud 1954, di Prisco 1998, Snyder & Sheafor 1999).

Conclusions: evolution of respiratory structures

The evolutionary interpretation of respiratory structures appears comparatively obvious. They are barely comparable between larger taxa, and often evolved within taxa in adaptation to either body size increase or to environmental changes. Terrestrialization, especially, requires fundamental changes of the respiratory system, as can be witnessed in molluscs, arthropods, and craniotes.

It appears plausible that the gill slits in the anterior region of the intestine are comparable between craniotes, acranians, tunicates, enteropneusts, and at least the cephalodiscid pterobranchs (genera *Cephalodiscus* and *Atubaria*), but it is debated when exactly such gill slits evolved. Due to the absence of gill slits in the pterobranch genus *Rhabdopleura*, Ax (2003) regards pterobranchs as paraphyletic and assumes that gill slits evolved in a common ancestor of Cephalodiscida + Enteropneusta + Chordata,

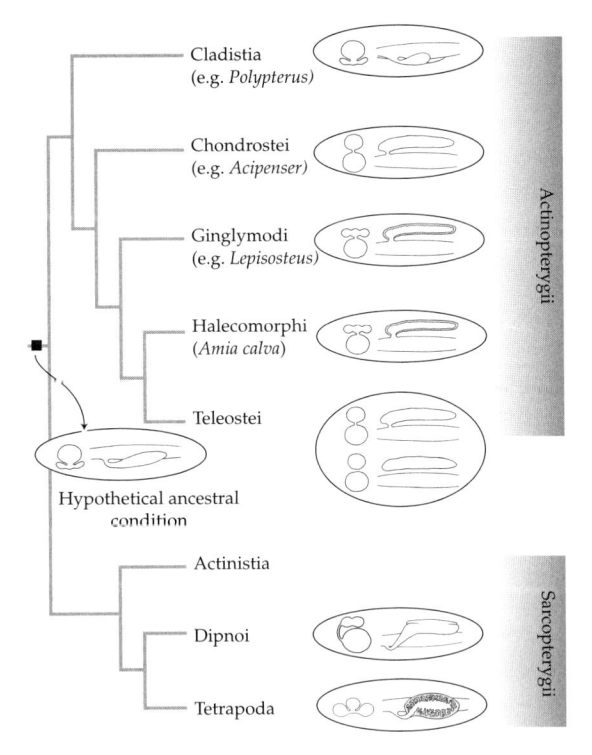

Fig. 11.9. Evolution of lungs in Osteognathostomata, see text for explanations.

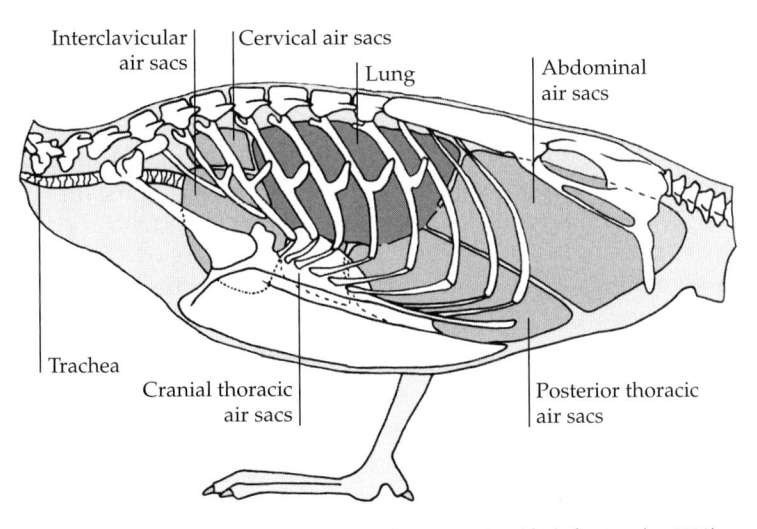

Fig. 11.10. Lung air sac – system in birds, showing the extension of the five air sacs (modified after Duncker 2000).

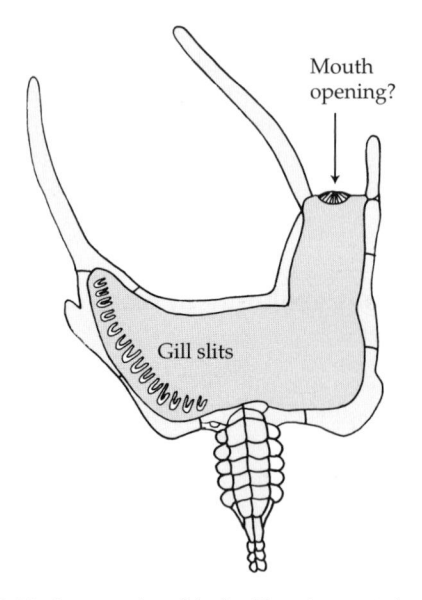

Fig. 11.11. Reconstruction of the fossil homalozoan *Cothurnocystis* with assumed gill slits (after Gee 1996).

this common taxon is named Pharyngotremata. The interpretation of the fossil Homalozoa either as stem lineage chordates or as echinoderms (see Chapter 2), and the probable presence of gill slits in at least the cornute *Cothurnocystis* (Fig. 11.11.; Jefferies 1968), make it possible that gills evolved even earlier. While the gills in craniotes undoubtely have a respiratory function, the ratio between nutritive and respiratory function in non-craniotes is unknown, and it may well be that the original function was not in respiration.

Evolution of respiratory pigments

Three types of respiratory pigments are used by animals: hemoglobins, hemerythrins, and hemocyanins. These are structurally different from each other and there is no indication that any of them shared a common ancestral protein. Therefore, respiratory pigments appear to have evolved at least three times independantly from each other.

Hemoglobin and related proteins
Several types of molecules involved in the transfer of electrons (e.g. cytochromes) or oxygen include a porphyrin ring with iron as the central structure. Those molecules that have adopted reversible binding of oxygen are called hemoglobins. Hemoglobin-like molecules are very widespread and have been found in bacteria, plants, and fungi (Riggs 1991, Hardison 1996). In animals, their occurrence is very scattered (see Fig. 11.12., Table 11.2.), but as all hemoglobins appear to be related to each other (Fig. 11.13.; Goodman et al. 1988, Moens et al. 1996, Burmester et al. 2000), a hemoglobin gene must be a plesiomorphy for metazoans, even if no hemoglobin has been reported from diploblastic metazoans.

Within Bilateria, hemoglobins became the most important molecules for oxygen transport and storage. They are very variable in size (they occur as monomers, dimers, tetramers, and larger polymers) and in their occurrence in animal tissues. They are present, for example, in muscle (e.g. radula muscle of certain gastropods, Bonaventura & Bonaventura 1983) and nerve tissue (e.g. in the polychaete *Aphrodite aculeata* and the gastropod *Aplysia californica*, Wittenberg et al. 1965, Dewilde et al. 1996), within mobile cells in body cavities such as in coelomocytes of annelids (Weber 1978) or erythrocytes of craniotes, and finally they occur extracellularly within body cavities, in particular in annelids. Often more than one type of hemoglobin is present and the differing properties of oxygen affinity and oxygen release together form effective systems for an optimal distribution of oxygen within the body (see as an example the probable pathway of oxygen from the gills to the circulatory system and then into the coelom in *Nephtys hombergi*, Weber 1971). From their distribution, it must be concluded that the expression of hemoglobin is very flexible, occurring wherever and whenever it is needed. This is nicely exemplified by platyhelminthes and nematodes, in which hemoglobin occurs only sporadically in the free-living forms but more regularly in the (in both cases derived) parasitic forms. Parasites are usually larger than free-living forms but live in environments with low oxygen content. Hemoglobin might be efficient to compete for oxygen from the host's tissue. In those highly specialized parasites which have an anaerobic

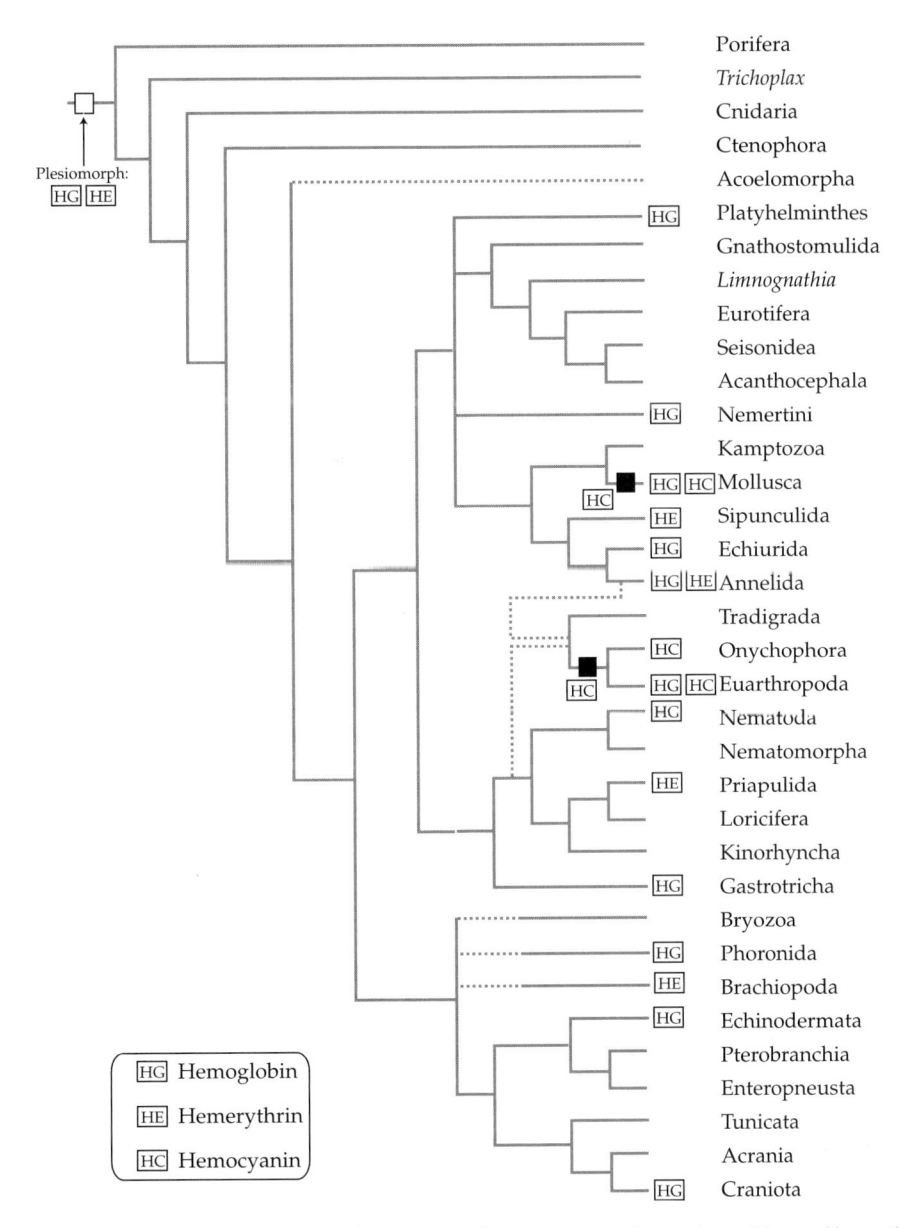

Fig. 11.12. Distribution of hemoglobin, hemerythrin, and hemocyanin on the metazoan tree. Because hemoglobin- and hemerythrin-related molecules occur outside metazoans, they are assumed to be plesiomorphic.

metabolism (see below), hemoglobin might immobilize oxygen and therefore protect the anaerobic pathways.

Besides hemoglobin, there are further related proteins. Craniote myoglobin can be distinguished from hemoglobin, but this is not the case in those invertebrates where hemoglobin has been found in the musculature and was sometimes termed myoglobin (e.g. van Holde 1997). Recently, two new kinds of globins have been found in craniotes. Cytoglobin is ubiquitously expressed in different tissues and is phylogenetically related to the craniote myoglobin (Burmester et al. 2002, Trent & Hargrove 2002,

Table 11.2 Respiratory pigments in metazoans. In the taxa not mentioned, respiratory pigments have so far not been detected.

	Respiratory pigment	Characterization	Reference
Gastrotricha *(Neodasys)*	Hemoglobin	Intracellular	Colacino & Kraus 1984
Nematoda	Hemoglobin	Intracellular, mono-, tetrameric in most species; extracellular, octomeric in *Ascaris suum*	Vinogradov 1985, Blaxter 1993
Priapulida	Hemerythrin	Intracellular	Fänge 1950, 1961, Fänge & Åkesson 1991, Schreiber et al. 1991
Platyhelminthes	Hemoglobin	Intracellular, monomeric	Phillips 1978, Tuchschmid et al. 1978, Jennings & Cannon 1985, 1987
Nemertini	Hemoglobin	In nerve and muscle tissue, sometimes extracellular	Wittenberg et al. 1965, Fänge 1969, Vandergon et al. 1998, Pesce et al. 2002
Mollusca	Hemocyanin	Extracellular	Terwilliger 1980, Bonaventura & Bonaventura 1983
	Hemoglobin	Intracellular (e.g. in muscles) and extracellular, mono-, di- or polymeric	
Sipunculida	Hemerythrin	Intracellular	Weber 1978, Klippenstein 1980
Echiurida	Hemoglobin	Intracellular, mono-, di- or tetrameric	Garey & Riggs 1984, Vinson & Bonaventura 1987
Annelida	Hemoglobin	Intracellular, monomeric, rarely polymeric Extracellular, polymeric	Weber 1978, Garlick 1980
	Chlorocruorin	Extracellular	
	Hemerythrin *(Magelona)*	Probably extracellular	Wells & Dales 1974
Onychophora	Hemocyanin-like	Extracellular	Kusche et al. 2002
Euarthropoda	Hemocyanin	Extracellular	Terwilliger 1980, Burmester 2002
	Hemoglobin	Monomeric, sometimes polymeric in insects, polymeric in branchiopods	Terwilliger 1980, Ilan et al. 1981, Vinogradov 1985
Phoronida	Hemoglobin	Intracellular, monomeric and dimeric	Garlick et al. 1979
Brachiopoda	Hemerythrin	Intracellular	Manwell 1960, Joshi & Sullivan 1973
Echinodermata	Hemoglobin	Dimeric and tetrameric	Binyon 1972, Terwilliger & Read 1972
Craniota	Hemoglobin	Intracellular, monomeric (Agnatha) and tetrameric	Andersen & Gibson 1971), Vinogradov 1985
	Myoglobin	Intracellular	Vinogradov 1985

Hankeln et al. 2005), while neuroglobin occurs only in the craniote brain (Burmester et al. 2000). It is possible that these new globins aid the oxygen supply for the cells, which is especially important in the brain (Sun et al. 2001, Pesce et al. 2002b) and the eye (Schmidt et al. 2003). While cytoglobin forms a group together with the craniote hemoglobin and myoglobin, neuroglobin branches off very early in the tree (Burmester et al. 2000) and is therefore probably a very ancient molecule (Fig. 11.13.). In invertebrates, hemoglobin-like molecules can also be present in nervous tissue, but while the 'neuroglobin'of *Aphrodite aculeata* (Annelida) is closely related to the craniote neuroglobin (Burmester et al. 2000, Hankeln et al. 2005), the 'neuroglobin' of molluscs is related to the 'usual' hemoglobins (T. Burmester, Hamburg, pers. commun.). From this distribution it seems likely that at least the bilaterian ancestor had already two types of globin, a hemoglobin and a neuroglobin.

There are several adaptations of hemoglobins to special environments. For example, neotropical camels (lama, guanaco, alpaca) have evolved a high oxygen affinity as an adaptation of living in high altitudes (Piccinini et al. 1990). In

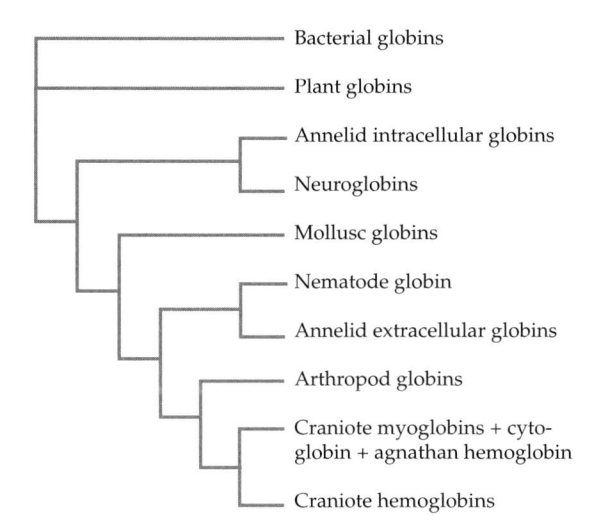

```
                    ─── Bacterial globins
                    ─── Plant globins
                ─── Annelid intracellular globins
                ─── Neuroglobins
            ─── Mollusc globins
            ─── Nematode globin
            ─── Annelid extracellular globins
        ─── Arthropod globins
        ─── Craniote myoglobins + cyto-
            globin + agnathan hemoglobin
        ─── Craniote hemoglobins
```

Fig. 11.13. Evolution of the globin superfamily, after Burmester et al. (2000).

humans, an adaptation to high altitudes is not performed by changes in the molecule structure, but by indirectly increasing the oxygen affinity through a higher pH or by differing hemoglobin concentrations (West 2006, Beall 2006). In the antarctic bivalve *Yoldia eightsi*, oxygen affinity of hemoglobin is adapted to the cold environment and is at 2°C comparable to the affinity of 'usual' hemoglobins at 25°C (Dewilde et al. 2003). In the deep-sea vestimentiferan *Riftia pachyptila*, hemoglobin is able to carry oxygen together with sulfide (Arp & Childress 1983, Arp et al. 1987). Under normal conditions, sulfide is toxic, because it blocks oxygen binding sites in the hemoglobin, but *Riftia* needs to transport sulfide without oxidizing it to its symbiotic bacteria in the trophosome.

Hemerythrin
Hemerythrin is a non-heme iron protein, which means it contains oxygen-binding iron, but this is not bound to a porphyr ring (Klippenstein 1980, Mangum 1992). Hemerythrin has been found as a respiratory pigment in four taxa: Priapulida, Sipunculida, Brachiopoda, and in the genus *Magelona* (Polychaeta, Annelida) (see Table 11.2.). These taxa are only distantly related, and the scattered distribution makes it likely that hemerythrin was present in the bilaterian (if brachiopods are related to deuterostomes) or protostome (if brachiopods are spiralians) ancestor. This interpretation, in contrast to a multiple convergent evolution, is supported by the finding that a hemerythrin-like domain has been found in a bacterium, *Desulphovibrio vulgaris* (Xiong et al. 2000).

The Hemocyanin superfamily
Hemocyanin and related proteins are very large molecules in which copper is the oxygen-binding metal (Bonaventura & Bonaventura 1980). Besides hemocyanin, the protein superfamily includes arthropod phenoloxidases, crustacean pseudohemocyanins, and insect storage hexamerins, as well as molluscan tyrosinases (Fig. 11.14.; Burmester 2001, 2002, van Holde et al. 2001). Hemocyanins occur as respiratory pigments in molluscs and arthropods (van Holde 1997). In arthropods, they have been found in chelicerates, crustaceans, diplopods, (Burmester 2001, Kusche & Burmester 2001) and occasionally in insects (Hagner-Holler et al. 2004). Phylogenetic analyses have shown that molecules of the hemocyanin superfamily in arthropods and molluscs are quite different from each other (Burmester 2001, van Holde et al. 2001) and probably evolved independantly from each other from phenoloxidases. Within arthropods, phenoloxidases form a cluster different from hemocyanins, pseudohemocyanins, and hexamerins. Phenoloxidases are oxygen-consuming molecules important in the melanin metabolism. They play, for example, a key role in cuticle sclerotization and it is therefore tempting to speculate whether they played a key role in the 'sudden' occurrence of hard exoskeletons at the Precambrian/Cambrian border (Burmester 2002). The suggestion that phenoloxidases conserved the ancestral function, and that hemocyanin is derived from a common ancestral molecule, comes from the finding of a hemocyanin-like protein with proposed phenoloxidase function in onychophorans (Kusche et al. 2002) and especially from the finding of two genes of the hemocyanin superfamily from tunicates (*Ciona intestinalis*) which also probably function as phenoloxidases (Immesberger &

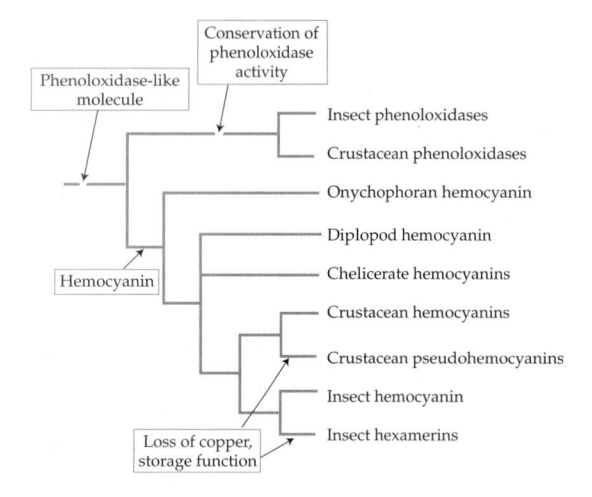

Fig. 11.14. Evolution of the hemocyanin protein superfamily in arthropods, after Burmester (2001, 2002).

Burmester 2004). Within euarthropods, crustacean pseudohemocyanins are closely related to crustacean hemocyanins and insect hexamerins are closely related to the insect hemocyanin (Burmester 2002). Both molecules are storage molecules without a central copper (Telfer & Kunkel 1991), but both are certainly derived from hemocyanins (Fig 11.14.).

Adaptations to low oxygen content

Although oxygen consumption appears to be characteristic for animal metabolisms, there are no few cases in which animals live in environments with a low oxygen content or even in absence of oxygen (anoxic). Such habitats occur in sediments, in particular in aquatic sediments, but also in terrestrial soil or in some caves (Fig. 11.15.; Nicholas 1991, Fenchel & Finlay 1995, Sarbu et al. 1996, Womersly et al. 1998) and for animals living parasitically in other organisms (Barrett 1991).

Examples of animals tolerating low oxygen contents come from different taxa. The gemmules of freshwater sponges (*Ephydatia muelleri*) survive up to 112 days of anoxic conditions (Reiswig & Miller 1998). Some gnathostomulid species (Schiemer 1973), the priapulid *Halycryptus spinulosus* (Oeschger et al. 1992), some annelids (Schöttler & Bennet 1991), some molluscs

Fig. 11.15. In sandy marine coastal sediments, a thin oxygenated layer (light colour) is followed by a (dark) layer with anoxia or low oxygen content. Only along burrows of larger animals (here from the lugworm *Arenicola marina*), can oxygen enter deeper layers. Photo by courtesy of Georg Mayer, Berlin.

(de Zwaan 1991), few arthropods (Zebe 1991), and particularly nematodes (e.g. Banage 1966, Por & Masry 1968, Wetzel et al. 1995, Womersly et al. 1998, van Voorhies & Ward 2000) survive low oxygen contents or anoxic conditions. Por and Masry (1968), for example, found large numbers of certain nematode species and one oligochaete species in Lake Tiberias, which is anoxic for eight months during the year. In freshwater and terrestrial sediments, anoxic conditions are usually temporary phenomena, but in marine sediments there is usually a sharp layer separating an oxygenized upper from an anoxic, sulphide-rich lower layer (Fig. 11.15.; Nicholas 1991, Fenchel & Finlay 1995). There are several meiofaunal species that have their highest abundances within this lower layer (which was termed the sulphide-biome; Fenchel & Riedl 1970, Boaden 1975, 1989, Fenchel & Finlay 1995, but see Reise & Ax 1979 for an alternative view).

From the physiological perspective it appears that the main strategy of free-living animals for survival of anoxic conditions is a reduction of the metabolism to very low rates (Schiemer 1973, Womersly et al. 1998). In the case of *Halicryptus spinulosus* this goes down to 2% of the normal rate (Oeschger et al. 1992). In some nematodes, this metabolic reduction includes cryptobiosis, a reversible state where no metabolism can be measured (Womersly et al. 1998). It has been stated by Nicholas (1991) that there is no evidence for anaerobic metabolism in organisms of the marine meiofauna, but that instead factors such as the abundance of mitochondria and the long and thin form of many species suggest an effective use of small amounts of oxygen. There are, however, few indications for the facultative use of metabolic pathways, for example, in the intertidal polychaete *Euzonus mucronata* (Ruby & Fox 1976) and in the soil nematode *Caenorhabditis elegans* (Föll et al. 1999). In summary it appears that anaerobic metabolism either plays no, or only a small, facultative role.

This assessment questions thoughts that meiofaunal animals living in the sulphide-biome are remnants of an early anoxic evolution of animals (Boaden 1975, 1989). This would require that these animals possess, at least in higher numbers, anaerobic metabolism. Furthermore, it must be questioned whether early animal evolution occurred in an anoxic environment. The early atmosphere was certainly oxygen-free and oxygen accumulated in the atmosphere as a product of photosynthesis during the proterozoic aeon (Cloud 1972). The appearance of macroscopic animal life towards the end of the Proterozoic and the 'explosion' in diversity in the Cambrian, can be correlated with the increase in atmospheric oxygen, following a phase of intensive tectonic activity and worldwide glaciation about 600 million years ago (Knoll 1991, 1992, Hoffmann et al. 1998). The scarcity of fossils from the earliest period could be explained by the delicacy or small size of the organisms. Small size or an epithelial organization, as in diploblastic animals, would also be congruent with a lower amount of oxygen needed in comparison to large bodied animals (Knoll 1991) and such amounts would have been present before 600 million years ago.

In contrast to free-living forms, intestinal parasites (neodermatan flatworms, nematodes, acanthocephalans) have evolved much more specific adaptations to their anoxic environment. They possess anaerobic metabolic pathways in which carbohydrates are taken up and a variety of substances, including acids (e.g. lactate) and alcohol, are excreted (see Barrett 1991 for an extensive review). This anaerobic pathway is a steady condition and is even maintained when oxygen is available. However, even parasites have oxic metabolic pathways, for example, functional cytochromes, and require at least some oxygen for normal growth and survival (Barrett 1991). In addition, the presence of three forms of hemoglobin in *Ascaris* (Nematoda; see above) suggests that there could be something like a differential supply of tissues with oxygen. While the peripheral tissues (e.g. musculature) may receive oxygen through the cuticle and body wall, the internal tissues might be protected from oxygen.

CHAPTER 12

Intestinal systems

The intestinal system is that part of the body into which food is ingested and in which nutrients are absorbed. This is not an exclusive characterization, because nutrients can also be absorbed by other tissues, in particular by the epidermis. The intestinal system is, when it is present, a sack-or tube-like invagination from the body wall. In a sense its lumen can be regarded as a continuation of the external into the body, but as it usually can be sealed off by sphincters or other structures, such characterization is not entirely exact. Typically, the intestinal system originates during development in the gastrulation as the 'archenteron'. The tissue forming the archenteron is named endoderm, but in adult animals, generally only the central part of the intestinal system is formed by endodermal tissue, whereas the terminal parts are of ectodermal origin. This is of particular importance, because it seems that, for example, cuticular structures can be formed only by tissue of ectodermal origin.

The food of metazoan animals can be very diverse, both in composition and in size. Three processes have to be distinguished: the capture of food, the breakdown into small particles, and the uptake of such particles into the body.

Endocytosis—the central process in intestinal systems

The uptake of the smallest particles into the body is the central process for the function of intestinal systems, and it is basically similar in all metazoan animals. It is performed by endocytosis, the enclosure of fluid or particles in small vesicles, which then detach from the cell membrane and migrate into the cell (see Griffiths 1996 for classification of different modes of endocytosis). When the vesicles fuse with other vesicles containing digestive enzymes (lysosomes), the content is further broken down to the molecular level.

In sponges, which do not have an intestinal system in the sense of a defined organ, endocytosis occurs in pinacocytes and choanocytes (Weissenfels 1976, Willenz & van de Vyver 1984, Langenbruch 1985, Imsiecke 1993). The content of the endocytotic vesicles is then passed to digestive lysosomes (Hahn-Keser & Stockem 1998) and may be transported through the body by archaeocytes (Leys & Reiswig 1998). In animals other than sponges, endocytosis occurs mainly in endodermal cells of the intestinal tract. Principally, other cells are also capable of endocytosis. For example, the uptake of nutrients is partly or completely taken over by epidermal cells in several parasitic animals living in the intestinal tract or in the body cavity of their host (e.g. acanthocephalans, cestodes, nematomorphs; see e.g. Jennings 1989 for the parasitic flatworm *Acholades asteris*).

Origin of the intestinal tract

From the three processes named in the introduction, sponges use only two: they capture food by creating a water current, which brings in food particles, and then absorb the smallest particles by endocytosis. Any utilization of larger particles requires a breakdown prior to endocytosis. Such breakdown can be performed mechanically or chemically, that is, by the secretion of digestive substances. While mechanical devices are present only in bilaterians, chemical breakdown was the first to evolve. As a secretion of digestive enzymes into the sea water would immediately dilute it to insignificant concentrations, it appears clear that such a process only makes sense if secretion is into a defined compartment separated from the external environment. This can in principle be the body of the prey, but there is also a tendency to form such a compartment as an integral part of the body, as the intestinal tract.

It should be noted here, that among sponges there is a fascinating exception to what is written in the paragraph above: sponges from the genus *Asbestopluma* live in the nutrient-poor deep sea or in caves, and capture larger prey (e.g. small crustaceans) with the aid of hook-like sclerites. This prey is completely overgrown by cells from the vicinity, so that digestion takes place in a temporary internal compartment of the body (Vacelet & Boury-Esnault 1995, Vacelet 2006).

An intestinal tract can be be observed in cnidarians and ctenophores and therefore evolved in the ancestor of Eumetazoa. There are two hypotheses about its origin, both starting from different reconstructions of the eumetazoan ancestor. The most common mode of the ontogenetic development of the intestinal tract is that the hollow blastula invaginates, creating the archenteron and the blastopore. When the eumetazoan ancestor was a blastula-like organism (or when its life cycle included such a stage), the evolutionary origin of the intestinal tract could be similar (Fig. 12.1.). One problem of this hypothesis is that other processes such as ingression of cells, delaminations, or combinations can lead to the same result (the presence of an archenteron). As all such processes are realized in cnidarians (see Chapter 3 and Fig. 3.8.), one can not exactly decide which of these is the ancestral mode.

The organization of *Trichoplax adhaerens* offers another hypothesis for the origin of the intestinal

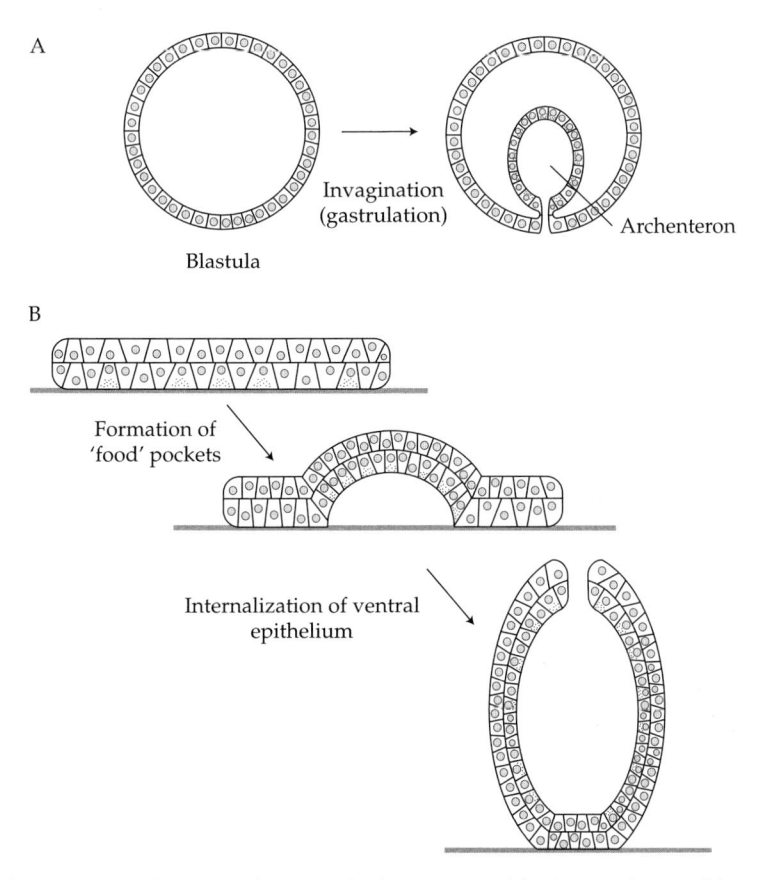

Fig. 12.1. Two models for the evolution of the intestinal system, either by invagination (A) or by internalization of the ventral epithelium (B).

tract. The ventral epithelium contains gland cells as well as endocytotically active cells and it can be observed that *Trichoplax* crawls over food particles with a subsequent formation of a 'feeding pocket', into which presumably products of the gland cells are secreted (Ruthmann et al. 1986, Grell & Ruthmann 1991). If this process is regarded as the creation of a compartment surrounded by an epithelium containing digestive gland cells, it is basically comparable to digestion in an intestine. If the ventral epithelium in *Trichoplax* can be homologized with the intestinal epithelium, then the crucial step in Eumetazoa would be to transfer this epithelium completely into the inside of the animal (Grell 1971a, b; Fig. 12.1.). This is probably supported by the expression of a gene, *Trox-2*, along the outer body margin (Schierwater 2005). *Trox-2* resembles *antennapedia*-like genes which are expressed in the anterior body region. It can be imagined that an internalization of the ventral epithelium would automatically lead to a closure of the margin to form the mouth opening, which would restrict the expression of *Trox-2* homologues to the anterior region. This hypothesis is based on the bilaterogastraea-hypothesis (Jägersten 1955), which requires a flat benthic stage in early metazoan evolution. What is problematic in this model is that a stage comparable to *Trichoplax*, that is, a flat arrangement of two layers, is never seen during development in other animals.

The evolution of an internalized intestinal tract offered the possibility to digest larger food particles and probably stimulated the evolution of a diversity of food capture mechanisms as well as targeted food requirements, which in turn stimulated changes in the sensory and locomotory apparatus.

From a sac-shaped intestine to a one-way gut

Within Bilateria, three principle forms of the intestinal system occur (neglecting its partial or complete reduction): a sac-shaped intestine with one opening to the exterior, a one-way gut with two openings (mouth and anus), and the presence of a digestive syncytium instead of an epithelially bordered intestine.

The last form, the lack of an epithelialized intestine but a digestion by a syncytial tissue, is present in acoels and was once regarded as an important transitory stage in metazoan evolution. Hadzi (1963), Hanson (1963), and Steinböck (1963), especially, advocated a view in which the metazoan ancestor originated from a ciliate-like organism by a multiplication of nuclei with subsequent cellularization. The syncytial nature of the central intestinal tract in acoels was regarded as proof for such a scenario. However, phylogenetic analyses have made this scenario unlikely and made it, in contrast, more likely that this special conformation of the intestinal system evolved within Acoela. The reasons are, on the one hand, that the sister group of Acoela, the Nemertodermatida, has an epithelialized intestine and therefore likely shows the ancestral condition. On the other hand, a basal acoel taxon, the genus *Paratomella*, shows a cellular organization of two layers around a central lumen. This has been interpreted as the following scenario: the ancestor of Acoelomorpha had a 'usual' epithelialized intestine, composed of digestive and gland cells. This condition was preserved in Nemertodermatida, whereas the ancestor of Acoela reduced the gland cells and separated the intestinal epithelial cells into two layers. While this condition was preserved in *Paratomella*, the proximal cells fused to form a syncytium in the ancestor of the remaining acoels (Euacoela) (Smith & Tyler 1985, Ehlers 1992a, Ax 1996; Fig. 12.2.).

The presence of a sac-shaped intestine in Platyhelminthes and Gnathostomulida has been taken in many cases as an indication of a basal position of these taxa among bilaterians. For example, Ax (1985) regards the Plathelminthomorpha (Platyhelminthes + Gnathostomulida) as a sister taxon of all other Bilateria (called Eubilateria), with the one-way gut being a central autapomorphy of Eubilateria. However, many subsequent analyses, morphological as well as molecular, make it more likely that Platyhelminthes and Gnathostomulida are part of the Spiralia (see Chapter 2). This implies two

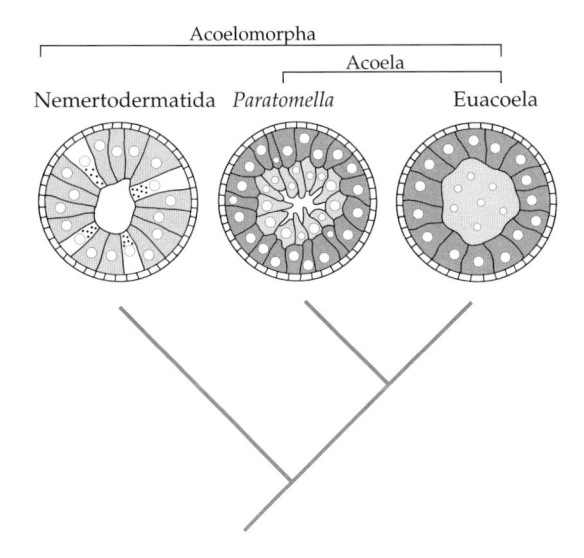

Fig. 12.2. Evolution of the intestine in Acoelomorpha. Nemertodermatida have epithelial (dark) and glandular (light) cells, *Paratomella* species have a bilayered intestinal system without gland cells and species of Euacoela have a central digestive syncytium. Modified after Ax (1996).

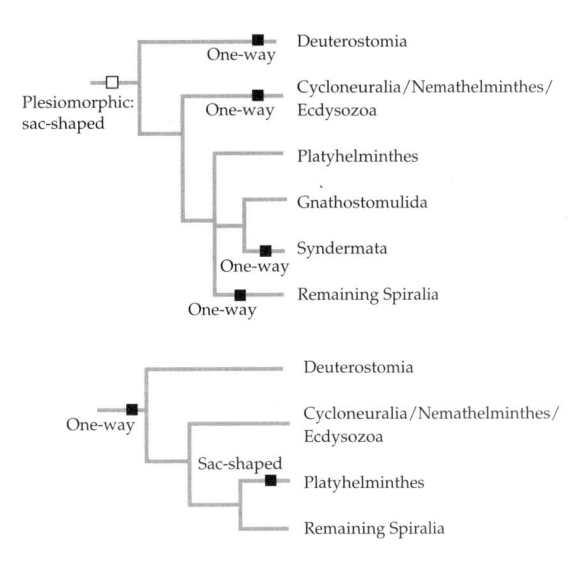

Fig. 12.3. Alternative scenarios for the evolution of the one-way gut, see text for explanation.

possible scenarios of intestine evolution. Either the sac-shaped intestine was conserved in Platyhelminthes and Gnathostomulida, then a one-way gut evolved four times independantly (in Deuterostomia, Nemathelminthes/Ecdysozoa, Syndermata + *Limnognathia* and in Euspiralia), or the one-way gut evolved in the bilaterian ancestor and a sac-shaped intestine evolved secondarily in Platyhelminthes and Gnathostomulida (Fig. 12.3.). Both scenarios appear to me to be difficult to weigh up against each other and a simple counting of evolutionary steps might be misleading. A one-way gut has advantages over a sac-shaped intestine and can develop in different ways (see below), therefore it might arise several times in parallel. On the other hand, the development of almost every organism goes through a gastrula, which is a stage with a sac-shaped intestine. If the development of the intestinal system arrests in this stage, adults may secondarily receive a sac-shaped intestine.

It should be mentioned here that there are few structures that have been indicated as *something like* an anus in flatworms and gnathostomulids. In two species of *Haplognathia* (Gnathostomulida) Knauss (1979) found a special region, in which epidermal and intestinal cells are interdigitating and are not separated by ECM. This structure might function as a temporary anus, but any further observation substantiating this interpretation is lacking. Some flatworms possess a connection between intestine and the bursa, which is a part of the female reproductive system. This bursa has been interpreted as a derivative of an ancestral anus (Steinböck 1924, Remane 1951). The main argument against this hypothesis is that the bursa always develops from the anlage of the reproductive system and gains contact to the intestine later in development, therefore there is no indication that the bursa originally did belong to the intestinal system (see Reisinger 1961). Additionally, the intestinal system in Ctenophora opens with two pores close to the apical pole. Whether these can be regarded as something like an anus is, however, questionable.

During development, a one-way gut can develop in three different ways. In addition to the blastopore, a second opening of the archenteron can break through and develop either as the anus ('protostomy') or as the mouth ('deuterostomy'), making the blastopore an anus or mouth, respectively. The blastopore can also elongate, its margins can approach and finally fuse in the centre, leaving an opening at either end (Fig. 12.4.). This process is called 'amphystomy' and occurs in representatives of several taxa (Nematoda, Onychophora, Annelida). Because of this scattered distribution and because amphystomy elegantly fits into some problems such as dorsoventral inversion (Arendt & Nübler-Jung 1997) or nervous system evolution (Nielsen & Nørrevang 1985), it is regarded as ancestral for protostomes (Nielsen 2001) or bilaterians (Arendt & Nübler-Jung 1997). I regard it as difficult to reconstruct which of the three modes of archenteron development can be regarded as ancestral. Even when amphystomy appears to be 'logically' ancient, parsimony would favour its convergent evolution. Reconstruction of the ancestral mode also depends on the questionable phylogenetic position of the tentaculate taxa which show protostomy.

The one-way gut has some advantages over the sac-shaped intestine. While in a sac-shaped intestine, the passage of food can be guided only in a limited way and food likely passes the same regions of the intestine more than once, the one-way gut allows a directed passage of food, where each 'station' is passed only once. This allows a regionalization with a much more specialized digestion than is possible in a simple sac. Consequently, in one-way guts we find a diversification of intestinal structures. It has to be added that sac-like intestines do not always have to be shaped like a 'simple' sac, but can be branched to differing degrees, this is realized in triclad, polyclad, and digenean Platyhelminthes (see Fig. 10.2.).

One-way guts are composed of three parts. While the central part is of endodermal origin, ectoderm invaginates at the anterior and posterior ends to a varying degree to form the anterior and posterior regions of the intestinal system. These parts differ considerably in structure. The anterior ectodermal part is often specialized for food uptake. Here we often find a strongly developed musculature, which is named the 'pharynx'. This can be understood as a further

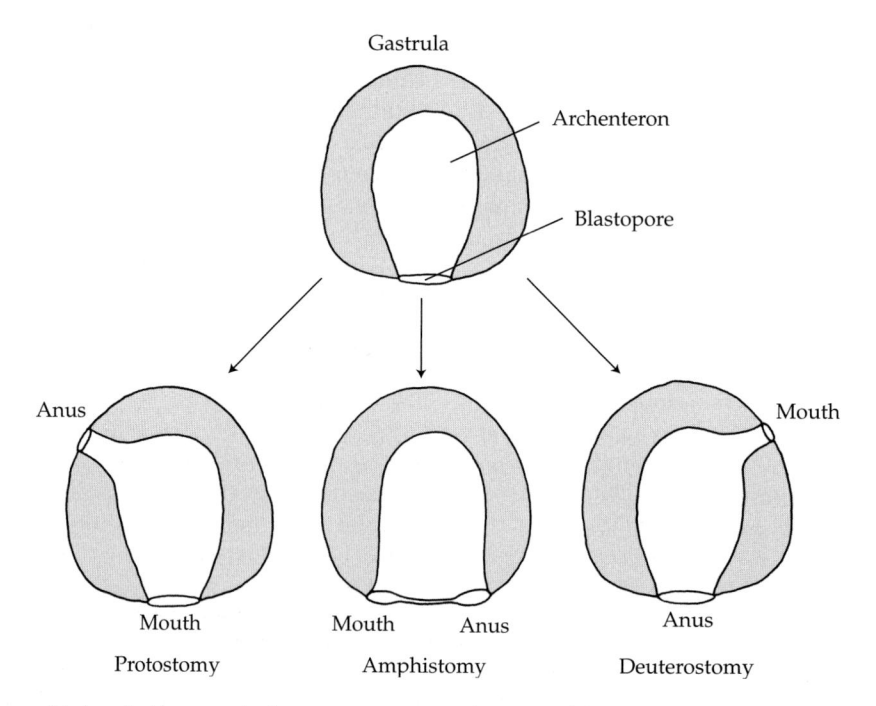

Fig. 12.4. Three possible fates for blastopore development: protostomy, amphistomy, and deuterostomy.

development of the subepidermal musculature underlying other parts of the ectoderm. Ectodermal cells are also capable of the secretion of extracellular material, and consequently hard structures can be formed in the anterior ectodermal part of the intestine. The central, endodermal part is specialized for digestion and nutrient uptake.

Many specializations of the intestinal system are unique features of certain taxa and I will only concentrate here on the discussion of three pharyngeal structures with potential for phylogenetic analyses: the structure of the pharynx, pharynges hard structures, and pharyngeal gill slits. A brief overview on the general structure of the intestine is given in Table 12.1.

Table 12.1 General structure of the intestinal system. The source of information is, if not otherwise indicated, taken from general textbooks (Westheide & Rieger 2004, 2007, Ruppert et al. 2004) and from the chapters in the series '*Microscopic Anatomy of Invertebrates*'. The terms 'pharynx' and 'esophagus' always indicate ectodermal structures, 'stomach' and 'intestine' are entodermal and 'hindgut' is ectodermal. When mouth or anus are located subterminal, the terms 'ventroterminal' or 'dorsoterminal' are used.

	Structure of the intestinal system
Cnidaria	Sac-shaped intestine with 1 opening. In polyps opening leads into a central gastrovascular cavity which may be divided by 4 (Scyphozoa) 6, 8 or a multiple of these (Anthozoa) septa into gastral pockets. Anthozoan polyps with ectodermal mouth tube ('pharynx'), including two ciliated grooves (siphonoglyphs). Medusae with stalked mouth opening (whole structure named manubrium) leading into a central 'stomach' and by radial canals to the periphery and into the tentacles. Radial canals connected by ring canal.
Ctenophora	Terminal mouth opening – ectodermal mouth tube ('pharynx'), which mechanically and chemically macerates food – stomach, from which 2 pharyngeal, 2 tentacular, and 2 transverse canals originate, the latter ones leading to 8 meridional canals running under the comb rows. Aboral canal leads from stomach to 4 short canals around the statocyst, 2 of which open by anal pores. The majority of food remnants are expelled through the mouth opening, only very little through anal pores.
Xenoturbella	Ventral mouth opening with short pharynx (musculature similar to subepidermal musculature, Raikova et al. 2000a), sac-shaped intestine without further specializations.
Acoelomorpha	Ventral mouth opening, simple pharynx is present in several acoels, its absence in some acoels is probably derived (Tyler 2001, Todt & Tyler 2006). Epithelial intestine in Nemertodermatida (lumen often occluded), bilayered intestine in Paratomella, central digestive syncytium and peripheral cells in Euacoela (Ax 1996).
Gastrotricha	Terminal mouth opening – pharynx with triradiate lumen (Y in Chaetonotida, inverted Y in Macrodasyida), composed of myoepithelial cells, pharyngeal pores in Macrodasyida – straight intestine – ventroterminal anus.
Nematoda	Terminal mouth opening – buccal tube, often with teeth or stylets – long pharynx with triradiate lumen (Y-shaped), composed of myoepithelial cells – straight intestine – ventroterminal anus (cloaca in males).
Nematomorpha	Ventroterminal mouth opening – esophagus – straight intestine, blindly ending in *Nectonema* – ventral (males) or terminal (females) cloacal opening in Gordiida.
Priapulida	Terminal mouth opening – pharynx (epithelium + muscular layer), with cuticular pharyngeal teeth, muscular polythyridium in *Tubiluchus* (Rothe et al. 2006) – straight intestine – terminal anus.
Kinorhyncha	Terminal mouth opening – pharynx (epithelium and muscular layer), lumen round (Cyclorhagida) or triradiate (inverted Y, Homalorhagida) – straight intestine – ventroterminal anus.
Loricifera	Terminal mouth opening – pharynx with triradiate lumen (Y), composed of myoepithelial cells – straight intestine – terminal anus.
Platyhelminthes	Mouth opening either anterior or ventral in midbody – muscular pharynx, which is eversible in some taxa (e.g. Tricladida). Shape of pharynx is phylogenetically informative (see Ehlers 1985a) – intestine sac-shaped, but often divided into few to numerous branches. Intestine completely reduced in e.g. cestodes.
Gnathostomulida	Ventroterminal mouth opening – muscular pharynx including specialized musculature of cuticular jaw apparatus (Sterrer 1969, Müller & Sterrer 2004) – straight intestine, anus lacking or probably temporal (Knauss 1979).
Limnognathia maerski	Ventroterminal mouth opening – muscular pharynx including specialized musculature of cuticular jaw apparatus – straight intestine, tapering towards end, probably forming a temporal anal pore (Kristensen & Funch 2000).

Table 12.1 (*Contd*)

	Structure of the intestinal system
Eurotifera	Mouth opening in center of wheel organ – muscular pharynx including specialized musculature of cuticular jaw apparatus – short esophagus – stomach – intestine – dorsoterminal cloacal opening.
Seisonida	Ventroterminal mouth opening – muscular pharynx including specialized musculature of cuticular jaw apparatus – long esophagus – stomach – short intestine only in *Seison nebaliae* (Ricci et al. 1993, Ahlrichs 1995). Protonephridial system and female reproductive system join the intestine in S. nebaliae. Cloaca (*S. nebaliae*) or anus (*S. annulatus*) dorsal.
Acanthocephala	Intestinal system completely absent.
Nemertini	Ventroterminal (Anopla) or almost terminal (Enopla) mouth opening in anterior end, joined by proboscis in Enopla – esophagus – stomach – intestine with serially arranged diverticula in most species – terminal anus.
Kamptozoa	U-shaped intestinal tract, mouth and anus inside tentacle crown. Mouth – esophagus – stomach – intestine – hindgut – anus.
Mollusca	Ventroterminal mouth opening – buccal cavity with radula (reduced in bivalves) – esophagus with salivary glands forming a mucous food string – stomach, in Eumollusca with a pair of digestive caeca (digestive glands), in Conchifera including a crystalline style (composed of enzymes), gastric shield, ciliated grooves, and other structures – intestine – hindgut – dorsoterminal anus. Digestion in stomach and digestive caeca, formation of fecal pellets in intestine.
Sipunculida	Terminal mouth opening – esophagus – long and coiled intestine – hindgut – middorsal anus.
Echiurida	Ventroterminal mouth opening at base of proboscis – esophagus (sometimes further divided into pharynx, esophagus, crop and gizzard based on diameter and surface structure) – long and coiled intestine – hindgut with pair of anal sacs – terminal anus. Bypass by a tube, the siphon, running parallel to the intestine.
Annelida	Variable structures due to different food sources. Ventroterminal mouth opening – pharynx – esophagus – midgut, often separated into stomach and 'intestine proper' – hindgut – anus in pygidium. Diversity of pharyngeal structures (ciliated folds, muscular parynges, eversible proboscis, jaws in polychaetes, sucking pharynx with radial musculature and triradiate lumen in leeches) (Tzetlin & Purschke 2005). Plesiomorphic condition are probably ciliated folds (Purschke & Tzetlin 1996).
Tardigrada	Terminal mouth opening – buccal tube with cuticular stylets, muscular pharynx (myoepithelial cells, triradiate, Y-shaped lumen) – short esophagus – intestine – hindgut receives malpighian tubules and, in Eutardigrada, the female gonoduct – cloacal opening (female eutardigrades) or anus (other tardigrades) ventroterminal.
Onychophora	Ventroterminal mouth – oral cavity with cuticular 'mandibles' – pharynx (at least in juveniles with triradiate lumen, Schmidt-Rhaesa et al. 1998) – esophagus – straight intestine – hindgut – ventroterminal anus.
Euarthropoda	Ventroterminal mouth opening – buccal cavity – pharynx – esophagus, often with specialized regions as crop (Insecta), proventriculus (Insecta, Xiphosura, Malacostraca) or pumping pharynx (Arachnida) – midgut with one or more pairs of caeca – hindgut – anus.
Chaetognatha	Ventral mouth – pharynx – straight intestine – hindgut – ventral anus.
Phoronida	Intestinal tract U-shaped. Mouth within tentacle lophophor – short esophagus – midgut divided into prestomach, stomach and intestine – hindgut – anus located close to, but outside the lophophor.
Brachiopoda	Mouth at base of lophophor – muscular pharynx – short esophagus – stomach with branching diverticula (caeca, this is the place for digestion) – intestine – anus in inarticulate brachiopods, intestine blindly ending in articulates.
Bryozoa	Intestinal tract U-shaped. Mouth within lophophor – sucking pharynx with triradiate lumen – esophagus – tripartite stomach (cardia, caecum, pylorus, cardia can be a strong crushing gizzard) – intestine – hindgut – anus outside of lophophor.
Echinodermata	Diverse feeding structures. **Crinoida**: mouth central on oral disc – esophagus – intestine, sometimes with diverticula – hindgut – anus excentrically on oral disc. **Asterioda**: mouth – stomach (eversible cardia, pylorus with diverticula into arms) – aboral anus. **Ophiuroida**: mouth with 5 teeth – esophagus – blindly ending stomach. **Echinoida**: either with complex jaw apparatus (Aristotle's lantern) and aboral anus (regular echinoids), with jaw apparatus and anus in 90° angle to mouth (irregular echinoids except Spatangoida) or without jaws and anus in 90° angle (Spatangoida). Intestinal tract coiled, divided into pharynx – esophagus – stomach – intestine – hindgut. Often with bypass, the siphon, branching from esophagus and rejoining at border of stomach and intestine. **Holothuroida**: terminal mouth – muscular pharynx – short esophagus – stomach only in some species – coiled intestine – terminal cloaca.

Table 12.1 (*Contd*)

	Structure of the intestinal system
Enteropneusta	Ventroterminal mouth opening – buccal cavity with anterior extension, the stomochord – pharynx with branchial pores, dorsal epibranchial ridge and ventral food channel – esophagus (in some species with esophageal pores) – straight intestine with diverticula ('hepatic caeca') – terminal anus. Food collection in pharynx, packing with mucous in esophagus, digestion in intestine.
Pterobranchia	Intestinal tract U-shaped. Mouth opening below oral shield – pharynx with anteriorly directed stomochord, pharynx in Cephalodiscida with 1 pair of pores – stomach, probably the main place for digestion – intestine – short hindgut – anus at base of tentacles (opposite side compared to mouth).
Tunicata	Intestinal tract U-shaped. Mouth apical – pharynx with gill slits, ventral endostyle and dorsal food groove – esphagus – stomach – intestine – hindgut – anus opening into peribranchial cavity.
Acrania	Ventroterminal mouth – buccal cavity – pharynx with gill slits, ventral endostyle and dorsal epipharyngeal groove – stomach with diverticle, the hepatic caecum – iliocolon – intestine – ventroterminal anus. Digestion primarily in stomach and iliocolon, absorption in the hepatic caecum.
Craniota	Ventroterminal mouth – buccal cavity, in Gnathostomata with teeth-bearing jaws – pharynx with gill slits used for nutrition only in larval Petromyzontida (ammocoetes-larva), respiratory gill slits in fish-like craniotes and larval amphibians – esophagus – stomach (mechanical, chemical and in some cases symbiotic digestion) – intestine (enzymatic digestion, absorption), divided in small intestine and colon – hindgut – anus (in some cases cloaca).

The evolution of pharynges

Musculature can be present everywhere along the intestinal tract, but the anterior ectodermal part appears to be especially well suited to form dominant muscular structures. The reason may be that this region is more or less an invaginated piece of body wall and as such already equipped with complex subepithelial musculature. Therefore, the presence of a muscular anterior part appears to be a common phenomenon which can already be found in cnidarians, more precisely in anthozoan polyps. 'Simple' pharynges, composed of musculature hardly more dominant than the adjacent body wall musculature, are present in *Xenoturbella* and Acoela (Doe 1981, Raikova et al. 2000a, Hooge 2001, Todt & Tyler 2006) and this can be interpreted in accordance with a potential basal position of these two taxa within Bilateria (see Chapter 2). Among other bilaterians, two trends in pharynx evolution are recognizable. The pharynx can become a large muscular bulbus, which is eversible in some cases. This trend is followed in platyhelminths and annelids, and in some species the everted pharynx can be of considerable length (Fig. 12.5.). Basal flatworms, such as species belonging to the taxa Catenulida and Macrostomida, have a

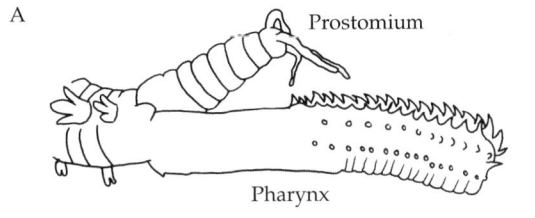

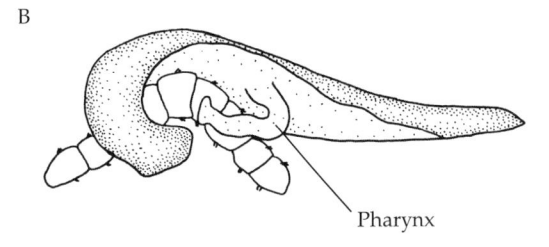

Fig. 12.5. Eversible pharynges in A. *Goniadides falcigera* (Polychaeta) and B. *Dugesia polychroa* feeding on a tubificid annelid. The everted pharynx of *Goniadides* measures about 700 μm. A. after micrograph in Tzetlin & Purschke (2005), B. after Odening in Gruner (1984).

'simple' pharynx (Doe 1981), showing that more complicated pharynges evolved within Platyhelminthes. Also within polychaetes, it can be assumed that dominant pharynges evolved within the taxon and that weakly muscular pharynges are ancestral (Purschke & Tzetlin

1996, Tzetlin & Purschke 2005). A similar kind of pharynx appears to be ancestral in sipunculids, which supports a closer relationship of sipunculids and annelids (Tzetlin & Purschke 2006). In Platyhelminthes, different types of pharynges can be recognized and used as phylogenetic markers (Ehlers 1985a, Ax 1996).

In many cases, the muscular wall of the pharynx creates a sucking force, which is essential for food uptake (especially liquid or small organisms) in a number of taxa. Although sucking can be performed with different pharyngeal constructions, the 'smartest' solution for a sucking pharynx is a radial orientation of muscle fibres in combination with a triradiate lumen (Fig. 12.6.). When muscles run from the border of the lumen (e.g. in myoepithelial cells from the cuticle bordering the lumen) to the pharyngeal periphery, a comparatively short contraction will create a triangular or almost round lumen rapidly (Fig. 12.7.). The rapid creation of a considerable lumen causes underpressure, by which liquid, small particles, and organisms can be sucked in. A sucking pharynx with a triradiate lumen can be found in a number of taxa: Gastrotricha, Nematoda, Kinorhyncha, Loricifera, in some polychaetes (*Microphthalmus*), leeches, tardigrades, juvenile onychophorans, some euarthropods (Pycnogonida, Amblypygi, Acari, Mystacocarida), and bryozoans (see Table 12.2.). Differences are found in the orientation of the lumen (Y or inverted-Y) and in the general composition of either myoepithelial cells or of a non-muscular epithelium and a separate muscular sheet (Table 12.2.). The distribution of triradiate sucking pharynges is (at least partially) so scattered that the functional constraints make it very likely that it evolved some times in parallel (this probably accounts for *Microphthalmus*, leeches, and bryozoans). The remaining taxa are,

Fig. 12.6. Cross section through a sucking pharynx showing the triradiate lumen and radially arranged musculature in the macrodasyid gastrotrich *Dactylopodola baltica*. Photo by Birgen H. Rothe & A. Schmidt-Rhaesa.

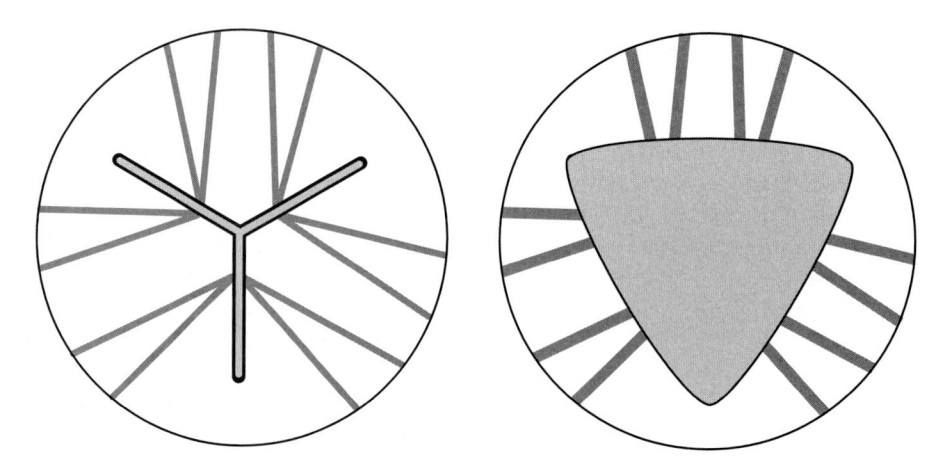

Fig. 12.7. Function of triradiate sucking pharynges: the contraction of radial musculature quickly creates a large lumen.

Table 12.2 Occurrence of muscular sucking pharynges with a triradiate lumen.

	Orientation of lumen	Myoepithelial (ME) versus epithelium + muscular layer	References
Gastrotricha Macrodasyida	Inverted Y	ME	Ruppert 1982, 1991b
Gastrotricha Chaetonotida	Y	ME	Ruppert 1982, 1991b
Nematoda	Y	ME	Wright 1991
Kinorhyncha Homalorhagida	Inverted Y	Epithelium + musculature	Kristensen & Higgins 1991
Loricifera	Y	ME	Kristensen 1991
Tardigrada	Y	ME	Dewel et al. 1993
Onychophora (juveniles)	Y	Epithelium + musculature	Schmidt-Rhaesa et al. 1998
Euarthropoda Pycnogonida	Y	Epithelium + musculature	Miyazaki 2002
Euarthropoda Acari (anactinotrichid mites)	Y	Epithelium + musculature	Alberti & Coons 1999, Coons & Alberti 1999
Euarthropoda Amblypygi	Y	?	Millot, J. 1968
Euarthropoda *Derocheilocaris*	Y	Epithelium + musculature[1]	Herrera-Alvarez et al. 1996
Polychaeta *Microphthalmus*	Inverted Y	?	Smith et al. 1986
Hirudinea Rhynchobdellida	Y	Epithelium + musculature (Sawyer 1986)	Sawyer 1986, Ax 1996
Hirudinea Arhynchobdellida	Inverted Y		Moser & Desser 1995, Ax 1996
Bryozoa	?	Epithelium + musculature	Bullivant & Bils 1968, Matricon 1973

[1] According to the investigation of Herrera–Alvarez et al. (1986), the triradiate lumen is surrounded only by circular musculature.

however, at least potentially, closely related (as Cycloneuralia, Nemathelminthes, or Ecdysozoa, see Chapter 2). Although the presence of triradiate sucking pharynges in further taxa and the functional constraints on its construction should raise some caution in using it as a phylogenetically informative character, I regard it as justified to assume the homology of sucking pharynges in Ecdysozoa, as long as such a relationship does not rest alone on the character 'sucking pharynx', but also on other characters or data. The differences may be easily explained. An evolution from a myoepithelium towards a separation of epithelium and musculature is a general trend that can also be found in the body wall (see Chapter 5) and in the coelom epithelium (see Chapter 8). A reversal in orientation of the lumen might not require a complicated scenaria, but rather a comparably simple change during pharynx development.

Pharyngeal hard structures

In contrast to the endodermal parts of the intestinal system, the ectodermal parts are capable of cuticle secretion. This can happen even when there is no body cuticle, as for example in gnathostomulids and eurotifers. Cuticular structures can form a variety of teeth, jaws, or stylets and can be important tools for prey capture, cell piercing, or algal scraping. Cuticular differentiations of the buccal cavity are found in nematodes and tardigrades, pharyngeal structures are present in priapulids, molluscs, polychaetes, and in Gnathifera. The pharyngeal teeth of macroscopic priapulids and the radula of molluscs are unique in structure and are likely to be autapomorphies of the respective taxa. In annelids, teeth or jaws occur within the taxa Phyllodocida, Eunicida (Fig. 12.8.), and in a few species of Ampharetidae (Tzetlin & Purschke 2005, Struck et al. 2006). The jaws, at least those in the eunicids *Ophryotrocha labronica* and *Diopatra aciculata*, are replaced during growth in a moult (Paxton 2004, 2005). Because the cuticular teeth and jaws in polychaetes are quite different in structure, it appears as if they all evolved within polychaetes.

In the taxon Gnathifera, however, the presence of jaws is the central character to unite gnathostomulids, *Limnognathia*, eurotifers (Fig. 12.9.), and seisonids (because acanthocephalans share other characters, they belong to Gnathifera

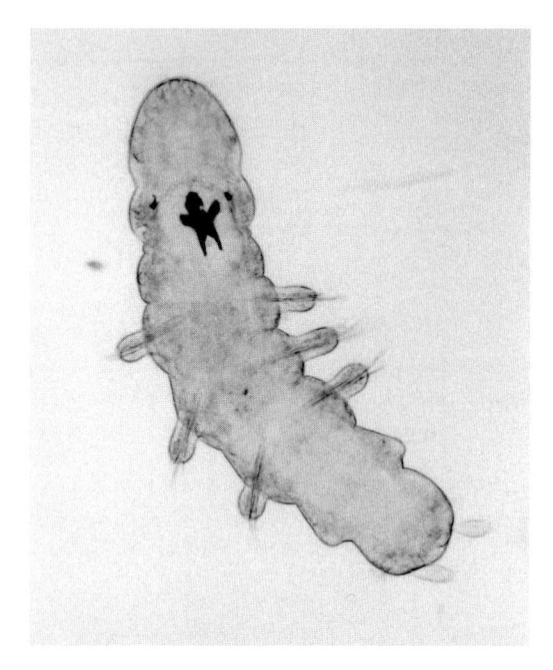

Fig. 12.8. The polychaete *Ophryotrocha* sp. with jaws in the pharyngeal region of the intestinal system.

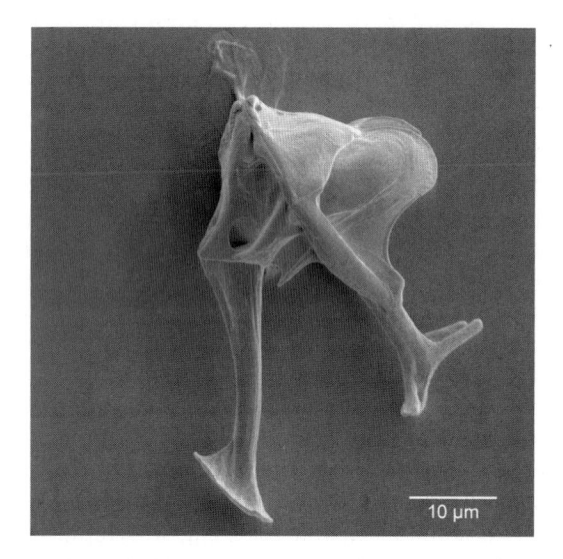

Fig. 12.9. Scanning electron micrograph of the cuticular jaw apparatus from the eurotifer *Cephalodella hyalina*. Photo by courtesy of Wilko Ahlrichs, Oldenburg.

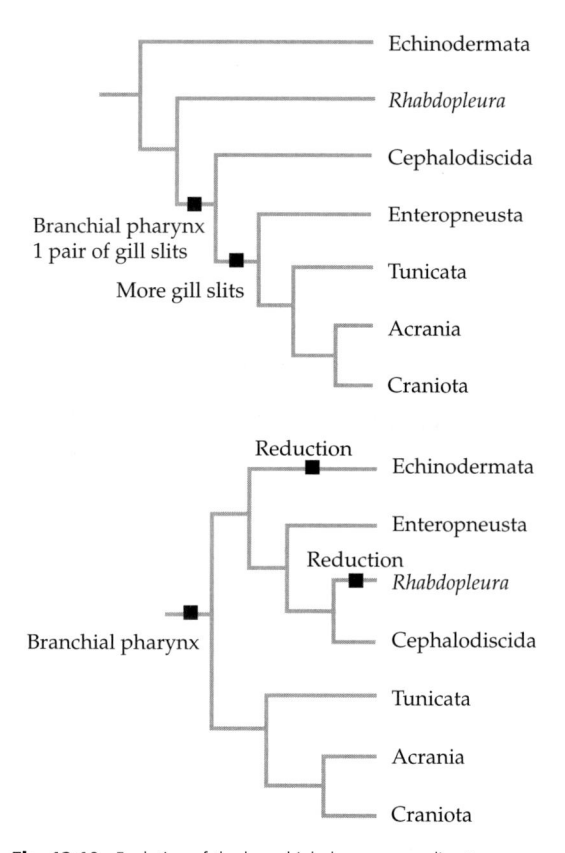

Fig. 12.10. Evolution of the branchial pharynx according to conflicting phylogenetic hypotheses, see text for details.

despite the absence of jaws, see Chapter 2). Although the jaws are variable concerning their composition of hard elements (Sørensen 2000, 2002, 2003, Sørensen & Sterrer 2002), the ultrastructure is similar in all cases (Ahlrichs 1995, Rieger & Tyler 1995, Herlyn & Ehlers 1997, Kristensen & Funch 2000).

Gill slits in deuterostomes

A pharynx with gill slits is found among deuterostome taxa, but this character has produced differing hypotheses on phylogeny. Among recent taxa, such a 'branchial pharynx' can be found in enteropneusts, in Cephalodiscida among pterobranchs, and in chordates (tunicates, acranians, and basal craniotes). On the one hand, Ax (2003) has hypothesized a sequential evolution of gill slits starting from one pair with the consequence that the taxa Pterobranchia and Hemichordata are paraphyletic (see Chapter 2; Fig. 12.10.). Regarding Pterobranchia and Hemichordata as monophyletic taxa, and hemichordates related to echinoderms (e.g. Dohle 2004), requires that the absence of gill pores in recent echinoderms and in *Rhabdopleura* is a secondary reduction or that gill slits evolved convergently in enteropneusts, cephalodiscid pterobranchs, and chordates (Fig. 12.10.).

It is quite likely that gill slits are also present in fossil deuterostomes. This is most evident in the cornute *Cothurnocystis* (Jefferies 1968; Fig. 11.11.). Gill slits have also been hypothesized for other fossil deuterostomes (see Gee 1996 for a summary and, more recently, Shu et al. 2001, Lacalli 2002 for vetulocolids and Shu et al. 2002 for vetulocystids), but in neither of these cases is the evidence as convincing as in *Cothurnocystis*. Because the fossils in question have a stereom, which is a particular three-dimensional structure of the endoskeleton present among recent taxa only in echinoderms, there are two hypotheses. Concentrating on the stereom as the important character, the fossil organisms would be stem lineage echinoderms, which in consequence means that gill slits were present in their common ancestor and therefore also in the deuterostome ancestor (see, e.g., Philip 1979). Concentrating on the presence of gill slits, the fossils would be stem lineage chordates and the stereom would be a character of the deuterostome ancestor (the 'calcichordate hypothesis', see Jefferies 1968, 1975, 1981, Cripps 1991; Fig. 12.11.).

Feeding in larvae

Many marine species have microscopically small pelagic larvae that usually secure dispersal of the species. In freshwater, there are often other larval types and swimming larva are of course reduced in terrestrial animals. The mode of nutrition in marine larvae as well as the method by which food particles are gathered, are both important characters from which phylogenetic conclusions have been drawn. Not all larvae feed, they may receive nutrients from internal yolk and therefore survive a certain time period in the water without additional nutrition. This mode is called 'lecithotrophy' and is contrasted by 'planctotrophy', in which larvae feed on dissolved particles

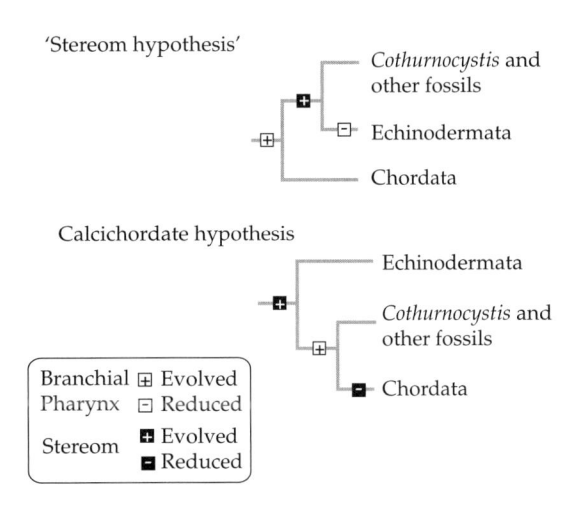

Fig. 12.11. Different interpretations of the evolution of stereom and branchial pharynx due to different positions of fossil organisms such as *Cothurnocystis*. Hemichordates are not included into these trees.

or other planctonic organisms. Planctotroph larvae usually collect food with the aid of their cilia, although other mechanisms such as capturing food with a string of mucous can also occur (Fenchel & Ockelmann 2002). Some larvae are completely ciliated, while many larvae develop a heterogeneous ciliation, in which particular ciliary bands are either especially pronounced among a complete ciliation (e.g. by being longer) or are the only ciliated cells of the larva. Cilia are used for locomotion and nutrition. For a summary of larval features see Table 12.3.

What is a larva? A larva is considered a stage during development that differs from the adult morphology (Hickman 1999). It transforms to the adult by metamorphosis. Development including larvae is also termed 'indirect development', in contrast to 'direct development', where embryos and juveniles resemble the adults in most respects. Larvae show particular larval features, which are adapted to the special requirements of this particular developmental stage.

Larval patterns are central in several hypotheses on metazoan evolution and therefore it is important to know whether the possession of a larval stage is comparable among animals. For example, Nielsen and Nørrevang (1985) have proposed a continuous evolution of larvae starting from a holopelagic blastea (in the metazoan ancestor) over a holopelagic gastraea with an archenteron (in the eumetazoan ancestor) to a holopelagic trochaea with one ciliary band (in the bilaterian ancestor, see also Nielsen 1985, 1995). In later publications, Nielsen (1998, 2001) assumes an early holopelagic evolution of metazoans, but is more cautious about a trochaea in the bilaterian ancestor. He assumes that both protostomes and deuterostomes have their own characteristic types of larvae. Such hypotheses imply that larvae (or at least one particular type of larva) evolved only once and became modified or lost during evolution. Is there evidence for this?

Most authors agree that two large groups of larvae can be distinguished, the 'trochophora' of Trochozoa (Mollusca, Kamptozoa, Annelida, Sipunculida, Echiurida) and the 'dipleurula' of echinoderms and enteropneusts (Figs. 12.12., 12.14.). As von Salvini-Plawen (1980b) and Rouse (1999a) explain, the name trochophore has often been used uncritically and has then been applied to a diversity of other taxa. A trochophore is regarded by von Salvini-Plawen (1980b) to have an apical plate, a pair of photoreceptors, protonephridia, an intestinal system with mouth and anus, and two ciliary bands called prototroch and metatroch. The prototroch is the characteristic feature of a trochophore according to Rouse (1999a). The prototroch is

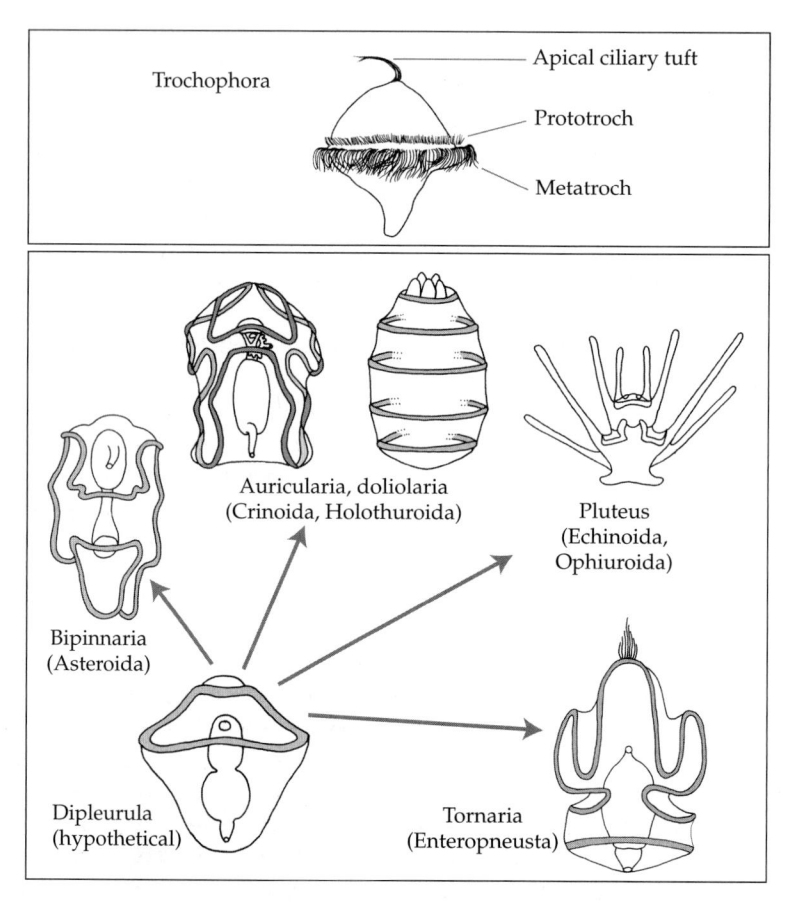

Fig. 12.12. The two basic larval types among protostomes trochophore and dipleurula. The figured trochophore is from the polychaete *Serpula vermicularis* (after photo in Westheide and Rieger 2007), the remaining figures are drawn after Westheide & Rieger (2007) and Young (2002).

characterized by a special cell lineage (Damen & Dictus 1994), starting in the third cleavage (Damen et al. 1996). Food particles are collected by the ciliary bands with a 'downstream collecting system'. This means that a water current is created by the cilia that brings in and accumulates particles on the downstream side of the ciliary band, from where they are transported by shorter cilia to the mouth (Fig. 12.13.). The dipleurula is a hypothetical larvae, but all larval forms of echinoderms and the tornaria of enteropneusts can be derived from it (e.g. Dohle 2004). Even derived larval morphologies such as the doliolaria of crinoids and holothurians can be derived from the auricularia, which resembles the dipleurula (Semon 1888, Lacalli 1988, Nakano et al. 2003). The dipleurula has a circumoral ciliary band that functions in a different way compared to the trochophore. The cilia create a water current away from the mouth and retain particles on the upstream side of the cilia, therefore this is called an 'upstream collecting system' (Fig. 12.13.). The more problematic aspect now is whether and how marine ciliated larvae from the remaining taxa relate to these two types of larvae.

Maslakova et al. (2004a) showed that in the completely ciliated larva of the palaeonemertean *Carinoma tremaphoros* several cells have a cell lineage corresponding to the prototroch cells of Trochozoa. This can be interpreted as an indication that larvae in Trochozoa and Nemertini are homologous, that these two taxa are closely related and that their common ancestor had a larva. Additionally, this observation supports the opinion that the ciliated early stages in Palaeonemertini and Enopla can be regarded as larvae (several authors call this a direct development), because the prototroch is a larval character. This evolutionary scenario (Fig. 12.14.) would imply that in the completely ciliated larvae of the ancestor of Nemertini + Trochozoa a certain subset of cells was created by a unique cell lineage. This stage was conserved in basal nemertean taxa, while another larva, the pilidium, evolved within Nemertini, in the taxon Heteronemertini. In the trochozoan ancestor, part of the cells lost their ciliation, but the particular subset of ciliated cells remained ciliated to form the prototroch.

The integration of further taxa of Protostomia makes the discussion of larval evolution more problematic. Most taxa are directly developing and have no larva (Gastrotricha, Nematoda, Kinorhyncha, Gnathostomulida, Eurotifera, Seisonida, *Limnognathia*) or have unique larvae not comparable to other forms (Nematomorpha, Loricifera, Priapulida). Only the interpretation of the presence of larvae in Platyhelminthes is problematic. While larval forms in parasitic flatworms are certainly derived, polyclads have ciliated larvae called Müller's or Götte's larva. All other flatworms are directly developing. Polyclads are not the basalmost taxon among Platyhelminthes (compare Ehlers 1985a, Littlewood & Olson 2001), making the possession of a larva in the platyhelminth ancestor not a parsimonious scenario. Additionally, the possession of larvae in polyclads appears to be explained by ecological constraints. While the microscopically small flatworms can leave the sediment to be dispersed with the water current

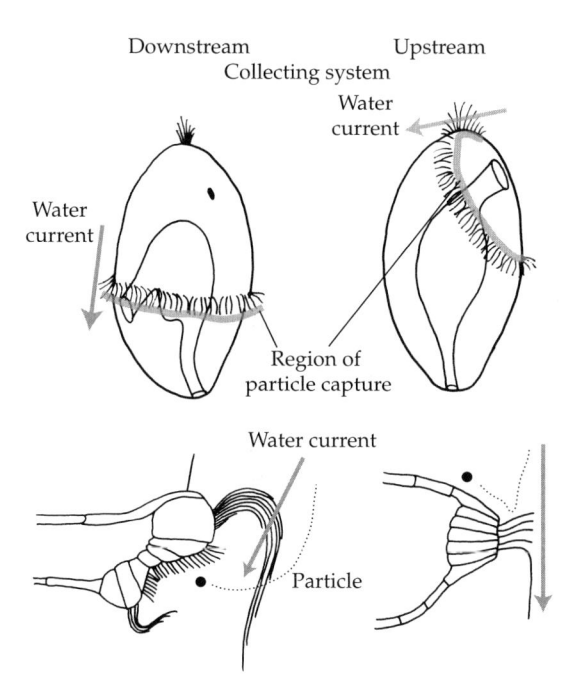

Fig. 12.13. Comparison of downstream and upstream collecting systems, figures after Nielsen (1987) and Lacalli (1996).

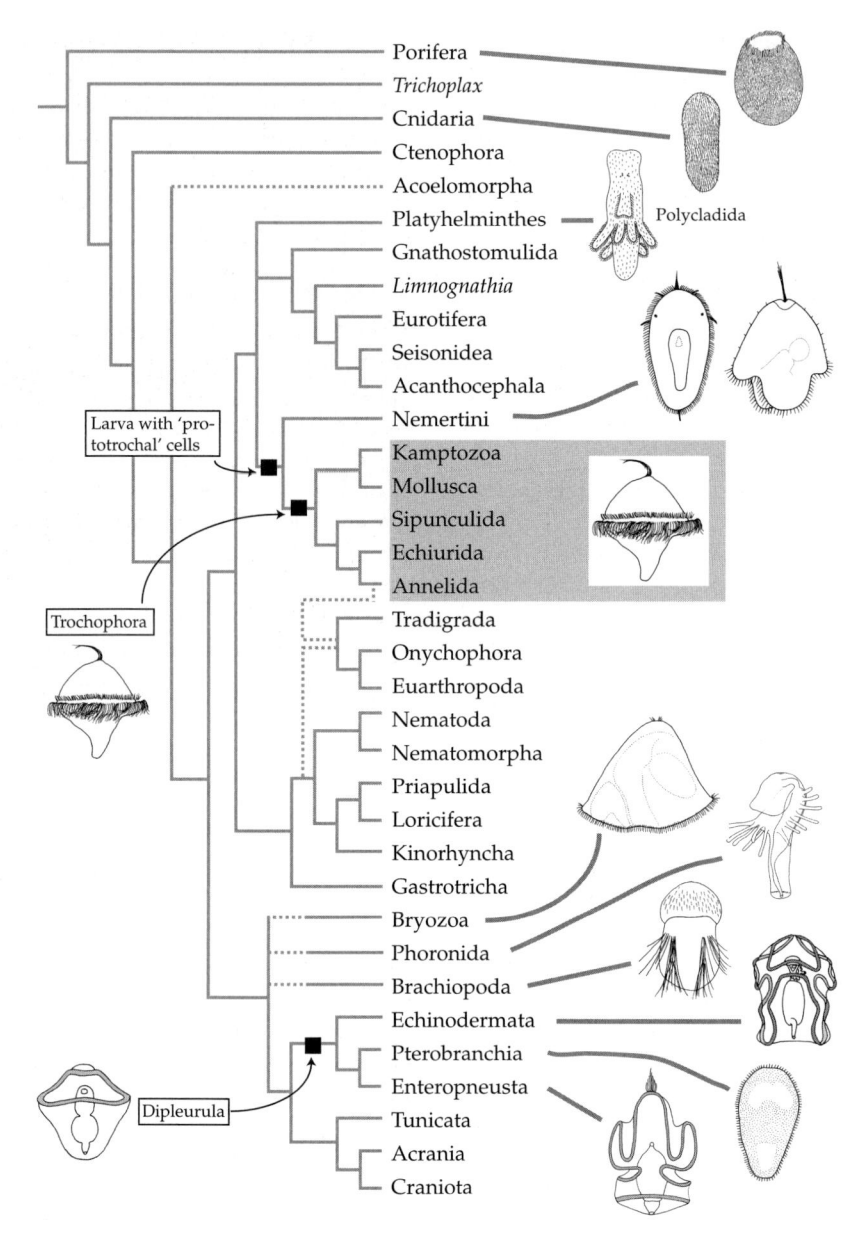

Fig. 12.14. Occurrence of marine ciliated larvae on the phylogenetic tree. As assumed ancestral states, only the dipleurula, the trochophor and a kind of 'pre-trochophora' in the common ancestor of Trochozoa and Nemertini are convincing. Images of larvae after figures in Dilly (1973), Ruppert (1978), Westheide and Rieger (2007), Young (2002) and cover photo of Invertebrate Biology 121 (3).

(Armonies 1989, 1994), polyclads are too large to colonize new habitats in this way. Therefore, they had to develop, comparable to many other taxa with large species, planctonic larvae. Polyclad larvae are completely ciliated, with the cilia along the 'arms' being longer. Therefore, the cells forming these cilia were called prototrochal cells (Ruppert 1978a) or a ciliary band (Lacalli 1982), but it is not yet clear whether they originate in a cell lineage comparable to that of the prototroch in Nemertini and Trochozoa. A common character in polyclad larvae and trochophores (as well as other larval types) is an apical organ, a tuft of longer cilia and sensory cells, which is directly above the larval brain in in polyclads (Lacalli 1983) and polychaetes (e.g. Lacalli 1981). Lacalli (1982) noted that the 'ciliary band' in Müller's larva is innervated by a peripheral nervous system originating from cells within the ciliary band and not from the central nervous system. This was also shown to be the case in a nemertean pilium larva (Lacalli & West 1985). It is hard to judge whether the similarites between polyclad larvae and nemertean and trochozoan larvae are homologies, but according to the position of polyclads within flatworms I would tend to regard them as convergences. Assuming that the platyhelminth ancestor did not have a larva, it becomes likely that ciliated marine larvae evolved within protostomes in the ancestor of Trochozoa + Nemertini and in polyclads, and that the protostome ancestor therefore had no larva. When, however, the polyclad larva and the trochophora are homologous, we must assume that a larva was present in the spiralian ancestor. As no (ciliated) larva is present in gastrotrichs, cycloneuralians, and arthropods, one protostome branch in this scenario has a larva while the other has not. Therefore, reconstruction of the bilaterian life cycle depends on the question of whether the dipleurula was present in the deuterostome ancestor, whether dipleurula, trochophora, and the larvae of the tentaculate taxa are homologous (see Fig. 12.14.).

The dipleurula is usually assumed to be the typical larva of Deuterostomia, but considering a possible close relationship between Echinodermata and Hemichordata (see Chapter 2), it might as well be an autapomorphy of the common ancestor of these two taxa, because chordates have no comparable larva (Fig. 12.14.). The larva of pterobranchs is an exception, it is a completely ciliated, lecithotrophic larva (Dilly 1973). The larvae of Bryozoa, Phoronida, and Brachiopoda cannot unambiguously be compared with either the trochophora or the dipleurula. Bryozoa have different larval forms; either as completely ciliated, lecithotrophic larvae in stenolaemates; or as planctotrophic cyphonautes

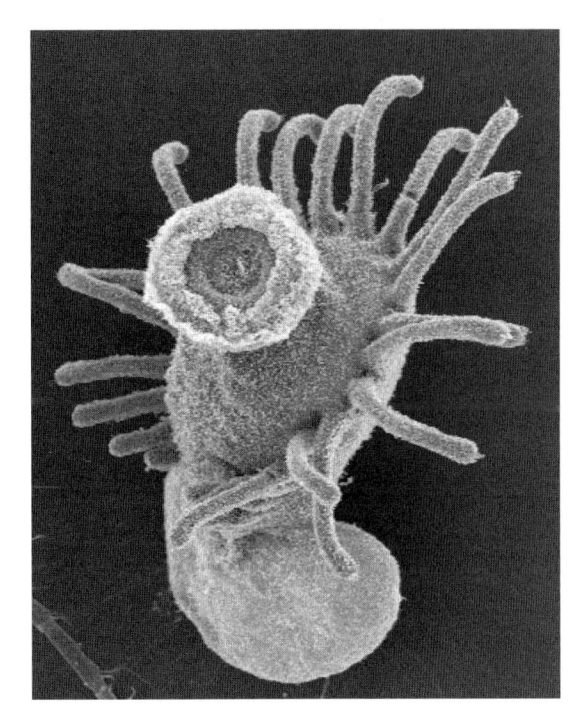

Fig. 12.15. Scanning electron micrograph of an actinotrocha, the larva of Phoronida.

or other forms in gymnolaemates (Woollacott 1999, Nielsen 2002c, Temkin & Zimmer 2002). The triangular cyphonautes has a ciliated lower ridge. Particles are captured by long laterofrontal cilia together with a varied water current created by lateral cilia and a flicking of the laterofrontal cilia towards the mouth (Strathmann 2006). Filtering by long and stiff cilia is also present in the phoronid actinotrocha (Riisgård 2002; Fig. 12.15.) and in planctotrophic brachiopod larvae (*Glottidia pyramidata*, Strathmann 2005). All three taxa have upstream collecting systems, which makes them comparable with echinoderms and hemichordates, but while in phoronids and brachiopods there are monociliary cells (as in echinoderms and hemichordates), cells are multiciliate in bryozoans (Nielsen 1987). Therefore, bryozoan larvae share characters of both protostomes and of deuterostomes. Larvae would support a relationship of at least phoronids and brachiopods to deuterostomes, and this would imply that a larva with an

Table 12.3 Summary of features concerning ciliated larvae in marine taxa. LE = lecitotrophic, PL = planctotrophic, DCS = downstream collecting system, UCS = upstream collecting system.

	Larval type	LE/PL	DCS/UCS	Ciliation	Reference
Porifera	'Blastea-larva'	LE	–	Complete, monociliate	Nielsen 1987, Maldonado 2004
Cnidaria	Planula	PL and LE	–	Complete, monociliate, rarely multiciliate	Nielsen 1987, 2001
Platyhelminthes	Müller's larva	Probably PL	?	Complete, bands of multiciliate cells along lobes	Lacalli 1982, Ballarin & Galleni 1987
Polycladida	Götte's larva				
Nemertini					
Enopla and Palaeonemertini	Ciliated larva	LE	–	Complete, with 'prae-prototroch'	Cantell 1989, Maslakova et al. 2004a
Heteronemertini	Pilidium	PL	?[1]	Ciliary band along lobes, partly double band, some cilia compound, multiciliate	Lacalli & West 1985, Nielsen 1987, Cantell 1989
Kamptozoa	Trochophora	PL	DCS	Proto- and metatroch with compound cilia and multiciliate cells	Nielsen 1971, 1990a, 2002c, see also Haszprunar et al. 1995
Mollusca					
Solenogastres	Trochophora or larva derived	LE	–	2 Ciliary rings, fine structure unknown	McFadien-Carter 1979, Pearse 1979, Nielsen 1987, Haszprunar et al. 1995
Polyplacophora	from trochophora	LE	–	Prototroch with compound cilia, multiciliate cells	
Gastropoda		LE and PL	DCS	Prototroch and metatroch with compound cilia, multiciliate cells	
Bivalvia		LE in protobranchs, PL in others	DCS	Prototroch and metatroch with compound cilia, multiciliate cells	
Scaphopoda		LE	–	?	
Sipunculida	Trochophora in egg envelope, sometimes followed by pelagosphaera	Trochophora LE, pelagosphaera PL	Trochophora –, pelagosphaera does not use cilia for feeding	Trochophora ?, pelagosphaera with compound cilia	Nielsen 1987, 2001, Rice 1989, Jaeckle & Rice 2002
Echiurida	Trochophora, derived larva in *Bonellia*	PL, LE in *Bonellia*	DCS	Probably multiciliate (Nielsen 1987)	Davis 1989, Pilger 2002
Annelida	Trochophora	LE and PL	DCS	Multiciliate, rarely monociliate (*Owenia*)	Nielsen 1987, Heimler 1988
Bryozoa	Cyphonautes and other larval types	cyphonautes PL other larvae LE	UCS	Ciliary band with multiciliate cells (not compound)	Nielsen 1987, Nielsen 1990b, Woollacott 1999, Temkin & Zimmer 2002
Phoronida	Actinotrocha	PL	UCS	Monociliary cells (rarely biciliary), not compound	Strathmann 1973, Nielsen 1987, Emig 1990, Johnson & Zimmer 2002
Brachiopoda					
Lingulacea, Discinacea	Tentaculate larva	PL	UCS	Monociliary cells, no compound cilia	Nielsen 1987, Chuang 1990, Pennington & Stricker 2002
Craniacea	'Lobed' larva	LE	–		
Articulata					
Pterobranchia	'Planula-like'	LE	UCS	Completely ciliated, no ciliary bands	Dilly 1973, Halanych 1993
Enteropneusta	Tornaria	PL	UCS	Monociliary cells, no compound cilia	Hadfield 1975, Nielsen 1987
Echinodermata	Doliolaria-type larvae	LE or PL	UCS	Monociliary cells, no compound cilia	Nielsen 1987

[1] Feeding has been assumed to be by DCS (Nielsen 1987), but only few cilia around the mouth contribute to feeding, the rest are involved in locomotion (Strathmann 1987, Haszprunar et al. 1995).

upstream collecting system of monociliate cells was present in a common ancestor.

Concluding from this brief review of larvae, it is not certain that larvae in general can be compared with each other. Probably the single character that is present in almost all types of larvae is an apical concentration of longer, sensory cilia. This, however, may reflect functional needs for orientation rather than a homologous character. This is supported by a different use of transcription factors in the formation of apical organs in a sea urchin and a gastropod (Dunn et al. 2007). There is evidence that several larval features are variable and depend on external constraints. For example, within the sea star genera *Asterina* and *Patiriella*, there are species with larvae as well as directly developing species, and lecithotrophic as well as planctotropic larvae (Byrne & Cerra 1996, Hart et al. 2004, Byrne 2006). Byrne (2006) estimates that lecithotrophy evolved independantly at least six times within Asterinidae (the taxon to which *Asterina* and *Patiriella* belong). Ciliation patterns are important for feeding as well as locomotion, and the final form of a larva is often a compromise between these two functions (Emlet 1994, Strathmann & Grünbaum 2006). During evolution, the constraints can change, that is when changing from feeding to non-feeding. Pernet (2003) showed that in echinoderms such changes usually cause rapid morphological changes in echinoderms, while in sabellid polychaetes the 'response' is slower.

Taking the uncertainties about the homology of larvae and the dynamics of changes in structure into account, the decision about whether planctotrophy (e.g. Strathmann 1978, 1985, Havenhand 1995, Nielsen 1998) or lecithotrophy (e.g. Haszprunar et al. 1995) is the ancestral mode, is almost impossible to make. It is easier, however, when considering particular groups of taxa. Let us start with diploblastic animals. While sponges have lecithotrophic larvae (Nielsen 1987), both lecithotrophic and planctotrophic larvae exist in cnidarians. There are comparatively few observations of feeding, but it appears that the planula larvae of anthozoans are planctotrophic and feed with the aid of mucous and cilia (Siebert 1974, Tranter et al. 1982, summary in Strathmann 1987). Because anthozoans are the basal cnidarian taxon, it appears probable that within cnidarians, lecithotrophy was derived from planctotrophy, whereas in metazoan evolution planctotrophy might be derived from lecithotrophy (as in sponge larvae).

In protostomes (at least in Trochozoa + Nemertini) and in Echinodermata + Hemichordata, larval feeding is performed by complex ciliary bands and particle capture mechanisms. This makes it likely that planctotrophy is ancient in these taxa. Indeed, there are no convincing examples, where planctotrophy is derived from lecithotrophy, but there are many examples of the reverse case (see Nielsen 1998). Downstream and upstream collecting systems appear to be restricted to particular taxa (Trochozoa and Deuterostomia, see, e.g., Nielsen 1987, 1994, 2001), but the potential position of the tentaculate taxa among protostomes would mean that upstream collecting systems evolved twice.

Feeding in adults

The ancestral mode of feeding in adult animals is filtration, in the broadest definition. This means that food is somehow extracted from a large volume of water. This is realized, with the exception of *Trichoplax adhaerens*, in all diploblastic taxa. However, filtration is performed in a variety of ways. While in sponges a water current produced by ciliary action brings in food particles, cnidarians and ctenophores 'comb' the water or use water currents to come into contact with food. Through their capture mechanisms, cnidocytes (Holstein & Tardent 1984, Mariscal 1984, Hessinger & Lenhoff 1988, Tardent 1995) or colloblasts (Benwitz 1978), are specialized to capture larger prey than sponges can (Table 12.4.).

Within Bilateria, filter feeding is also widely distributed. Some taxa such as molluscs, annelids, and echinoderms show a wide variety of modes of nutrition, among which filtration evolves secondarily (in bivalves, sedentary polychaetes, and in some ophiuroids and holothurians). In many cases, cilia are important

Table 12.4 Feeding and food source in adults. The source of information is, if not otherwise indicated, taken from general textbooks (Westheide & Rieger 2004, 2007, Ruppert et al. 2004) and from the chapters in the series '*Microscopic Anatomy of Invertebrates*'.

	Food source, capture of food and transport to mouth
Porifera	Food: fine, suspended particles, small planctonic organisms (< 50 m), captured by choanocytes from water incurrent. Exception: carnivorous sponges capture zooplanctonic organisms (Vacelet & Boury-Esnault 1995).
Trichoplax adhaerens	Food: flagellates (*Cryptomonas*) in laboratory cultures, in sea presumably protozoans and bacteria colonizing hard substrates. Food is captured in pockets between substrate and ventral epithelium.
Cnidaria	Food: zooplanctonic organisms. Prey captured by nematocyst explosion and moved to mouth by tentacles.
Ctenophora	Food: zooplanctonic organisms. Prey captured by contact with sticky colloblasts and moved to mouth by contraction of tentacles.
Xenoturbella	Food: not known with certainty. Probably bivalves (*Nucula*, see Bourlat et al. 2003), probably also dead animals (Westblad 1950). Prey ingested through ventral mouth.
Acoelomorpha	Food: no data are available for Nemertodermatida, very little is known for Acoela, but probably a range of very small (diatoms, protozoans, algal spores) to larger (small annelids and crustaceans) food can be used. According to observations on *Convoluta convoluta* (Jennings 1957), small food is engulfed by syncytial intestinal tissue protruded through the mouth while large food is trapped by lifting the anterior end and moving it over approaching prey.
Gastrotricha	Food: small organisms (bacteria, diatoms, small protozoans). Prey is sucked in by pharyngeal action.
Nematoda	Food (of free-living forms): bacteria, larger organisms, content of plant or fungal cells. Food is sucked in by pharyngeal action, in nematodes with stylets or teeth after or with simultaneous mechanical treatment.
Nematomorpha	No feeding as adults. Only in the (juvenile) parasitic phase, nutrients are absorbed, mainly over body surface (Hanelt et al. 2005).
Priapulida	Food incompletely known. Macroscopic forms probably predators of polychaetes and other prey, microscopic forms deposit-feeders or microcarnivoures. In macroscopic species, prey may be captured and swallowed with the aid of pharyngeal teeth.
Kinorhyncha	Food: diatoms, bacteria which are, at least in Homalorhagida, sucked in by pharyngeal action.
Loricifera	Food: bacteria, sucked in by pharyngeal action.
Platyhelminthes	Food: diversity of prey organisms, ranging in size almost up to the flatworm's body size. Pharyngeal action is central in prey capture and uptake. According to the flatworm taxon, small prey is swallowed as a whole, large prey is damaged by pharyngeal action and sucked out (Jennings 1968). Parasites feed on host's blood or tissue (e.g. Digenea) or on host's intestinal content, taken up over the body surface (e.g. tapeworms).
Gnathostomulida	Food: probably algal or bacterial films on sand grains, scraped off with jaw apparatus (Sterrer 1971).
Limnognathia maerski	Food: probably bacteria or diatoms, collected by preoral ciliary field and taken over by jaw apparatus (Kristensen & Funch 2000).
Eurotifera	Food: suspended particles (in suspension feeders), algae or zooplanctonic organisms (in predatory eurotifers). Suspension feeders (e.g. bdelloids, monogononts such as *Brachionus*, *Keratella*) create water current by action of wheel organ cilia, particles are sucked in by pharynx and ground by jaw (mastax). Selective feeders capture prey (algal cells, animal prey) by a combination of sucking and piercing (e.g. *Notommata*, *Polyarthra*, *Synchaeta*) or by grasping with forceps-like jaws (*Asplanchna*) (Wallace & Ricci 2002). Jaws are adopted to respective modes of feeding and can be distinguished into several types (e.g. ramate, virgate, malleate, forcipate; see e.g. Wallace & Snell 1991).
Seisonida	Food: according to stomach contents, *Seison nebaliae* feeds on bacteria and *S. annulatus* on hemolymph of the host (Ahlrichs 1995). Bacteria are probably picked up with jaw.
Acanthocephala	Food: nutrients from host's intestine, absorbed over the epidermis.
Nemertini	Food: prey organisms, particularly annelids and crustaceans, in some cases also dead animals. Prey is captured with proboscis, either by wrapping around prey or, in Enopla, with the aid of proboscidal stylet. Toxins are applied in both cases.
Kamptozoa	Food: suspended particles. Particle capture with tentacles (downstream collecting system), ciliary transport to mouth.
Mollusca	**Caudofoveata**: selective deposit feeders. **Solenogastres**: carnivores on cnidarians. **Polyplacophora**: algal scrapers. **Neopilinida**: deposit feeders. **Gastropoda**: variety of food sources, abundantly redula scraping on plant material, but also carnivory. **Cephalopoda**: carnivores, prey capture with tentacles, maceration with beak and radula. **Bivalvia**:

Table 12.4 (*Contd*)

	Food source, capture of food and transport to mouth
	deposit or suspension feeders with targeted (in deposit feeders) or untargeted (in suspension feeders) water incurrent, trapping of particles by mucous on gills, transport with labial papilla and cilia to mouth. **Scaphopoda**: microcarnivors or microomnivors, prey or food captured and transported to mouth by captacula. (See also von Salvini-Plawen 1988).
Sipunculida	Food: fine particles, either suspended in water, within or on sediment. Suspension and deposit feederscapture particles with tentacles by mucous and transport them by cilia to the mouth, sediment feeders (e.g. *Sipunculus nudus*) ingest entire sediment.
Echiurida	Food: Fine particles on sediment. Deposit feeders capture particles in mucous on extended proboscis and transport it by cilia to the mouth. Exception is *Urechis caupo* who captures fine particles and plancton by a water current through a mucous net, which is ingested periodically.
Annelida	Food: diversity ranging from fine organic particles (suspended or on sediment) to plant material, animal prey, entire sediment and blood (gnathobdellid leeches). Food is either collected with anterior appendages (e.g. in sedentary polychaetes), grabbed with the pharynx (or jaws) or directly taken up into the mouth.
Tardigrada	Food: content of plant cells or animal prey. Plant cells (algae etc.) as well as animals (nematodes, eurotifers, sometimes even other tardigrades) are pierced with stylets and either sucked out or ingested completely by pharyngeal action.
Onychophora	Food: smaller euarthropods. Food capture by squirting of slime gland secretions over the prey, followed by maceration with 'mandibles'. Prey is partly digested externally with saliva, which is then sucked in.
Euarthropoda	Food: diversity of food sources, many euarthropods are carnivorous, but also herbivory, filtration, blood sucking and other utilizations of food sources exist. Food uptake and manipulation with specialized mouth parts, arachnids have extraintestinal digestion.
Chaetognatha	Food: zooplanctonic organisms. Grasping of prey with hooks, teeth may help in maceration.
Phoronida	Food: fine, suspended particles. Particle capture with tentacles of lophophor (upstream collecting system), ciliary-mucous capture and transport to mouth.
Brachiopoda	Food: fine, suspended particles. Particle capture with tentacles of lophophor (upstream collecting system), ciliary transport to mouth (probably without mucous).
Bryozoa	Food: suspended particles, phytoplankton, exceptionally zooplancton. Particle capture with tentacles of lophophor (upstream collecting system), ciliary transport to mouth.
Echinodermata	**Crinoida**: suspension feeders, particle capture with mucous on tentacular ambulacral feet, ciliary transport to mouth. **Asteroida**: microvorous or macrovorous predators. **Ophiuroida**: microphagous or macrophagous carnivores, suspension feeders or mixture of mechanisms. When suspension feeding, particles are trapped by mucous. **Echinoida**: algal scrapers (regular echinoids) or deposit feeders using ambulacral feet for selective uptake (irregular echinoids). **Holothuroida**: suspension feeders (with branched, filtering buccal podia), deposit feeders (selective uptake of particles with buccal podia) or endobenthic sediment feeders (Apodida, Molpadiida).
Enteropneusta	Food: fine particles (detritus, suspended particles). Species are either suspension, deposit, or sediment feeders. Suspension feeders create water current with proboscial and pharyngeal cilia, particles are trapped by mucous on proboscis and in pharynx. Deposit feeders (e.g. *Saccoglossus* species) trap particles on the sediment with mucous and transport this to the mouth. Sediment feeders (e.g. *Balanoglossus* species) feed on sediment rich in organic particles.
Pterobranchia	Food: fine, suspended particles. Particle capture with tentacles (upstream collecting system), ciliary transport to mouth.
Tunicata	Food: planctonic organisms. Cilia create a water current through the mouth (buccal siphon) and through the gill slits. Particles trapped in pharynx by mucous net created in the ventral endostyle, this net is collected by the dorsal groove, concentrated into a cord and transported posteriorly.
Acrania	Food: fine, suspended particles. Cilia create a water current into the mouth, water leaves through the branchial slits, atrium, and atriopore. Food capture similar to tunicates.
Craniota	Food: diversity of food. Originally presumably fine suspended particles captured with branchial pharynx similar to Acrania (realized in larva of Petromyzontida). Ancestor of Gnathostomata likely switched to large animal prey, specializations on other food sources (e.g. plant material) are secondary.

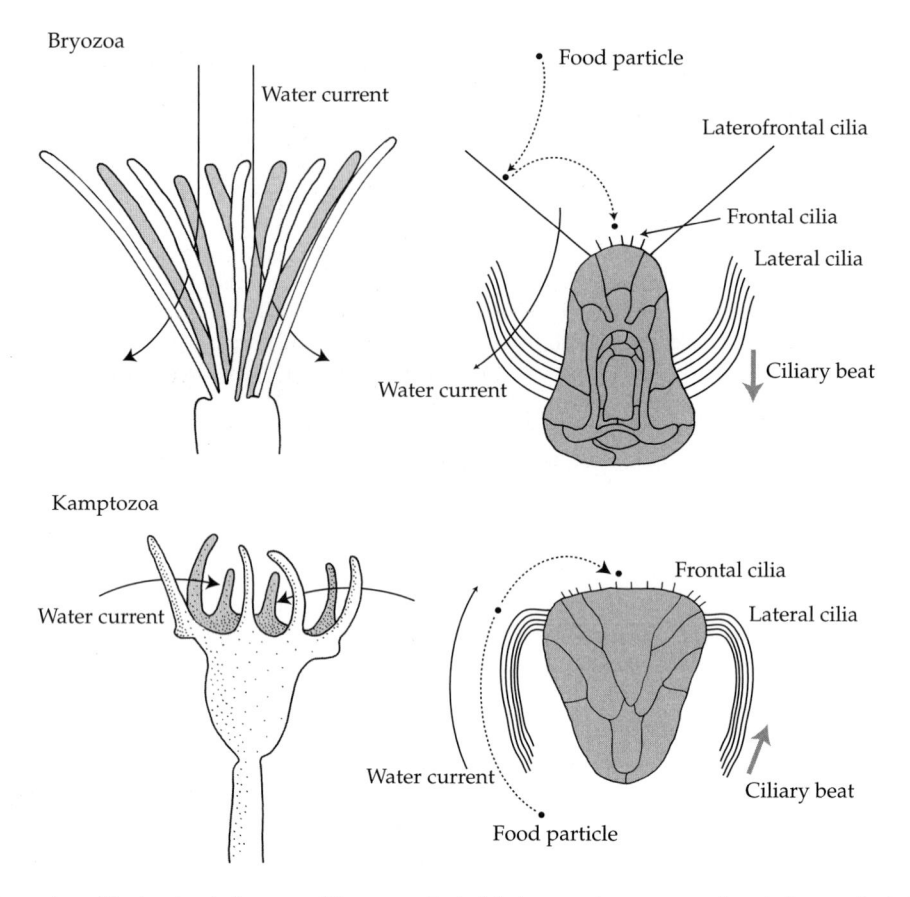

Fig. 12.16. Comparison of food capture in Bryozoa and Kamptozoa. To the left the general water current through the tentacles is shown. To the right are cross sections of single tentacles, with indicated direction of ciliary beat, water current, and particle capture. The frontal cilia direct towards the internal of the tentacular crown. Figures modified after Ryland (1970), Nielsen (1998) and Ruppert et al. (2004).

to create water currents and transport captured particles to the mouth, but the actual capture is often performed with mucous, to which dissolved food particles adhere. In some taxa, cilia play the dominant role in filtration and here we can distinguish, similarly to the larvae, downstream and upstream collecting systems (see Table 12.4.). For example, despite having quite similar tentacular crowns on first view, kamptozoans and bryozoans use completely different mechanisms for particle capture (Fig. 12.16.). Kamptozoans have two sets of cilia along their tentacles: longer lateral cilia creating a water current from outside into the tentacular crown (atrium), and shorter frontal cilia that transport the particles along the tentacle to the mouth (Nielsen & Rostgaard 1976, Emschermann in Westheide & Rieger 2007). Particles can be captured because turbulences occur in the water current while it passes the tentacles, and particles are caught in the 'current shadow' on the frontal (atrial) side of the tentacles (i.e. downstream). Bryozoans (Fig. 12.17.) also have lateral and frontal cilia on their tentacles, but they additionally have long and stiff laterofrontal cilia (Nielsen 2001). The lateral cilia create a water current from the atrium to the outside, that is, in the opposite direction as in kamptozoans. The stiff laterofrontal cilia mechanically filter particles from the water current (Riisgård & Manriquez 1997, Nielsen & Riisgård 1998, Riisgård et al. 2004), upstream of the filtering structure. These are 'kicked' by a flicking of the laterofrontal cilia and/or of the tentacle towards

Fig. 12.17. Colony of *Flustrellidra hispida* (Bryozoa, Gymnolaemata), showing lophophor with tentacles in circular arrangement. Photo by courtesy of Georg Mayer, Berlin.

the mouth. The tentacles (lophophor) of brachiopods and phoronids function in a similar way. The presence of upstream collecting systems in adults is, as in larvae, an argument for a closer relationship of the tentaculate taxa to deuterostomes. In this case, the tentacle apparatus of phoronids, brachiopods, and bryozoans might be homologous to the pterobranch tentacles. If, however, bryozoans or all tentaculate taxa belong to the protostomes, then such a tentacular collecting system would be of convergent origin (see Halanych 1996b).

Conclusions

There are three basic steps in nutrition: the capture of food, its breakdown (when the captured particles are not already very small), and endocytosis—the uptake of particles into the body. It appears that only capture and uptake are ancestral patterns, and that an intestinal tract evolved particularly to extend the range of food to larger items. These are treated in a compartment separated from the environment mechanically and chemically. The intestinal tract is originally sac-shaped, with only one opening, but within bilaterians it becomes, probably several times convergently, an unidirectional one-way-gut with two openings. This allows a more specialized treatment of food and therefore more effective nutrition. With a diversification of bilaterian animals goes a diversification in the utilization of food sources and a diversification of feeding structures adopted to sucking, swallowing, scraping, piercing, and so on (see Table. 12.4.).

In marine, ciliated larvae, some patterns with some phylogenetic impact can be realized in the mode of nutrition (lecithotrophic versus planctotrophic) and the presence as well as the function of ciliary bands (downstream versus upstream collecting bands). While these patterns are very informative and helpful when considering particular taxa, broader hypotheses (e.g. for bilaterian or metazoan evolution) depend on several ambiguous considerations and are therefore not well substantiated.

CHAPTER 13

Reproductive organs

Reproduction in metazoans is extremely diverse. It can be asexual or sexual. In the case of sexual reproduction, gametes are produced, often, but not always, in special organs called gonads. Sexes can be separate or both kinds of gametes can be produced within one organism. The structures to transfer or receive gametes are very diverse and relate to a diversity of reproductive behaviours. It is impossible to cover any aspect concerning reproductive organs within the frame of this book, but I will try to touch on some in the following.

Asexual versus sexual reproduction

While sexual reproduction always describes the development of new individuals from a zygote, that is, fused male and female gametes, asexual reproduction summarizes two different processes that must be distinguished from each other. New organisms can be produced entirely without gametes being involved (called here 'agametic-asexual'). Offspring can also develop from unfertilized oocytes (this 'gametic-asexual' reproduction is called 'parthenogenesis').

Both asexual and sexual modes of reproduction have certain advantages and disadvantages and their evolutionary comparison has received much attention (see, e.g., Ghiselin 1974, Maynard Smith 1978, Bell 1982, Stearns 1987). The main advantage of sexual reproduction is the combination of genetic information from two individuals in one offspring, which leads to an increased genetic diversity. This is assumed to be advantageous in the long term and therefore important for the existence of species through time. Asexual reproduction generally requires less energy and is faster compared to sexual reproduction. It has therefore short term advantages.

Both agametic-asexual and sexual reproduction are very old characteristics, and were likely present in the metazoan ancestor and predate the Metazoa. A metazoan autapomorphy is a characteristic lineage of oogensis and spermiogenesis (Ax 1996). Agametic-asexual reproduction is present in every diploblastic taxon. In sponges and cnidarians (excluding anthozoans), both modes of reproduction are present in the life cycle of the same organism (called 'metagenesis'). In these cases, the sexually produced stages in the life cycle are vagile larvae, while asexual reproduction is present in the stationary phases of the life cycle (as well as in the production of a second vagile stage, the medusa of Tesserazoa) (Fig. 13.1.). This may be generalized in stating that animals tend to exploit new habitats with genetically diverse, sexually reproduced stages, while they spread locally with clonal stages, which are produced comparably 'easy' and fast by asexual reproduction.

In bilaterian animals, both asexual and sexual reproduction are broadly distributed. Asexual modes of development include here parthenogenesis, which is clearly derived from sexual reproduction, because gametes are formed. Compared to sexual reproduction, each parthenogenetic individual (and not every second as in sexual reproduction) contributes to the production of offspring, which is therefore, theoretically, twice as much as in sexually reproducing animals. It is likely that the bilaterian ancestor was capable of agametic-asexual as well as sexual reproduction, but within bilaterians the

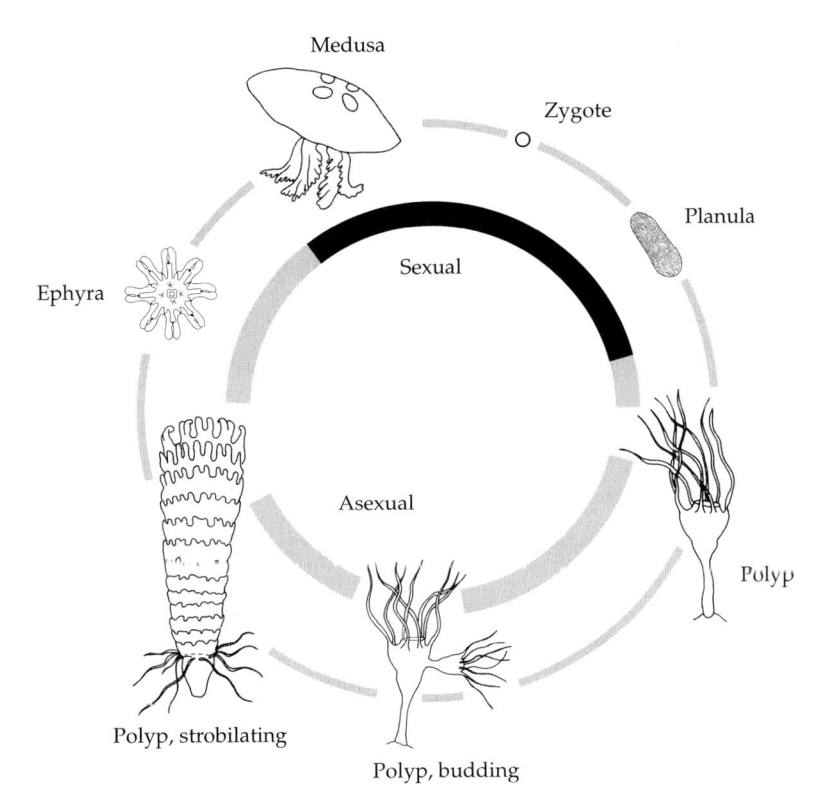

Fig. 13.1. The life cycle of *Aurelia aurita* (Scyphozoa, Cnidaria) as an example for metagenesis. The medusa reproduces sexually, the polyp is the asexual stage. After Westheide and Rieger (2007).

patterns changed very often and in very diverse ways:

• Agametic-asexual reproduction is an important way to form colonies (Fig. 12.17.). Colonies are present in several taxa, and the growing by asexual reproduction (budding) from one single individual ensures that the individual modules within a colony remain in close contact (sometimes even sharing certain organ systems). A general advantage of colonies is that it allows much more growth compared to single specimens in those species, where the body organization prevents growing to large sizes. Wasson (1999a) compared colonial and solitary anthozoans, showing that the surface:volume ratio sets a limit to the growth of solitary, but not to colonial, anthozoans.

• Budding occurs also in non-colonial bilaterians, for example in the acoel *Convoluta retrogemma* (Hendelberg & Åkesson 1991), where buds are formed at the posterior end of the animal. In catenulid and macrostomid platyhelminths and some annelids, in particular syllids (Fig. 13.2.; Franke 1999), terminal budding leads to chains of individuals in different stages of maturity, this is called paratomy. (In syllids, the 'daughter' individuals do not have to form chains, but can also form other clusters at the terminal end of the 'mother', see Franke 1999).

• Agametic-asexual reproduction by fragmentation or fission is performed in one acoel (*Convolutriloba longifissura*, Bartolomaeus & Balzer 1997), some platyhelminths, annelids, nemerteans, phoronids, echinoderms, and hemichordates (see Table 13.1.).

• The combination of sexual and asexual reproduction in a metagenetic life cycle can be advantageous for parasites, which usually need a high

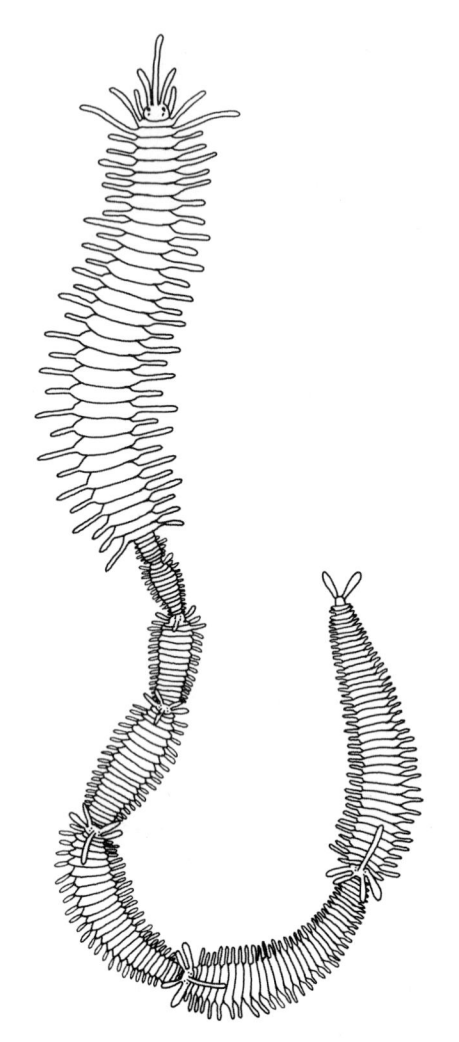

Fig. 13.2. Paratomy in the syllid *Autolytus purpureomaculatus* (Polychaeta, Annelida): new individuals bud off from the posterior end of the 'mother' and form a chain of differently developed offspring. After Westheide and Rieger (2007).

number of offspring to ensure that at least some of them enter a new host. Digeneans and some cestodes (*Echinococcus* species) are well known for their asexually produced larvae, either as sporocysts (Digenea; Fig. 13.3.) or as proliferating hydatids (*Echinococcus*).
• Parthenogenesis has evolved in several taxa (see Table 13.1.), probably many times convergently. Normark (2003) assumes about 900 origins within insects alone. The switch from sexual reproduction to parthenogenesis (and back?) therefore appears to be comparatively easy. In some asteroids (*Asterias* and *Marthasterias*) and echinoids, parthenogenesis can be induced experimentally, while it is absent in natural populations (Chia & Walker 1991, Pearse & Cameron 1991). In some groups (especially gastrotrichs and tardigrades), parthenogenetic species are found in freshwater, while marine species are all sexually reproducing. This appears to be related to the fact that many freshwater habitats are unstable and require quick and efficient responses to ecological changes. Freshwater chaetonotoid gastrotrichs and bdelloid eurotifers, for example, form two kinds of parthenogenetic eggs. The first one is to rapidly obtain high population densities, the second one is a resistant resting egg (Strayer & Hummon 1991, Mark Welch & Meselson 2000). Parthenogenetic reproduction is also helpful in colonizing islands. For example, it has recently been shown that the Komodo dragon has the potential to reproduce parthenogenetically (Watts et al. 2006).
• Sexual reproduction and parthenogenesis can also alter within one life cycle, this is called 'heterogony'. Heterogony occurs in several cases: Cladocera (Crustacea), Monogononta (Eurotifera), and Aphidina (Insecta). The life cycles of cladocerans and monogononts (Fig. 13.4.) are comparable to that described above for freshwater chaetonotid gastrotrichs, and they are also related to ecological factors (e.g. Ricci 1992 for monogononts). Compared to marine environments, freshwater habitats are usually smaller, younger, and less stable. Therefore, ecological conditions are likely to vary considerably and monogononts and cladocerans quickly develop high population densities following spring algal blooms (asexual reproduction), and form resting stages for inconvenient conditions later in the year by sexual reproduction.
• Parthenogenesis is certainly derived from sexual reproduction, and there are some cases in which parthenogenesis needs a heterosexual stimulus. New Mexico whiptails (15 species in the genus *Cnemidophorus*) reproduce exclusively by parthenogenesis, but need stimuli from a sexual behaviour (Maslin 1971). There are some

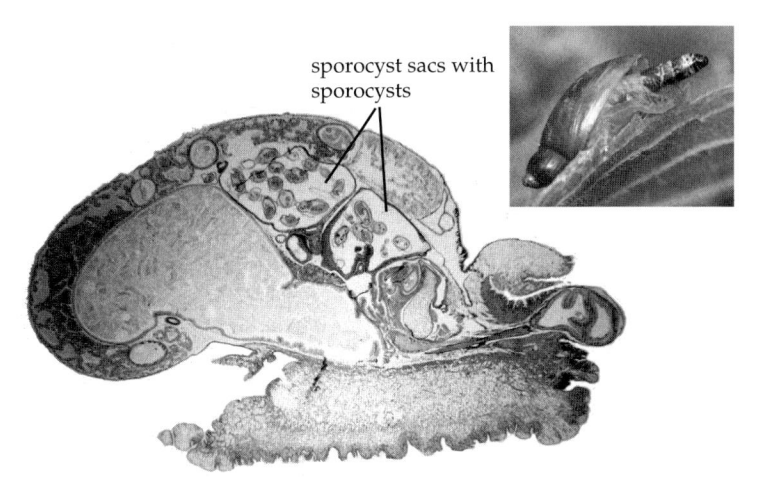

Fig. 13.3. Asexual reproduction in sporocysts of the digenean *Leucochloridium* sp., which creates swollen tentacles with oscillating ring patterns in its intermediate host, gastropods from the genus *Succinea* (top right). A section through the snail shows sacs with asexually produced sporocysts in the visceral mass of the body.

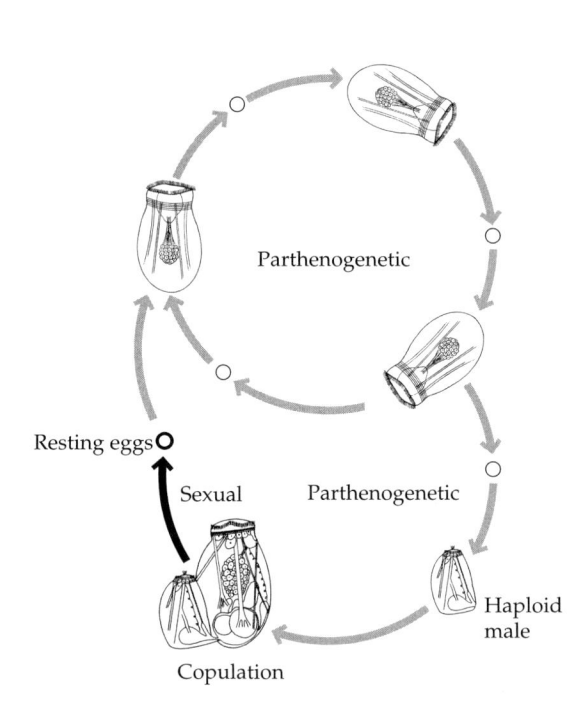

Fig. 13.4. Life cycle of the monogonont eurotifer *Asplanchna* sp. as an example for heterogony. During the year, *Asplanchna* reproduces parthenogenetically several times with diploid oocytes. Towards the end of the year, haploid males are formed, which copulate with a female producing haploid oocytes. The resulting diploid zygotes are resting eggs that survive winter conditions. Life cycle incudes figures after Thane (1974) and Westheide and Rieger (2007).

cases among craniotes, in which the parthenogenetic oocyte even needs the stimulus of a sperm cell, although the gametes do not fuse. This phenomenon is called 'gynogenesis' and occurs, for example, in the fish genus *Poeciliopsis* and the salamander *Ambystoma* (Schlupp 2005). There are further, strange examples for unisexual reproduction involving the other sex, such as the possible elimination of female genetic material from the zygote, which occurs in the fire ant *Wasmannia auropunctata* (Fournier et al. 2005).

Parthenogenesis is more diverse and further distributed in bilaterians than agametic-asexual reproduction. This may be linked to a trend to reduce the capacity to regenerate (Steinböck 1963b). This capacity is very high in non-bilaterian taxa. Within bilaterian taxa, it is very diverse. There are some taxa which have a high regeneration potential (and therefore the capacity for agametic-asexual reproduction, see Table 13.1.). To give just one example, the polychaete *Dorvillea bermudensis* is able to regenerate completely from fragments comprising only two body segments (Åkesson & Rice 1992, Paulus & Müller 2006). In several animal taxa, the regeneration capacity is very low or even absent. Examples are Gnathifera and Nemathelminthes. For example, Lorenzen (1985) assumed the lack

Table 13.1 Occurrence of sexual (SR) and asexual reproduction (AR), gonochorism (GC) and hermaphroditism (HE) in metazoan taxa. The references cited are usually from the four-volume '*Encyclopedia of Reproduction*', for further references see volumes of the series '*Reproductive Biology of Invertebrates*' and '*Reproduction of Marine Invertebrates*'.

	Sexual – asexual reproduction	Gonochorism – hermaphroditism	Selected references
Porifera	Both present, usually together in one individual. AS by various modes (fragmentation, budding, gemmules = resting stages)	HE, some species GC	Fell 1999
Trichoplax adhaerens	SR likely, but not completely known, AR by fission or 'swarmers' formed by budding	Unknown (gametes were rarely observed)	Grell & Ruthmann 1991
Cnidaria	Both present, in Tesserazoa often combined in metagenetic life cycles	Both present, GC occurs more often than HE	Fautin 1999
Ctenophora	SR is the regular mode, AR by fission or budding, followed by regeneration, occurs in some species	Most species HE, *Ocyropsis* species are dioecious	Harbison & Miller 1986, Matsumoto 1999
Xenoturbella	Probably only SR	HE (protandrous)	Westblad 1950
Acoelomorpha	SR and AR (fragmentation, fission, budding)	HE, rarely GC	Tyler 1999
Gastrotricha	SR in marine species, AR in freshwater chaetonotids (as parthenogenesis)	HE or females with rudimentary male reproductive organs (freshwater chaetonotids)	Hummon & Hummon 1992, 1993, Weiss 2001
Nematoda	Mainly SR, sometimes AR (parthenogenesis)	GC, rarely HE	Poinar & Hansen 1983, Wharton 1986, Wright 1999
Nematomorpha	Only SR	Only GC	Schmidt-Rhaesa 1999
Priapulida	Only SR	Only GC	Lemburg & Schmidt-Rhaesa 1999
Kinorhyncha	only SR	Only GC	Neuhaus 1999
Loricifera	SR and AR (parthenogenesis), both occurring in complex life cycles	Only GC	Gad 2005a, b
Platyhelminthes	SR and AR (fragmentation, fission)	HE, few species with GC (e.g. *Schistosoma mansoni*)	Tyler 1999
Gnathostomulida	Only SR	Only HE	Sterrer 1999
Limnognathia maerski	Probably AS (parthenogenesis)	Only females are known	Kristensen & Funch 2000
Eurotifera	SR and AR (parthenogenesis) in Monogononta, only parthenogenesis in Bdelloida	Only GC	Wallace 1999
Seisonidea	only SR	Only GC	Wallace 1999
Acanthocephala	only SR	Only GC	Crompton 1999
Cycliophora	SR and AR (internal budding)	GC	Funch & Kristensen 1997
Nemertini	most species SR, AR (fragmentation) possible in some *Lineus* species	Most species GC, some species HE	Friedrich 1979, Bierne 1983
Kamptozoa	both SR and AR (budding)	Both GC and HE (often sequential with protandry)	Mariscal 1975, Wasson 1999b
Mollusca	Only SR	GC (Caudofoveata, Polyplacophora, Monoplacophora, Scaphopoda and Cephalopoda), other groups GC and HE	Heller 1993, Saleuddin 1999
Sipunculida	SR, only one species (*Themiste lageniformis*) can reproduce parthenogenetically	GC, only one species (*Golfingia minuta*) HE	Rice 1999
Echiurida	Only SR	Only GC	McHugh 1999
Annelida	SR and AS (fragmentation, budding as paratomy)	Polychaetes and pogonophorans usually GC, clitellates HE	Fischer 1999, McHugh 1999

Table 13.1 (*Contd*)

	Sexual – asexual reproduction	Gonochorism – hermaphroditism	Selected references
Tardigrada	SR and AR (parthenogenesis in many non-marine tardigrades)	Marine species always GC, freshwater and terrestrial species GC or HE	Bertolani & Rebecchi 1999
Onychophora	SR, one population of *Epiperipatus imthurni* AR (parthenogenesis)	GC	Storch & Ruhberg 1993
Euarthropoda	SR, some species AR by parthenogenesis (several species of Acari, Crustacea, Chilopoda, Progoneata, apterygote insects, aphids; unisexual parthenogenesis in some Hymenoptera)	GC, rarely HE (some crustaceans and insects)	Davey 1999, Kaufman 1999
Chaetognatha	Only SR (parthenogenesis may rarely be present, but has not been shown with certainty)	HE	Shinn 1999
Bryozoa	Both SR and AR (budding)	Single zooids GC or HE	Reed 1991, Woollacott 1999
Phoronida	Both SR and AR (fragmentation occurs regularly in *Phoronis ovalis*)	Most species HE (protandry), *Phoronopsis harmeri* GC	Zimmer 1991, Herrmann 1997
Brachiopoda	Only SR	Most species GC, few HE (genus *Argyrotheca*)	Stricker 1999
Echinodermata	SR, sometimes AR in asteroids, ophiuroids, and holothurians (division or fission)	Most species GC, small species often HE	Byrne 1999
Enteropneusta	SR and AR (fission in some species, natural parthenogenesis only in the sea star *Ophidiaster granifer*)	GC	Hadfield 1975, King 1999
Pterobranchia	SR and AR (budding or fission)	GC	King 1999
Tunicata	SR and AR (budding)	HE, only *Oikopleura dioica* (Appendicularia) GC	Newberry 1999
Acrania	Only SR	GC, some species can have a small percentage of hermaphroditic specimens	Stokes 1999
Craniota	SR, rarely AR (by parthenogenesis in some teleost fishes, amphibians, and squamatans)	Most species GC, several teleost fishes HE	Westheide & Rieger 2004

of asexual reproduction, and the little or no ability to regenerate, as characteristic for 'pseudo-coelomates' (now recognized as an artificial taxon composed of Gastrotricha, Cycloneuralia, and Syndermata). The absence of regeneration is closely related to eutely, the presence of a constant number of somatic cells. This has been shown in some nematodes (e.g. *Caenorhabditis elegans*) and and syndermatans (Wallace 2002 for eurotifers). However, this character is probably over-estimated and not as broadly distributed as sometimes thought. Less-derived nematodes than *C. elegans* show no eutely (Malakhov 1998) and gastrotrichs possess the ability for regeneration of wounds (Manylov 1995). The regeneration capacity, nevertheless, remains low compared to other bilaterians.

Hermaphroditism versus gonochorism

Gonochorism and hermaphroditism are widespread in animals and there are many taxa in which species with both kinds of reproduction occur. In some taxa only single or few species show a differing mode. For example, among the gonochoristic nematodes, *Caenorhabditis elegans* is (largely) hermaphroditic and among the hermaphroditic flatworms, *Schistosoma mansoni* (Fig. 2.25.) is gonochoristic. Both species are certainly derived within their groups, making it likely that their reproductive mode is also derived. While the ancestral mode of reproduction can be reconstructed in such cases convincingly, there are other taxa in which this is not so easy. This is the case, unfortunately, in sponges and cnidarians. Gonochorism is more abundant in cnidarians than hermaphroditism, but this does not have to mean that it is the ancestral mode, in particular because several anthozoans (as the basal branch of cnidarians) are hermaphroditic. Sponges can also be gonochoristic or hermaphroditic, but according to Fig. 1 in Sarà (1992), hermaphroditism appears more likely to be ancestral (if sponges are monophyletic).

Hermaphroditism is sometimes taken as a phylogenetically informative character (e.g. supporting a sister-group relationship of Platyhelminthes and Gnathostomulida, Ax 1996) or assumed to be the ancestral mode in (at least) bilaterian animals (Balsamo 1992). By looking at the distribution on the phylogenetic tree (Fig. 13.5.), it appears that going back from the taxa to their common ancestors adds more and more uncertainty to the reconstruction of the ancestral modes of reproduction. Taking the mode of reproduction as a phylogenetically informative character therefore requires some caution.

The origin of gametes

Gametes are cells that have the capacity to develop into all the cells of a newly developing organism. The time in which cells are determined to become gametes, varies from very early to quite late in development. In some animals, maternally inherited determinants for germ cells are segregated before or immediately following fertilization to become the germ cells. This is called 'preformation'. In other animals, germ cells are recognized later, either during certain cleavage stages or even only when the gonads are formed. In these cases it appears that the prospective germ cells receive an inductive signal from neighbouring cells. This is called 'epigenesis' (see, e.g., Nieuwkoop & Sutasurya 1979, 1981, Extavour & Akam 2003). Recognizing a cell as a germ cell requires either special cellular characteristics or the presence of particular markers. In most cases, such cells contain aggregates of electron dense, basophilic structures in the cytoplasm, and many express the gene *vasa* or its homologue (Extavour & Akam 2003). As soon as a cell is recognizable as a prospective germ cell it is called a 'primordial germ cell' (PGC).

In discussing whether preformation or epigenesis is the ancestral way of PGC determination, matters are sometimes biased by the fact that we have quite detailed knowledge on some species with preformation (e.g. the so-called model organisms *Caenorhabditis elegans*, Nematoda and *Drosophila* spp., Insecta), but only superficial knowledge in many other taxa. Nevertheless, epigenesis appears to be more widely distributed than preformation and is also the ancestral mode (Table 13.3.; Extavour & Akam 2003).

In sponges, germ cells originate from totipotent cells (archaeocytes) or from already differentiated cells such as choanocytes (Fell 1999). In cnidarians, gametes develop from either endodermal (Anthozoa, Scyphozoa) or ectodermal (several Hydrozoa) cells that migrate into the ECM (mesogloea) between both cell layers (Larkman 1983, 1984, Schäfer 1983, Fautin 1999). In hydrozoans, these are pluripotent stem cells, the I-cells (interstitial cells) (see Thomas & Edwards 1991 for summary). Therefore, it is likely that all diploblastic animals form germ cells by epigenesis. Within bilaterians, epigenesis is also broadly distributed (see Table 13.3.). Platyhelminths and acoelomorphs derive germ cells from totipotent stem cells (neoblasts) (e.g. Ladurner et al. 2000). Preformation occurs in trematodes (Bednarz 1973), but this is derived within Platyhelminthes.

In other spiralians, the PGCs appear to derive from a subpopulation of mesodermal cells, which themselves are derived from the 4d cell (e.g. Dohmen 1983 for molluscs). With few exceptions (see Extavour & Akam 2003), PGC determination is by epigenesis. The sources and timing of PGC differentiation are variable. PGCs may also derive from differentiated epithelial cells, as has been reported, for example, from the pulmonate slug *Deroceras reticulatum*. Here, Hogg and Wijdens (1979) observed that the reproductive system at hatching consists of only one cell type (named the gonadal stem cell) that later differentiates into germinal and supporting cells.

In nematodes, more precisely in *Caenorhabditis elegans*, one cell called P4 is set aside very early in development. It divides to form four cells in a

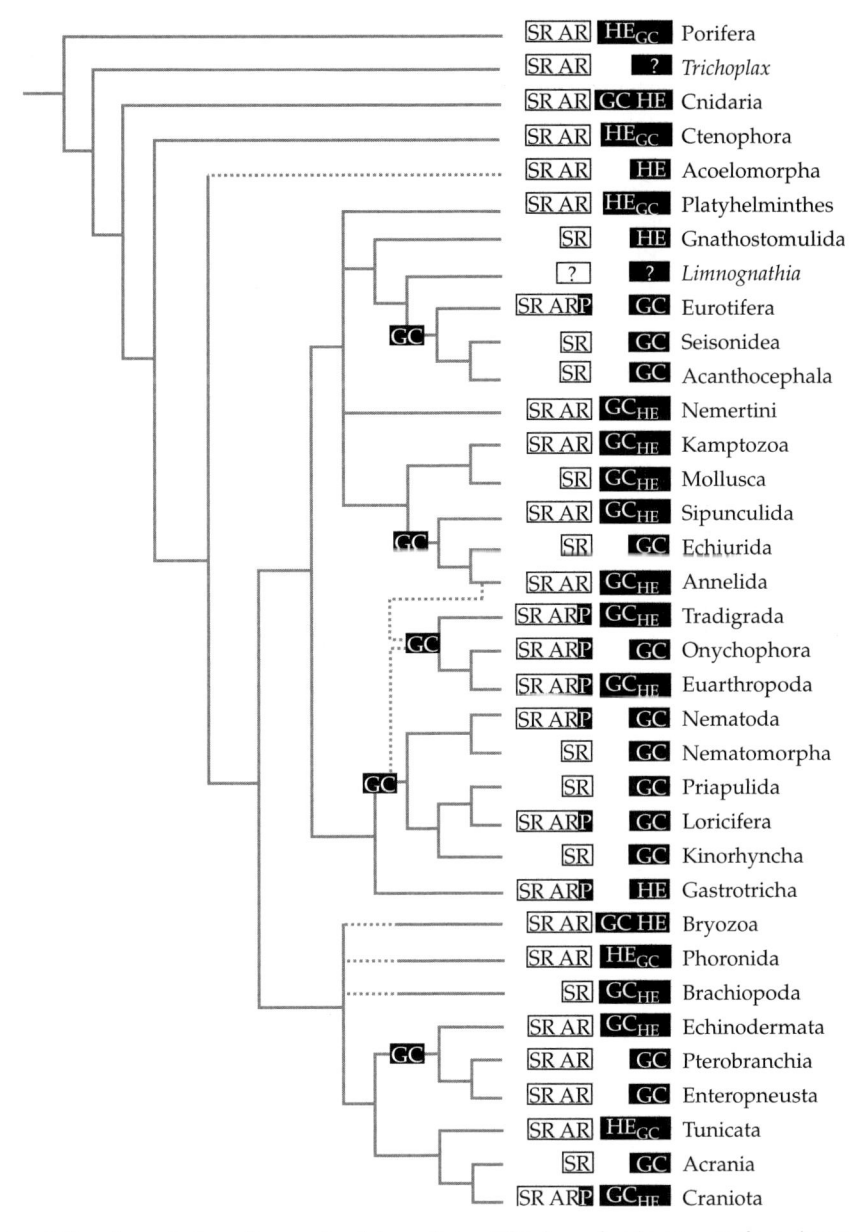

Fig. 13.5. Mapping of different reproductive patterns on the phylogenetic tree. White boxes describe presence of sexual reproduction (SR), asexual reproduction (AR), and parthenogenesis (ASP). Black boxes describe the presence of gonochorism (GC) and hermaphroditism (HE). When both patterns occur in one taxon and the ancestral condition is not clear, both patterns have the same font. When one pattern is clearly derived, it is written in a smaller font. Only a few patterns in ancestors of larger taxa can be reconstructed unequivocally.

row, Z1 to Z4. The first and the last, Z1 and Z4, proliferate and form the gonad, while the central ones, Z2 and Z3, are the precursors of a syncytial cluster of germline cells, which become surrounded by the growing gonad wall (Schedl 1997, Pilgrim 1999). This appears to account for all nematodes investigated so far (Extavour & Akam 2003).

In euarthropods, however, knowledge of the *Drosophila* germ line cannot be generalized.

Germline cells are set apart very early in development and later become surrounded by the gonad (Huebner & Diehl-Jones 1992). In crustaceans (e.g. Pochon-Masson 1983, Scanabissi et al. 2005) and chelicerates (e.g. Coons & Alberti 1999, Fahrenbach 1999) PGCs are derived from the germinal epithelium of the gonad and this (and therefore epigenesis) appears to be the ancestral mode in euarthropods.

In deuterostomes, there is a mixture of epigenesis and preformation, but probably epigenesis is also the ancestral mode (Extavour & Akam 2003). Among craniotes, several groups including agnathans use epigenesis, but in several (all?) teleost fishes, frogs, and birds, preformation is present (Extavour & Akam 2003, Johnson et al. 2003). In amniotes, PGCs are found in extraembryonic tissue, regardless of whether they originate by preformation or epigenesis. They later migrate into the embryo. In mammals, they migrate through the allantois; while in birds, they migrate through the circulatory system (see Gilbert 2000). As has been shown in chicken embryos, the PGCs then pass the epithelium of a blood vessel as well as the epithelium of the developing gonad (Fig. 13.6.; Tsunekawa et al. 2000).

For further development it is important that gradients of molecules form within the oocyte. In holothurians and acranians it has been shown that the polarity of germ cells in their epithelial stage is conserved to form the polarity of the oocyte (Frick & Ruppert 1996, 1997, Frick et al. 1996).

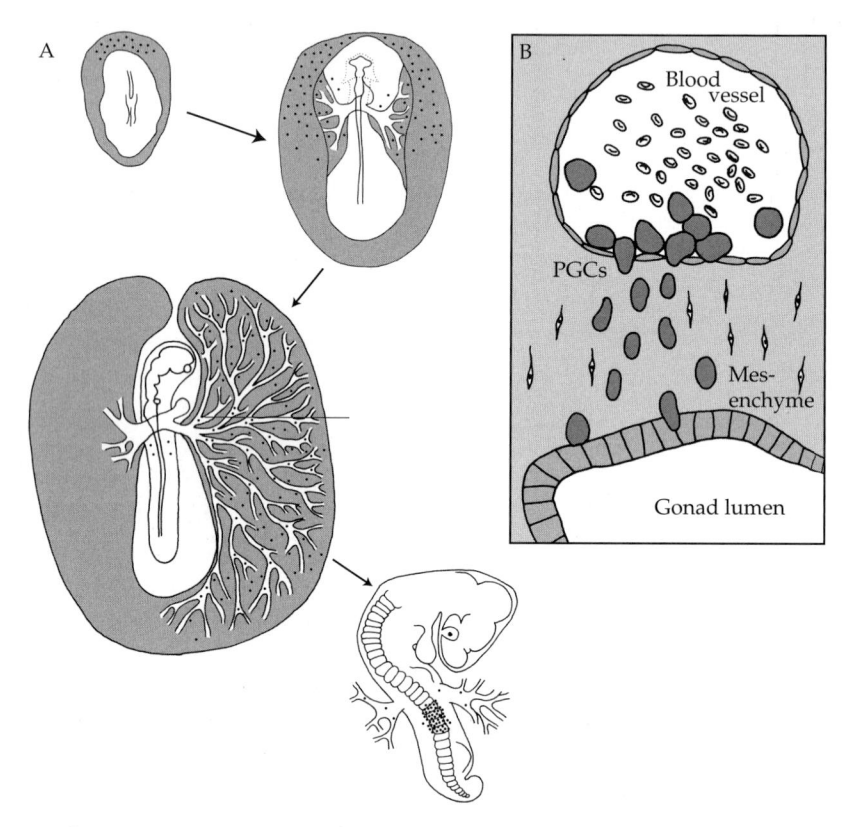

Fig. 13.6. The migration of primary germ cells (PGCs) in the avian embryo. A. PGCs (black dots) are first visible in a cresent within the 'aera opaca' anterior of the early embryo in the primitive streak stage. They distribute and then migrate through the extraembryonic circulatory system into the embryo, where they concentrate in the region of the gonads. In the third stage, only one half is drawn completely. B. PGCs penetrate the epithelium of the blood vessels, migrate through the mesenchyme and through the gonadal epithelium into the gonad. A after Nieuwkoop and Sutasurya (1979), B after Gilbert (2000).

In their review, Extavour and Akam (2003) suggested that PGCs can be regarded as homologous across all metazoans. The similar function of these cells, as well as similar structural and molecular characteristics, support this assumption. However, homology implies common origin also, and this is not the case in PGCs.

Evolution of gonads

The possible presence of PGCs very early in development, and long before gonads are formed, shows that gonads do not have to be the site of origin for the gametes. In these cases, gonads are just the organs, in which gametes mature. Gonads are defined here as epithelially bordered organs surrounding the gametes. This definition excludes cases in which gametes are present between epithelia (as in cnidarians) or between parenchymal cells (as in acoels). It also excludes cases where gametes derive directly from the coelom epithelium, and therefore from particular regions of a different organ system. This definition includes, however, cases, where an epithelium disintegrates up to a stage where only ECM surrounds the gametes.

Gametes must have been present already in the metazoan ancestor, because they are present in every metazoan taxon. Gonads, however, evolved later. In sponges, gametes can originate almost everywhere in the body (Boury-Esnault & Jamieson 1999, Fell 1999). Gametes are not clustered in particular body regions, but they can be associated with further cells. Both spermatozoa and oocytes can be surrounded by accessory cells, which form the 'sperm cyst' or a 'follicular epithelium'. The follicular cells likely pass yolk material to the growing oocyte (Fell 1999). Cnidarians and ctenophores possess particular locations where gametes are concentrated. As these are more or less only concentrations of gametes, the term 'gonad' is inappropriate and many authors prefer to talk about 'gametogenic areas', although these areas can show some forms of epithelial elaborations (see below). When they are present, medusae are always the stages in which gametes mature in cnidarians. In taxa without medusae, this is also possible in the polyp. In general, germ cells develop from ectodermal or endodermal cells and migrate into the ECM (mesogloea) between both cell layers (see above). In scyphozoans, an additional epithelium can be integrated into the gonadal tissue, separating a genital sinus (Fig. 13.7a.; Widersten 1965, Lesh-Laurie & Suchy 1991). The gonads in Ctenophora are associated with the endodermal epithelium and develop close to the canal system (Fig. 13.7b.; Martindale & Henry 1997, Matsumoto 1999). They are bordered on one side by the endodermal cells of the meridional canal and on the other side by ECM. In the female part, glandular and some epithelially organized flat cells form a canal, the oviduct. The testicular portion of the gonad contains clusters of different stages of spermiogenesis, while the female part contains oocytes as well as numerous nurse cells which supply the oocytes with nutrients and yolk.

In some bilaterian animals, the gametes are not enclosed by structures that can be called a gonad, but are present between parenchymal cells or between intestinal and parenchymal cells. Such cases occur in *Xenoturbella* (Israelsson 1999), several acoelomorphs (Ehlers 1985a), and catenulids (Platyhelminthes; Ehlers 1985a). In comparison with diploblastic taxa, this may be taken as an ancestral pattern, while in most other bilaterians, an epithelialized gonad is present. There are, however, some acoels and nemertodermatids (Rieger et al. 1991b, data matrix in Lundin & Sterrer 2001), where epithelial cells have been found in the gonad. Therefore, epithelialized gonads may either have evolved more than one time in parallel, or they are ancient and were reduced some times.

In bilaterians, the gonads are often epithelialized sacs. In many cases, this epithelium is a 'germinal epithelium', from which gametes originate. In several cases this epithelium has been found to be discontinuous or even absent (Ahlrichs 1995 for Seisonidea), so that a layer of ECM is the only border of the gonad. I suspect that such observations are due to the dissociation of maturing germ cells from the germinal epithelium, and it should be tested whether earlier developmental stages of the gonad possess a more complete epithelium.

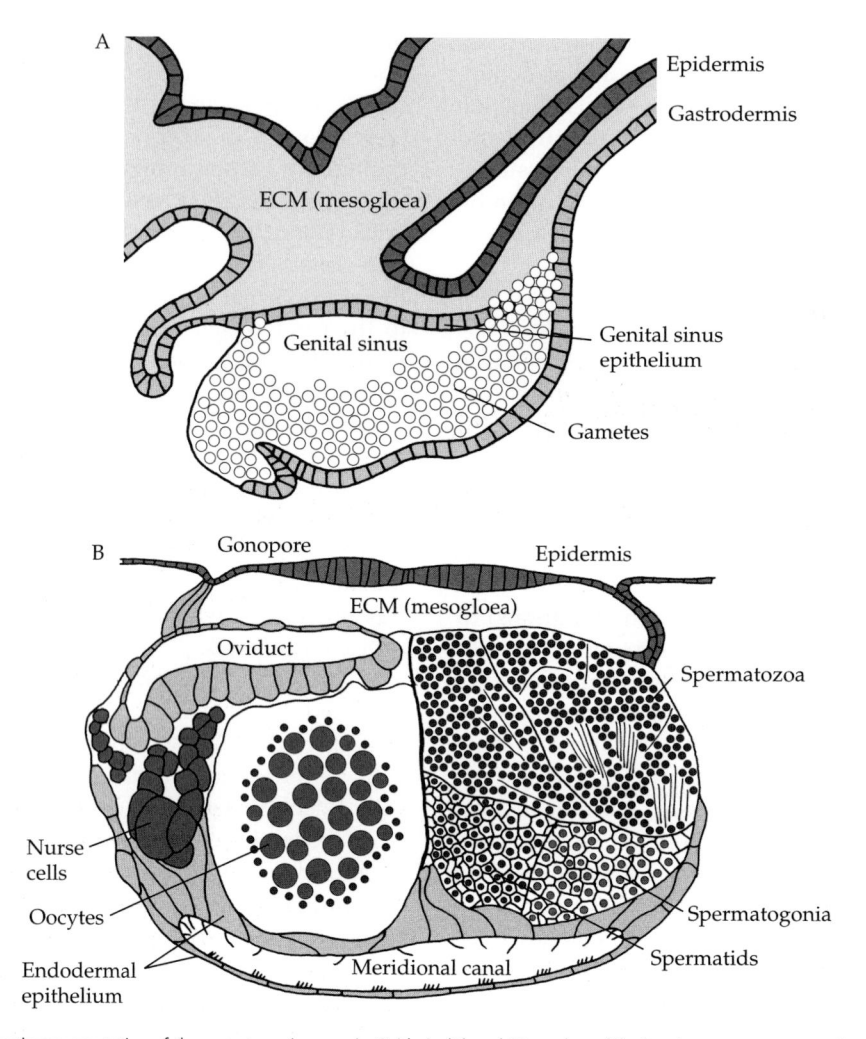

Fig. 13.7. Schematic representation of the gametogenic areas in Cnidaria (A) and Ctenophora (B), showing gametes accumulated between epidermis and gastrodermis, with some epithelial components of endodermal origin. A after Widersten (1965), B after Hernandez-Nicaise (1991).

In hermaphrodites, male and female gametes can mature within one gonad (Fig. 13.8.) or both male and female gonads are present. When gametes mature in the germinal epithelium, they have to leave this epithelium at some point of time. In some cases this is comparably early, but in other cases the maturing oocytes remain connected to the epithelium for a longer time. Because this creates spatial problems, such oocytes become dislocated from the epithelium and connected by a small stalk. In Peripatopsidae (Onychophora) and ticks (Acari, Chelicerata), they bulge to the basal side of the epithelium (Fig. 13.9.; Diehl et al. 1982, Evans 1992, Storch & Ruhberg 1993). This is also the case in echinoderms (Fig. 13.10.), where they bulge into the hemal sinus surrounding the gonad itself (e.g. Heinzeller & Welsch 1994). There often is a maturation grade of gametes within the gonads, in some cases it is from peripheral (young stages) to central (old stages). In insects and nematodes, gonads are long tubes with a proximal (young stage) to distal (old stage) gradient (Fig. 13.11.).

REPRODUCTIVE ORGANS   251

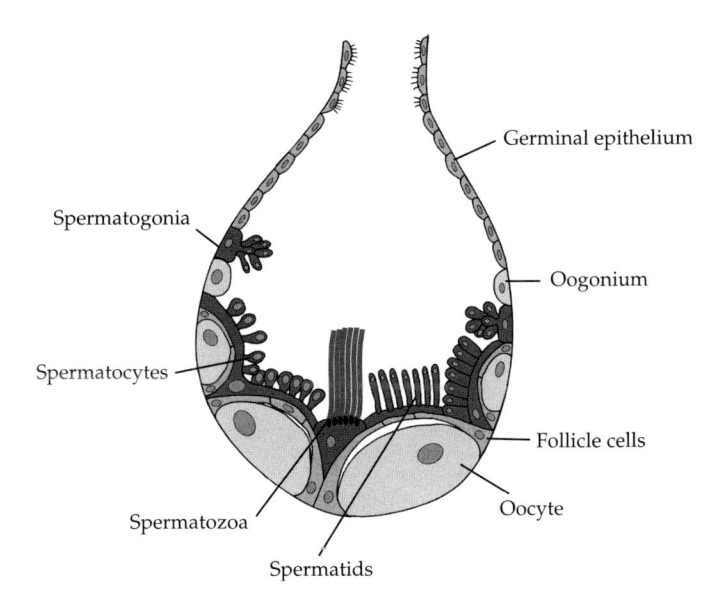

Fig. 13.8. Schematic drawing of a hermaphroditic gonad in a pulmonate gastropod, where male and female gametes develop in very close proximity. Lighter grey symbolizes female components (oogenesis stages and supportive cells), darker grey male components (spermatogenesis stages). After Saleuddin (1999).

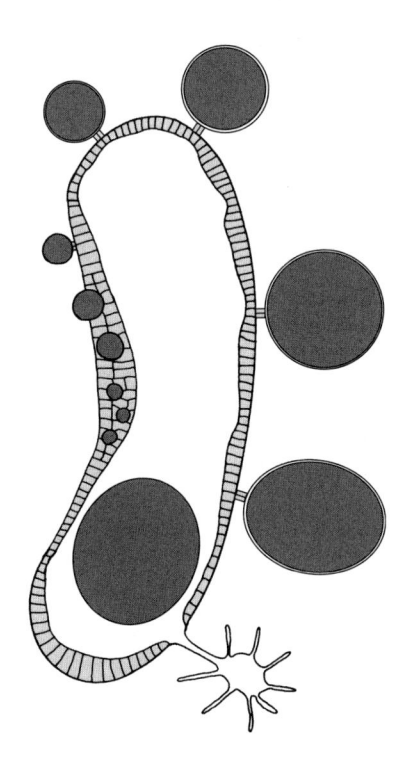

Fig. 13.9. Development of oocytes in the tick gonad (Acari, Chelicerate). Oocytes originate in the epithelium and lose the epithelial context when growing (maturation cycle in clockwise direction). They remain connected to the epithelium the whole time. ECM surrounds the entire structure (gonad + oocytes; ECM is not shown here). Mature oocytes are released into the gonad lumen. After Diehl et al. (1982).

Bilaterian gonads can originate in different ways. Unfortunately, developmental data on gonads are comparably rare, but several patterns emerge:

- **Absence of gonads**: gametes derive from the coelom epithelium and mature in the coelom. This is the case in the tentaculate taxa Bryozoa, Phoronida, and Brachioloda and likely also the case in annelids and probably sipunculans.
- **Gonad derived from coelom**: in some molluscs (some gastropods and bivalves, Moor 1983; cephalopods, Budelmann et al. 1997), the gonad epithelium is reported to bud off from the pericard epithelium. In echinoderms, the PGCs proliferate and form an extending structure, the rhachis, which is surrounded by an epithelium of coelomic origin according to Houk & Hinegardner 1980).
- **Gonads go from compact to epithelialized**: such a process is comparable to schizocoely (see Chapter 8). In nematomorphs, the first stage of the gonad is a solid strand of cells, which later separate into peripheral and central cells (Schmidt-Rhaesa 2005). However, it is not certain whether an epithelium is already preformed in the solid strand.
- **The epithelium grows around the germ cells**: in species with an early specification of PGCs,

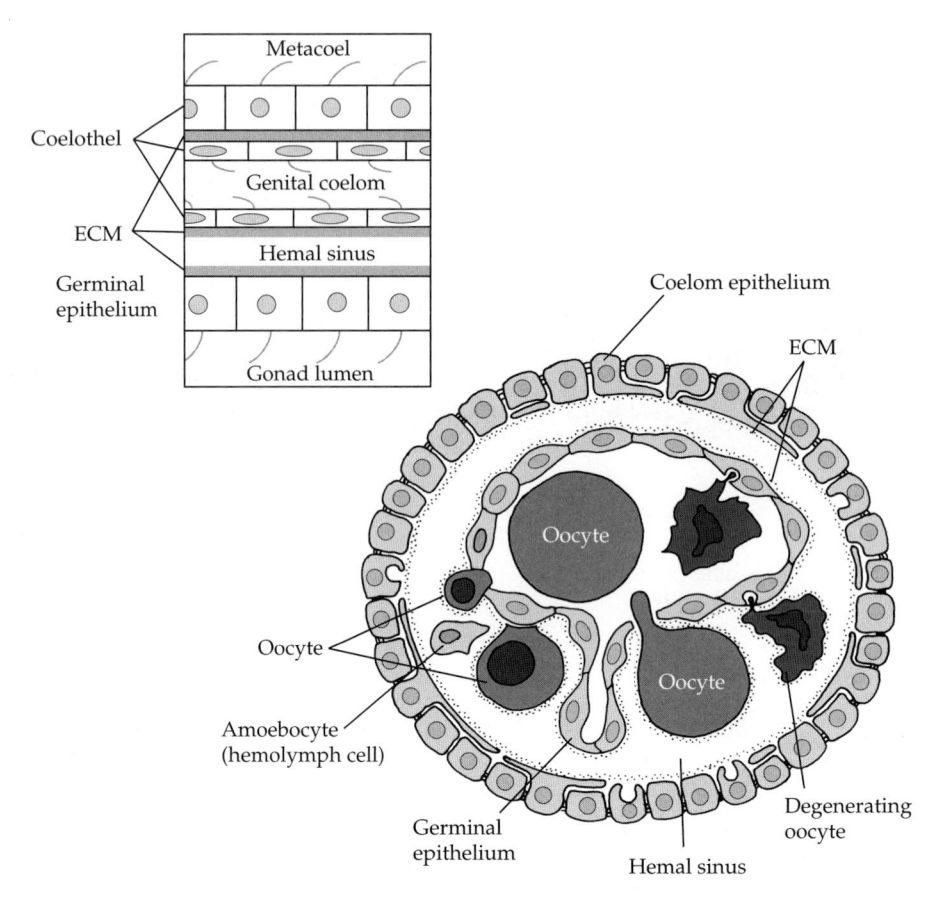

Fig. 13.10. Schematic representation of the structures surrounding the gonad in echinoderms (top) and a crinoid ovary as one example. The gonad itself is surrounded by a hemal sinus (primary body cavity), a genital coelom, and the metacoel (only innermost coelom epithelium shown in the crinoid ovary). Crinoid ovary after Heinzeller and Welsch (1994).

these often migrate to the prospective site of the gonads and induce there either somatic cells (e.g. insects) or cells from their cluster (nematodes) to grow epithelially around the PGCs (Bünting 1992, Huebner & Diehl-Jones 1992).

The diversity of modes in which gonads are formed makes it hard to substantiate a common origin of gonads. It appears to be more likely that gonads evolved independantly several times. This interpretation would argue against the gonocoel theory (Goodrich 1945, see Chapter 8), in which the presence of an epithelialized gonad is the central starting point for the evolution of coeloms. For Goodrich (1945), the gonad of platyhelminths is of central importance, because the evolution of the coelom is assumed to start from a state comparable to flatworms. However, as in catenulids, the gonad can be non-epithelialized (see above), it is not clear whether the platyhelminth ancestor had epithelialized gonads.

Another questionable matter is whether ducts and pores to release the gametes are ancestral features. This is very likely for the male system: spermatozoa are always released to the external environment or transferred to a partner through a duct and a pore. In the female system of several acoelomorphs, gastrotrichs, platyhelminths, and gnathostomulids, however, duct and pore are lacking. The hypodermal injection of spermatozoa (either undirected or into the region of the female gametes) brings spermatozoa to the oocytes. In

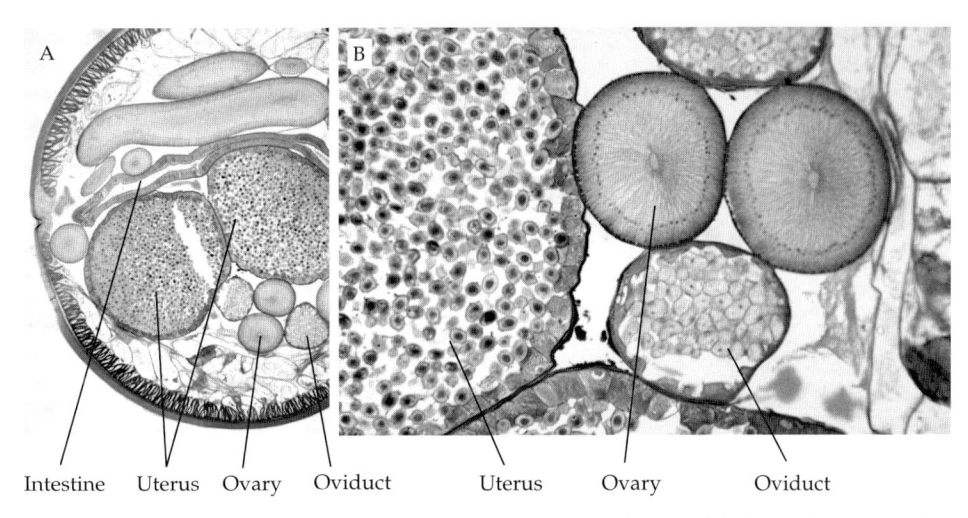

Intestine Uterus Ovary Oviduct   Uterus Ovary Oviduct

Fig. 13.11. The female reproductive system of the large nematode *Ascaris suum* is a very long tube, folded several times within the worm body. Therefore, sections show different parts of the tube: the proximal ovary, the intermediate oviduct, and the distal uterus. In the ovary, oogenesis stages are arranged radially around a central rhachis.

some gastrotrichs (Teuchert 1968) and gnathostomulids (Riedl 1969), oocytes are released through ruptures in the body wall. As oocytes in both cases are very large, they are squeezed through the much smaller pores. It is sometimes assumed that such an absence of female ducts and pores may be the ancestral pattern, but this interpretation is made difficult by the fact that within both taxa oviducts and pores do also exist.

Gamete transfer and receiving

It is quite evident that, within metazoan animals, there is an evolutionary trend from external to internal fertilization (see Chapter 14). The release of gametes to the exterior requires a comparatively simple structure of the reproductive system, usually only the gonad, a duct, and a pore to the external environment. As the probability of gametes to meet externally is comparatively low, high numbers of gametes have to be produced, which is physiologically 'expensive'. In internal fertilization, the number of gametes can be reduced dramatically, but systems have to evolve for the directed transfer of gametes. The fertilization of oocytes within the animal body or in its close vicinity makes it possible to invest in the development of the offspring, in particular in giving it protection and in providing more nutrition than is possible in externally fertilizing animals. This leads to a great diversification of the reproductive system and to the evolution of several structures. Some examples are:

• Eversible copulatory structures (penes, cirri): these are tube-like structures with an apical pore, through which spermatozoa can be transferred (Fig. 13.12.). The eversion is always by fluid pressure. Copulatory structures can have very diverse shapes. They are either used for an injection of spermatozoa into the partner's body (hypodermal injection) or they are inserted into the female genital opening in copulation. Hard structures of diverse shapes are sometimes present, for example in some platyhelminths (Brüggemann 1985) and all nematodes (Wright 1991). There is no restriction of hard structures to hypodermal injection, both hypodermal injection and copulation can be performed using penes with and without hard structures. In hypodermal injection without hard structures, the penis is muscular-glandular. Gland products are used to attach the semi-everted penis to the integument of the partner, which is penetrated by applying further pressure to evert the penis completely.

- In several cases, 'spermatopores' are formed. These are variously shaped cases, in which numerous spermatozoa are packed together in a case and transferred to the partner (Mann 1984). They are either deposited on the partner's surface, onto sediment, or transferred into the partner's genital opening. The production of a spermatophore requires a special elaboration of the male reproductive tract, which contains glands to form the spermatophore wall.
- Further glands can be present in the male reproductive tract. To give one example, male acanthocephalans have cement glands, the secretion of which is transferred to the female genital opening after copulation and hardens to form a plug (Crompton 1985). This plug needs a few days to dissolve and prevents the female from mating with another male, therefore raising the chance that oocytes are fertilized by the male which applied the plug. Interestingly, males also apply their cement to the posterior end of other males, thereby preventing them copulating for a while (Abele & Gilchrist 1977).
- In very many cases, spermatozoa can be stored in the female reproductive system after copulation, leading to a varying time period between sperm transfer and fertilization. Such storage structures are usually called 'seminal receptacles' and may be more or less 'simple' extensions from the female genital duct.

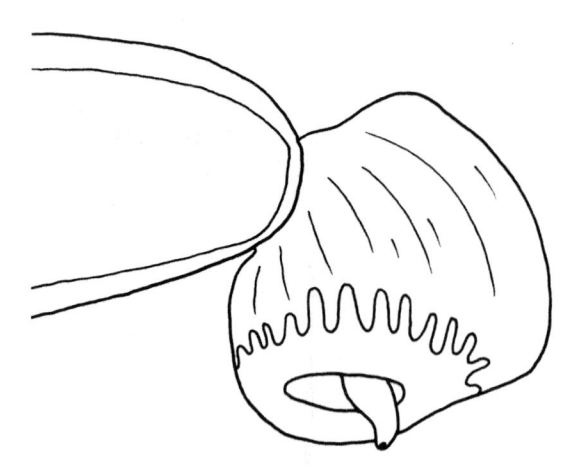

Fig. 13.12. Posterior end of a male acanthocephalan (*Pseudoacanthocephalus bufonis*) with everted 'bursa' and penis. The bursa is wrapped around the female posterior end during copulation. After Crompton (1985).

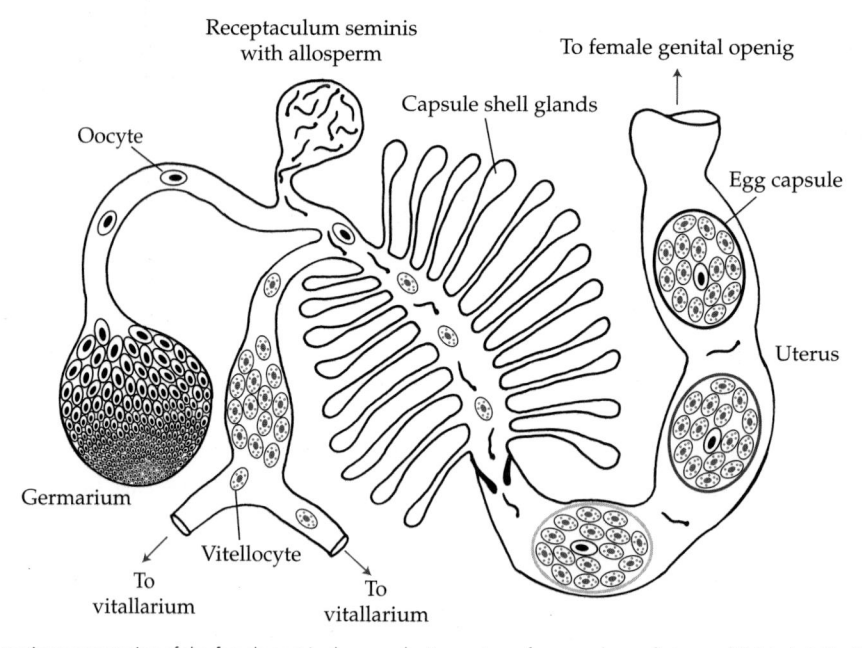

Fig. 13.13. Schematic representation of the female part in the reproductive system of a neoophoran flatworm (Platyhelminthes). Germarium and vitellarium are separated, oocytes are fertilized with allosperm stored in the seminal receptacle. One oocyte and several vitallocytes are surrounded by a capsule and leave the system as compound, 'ectolecithal' eggs. Modified after Ehlers (1985a).

- Oocytes are those gametes which contain almost all cytoplasm compared to spermatozoa. This cytoplasm contains nutrients used for furher development. In externally as well as internally fertilizing animals, oogenesis often includes supportive cells, which transfer nutrients to the developing oocyte. In some internally fertilizing taxa, the female gonad is divided in special germinal and vitellar areas (e.g. platyhelminths and eurotifers). In some cases (Neoophora among Platyhelminthes), the vitallaria even produce nutritive cells (vitellocytes) that are packed together with one oocyte into a capsule (Fig. 13.13.; Ehlers 1985a).
- As a certain protection for the oocyte or the embryo, or to hold together oocytes and vitellocytes, several animals with internal fertilization produce 'capsules' or 'shells'. Such shell material is produced by special glands in the female reproductive system.
- Apart from laying eggs (ovipary), offspring can leave the mother in a more or less developed stage (ovovivipary and vivipary). This retention of embryos is called 'brood protection' and requires special structures, either in the reproductive tract or elsewhere in the body in the form of protective pouches (see, e.g., Clutton-Brock 1991).
- An extreme case of brood protection is the nutrition of the developing embryo by maternal tissue. This is possible when maternal and embryonic tissue are closely associated. Such elaborations within the female reproductive system are called 'placenta'. They are well known from ourselves and are name-giving for the mammalian subtaxon Placentalia, but placentas can also be found in other taxa. These are peripatid Onychophora (Walker & Campiglia 1988, Ruhberg 1990), some sharks and rays (Wourms 1981, Hamlett 1993), and some teleost fishes such as members of the Poeciliidae (Reznick et al. 2002).

Many of these specializations evolved more than once and are not characteristic for larger taxa, but may be important for phylogenetic considerations within taxa. One thing that appears fascinating to me is the fact that the reproductive system and the reproductive biology of presumably basal bilaterian taxa are amazingly diverse and complicated. I want to illustrate this with a brief overview on what is known from gastrotrichs and gnathostomulids.

Gastrotrichs are cross-fertilizing hermaphrodites, but freshwater chaetonotid gastrotrichs reproduce parthenogenetically, as far as is known. The discovery of sperm in such species (Weiss & Levy 1979, Weiss 2001) is puzzling, and it is not yet known if fertilization (by autosperm or even allosperm) is possible or if these structures are only nonfunctional rudiments. Macrodasyid gastrotrichs can have, in addition to ovary and tesis, two accessory structures called the frontal and the caudal organ. When present, the frontal organ appears to be a seminal vesicle and the caudal organ a copulatory organ or a

Table 13.2 Diversity of structures in the reproductive system of Gnathostomulida. After Graebner & Adam (1970), Sterrer (1974), Knauss & Rieger (1979), Mainitz (1979, 1983, 1989), Alvestad-Graebner & Adam (1983), and Lanfranchi & Falleni (1998).

	Filospermoidea	Bursovaginoidea	
		Scleroperalia	Conophoralia
Testes	Paired	Paired	Unpaired
Penis	Glandular	Muscular, often with stylet	Muscular, without stylet
Female genital opening	Absent	Present	Present
Bursa (sperm-receiving organ)	Absent	Present, sclerotized	Present, unsclerotized
Reproduction	Hypodermal injection with glandular penis into various body regions	Hypodermal injection with stylet-penis into bursa tissue	Probably transfer of 1–2 giant spermatozoa into bursa
Spermatozoa	Filiform	Aciliary, dwarf	Aciliary, giant

Table 13.3 Summary of selected data from the reproductive system.

	General structure	Gonad structure	Gamete origin	Fertilization	References
Porifera	No gonads	–	From totipotent cells anywhere in sponge body	External or internal (by spermatozoa in water current)	Fell 1999
Trichoplax adhaerens	No gonads	–	Oocytes derived from ventral epithelium, origin of probable spermatogenesis-stages unknown	Unknown	Grell & Ruthmann 1991
Cnidaria	Gonad-like regions are accumulations of gametes (=gametogenic areas)	No gonads, but gametogenic areas in ECM	From entodermal (Anthozoa, Scyphozoa) or ectodermal (several Hydrozoa) cells	External or internal (by spermatozoa in water current)	Fautin 1999
Ctenophora	Gametes are accumulated in particular regions close to meridional canals	Gametes are incompletely bordered by entodermal and further cells	From entodermal epithelium	External	Hernandez-Nicaise 1991, Matsumoto 1999
Xenoturbella	No gonads	No gonads, gametes basal of entoderm or in parenchyma	Unknown	Unknown, probably internal	Israelsson 1999, Israelsson & Budd 2005
Acoelomorpha	Gonads as ovotestis or separate, ducts lacking, but gonopores present. Bursa or bursal tissue to receive allosperm	Gonads diffuse, follicular or compact, epithelial cells in some species, but most species without	Probably from pluripotent cells (neoblasts) in parenchyma	Internal, transferred by copulation with either inconspicuous structures (Nemertodermatida) or muscular-glandular copulatory organs (Acoela)	Rieger et al. 1991b, Lundin & Sterrer 2001
Gastrotricha	Paired or unpaired gonads, oviduct and female pore only in few species, male duct and pore present. Often two accessory structures present (frontal and caudal organ)	Ovotestis or separate ovary and testis, epithelially lined	Spermatocytes originate from testicular epithelium, origin of oocytes is unknown	(where present:) internal, sperm transfer by spermatophores, hypodermal injection or copulation	Teuchert 1976b, Hummon & Hummon 1983a, b, 1989, Ruppert 1991b
Nematoda	Paired tubular gonads, continuing into gonoducts. Female pore separate, male gonoduct leads into hindgut	Epithelially lined gonads, gametogenesis either in entire gonad or in proximal tip	Preformation, going back to P4 cell in early embryogenesis	Internal, sperm transfer in copulation using copulatory spines	Wright 1991, 1999
Nematomorpha	Paired tubular gonads, continuing to gonoducts, gonopores in *Nectonema*, cloaca in Gordiida	Epithelially-lined tubes in Gordiida¹, probably absent in female *Nectonema* and as imperfectly known sperm sacs in male *Nectonema*	Unknown, oogenesis in *Nectonema* probably from parenchymal cells	Internal, sperm transfer onto female posterior end in pseudocopulation, fertilization in female genital tract	Schmidt-Rhaesa 1997b, 1999, 2005, de Villalobos et al. 2005

Taxon	Gonads	Gametogenesis	Reproduction	References
Priapulida	Paired gonads, gonoducts fuse with excretory system	From gonad epithelium	External in macroscopic species, internal fertilization in meiobenthic species[2]	Nørrevang 1965, Lemburg & Schmidt-Rhaesa 1999
Kinorhyncha	Paired gonads, gonoducts, and single pair of gonopores, seminal receptacle in females	Sac-like, epithelialized gonads in both sexes	Internal, copulation likely	Kristensen & Higgins 1991, Neuhaus 1999
Loricifera	Paired dorsal gonads, gonoducts, and pores	Fine structure not completely clear, probably with epithelial cells	Internal, copulation likely	Kristensen 1991
Platyhelminthes	Gonads as ovotestis or separate, ducts present or lacking, gonopores present. Bursa and copulatory organs often present	Gonads diffuse, follicular or compact, epithelial cells present or absent	Internal (by hypodermal injection or copulation), copulatory organs muscular-glandular or with hard structures	Rieger et al. 1991b, Tyler 1999
Gnathostomulida	Gonads paired or unpaired, female gonopore absent or present, bursa absent or present, males with spermioducts and copulatory organs	Few details known, an epithelium may be present	Internal with copulation, copulatory organ glandular, muscular or with hard structures	Alvestad-Graebner & Adam 1983, Mainitz 1983, Sterrer 1999
Limnognathia maerski	Paired ovary, eggs probably laid through ventral pore	Compact ovaries, cells other than developing oocytes are unknown	Probably absent	Kristensen & Funch 2000
Eurotifera	Bdelloida: paired ventral ovary, oviducts lead to hindgut. Monogononta: unpaired ventral ovary and testis, oviducts lead to hindgut, spermioducts lead to copulatory organ	Paired or unpaired gonads, ovary as syncytial germovitellar, in some species surrounded by syncytial epithelium	Monogononta: internal, copulation	Clément & Wurdak 1991
Seisonidea	Paired gonads, gonoducts lead to hindgut	Paired gonads without epithelium, surrounded by ECM and containing oocytes or spermatozoa	Unknown	Internal, deposition of spermatophores on host gills, uptake by female	Ricci et al. 1993, Ahlrichs 1995
Acanthocephala	Ovaries dissolve early in development, gametes in ligament sacs or primary body cavity, release of embryos through uterine bell, testes with spermioducts, they receive excretory ducts and lead into hindgut	no epithelialized gonads[4]	Oocytes can be traced back to a 'central nuclear mass' in the acanthor larva	internal, by copulation	Crompton 1999
Nemertini	Bilateral, serial gonads with duct and gonopore	Gonads with epithelium and musculature	From epithelium	External or internal, either in pseudocopulation (within mucous cocoon) or, rarely, with copulatory penes	Bierne 1983, Turbeville 1991, von Döhren & Bartolomaeus 2006

Table 13.3 (*Contd*)

	General structure	Gonad structure	Gamete origin	Fertilization	References
Kamptozoa	Paired gonads with short gonoducts and pores in the atrium	Paired sac-like gonads with epithelium	From epithelium	Internal, by collection of spermatozoa from water current	Wasson 1999b
Mollusca	Gonads in gonocoel, either separate gonoducts (Eumollusca) or ducts lead into pericard (Aplacophora)	Epithelialized gonad	From primordial germ cells or gonad epithelium	External or internal	Dohmen 1983, Moor 1983
Sipunculida	Small paired gonads at base of retractor muscle, gametes break free into coelom, where further development takes place. Release through nephridia	Compact gonads surrounded by coelomic epithelium	Unknown, possibly from peritoneal cells	External	Rice 1993, 1999
Echiurida	Unpaired gonads in association with coelom epithelium, close to ventral hemolymph vessel. Gametes break free into coelom, where further development takes place. Release through nephridia	Fine structure unknown, probably as in Sipunculida	Unknown, possibly from peritoneal cells	External ('internal' in *Bonellia*, where male lives in female nephridia)	Davis 1989, Pilger 1993
Annelida	Serially arranged 'gonads' probably as thickened areas in coelom epithelium, often with hemolymph vessels. Gamete maturation in coelom. Release through coelom and nephridia	Coelom epithelium[5]	From coelom epithelium	Abundantly external in polychaetes, internal with copulation in meiofaunal polychaetes, sperm transfer without copulatory organs in 'oligochaetes', hypodermal injection or copulation, each with spermatophore transfer, in hirudineans	Eckelbarger 1988, 1992, 2005, Rice 1992, McHugh 1999
Tardigrada	Unpaired dorsal gonads, oviduct leads to gonopore (Heterotardigrada) or into hindgut (Eutardigrada)	Gonads with discontinuous epithelium	Not exactly known, oocytes may arise from syncytial mass central in the ovary	Probably external in some heterotardigrades, internal in eutardigrades	Dewel et al. 1993
Onychophora	Paired gonads (ovaries often fuse during development), gonoducts, often seminal vesicles in females, copulatory organs in few males	Gonads with germinal epithelium; in females of Peripatopsidae, maturing oocytes bulge into the surrounding body cavity (exogeneous development), in Peripatidae, they develop within the ovary (endogeneous dev.)	From gonad epithelium	Internal with spermatophores, either attached to female skin (Peripatopsidae) or transferred to female gonopore (Peripatopsidae), but only few species have a penis	Herzberg et al. 1980, Storch & Ruhberg 1993

Euarthropoda	Usually paired gonads, gonoducts and pores	Gonads with non-germinative epithelium, surrounding developing gametes	Gametes originate by preformation in several, but not all insects. In chelicerates, crustaceans and 'myriapods' from gonad epithelium	External only in Xiphosura and Pycnogonida, all others internal by copulation	Bünting 1992, Huebner & Diehl-Jones 1992, Krol et al. 1992
Chaetognatha	Ovaries in trunk coelom, with seminal receptacle and female pore; testes in tail coelom, spermatozoa released to coelom and packed into spermatophore in seminal vesicle	Testes: spermatogonia surrounded by coelomic epithelium; ovary: myoepithelial ovarian wall surrounds oocytes	Probably preformed from a particular cell expressing vasa-like protein in the 32-cell stage	Internal, sperm or spermatophore deposited on partner, no copulatory structures	Shinn 1997, 1999, Carré et al. 2002
Bryozoa	No gonads, gametes mature in epthelium of metacoel	Gonads are parts of the coelom epithelium of metacoel	From coelom epithelium	External or internal, when spermatozoa are "caught" from the water by tentacles and transferred to pore of metacoel	Woollacott 1999
Phoronida	No gonads, thickened area in peritoneum close to BVS are gametogenic. Gametes released to metacoel and to the external through coelomopores	Gonads are thickened regions of the metacoel epithelium	From coelom epithelium	External in most species, in some internal fertilization by transfer of spermatophores with tentacles to partner	Herrmann 1997
Brachiopoda	No gonads, gametes mature in epthelium of metacoel, which often forms a tubular system in the mantle. Release of gametes into metacoel and to the external by coelomopores	Coelom epithelium of metacoel	From coelom epithelium	External	James 1997
Echinodermata	Epithelialized gonads, surrounded by a hemal sinus, genital coelom (absent in holothurians) and part of the metacoel, genital ducts and pores	Cluster of primordial germ cells surrounded by epithelium probably derived from coelom. Often surrounded by hemal sinus, genital coelom and metacoel	From germinal epithelium	External or internal by collection of spermatozoa from water current	Chia & Walker 1991, Holland 1991, Pearse & Cameron 1991, Smiley et al. 1991, Byrne 1999
Enteropneusta	Serially arranged sac-like gonads, each with own gonoduct and gonopore. Gonads bulge into metacoel, but remain separated from it	Sac-like gonads with epithelium	From gonad epithelium, which appears to be formed by primordial germ cells	External	Benito & Pardos 1997
Pterobranchia	Single or paired gonad close to metacoel	Sac-like gonads with epithelium	From gonad epithelium	External	Benito & Pardos 1997

Table 13.3 (*Contd*)

	General structure	Gonad structure	Gamete origin	Fertilization	References
Tunicata	Unpaired or more than one hermaphroditic gonads with separate gonoducts for sperm and eggs	Epithelialized gonads of various shape and size	Exact origin unknown, probably determination of PGCs by a "centrosome-attracting body" Oocytes are reported to be derived from hemolymph cells that float to the gonads early in development.	External	Burighel & Cloney 1997, Extavour & Akam 2003
Acrania	Serially arranged gonads, surrounded by perigonadal coelom, close proximity to atrium	Sac-like gonad with gametogenic epithelium	From gonad epithelium	External	Ruppert 1997
Craniota	Paired gonads with gonoducts leading to the excretory system	Gonad separated in cortex (derived coelom epithelium and primordial germ cells) and medulla	Primordial germ cells originate as mesodermal cells early in development and migrate to the gonad anlagen	External (in agnathans, many teleost fishes and several amphibians), or internal, by copulation	Frye 1977, Liem et al. 2000, Kardong 2005

[1] It is likely that the gonads in both sexes originate as solid longitudinal strands. In males, the cells separate into epithelial and central cells, in females, the developing oocytes appear to burst the gonad epithelium and proliferate into the body cavity (Lanzavecchia et al. 1995, Schmidt-Rhaesa 2005).

[2] There is indirect evidence for internal fertilization: sperm was found in the female reproductive system in *Tubiluchus* (Alberti & Storch 1988) and an embryo was found in the state of birth in *Meiopriapulus* (Higgins & Storch 1991).

[3] Alvestad-Graebner & Adam (1983) describe a 'follicular wall' in the testes and a developmental gradient in the spermatozoa within the testis. As this gradient is from peripheral (young) to central (old), an origin from the epithelium may be suspected, but remains to be shown. In the ovary, the germinal zone is described to be in the anterior part (Mainitz 1983, Sterrer 1999).

[4] Gonadal primordia arise from a central nuclear mass in the acanthor larva. Further development is in association with the development of ligament sacs. The ovary divides further and further, forming numerous ovarian fragments that float as ovarian balls in the ligament sacs or in the body cavity. In the testes, primordial germ cells and supportive cells develop (Crompton 1985, 1999).

[5] Despite the presence of several excellent compilations of annelid, especially polychaete, anatomy, the gonads are often only touched on and good descriptions are rare. Some authors speak of a germinal epithelium, others assume an origin from plutipotent cells early in development. According to T. Bartolomaeus (Berlin, pers. commun.), gametes originate in the epithelium of the coelom.

spermatophore-forming organ. There are very few observations on sperm transfer, but these already show diversity. *Dactylopodola* species transfer spermatophores, either onto the surface of the partner (*D. baltica*, Teuchert 1968) or into the female part of the reproductive system (*D. typhle*; A. Kienecke, Oldenburg, pers. comm.). In *Turbanella cornuta*, spermatozoa appear to be transferred directly from the male pore to the female system without spermatophores or elaborated copulatory structures (Teuchert 1968). *Macrodasys* sp. uses its caudal organ for sperm transfer. During copulation, spermatozoa are first transferred from the male genital pore to the more posterior caudal organ and then from there to a partner (Ruppert 1978b). Ducts and pores in the female system may be present or absent, in the latter case the eggs are released by rupture of the body wall.

Gnathostomulida include about 100 known species and three subtaxa are recognized: Filospermoidea, Scleroperalia and Conophoralia. The latter two are united as Bursovaginoidea. These names already refer to several characters of the reproductive system. Gnathostomulids are, like gastrotrichs, cross-fertilizing hermaphrodites. The three groups vary in all aspects of the reproductive system (see Table 13.2.), reflecting an enormous dynamic in the evolution of this system.

Conclusions

Reproductive patterns are very diverse. Both sexual and asexual reproduction are ancient modes of reproduction and were present in the metazoan ancestor. Within bilaterians, a special mode of asexual reproduction, nevertheless derived from sexual reproduction, evolved several times in parallel: parthenogenesis. Asexual reproduction without gametes is correlated with the potential of organisms to regenerate, and there appears to be a trend to reduce this potential. Hermaphroditism and gonochorism change many times during evolution, making it very hard to reconstruct ancestral states for larger taxa. Gametes can originate very early (preformation) or comparatively late (epigenesis) during development. It is quite evident that epigenesis is the ancestral mechanism and epigenesis is derived. The first recognizable germ cells, the primary germ cells (PGCs), appear to be comparable at the cellular level. This means that the genetic programme that specifies PGCs is homologous, but can be activated in different types of cells and at different times during development. Gamete development can be independant from gonads, but many bilaterian animals have such gonads, in which gametes either only mature or in which they originate from the gonad epithelium. Gonads have different origins and it is possible that gonads evolved several times in parallel. The transition from external to internal fertilization caused elaborations of the reproductive system in many ways. A diversity of copulatory structures evolved, as well as mechanisms to pack spermatozoa into a spermatophore or oocytes and vitellocytes into a compound egg. Sperm is often stored in a seminal receptacle. The investment in protection and nutrition for the embryo is quite high in many cases, and involves structures for embryonic nutrition (e.g. a placenta) or for brood protection.

CHAPTER 14

Gametes (Spermatozoa)

As gametes are unicellular structures, they cannot be regarded as organs; but they are included here for two reasons. First, they are (often) the products of an organ system, the gonads, and second gametes, especially spermatozoa, have been used abundantly in systematics (e.g. Franzén 1956, 1977a, Baccetti 1970, 1991, Wirth 1984, Jamieson et al. 1995). For this last reason, I will concentrate here exclusively on spermatozoa.

Spermatozoa generally have very little cytoplasm and a minimum amount of organelles. There are extreme cases (see gastrotrichs), where a spermatozoon only contains the nucleus, but more often we find additional mitochondria, an acrosome (for the entry reaction with the oocyte), and a cilium (sometimes more than one). There are two 'problematic' terms. In most spermatological publications, the term 'flagellum' is used instead of 'cilium'. There is no morphological difference between these structures, but cilium is usually used when a large number is present, while flagellum usually refers to single or small numbers. However, such use is not consistent as, for example, single sensory cilia are not called 'sensory flagellum'. For reasons of consistency, I will use the name 'cilium' (biciliary, aciliary ...) here. The second problematic term concerns the centrioles. Centrioles (or, more generally, microtubule-organizing centres, MTOC) are central organelles for cilia formation and cell division. They always have nine peripheral triplets of microtubuli and represent the basal structure of a cilium. Often, but not always, a second centriole with an orientation perpendicular to the ciliary base is present, the accessory centriole. In somatic cells, the terms 'basal body' and 'accessory centriole' are commonly used, while in spermatozoa the terms 'proximal centriole' (for the accessory centriole) and 'distal centriole' (for the basal body) are often used. I will use in the following the same terms for somatic and germ cells: basal body and accessory centriole.

Spermatozoa are extremely variable in structure, but can be roughly divided into some types. There appears to be a strong correlation between sperm type and mode of fertilization (Franzén 1956, 1970, 1977a, Rouse & Jamieson 1987), therefore it is important to note how spermatozoa are transferred to, and where they fertilize, the oocytes. There are several terms applied to spermatozoa (see discussion at the end of this chapter), but for the following review of spermatozoa I regard it most helpful to use only descriptive terms:

- **round-headed spermatozoon**: spermatozoon with a more or less spherical nucleus, an anterior acrosome, mitochondria behind the nucleus and a terminal cilium.
- **filiform spermatozoon**: very elongated, thin spermatozoa.
- **aciliary spermatozoon**: spermatozoa lacking a cilium.

Particular attention is placed in the following on the general shape of the spermatozoon, absence or presence of an acrosome and whether it is composed of one vesicle or not, the form of nucleus and mitochondria, and the position and structure of the cilium. The distribution of microtubules in the cross section of a spermatozoon is given in the standard formula (number of peripheral + number of central tubules) where 9×2 means nine doublets and 9×3 means nine triplets of tubules (example: $9 \times 2 + 2$).

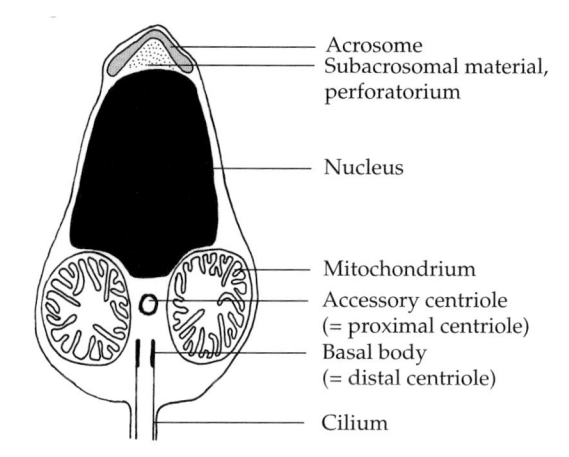

Figure 14.1 General explanation for all schematic figures in Chapter 14. The nucleus is always black, the acrosome grey. The basal body of the cilium (= distal centriole) is represented by bold lines, the accessory centriole (= proximal centriole), when present, as a circle. The free part of the cilium is not shown, only its proximal insertion on the cell. Additional, special features of certain sperm cells are shown in only a few cases.

Porifera

Spermatozoa are released in all sponges to the sea water, but oocytes are often retained in the sponge body (Fell 1989). In those cases where both gametes are released, fertilization is external, in the second case fertilization is internal. Spermatozoa follow the water-flow into the sponge and then make contact with choanocytes, which can transform into carrier cells to transport the spermatocytes to the oocytes (Fell 1989). Spermatozoa are derived from choanocytes (Tuzet et al. 1970, Gaino et al. 1986, Paulus & Weissenfels 1986). They are more or less round-headed (Boury-Esnault & Jamieson 1999; Fig. 14.2.), but can also vary somewhat in shape, for example with a laterally inserting cilium in the spermatozoon of *Oscarella lobularis* (Homoscleromorpha, see Bacetti et al. 1986, Gaino et al. 1986; Fig. 14.2.). An acrosome is lacking in most species, but it can be present as

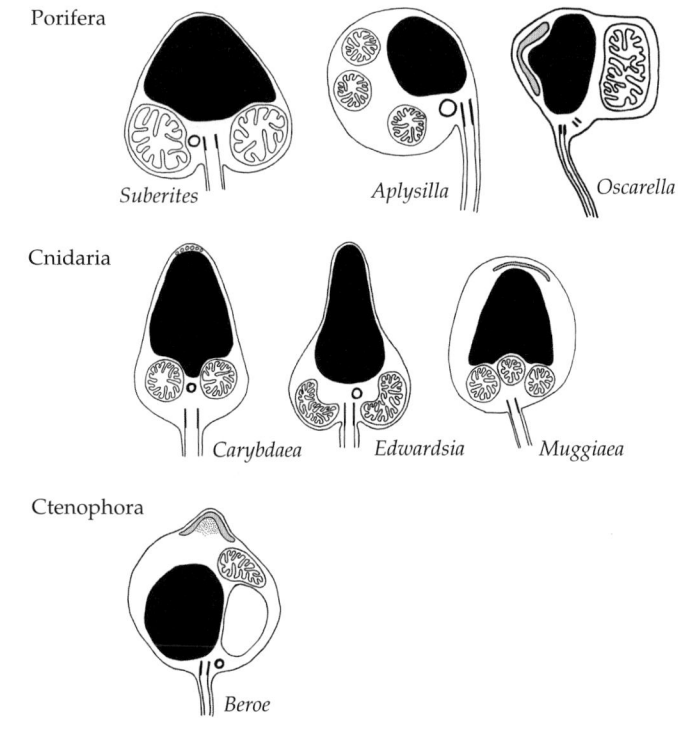

Fig. 14.2. Schematic representation of spermatozoa from **Porifera**: *Suberites massa* (Demospongia), after micrograph in Simpson (1984); *Aplysilla rosea* (Demospongia), after micrograph in Tuzet et al. (1970); *Oscarella lobularis* (Homoscleromorpha), after Baccetti et al. (1986). **Cnidaria**: *Carybdea marsupialis* (Cubozoa), after Corbelli et al. (2003); *Edwardsia* sp. (Anthozoa, Hexacorallia), after Schmidt and Zissler (1979); *Muggiaea kochi* (Hydrozoa, Siphonophora), after Carré (1984). **Ctenophora**: *Beroe ovata*, after Franc (1973).

numerous small vesicles or as one large vesicle. This has been stated for few species after light microscopical investigation (Boury-Esnault & Jamieson 1999), but could be confirmed ultrastructurally only in *Oscarella lobularis* (Homoscleromorpha; Bacetti et al. 1986, Gaino et al. 1986; Fig. 14.2.).

Trichoplax
Aciliary cells can be present freely between the dorsal and ventral epithelium during the sexual phase (i.e. in the phase where no asexual reproduction takes place and where an oocyte is found). Because of their occurrence in the sexual phase and because of their possession of a larger vesicle that may represent an acrosome, they are thought to represent spermatozoa (Grell & Benwitz 1974, 1981, Grell & Ruthmann 1991). Fertilization, however, has never been observed in *Trichoplax*.

Cnidaria
Most cnidarians fertilize oocytes externally, but within all subtaxa, some species have changed towards internal fertilization (Fautin et al. 1989, Fautin 1999, Harrison & Jamieson 1999). Spermatozoa are then ingested from the surrounding water, it is unknown whether this is a chance or targeted process. Oocytes are fertilized within the 'gonad' or in the gastrovascular cavity. All gametes found so far in cnidarians are in principle round-headed spermatozoa (Fig. 14.2.). The restriction 'in principle' refers to the fact that the head of spermatozoa is in many cases cone- or pear-shaped and in very few cases slightly elongated (Schmidt & Zissler 1979, Harrison & Jamieson 1999). In anthozoans, it was found that brooding species (with internal fertilization) possess spermatozoa with spherical nuclei, whereas externally fertilizing species possess spermatozoa with pear-shaped nuclei (Harrison & Jamieson 1999). Spermatozoa possess a variable number of mitochondria and a cilium with a 9 × 2 +2 pattern of microtubules (Harrison & Jamieson 1999). The possession of an acrosome is controversial. Several species do not possess an acrosome, but others possess either a so-called anterior process (which is probably not homologous to an acrosome) or small vesicles (e.g. scleractinian species: Steiner 1993; the hydrozoan *Halammohydra schulzei*: Ehlers 1993; the scyphozoan *Nausithoë* sp.: Afzelius & Franzén 1971; the cubozoan *Carybdea marsupialis*: Corbelli et al. 2003; Fig. 14.2.). In one case, in the siphonophore *Muggiaea kochi*, one single acrosomal vesicle was found (Carré 1984; Fig. 14.2.), but Ehlers (1993) regards this as a derived condition (see below for discussion of acroscomal evolution). An acrosomal function of these vesicles is likely because O'Rand & Miller (1974) observed that they are lost during fertilization in the hydrozoan *Campanularia flexuosa*.

Ctenophora
Most ctenophores release their spermatozoa into the water, where fertilization occurs (Martindale & Henry 1997, Matsumoto 1999). Only a few species (platyctenids) have internal fertilization with subsequent brood protection. There are few data on spermatozoa, but the investigation of Franc (1973) shows that a round-headed spermatozoon is present (Fig. 14.2.). There is one acrosomal complex, followed by a perforatorium. Several mitochondria fuse during spermiogenesis to form one large mitochondrium in the mature spermatozoon. This, and a so-called paranuclear body, is positioned laterally of the nucleus. The cilium has a 9 × 2 + 2 pattern of microtubules.

'Mesozoa'
Spermatozoa have been detected in both Dicyemida and Orthonectida. In dicyemids, the few data indicate that they are small and aciliary (Ridley 1969). In the orthonectid *Rhopalura littoralis*, Slyusarev & Ferraguti (2002) found a round-headed spermatozoon lacking an acroscome and with an oval nucleus, a single mitochondrium displaced at the side of the nucleus, and two serially arranged centrioles (Fig. 14.3.). A striated rootlet is attached to the anterior centriole while the posterior one gives rise to a cilium (9 × 2 + 2 pattern).

Xenoturbella
There has been no ultrastructural investigation of the spermatozoa, but according to

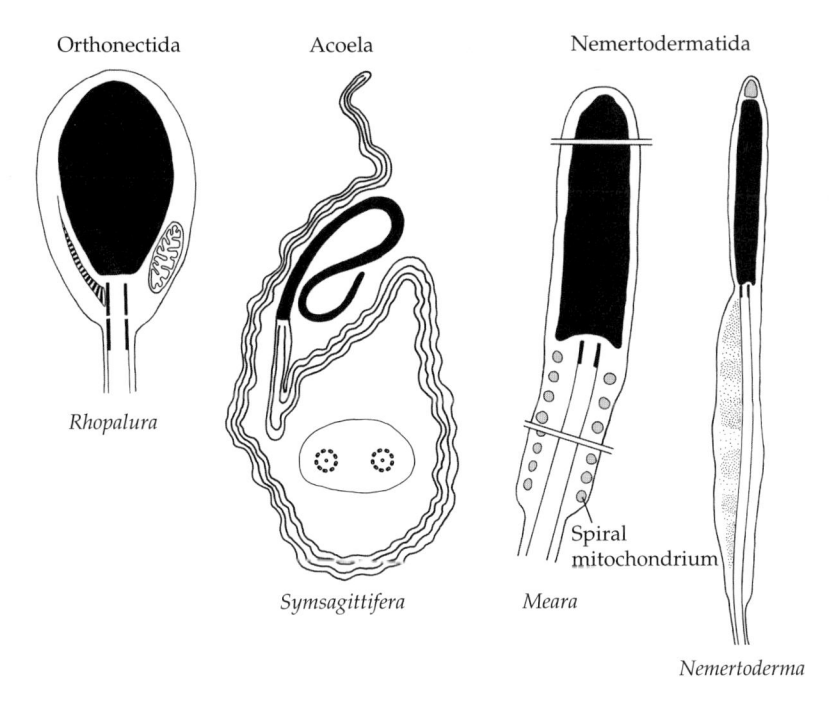

Fig. 14.3. Schematic representation of spermatozoa from **Orthonectida**: *Rhopalura littoralis*, after Slyusarev and Ferraguti (2002). **Acoela**: *Symsagittifera schultzei*, after images in Raikova et al. 1998. **Nemertodermatida**: *Meara stichopi*, after micrograph in Hendelberg 1983; *Nemertoderma* sp., after micrograph in Tyler and Rieger (1975).

light-microscopical investigations by Westblad (1950) and Reisinger (1960) they are round-headed spermatozoa. The transfer of spermatozoa is unknown.

Acoelomorpha

All acoelomorphs transfer sperm for an internal fertilization. Acoels may copulate or perform hypodermal injection, the mode of sperm transfer has been corellated to the population structure by Apelt (1969). Acoels have filiform spermatozoa with two cilia that are incorporated into the sperm cytoplasm (Fig. 14.3.). These cilia can have $9 \times 2 + 2$, $9 \times 2 + 1$ or $9 \times 2 + 0$ microtubules, the occurrence of these patterns corresponds to phylogeny of acoels (Petrov et al. 2004, Tekle et al. 2007). In contrast to acoels, the nemertodermatid spermatozoa have only one cilium (Tyler & Rieger 1975, Lundin & Hendelberg 1998, Lundin & Sterrer 2001). In *Nemertoderma* sp., spermatozoa are filiform, have an acrosome, an elongate nucleus, an unknown structure in the mid-region (probably a mitochondrial derivate), and a cilium ($9 \times 2 + 2$ pattern) originating without a basal body (i.e. without a $9 \times 3 + 0$ pattern) (Tyler & Rieger 1975; Fig. 14.3.). In *Meara stichopi*, the sperm is elongated (Hendelberg 1983, Lundin & Hendelberg 1998, Sterrer 1998; Fig. 14.3.). During spermiogenesis, the cylindrical nucleus elongates, and the two mitochondria fuse and coil around the proximal part of the single cilium. The proximal part of the cilium is incorporated into the sperm cell while the distal part is free. Vesicles at the anterior end may represent acrosomal vesicles (Lundin & Hendelberg 1998).

Gastrotricha

The structure of spermatozoa is very diverse among gastrotrichs. Macrodasyids always have an internal fertilization, but spermatozoa appear to be transferred to the (hermaphroditic) partner in different ways. This has been observed for only a small number of species. In *Dactylopodola baltica*, spermatozoa are packed into spermatophores which are deposited on the individual partner (Teuchert 1968). In *Mesodasys* sp., spermatozoa are injected hypodermally into the partner's body (Ruppert 1991b). In *Turbanella* and *Paraturbanella*, a direct transfer from the male genital pore to the female pore is likely (Teuchert 1968, Balsamo et al. 2002). In *Macrodasys* sp., Ruppert (1978b) showed that spermatozoa are transferred first into the caudal organ and then into the partner. All spermatozoa are filiform (Fig. 14.4.). The acrosome is spirally coiled, the nucleus is elongate and winds spirally around one long or few smaller mitochondria, and there is a terminal cilium (Fischer 1994, Ferraguti & Balsamo 1995, Fregni et al. 1999, Guidi et al. 2003a, 2004, Marotta et al. 2005a; Fig. 14.4.). In chaetonotids, marine species are assumed to cross-fertilize, while freshwater species reproduce by parthenogenesis (Balsamo 1992). The sperm structure is more diverse compared to Macrodasyida and may reflect the transition from cross-fertilization to parthenogenesis within this group. In *Neodasys*, the spermatozoon is filiform, with an anterior conical acrosome, an elongate nucleus and small mitochondria distributed (but not wound) around the nucleus, and a posteriorly inserting, but perpendicular oriented, cilium (Guidi et al. 2003b; Fig. 14.4.). Spermatozoa in *Musellifer delamarei* and *Xenotrichula intermedia* are filiform with an anteroposterior succession of an elongate acrosome (only in *Musellifer*), elongate nucleus, one (*Xenotrichula*) or four (*Musellifer*) small mitochondria, and a cilium (Ferraguti et al. 1995, Guidi et al. 2003b; Fig. 14.4.). Other species of Xenotrichulidae have an acrosome and two mysterious appendages, the so-called paraacrosomal bodies (Ferraguti et al. 1995; Fig. 14.4). The hermaphroditic freshwater species can also have spermatozoa (Weiss & Levy 1979, Weiss 2001), but they are probably not used in reproduction and seem to be reduced to a nuclear rod only (Ferraguti & Balsamo 1995, Balsamo et al. 1999).

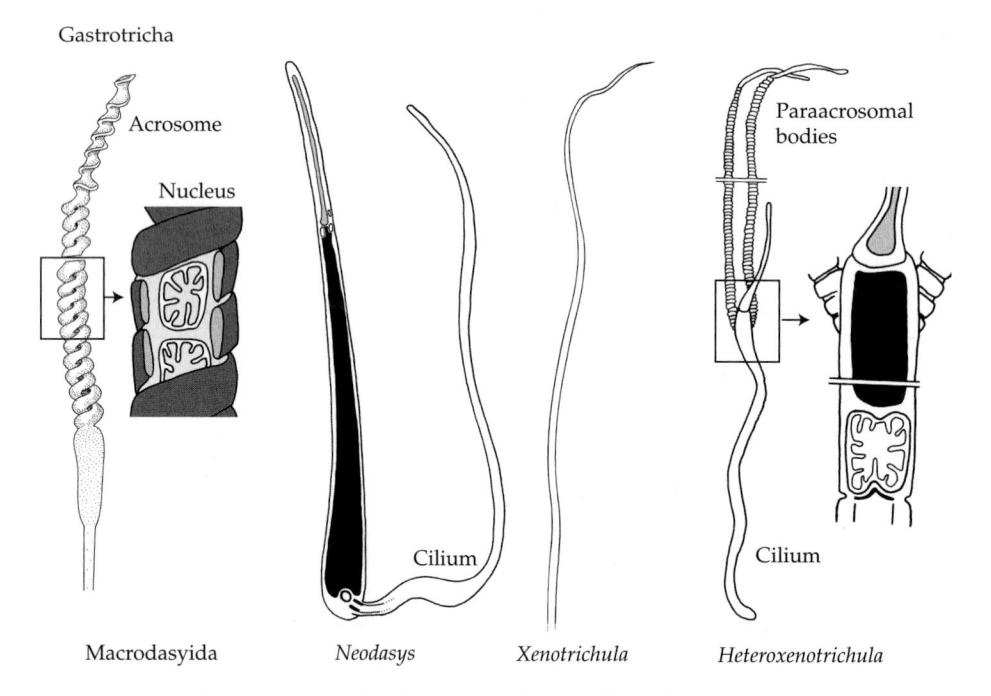

Fig. 14.4. Schematic representation of spermatozoa from **Gastrotricha**: Macrodasyida combined after figures from *Pseudostomella etrusca* and *Mesodasys laticaudatus*, after Ferraguti and Balsamo (1995); *Neodasys ciritus* (Chaetonotida, Multitubulatina), after Guidi et al. (2003); *Xenotrichula intermedia* (Chaetonotida, Paucitubulatina), after Ferraguti et al. (1995); *Heteroxenotrichula squamosa* (Chaetonotida, Paucitubulatina), after Ferraguti and Balsamo (1995).

Nematoda

Nematodes transfer spermatozoa from males to females in copulation using male copulatory stylets, and fertilization is internal. All spermatozoa are aciliary (Fig. 14.5.) and it has been shown in several cases that they are capable of amoeboid movement within the female uterus. Recent reviews of spermatozoa by Justine and Jamieson (2000) and Justine (2002) show that spermatozoa of most species contain so-called membraneous organelles (these are lacking, for example, in Chromadorida, see Yushin & Coomans 2000, 2005). In several cases, the membraneous organelles form a complex together with a fibrous body that contains a protein typical for nematode spermatozoa, the major-sperm protein (MSP). This MSP appears to be responsible for the amoeboid movement (King et al. 1992). In many nematodes, the nucleus lacks a membrane, but this is present in the enoplids (Yushin & Malakhov 1994, 1998, Yushin et al. 2002); its absence may therefore be a derived feature within nematodes. Although aciliary, several spermatozoa contain centrioles which in most cases have a $9 \times 1 + 0$ pattern. Exceptions have been found in *Gastromermis* sp. (Mermithida, $9 \times 2 + 0$, Poinar & Hess-Poinar 1993) and *Heligmosoides polygyrus* (Strongylida, $10 \times 1 + 0$, Mansir & Justine 1995).

Nematomorpha

While marine nematomophs (*Nectonema*) copulate, males of the freshwater Gordiida deposit sperm on the posterior end of the female (Schmidt-Rhaesa 1999). At least part of this sperm reaches the female seminal receptacle and

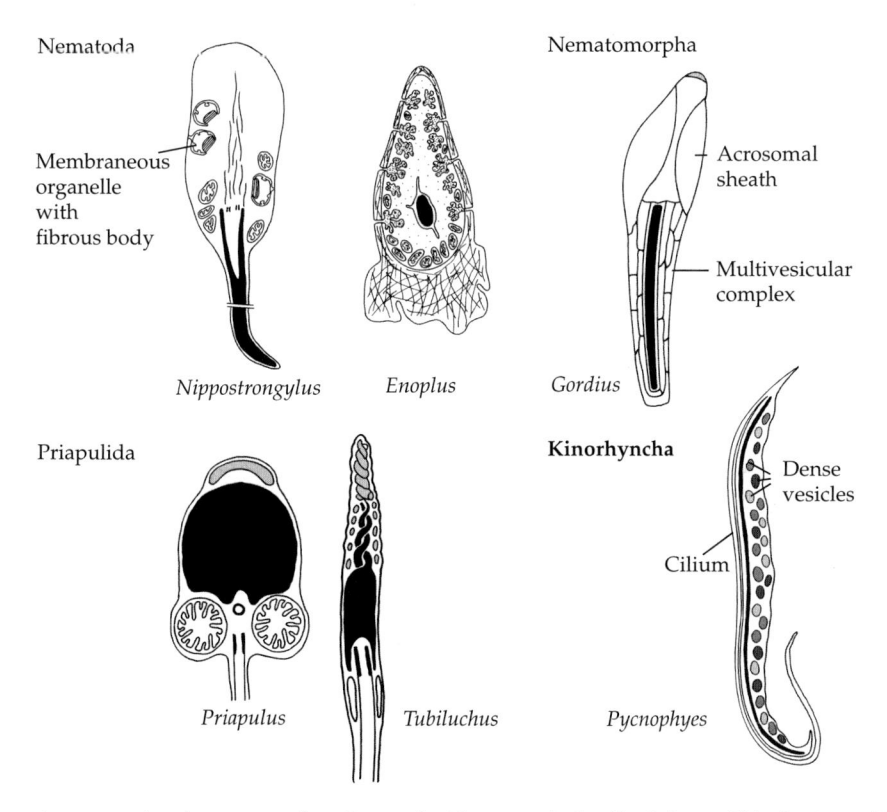

Fig. 14.5. Schematic representation of spermatozoa from **Nematoda**: *Nippostrongylus brasiliensis* (Strongylida), after Jamuar (1966); *Enoplus demani* (Enoplida), after Yushin & Malakhov (1994). **Nematomorpha**: *Gordius aquaticus* (Gordiida), after Schmidt-Rhaesa (1997). **Priapulida**: *Priapulus caudatus*, after Afzelius & Ferraguti (1978); *Tubiluchus corallicola*, after Storch and Higgins (1989). **Kinorhyncha**: *Pycnophyes kielensis* (Homalorhagida), after Adrianov and Malakhov (1999).

fertilization is internal (Schmidt-Rhaesa 1997c). Spermatozoa of Gordiida are aciliary, with a rod-like nucleus, an acrosome, a perforatorium, and further unique structures named the multivesicular complex and acrosomal sheath (Lora Lamia Donin & Cotelli 1977, Schmidt-Rhaesa 1997c, Valvassori et al. 1999, de Villalobos et al. 2005; Fig. 14.5.). They modify into thin rods when transferred to the female (Schmidt-Rhaesa 1997c). In *Nectonema*, only undifferentiated cells have been found that likely represent early stages of spermiogenesis (Schmidt-Rhaesa 1999).

Priapulida
Fertilization and sperm structure in priapulids are diverse. The macroscopic species (*Priapulus*, *Priapulopsis*, and *Halicryptus* have been investigated; Afzelius & Ferraguti 1978a, Storch et al. 2000a) correspond in having an external fertilization and round-headed spermatozoa with an acrosome, a spherical nucleus, four (rarely five) mitochondria, a cilium ($9 \times 2 + 2$ pattern), and an accessory centriole (Fig. 14.5.). A round-headed spermatozoon is also present in *Meiopriapulus fijiensis* (Storch et al. 1989), but judging from the paucity of female gametes and the finding of a larva in the state of birth (Higgins & Storch 1991), internal fertilization is likely to be present. In *Tubiluchus* species, internal fertilization has been shown by the presence of spermatozoa in the female reproductive system (Alberti & Storch 1988). During spermiogenesis, the spermatids of *Tubiluchus* resemble round-headed spermatozoa, but subsequently the cell elongates, the acrosome becomes bipartite, and two processes grow out from the anterior region of the nucleus (Storch & Higgins 1989, Ferraguti & Garbelli 2006; Fig. 14.5.). In the mature spermatozoon these two processes wind spirally around each other and the two parts of the acrosome are spirally coiled around the processes (Fig. 14.5.). The three mitochondria elongate and surround the anterior part of the cilium. Accessory centrioles are absent in both *Meiopriapulus* and *Tubiluchus*, but there are several accessory microtubules in *Tubiluchus* (Ferraguti & Garbelli 2006).

Loricifera
Few data are known from loriciferans. The presence of seminal receptacles in the females at least of *Nanaloricus* suggests an internal fertilization (Kristensen 1991). Fine structural data on spermatozoa require additional investigations but - indicate that the structure can be variable. In *Nanaloricus mysticus*, the general shape is spherical, but the nucleus is rod-like and a large vesicle is present in addition to the acrosome (Kristensen 1991). In *Pliciloricus enigmaticus*, the nucleus is (at least anteriorly) spirally coiled (Kristensen & Brooke 2002, R.M. Kristensen, Copenhagen, pers. commun.).

Kinorhyncha
Internal fertilization is likely in kinorhynchs from the presence of spermatozoa in the female seminal receptacle (Nyholm 1977), probably transferred by a spermatophore (Brown 1983 for *Kinorhynchus phyllotropis*). In contrast, Neuhaus and Higgins (2002) observed copulation, including the transfer of a spermatozoa-containing mass which should not be called a spermatophore. The exact procedure of copulation, especially the role of the 'penile spines' is not completely clear. The spermatozoa of all kinorhynchs are elongated (Adrianov & Malakhov 1999). Ultrastructural data exist only for homalorhagid species. Here the nucleus is long and thin. It is surrounded by three types of vesicles (named platelets, Nyholm & Nyholm 1982, Adrianov & Malakhov 1999). An acrosome is lacking, but a perforatorium may be present. The cilium attaches anteriorly, has a $9 \times 2 + 2$ pattern and runs parallel to the elongated sperm cell, being only slightly longer than it (Fig. 14.5.). The spermatozoon develops from a spherical spermatid with a long mobile cilium (Nyholm & Nyholm 1982).

Platyhelminthes
All platyhelminths have internal fertilization and transfer their spermatozoa either through hypodermal injection or in copulation. Sterrer and Rieger (1974) speculate that in marine catenulids (and the acoel *Childia groenlandica*) fertilization may take place after cannibalistically ingesting a 'partner'. Spermatozoa of

platyhelminths can be either aciliary (e.g. Catenulida, Macrostomida, Prolecithophora, some Rhabdocoela), monociliary (some species among Rhabdocoela; Fig. 14.6.) or biciliary (Polycladida, Lecithoepitheliata, Seriata, most Rhabdocoela; Fig. 14.6.) (Hendelberg 1983, Ehlers 1985a, Watson & Rohde 1995, Watson 1999, Culioli et al. 2006a,b). When present, cilia show in almost all cases a 9 × 2 + '1' pattern, only in very few cases is a 9 × 2 + 0 pattern present (Ehlers 1985a). In the 9 × 2 + '1' pattern, '1' means that the central microtubulus differs from the peripheral doublets in its immunocytological staining (both react with antibodies against α- and β-tubulin, the peripheral doublets additionally react against anti-acetylated α-tubulin; Iomini et al. 1995). Cilia can be free or incorporated into the sperm cell. The sperm structure in catenulids, as the first branch of Platyhelminthes, is interesting, but comparatively few data are available. The mature spermatozoon is aciliary and distinct from other flatworm spermatozoa in, for example, possessing lamellar bodies (Rieger 1978, Schuchert & Rieger 1990b). In spermatids of *Retronectes atypica*, Rieger (1978) found short ciliary structures in association with these lamellar bodies, consisting of a transition from a 9 × 3 + 0 to a 9 × 2 + 0 pattern. Schuchert and Rieger (1990b) speculated that these structures were more likely to be intercellular bridges. Ehlers (1985a) found a single ciliary structure in the periphery of a spermatid of *Retronectes* cf. *sterreri*, which he interpreted as a transitory monociliary stage (see also Schuchert & Rieger 1990b). Ehlers (1985a) interprets this as an indication that the platyhelminth ancestor had monociliary spermatozoa (the assumed plesiomorphic condition for Metazoa). The basal taxa Catenulida (in their adult stage) and Macrostomida shifted to aciliary spermatozoa,

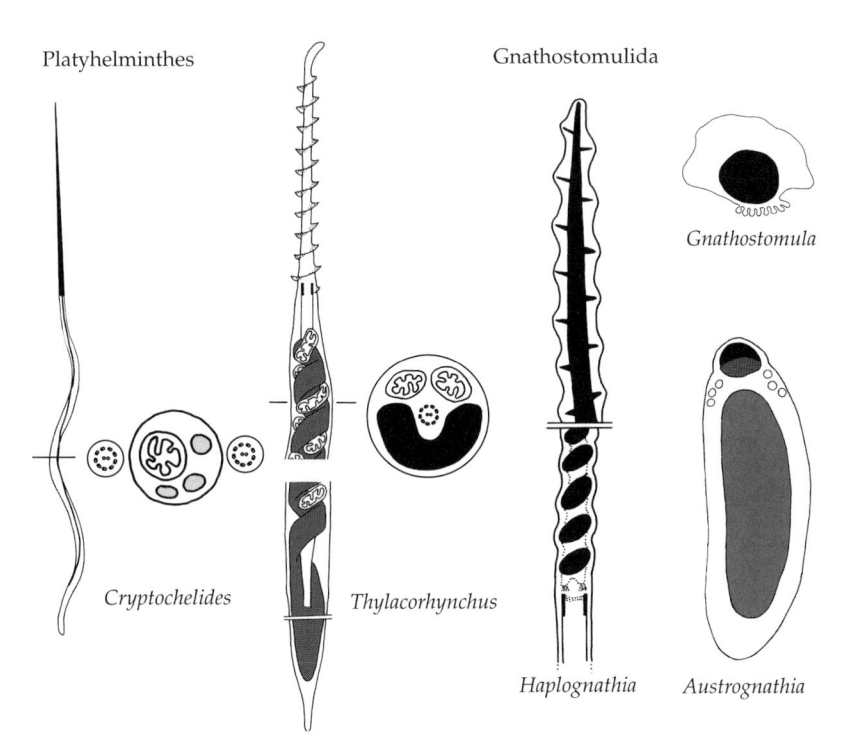

Fig. 14.6. Schematic representation of spermatozoa from **Platyhelminthes**: *Cryptochelides loweni* (Polycladida), after Hendelberg (1983); *Thylacorhynchus ambronensis* (Kalyptorhynchia, Rhabdocoela), after Watson and Schockaert (1996). **Gnathostomulida**: *Haplognathia rosacea* (Filospermoida), after micrograph in Sterrer et al. (1985); *Gnathostomula paradoxa* (Scleroperalia), after micrograph in Graebner (1969); *Austrognathia* sp. (Conophoralia), after micrograph in Lanfranci and Falleni (1998).

and the ancestor of all remaining flatworms (Trepaxonemata) acquired biciliary spermatozoa, which were secondarily altered several times (Justine 1991a, 2001, Watson 1999, 2001).

Gnathostomulida
Gnathostomulids are variable in their sperm morphology and show three distinct types of sperm in three subtaxa. All appear to transfer their spermatozoa to the partner, probably by hypodermal injection in Filospermoida and Scleroperalia, but by a transfer (copulation?) to the (female) bursa in Conophoralia (Mainitz 1989). The Filospermoida have filiform spermatozoa with a spirally coiled nucleus and a single mitochondrium surrounding the anterior part of the cilium (Sterrer et al. 1985, Lammert 1991; Fig. 14.6.). The Scleroperalia have tiny aciliary spermatozoa, in which only the spherical nucleus is recognizable, and which have feet-like extensions (Graebner 1969, Graebner & Adam 1970; Fig. 14.6.). The Conophoralia have a few giant aciliary spermatozoa, in which the mushroom-shaped nucleus consists of three regions with differing condensation of chromatin (Lanfranchi & Falleni 1998; Fig. 14.6.).

Eurotatoria
As males are present only in Monogononta, spermatozoa are found only in this eurotatorian taxon. They are transferrred from the dwarf male to the female in copulation (Thane 1974). There is comparatively sparse information on the ultrastructure of the spermatozoon, only *Asplanchna brightwelli* (Gilbert 1983), *Brachionus plicatilis*, and *Epiphanes senta* (Melone & Ferraguti 1994) have been investigated in detail. The spermatozoon is filiform, the cilium inserts anteriorly and runs inside the sperm cell throughout its length (Fig. 14.7.). An acrosome appears to be lacking. The additional light microscopical observations make it likely that this type of spermatozoon is present in other monogononts as well (Ahlrichs 1995).

Seisonidea
The spermatozoa of *Seison* are transferred in a spermatophore to the females (Ricci et al. 1993,

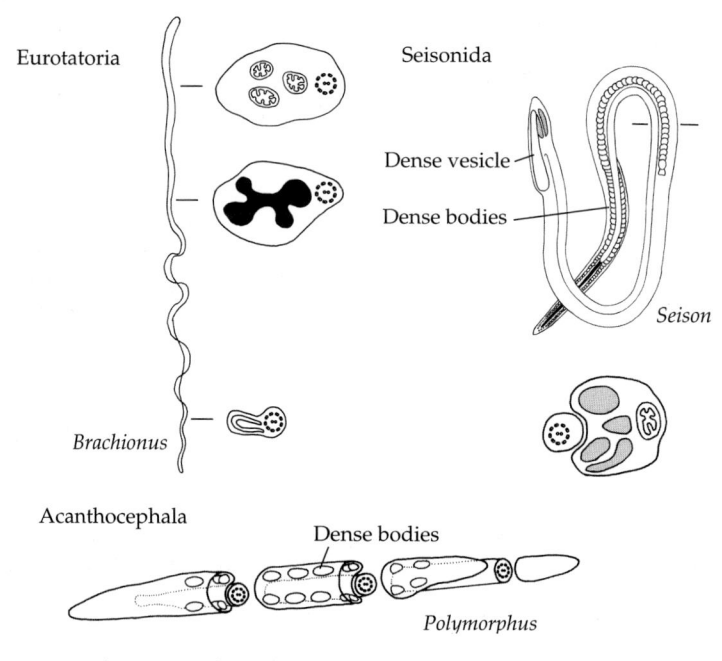

Fig. 14.7. Schematic representation of spermatozoa from **Eurotatoria**: *Brachionus plicatilis* (Monogononta), after Melone and Ferraguti (1994). **Seisonida**: *Seison nebaliae*, after Ahlrichs (1998). **Acanthocephala**: *Polymorphus minutus*, after Whitfield (1971).

Ferraguti & Melone 1999). They are filiform, contain an acrosome, an anteriorly inserting cilium, and a large number of so-called dense bodies (Ahlrichs 1998, Ferraguti & Melone 1999; Fig. 14.7.).

Acanthocephala

All acanthocephalans have internal fertilization (Crompton 1989). Spermatozoa are filiform, they lack (at least in the mature stage) an acrosome and mitochondria, but possess numerous aligned dense bodies (Whitfield 1971, Zhao & Liu 1992; summaries in Crompton 1985, Carcupino & Dezfuli 1999; Fig. 14.7.). The cilium inserts anteriorly, and runs parallel to the remaining sperm cell, obviously attached by extracellular material but not incorporated into the cell. In most cases, the cilium has a $9 \times 2 + 2$ pattern of microtubules, but there is a great intraspecific variation in the number of the central microtubules, with patterns ranging from $9 \times 2 + 0$ to $9 \times 2 + 5$ (see summary in Crompton 1985). The nuclear membrane breaks up during spermiogenesis.

Nemertini

Nemerteans release their gametes either to the external sea water or into mucous masses, which allows an external fertilization in a spatially restricted environment (Friedrich 1979, Franzén & Sensenbaugh 1988, Cantell 1989, Stricker & Folsom 1998). This last mode is called pseudocopulation and occurs in terrestrial and among marine species. External fertilization is (at least sometimes) synchronized, and pelagic nemerteans spawn in close proximity (Norenburg & Roe 1998). Few species, for example the hoplonemertean *Carinonemertes epialti*, have internal fertilization, as is evidenced by the presence of zygotes and early embryos in the female reproductive tract (Stricker 1986). The general sperm morphology differs between round-headed spermatozoa, those with slightly elongated heads, and almost filiform ones with very long heads (Turbeville 1991, Franzén & Afzelius 1999, von Döhren & Bartolomaeus 2006; Fig. 14.8.). There is a correlation between sperm structure and mode of fertilization in some species, with round-headed spermatozoa occurring in external fertilization, and elongated-head spermatozoa occurring in pseudocopulation and internal fertilization (Stricker & Folsom 1998, Franzén & Afzelius 1999), but there are also cases in which spermatozoa with elongated heads are used in external fertilization (among others, *Cerebratulus lacteus* and pelagic species; Norenburg & Roe 1998, Stricker & Folsom 1998).

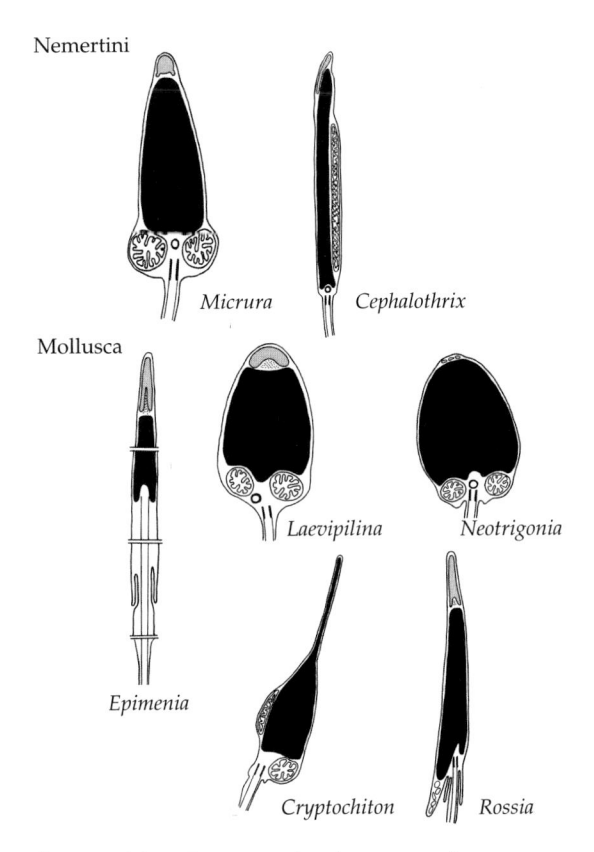

Fig. 14.8. Schematic representation of spermatozoa from **Nemertini**: *Micrura fasciolata* (Heteronemertini) and *Cephalothrix rufifrons* (Palaeonemertini), both after Franzén & Afzelius (1999). **Mollusca**: *Epimenia australis* (Aplacophora, Caudofoveata), after Buckland-Nicks and Scheltema (1995); *Laevipilina antarctica* (Neopilinida), after Healy et al. (1995); *Cryptochiton stelleri* (Polyplacophora), after Buckland-Nicks et al. (1990); *Neotrigonia bednalli* (Bivalvia), after Healy (2000); *Rossia* sp. (Cephalopoda), after Healy (2000).

Kamptozoa

As far as is known, spermatozoa are released into the water by males, but fertilization takes place within the females (Wasson 1999b). Mature spermatozoa are filiform, with an elongate nucleus, one elongate mitochondrion, and a cilium, from which the anterior part runs inside the sperm cell and only the posterior part is free (Franzén 1979, Nielsen & Jespersen 1997; Fig. 14.9.).

Mollusca
There is a wide variability of fertilization modes and sperm types in molluscs (as long as not otherwise indicated, information on molluscs is taken from the summary by Healy 2000). Solenogastres (Neomeniomorpha) have internal fertilization and their spermatozoa are filiform (Buckland-Nicks 1995, Buckland-Nicks & Scheltema 1995; Fig. 14.8.). Caudofoveata (Chaetodermomorpha) have external fertilization, their mature spermatozoa have some similarities with round-headed spermatozoa, but possess special features such as an anterior extension (Buckland-Nicks 1995). The spermiogenesis is very unusual in passing through a stage with two internal cilia (Buckland-Nicks 1995). In Monoplacophora, spermatozoa are likely released externally and fertilize the eggs in the mantle cavity—that means in a restricted, but 'external' environment (the only exception appears to be *Micropilina arntzi* with probable internal fertilization). The spermatozoa of *Laevipilina antarctica* (the only species investigated ultrastructurally) are round-headed. Acrosome, perforatorium, sphaerical nucleus, five mitochondria, and a cilium with a $9 \times 2 + 2$ pattern are present (Healy et al. 1995; Fig. 14.8.). Fertilization in polyplacophorans is external, their spermatozoa are unique in having an ovoid nucleus with an anterior filiform projection, they also contain an acrosome, a variable number of mitochondria, and a cilium (Buckland-Nicks et al. 1990; Fig. 14.8.). Basal polyplacophores probably have round-headed spermatozoa (Buckland-Nicks 1995). Gastropods have external as well as internal fertilization. It appears that the basal taxa are externally fertilizing and have an almost round-headed spermatozoon with an ovoid nucleus and therefore a slightly elongated shape (see also Fretter 1984). In pulmonates (Tompa 1984) and opisthobranchs (Hadfield & Switzer-Dunlap 1984) internal fertilization with filiform spermatozoa occurs. Cephalopods transfer spermatophores to their partners (Mann 1984); their spermatozoa are all filiform and include an acrosome, an elongated nucleus, several mitochondria, and a cilium with a $9 \times 2 + 2$ pattern (Arnold 1984; Fig. 14.8.). In Bivalvia, spermatozoa are usually round-headed (Fig. 14.8.), but transitions to ovoid or elongated forms do also occur (Morse & Zardus 1997). Scaphopods are broadcast spawners with external fertilization (McFadien-Carter 1979, Steiner 1993). The sperm structure is not very well known. Spermatozoa have a pointed head, and the acrosome is lacking in some species (Shimek & Steiner 1997). The reconstruction of the mode of fertilization and the sperm type in the mollusc ancestor has been controversially discussed. Because external fertilization with round-headed spermatozoa is generally regarded as being the ancient pattern, it may be assumed that internal fertilization with derived sperm types occurred in parallel in aplacophorans, cephalopods, scaphopods, and within gastropods (Franzén 1955). Haszprunar (1992) and Buckland-Nicks and Scheltema (1995), however, hypothesize that the molluscan ancestor was microscopically small in size, which reduces the probability of external fertilization. Therefore, these authors assume fertilization in the mantle cavity (Haszprunar 1992) or internal fertilization (Buckland-Nicks 1995, Buckland-Nicks & Scheltema 1995) as the ancestral modes in molluscs.

Sipunculida
Fertilization is external in all sipunculids (Rice 1989). As far as has been investigated (see summary of data in Rice 1993), all spermatozoa are round-headed or slightly oval (e.g. in *Golfingia ikedai*; Sawada 1980; Fig. 14.9.).

Echiurida
With the exception of Bonelliidae (e.g. *Bonellia viridis*), fertilization is external (Davis 1989). Because the dwarf male of Bonelliidae lives inside the female metanephridia, gametes are

not released to the exterior, but fertilization is internal. The few available descriptions of mature spermatozoa indicate that echiurids with external fertilization have almost round-headed spermatozoa (see Pilger 1993; Fig. 14.9.), but in Bonelliidae (*Bonellia viridis* and *Hamingia arctica*), spermatozoa are filiform (Franzén & Ferraguti 1992; Fig. 14.9.). Apart from the differences in the shape, spermatozoa correspond in having an acrosome, an accessory centriole, and a 'usual' cilium with the $9 \times 2 + 2$ pattern. The number of mitochondria appears to be reduced during spermiogenesis, to four in *Echiurus echiurus*, and one or two in *Listeriolobus*, *Ikedosoma*, *Urechis*, *Bonellia*, and *Hamingia* (Franzén & Ferraguti 1992, Pilger 1993).

Annelida

The mode of fertilization and sperm characters are so diverse in annelids that this group has almost developed into a model group to discuss sperm evolution. However, there are problems in reconstructing the sperm morphology and mode of fertilization in the annelid ancestor, because of uncertainties concerning phylogenetic relationships within annelids. It is quite clear that Clitellata are a monophyletic taxon that form cocoons with secretions of their clitellum. Fertilization takes place within this cocoon in most 'oligochaetes' and branchiobdellids, following a transfer of sperm to the (hermaphroditic) partner's seminal receptacle (see summary of Ferraguti 2000). In leeches and eudrilid 'oligochaetes', spermatozoa are either hypodermally injected into the partner or transferred in copulation. This means that internal fertilization in clitellates is derived from a strongly modified form of external fertilization (i.e. within a spatially restricted compartment). Clitellate spermatozoa are always filiform, including an acrosome, elongated nucleus, mitochondria, and a cilium with usually a $9 \times 2 + 2$ pattern (Ferraguti 2000; Fig. 14.10.). Other characters of spermatozoa are variable and can be used as phylogenetic characters in several cases (e.g. Ferraguti 1984, Ferraguti & Erséus 1999, Cardini & Ferraguti 2004). In tubificids, two kinds of spermatozoa, eusperm and parasperm, are formed, the parasperm protects and carries the eusperm (Ferraguti et al. 2002). In polychaetes,

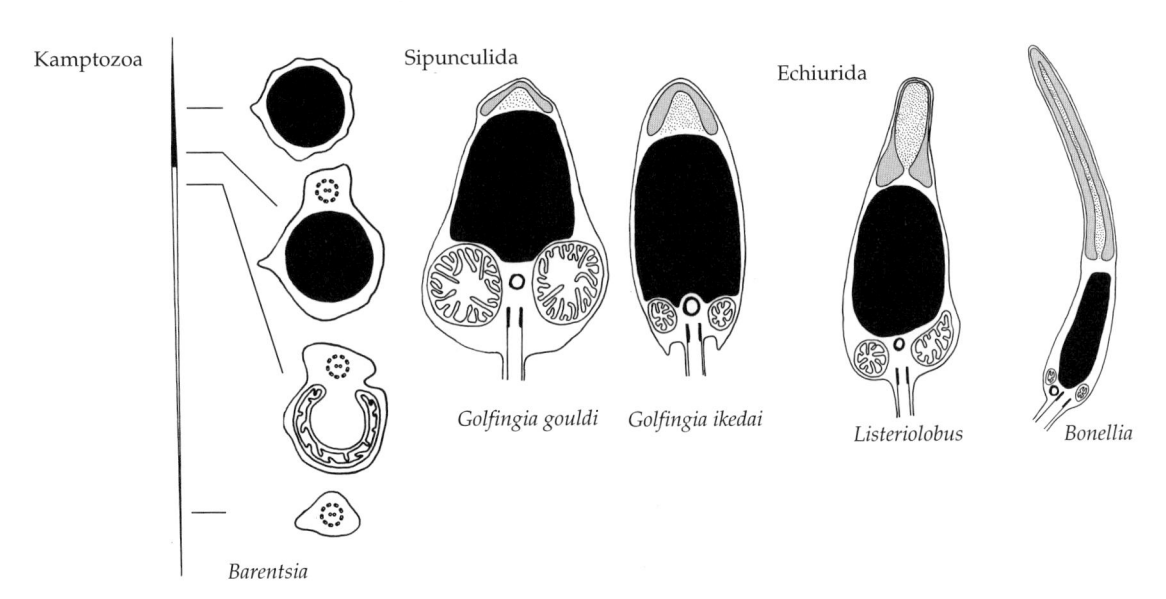

Fig. 14.9. Schematic representation of spermatozoa from **Kamptozoa**: *Barentsia laxa*, after Franzén (1979). **Sipunculida**: *Golfingia gouldi*, after Baccetti (1979b); *G.ikedai*, after Sawada (1980). **Echiurida**: *Listeriolobus pelodes*, after micrograph in Pilger (1993); *Bonellia viridis*, after Franzén & Ferraguti (1992).

the situation is much more complex (see Franzén & Rice 1988, Rouse 1999b, 2000, 2005 for summaries). Both external and internal (hypodermal injection, copulation, spermatophore transfer, see Westheide 1984) fertilization occur and spermatozoa range from round-headed to filiform (Fig. 14.10.). A few spermatozoa differ more drastically in, for example, being aciliary (e.g. *Ophryotrocha species*, Berruti et al. 1978, but see Troyer & Schwager 1979 for the simultaneous presence of aciliary and ciliary spermatozoa). Usually, there is a good correlation between external fertilization and round-headed spermatozoa, or internal fertilization and filiform spermatozoa, but there are exceptions, in which almost round-headed spermatozoa (oval nucleus and projecting acrosome) are used in internal fertilization, for example in the terrestrial polychaete *Parergodrilus heideri* (Purschke 2002b). Small, meiofaunal polychaetes often have filiform spermatozoa (Franzén 1977b, Franzén & Sensenbaugh 1984), but for example, *Polygordius lacteus* has round-headed spermatozoa (Franzén 1977b). The situation is extremely complex in polychaetes, because not only are external and internal fertilization present, but several intermediate steps where spermatozoa are transferred to cocoons or into tubes, deposited onto externally deposited oocytes, or packed into bundles (spermatozeugmata) or spermatophores, which then are transferred to the partner or released as buoyant bodies into

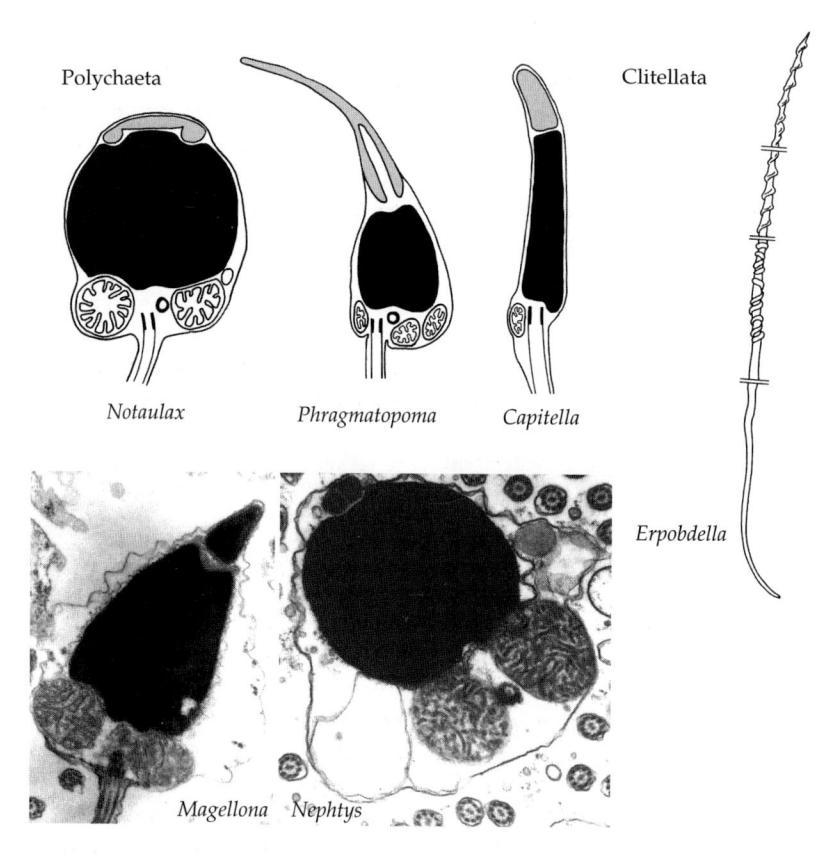

Fig. 14.10. Schematic representation of spermatozoa from **Annelida**: *Notaulax nudicollis* (Polychaeta, Sabellida), after micrograph in Rouse (2000); *Phragmatopoma lapidosa* (Polychaeta, Sabellida), after Eckelbarger (1984); *Capitella capitata* (Capitellidae), after micrograph in Franzén and Rice (1988); *Erpobdella octoculata* (Clitellata, Hirudinea), after Ferraguti (2000). Micrographs show spermatozoa from *Magelona papillicornis* (pear-shaped) and *Nephtys* sp. (round-headed). The cell membrane of the Magelona spermatozoon is artificially swollen. Both photos by courtesy of Marco Ferraguti, Milano.

the water where they are collected by females (see Westheide 1984, Rouse 2000). There is such a diversity of fertilization modes and sperm types in polychaetes that one has to assume several evolutionary transitions among polychaetes, even within smaller taxa such as Sabellida. Within this taxon, Rouse & Fitzhugh (1994) have hypothesized that filiform spermatozoa are ancestral and that round-headed spermatozoa are derived. Whether this is an exception, or just one out of several examples in polychaetes, remains unresolved and depends on the reconstruction of the ancestral annelid. In addition, it should be mentioned that myzostomids have filiform spermatozoa with a long cilium and a strange nucleus containing numerous chromatin condensations and probably lacking a nuclear membrane (Afzelius 1983, Mattei & Marchant 1988, Eeckhaut & Lanterbecq 2005). Furthermore, pogonophorans have also elongate spermatozoa, which are probably transferred in internal fertilization (Hilário et al. 2005) and consist of an acrosome, an elongate nucleus, two mitochondria—all of which are helically coiled—and a long cilium (Franzén 1973, Marotta et al. 2005b).

Tardigrada
Fertilization in tardigrades has not been observed for many species, but available data suggest a diversity of modes. Kristensen (1979) reports that heterotardigrades from the genera *Echiniscoides* and *Pseudechiniscus* have external fertilization in the water. In the heterotardigrade *Batillipes noerrevangi*, fertilization is also external, but in close proximity: following stimulation by the male, the female lays eggs on a sand grain, which are then fertilized by the male (Kristensen 1979). In eutardigrades, fertilization has been observed in *Pseudobiotus megalonyx* (see Rebecchi et al. 2000). Males penetrate and inject sperm into the female exuvia, while the female is not completely emerged from it. Spermatozoa then move from the interior of the exuvia to the female genital opening and fertilize oocytes internally. Spermatozoa vary considerably in tardigrades. Species with external fertilization in the water (see above) probably have round-headed spermatozoa, but this is not documented by micrographs (see Kristensen 1979). All tardigrade spermatozoa have an acrosome, mitochondria are present in most, but not all, species and the cilium has the $9 \times 2 + 2$ pattern (Rebecchi et al. 2000). The spermatozoa of heterotardigrades such as *Batillipes noerrevangi* (Kristensen 1979) and Echiniscidae (Rebecchi et al. 2003) correspond in having a sack-like extension from the mid-region containing the mitochondria (Fig. 14.11.). Among Eutardigrada, there are species with slightly elongated (e.g. *Amphibolus* species; Rebecchi & Guidi 1995; Fig. 14.11.) or filiform heads (e.g. *Macrobiotus hufelandi* and *Pseudobiotus megalonyx*; Baccetti et al. 1971, Rebecchi & Bertolani 1999; Fig. 14.11.), or species in which the acrosome and nuclear region turn 180° into the midpiece and ciliar region (e.g. *Xerobiotus pseudohufelandi*; Rebecchi 1997; Fig. 14.11.) (see Rebecchi et al. 2000 for summary). The eutardigrade spermatozoa appear to correspond in a terminal tuft of filaments on the cilium. In most, but not all eutardigrades, the nucleus is spirally coiled.

Onychophora
Spermatozoa are transferred in a spermatophore. They are filiform and most components are helically wound, for example, the acrosome and the elongated nucleus. One to five mitochondria are present in the midpiece region. An accessory centriole is lacking and the cilium has the $9 \times 2 + 2$ pattern (Storch et al. 2000b, Marotta & Ruhberg 2004; Fig. 14.11.).

Euarthropoda
Among euarthropods, there are only two taxa with external fertilization, Pycnogonida and Xiphosura. Typically round-headed spermatozoa are only present in Xiphosura (Fahrenbach 1973, Alberti & Janssen 1986; Fig. 14.11.). Within Pycnogonida, some species such as *Nymphon* spp. have elongated spermatozoa which show the succession of nucleus, mitochondria, and a free cilium (an acrosome is lacking), therefore they roughly resemble round-headed spermatozoa (van Deurs 1974a, Alberti 2000; Fig. 14.11.). Other species such as *Pycnogonum littorale* have

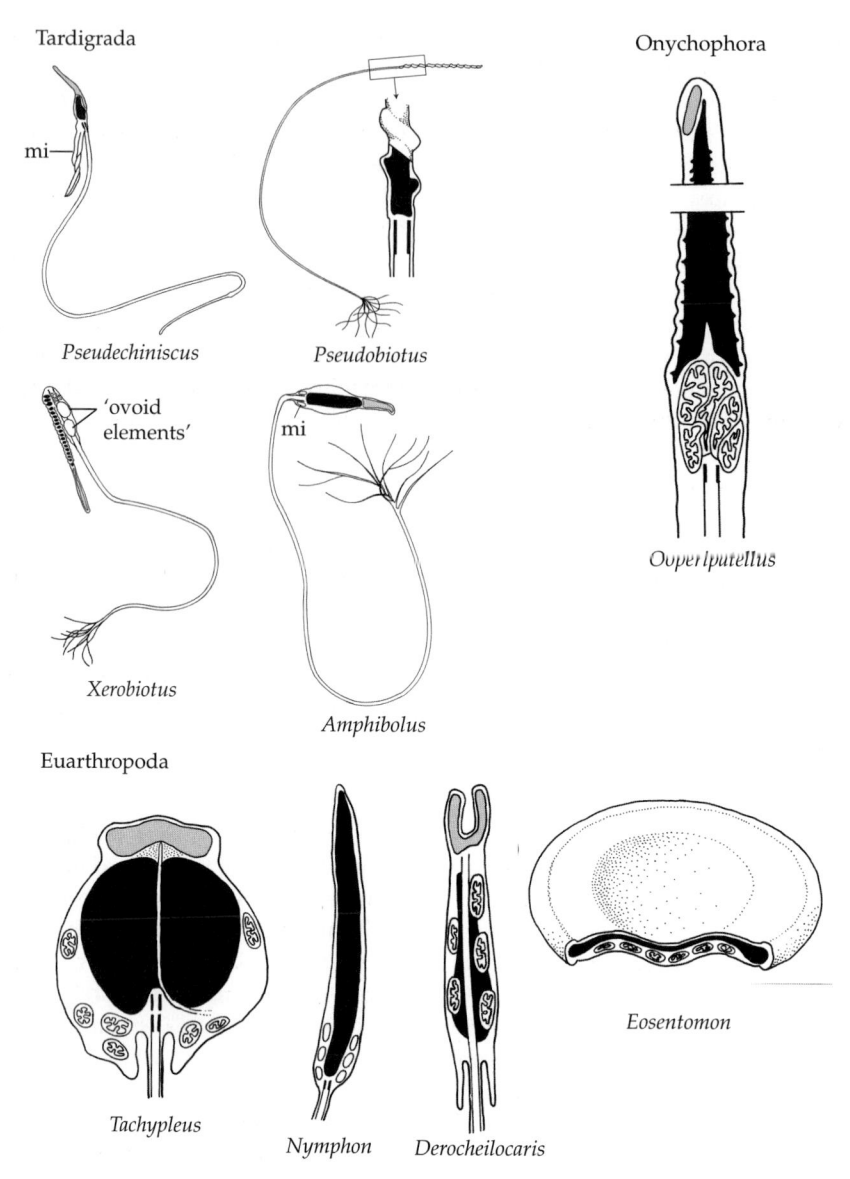

Fig. 14.11. Schematic representation of spermatozoa from **Tardigrada**: *Pseudechiniscus juanitae* (Heterotardigrada), after Rebecchi et al. (2003); *Pseudobiotus megalonyx* (Eutardigrada), after Rebecchi and Bertolani (1999); *Xerobiotus pseodohufelandi* (Eutardigrada), after Rebecchi (1997); *Amphibolus volubilis* (Eutardigrada), after Rebecchi & Guidi (1995). **Onychophora**: *Ooperipatellus insignis* (Peripatopsidae), after Marotta and Ruhberg (2004). **Euarthropoda**: *Tachypleus gigas* (Chelicerata, Xiphosura), after Alberti and Janssen (1986); *Nymphon* sp. (Pycnogonida), after Alberti (2000); *Derocheilocaris typicus* (Crustacea, Mystacocarida), after Brown and Metz (1967); *Eosentomon transitorium* (Protura), after Baccetti et al. (1973).

modified, filiform spermatozoa in which, other than a large amount of microtubules, neither the nucleus, mitochondria, or a cilium can be clearly recognized (van Deurs 1974a). The cilia in the *Nymphon* species show a $9 \times 2 + 0$ or $12 \times 2 + 0$ pattern with considerable intraspecific variation (van Deurs 1973, 1974b). In all other euarthropods there has been a shift to internal fertilization, either in the aquatic environment or in connection with terrestrialization (Proctor 1998).

Connected with this, there are extensive alterations of spermatozoal morphology (Baccetti 1979a). Some plesiomorphic features—such as an acrosome followed by nucleus, mitochondria, and cilium—can still be found in basal crustacean taxa such as Mystacocarida (Brown & Metz 1967; Fig. 14.11.) and Remipedia (Felgenhauer et al. 1992) and in basal insects such as japygid Diplura and collemboles (see Baccetti 1979a, 1998). In other taxa, there is a strong tendency to modify spermatozoa into filiform, aciliary, or otherwise derived forms (see Alberti 2000 for Chelicerata; Mazzini et al. 2000 for Chilopoda and Progoneata; contributions in Harrison & Humes 1992 and Jamieson & Tudge 2000 for Crustacea; Baccetti 1998 for Insecta). A well-known case, where sperm morphology has pointed the way for recovering phylogenetic relationships, is the association of pentastomids with branchiuran crustaceans on the basis of comparison of their spermatozoa (Wingstrand 1972, Storch & Jamieson 1992), which was confirmed, by DNA sequence comparisons (Abele et al. 1989).

Chaetognatha
Chaetognaths transfer spermatozoa as sperm masses or spermatophores directly onto the partner, either close to the female opening or elsewhere on the partner's body (Bergey et al. 1994, Shinn 1999). The mature spermatozoa are filiform. They contain an intracellular, anteriorly inserting cilium, paralleled by one long mitochondrion, and in the posterior part by the elongated nucleus (van Deurs 1972, Shinn 1997; Fig. 14.12.).

Bryozoa
Spermatozoa are released into the water in all bryozoans, but they appear to be collected by the tentacles of other individuals to fertilize the oocytes either inside an intertentacular organ or even in the ovary (Nielsen 1990b, Reed 1991, Temkin 1994). The spermatozoa range from those roughly resembling the round-headed type (with slightly elongated nucleus, no acrosome, but mitochondria in the midpiece and a cilium)— as in the gymnolaemate *Bugula* sp. (Reger 1971; Fig. 14.12)— to filiform (see, e.g., Franzén 1976, 1982, 1984, Jamieson 2000a; Fig. 14.12.). An acrosome is present in phylactolaemate bryozoans, but is lacking in gymnolaemates and stenolaemates.

Phoronida
Both external and internal fertilization appears to be present in phoronids. Internal fertilization follows the release of spermatozoa in spermatophores, which are then collected by the tentacles of another individual and are probably able to lyse the epidermis, so that spermatozoa can fertilize oocytes within the metacoel (Emig 1990). The spermatozoa, as far as has been observed, are composed of two parts that are connected in a sharp angle, one mitochondrial-nuclear part and one ciliary part (Franzén & Ahlfors 1980, Jamieson 2000a; Fig. 14.12.). An acrosome is present in an apical extension of the connection between the two parts. The two mitochondria are not intermediate between nucleus and cilium, but apical to the nucleus.

Brachiopoda
All brachiopods have external fertilization (Chuang 1990, Long & Stricker 1991), the spermatozoa are either round-headed as in, for example, *Terebratulina caputserpentis* (Afzelius & Ferraguti 1978b; Fig. 14.12.) or pear-shaped due to a pointed acrosome as in, for example, *Crania anomala* (Afzelius & Ferraguti 1978b; Fig. 14.12.).

Echinodermata
Echinoderms generally release spermatozoa and oocytes into the sea water, where fertilization takes place. However, in all taxa there are some species in which brood protection occurs and fertilization is likely to be internal (in intraovarian brood protection), or external but in close proximity of the female (in external brood protection). In many such cases, the fertilization is unknown but, for example, several deep-sea echinoids are known to release their spermatozoa in mucous strands to avoid their floating away (Young 1994). The spermatozoa are almost always round-headed, they contain an acrosome with periacrosomal material, a spherical nucleus,

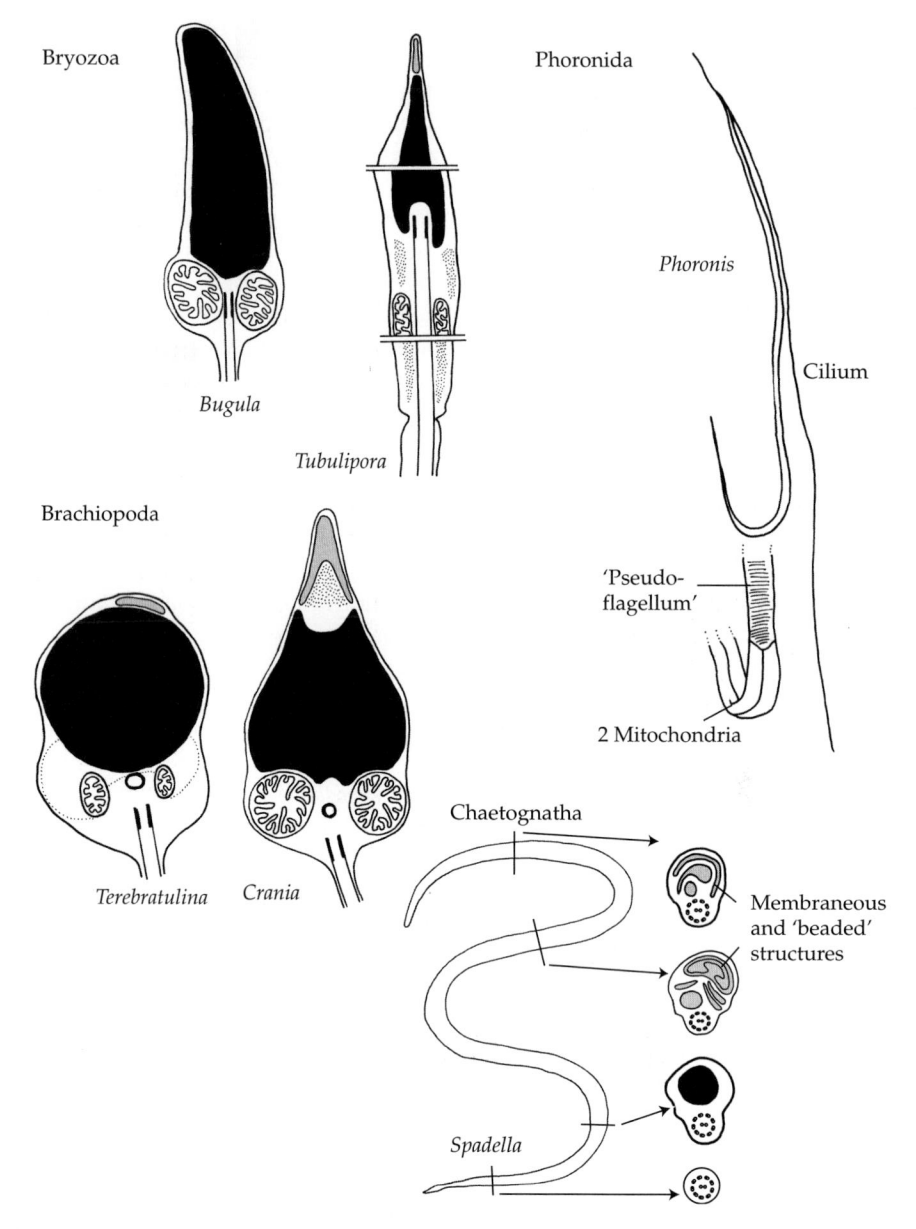

Fig. 14.12. Schematic representation of spermatozoa from **Bryozoa**: *Bugula* sp. (Cheilostomata), after micrograph in Reger (1971); *Tubulipora littacea* (Cyclostomata), after Franzén (1984). **Phoronida**: *Phoronis pallida*, after Franzén and Ahlfors (1980). **Brachiopoda**: *Terebratulina caputserpentis* (Articulata); *Crania anomala* (Inarticulata), both after Afzelius and Ferraguti (1978). **Chaetognatha**: *Spadella cephaloptera*, after van Deurs (1972).

and one single mitochondrion (Jamieson 2000b; Fig. 14.13.). Exceptions with conical to elongated spermatozoa are present in a few crinoids and holothuroids and especially in echinoids (Jamieson 2000b; Fig. 14.13.). A big exception are the spermatozoa of Concentricycloidea, which have some similarities with the spermatozoa of phoronids and pterobranchs. Two 'arms' of the filiform spermatozoon are arranged in the shape of a 'v'. One arm contains the distal

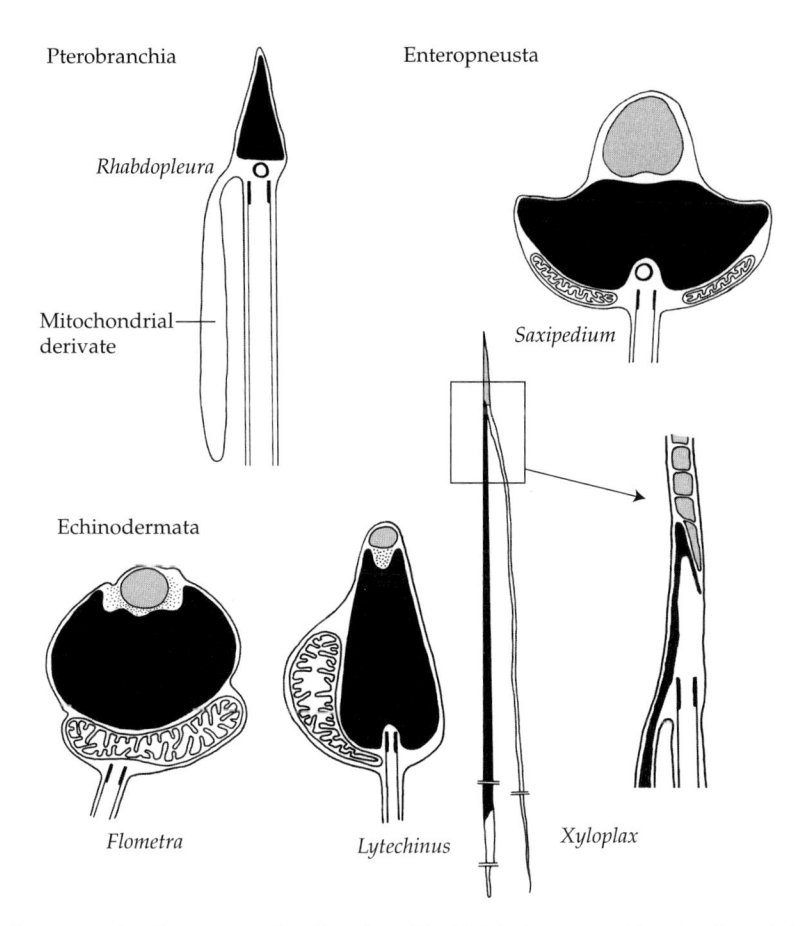

Fig. 14.13. Schematic representation of spermatozoa from **Pterobranchia**: *Rhabdopleura normani*, based on figures in Lester (1988) and Jamieson (2000d). **Enteropneusta**: *Saxipedium coronatum*, after Franzén et al. (1985). **Echinodermata**: *Flometra serratissima* (Crinoida), after Jamieson (2000b); *Lytechinus variegatus* (Echinoida), after Jamieson (2000b); *Xyloplax turnerae* (Concentricycloida), after Healy et al. (1988).

mitochondrion and the elongated nucleus, the other the cilium (9 × 2 + 2 pattern). In the pointed connecting region is the acrosome (Healy et al. 1988, Rowe et al. 1994; Fig. 14.13.). Chia et al. (1975) noted some cases in which the correlation between round-headed spermatozoa and external fertilization does not hold in echinoderms. These are the holothurians *Cucumaria lubrica* and *C. pseudocurata*, which have external fertilization but derived spermatozoa, and *Leptosynapta clarki* with internal fertilization but round-headed spermatozoa. Some holothurians, such as *Stichopus californicus* and *Leptosynapta clarki*, have been observed to aggregate in pairs or groups of up to four individuals for external fertilization in close proximity (Smiley et al. 1991). Such 'pseudocopulation' has also been observed in the asteroid *Archaster typicus* (see Chia & Walker 1991).

Pterobranchia
Fertilization and sperm structure are unknown in *Cephalodiscus*, but have been described for *Rhabdopleura normani*. As oocytes develop externally, but within the tube, this is probably also the place where fertilization takes place. Spermatozoa may be released externally and then collected by the females. The spermatozoon is filiform and v-shaped (Lester 1988; Fig. 14.13.). The two arms of the 'v' are composed of a mitochondrial derivative on the one side, and the cilium (9 × 2 + 2 pattern) on the other side. The

pointed nucleus is positioned in the connecting region, and an acrosome is absent.

Enteropneusta
There is external fertilization in enteropneusts, and spermatozoa are either round-headed (e.g. *Glossobalanus sarniensis*, Franzén 1956) or pear-shaped (e.g. *Saxipedium coronatum*, Franzén et al. 1985; Fig. 14.13.) due to a distinctly protruding acrosome. The nucleus is always spherical, an accessory centriole is present, and the cilium has a $9 \times 2 + 2$ pattern. The number of mitochondria is unclear, it may range from one which is large and ring-shaped, up to five.

Tunicata
Both external and internal fertilization exist in tunicates: solitary ascidians and appendicularians usually fertilize externally, colonial ascidians and thaliaceans internally (Burighel & Martinucci 2000). In internal fertilization, spermatozoa are probably gathered from the water with the incurrent. Despite external fertilization, spermatozoa are never typically round-headed. In species of Ascidiacea and Thaliacea, they are filiform, containing an elongated nucleus, flanked by a single mitochondrium (Franzén 1983, Holland 1991, Burighel & Martinucci 2000; Fig. 14.14.). The cilium has a $9 \times 2 + 2$ pattern, and an accessory centriole is present in only few species (e.g. *Perophora*, see Burighel & Martinucci 2000). An acrosome is present in some, but not all species. The mitochondrium is lost prior to the fusion with the oocyte (Ursprung & Schabtach 1965). In some species, some structures are spirally coiled. For example, in the spermatozoon of *Thalia democratica*, the mitochondrium is wound around the nucleus (Holland 1988); in *Pyrosoma atlanticum*, the single mitochondrium is slightly wound around the posterior part of the nucleus (Holland 1990; Fig. 14.14.); in *Boltenia*, the anterior part of the nucleus is spirally coiled (Cavey 1994); and in *Trididemnum*, the anterior part of the nucleus has corkscrew-like extensions (Burighel & Martinucci 2000; Fig. 14.14.). In Appendicularia, as investigated for *Oikopleura dioica*, spermatozoa are ovoid and have an acrosome, surrounded by an apically positioned nucleus and by a mitochondrium (Flood & Afzelius 1978, Holland et al. 1988; Fig. 14.14.). The cilium inserts directly posterior of the acrosome and crosses the whole cell before running free.

Acrania
There is external fertilization in acranians and the spermatozoa are round-headed (see Jamieson 2000c for summary and further references). An acrosome is followed by a spherical nucleus and a single mitochondrium. An accessory centriole is present and the cilium has a $9 \times 2 + 2$ pattern (Fig. 14.14.). There is a canal, which starts from the subacrosomal material and traverses the nucleus.

Craniota
Cyclostomes, lungfishes, and actinopterygian fishes have external fertilization, mostly in close proximity of their partners; while chondrichthyes, actinistians, and tetrapods have internal fertilization (Mattei 1991a,b). The structure of the spermatozoa varies between round-headed and filiform. Lampreys (Petromyzontida) spawn in close proximity. Although the males develop a penis-like urogenital papilla in the spawning season (Hardisty 1986), this is not used as a penis. Oocytes and spermatozoa are spawned to the exterior, where fertilization takes place. The spermatozoa of lampreys (Follenius 1965, Stanley 1967) and hagfishes (Myxinoida; Morisawa 2005) are elongated, have an acrosome, an elongated nucleus—which is crossed by a canal starting from the subacrosomal material—a few mitochondria, and a cilium with a $9 \times 2 + 2$ pattern (Fig. 14.14.). Chondrichthyes have internal fertilization with filiform spermatozoa (Mattei 1991a). In actinopterygian fishes, there is external fertilization, and spermatozoa range from being round-headed (Fig. 14.14.) to elongated or filiform (Mattei 1991a,b). The exceptions are aflagellate spermatozoa in the Mormyridae (Mattei et al. 1972, Mattei 1991a, Morrow 2004). Lungfishes (Dipnoi) and Actinistia have elongated spermatozoa (Mattei 1991a; Fig. 14.14.); spermatozoa with two cilia occur in the lungfish *Polypterus* (Mattei 1970). Among tetrapods, several frogs have

external fertilization, while a few frogs, urodeles, gymnophiones, and amniotes have internal fertilization. Spermatozoa are generally elongated, regardless of the mode of fertilization (Jamieson 1995a, b, Scheltinga & Jamieson 2003, van der Horst et al. 1995), but in many cases (e.g. mammals with the exception of rodents), the spermatozoon resembles the round-headed type (Guraya 1987, Roldan et al. 1991). An acrosome and a perforatorium are generally present, but can also be reduced (acrosome reduction in teleost fishes, perforatorium absent in frogs, passerine birds, and mammals; Nicander 1970, Jamieson 1995a, Jamieson et al. 2006), the

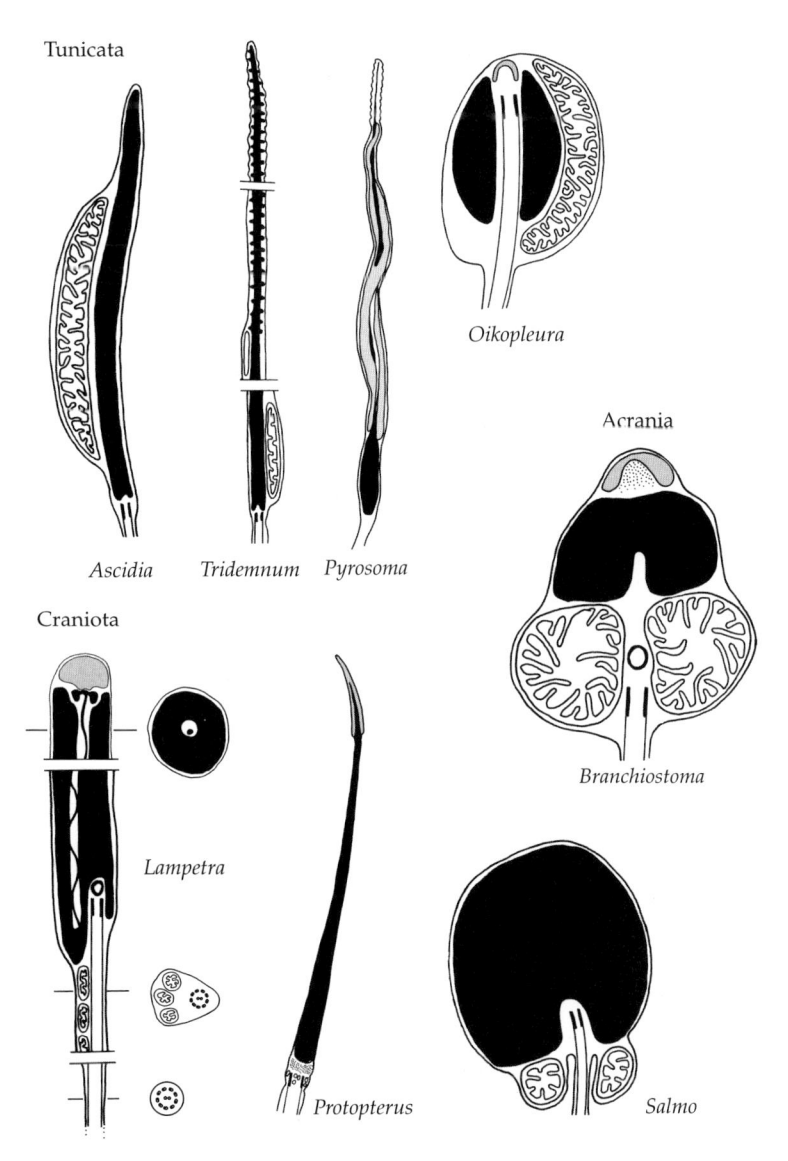

Fig. 14.14. Schematic representation of spermatozoa from **Tunicata**: *Ascidia* sp. (Ascidiacea); *Tridemnum* sp. (Ascidiacea), both after Burighel and Martinucci (2000); *Pyrosoma atlanticum* (Thaliacea), after Holland (1990); *Oikopleura dioica* (Appendicularia), after Burighel and Martinucci (2000). **Acrania**: *Branchiostoma moretonensis*, after Jamieson (2000c). **Craniota**: *Lampetra fluviatilis* (Petromyzontida), after micrographs in Stanley (1967); *Salmo salar* (Teleostei), after micrograph in Nicander (1970); *Protopterus annectens* (Dipnoi), after Mattei (1970).

cilium has a 9 × 2 +2 pattern of microtubules (Nicander 1970). In passerine birds the spermatozoon is helically wound (Asa & Phillips 1987, Koehler 1995).

The correlation between sperm type and mode of fertilization

The recognition of a strong correlation between the sperm type and the mode of fertilization goes back to Retzius (1904, 1905) and was prominently advocated by Franzén (1956). In addition to the discovery that the spermatozoa used in external fertilization are round-headed, motile spermatozoa, Retzius's and Franzén's statements include a phylogenetic reading direction: round-headed spermatozoa are 'primitive' and all other types of spermatozoa are derived from the round-headed type. These two issues, the correlation between sperm type and mode of fertilization as well as the validity of the phylogenetic reading direction from round-headed to other sperm types, have to be tested by the amount of current data on spermatozoa.

There are numerous cases in which the correlation between sperm type and mode of reproduction is in accord with Franzén (1956), so that it is usually correct to speculate on the mode of fertilization when only the sperm morphology is known. However, there are several cases in which this correlation is not correct. There are examples in which externally fertilizing species have spermatozoa that are not typically round-headed but ovoid, elongated, or pear-shaped (see, e.g., Cnidaria, Nemertini, Echinodermata, Tunicata) or have other derived morphological features such as the anterior process in chitons (Polyplacophora, Mollusca). Also remarkable are the few cases in which only slightly elongated spermatozoa (as in the polychaete *Parergodrilus heideri*, Purschke 2002b), or round-headed spermatozoa (as in the priapulid *Meiopriapulus fijiensis*, Storch et al. 1989), are found in species with internal fertilization.

The first case, the sperm variability in external fertilization, can probably best be explained by the notion that external fertilization is not the same in all cases. External fertilization is, simply expressed, convenient for organisms because it requires a minimal organization of reproductive components—including morphology of genital structures, reproductive behaviour, and physiology. It is, however, disadvantageous because gametes are diluted in the water which reduces the probability that gametes meet and fuse. Therefore, external fertilization requires the production of large amounts of gametes, which in turn is quite energy expensive for an organism, and usually requires a large body size to produce and store large amounts of gametes (Baccetti & Afzelius 1976). Therefore, a tendency towards strategies that optimize reproductive success, that is, the probability that gametes meet in the external environment, is often observed. One can recognize a diversity of strategies, starting from the synchronization of spawning, to the tendency to spatially reduce the external environment by close proximity of the partners, or the bringing of spermatozoa into tubes, mantle cavities, exuvia, cocoons, or else (see Fig. 14.15.). Additionally, spermatozoa may be released to the exterior in a broadcast spawning (as isolated spermatozoa, bundles, or in spermatophores) and be collected by the female using filtration or selective collection. Such cases of modified external fertilization, or external spawning with subsequent internal fertilization, produce altered requirements of the spermatozoa—for example, reduced motility—and therefore open the gates for evolutionary alterations. It has been documented, in particular in nemerteans and polychaetes, how such modified external fertilizations result in altered sperm morphology, and this has led Rouse & Jamieson (1987) to distinguish between not only round-headed spermatozoa (which they name ect-aquasperm) and derived spermatozoa (named introsperm), but also the 'ent-aquasperm' which are released into the water but somehow reach the female.

The problem with Rouse and Jamieson's definition of terms is that it requires knowledge of both the sperm type and the mode of reproduction. Such complete knowledge is not commonly available, because the sperm type is much easier to investigate than the mode of fertilization. For example, it is not known if the examples cited

Spermatozoa released	Into the water	External fertilization
	In close proximity to partner Attached to externally deposited eggs In mucus In cocoon Into partner´s tube Into exuvia of female	Modified form of external fertilization
	In bundles into water, collected or filtered by female In spermatophore into water, collected or filtered by female On partner´s surface, then migration into female	External spawning, internal fertilization
	Into female genital opening (copulation) Into female seminal receptacles Hypodermal injection In spermatophore on partner In spermatophore on ground	Internal fertilization

Fig. 14.15. Summary of different modes of how released spermatozoa can be transferred to the oocytes. There are several modes which are intermediate between external and internal fertilization.

above as oval, elongated, or pear-shaped spermatozoa used in external fertilization, are in fact modified because some kind of 'restricted' external fertilization occurs, or if a 'real' external fertilization is present. Additionally, the modification of spermatozoa is not moving in one particular direction, but results in numerous derived sperm types.

The phylogenetic reading direction, from round-headed to other types of spermatozoa, is surely correct for diploblastic animals, because, on the one hand, round-headed spermatozoa predominate and exceptions are very likely to be secondary. On the other hand, deviations from the round-headed pattern are, when they are present, not dramatic. Therefore, the round-headed spermatozoon and external fertilization are surely early traits in Metazoa, although calling the round-headed spermatozoon 'primitive' should be avoided. Among Bilateria, matters become more complicated.

Buckland-Nicks and Scheltema (1995) investigated the spermatozoon of the caudofoveate mollusc *Epimenia australis*, which has internal fertilization and filiform spermatozoa. In correspondance with hypotheses that the mollusc ancestor was microscopically small (Haszprunar 1992), they conclude that the mollusc ancestor also had internal fertilization with a derived sperm type (see also Buckland-Nicks 1995). Furthermore, they refer to hypotheses that the bilaterian ancestor was also small, and conclude that the bilaterian ancestor also had internal fertilization. This means that round-headed spermatozoa within Bilateria are derived. Buckland-Nicks and Scheltema (1995) offer an elegant explanation for

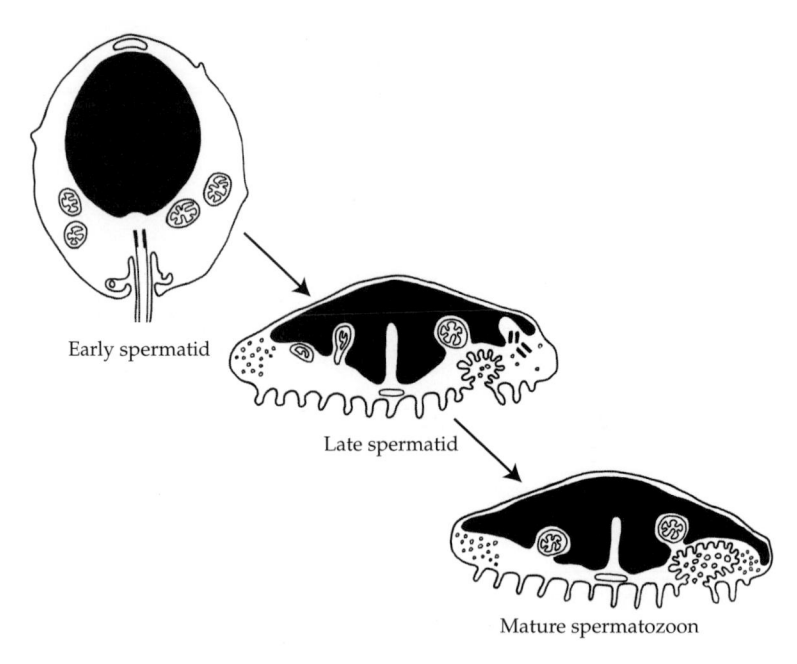

Fig. 14.16. Spermiogenesis in *Siro rubens* (Opiliones, Chelicerata). Early spermatids resemble round-headed spermatozoa, but become strongly modified during spermiogenesis. After Juberthie et al. (1976).

this: during spermiogenesis, many derived sperm types pass through a spherical stage. This has been documented in a number of cases, for example in priapulids (*Tubiluchus corallicola*: Storch & Higgins 1989), chitons (*Cryptochiton stelleri*: Buckland-Nicks et al. 1990), opiliones (*Siro rubens*, Chelicerata; Juberthie et al. 1976; Fig. 14.16.), and concentricycloid echinoderms (*Xyloplax turnerae*: Healy et al. 1988), to give just a few examples. In fact, spermatids are generally spherical cells with spherical nuclei, and modifications occur during sperm maturation. When such early stages become the mature spermatozoa, a secondarily derived round-headed spermatozoon results (Fig. 14.17.). Such a progress has been proposed for sabellid polychaetes, in which a phylogenetic analysis makes it likely that external fertilization and round-headed spermatozoa are derived (Rouse & Fitzhugh 1994). This process has been named 'progenetic spermiogenesis' by Justine (1991b) (see also Buckland-Nicks & Scheltema 1995).

I would like make things more complicated by discussing a further example. Within the Nemathelminthes, there are only a handful of species (the macroscopic priapulids) that have external fertilization and round-headed spermatozoa. All other taxa have internal fertilization and some kind of derived spermatozoa. Although priapulids are, as members of the Scalidophora, not in a basal position among Nemathelminthes, Lorenzen (1985) assumes that the cycloneuralian ancestor (in his context the 'pseudocoelomate' ancestor) was large and had external fertilization. Because the nemathelminth taxa differ considerably in their spermatozoa, it can be assumed that these evolved independantly from each other (the only exception might be that the common ancestor of Nematoda and Nematomorpha had aciliary spermatozoa). This makes it likely that the macroscopic priapulids retained an ancestral round-headed spermatozoon. On the other hand, the microscopic size in most Nemathelminthes makes it likely that their common ancestor was also small in size. Based on the reproduction of the priapulid *Meiopriapulus*, I regard it as possible that small organisms can also have round-headed spermatozoa. This may be true for the nemathelminth ancestor as well as for the bilaterian ancestor.

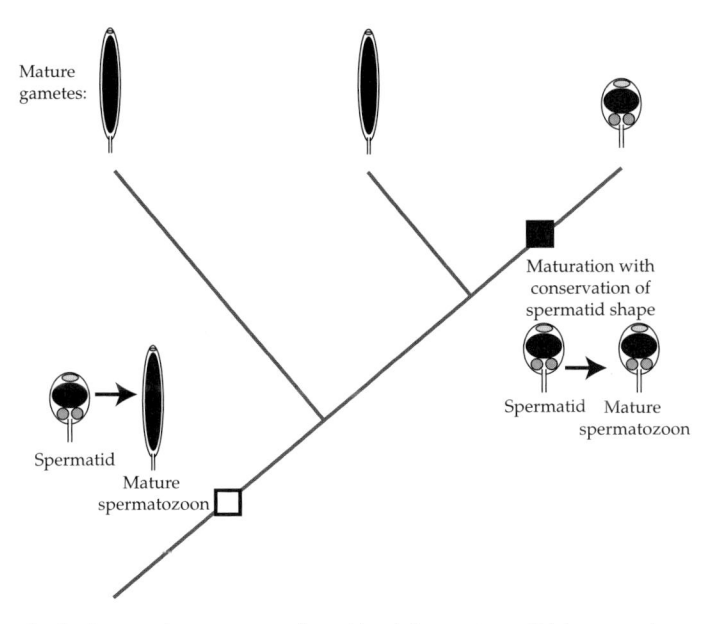

Fig. 14.17. Possible explanation for the secondary occurrence of round-headed spermatozoa. This happens when round-headed spermatids mature without modifying their shape.

Function and phylogeny of derived spermatozoa

It appears that external fertilization puts more constraints onto sperm structure than internal fertilization. The cilium is necessary for sperm mobility, this in turn requires the presence of nearby mitochondria in the sperm midpiece. Spermatozoa used in internal fertilization vary considerably in structure. Even within one taxon, a diversity of structures can be found (see, e.g., Gastrotricha, Tardigrada, Gnathostomulida, Insecta). It appears that in most cases derived spermatozoa are so different in structure that one has to assume their multiple derivation from round-headed spermatozoa. Nevertheless, some trends appear to occur several times.

In internal fertilization, spermatozoa can deal with reduced motility compared to external fertilization. This results either in changes of the microtubular pattern within the cilium (see, e.g., above under Acanthocephala and Evarthropoda: Pycnogonida) or in aciliary spermatozoa (Morrow 2004). Such spermatozoa can change to different modes of motility, for example an amoeboid movement with help of the major sperm protein (MSP) in nematodes (King et al. 1992), or with the help of peripheral microtubules in flatworms (Bedini & Papi 1970, Schmidt-Rhaesa 1993; Fig. 14.18.). Other spermatozoa are immobile (e.g. among insects, see Baccetti 1998).

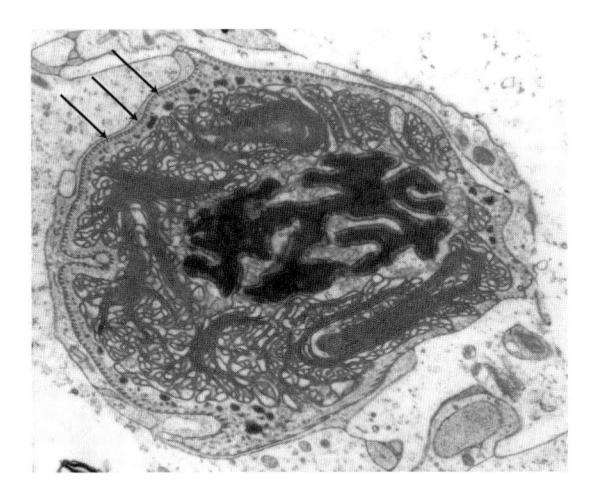

Fig. 14.18. Aflagellate spermatozoon of *Multipeniata* sp. (Prolecithophora, Platyhelminthes), showing peripheral microtubuli (some indicated by arrows). Additionally, the spermatozoa contain a strongly lobed nucleus and an intensive membraneous system.

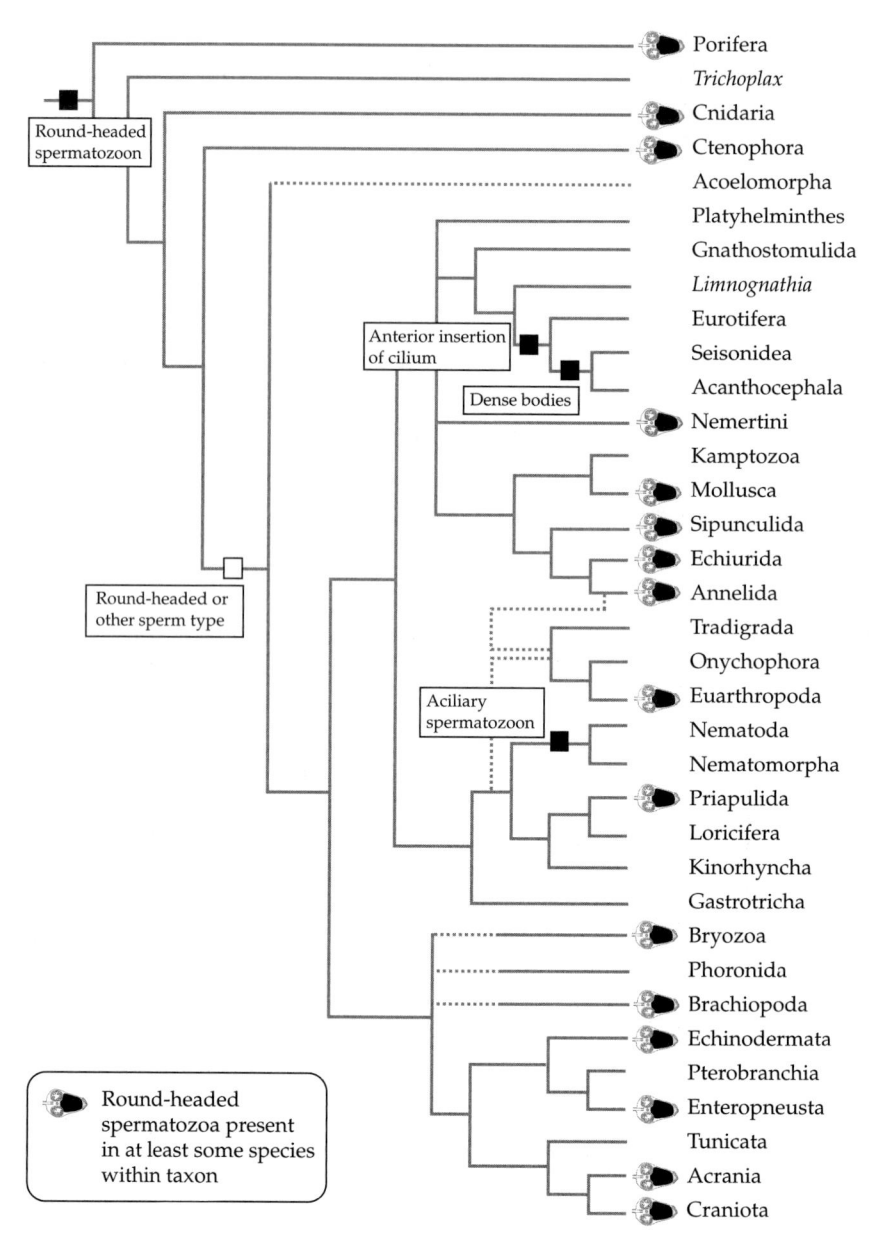

Fig. 14.19. Mapping of the occurrence of round-headed spermatozoa on the phylogenetic tree: round-headed spermatozoa occur in the majority of taxa, at least in some representatives. The sperm type in the bilaterian ancestor remains under debate. Few further autapomorphies concerning spermatozoa are indicated.

In filiform spermatozoa, there also is a considerable trend towards the spiralization of structures. This can occur in the acrosome, nucleus, or mitochondria. Examples can be found among Gastrotricha (Macrodasyida), Priapulida (*Tubiluchus*), Gnathostomulida (Filospermoida), Annelida (Pogonophora, Clitallata), Tardigrada (Eutardigrada), Onychophora, and Tunicata. It may be speculated that spiral structures or a corkscrew-like organization (i.e. with a solid core and a spirally organized edge) are also helpful in mobility.

Table 14.1 Summary of characters of the mature spermatozoa. ac = accessory centriole.

	Fertilization	General shape	Acrosome	Nucleus	Mitochondria	Cilium	Miscellaneous	Reference
Porifera	External or internal	Round-headed	Present as small vesicles (exception *Oscarella*)	Spherical	1–3	$9 \times 2 + 2$ ac present		Boury-Esnault & Jamieson 1999
Cnidaria	External or internal	Round-headed	Present as small vesicles (exception *Muggiaea*)	Spherical, slightly elongated or pear-shaped	variable number	$9 \times 2 + 2$ ac present		Harrison & Jamieson 1999
Ctenophora	External	Round-headed	Present, with perforatorium	Spherical	1	$9 \times 2 + 2$ ac present		Franc 1973
Acoelomorpha Acoela	Internal	Filiform	Absent	Elongated	numerous	2 Intracellular cilia, $9 \times 2 + 2$, $9 \times 2 + 1$ or $9 \times 2 + 0$, ac absent		Watson 1999
Nemertodermatida			Present		Spirally wound (*Meara*) or derivative (*Nemertoderma*)	$9 \times 2 + 2$ ac absent		Hendelberg 1983, Watson 1999
Gastrotricha	Internal	Filiform	Present, coiled in Macrodasyida	Elongated, coiled in Macrodasyida	Variable number	$9 \times 2 + 2$ ac absent	Nucleus, acrosome and mitochondria can be spirally coiled or corkscrew-shaped, paraacrosomal bodies in Xenotrichulidae	Balsamo et al. 1999
Nematoda	Internal	Compact	Absent	Variable in shape	Absent	Absent	Membraneous organelles, fibrous body often present	Justine & Jamieson 1999
Nematomorpha	Internal	Rod-shaped	Present, with perforatorium	Elongated	Absent	Absent	Acrosomal sheath, multivesicular complex	Schmidt-Rhaesa 1997c

Table 14.1 (*Contd*)

	Fertilization	General shape	Acrosome	Nucleus	Mitochondria	Cilium	Miscellaneous	Reference
Priapulida	External	Round-headed	Present	Spherical	4–5	$9 \times 2 + 2$ ac present		Storch et al. 2000
	Internal	Filiform or round-headed	Present, coiled in *Tubiluchus*	Elongated, anteriorly coiled in *Tubiluchus* or spherical	3 (*Tubiluchus*) or 4 (*Meiopriapulus*)	$9 \times 2 + 2$ ac absent		
Loricifera	Internal	Spherical	Present	Elongated, coiled in *Pliciloricus*	Unknown number of modified mitochondria	$9 \times 2 + 2$ ac present		Kristensen 1991
Kinorhyncha	Internal	Filiform	Absent	Elongated	Absent	$9 \times 2 + 2$ ac absent, anterior insertion	Perforatorium probably present	Adrianov & Malakhov 1999
Platyhelminthes	Internal	Filiform	Absent	Elongated	Several (absent in Eucestoda)	Mostly 2 cilia, also 1 or no cilia, $9 \times 2 + 2$ ac absent, cilia free or intracellular		Watson 1999
Gnathostomulida								
Filospermoida	Internal	Filiform	Probably present	Elongated, spirally coiled	1, wrapped around cilium	$9 \times 2 + 2$ ac present		Sterrer et al. 1985, Lammert 1991
Scleroperalia		Compact, dwarf	Absent	Spherical	Absent	Absent		Graebner 1969
Conophoralia		Compact, giant		Mushroom-shaped, zones with different condensation of chromatin	Several small			Lanfranchi & Falleni 1998
Eurotatoria (Monogononta)	Internal	Filiform	Absent	Lobate	Several	$9 \times 2 + 2$, ac absent, anteriorly inserting, intracellular		Melone & Ferraguti 1994

Taxon	Fertilization	Sperm shape	Acrosome	Nucleus shape	Mitochondria	Flagellum	Other	References
Seisonida	Internal	Filiform	Present, with perforatorium	Elongated	1 Elongated	9 × 2 + 2, ac absent, anteriorly inserting	Dense bodies	Ahlrichs 1998, Ferraguti & Melone 1999
Acanthocephala	Internal	Filiform	Absent	Elongated	Absent in mature spermatozoa	9 × 2 + 2, ac absent, anteriorly inserting	Dense bodies	Carcupino & Dezfuli 1999
Nemertini	External Pseudocopulation, internal	Round-headed Elongate head	Present, with perforatorium	Conical Elongated	4–5 1	9 × 2 + 2 ac present		Franzén & Afzelius 1999
Kamptozoa	Internal	Filiform	Absent	Elongated	1 elongated	9 × 2 + 2, ac absent		Franzén 1979
Mollusca	External Internal	Round-headed Elongated or with anterior process	Present, with perforatorium Present	Spherical Elongated	Variable number Variable number	9 × 2 + 2 ac present 9 × 2 + 2 ac mostly absent		Buckland-Nicks & Scheltema 1995, Healy 2000
Sipunculida	External	Round-headed	Present, with perforatorium	Spherical or oval	4–5	9 × 2 + 2 ac present		Rice 1993
Echiurida	External Internal	Round-headed Filiform	Present, with perforatorium	Spherical Elongated	1–4 1–2	9 × 2 + 2 ac present		Franzén & Ferraguti 1992, Pilger 1993
Annelida								
Polychaeta	External	Round-headed	Present	Spherical	often 4–5	9 × 2 + 2 ac present		Rouse 2000, 2005
	Modified external or internal	Elongate, filiform or other	Present or absent	Elongated	Variable	9 × 2 + 2 ac present or aciliary		
Clitellata	Internal	Filiform	Present	Elongated, sometimes corcscrew-shaped or coiled	Several, sometimes coiled	9 × 2 + 2 ac present		Ferraguti 2000

Table 14.1 (*Contd*)

	Fertilization	General shape	Acrosome	Nucleus	Mitochondria	Cilium	Miscellaneous	Reference
Myzostomida	Internal	Filiform	Absent	Elongated, no nuclear membrane	2	$9 \times 2 + 0$ ac absent		Afzelius 1983, Eeckhaut & Lanterbecq 2005
Pogonophora	Probably internal	Filiform	Present	Elongated	2	$9 \times 2 + 2$ ac present		Franzén 1973, Marotta et al. 2005
Tardigrada	(external?) internal	Filiform or V-shaped	Present	Elongated	Variable number or absent	$9 \times 2 + 2$, ac absent		Rebecchi et al. 2000
Onychophora	Internal	Filiform	Present	Elongated	1–5	$9 \times 2 + 2$, ac absent		Storch et al. 2000b, Marotta & Ruhberg 2004
Euarthropoda	External: Xiphosura	Rround-headed	Present, with perforatorium	Spherical	high number, small	$9 \times 2 + 2$, ac present		Alberti 2000
	External: Pycnogonida	Elongated	Absent	Elongated	High number, small	Variable microtubular pattern, ac absent		
	Internal	Variable forms	Absent or present	Variable shapes	Variable	Absent or present		Baccetti 1998, Alberti 2000, Mazzini et al. 2000
Bryozoa	Internal or external in close proximity	Round-headed or filiform	present in Gymnolaemata, absent in other bryozoan taxa	Spherical to elongated	Few to several, in some mature spermatozoa mitochondrial derivatives	$9 \times 2 + 2$, ac absent		Jamieson 2000a
Phoronida	External or internal	Filiform, V-shaped	Present	Elongated	2	$9 \times 2 + 2$, ac absent		Franzén & Ahlfors 1980
Brachiopoda	External	Round-headed	Present	Spherical	1 (Articulata) or 4–5 (Inarticulata)	$9 \times 2 + 2$ ac present		Jamieson 2000a

Echinodermata	External or internal	Round-headed	Present	Spherical to slightly elongate	1	$9 \times 2 + 2$ ac present in most species	Jamieson 2000b
Pterobranchia	External	Filiform, V-shaped	Absent	Elongate, pointed	'Mitochondrial filament', probably mitochondrial derivative	$9 \times 2 + 2$ ac present	Lester 1988
Enteropneusta	External	Round-headed or pear-shaped	Present	spherical	Presumably small number (4–5)	$9 \times 2 + 2$ ac present	Jamieson 2000d
Tunicata	External or internal	Elongated	Present or absent	Elongated, sometimes anteriorly corcscrew-shaped or coiled	1, sometimes coiled	$9 \times 2 + 2$ ac rarely present, absent in most species	Burighel & Martinucci 2000
Acrania	External	Round-headed	Present, with perforatorium	Spherical	1	$9 \times 2 + 2$ ac present	Jamieson 2000c
Craniota	External or internal	Round-headed or elongated	Present, with perforatorium (absent in Teleostei)	Spherical or elongated	Variable number	$9 \times 2 + 2$ ac present, rare exceptions	Nicander 1970

Filiform, V-shaped spermatozoon in Concentricycloidea

Acrosome evolution

The proposal that one cup-shaped acrosomal vesicle with a subacrosomal perforatorium is a common feature of ctenophores and bilaterians goes back to Baccetti (1979b), and it led Ehlers (1993) to evaluate this feature as a synapomorphy of both taxa. The taxon (Ctenophora + Bilateria) is consequently named Acrosomata. The discovery of one large acrosomal vesicle in the sponge *Oscarella lobularis* (Bacetti et al. 1986, Gaino et al. 1986) and the siphonophore *Muggiaea kochi* (Carré 1984) makes it a bit more complicated to evaluate the presence of one acrosomal vesicle. However, the isolated occurrence of one acrosomal vesicle in species not belonging to basal taxa within Porifera and Cnidaria, makes it likely that this character evolved in parallel at least three times. The presence of the subacrosomal perforatorium still supports the monophyly of Acrosomata.

Conclusions

It is very likely that the metazoan ancestor had spermatozoa and oocytes, which developed in a characteristic spermiogenesis and oogenesis. It is also likely that the spermatozoon was a round-headed spermatozoon used in external fertilization in the sea water. This basic type of spermatozoon was maintained by diploblastic animals, but several shifts towards internal fertilization following the release of spermatozoa into the water resulted in some variation of the round-headed spermatozoon, that is, slight elongation or a pear-shaped form.

From the broad distribution among Bilateria (Fig. 14.19.), I assume that the round-headed spermatozoon was also present in the bilaterian ancestor. This does not, however, have to be automatically correlated to a particular body size and an exact mode of fertilization. Especially within the Protostomia, there were several shifts from external to internal fertilization, which resulted in many different lines of sperm derivations. Derived spermatozoa are rarely comparable between taxa and are therefore of limited help in reconstructing large-scale phylogenetic relationships, although sperm structure is of great help to reconstruct phylogeny within certain taxa. Exceptions are the basically similar spermatozoa of Seisonida and Acanthocephala, which support a sister-group relationship between these two, the anterior insertion of the cilium in Syndermata, and the similarity of pentastomid and branchiuran spermatozoa which argues for an inclusion of Pentastomida in the Crustacea. Because Nematoda and Nematomorpha are likely to be sister-taxa, based on several synapomorphies, the aciliarity of their spermatozoa has to be regarded as a further synapomorphy.

Although derived spermatozoa are rarely directly comparable, there are several parallel trends such as becoming filiform, the loss of the cilium and acrosome, or the spiral organization of components. The diversity of forms makes it doubtful, whether 'filiform spermatozoa' should be used as a character in phylogenetic matrices (see, e.g., character 117 in Zrzavý et al. 1998). Even such a strongly derived sperm type, in which two filiform branches meet in an acute angle, have evolved five times in parallel within tardigrades, in isopods, in phoronids, in concentricycloid echinoderms, and in pterobranchs.

CHAPTER 15

Final conclusions

The sources of information to draw on, in order to gain a picture of the diversity and evolution of organ systems, are fascinatingly broad. For a long time, the main information available was on the 'design' of organs. More recent is the information on the molecular components and the underlying regulatory networks.

In several cases, particular components typical for an organ can be identified and traced back in time. In several organ systems, we can follow an evolution from the molecular level to the cellular level to the tissue level. The 'first' thing to find are often genes orthologous to the ones characteristically involved in a particular organ system. These may code for molecules with a different function. Functional changes in gene evolution lead to the expression of characteristic molecules. These are further integrated into simple functional and structural contexts, and then into more and more complex systems. An example is the musculature, where the characteristic molecules actin and myosin have a long history with functional contexts different from musculature. At some point in time, they formed the functional actin–myosin context, first as one of several functional components in a cell (epithelio muscle cell), then as the only component (fibre muscle cell). Finally, fibre muscle cells formed larger tissues, the muscles.

This sequence from gene to tissue makes it clear how important different approaches are for the subject. It also makes it clear that all these approaches are needed *in combination*. In biology, one particular approach often appears to be fashionable, implying that other approaches are redundant. While it is understood that new approaches need to gain a foothold, other approaches are still necessary. It is not 'modern' to regard one approach as the main key to understanding nature, but it is modern to recognize that only the integration of different approaches brings us close to understanding.

It also becomes clear that we should try to bridge the gaps between morphology and molecular biology. It is true that there are conceptual differences in the methods, with which we approach both directions, but if one is interested in the evolution of structures, organs, and organisms, the aim must be to develop an evolutionary scenario that is as complete as possible. Recent approaches using gene expression patterns and working with transcriptomes (the expressed sequence tag approach) look very promising to narrow the gap between molecules and morphology. However, this does not mean that we should neglect, for example, morphology. A look at the reference section shows that important morphological information comes from the 1960s and 1970s. This was the prime time for electron microscopy and one could get the impression that the main questions were solved during this period. This is not the case! For many organisms and many organ systems, very isolated and sketchy information is available and it is very important to gain, for example, more information on the diversity of structures within particular taxa. Comparative morphology is one important part and should not be neglected in current teaching and research (see also Budd & Olsson 2007).

Looking at evolution from the perspective of organ systems offers an intriguing view on how organisms face their environment. There is a fascinating combination of conservation and innovation, of diversity and unity!

References

Abele, L.G. & Gilchrist, S. (1977): Homosexual rape and sexual selection in acanthocephalan worms. *Science* **197**: 81–83.

Abele, L.G., Kim, W. & Felgenhauer, B.E. (1989): Molecular evidence for inclusion of the phylum Pentastomida in the Crustacea. *Mol. Biol. Evol.* **6**: 685–91.

Aboobaker, A. & Blaxter, M.L. (2003a): Hox gene evolution in nematodes: novelty conserved. *Curr. Opin. Gen. Dev.* **13**: 593–98.

Aboobaker, A. & Blaxter, M.L. (2003b): Hox gene loss during dynamic evolution of the nematode cluster. *Curr. Biol.* **13**: 37–40.

Ache, B.W. (1982): Chemoreception and thermoreception. In: *The Biology of Crustacea*, Atwood, H.L. & Sandeman, D.C. (eds.). Vol. 3: Neurobiology: structure and function. Academic Press, New York: 369–98.

Adams, C.L., McInerney, J.O. & Kelly, M. (1999): Indications of relationships between poriferan classes using full-length 18S rRNA gene sequences. *Mem. Queensl. Mus.* **44**: 33–43.

Adams, R.J. & Pollard, T D. (1986): Propulsion of organelles isolated from *Acanthamoeba* along actin filaments by myosin-1. *Nature* **322**: 754–56.

Adrianov, A.V. & Malakhov, V.V. (1995): The phylogeny and classification of the phylum Cephalorhynchia. *Zoosyst. Rossica* **3**: 181–201.

Adrianov, A.V. & Malakhov, V.V. (1999): Kinorhyncha. In: *Reproductive Biology of Invertebrates*, Adiyodi, K.G., Adiyodi, R.G. & Jamieson, B.G.M. (eds.). Vol. 9, part A: Progress in male gamete ultrastructure and phylogeny. John Wiley & Sons, Chichester: 193–211.

Adrianov, A.V. & Malakhov, V.V. (2001a): Symmetry of priapulids (Priapulida). 1. Symmetry of adults. *J. Morphol.* **247**: 99–110.

Adrianov, A.V. & Malakhov, V.V. (2001b): Symmetry of priapulids (Priapulida). 2. Symmetry of larvae. *J. Morphol.* **247**: 111–21.

Afzelius, B.A. (1983): The spermatozoon of *Myzostomum cirriferum* (Annelida, Myzostomida). *J. Ultrastruct. Res.* **83**: 58–68.

Afzelius, B.A. & Ferraguti, M. (1978a): The spermatozoon of *Priapulus caudatus* Lamarck. *J. Submicrosc. Cytol.* **10**: 71–79.

Afzelius, B.A. & Ferraguti, M. (1978b): Fine structure of the brachiopod spermatozoa. *J. Ultrastruct. Res.* **63**: 308–15.

Afzelius, B.A. & Franzén, Å. (1971): The spermatozoon of the jellyfish *Nausithoë*. *J. Ultrastruct. Res.* **37**: 186–99.

Aguinaldo, A.M.A., Turbeville, J.M., Linford, L.S., Rivera, M.C., Garey, J.R., Raff, R.A. & Lake, J.A. (1997): Evidence for a clade of nematodes, arthropods and other moulting animals. *Nature* **387**: 489–93.

Ahlrichs, W.H. (1993a): On the protonephridial system of the brackish water rotifer *Proales reinhardti* (Rotifera, Monogononta). *Microfauna Marina* **8**: 39–53.

Ahlrichs, W.H. (1993b): Ultrastructure of the protonephridia of *Seison annulatus* (Rotifera). *Zoomorphology* **113**: 245–51.

Ahlrichs, W.H. (1995): Ultrastruktur und Phylogenie von *Seison nebaliae* (Grube 1859) und *Seison annulatus* (Claus 1876). Hypothesen zu phylogenetischen Verwandtschaftsverhältnissen innerhalb der Bilateria. Cuvillier Verlag, Göttingen.

Ahlrichs, W.H. (1997): Epidermal ultrastructure of *Seison nebaliae* and *Seison annulatus*, and a comparison of epidermal structures within the Gnathifera. *Zoomorphology* **117**: 41–48.

Ahlrichs, W.H. (1998): Spermatogenesis and ultrastructure of the spermatozoa of *Seison nebaliae* (Syndermata). *Zoomorphology* **118**: 255–61.

Aizenberg, J., Tkachenko, A., Weiner, S., Addadi, L. & Hendler, G. (2001): Calcitic microlenses as part of the photoreceptor system in brittlestars. *Nature* **412**: 819–22.

Åkesson, B. & Rice, S.A. (1992): Two new *Dorvillea* species (Polychaeta, Dorvilleidae) with obligate asexual reproduction. *Zool. Scr.* **21**: 351–62.

Akiyama, S.K. & Johnson, M.D. (1983): Fibronectin in evolution: presence in invertebrates and isolation from *Microciona prolifera*. *Comp. Biochem. Physiol. B* **76**: 687–94.

Alberti, G. (2000): Chelicerata. In: *Reproductive Biology of Invertebrates*, Adiyodi, K.G., Adiyodi, R.G. & Jamieson, B.G.M. (eds.). Vol. 9, part B: Progress in male gamete ultrastructure and phylogeny. John Wiley & Sons, Chichester: 311–88.

Alberti, G. & Coons, L.B. (1999): Acari: mites. In: *Microscopic Anatomy of Invertebrates*, Harrison, F.W. & Foelix, R.F. (eds.), Vol. 8C: Chelicerate Arthropoda. Wiley-Liss, New York: 515–1215.

Alberti, G. & Janssen, H.H. (1986): On the fine structure of spermatozoa of *Tachypleus gigas* (Xiphosura, Merostomata). *Int. J. Invertebr. Reprod. Dev.* **9**: 309–19.

Alberti, G. & Storch, V. (1986): Zur Ultrastruktur der Protonephridien von *Tubiluchus philippinensis* (Tubiluchidae, Priapulida). *Zool. Anz.* **217**: 259–71.

Alberti, G. & Storch, V. (1988): Internal fertilization in a meiobenthic priapulid worm: *Tubiluchus philippinensis* (Tubiluchidae, Priapulida). *Protoplasma* **143**: 193–96.

Alberts, B., Johnson, A., Lewis, J., Raff, M., Roberts, K. & Walter, P. (2002): *Molecular Biology of the Cell*. 4th ed., Garland Science, New York.

Altner, H. (1968): Die Ultrastruktur der Labialnephridien von *Onychiurus quadriocellatus* (Collembola). *J. Ultrastruct. Res.* **24**: 349–66.

Alvestad-Graebner, I. & Adam, H. (1983): Gnathostomulida. II: Spermatogenesis and sperm function. In: *Reproductive Biology of Invertebrates*, Adiyodi, K.G. & Adiyodi, R.G. (eds.). Vol. 2: Spermatogenesis and sperm function. John Wiley & Sons, Chichester: 171–180.

Amor, A. (1973): Modelos de sistema vascular en Echiura. *Physis, Sec. A* **32**: 115–20.

Anctil, M. & Bouchard, C. (2004): Biogenic amine receptors in the sea pansy: activity, molecular structure, and physiological significance. *Hydrobiologia* **530/531**: 35–40.

Anctil, M., Hurtubise, P. & Gillis, M.-A. (2002): Tyrosine hydrolase and dopamine-beta-hydroxylase immunoreactivities in the cnidarian *Renilla koellikeri*. *Cell Tiss. Res.* **310**: 109–17.

Anderson, C.L., Canning, E.U. & Okamura, B. (1998) A Tripolblast origin for Myxozoa? *Nature* **392**: 346–47.

Anderson, D.T. (1973): Embryology and phylogeny in annelids and arthropods. *Int. Ser. Monogr. Pure Appl. Biol., Zool. Div.* **50**, Pergamon Press, Oxford.

Anderson, F.E. & Swofford, D.L. (2004): Should we be worried about long-branch attraction in real data sets? Investigations using metazoan 18S rDNA. *Mol. Phyl. Evol.* **33**: 440–51.

Anderson, P.A.V. (1985): Physiology of a bidirectional, excitatory, chemical synapse. *J. Neurophysiol.* **53**: 821–35.

Anderson, P.A.V. & Schwab, W.E. (1981): The organization and structure of nerve and muscle in the jellyfish *Cyanea capillata* (Coelenterata; Scyphozoa). *J. Morphol.* **170**: 383–99.

Anderson, P.A.V. & Schwab, W.E. (1982): Recent advances and model systems in coelenterate neurobiology. *Progr. Neurobiol.* **19**: 213–36.

Anderson, P.A.V. & Schwab, W.E. (1983): Action potential in neurons of motor nerve net of *Cyanea* (Coelenterata). *J. Neurophysiol.* **50**: 671–83.

Andrews, E.B. (1988): Excretory systems of molluscs. In: *The Mollusca*, Wilbur, K.M. (ed.). Vol. 11: Form and function, Trueman, E.R. & Clarke, M.R. (volume eds.). Academic Press, San Diego: 381–448.

Aouacheria, A., Geourjon, C., Aghajari, N., Navratil, V., Deléage, G., Lethias, C. & Exposito, J.-Y. (2006): Insights into early extracellular matrix evolution: spongin short chain collagen-related proteins are homologous to basement membrane type IV collagens and form a novel family widely distributed in invertebrates. *Mol. Biol. Evol.* **23**: 2288–302.

Apelt, G. (1969): Fortpflanzungsbiologie, Entwicklungszyklen und vergleichende Frühentwicklung acoeler Tuebellarien. *Mar. Biol.* **4**: 267–325.

Arendt, D. (2003): Evolution of eyes and photoreceptor cell types. *Int. J. Dev. Biol.* **47**: 563–571.

Arendt, D. & Nübler-Jung, K. (1994): Inversion of dorsoventral axis? *Nature* **371**: 26.

Arendt, D. & Nübler-Jung, K. (1996): Common ground plans in early brain development in mice and flies. *BioEssays* **18**: 255–59.

Arendt, D. & Nübler-Jung, K. (1997): Dorsal or ventral: similarities in fate maps and gastrulation patterns in annelids, arthropods and chordates. *Mech. Dev.* **61**: 7–21.

Arendt, D., Tessmar-Raible, K., Snyman, H., Dorresteijn, A.W. & Wittbrodt, J. (2004): Ciliary photoreceptors with a vertebrate-type opsin in an invertebrate brain. *Nature* **306**: 869–71.

Arendt, D. & Wittbrodt, J. (2001): Reconstructing the eyes of Urbilateria. *Phil. Trans. R. Soc. London B* **356**: 1545–63.

Armonies, W. (1989): Semiplanctonic Plathelminthes in the wadden sea. *Mar. Biol.* **101**: 521–27.

Armonies, W. (1994): Drifting meio- and macrobenthic invertebrates on tidal flats in Königshafen: a review. *Helgol. Meeresunt.* **48**: 299–320.

Arnold, J.M. (1984): Cephalopods. In: The Mollusca, Wilbur, K.M. (ed.), *Vol. 7: Reproduction*, Tompa, A.S., Verdonk, N.H. & van den Biggelaar, J.A.M. (volume eds.). Academic Press, Orlando: 419–54.

Aronson, J.M. (1965): The cell wall. In: *The Fungi. An Advanced Treatise*, Ainsworth, G.C. & Sussman, A.S. (eds.), Vol. 1: The fungal cell. Academic Press, New York: 49–76.

Arp, A.J. & Childress, J.J. (1983): Sulfide binding by the blood of the hydrothermal vent tube worm *Riftia pachyptila*. *Science* **219**: 295–97.

Arp, A.J., Childress, J.J. & Vetter, R.D. (1987): The sulphide-binding protein in the blood of the vestimentiferan tube-worm, *Riftia pachyptila*, is the extracellular haemoglobin. *J. Exp. Biol.* **128**: 139–58.

Asa, C.S. & Philipps, D.M. (1987): Ultrastructure of avian spermatozoa: a short review. In: *New Horizons in Sperm Cell Research*, Mohri, H. (ed.). Japan Scientific Society Press, Tokyo & Gordon and Breach Scientific Publishers, New York: 365–73.

Asano, A., Asano, K., Sasaki, H., Furuse, M. & Tsukita, S. (2003): Claudins in *Caenorhabditis elegans*: their distribution and barrier function in the epithelium. *Curr. Biol.* **13**: 1042–46.

Atkinson, H.J. (1976): The respiratory physiology of nematodes. In: *The Organization of Nematodes*, Croll, N.A. (ed.). Academic Press, London: 243–72.

Atkinson, H.J. (1980): Respiration in nematodes. In: *Nematodes as Biological Models*, Zuckerman, B.N. (ed.). Academic Press, New York: 101–42.

Ax, P. (1958): Vervielfachung des männlichen Kopulationsapparates bei Turbellarien. *Verh. Dtsch. Zool. Ges.* 1957 (*Zool. Anz. Suppl.* **21**): 227–49.

Ax, P. (1966): Die Bedeutung der interstitiellen Sandfauna für allgemeine Probleme der Systematik, Ökologie und Biologie. *Veröff. Inst. Meeresforsch. Bremerhaven*, Spec. Iss. **2**: 15–66.

Ax, P. (1985): The position of the Gnathostomulida and Platyhelminthes in the phylogenetic system of the Bilateria. In: *The Origins and Relationships of Lower Invertebrates*, Conway Morris, S., George, J.D., Gibson, R. & Platt, H.M. (eds.). *Syst. Assoc. Spec. Vol.* **28**: 168–80.

Ax, P. (1987): *The Phylogenetic System. The Systematization of Organisms on the Basis of Their Phylogenesis*. John Wiley & Sons, Chichester.

Ax, P. (1989): Basic phylogenetic systematization of the Metazoa. In: *The Hierarchy of Life*, Fernholm, B., Bremer, K. & Jörnvall, H. (eds.). Elsevier, Amsterdam: 229–45.

Ax, P. (1996): *Multicellular Animals I. A New Approach to the Phylogenetic Order in Nature*. Springer, Berlin.

Ax, P. (2000): *Multicellular Animals II. The Phylogenetic System of the Metazoa*. Springer, Berlin.

Ax, P. (2003): *Multicellular Animals III. Order in Nature – System Made by Man*. Springer, Berlin.

Babu, K.S. (1985): Patterns of arrangement and connectivity in the central nervous system of arachnids. In: *Neurobiology of Arachnids*, Barth, F.G. (ed.). Springer, Berlin: 3–19.

Baccetti, B. (1970): *Comparative Spermatology*. Academic Press, New York.

Baccetti, B. (1979a): Ultrastructure of sperm and its bearing on arthropod phylogeny. In: *Arthropod Phylogeny*, Gupta, A.P. (ed.). Van Nostrand Reinhold Company, New York: 609–44.

Baccetti, B. (1979b): The evolution of the acrosomal complex. In: *The Spermatozoon*, Fawcett, D.W. (ed.). International Symposium on the Spermatozoon, Baltimore: 305–329.

Baccetti, B. (1991): Comparative spermatology 20 years after. Serono Symp. Publ. Raven Press 75. Raven Press, New York.

Baccetti, B. (1998): Spermatozoa. In: *Microscopic Anatomy of Invertebrates*, Harrison, F.W. & Locke, M. (eds.). Vol. 11C: Insecta. Wiley-Liss, New York: 843–94.

Baccetti, B. & Afzelius, B.A. (1976): The biology of the sperm cell. *Monogr. Dev. Biol.* **10**. S. Karger, Basel.

Baccetti, B., Dallai, R. & Fratello, B. (1973): The spermatozoon of Arthropoda. XXII. The '12+0', '14+0' or aflagellate sperm of Protura. *J. Cell Sci.* **13**: 321–35.

Bacetti, B., Gaino, E. & Sará, M. (1986): A sponge with acrosome: *Oscarella lobularis*. *J. Ultrastruct. Mol. Struct. Res.* **94**: 195–98.

Baccetti, B. & Rosati, F. (1971): Electron microscopy on tardigrades. III. The integument. *J. Ultrastruct. Res.* **34**: 214–43.

Baccetti, B., Rosati, F. & Selmi, G. (1971): Electron microscopy on tardigrades. 4. The spermatozoon. *Monit. Zool. Ital.* **5**: 231–40.

Bagby, R.M. (1966): The fine structure of myocytes in the sponges *Microciona prolifera* (Ellis and Solander) and *Tedania ignis* (Duchassaing and Michelotti). *J. Morphol.* **118**: 167–82.

Bagby, R.M. (1970): The fine structure of pinacocytes in the marine sponge *Microciona prolifera* (Ellis and Solander). *Z. Zellforsch.* **105**: 579–94.

Baguñà, J. & Riutort, M. (2004a): Molecular phylogeny of the platyhelminthes. *Can. J. Zool.* **82**: 168–93.

Baguñà, J. & Riutort, M. (2004b): The dawn of bilaterian animals: the case of acoelomorph flatworms. *BioEssays* **26**: 1046–57.

Baguñà, J., Ruiz-Trillo, I., Paps, J., Loukota, M., Ribera, C., Jondelius, U. & Riutort, M. (2001): The first bilaterian organisms: simple or complex? New molecular evidence. *Int. J. Dev. Biol.* **45**: S133–S134.

Baker, R.H. & Gatesy, J. (2002): Is morphology still relevant? In: *Molecular Systematics and Evolution: Theory and Practice*, De Salle, R., Giribet, G. & Wheeler, W. (eds.). Birkhäuser Verlag, Basel: 163–74.

Balavoine, G. (1998): Are Platyhelminthes coelomates without a coelom? An argument based on the evolution of HOX genes. Am. Zool. 38: 843–858.

Balavoine, G. & Adoutte, A. (2003): The segmented *Urbilateria*: a testable scenario. *Integr. Comp. Biol.* **43**: 137–47.

Balavoine, G., de Rosa, R. & Adoutte, A. (2002): Hox clusters and bilaterian phylogeny. *Mol. Phyl. Evol.* **24**: 366–73.

Baldwin, D., McCabe, M. & Thomas, F. (1984): The respiratory gas carrying capacity of ascidian blood. *Comp. Biochem. Physiol. A* **79**: 479–82.

Ballarin, L. & Galleni, L. (1987): Evidence for planktonic feeding in Götte's larva of *Stylochus mediterraneus* (Turbellaria – Polycladida). *Boll. Zool.* **54**: 83–85.

Balsamo, M. (1992): Hermaphroditism and parthenogenesis in lower Bilateria: Gnathostomulida and Gastrotricha. In: Sex origin and evolution, Dallai, R. (ed.). Selected Symposia and Monographs U.Z.I. 6, Mucchi, Modena: 309–327.

Balsamo, M., Ferraguti, M., Guidi, L., Todaro, A. & Tongiorgi, P. (2002): Reproductive system and spermatozoa of *Paraturbanella teissieri* (Gastrotricha, Macrodasyida): implications for sperm transfer modality in Turbanellidae. *Zoomorphology* **121**: 235–41.

Balsamo, M., Fregni, E. & Ferraguti, M. (1999): Gastrotricha. In: *Reproductive Biology of Invertebrates*, Adiyodi, K.G., Adiyodi, R.G. & Jamieson, B.G.M. (eds.). Vol. 9, part A: Progress in male gamete ultrastructure and phylogeny. John Wiley & Sons, Chichester: 171–91.

Balser, E.J. & Ruppert, E.E. (1990): Structure, ultrastructure, and function of the preoral heart-kidney in *Saccoglossus kowalevskii* (Hemichordata, Enteropneusta) including data on the stomochord. *Acta Zool.* **71**: 235–49.

Balser, E.J. & Ruppert, E.E. (1993): Ultrastructure of axial vascular and coelomic organs in comasterid featherstars (Echinodermata: Crinoidea). *Acta Zool.* **74**: 87–101.

Balser, E.J., Ruppert, E.E. & Jaeckle, W.B. (1993): Ultrastructure of the coeloms of auricularia larvae (Holothuroidea: Echinodermata): evidence for the presence of an axocoel. *Biol. Bull.* **185**: 86–96.

Baltzer, F. (1931): Echiurida. In: *Handbuch der Zoologie*, Krumbach, T. (ed.). Vol. 2, 2nd half: Vermes Polymera: Priapulida, Sipunculida, Echiurida: (9)62–(9)168. (pp. 62–160 published in 1931, pp. 161–168 in 1934).

Banage, W.B. (1966): Survival of a swamp nematode (*Dorylaimus* sp.) under anaerobic conditions. *Oikos* **17**: 113–20.

Barber, V.C. (1968): The structure of mollusc statocysts, with particular reference to cephalopods. *Symp. Zool. Soc. London* **23**: 37–62.

Barber, V.C. & Wright, D.E. (1969): The fine structure of the eye and optic tentacle of the mollusc *Cerastoderma edule*. *J. Ultrastruct. Res.* **26**: 515–28.

Barber, V.C., Evans, E.M. & Land, M.F. (1967): The fine structure of the eye of the mollusc *Pecten maximus*. *Z. Zellforsch.* **76**: 295–312.

Barnes, S.N. (1971): Fine structure of the photoreceptor and cerebral ganglion of the tadpole larva of *Amaroucium constellatum* (Verrill) (subphylum: Urochordata; class: Ascidiacea). *Z. Zellforsch.* **117**: 1–16.

Barrett, J. (1991): Parasitic helminths. In: *Metazoan Life Without Oxygen*, Byrant, C. (ed.). Chapman and Hall, London: 146–164.

Barth, F.G. (1986): Zur Organisation sensorischer Systeme: die cuticularen Mechanorezeptoren der Arthropoden. *Verh. Dtsch. Zool. Ges.* **79**: 69–90.

Barth, F.G. (2002): Spider senses – technical perfection and biology. *Zoology* **105**: 271–85.

Barth, F.G. & Blickhan, R. (1984): Mechanoreception. In: *Biology of the Integument*, Bereiter-Hahn, J., Matoltsy, A.G. & Sylvia Richards, K. (eds.). Vol. 1: Invertebrates. Springer, Berlin: 554–82.

Bartnik, E. & Weber, K. (1989): Widespread occurrence of intermediate filaments in invertebrates; common principles and aspects of division. *Europ. J. Cell Biol.* **50**: 17–33.

Bartolomaeus, T. (1985): Ultrastructure and development of the protonephridia of *Lineus viridis* (Nemertini). *Microfauna Marina* **2**: 61–83.

Bartolomaeus, T. (1988): No direct contact between the excretory system and the circulatory system in *Prostomatella arenicola* Friedrich (Hoplonemertini). *Hydrobiologia* **156**: 175–81.

Bartolomaeus, T. (1989a): Larvale Nierenorgane bei *Lepidochiton cinereus* (Polyplacophora) und *Aeolidia papillosa* (Gastropoda). *Zoomorphology* **108**: 297–307.

Bartolomaeus, T. (1989b): Ultrastructure and development of the nephridia in *Anaitides mucosa* (Annelida, Polychaeta). *Zoomorphology* **109**: 15–32.

Bartolomaeus, T. (1989c): Ultrastructure and relationship between protonephridia and metanephridia in *Phoronis muelleri* (Phoronida). *Zoomorphology* **109**: 113–22.

Bartolomaeus, T. (1992a): Ultrastructure of the photoreceptor in the larva of *Lepidochiton cinereus* (Mollusca, Polyplacophora) and *Lacuna divaricata* (Mollusca, Gastropoda). *Microfauna Marina* **7**: 215–36.

Bartolomaeus, T. (1992b): Ultrastructure of photoreceptors in certain larvae of the Annelida. *Microfauna Marina* **7**: 191–214.

Bartolomaeus, T. (1993a): Die Leibeshöhlenverhältnisse und Nephridialorgane der Bilateria – Ultrastruktur, Entwicklung und Evolution. Unpublished habilitation thesis, University Göttingen.

Bartolomaeus, T. (1993b): Die Leibeshöhlenverhältnisse und Verwandtschaftsbeziehungen der Spiralia. *Verh. Dtsch. Zool. Ges.* **86.1**: 42.

Bartolomaeus, T. (1994): On the ultrastructure of the coelomic lining in the Annelida, Sipuncula and Echiura. *Microfauna Marina* **9**: 171–220.

Bartolomaeus, T. (1995): Secondary monociliarity in the Annelida: monociliated epidermal cells in larvae of *Magelona mirabilis* (Magelonida). *Microfauna Marina* **10**: 327–32.

Bartolomaeus, T. (1996): Ultrastructure of the renopericardial complex of the interstitial gastropod *Philinoglossa helgolandica* Hertling, 1932 (Mollusca: Opisthobranchia). *Zool. Anz.* **235**: 165–76.

Bartolomaeus, T. (1997a): Chaetogenesis in polychaetous Annelida – significance for annelid systematics and the position of the Pogonophora. *Zoology* **100**: 348–64.

Bartolomaeus, T. (1997b): Structure and development of the nephridia of *Tomopteris helgolandica* (Annelida). *Zoomorphology* **117**: 1–11.

Bartolomaeus, T. (1999): Structure, function and development of segmental organs in Annelida. *Hydrobiologia* **402**: 21–37.

Bartolomaeus, T. (2001): Ultrastructure and formation of the body cavity lining in *Phoronis muelleri* (Phoronida, Lophophorata). *Zoomorphology* **120**: 135–48.

Bartolomaeus, T. & Ax, P. (1992): Protonephridia and metanephridia – their relation within the Bilateria. *Z. Zool. Syst. Evolutionsforsch.* **30**: 21–45.

Bartolomaeus, T. & Balzer, I. (1997): *Convolutriloba longifissura*, nov. spec. (Acoela) – first case of longitudinal fission in Plathelminthes. *Microfauna Marina* **11**: 7–18.

Bartolomaeus, T. & Quast, B. (2005): Structure and development of nephridia in Annelida and related taxa. In: Morphology, Molecules, Evolution and Phylogeny in Polychaeta and Related Taxa, Bartolomaeus, T. & Purschke, G. (eds.). *Hydrobiologia* **535/536**: 139–165.

Bartolomaeus, T. & Ruhberg, H. (1999): Ultrastructure of the body cavity lining in embryos of *Epiperipatus biolleyi* (Onychophora, Peripatidae) – a comparison with annelid larvae. *Invertebr. Biol.* **118**: 165–174.

Bateson, W. (1894): *Materials for the Study of Variation Treated with Especial Regard to Discontinuity in the Origin of Species*. John Hopkins University Press, Baltimore.

Battelle, B.-A. (2006): The eyes of *Limulus polyphemus* (Xiphosura, Chelicerata) and their afferent and efferent projections. *Arthropod Struct. Dev.* **35**: 261–74.

Bauer-Nebelsick, M., Blumer, M., Urbancik, W. & Ott, J.A. (1995): The glandular sensory organ of Desmodoridae (Nematoda) – ultrastructure and phylogenetic importance. *Invertebr. Biol.* **114**: 211–19.

Baxter, J.M., Jones, A.M. & Sturrock, M.G. (1987): The ultrastructure of aesthetes in *Tonicella marmorea* (Polyplacophora; Ischnochitonina) and a new functional hypothesis. *J. Zool.* **211**: 589–604.

Baxter, J.M., Sturrock, M.G. & Jones, A.M. (1990): The structure of the intrapigmented aesthetes and the properiostracum layer in *Callochiton achatinus* (Mollusca: Polyplacophora). *J. Zool.* **220**: 447–68.

Beall, C.M. (2006): Andean, Tibetan, and Ethiopian patterns of adaptation to high-altitude hypoxia. *Integr. Comp. Biol.* **46**: 18–24.

Bedini, C., Ferrero, E. & Lanfranchi, A. (1973): The ultrastructure of ciliary sensory cells in two Turbellaria Acoela. *Tiss. Cell* **5**: 359–72.

Bedini, C. & Papi, F. (1970): Peculiar patterns of microtubular organization in spermatozoa of lower Turbellaria. In: *Comparative Spermatology*, Baccetti, B. (ed.). Academic Press, New York: 363–66.

Bednarz, S. (1973): The developmental cycle of the germ cells in several representatives of Trematoda (Digenea). *Zool. Pol.* **23**: 279–326.

Beer, A.-J., Moss, C. & Thorndyke, M. (2001): Development of serotonin-like and SALMFamide-like immunoreactivity in the nervous system of the sea urchin *Psammechinus miliaris*. *Biol. Bull.* **200**: 268–80.

Behr, M., Riedel, D. & Schuh, R. (2003): The claudin-like megatrachea is essential in septate junctions fort he epithelial barrier function in *Drosophila*. *Dev. Cell* **5**: 611–620.

Behrendt, G. & Ruthmann, A. (1986): The cytoskeleton of the fiber cells of *Trichoplax adhaerens* (Placozoa). *Zoomorphology* **106**: 123–30.

Beinbrech, G. (1998): Muscle structure. In: *Microscopic Anatomy of Invertebrates*, Harrison, F.W. & Locke, M. (eds.), Vol. 11B: Insecta. Wiley-Liss, New York: 553–72.

Bell, G. (1982): *The Masterpiece of Nature. The Evolution and Genetics of Sexuality*. University of California Press, Berkeley.

Bengtson, S. & Budd, G. (2004): Comment on 'Small bilaterian fossils from 40 to 55 million years before the Cambrian'. *Science* **306**: 1291a.

Bengtson, S. & Zhao, Y. (1997): Fossilized metazoan embryos from the earliest Cambrian. *Science* **277**: 1645–48.

Benito, J. & Pardos, F. (1997): Hemichordata. In: *Microscopic Anatomy of Invertebrates*, Harrison, F.W. & Ruppert, E.E. (eds.), Vol. 15: Hemichordata, Chaetognatha, and the invertebrate chordates. Wiley-Liss, New York: 15–101.

Bennett, H.S. (1963): Morphological aspects of extracellular polysaccharides. *J. Histochem. Cytochem.* **11**: 14–23.

Benton, M.J. (2002): Cope's rule. In: *Encyclopedia of Evolution*, Pagel, M. (ed.). Oxford University Press, Oxford: 209–10.

Benton, M.J. & Ayala, F.J. (2003): Dating the tree of life. *Science* **300**: 1698–1700.

Benwitz, G. (1978): Elektronenmikroskopische Untersuchung der Colloblasten-Entwicklung bei der Ctenophore *Pleurobrachia pileus* (Tentaculifera, Cydippea). *Zoomorphologie* **89**: 257–78.

Berchtold, J.-P., Sauber, F. & Reuland, M. (1985): Etude ultrastructurale de l'évolution du tégument de la sangsue *Hirudo medicinalis* L. (Annélide, Hirudinée) au cours d'un cycle de mue. *Int. J. Invertebr. Reprod. Dev.* **8**: 127–38.

Berg, G. (1985): *Annulonemertes* gen. nov., a new segmented hoplonemertean. In: The Origins and

Relationships of Lower Invertebrates, Conway Morris, S., George, J.D., Gibson, D.I. & Platt, H.M. (eds.). Clarendon Press, Oxford. *Syst. Assoc. Spec.* **28**: 200–209.

Berg, J.S., Powell, B.C. & Cheney, R.E. (2001): A millenial myosin census. *Mol. Biol. Cell* **12**: 780–94.

Bergey, M.A., Crowdner, R.J. & Shinn, G.L. (1994): Morphology of the male system and spermatophores of the arrowworm *Ferosagitta hispida* (Chaetognatha). *J. Morphol.* **221**: 321–41.

Bergquist, P.R. (1978): *Sponges*. Hutchinson, London.

Bergsten, J. (2005): A review of long-branch attraction. *Cladistics* **21**: 163–193.

Bergström, J. & Hou, X.-G. (2001): Cambrian Onychophora or Xenusians. *Zool. Anz.* **240**: 237–45.

Berney, C., Pawlowski, J. & Zaninetti, L. (2000): Elongation factor 1-α sequences do not support an early divergence of the Acoela. *Mol. Biol. Evol.* **17**: 1032–39.

Berruti, G., Ferraguti, M. & Lora Lamia Donin, C. (1978): The aflagellate spermatozoon of *Ophryotrocha*: a line of evolution of fertilization among polychaetes. *Gamete Res.* **1**: 287–92.

Bertolani, R. & Rebecchi, L. (1999): Tardigrada. In: *Encyclopedia of Reproduction*, Knobil, E. & Neill, J.D. (eds.). Vol. 4. Academic Press, San Diego: 703–17.

Bharathan, G., Janssen, B.-J., Kellogg, E. & Sinha, N. (1997): Did homeodomain proteins duplicate before the origin of angiosperms, fungi, and Metazoa? *Proc. Natl. Acad. Sci. USA* **94**: 13749–53.

Bierne, J. (1983): Nemertina. In: *Reproductive Biology of Invertebrates*, Adiyodi, K.G. & Adiyodi, R.G. (eds.). Vol. 1: Oogenesis, oviposition, and oosorption. John Wiley & Sons, Chichester: 147–67.

Bird, A.F. & Bird, J. (1991): *The Structure of Nematodes*. 2nd Edn., Academic Press, San Diego.

Bitsch, C. & Bitsch, J. (2005): Evolution of eye structure and arthropod phylogeny. In: Crustacea and Arthropod Relationships, Koenemann, S. & Jenner, R.A. (eds.). *Crust. Issues* **16**: 185–214.

Bittner, K., Ruhberg, H. & Storch, V. (1998): Ultrastructural analysis of the sensilla of *Austroperipatus aequabilis* Reid, 1996 (Onychophora, Peripatopsidae). *Acta Zool.* **79**: 267–75.

Black, M.B., Halanych, K.M., Maas, P.A.Y., Hoeh, W.R., Hashimoto, J., Desbruyéres, D., Lutz, R.A. & Vrijenhoek, R.C. (1997): Molecular systematics of vestimentiferan tubeworms from hydrothermal vents and cold-water seeps. *Mar. Biol.* **130**: 141–49.

Blair, J.E. & Hedges, S.B. (2005): Molecular phylogeny and divergence times of deuterostome animals. *Mol. Biol. Evol.* **22**: 2275–84.

Blair, J.E., Ikeo, K., Gojobori, T. & Hedges, S.B. (2002): The evolutionary position of nematodes. *BMC Evol. Biol.* **2**: 7–13.

Blaxter, M.L. (1993): Nemoglobins: divergent nematode globins. *Parasitol. Today* **9**: 353–60.

Blaxter, M.L., De Ley, P., Garey, J.R., Liu, L.X., Scheldeman, P., Vierstraete, A., Vanfleteren, J.R., Mackey, L.M., Dorris, M., Frisse, L.M., Vida, J.T. & Kelly Thomas, W. (1998): A molecular framework for the phylum Nematoda. *Nature* **392**: 71–75.

Bleidorn, C., Podsiadlowski, L. & Bartolomaeus, T. (2006): The complete mitochondrial genome of the orbiniid polychaete *Orbinia latreillii* (Annelida, Orbiniidae) – a novel gene order for Annelida and implications for annelid phylogeny. *Gene* **370**: 96–103.

Bleidorn, C., Schmidt-Rhaesa, A. & Garey, J.R. (2002): Systematic relationships of Nematomorpha based on molecular and morphological data. *Invertebr. Biol.* **121**: 357–64.

Bleidorn, C., Vogt, L. & Bartolomaeus, T. (2003): New insights into polychaete phylogeny (Annelida) inferred from 18S rDNA sequences. *Mol. Phyl. Evol.* **29**: 279–88.

Bleve-Zacheo, T., Grimaldi de Zio, S., Lamberti, F. & Morone de Lucia, M.R. (1975): Nematodes do have a coelomic cavity. *Nematol. Mediterranea* **3**: 109–12.

Blumer, M. (1994): The ultrastructure of the eyes in the veliger-larvae of *Aporrhais* sp. and *Bittium reticulatum* (Mollusca, Caenogastropoda). *Zoomorphology* **114**: 149–59.

Blumer, M.J.F., von Salvini-Plawen, L., Kikinger, R. & Büchinger, T. (1995): Ocelli in a Cnidaria polyp: the ultrastructure of the pigment spots in *Stylocoronella riedli* (Scyphozoa, Stauromedusae). *Zoomorphology* **115**: 221–27.

Boaden, P.J.S. (1975): Anaerobiosis, meiofauna and early metazoan evolution. *Zool. Scr.* **4**: 21–24.

Boaden, P.J.S. (1989): Meiofauna and the origins of the Metazoa. *Zool. J. Linn. Soc.* **96**: 217–27.

Bodmer, R. & Venkatesh, T.V. (1998): Heart development in *Drosophila* and vertebrates: conservation of molecular mechanisms. *Dev. Genetics* **22**: 181–86.

Boelsterli, U. (1977): An electron microscopic study of early developmental stages, myogenesis, oogenesis and cnidogenesis in the anthomedusa, *Podocoryne carnea* M. Sars. *J. Morphol.* **154**: 259–90.

Boero, F., Gravili, C., Pagliara, P., Piraino, S., Bouillon, J. & Schmid, V. (1998): The cnidarian premises of metazoan evolution: from triploblasty, to coelom formation, to metamery. *Ital. J. Zool.* **65**: 5–9.

Böger, H. (1988): Versuch über das phylogenetische System der Porifera. *Meyniana* **40**: 143–54.

Bonaventura, C. & Bonaventura, J. (1983): Respiratory pigments: structure and function. In: *The Mollusca*, Wilbur, K.M. (ed.). Vol. 2: Environmental biochemistry and physiology, Hochachka, P.W. (volume ed.). Academic Press, New York: 1–50.

Bonaventura, J. & Bonaventura, C. (1980): Hemocyanins: relationships in their structure, function and assembly. *Am. Zool.* **20**: 7–17.

Bone, Q. & Mackie, G.O. (1982): Urochordata. In: *Electrical Conduction and Behaviour in "Simple" Invertebrates*, Sheldon, G.A.B. (ed.). Clarendon Press, Oxford: 473–35.

Bone, Q. & Ryan, K.P. (1973): The structure and innervation of the locomotor muscles of salps (Tunicata: Thaliacea). *J. Mar. Biol. Assoc. UK* **53**: 873–83.

Bone, Q. & Ryan, K.P. (1974): On the structure and innervation of the muscle bands of *Doliolum* (Tunicata: Cyclomyaria). *Proc. R. Soc. London B* **187**: 315–27.

Bone, Q. & Ryan, K.P. (1978): Cupular sense organs in *Ciona* (Tunicata: Ascidiacea). *J. Zool. London* **186**: 417–29.

Bonner, J.T. (1998): The origins of multicellularity. *Integr. Biol.* **1**: 27–36.

Boore, J.L. & Staton, J.L. (2002): The mitochondrial genome of the sipunculid *Phascolopsis gouldii* supports its association with Annelida rather than Mollusca. *Mol. Biol. Evol.* **19**: 127–37.

Borchiellini, C., Boury-Esnault, N., Vacelet, J. & Le Parco, Y. (1998): Phylogenetic analysis of the Hsp70 sequences reveals the monophyly of Metazoa and specific phylogenetic relationships between animals and fungi. *Mol. Biol. Evol.* **15**: 647–55.

Borchiellini, C., Chombard, C., Lafay, B. & Boury-Esnault, N. (2000): Molecular systematics of sponges (Porifera). *Hydrobiologia* **420**: 15–27.

Borchiellini, C., Chombard, C., Manuel, M., Alivon, E., Vacelet, J. & Boury-Esnault, N. (2004): Molecular phylogeny of Demospongiae: implications for classification and scenarios of character evolution. *Mol. Phyl. Evol.* **32**: 823–37.

Borchiellini, C., Manuel, M., Alivon, E., Boury-Esnault, N., Vacelet, J. & Le Parco, Y. (2001): Sponge paraphyly and the origin of Metazoa. *J. Evol. Biol.* **14**: 171–79.

Bosch, C. (1975): Sur l'ultrastructure de la paroi vaisseaux latéraux de la trompe de la Bonellie (*Bonellia viridis*, Echiuridae). *C.R. Acad. Sci. Paris* **281**: 803–6.

Bosch, C. (1984): Les vaisseaux sanguins de la trompe chez *Bonellia viridis* (Echiurien) et leur musculature à fonction multiple. *Ann. Sci. Nat., Zool., 13e serie* **6**: 3–32.

Bourlat, S.J., Juliusdottoir, T., Lowe, C.J., Freeman, R., Aronowicz, J., Kirschner, M., Lander, E.S., Thorndyke, M., Nakano, H., Kohn, A.B., Heyland, A., Moroz, L.L., Copley, R.R. & Telford, M.J. (2006): Deuterostome phylogeny reveals monophyletic chordates and the new phylum Xenoturbellida. *Nature* **444**: 85–88.

Bourlat, S.J., Nielsen, C., Lockyer, A.E., Littlewood, D.T.J. & Telford, M. (2003): *Xenoturbella* is a deuterostome that eats molluscs. *Nature* **424**: 925–28.

Boury-Esnault, N., Efremova, S., Bezac, C. & Vacelet, J. (1999): Reproduction of a hexactinellid sponge: first description of gastrulation by cellular delamination in the Porifera. *Invertebr. Reprod. Dev.* **35**: 187–201.

Boury-Esnault, N., Ereshovsky, A., Bezac, C. & Tokina, D. (2003): Larval development in the Homoscleromorpha (Porifera: Demospongiae). *Invertebr. Biol.* **122**: 187–202.

Boury-Esnault, N. & Jamieson, B.G.M. (1999): Porifera. In: *Reproductive Biology of Invertebrates*, Adiyodi, K.G., Adiyodi, R.G. & Jamieson, B.G.M. (eds.). Vol. 9, part A: Progress in male gamete ultrastructure and phylogeny. John Wiley & Sons, Chichester: 1–20.

Boute, N., Exposito, J.-Y., Boury-Esnault, N., Vacelet, J., Noro, N., Miyazaki, K., Yoshizato, K. & Garrone, R. (1996): Type IV collagen in sponges, the missing link in basement membrane ubiquity. *Biol. Cell* **88**: 37–44.

Boyer, B.C. & Henry, J.Q. (1998): Evolutionary modifications of the spiralian developmental program. *Am. Zool.* **38**: 621–33.

Boyer, B.C., Henry, J.Q. & Martindale, M.Q. (1996): Dual origins of mesoderm in a basal spiralian: cell lineage analyses in the polyclad turbellarian *Hoploplana inquilina*. *Dev. Biol.* **179**: 329–38.

Bradley, T.J. (1998): Malpighian tubules. In: *Microscopic Anatomy of Invertebrates*, Harrison, F.W. & Locke (eds.). Vol. 11B: Insecta. Wiley-Liss, New York: 809–29.

Brandenburg, J. (1962): Elektronenmikroskopische Untersuchung des Terminalapparates von *Chaetonotus* sp. (Gastrotricha) als ersten Beispiels einer Cyrtocyte bei Askhelminthen. *Z. Zellforsch.* **57**: 136–44.

Brandenburg, J. (1966): Die Reusenform der Cyrtocyten. *Zool. Beitr.* **12**: 345–417.

Brandenburg, J. (1970): Die Reusenzelle (Cyrtocyte) des *Dinophilus* (Archiannelida). *Z. Morphol. Tiere* **68**: 83–92.

Brandenburg, J. (1975): The morphology of the protonephridia. *Fortschr. Zool./Progr. Zool.* **23**: 1–17.

Brandenburg, J. & Kümmel, G. (1961): Die Feinstruktur der Solenocyten. *J. Ultrastruct. Res.* **5**: 437–52.

Brandenburger, J.L., Woollacott, R.M. & Eakin, R.M. (1973): Fine structure of eyespots in tornarian larvae (phylum: Hemichordata). *Z. Zellforsch.* **142**: 89–102.

Bridge, D., Cunningham, C.W., DeSalle, R. & Buss, L.W. (1995): Class-level relationships in the phylum Cnidaria: molecular and morphological evidence. *Mol. Biol. Evol.* **12**: 679–89.

Brivio, M.F., de Eguileor, M., Grimaldi, A., Vigetti, D., Valvassori, R. & Lanzavecchia, G. (2000): Structural and biochemical analysis of the parasite *Gordius villoti* (Nematomorpha, Gordiacea) cuticle. *Tiss. Cell* **32**: 366–76.

Bromham, L. (2003): What can DNA tell us about the Cambrian explosion? *Integr. Comp. Biol.* **43**: 148–56.

Bromham, L.D. & Degnan, B.M. (1999): Hemichordates and deuterostome evolution: robust molecular phylogenetic support for a hemichordate + echinoderm clade. *Evol. Dev.* **1**: 166–71.

Brooke, N.M., Garcia- Fernàndez, J. & Holland, P.W.H. (1998): The ParaHox gene cluster is an evolutionary sister of the Hox gene cluster. *Nature* **392**: 920–22.

Brooker, B.E. (1972): The sense organs of trematode miracidia. In: *Behavioural Aspects of Parasite Transmission*, Canning, E.U. (ed.). Academic Press, London: 171–80.

Brower, D.L., Brower, S.M., Hayward, D.C. & Ball, E.E. (1997): Molecular evolution of integrins: genes encoding integrin β subunits from a coral and a sponge. *Proc. Natl. Acad. Sci. USA* **94**: 9182–87.

Brown, G.G. & Metz, C.B. (1967): Ultrastructural studies on the spermatozoa of two primitive crustaceans *Hutchinsoniella macracantha* and *Derocheilocaris typicus*. *Z. Zellforsch.* **80**: 78–92.

Brown, L.S. & Jung, K.-H. (2006): Bacteriorhodopsin like proteins of eubacteria and fungi: the extent of conservation of the haloarchaeal proton-pumping mechanism. *Photochem. Photobiol. Sci.* **5**: 538–46.

Brown, R. (1983): Spermatophore transfer and subsequent sperm development in a homalorhagid kinorhynch. *Zool. Scr.* **12**: 257–66.

Brown, W.I. & Shick, J.M. (1979): Bimodal gas exchange and the regulation of oxygen uptake in holothurians. *Biol. Bull.* **156**: 272–88.

Brüggemann, J. (1985): Ultrastruktur und Bildungsweise penialer Hartstrukturen bei freilebenden Plathelminthen. *Zoomorphology* **105**: 143–189.

Brüggemann, J. & Ehlers, U. (1981): Ultrastruktur der Statocyste von *Ototyphlonemertes pallida* (Keferstein, 1862) (Nemertini). *Zoomorphology* **97**: 75–87.

Brusca, R.C. & Brusca, G.J. (2003): *Invertebrates*, 2nd Edn., Sinauer, Sunderland.

Bubel, A. (1984): Epidermal cells. In: *Biology of the Integument*, Bereiter-Hahn, J., Matoltsy, A.G. & Sylvia Richards, K. (eds.). Vol. 1: Invertebrates. Springer, Berlin: 400–47.

Bubel, A., Stephens, R.M., Fenn, R.H. & Fieth, P. (1983): An electron microscope, X-ray diffraction and amino acid analysis study of the opercular filament cuticle, calcareous opercular plate and habitation tube of *Pomatoceros lamarckii* Quatrefages (Polychaeta: Serpulidae). *Comp. Biochem. Physiol. B* **74**: 837–50.

Buckland-Nicks, J. (1995): Ultrastructure of sperm and sperm-egg interaction in Aculifera: implications for molluscan phylogeny. In: Advances in Spermatozoal Phylogeny and Taxonomy, Jamieson, B.G.M., Ausio, J. & Justine, J.-L. (eds.). *Mém. Mus. Natn. Hist. Nat.* **166**: 129–53.

Buckland-Nicks, J., Chia, F.-S. & Koss, R. (1990): Spermiogenesis in Polyplacophora, with special reference to acrosome formation (Mollusca). *Zoomorphology* **109**: 179–88.

Buckland-Nicks, J. & Scheltema, A. (1995): Was internal fertilization an innovation of early Bilateria? Evidence from sperm structure of a mollusc. *Proc. R. Soc. London B* **261**: 11–18.

Budd, G.E. (1993): A Cambrian gilled lobopod from Greenland. *Nature* **364**: 709–11.

Budd, G.E. (1996): The morphology of *Opabinia regalis* and the reconstruction of the arthropod stem-group. *Lethaia* **29**: 1–14.

Budd, G.E. (1997): Stem group arthropods from the Lower Cambrian Sirius Passet fauna of North Greenland. In: Arthropod Relationships, Fortey, R.A. & Thomas, R.H. (eds.). *Syst. Assoc. Spec.* Vol. **55**, Chapman & Hall, London: 125–38.

Budd, G.E. (1998): Arthropod body-plan evolution in the cambrian with an example from anomalocaridid muscle. *Lethaia* **31**: 197–210.

Budd, G.E. (1999): The morphology and phylogenetic significance of *Kerygmachela kierkegaardi* Budd (Buen formation, Lower Cambrian, N Greenland). *Trans. R. Soc. Edinburgh Earth Sci.* **89**: 249–90.

Budd, G.E. (2001a): Tardigrades as 'stem-group arthropods': the evidence fropm the Cambrian fauna. *Zool. Anz.* **240**: 265–79.

Budd, G.E. (2001b): Why are arthropods segmented? *Evol. Dev.* **3**: 332–42.

Budd, G.E. (2002): A palaeontological solution to the arthropod head problem. *Nature* **417**: 271–75.

Budd, G.E. (2004): Lost children of the Cambrian. *Nature* **427**: 205–7.

Budd, G.E. & Jensen, S. (2000): A critical reappraisal of the fossil record of the bilaterian phyla. *Biol. Rev.* **75**: 253–95.

Budd, G.E. & Olsson, L. (2007): Editorial: a renaissance for evolutionary morphology. *Acta Zool.* **88**: 1.

Budelmann, B.U. (1994): Cephalopod sense organs, nerves and the brain: adaptations for high performance and life style. *Mar. Fresh. Behav. Physiol.* **25**: 13–33.

Budelmann, B.U., Schipp, R. & von Boletzky, S. (1997): Cephalopoda. In: *Microscopic Anatomy of Invertebrates*, Harrison, F.W. & Kohn, A.J. (eds.), Vol. 6A: Mollusca II. Wiley-Liss, New York: 119–414.

Büning, J. (1992): The ovariole: structure, type, and phylogeny. In: *Microscopic Anatomy of Invertebrates*, Harrison, F.W. & Locke, M. (eds.). Vol. 11C: Insecta. Wiley-Liss, New York: 897–932.

Bullard, B., Luke, B. & Winkelman, L. (1973): The paramyosin of insect flight muscle. *J. Mol. Biol.* **75**: 359–67.

Bullivant, J.S. & Bils, R.F. (1968): The pharyngeal cells of *Zoobotryon verticillatum* (Delle Chiaje) a gymnolaemate bryozoan. *New Z. J. Mar. Freshwater Res.* **2**: 438–46.

Bullock, T.H. (1940): The functional organization of the nervous system of Enteropneusta. *Biol. Bull.* **79**: 91–113.

Bullock, T.H. & Horridge, G.A. (1965): *Structure and Function in the Nervous System of Invertebrates*. 2 volumes. Freeman, San Francisco.

Burdon-Jones, C. (1952): Development and biology of the larva of *Saccoglossus horsti* (Enteropneusta). *Philos. Trans. R. Soc. London B* **236**: 553–90.

Burger, G., Forget, L., Zhu, Y., Gray, M.W. & Lang, F. (2003): Unique mitochondrial genome architecture in unicellular relatives of animals. *Proc. Nat. Acad. Sci. USA* **100**: 892–97.

Burighel, P. & Cloney, R.A. (1997): Urochordata: Ascidiacea. In: *Microscopic Anatomy of Invertebrates*, Harrison, F.W. & Ruppert, E.E. (eds.), Vol. 15: Hemichordata, Chaetognatha, and the invertebrate chordates. Wiley-Liss, New York: 221–347.

Burighel, P. & Martinucci, G.B. (2000): Urochordata. In: *Reproductive Biology of Invertebrates*, Adiyodi, K.G., Adiyodi, R.G. & Jamieson, B.G.M. (eds.). Vol. 9, part C: Progress in male gamete ultrastructure and phylogeny. John Wiley & Sons, Chichester: 261–98.

Burmester, T. (2001): Molecular evolution of the arthropod hemocyanin superfamily. *Mol. Biol. Evol.* **18**: 184–95.

Burmester, T. (2002): Origin and evolution of arthropod hemocyanins and related proteins. *J. Comp. Physiol. B* **172**: 95–107.

Burmester, T., Ebner, B., Weich, B. & Hankeln, T. (2002): Cytoglobin: a novel globin type ubiquitously expressed in vertebrate tissues. *Mol. Biol. Evol.* **19**: 416–21.

Burmester, T., Weich, B., Reinhardt, S. & Hankeln, T. (2000): A vertebrate globin expressed in the brain. *Nature* **407**: 520–22.

Burr, A.H.J., Hunt, P., Wagar, D.R., Dewilde, S., Blaxter, M., Vanfleteren, J. & Moens, L. (2000): A hemoglobin with an optical function. *J. Biol. Chem.* **275**: 4810–15.

Bush, A.O., Fernández, J.C., Esch, G.W. & Seed, J.R. (2001): *Parasitism. The Diversity and Ecology of Animal Parasites*. Cambridge University Press, Cambridge.

Bush, B.M.H. & Laverack, M.S. (1982): Mechanoreception. In: *The Biology of Crustacea*, Atwood, H.L. & Sandeman, D.C. (eds.). Vol. 3: Neurobiology: structure and function. Academic Press, New York: 399–468.

Butterfield, N.J. (2003): Exceptional fossil preservation and the Cambrian explosion. *Integr. Comp. Biol.* **43**: 166–77.

Byrne, M. (1994): Ophiuroidea. In: *Microscopic Anatomy of Invertebrates*, Harrison, F.W. & Chia, F.-S. (eds.), Vol. 14: Echinodermata. Wiley-Liss, New York: 247–343.

Byrne, M. (1999): Echinodermata. In: *Encyclopedia of Reproduction*, Knobil, E. & Neill, J.D. (eds.). Vol. 1. Academic Press, San Diego: 940–54.

Byrne, M. (2006): Life history diversity and evolution in the Asterinidae. *Integr. Comp. Biol.* **46**: 243–254.

Byrne, M. & Cerra, A. (1996): Evolution of intragonadal development in the diminuitive asterinid sea stars *Patiriella vivipara* and *P. parvipara* with an overview of development in the Asterinidae. *Biol. Bull.* **191**: 17–26.

Byrne, M., Cisternas, P., Elia, L. & Relf, B. (2005): *Engrailed* is expressed in larval development and in the radial nervous system of *Patiriella* sea stars. *Dev. Genes Evol.* **215**: 608–17.

Cabib, E., Shaw, J.A., Mol, P.C., Bowers, B. & Choi, W.J. (1996): Chitin biosynthesis and morphogenetic processes. In: *The Mycota*, Brambl, R. & Marzluf, G.A. (eds.). Vol. III. Biochemistry and molecular biology. Springer, Berlin: 243–67.

Camatini, M., Ceresa Castellani, L., Franchi, E., Lanzavecchia, G. & Paoletti, L. (1976): Thick filaments and paramyosin of annelid muscles. *J. Ultrastruct. Res.* **55**: 433–47.

Camatini, M., Franchi, E. & Lanzavecchia, G. (1979): The body muscles of Onychophora: an atypical contractile system. In: *Myriapod Biology*, Camatini, M. (ed.), Academic Press, London: 419–31.

Cameron, C.B., Garey, J.R. & Swalla, B.J. (2000): Evolution of the chordate body plan: new insights from phylogenetic analyses of deuterostome phyla. *Proc. Natl. Acad. Sci. USA* **97**: 4469–74.

Cameron, C.B. & Mackie, G.O. (1996): Conduction pathways in the nervous system of *Saccoglossus* sp. (Enteropneusta). *Can. J. Zool.* **74**: 15–19.

Campbell, R.D. (1972): Statocyst lacking cilia in the coelenterate *Corymorpha palma*. *Nature* **238**: 49–51.

Candia Carnevali, M.D. & Ferraguti, M. (1979): Structure and ultrastructure of muscles in the priapulid *Halycryptus spinulosus*: functional and phylogenetic remarks. *J. Mar. Biol. Assoc. UK* **59**: 737–44.

Candia Carnevali, M.D. & Saita, A. (1976): Correlated structural and contractile properties in specialized fibers of a woodlouse *Armadillidium vulgare* (Latr.). *J. Exp. Zool.* **198**: 241–52.

Candia Carnevali, M.D. & Valvassori, R. (1981): Z-line in insect muscles: structural and functional diversities. *Boll. Zool.* **48**: 1–9.

Candia Carnevali, M.D. & Valvassori, R. (1982): Active supercontraction in rolling-up muscles of *Glomeris marginata* (Myriapoda, Diplopoda). *J. Morphol.* **172**: 75–82

Candia Carnevali, M.D., Saita, A. & Fedrigo, A. (1986): An unusual Z-system in the obliquely striated muscles of crinoids: three-dimensional structure and computer simulations. *J. Muscle Res. Cell Motil.* **7**: 568–78.

Cantell, C.-E. (1989): Nemertina. In: *Reproductive Biology of Invertebrates*, Adiyodi, K.G. & Adiyodi, R.G. (eds.). Vol. 4, part A: fertilization, development, and parental care. John Wiley & Sons, Chichester: 147–65.

Carcupino, M. & Dezfuli, B.S. (1999): Acanthocephala. In: *Reproductive Biology of Invertebrates*, Adiyodi, K.G., Adiyodi, R.G. & Jamieson, B.G.M. (eds.). Vol. 9, part A: Progress in male gamete ultrastructure and phylogeny. John Wiley & Sons, Chichester: 229–42.

Cardini, A. & Ferraguti, M. (2004): The phylogeny of *Branchiobdella* (Annelida, Clitallata) assessed by sperm characters. *Zool. Anz.* **243**: 37–46.

Carle, K.J. & Ruppert, E.E. (1983): Comparative ultrastructure of the bryozoan funiculus: a blood vessel homologue. *Z. Zool. Syst. Evolutionsforsch.* **21**: 181–93.

Carré, D. (1984): Existence d'un complexe acrosomal chez les spermatozoides du cnidaire *Muggiaea kochi* (Siphonophore, Calycophore): differenciation et réaction acrosomale. *Int. J. Invertebr. Reprod. Dev.* **7**: 95–103.

Carré, D., Djediat, C. & Sardet, C. (2002): Formation of a large *vasa*-positive granule and its inheritance by germ cells in the enigmatic chaetognaths. *Development* **129**: 661–70.

Carroll, S.B., Grenier, J.K. & Weatherbee, S.D. (2005): *From DNA to Diversity. Molecular Genetics and the Evolution of Animal Design.* 2nd Edn., Blackwell, Malden.

Casanova, J.-P. & Duvert, M. (2002): Comparative studies and evolution of muscles in chaetognaths. *Mar. Biol.* **141**: 925–38.

Castellani, L.C. & Saita, A. (1974): Ultrastructural analysis of muscle fibres in *Glossobalanus minutus* (Kowalewski, 1986) (Enteropneusta). *Monit. Zool. Ital.* **8**: 117–32.

Cavalier-Smith, T. (1998): A revised six-kingdom system of life. *Biol. Rev.* **73**: 203–66.

Cavalier-Smith, T. & Chao, E.E.Y. (2003): Phylogeny of Choanozoa, Apusozoa, and other Protozoa and early eucaryote megaevolution. *J. Mol. Evol.* **56**: 540–63.

Caveney, S. (1998): Compound eyes. In: *Microscopic Anatomy of Invertebrates*, Harrison, F.W. & Locke, M. (eds.). Vol. 11B: Insecta. Wiley-Liss, New York: 423–45.

Cavey, M.J. (1994): Spermatogenesis in the ovotestes of the solitary ascidian *Boltenia villosa*. In: *Reproduction and Development of Marine Invertebrates*, Wilson, W.H. (eds.). Friday Harbor Laboratories, Baltimore: 64–76.

Cavey, M.J. (2006): Organization of the coelomic lining and a juxtaposed nerve plexus in the suckered tube feet of *Parastichopus californicus* (Echinodermata: Holothuroida). *J. Morphol.* **267**: 41–49.

Cavey, M.J. & Märkel, K. (1994): Echinoidea. In: *Microscopic Anatomy of Invertebrates*, Harrison, F.W. & Chia, F.-S. (eds.), Vol. 14: Echinodermata. Wiley-Liss, New York: 345–400.

Cavey, M.J. & Wilkens, J.L. (1982): Ultrastructure of putative statocysts in the mantle of an articulate brachiopod. *Am. Zool.* **22**: 940.

Celerin, M., Ray, J.M., Schisler, N.J., Day, A.W., Stetler-Stevenson, W.G. & Laudenbach, D.E. (1996): Fungal fimbriae are composed of collagen. *EMBO J.* **15**: 4445–53.

Chalfie, M. & Sulston, J. (1981): Developmental genetics of the mechanosensory neurons of *Caenorhabditis elegans*. *Dev. Biol.* **82**: 358–70.

Chapman, D.M. (1978): Microanatomy of the cubopolyp, *Tripedalia cystophora* (class Cubozoa). *Helgol. Wiss. Meeresunters.* **31**: 128–68.

Chapman, D.M. (1985): X-ray microanalysis of selected coelentherat statoliths. *J. Mar. Biol. Assoc. UK* **65**: 617–27.

Chapman, D.M., Pantin, C.F.A. & Robson, E.A. (1962): Muscle in coelenterates. *Rev. Can. Biol.* **21**: 267–78.

Chapman, H.D. (1973): The functional organization and fine structure of the tail musculature of the cercariae of *Cryptocotyle lingua* and *Himasthla secunda*. *Parasitology* **66**: 487–97.

Chen, J.-Y., Bottjer, D.J., Oliveri, P., Dornbos, S.Q., Gao, F., Ruffins, S., Chi, H., Li, C.-W. & Davidson, E.H. (2004): Small bilaterian fossils from 40 to 55 million years before the Cambrian. *Science* **305**: 218–22.

Chen, J.-Y., Ramsköld, L. & Zhou, G.-Q. (1994): Evidence for monophyly and arthropod affinity of Cambrian giant predators. *Science* **264**: 1304–8.

Chen, J.-Y., Zhou, G.-Q. & Ramsköld, L. (1995): A new early Cambrian onychophoran-like animal, *Paucipodia* gen. nov., from the Chengjiang fauna, China. *Trans. R. Soc. Edinburgh Earth Sci.* **85**: 275–82.

Chen, L., DeVries, A.L. & Cheng, C.-H.C. (1997): Convergent evolution of antifreeze glycoproteins in Antarctic notothenoid fish and arctic cod. *Proc. Natl. Acad. Sci. USA* **94**: 3811–16.

Chia, F.-S. & Koss, R. (1979): Fine structural studies of the nervous system and the apical organ in the planula larva of the sea anemone *Anthopleura elegantissima*. *J. Morphol.* **160**: 275–98.

Chia, F.-S. & Koss, R. (1994): Asteroidea. In: *Microscopic Anatomy of Invertebrates*, Harrison, F.W. & Chia, F.-S. (eds.), Vol. 14: Echinodermata. Wiley-Liss, New York: 169–245.

Chia, F.S. & Walker, C.W. (1991): Echinodermata: Asteroidea. In: *Reproduction of Marine Invertebrates*, Giese, A.C., Pearse, J.S. & Pearse, V.B. (eds.). Vol. 6: Echinoderms and lophophorates. The Boxwood Press, Pacific Grove: 301–53.

Chia, F.-S., Amerongen, H.M. & Peteya, D.J. (1984): Ultrastructure of the neuromuscular system of the polyp of *Aurelia aurita* L., 1758 (Cnidaria, Scyphozoa). *J. Morphol.* **180**: 69–79.

Chia, F.-S., Atwood, D. & Crawford, B. (1975): Comparative morphology of echinoderm sperm and possible phylogenetic implications. *Am. Zool.* **15**: 553–65.

Chitwood, B.G. & Chitwood, M.B. (1950): *Introduction to Nematology*. University Park Press, Baltimore.

Chuang, S.H. (1990): Brachiopoda. In: *Reproductive Biology of Invertebrates*, Adiyodi, K.G. & Adiyodi, R.G. (eds.). Vol. 4, part B: fertilization, development, and parental care. John Wiley & Sons, Chichester: 211–54.

Cisternas, P.A. & Byrne, M. (2003): Peptidergic and serotinergic immunoreactivity in the metamorphosing ophiopluteus of *Ophiactis resiliens* (Echinodermata, Ophiuroidea). *Invertebr. Biol.* **122**: 177–85.

Clark, A.W. (1967): The fine structure of the eye of the leech, *Helobdella stagnalis*. *J. Cell Sci.* **2**: 341–48.

Clark, R.B. (1964): *Dynamics in Metazoan Evolution. The Origin of the Coelom and Segments*. Clarendon Press, Oxford.

Clark, R.B. (1980): The nature and origins of metameric segmentation. *Zool. Jahrb. Anat.* **103**: 169–95.

Clarkson, E., Levi-Setti, R. & Horváth, G. (2006): The eyes of trilobites: the oldest preserved visual system. *Arthropod Struct. Dev.* **35**: 247–59.

Clausen, C. (1986): *Microphthalmus ephippiophorus* sp.n. (Polychaeta: Hesionidae) and two other *Microphthalmus* species from the Bergen area, Western Norway. *Sarsia* **71**: 177–91.

Clément, P. (1968): Ultrastructures d'un rotifère: *Notommata copeus*. I. La cellule-flamme. Hypothèses physiologiques. *Z. Zellforsch.* **89**: 478–98.

Clément, P. & Amsellem, J. (1989): The skeletal muscles of rotifers and their innervation. *Hydrobiologia* **186/187**: 255–78.

Clément, P. & Wurdak, E. (1984): Photoreceptors and photoreception in rotifers. In: *Photoreception and Vision in Invertebrates*, Ali, M.A. (ed.). Plenum Press, New York: 241–88.

Clément, P. & Wurdak, E. (1991): Rotifera. In: *Microscopic Anatomy of Invertebrates*, Harrison, F.W. & Ruppert, E.E. (eds.), Vol. 4: Aschelminthes. Wiley-Liss, New York: 219–97.

Cloud, P. (1972): A working model of the primitive earth. *Am. J. Sci.* **272**: 537–48.

Clutton-Brock, T.H. (1991): *The Evolution of Parental Care. Monographs in Behavior and Ecology*. Princeton University Press, Princeton.

Cobb, J.L.S. (1995): The nervous systems of Echinodermata: recent results and new approaches. In: *The Nervous Systems of Invertebrates: An Evolutionary and Comparative Approach*, Breidbach, O. & Kutsch, W. (eds.). Birkhäuser Verlag, Basel: 407–24.

Cobb, J.L.S. & Laverack, M.S. (1967): Neuromuscular systems in echinoderms. In: Echinoderm Biology, Millott, N. (ed.), *Symp. Zool. Soc. London* **20**: 25–51.

Cohen, B.L. (2000): Monophyly of brachiopods and phoronids: reconciliation of molecular evidence with Linnean classification (the subphylum Phoroniformea nov.). *Proc. R. Soc. London B* **267**: 225–31.

Cohen, B.L. & Weydmann, A. (2005): Molecular evidence that phoronids are a subtaxon of brachiopods (Brachiopoda: Phoronata) and that genetic divergence of metazoan phyla began long before the early Cambrian. *Org. Div. Evol.* **5**: 253–73.

Cohen, B.L., Gawthorp, A. & Cavalier-Smith, T. (1998): Molecular phylogeny of brachiopods and phoronids based on nuclear-encoded small subunit ribosomal RNA gene sequences. *Phil. Trans. R. Soc. London B* **353**: 2039–61.

Cohen, B.L., Holmer, L.E. & Lüter, C. (2003): The brachiopod fold: a neglected body plan hypothesis. *Palaeontology* **46**: 59–65.

Cohen, B.L., Stark, S., Gawthorp, A.B., Burke, M.E. & Thayer, C.W. (1998): Comparison of articulate brachiopod nuclear and mitochondrial gene trees lead to a clade-based redefinition of protostomes (Protostomozoa) and deuterostomes (Deuterostomozoa). *Proc. R. Soc. London B* **265**: 475–82.

Colacino, J.M. & Kraus, D.W. (1984): Hemoglobin-containing cells of *Neodasys* (Gastrotricha, Chaetonotoida). II. Respiratory significance. *Comp. Biochem. Physiol.* **79A**: 363–69.

Cole, A.G. & Hall, B.K. (2004): The nature and significance of invertebrate cartilages revisited: distribution and histology of cartilage and cartilage-like tissues within Metazoa. *Zoology* **107**: 261–73.

Colgan, D.J., Hutchings, P.A. & Braune, M. (2006): A multigene framework for polychaete phylogenetic studies. *Org. Div. Evol.* **6**: 220–35.

Collins, A.G. (1998): Evaluating multiple alternative hypotheses for the origin of Bilateria: an analysis of 18S rRNA molecular evidence. *Proc. Natl. Acad. Sci. USA* **95**: 15458–63.

Collins, A.G. (2002): Phylogeny of Medusozoa and the evolution of cnidarian life cycles. *J. Evol. Biol.* **15**: 418–32.

Collins, A.G., Schuchert, P., Marques, A.C., Jankowski, T., Medina, M. & Schierwater, B. (2006): Medusozoan phylogeny and character evolution clarified by new large and small subunit rDNA data and an assessment of the utility of phylogenetic mixture models. *Syst. Biol.* **55**: 97–115.

Conway Morris, S. (1977): Fossil priapulid worms. *Spec. Papers Palaeont.* **20**: 1–95.

Conway Morris, S. (1997): The cuticular structure of the 495-Myr-old type species of the fossil worm *Palaeoscolex*, *P. piscatorum* (?Priapulida). *Zool. J. Linn. Soc.* **119**: 69–82.

Conway Morris, S. (1998): *The Crucible of Creation. The Burgess Shale and the Rise of Animals*. Oxford University Press, Oxford.

Conway Morris, S. (2003): *Life's Solution. Inevitable Humans in a Lonely Universe*. Cambridge University Press, Cambridge.

Conway Morris, S. & Peel., J.S. (1995): Articulated halkieriids from the Lower Cambrian of North Greenland and their role in early protostome evolution. *Phil. Trans. R. Soc. London B* **347**: 305–58.

Cook, C.E., Jiménez, E., Akam, M. & Saló, E. (2004): The hox gene complement of acoel flatworms, a basal bilaterian clade. *Evol. Dev.* **6**: 154–63.

Cook, C.E., Smith, M.L., Telford, M.J., Bastianello, A. & Akam, M. (2001): Hox genes and the phylogeny of the arthropods. *Curr. Biol.* **11**: 759–63.

Coomans, A. (1979): The anterior sensilla of nematodes. *Rev. Nematol.* **2**: 259–83.

Coons, L.B. & Alberti, G. (1999): Acari: ticks. In: *Microscopic Anatomy of Invertebrates*, Harrison, F.W. & Foelix, R.F. (eds.), Vol. 8B: Chelicerate Arthropoda. Wiley-Liss, New York: 267–514.

Copley, R.R., Aloy, P., Russell, R.B. & Telford, M.J. (2004): Systematic searches for molecular synapomorphies in model metazoan genomes give some support for Ecdysozoa after accounting for the idiosyncrasies of *Caenorhabditis elegans*. *Evol. Dev.* **6**: 164–69.

Coppi, A., Merali, S. & Eichinger, D. (2002): The enteric parasite *Entamoeba* uses an autocrine catecholamine system during differentiation into the infectious cyst stage. *J. Biol. Chem.* **277**: 8083–90.

Corbelli, A., Avian, M., Marotta, R. & Ferraguti, M. (2003): The spermatozoon of *Carybdaea marsupialis* (Cubozoa, Cnidaria). *Invertebr. Reprod. Dev.* **43**: 95–104.

Cormier, S.M. & Hessinger, D.A. (1980): Cnidocil apparatus: sensory receptor of *Physalia* nematocytes. *J. Ultrastruct. Res.* **72**: 13–19.

Corrêa, D.D. (1948): A embriologia de *Bugula flabellata* (J.V. Thompson) (Bryozoa Ectoprocta). *Bolm Fac. Filos. Ciênc. Univ. Sao Paolo, Zool.* **13**: 7–71.

Cracraft, J. & Donoghue, M.J. (2004): *Assembling the Tree of Life*. Oxford University Press, Oxford.

Cripps, A.P. (1991): A cladistic analysis of the cornutes (stem chordates). *Zool. J. Linn. Soc.* **102**: 333–66.

Croll, N.A., Evans, A.A.F. & Smith, J.M. (1975): Comparative nematode photoreceptors. *Comp. Biochem. Physiol. A* **51**: 139–43.

Crompton, D.W.T. (1985): Reproduction. In: *Biology of the Acanthocephala*, Crompton, D.W.T. & Nickol, B.B. (eds.). Cambridge University Press, Cambridge: 213–71.

Crompton, D.W.T. (1989): Acanthocephala. In: *Reproductive Biology of Invertebrates*, Adiyodi, K.G. & Adiyodi, R.G. (eds.). Vol. 4, part A: Fertilization, development, and parental care. John Wiley & Sons, Chichester: 251–58.

Crompton, D.W.T. (1999): Acanthocephala. In: *Encyclopedia of Reproduction*, Knobil, E. & Neill, J.D. (eds.). Vol. 1. Academic Press, San Diego: 6–16.

Crompton, D.W.T. & Lee, D.L. (1965): The fine structure of the body wall of *Polymorphus minutus* (Goetze, 1782) (Acanthocephala). *Parasitology* **55**: 357–64.

Culioli, J.-L., Foata, J., Mori, C. & Marchant, B. (2006a): Spermiogenesis and sperm in *Mesostoma viareggimum* (Plathelminthes, Rhabdocoela): an ultrastructural study to infer sperm orientation. *Zoomorphology* **125**: 47–56.

Culioli, J.-L., Foata, J., Mori, C., Orsini, A. & Marchant, B. (2006b): Ultrastructure of spermiogenesis and spermatozoa in *Mesocastrada fuhrmanni* Voltz, 1898 (Platyhelminthes, Rhabdocoela, Typhloplanoida). *Zool. Anz.* **245**: 3–12.

Czaker, R. (2000): Extracellular matrix (ECM) components in a very primitive multicellular animal, the dicyemid mesozoan *Kantharella antarctica*. *Anat. Rec.* **259**: 52–59.

Czaker, R. & Janssen, H.H. (1998): Outer extracellular matrix (ECM) in the dicyemid mesozoan *Kantharella antarctica*. *J. Submicrosc. Cytol. Pathol.* **30**: 349–53.

Damen, W.G.M. (2002): Parasegmental organization of the spider embryo implies that the paragegment is an evolutionary conserved entity in arthropod embryogenesis. *Development* **129**: 1239–50.

Damen, P. & Dictus, W.J.A.G. (1994): Cell lineage of the prototroch of *Patella vulgata* (Gastropoda, Mollusca). *Dev. Biol.* **162**: 364–83.

Damen, W.G.M., Hausdorf, M., Seyfarth, E.-A. & Tautz, D. (1998): A conserved mode of head segmentation in arthropods revealed by the expression pattern of Hox genes in a spider. *Proc. Nat. Acad. Sci. USA* **95**: 10665–70.

Damen, W.G.M., Klerkx, A.H.E.M. & van Loon, A.E. (1996): Micromere formation at third cleavage is decisive for trochoblast specification in the embryogenesis of *Patella vulgata*. *Dev. Biol.* **178**: 238–50.

Dautov, S.S. & Nezlin, L.P. (1992): Nervous system of the tornaria larva (Hemichordata: Enteropneusta). A histochemical and ultrastructural study. *Biol. Bull.* **183**: 463–75.

Davey, K.G. (1999): Insect reproduction, overview. In: *Encyclopedia of Reproduction*, Knobil, E. & Neill, J.D. (eds.). Vol. 2. Academic Press, San Diego: 845–52.

Davidson, E.H. (2001): *Genomic Regulatory Systems. Development and Evolution*. Academic Press, San Diego.

Davis, F.C. (1989): Echiura. In: *Reproductive Biology of Invertebrates*, Adiyodi, K.G. & Adiyodi, R.G. (eds.). Vol. 4, part A: fertilization, development, and parental care. John Wiley & Sons, Chichester: 349–81.

Davis, G.K. & Patel, N.H. (1999): The origin and evolution of segmentation. *Trends Gen.* **15**: M68-M72.

De Eguileor, M. & Valvassori, R. (1977): Studies on the helical and paramyosinic muscles. VII. Fine structure of body wall muscles in *Sipunculus nudus*. *J. Submicrosc. Cytol.* **9**: 363–72.

De Ley, P. & Blaxter, M. (2002) : Systematic position and phylogeny. In: *The Biology of Nematodes*, Lee, D.L. (ed.). Taylor & Francis, London: 1–30.

De Robertis, E.M. (1997): The ancestry of segmentation. *Nature* **387**: 25–26.

De Robertis, E.M. & Sasai, Y. (1996): A common plan for dorsoventral patterning in Bilateria. *Nature* **380**: 37–40.

De Rosa, R. (2001): Molecular data indicate the protostome affinity of brachiopods. *Syst. Biol.* **50**: 848–59.

De Rosa, R., Grenier, J.K., Andreeva, T., Cook, C.E., Adoutte, A., Akam, M., Carroll, S.B. & Balavoine, G. (1999): Hox genes in brachiopods and priapulids and protostome evolution. *Nature* **399**: 772–76.

De Villafranca, G.W. & Haines, V.E. (1974): Paramyosin from arthropod cross-striated muscle. *Comp. Biochem. Physiol. B* **47**: 9–26.

De Villalobos, C., Restelli, M., Schmidt-Rhaesa, A. & Zanca, F. (2005) : Ultrastructural observations of the testicular epithelium and spermatozoa of *Pseudochordodes bedriagae* (Gordiida, Nematomorpha). *Cell Tiss. Res.* **321**: 251–55.

De Zwaan, A. (1991): Molluscs. In: *Metazoan Life Without Oxygen*, Bryant, C. (ed.). Chapman and Hall, London: 186–217.

Degnan, B.M., Degnan, S.M., Giusti, A. & Morse, D.E. (1995): A hox/hom homeobox gene in sponges. *Gene* **155**: 175–77.

Dehal, P. and 86 other authors (2002): The draft genome of *Ciona intestinalis*: insights into chordate and vertebrate origins. *Science* **298**: 2157–67.

Del Cacho, E., Lopez-Bernad, F., Quilez, J. & Sanchez-Acedo, C. (1997): A fibronectin-like molecule expressed by *Eimeria tenella* as a potential coccidial vaccine. *Parasitol. Today* **13**: 405–6.

Delage, Y. (1886): Études histologiques sur les planaires rhabdocoeles acoeles (*Convoluta schultzii* [O. Schm.]). *Arch. Zool. Exper. Gen.* **4**: 109–160.

Delsuc, F., Brinkmann, H., Chourrout, D. & Philippe, H. (2006): Tunicates and not cephalochordates are the closest living relatives of vertebrates. *Nature* **439**: 965–68.

Dennis, R.D. (1976): Insect morphogenetic hormones and developmental mechanisms in the nematode, *Nematospiroides dubius*. *Comp. Biochem. Physiol. A* **53**: 53–56.

Dewel, R.A. (2000): Colonial origin for Eumetazoa: major morphological transitions and the origin of bilaterian complexity. *J. Morphol.* **243**: 35–74.

Dewel, R.A. & Dewel, W.C. (1979): Studies on the tardigrades. IV. Fine structure of the hindgut of *Milnesium tardigradum* Doyere. *J. Morphol.* **161**: 79–109.

Dewel, R.A. & Dewel, W.C. (1996): The brain of *Echiniscus viridissimus* Peterfi, 1956 (Heterotardigrada): a key to understanding the phylogenetic position of the arthropod head. *Zool. J. Linn. Soc.* **116**: 35–49.

Dewel, R.A. & Dewel, W.C. (1997): The place of tardigrades in arthropod evolution. In: Arthropod Relationships, Fortey, R.A. & Thomas, R.H. (eds.). *Syst. Assoc. Spec.* Vol. 55, Chapman & Hall, London: 109–123.

Dewel, R.A., Nelson, D.R. & Dewel, W.C. (1993): Tardigrada. In: *Microscopic Anatomy of Invertebrates*, Harrison, F.W. & Rice, M.E. (eds.), Vol. 12: Onychophora, Chilopoda, and lesser Protostomata. Wiley-Liss, New York: 143–83.

Dewilde, S., Angelini, E., Kiger, L., Marden, M.C., Beltramini, M., Salvato, B. & Moens, L. (2003): Structure and function of the globin and globin genes from the Antarctic mollusc *Yoldia eightsi*. *Biochem. J.* **370**: 245–53.

Dewilde, S., Blaxter, M., van Hauwaert, M.-L., Vanfleteren, J., Esmans, E.L., Marden, M., Griffon, N. & Moens, L. (1996): Globin and globin gene structure of the nerve myoglobin of *Aphrodite aculeata*. *J. Biol. Chem.* **271**: 19865–70.

Di Prisco, G. (1998): Molecular adaptations of Antarctic fish hemoglobins. In: *Fishes of Antarctica*, Di Prisco, G., Pisano, E. & Clarke, A. (eds.). Springer, Berlin: 339–53.

Diehl, P.A., Aeschlimann, A. & Obenchain, F.D. (1982): Tick reproduction: oogenesis and oviposition. In: *Physiology of Ticks*, Obenchain, F.D. & Galun, R. (eds.). Pergamon Press, New York: 277–350.

Dilly, P.N. (1969): The structure of a photoreceptor organelle in the eye of *Pterotrochea mutica*. *Z. Zellforsch.* **99**: 420–29.

Dilly, P.N. (1973): The larva of *Rhabdopleura compacta* (Hemichordata). *Mar. Biol.* **18**: 69–86.

Dilly, P.N. (1975): The pterobranch *Rhabdopleura compacta*: its nervous system and phylogenetic position. In: *Protochordates*, Barrington, E.J.W. & Jefferies, R.P.S. (eds.). *Symp. Zool. Soc. London* **36**: 1–16.

Dilly, P.N. & Wolken, J.J. (1973): Studies on the receptors in *Ciona intestinalis*. IV. The ocellus in the adult. *Micron* **4**: 11–29.

Dilly, P.N., Welsch, U. & Rehkämper, G. (1986): Fine structure of heart, pericardium and glomerular vessel in *Cephalodiscus gracilis* M'Intosh, 1882 (Pterobranchia, Hemichordata). *Acta Zool.* **67**: 173–79.

Dilly, P.N., Welsch, U. & Storch, V. (1970): The structure of the nerve fibre layer and neurocord in the enteropneusts. *Z. Zellforsch.* **103**: 129–48.

Dodge, J.D. (1991): Photosensory systems in eukaryotic algae. In: Vision and visual dysfunction, Cronly-Dillon, J.R. & Gregory, R.L. (eds.). Vol. 2: *Evolution of the Eye and Visual System.* Macmillan Press, Houndmills: 323–40.

Doe, D.A. (1981): Comparative ultrastructure of the pharynx simplex in Turbellaria. *Zoomorphology* **97**: 133–93.

Dohle, W. (1967): Zur Morphologie und Lebensweise von *Ophryotrocha gracilis* Huth 1934 (Polychaeta, Eunicidae). *Kieler Meeresforsch.* **23**: 68–74.

Dohle, W. (2001): Are the insects terrestrial crustaceans? A discussion of some new facts and arguments and the proposal of a proper name 'Tetraconata' for the monophyletic unit Crustacea + Hexapoda. In: Origin of the Hexapoda, Deuve, T. (ed.). *Ann. Soc. Entomol. Fr.* **37**: 85–103.

Dohle, W. (2004): Die Verwandtschaftsbeziehungen der Großgruppen der Deuterostomier: Alternative Hypothesen und ihre Begründungen. In: Kontroversen in der Phylogenetischen Systematik der Metazoa, Richter, S & Sudhaus, W. (eds.). *Sitzungsber. Ges. Naturforsch. Freunde Berlin* **43**: 123–62.

Dohmen, M.R. (1983): Gametogenesis. In: *The Mollusca*, Wilbur, K.M. (ed.). Vol. 3: Development, Verdonk, N.H., van den Biggelaar, J.A.M. & Tompa, A.S. (volume eds.). Academic Press, New York: 1–48.

Donaldson, S., Mackie, G.O. & Roberts, A. (1980): Preliminary observations on escape swimming and giant neurons in *Aglantha digitale* (Hydromedusae: Trachylina). *Can. J. Zool.* **58**: 549–52.

Dong, X.-P., Donoghue, P.C.J., Cheng, H. & Liu, J.-B. (2004): Fossil embryos from the middle and late Cambrian period of Hunan, South China. *Nature* **427**: 237–240.

Dong, X.-P., Donoghue, P.C.J., Cunningham, J.A., Liu, J.-B. & Cheng, H. (2005): The anatomy, affinity, and phylogenetic significance of *Markuelia*. *Evol. Dev.* **7**: 468–82.

Donoghue, P.C.J., Kouchinsky, A., Waloszek, D., Bengtson, S., Dong, X.-P., Val'kov, A.K., Cunningham, J.A. & Repetski, J.E. (2006): Fossilized embryos are widespread but the record is temporally and taxonomically biased. *Evol. Dev.* **8**: 232–38.

Dopazo, H. & Dopazo, J. (2005): Genome-scale evidence of the nematode-arthropod clade. *Genome Biol.* **6**: R41.1–10.

Ducret, F. (1978): Particularités structurales du système optique chez deux chaetognathes (*Sagitta tasmanica* et *Eukrohnia hamata*) et incidentces phylogénétiques. *Zoomorphologie* **91**: 201–15.

Dunagan, T.T. & Miller, D.M. (1983): Apical sense organ of *Macracanthorhynchus hirudinaceus* (Acanthocephala). *J. Parasitol.* **69**: 897–902.

Dunagan, T.T. & Miller, D.M. (1986a): Ultrastructure of flame bulbs in male *Macracanthorhynchus hirudinaceus* (Acanthocephala). *Proc. Helminthol. Soc. Washington* **53**: 102–9.

Dunagan, T.T. & Miller, D.M. (1986b): A review of protonephridial excretory systems in Acanthocephala. *J. Parasitol.* **72**: 621–32.

Dunagan, T.T. & Miller, D.M. (1991): Acanthocephala. In: *Microscopic Anatomy of Invertebrates*, Harrison, F.W. & Ruppert, E.E. (eds.), Vol. 4: Aschelminthes. Wiley-Liss, New York: 299–332.

Dunagan, T.T. & Rashed, R.-M.A. (1988): Capsular protonephridia in male *Oligacanthorhynchus atrata* (Acanthocephala). *J. Parasitol.* **74**: 180–185.

Duncker, H.-R. (1971): The lung air sac system of birds. *Adv. Anat. Embryol. Cell Biol.* **45**: 1–171.

Duncker, H.-R. (2000): Der Atemapparat der Vögel und ihre lokomotorische und metabolische Leistungsfähigkeit. *J. Ornithologie* **141**: 1–67.

Duncker, H.-R. (2001): The emergence of macroscopic complexity. An outline of the history of the respiratory apparatus of vertebrates from diffusion to language production. *Zoology* **103**: 240–59.

Duncker, H.-R. (2004): Vertebrate lungs: structure, topography and mechanics. A comparative perspective of the progressive integration of respiratory system, locomotor apparatus and ontogenetic development. *Respir. Physiol. Neurobiol.* **144**: 111–24.

Dunlop, J. & Arango, C.P. (2004): Pycnogonid affinities: a review. *J. Zool. Syst. Evolut. Res.* **43**: 8–21.

Dunn, E.F., Moy, V.N., Angerer, L.M., Angerer, R.C., Morris, R.L. & Peterson, K.J. (2007): Molecular paleoecology: using gene regulatory analysis to address the origins of complex life cycles in the late Precambrian. *Evol. Dev.* **9**: 10–24.

Duvert, M. (1991): A very singular muscle: the secondary muscle of chaetognaths. *Phil. Trans. R. Soc. London B* **332**: 245–60.

Duvert, M. & Savineau, J.-P. (1986): Ultrastructural and physiological studies of the contraction of the trunk musculature of *Sagitta setosa* (Chaetognath). *Tiss. Cell* **18**: 937–52.

Dybas, L. (1976): A light an electron microscopic study of the ciliated urn of *Phascolosoma agassizii* (Sipunculida). *Cell Tiss. Res.* **169**: 67–75.

Dybas, L. (1981): Sipunculans and echiuroids. In: *Invertebrate Blood Cells*, Ratcliff, N.A. & Rowley, A.F. (eds.). Academic Press, New York: 161–88.

Eakin, R.M. (1963): Lines of evolution of photoreceptors. In: *General Physiology of Cell Specialization*, Mazia, D. & Tyler, A. (eds.). McGraw Hill, New York: 393–425.

Eakin, R.M. (1965): Evolution of Photoreceptors. *Cold Spring Harbor Symp. Quant. Biol.* **30**: 363–70.

Eakin, R.M. (1968): Evolution of photoreceptors. *Evol. Biol.* **2**: 194–242.

Eakin, R.M. (1973): *The Third Eye*. University of California Press, Berkeley.

Eakin, R.M. (1979): Evolutionary significance of photoreceptors: in retrospect. *Am. Zool.* **19**: 647–53.

Eakin, R.M. (1982): Continuity and diversity in photoreceptors. In: *Visual Cells in Evolution*, Westfall, J.A. (ed.). Raven Press, New York: 91–105.

Eakin, R.M. & Brandenburger, J.L. (1979): Effects of light on ocelli of seastars. *Zoomorphologie* **92**: 191–200.

Eakin, R.M. & Brandenburger, J.L. (1981a): Unique eye of probable evolutionary significance. *Science* **211**: 1189–90.

Eakin, R.M. & Brandenburger, J.L. (1981b): Fine structure of the eyes of *Pseudoceros canadensis* (Turbellaria, Polycladida). *Zoomorphology* **98**: 1–16.

Eakin, R.M. & Hermans, C.O. (1988): Eyes. In: The Ultrastructure of Polychaeta, Westheide, W. & Hermans, C.O. (eds.). *Microfauna Marina* **4**: 135–56.

Eakin, R.M. & Kuda, A. (1971): Ultrastructure of sensory receptors in ascidian tadpoles. *Z. Zellforsch.* **112**: 287–312.

Eakin, R.M. & Westfall, J.A. (1964): Fine structure of the eye of a chaetognath. *J. Cell Biol.* **21**: 115–132.

Eakin, R.M. & Westfall, J.A. (1965): Fine structure of the eye of *Peripatus* (Onychophora). *Z. Zellforsch.* **68**: 278–300.

Eastman, J.T., DeVries, A.L., Coalson, R.E., Nordquist, R.E. & Boyd, R.B. (1979): Renal conservation of antifreeze peptide in Antarctic eelpout, *Rhigophila dearborni*. *Nature* **282**: 217–18.

Ebner, B., Burmester, T. & Hankeln, T. (2003): Globin genes are present in *Ciona intestinalis*. *Mol. Biol. Evol.* **20**: 1521–25.

Ebnet, E., Fischer, M., Deininger, W. & Hegemann, P. (1999): Volvoxrhodopsin, a light-regulated sensory photoreceptor of the spheroidal green alga *Volvox carteri*. *Plant Cell* **11**: 1473–84.

Eckelbarger, K.J. (1984): Ultrastructure of spermatogenesis in the reef-building polychaete *Phragmatopoma lapidosa* (Sabellariidae) with special reference to acrosome morphogenesis. *J. Ultrastruct. Res.* **89**: 146–64.

Eckelbarger, K.J. (1988): Oogenesis and female gametes. In: The Ultrastructure of Polychaeta, Westheide, W. & Hermans, C.O. (eds.). *Microfauna Marina* **4**: 281–307.

Eckelbarger, K.J. (1992): Polychaeta: oogenesis. In: *Microscopic Anatomy of Invertebrates*, Harrison, F.W. & Gardiner, S.L. (eds.), Vol. 7: Annelida. Wiley-Liss, New York: 109–127.

Eckelbarger, K.J. (2005): Oogenesis and oocytes. In: Morphology, Molecules, Evolution and Phylogeny in Polychaeta and Related Taxa, Bartolomaeus, T. & Purschke, G. (eds.). *Hydrobiologia* **535/536**: 179–98.

Edgecombe, G.D., Wilson, G.D.F., Colgan, D.J., Gray, M.R. & Cassis, G. (2000): Arthropod cladistics: combined analysis of histone H3 and U2 snRNA sequences and morphology. *Cladistics* **16**: 155–203.

Eeckhaut, I., Fievez, L. & Müller, M.C.M. (2003): Larval development of *Myzostoma cirriferum* (Myzostomida). *J. Morphol.* **258**: 269–83.

Eeckhaut, I. & Jangoux, M. (1993): Life cycle and mode of infestation of *Myzostoma cirriferum* (Annelida), a symbiotic myzostomid of the comatulid crinoid *Antedon bifida* (Echinodermata). *Dis. Aquat. Org.* **15**: 207–17.

Eeckhaut, I. & Lanterbecq, D. (2005): Myzostomida: a review of the phylogeny and ultrastructure. In: Morphology, Molecules, Evolution and Phylogeny in Polychaeta and Related Taxa, Bartolomaeus, T. & Purschke, G. (eds.). *Hydrobiologia* **535/536**: 253–75.

Eeckhaut, I., McHugh, D., Mardulyn, P., Tiedemann, R., Monteyne, D., Jangoux, M. & Milinkovitch, M.C. (2000): Myzostomida: a link between trochozoans and flatworms? *Proc. R. Soc. London B* **267**: 1383–92.

Eernisse, D.J. (1997): Arthropod and annelid relationships re-examined. In: Arthropod Relationships, Fortey, R.A. & Thomas, R.H. (eds.). *Syst. Assoc. Spec. Vol.* **55**, Chapman & Hall, London: 43–56.

Eernisse, D.J. & Peterson, K.J. (2004): The history of animals. In: *Assembling the Tree of Life*, Cracraft, J. & Donoghue, M.J. (eds.). Oxford University Press, Oxford: 197–208.

Eernisse, D.J. & Reynolds, P.D. (1994): Polyplacophora. In: *Microscopic Anatomy of Invertebrates*, Harrison, F.W. & Kohn, A.J. (eds.), Vol. 5: Mollusca I. Wiley-Liss, New York: 55–110.

Eernisse, D.J., Albert, J.S. & Anderson, F.E. (1992): Annelida and Arthropoda are not sister taxa: a phylogenetic analysis of spiralian metazoan morphology. *Syst. Biol.* **41**: 305–30.

Ehlers, U. (1985a): *Das Phylogenetische System der Plathelminthes*. Gustav Fischer Verlag, Stuttgart.

Ehlers, U. (1985b): Cytoskelette bei freilebenden Plathelminthen. *Verh. Dtsch. Zool. Ges.* **78**: 161.

Ehlers, U. (1985c): Organisation der Statocyste von *Retronectes* (Catenulida, Plathelminthes). *Microfauna Marina* **2**: 7–22.

Ehlers, U. (1991): Comparative morphology of statocysts in the Plathelminthes and the Xenoturbellida. *Hydrobiologia* **227**: 263–71.

Ehlers, U. (1992a): On the fine structure of *Paratomella rubra* Rieger & Ott (Acoela) and the position of the taxon *Paratomella* Dörjes in a phylogenetic system of the Acoelomorpha (Plathelminthes). *Microfauna Marina* **7**: 265–93.

Ehlers, U. (1992b): Dermonephridia – modified epidermal cells with a probable excretory function in *Paratomella rubra* (Acoela, Plathelminthes). *Microfauna Marina* **7**: 253–64.

Ehlers, U. (1993): Ultrastructure of the spermatozoa of *Halammohydra schulzei* (Cnidaria, Hydrozoa): the significance of acrosomal structures for the systematization of the Eumetazoa. *Microfauna Marina* **8**: 115–30.

Ehlers, U. (1994a): Absence of a pseudocoel or pseudocoelom in *Anoplostoma vivipara* (Nematodes). *Microfauna Marina* **9**: 345–50.

Ehlers, U. (1994b): Ultrastructure of the unusual bodywall musculature of *Anaperus tvaerminnensis* (Acoela, Plathelminthes). *Microfauna Marina* **9**: 291–301.

Ehlers, U. (1994c): On the ultrastructure of the protonephridium of *Rhynchoscolex simplex* and the basic systematization of the Catenulida (Plathelminthes). *Microfauna Marina* **9**: 157–69.

Ehlers, U. & Ehlers, B. (1977): Monociliary receptors in interstitial Proseriata and Neorhabdocoela (Turbellaria Neoophora). *Zoomorphologie* **86**: 197–222.

Ehlers, U. & Sopott-Ehlers, B. (1997): Ultrastructure of the subepidermal musculature of *Xenoturbella bocki*, the adelphotaxon of the Bilateria. *Zoomorphology* **117**: 71–79.

Ehlers, U., Ahlrichs, W., Lemburg, C. & Schmidt-Rhaesa, A. (1996): Phylogenetic systematization of the Nemathelminthes (Aschelminthes). *Verh. Dtsch. Zool. Ges.* **89.1**: 8.

Eibye-Jacobsen, J. (1996): New observations on the embryology of the Tardigrada. *Zool. Anz.* **235**: 201–16.

Eibye-Jacobsen, D. & Kristensen, R.M. (1994): A new genus and species of Dorvilleidae (Annelida, Polychaete) from Bermuda, with a phylogenetic analysis of Dorvilleidae, Iphitimidae and Dinophilidae. *Zool. Scr.* **23**: 107–31.

Eichelberg, D. & Wessing, A. (1975): Morphology of the malpighian tubules of insects. *Fortschr. Zool./Progr. Zool.* **23**: 124–47.

Eichinger, D., Coppi, A., Frederick, J. & Merali, S. (2002): Catecholamines in Entamoebae: recent (re)discoveries. *J. Biosci.* **27**, Suppl. **3**: 589–93.

Ellenby, C. & Smith, L. (1966): Haemoglobin in *Mermis subnigrescens* (Cobb), *Enoplus brevis* (Bastian) and *E. communis* (Bastian). *Comp. Biochem. Physiol.* **19**: 871–77.

Elofsson, R. (2006): The frontal eyes of crustaceans. *Arthropod Struct. Dev.* **35**: 275–91.

Emig, C.C. (1990): Phoronida. In: *Reproductive Biology of Invertebrates*, Adiyodi, K.G. & Adiyodi, R.G. (eds.). Vol. 4, part B: fertilization, development, and parental care. John Wiley & Sons, Chichester: 165–184.

Emlet, R.B. (1994): Body form and patterns of ciliation in nonfeeding larvae of echinoderms: functional solutions to swimming in the plancton? *Am. Zool.* **34**: 570–585.

Emschermann, P. (1965): Das Protonephridialsystem von *Urnatella gracilis* Leidy (Kamptozoa). *Z. Morphol. Ökol. Tiere* **55**: 859–914.

Emschermann, P. (1969): Ein Kreislauforgan bei Kamptozoen. *Z. Zellforsch.* **97**: 576–607.

Engel, J., Efimov, V.P. & Maurer, P. (1994): Domain organizations of extracellular matrix proteins and their evolution. *Development* **1994 Suppl.**: 35–42.

Epstein, H.F., Miller, D.N., Ortiz, I. & Berliner, G.C. (1985): Myosin and paramyosin are organized about a newly identified core structure. *J. Cell Biol.* **100**: 904–15.

Erber, A., Riemer, D., Bovenschulte, M. & Weber, K. (1998): Molecular phylogeny of metazoan intermediate filament proteins. *J. Mol. Evol.* **47**: 751–62.

Ereskovsky, A.V. & Dondua, A.K. (2006): The problem of germ layers in sponges (Porifera) and some issues concerning early metazoan evolution. *Zool. Anz.* **245**: 65–76.

Eriksson, B.J. & Budd, G.E. (2000): Onychophoran cephalic nerves and their bearing on our understanding of head segmentation and stem-group evolution of Arthropoda. *Arthr. Struct. Dev.* **29**: 197–209.

Erwin, D.H. & Davidson, E.H. (2002): The last common bilaterian ancestor. *Development* **129**: 3021–32.

Eshleman, W.P., Wilkens, J.L. & Cavey, M.J. (1982): Electrophoretic and electron microscopic examination of the adductor and diductor muscles of an articulate brachiopod, *Terebratulina transversa*. *Can. J. Zool.* **60**: 550–59.

Espeel, M. (1985): Fine structure of the statocyst sensilla of the mysid shrimp *Neomysis integer* (Leach, 1814) (Crustacea, Mysidacea). *J. Morphol.* **186**: 149–65.

Evans, G.O. (1992): *Principles of Acarology*. CABI International, Wallingford.

Exposito, J.-Y., Le Guellec, D., Lu, Q. & Garrone, R. (1991): Short chain collagens in sponges are encoded by a family of closely related genes. *J. Biol. Chem.* **266**: 21923–28.

Extavour, C.G. & Akam, M. (2003): Mechanisms of germ cell specification across the metazoans: epigenesis and preformation. *Development* **130**: 5869–84.

Fänge, R. (1950): Haemerythrin in *Priapulus caudatus* Lam. *Nature* **165**: 613–14.

Fänge, R. (1969): Gastrotricha, Kinorhyncha, Rotatoria, Kamptozoa, Nematomorpha, Nemertinea, Priapuloidea. In: *Chemical Zoology*, Florkin, M. & Scheer, B.T. (eds.). Vol. 3. Academic Press, New York: 593–609.

Fänge, R. & Åkesson, B. (1951): The cells of the coelomic fluid of priapulides and their content of haemerythrin. *Ark. Zool.* **3**: 25–31.

Fänge, R. & Mattisson, A. (1961): Function of the caudal appendage of *Priapulus caudatus*. *Nature* **190**: 1216–17.

Fahrenbach, W.H. (1969): The morphology of the eyes of *Limulus*. II. Ommatidia of the compound eye. *Z. Zellforsch.* **93**: 451–83.

Fahrenbach, W.H. (1973): Spermiogenesis in the horseshoe crab, *Limulus polyphemus*. *J. Morphol.* **140**: 31–52.

Fahrenbach, W.H. (1999): Merostomata. In: *Microscopic Anatomy of Invertebrates*, Harrison, F.W. & Foelix, R.F. (eds.), Vol. 8A: Chelicerate Arthropoda. Wiley-Liss, New York: 21–115.

Fain, G.L. (2003): *Sensory Transduction*. Sinauer, Sunderland.

Fanenbruck, M., Harzsch, S. & Wägele, J.W. (2004): The brain of the Remipedia (Crustacea) and an alternative hypothesis on their phylogenetic relationships. *Proc. Natl. Acad. Sci. USA* **101**: 3868–73.

Farley, R.D. (1999): Scorpiones. In: *Microscopic Anatomy of Invertebrates*, Harrison, F.W. & Foelix, R.F. (eds.), Vol. 8A: Chelicerate Arthropoda. Wiley-Liss, New York: 117–222.

Farquhar, M.G. (1982): The glomerular basement membrane – a selective macromolecular filter. In: *Cell Biology of Extracellular Matrix*, Hay, E.D. (ed.). Plenum Press, New York: 335–78.

Fautin, D.G. (1999): Cnidaria. In: *Encyclopedia of Reproduction*, Knobil, E. & Neill, J.D. (eds.). Vol. 1. Academic Press, San Diego: 645–53.

Fautin, D.G. & Mariscal, R.N. (1991): Cnidaria: Anthozoa. In: *Microscopic Anatomy of Invertebrates*, Harrison, F.W. & Westfall, J.A. (eds), Vol. 2: Placozoa, Porifera, Cnidaria, and Ctenophora. Wiley-Liss, New York: 267–358.

Fautin, D.G., Spaulding, J.G. & Chia, F.-S. (1989): Cnidaria. In: *Reproductive Biology of Invertebrates*, Adiyodi, K.G. & Adiyodi, R.G. (eds.). Vol. 4, part A: Fertilization, development, and parental care. John Wiley & Sons, Chichester: 43–62.

Fei, K., Yan, L., Zhang, J. & Sarras, M.P. (2000): Molecular and biological characterization of a zonula occludens-1 homologue in *Hydra vulgaris*, named HZO-1. *Dev. Genes Evol.* **210**: 611–16.

Felgenhauer, B.E. (1999): Araneae. In: *Microscopic Anatomy of Invertebrates*, Harrison, F.W. & Foelix, R.F. (eds.), Vol. 8A: Chelicerate Arthropoda. Wiley-Liss, New York: 223–66.

Felgenhauer, B.E., Abele, L.G. & Felder, D.L. (1992) Remipedia. In: *Microscopic Anatomy of Invertebrates*. Harrison, F.W. & Humes, A.G. (eds.), Vol. 9: Crustacea. Wiley-Liss, New York: 225–47.

Fell, P.E. (1989): Porifera. In: *Reproductive Biology of Invertebrates*, Adiyodi, K.G. & Adiyodi, R.G. (eds.). Vol. 4, part A: Fertilization, development, and parental care. John Wiley & Sons, Chichester: 1–41.

Fell, P.E. (1997): Porifera, the sponges. In: *Embryology – Constructing the Organism*, Gilbert, S.F. & Raunio, A.M. (eds.). Sinauer, Sunderland: 39–54.

Fell, P.E. (1999): Porifera. In: *Encyclopedia of Reproduction*, Knobil, E. & Neill, J.D. (eds.). Vol. 1. Academic Press, San Diego: 938–45.

Felsenstein, J. (1978): Cases in which parsimony or compatibility methods will be positively misleading. *Syst. Zool.* **27**: 401–10.

Fenchel, T. & Finlay, B.J. (1995): *Ecology and Evolution in Anoxyc Worlds*. Oxford University Press, Oxford.

Fenchel, T. & Ockelmann, K.W. (2002): Larva on a string. *Ophelia* **56**: 171–78.

Fenchel, T. & Riedl, R.J. (1970): The sulfide system: a new biotic community underneath the oxidized layer of marine sand bottoms. *Mar. Biol.* **7**: 255–68.

Fernández, J., Téllez, V. & Olea, N. (1992): Hirudinea. In: *Microscopic Anatomy of Invertebrates*, Harrison, F.W. & Gardiner, S.L. (eds.), Vol. 7: Annelida. Wiley-Liss, New York: 323–394.

Ferraguti, M. (1984): The comparative ultrastructure of sperm flagella central sheath in Clitellata reveals a new autapomorphy of the group. *Zool. Scr.* **13**: 201–7.

Ferraguti, M. (2000): Euclitellata. In: *Reproductive Biology of Invertebrates*, Adiyodi, K.G., Adiyodi, R.G. & Jamieson, B.G.M. (eds.). Vol. 9, part B: Progress in male gamete ultrastructure and phylogeny. John Wiley & Sons, Chichester: 125–82.

Ferraguti, M. & Balsamo, M. (1995): Comparative spermatology of Gastrotricha. In: Advances in Spermatozoal Phylogeny and Taxonomy, Jamieson, B.G.M., Ausio, J. & Justine, J.-L. (eds.) *Mém. Mus. Natn. Hist. Nat.* **166**: 105–17.

Ferraguti, M. & Erséus, C. (1999): Sperm types and their use for a phylogenetic analysis of aquatic clitellates. *Hydrobiologia* **402**: 225–37.

Ferraguti, M., Fascio, U. & Boi, S. (2002): Mass production of basal bodies in paraspermiogenesis of Tubificinae (Annelida, Oligochaeta). *Biol. Cell* **94**: 109–15.

Ferraguti, M. & Garbelli, C. (2006): The spermatozoon of a 'living fossil': *Tubiluchus troglodytes* (Priapulida). *Tiss. Cell* **38**: 1–6.

Ferraguti, M. & Melone, G. (1999): Spermiogenesis in *Seison nebaliae* (Rotifera, Seisonidea): further evidence of a rotifer-acanthocephalan relationship. *Tiss. Cell* **31**: 428–40.

Ferraguti, M., Balsamo, M. & Fregni, E. (1995): The spermatozoa of three species of Xenotrichulidae (Gastrotricha, Chaetonotoida): the two 'dünne Nebengeisseln' of spermatozoa in *Heteroxenotrichula squamosa* are peculiar para-acrosomal bodies. *Zoomorphology* **115**: 151–59.

Ferraris, J.D. (1985): Putative neuroendocrine devices in the Nemertinea – an overview of structure and function. *Am. Zool.* **25**: 73–85.

Ferrero, E. (1973): A fine structural analysis of the statocyst in Turbellaria Acoela. *Zool. Scr.* **2**: 5–16.

Ferrero, E.A. & Bedini, C. (1991): Ultrastructural aspects of nervous system and statocyst morphogenesis during embryonic development of *Convoluta psammophila* (Turbellaria, Acoela). *Hydrobiologia* **227**: 131–37.

Ferrero, E.A., Bedini, C. & Lanfranchi, A. (1985): An ultrastructural account of otoplanid Turbellaria neuroanatomy. II. The statocyst design: evolutionary and functional implications. *Acta Zool.* **66**: 75–87.

Ferrier, D.E.K. & Holland, P.W.H. (2001): Ancient origin of the Hox gene cluster. *Nature Rev. Gen.* **2**: 33–38.

Field, K.G., Olsen, G.J., Lane, D.J., Giovannoni, J., Ghiselin, M.T., Raff, E.C., Pace, N.R. & Raff, R.A. (1988): Molecular phylogeny of the animal kingdom. *Science* **239**: 748–53.

Filippova, A., Purschke, G., Tzetlin, A.B. & Müller, M.C.M. (2005): Reconstruction of the musculature of *Magelona* cf. *mirabilis* (Magelonidae) and *Prionspio cirrifera* (Spionidae) (Polychaeta, Annelida) by phalloidin labelling and cLSM. *Zoomorphology* **124**: 1–8.

Finnerty, J.R. (2003): The origins of axial patterning in the Metazoa: how old is bilateral symmetry. *Int. J. Dev. Biol.* **47**: 523–29.

Finnerty, J.R. & Martindale, M.Q. (1999): Ancient origins of axial patterning genes: Hox genes and ParaHox genes in the Cnidaria. *Evol. Dev.* **1**: 16–23.

Finnerty, J.R., Master, V.A., Irvine, S., Kourakis, M.J., Warriner, S. & Martindale, M.Q. (1996): Homeobox genes in the Ctenophora: identification of paired-type and Hox homologues in the atentaculate ctenophore, *Beroe ovata*. *Mol. Mar. Biol. Biotech.* **5**: 249–58.

Finnerty, J.R., Pang, K., Burton, P., Paulson, D. & Martindale, M.Q. (2004): Origins of bilateral symmetry: *Hox* and *Dpp* expression in a sea anemone. *Science* **304**: 1335–37.

Firestein, S. (2001): How the olfactory system makes sense of scents. *Nature* **413**: 211–18.

Fischer, A. (1999): Reproductive and developmental phenomena in annelids: a source of exemplary research problems. *Hydrobiologia* **402**: 1–20.

Fischer, F.P. (1978): Photoreceptor cells in chiton aesthetes. *Spixiana* **1**: 209–13.

Fischer, F.P. (1979): Die Ästheten von *Acanthochiton fascicularis* (Mollusca, Polyplacophora). *Zoomorphologie* **92**: 95–106.

Fischer, F.P. (1980): Fine structure of the larval eye of *Lepidochiton cinerea* L. (Mollusca, Polyplacophora). *Spixiana* **3**: 53–57.

Fischer, F.P. & Renner, M. (1978): Die Feinstruktur der Ästheten von *Chiton olivaceus* (Mollusca, Polyplacophora). *Helgol. Wiss. Meeresunters.* **31**: 425–43.

Fischer, F.P., Maile, W. & Renner, M. (1980): Die Mantelpapillen und Stacheln von *Acanthochiton fascicularis* L. (Mollusca, Polyplacophora). *Zoomorphologie* **94**: 121–31.

Fischer, U. (1994): Ultrastructure of spermatogenesis and spermatozoa of *Cephalodasys maximus* (Gastrotricha, Macrodasyida). *Zoomorphology* **114**: 213–25.

Fleming, T.P. & Johnson, M.H. (1988): From egg to epithelium. *Ann. Rev. Cell Biol.* **4**: 459–85.

Flood, P.R. (1966): A peculiar mode of muscular innervation in amphioxus. Light and electron microscopic studies of the so-called ventral roots. *J. Comp. Neurol.* **126**: 181–218.

Flood, P.R. (1968): Structure of the segmental trunk muscle in *Amphioxus*. *Z. Zellforsch.* **84**: 389–416.

Flood, P.R. & Afzelius, B. (1978): The spermatozoon of *Oikopleura dioica* Fol (Larvacea, Tunicata). *Cell Tiss. Res.* **191**: 27–37.

Föll, R.L., Pleyers, A., Lewandovski, G.J., Wermter, C., Hogemann, V. & Paul, R.J. (1999): Anaerobiosis in the nematode *Caenorhabditis elegans*. *Comp. Biochem. Physiol. B* **124**: 269–80.

Follenius, E. (1965): Particularités de structure des spermatozoïdes de *Lampetra planeri*. *J. Ultrastruct. Res.* **13**: 459–68.

Fortey, R. (2000): *Trilobite! Eyewitness to Evolution*. Alfred Knopf, New York.

Fournier, A. & Combes, C. (1978): Structure of photoreceptors of *Polystoma integerrimum* (Platyhelminthes, Monogenea). *Zoomorphologie* **91**: 147–55.

Fournier, D., Estoup, A., Orivel, J., Foucaud, J., Jourdan, H., Le Breton, J. & Keller, L. (2005): Clonal reproduction by males and females in the little fire ant. *Nature* **435**: 1230–34.

Fowler, V.M., McKeown, C.R. & Fischer, R.S. (2006): Nebulin: does it measure up as a ruler? *Curr. Biol.* **16**: R18–R20.

Franc, J.-M. (1973): Etude ultrastructurale de la spermatogenèse du cténaire *Beroe ovata*. *J. Ultrastruct. Res.* **42**: 255–67.

Francis, R. & Waterston, R.H. (1991): Muscle cell attachment in *Caenorhabditis elegans*. *J. Cell Biol.* **114**: 465–79.

François, J. (1998): Labial kidney. In: *Microscopic Anatomy of Invertebrates*, Harrison, F.W. & Locke, M. (eds.). Vol. 11B: Insecta. Wiley-Liss, New York: 831–40.

Franke, H.-D. (1999): Reproduction of the Syllidae (Annelida: Polychaeta). *Hydrobiologia* **402**: 39–55.

Franke, M. (1993): Ultrastructure of the protonephridia in *Loxosomella fauveli*, *Barentsia matushimana* and *Pedicellina cernua*. Implications for the protonephridia in the ground pattern of the Entoprocta (Kamptozoa). *Microfauna Marina* **8**: 7–38.

Fransen, M.E. (1980): Ultrastructure of coelomic organization in annelids. I. Archiannelids and other small polychaetes. *Zoomorphologie* **95**: 235–49.

Fransen, M.E. (1988): Coelomic and vascular systems. In: The Ultrastructure of Polychaetes, Westheide, W. & Hermans, C.O. (eds.). *Microfauna Marina* **4**: 199–213.

Franzén, Å. (1955): Comparative morphological investigations into the spermiogenesis among *Mollusca. Zool. Bidr. Uppsala* **30**: 399–456.

Franzén, Å. (1956): On spermiogenesis, morphology of the spermatozoon, and biology of fertilization among invertebrates. *Zool. Bidr. Uppsala* **31**: 355–482.

Franzén, Å. (1970): Phylogenetic aspects of the morphology of spermatozoa and spermiogenesis. In: *Comparative Spermatology*, Baccetti, B. (ed.). Academic Press, New York: 29–46.

Franzén, Å. (1973): The spermatozoon of *Siboglinum* (Pogonophora). *Acta Zool.* **54**: 179–92.

Franzén, Å. (1976): On the ultrastructure of spermiogenesis of *Flustra foliacea* (L.) and *Triticella korenii* G.O. Sars (Bryozoa). *Zoon* **4**: 19–29.

Franzén, Å. (1977a): Sperm structure with regard to fertilization biology and phylogenetics. *Verh. Dtsch. Zool. Ges.* **1977**: 123–38.

Franzén, Å. (1977b): Ultrastructure of spermatids and spermatozoa in Archiannelida. *Zoon* **5**: 97–105.

Franzén, Å. (1979): A fine structure study on spermiogenesis in the Entoprocta. *J. Submicrosc. Cytol.* **11**: 73–84.

Franzén, Å. (1982): Ultrastructure of spermatids and spermatozoa in the fresh-water bryozoan *Plumatella* (Bryozoa, Phylactolaemata). *J. Submicrosc. Cytol.* **14**: 323–36.

Franzén, Å. (1983): Urochordata. In: *Reproductive Biology of Invertebrates*, Adiyodi, K.G. & Adiyodi, R.G. (eds.). Vol. 2: Spermatogenesis and sperm function. John Wiley & Sons, Chichester: 621–32.

Franzén, Å. (1984): Ultrastructure of spermatids and spermatozoa in the cyclostomatous bryozoan *Tubulipora* (Bryozoa, Cyclostomata). *Zoomorphology* **104**: 140–46.

Franzén, A. & Afzelius, B.A. (1987): The ciliated epidermis of *Xenoturbella bocki* (Platyhelminthes, Xenoturbellida) with some phylogenetic considerations. *Zool. Scr.* **16**: 9–17.

Franzén, Å. & Afzelius, B.A. (1999): Nemertea. In: *Reproductive Biology of Invertebrates*, Adiyodi, K.G., Adiyodi, R.G. & Jamieson, B.G.M. (eds.). Vol. 9, part A: Progress in male gamete ultrastructure and phylogeny. John Wiley & Sons, Chichester: 143–56.

Franzén, Å. & Ahlfors (1980): Ultrastructure of spermatids and spermatozoa in *Phoronis*, phylum Phoronida. *J. Submicrosc. Cytol.* **12**: 585–97.

Franzén, Å. & Ferraguti, M. (1992): Ultrastructure of spermatozoa and spermatids in *Bonellia viridis* and *Hamingia arctica* (Echiura) with some phylogenetic considerations. *Acta Zool.* **73**: 25–31.

Franzén, Å. & Rice, S. (1988): Spermatogenesis, male gametes and gamete interactions. In: The Ultrastructure of Polychaeta, Westheide, W. & Hermans, C.O. (eds.). *Microfauna Marina* **4**: 309–33.

Franzén, Å. & Sensenbaugh, T. (1984): Fine structure of spermiogenesis in the archiannelid *Nerilla antennata* Schmidt. *Vidensk. Meddr. Dansk. Naturh. Foren.* **145**: 23–36.

Franzén, Å. & Sensenbaugh, T. (1988): The spermatozoon of *Geonemertes parasita* Nemertea, (Hoplonemertea) with a note on sperm evolution in the nemerteans. *Int. J. Invertebr. Reprod. Dev.* **14**: 25–36.

Franzén, Å., Woodwick, K.H. & Sensenbaugh, T. (1985): Spermiogenesis and ultrastructure of spermatozoa in *Saxipendium coronatum* (Hemichordata, Enteropneusta), with consideration of their reproduction and dispersal. *Zoomorphology* **105**: 302–7.

Fregni, E., Balsamo, M. & Ferraguti, M. (1999): Morphology of the reproductive system and spermatozoa of *Mesodasys adenotubulatus* (Gastrotricha: Macrodasyida). *Mar. Biol.* **135**: 515–20.

Fretter, V. (1984): Prosobranchs. In: *The Mollusca*, Wilbur, K.M. (ed.), Vol. 7: Reproduction, Tompa, A.S., Verdonk, N.H. & van den Biggelaar, J.A.M. (volume eds.). Academic Press, Orlando: 1–45.

Frick, J.E. & Ruppert, E.E. (1996): Primordial germ cells of *Synaptula hydriformis* (Holothuroidea; Echinodermata) are epithelial flagellated-collar cells: their apical-basal polarity becomes primary egg polarity. *Biol. Bull.* **191**: 168–77.

Frick, J.E. & Ruppert, E.E. (1997): Primordial germ cells and oocytes of *Branchiostoma virginiae* (Cephalochordata, Acrania) are flagellated epithelial cells: relationship between epithelial and primary egg polarity. *Zygote* **5**: 139–51.

Frick, J.E., Ruppert, E.E. & Wourms, J.P. (1996): Morphology of the ovotestis of *Synaptula hydriformis* (Holothuroidea, Apoda): an evolutionary model of oogenesis and the origin of egg polarity in echinoderms. *Invertebr. Biol.* **115**: 46–66.

Fried, B. & Haseeb, M.A. (1991): Platyhelminthes: Aspidogastrea, Monogenea, and Digenea. In: *Microscopic Anatomy of Invertebrates*, Harrison, F.W. & Bogitsh, B.J. (eds.), Vol. 3: Platyhelminthes and Nemertinea. Wiley-Liss, New York: 141–209.

Friedlander, T.P., Regier, J.C. & Mitter, C. (1992): Nuclear gene sequences for higher level phylogenetic analysis: 14 promising candidates. *Syst. Biol.* **41**: 483–90.

Friedlander, T.P., Regier, J.C. & Mitter, C. (1994): Phylogenetic information content of five nuclear gene sequences in animals: initial assessment of character sets from concordance and divergence studies. *Syst. Biol.* **43**: 511–25.

Friedrich, H. (1979): Nemertini. In: *Morphogenese der Tiere*, Seidel, F. (ed.). Gustav Fischer, Jena.

Friedrich, M. & Tautz, D. (1995): Ribosomal DNA phylogeny of the major extant arthropod classes and the evolution of myriapods. *Nature* **376**: 165–67.

Friedrich, S., Wanninger, A., Brückner, M. & Haszprunar, G. (2002): Neurogenesis in the mossy chiton, *Mopalia muscosa* (Gould) (Polyplacophora): evidence against molluscan metamerism. *J. Morphol.* **258**: 109–17.

Frisch, S.M. (1997): The epithelial cell default-phenotype hypothesis and its implications for cancer. *BioEssays* **19**: 705–9.

Fritzenwanker, J.H., Saina, M. & Technau, U. (2004): Analysis of *forkhead* and *snail* expression reveals epithelial-mesenchymal transitions during embryonic and larval development of *Nematostella vectensis*. *Dev. Biol.* **275**: 389–402.

Frye, B.F. (1977): Reproduction. In: *Chordate Structure and Function*, Kluge, A.G. (ed.). 2nd edition. Macmillan Publishing, New York: 554–97.

Fuchs, J., Bright, M., Funch, P. & Wanninger, A. (2006): Immunocytochemistry of the neuromuscular systems of *Loxosomella vivipara* and *L. parguerensis* (Entoprocta: Loxosomatidae). *J. Morphol.* **267**: 866–83.

Funch, P. (1996): The chordoid larva of *Symbion pandora* (Cycliophora) is a modified trochophore. *J. Morphol.* **230**: 231–63.

Funch, P. & Kristensen, R.M. (1995): Cycliophora is a new phylum with affinities to Entoprocta and Ectoprocta. *Nature* **378**: 711–14.

Funch, P. & Kristensen, R.M. (1997): Cycliophora. In: *Microscopic Anatomy of Invertebrates*, Harrison, F.W. & Woollacott, R.M. (eds.), Vol. 13: Lophophorates, Entoprocta, and Cycliophora. Wiley-Liss, New York: 409–74.

Funch, P., Sørensen, M.V. & Obst, M. (2005): On the phylogenetic position of Rotifera – have we come any further? *Hydrobiologia* **546**: 11–28.

Furuya, H., Tsuneki, K. & Koshida, Y. (1997): Fine structure of dicyemid mesozoans, with special reference to cell junctions. *J. Morphol.* **231**: 297–305.

Gad, G. (2004a): A new genus of Nanaloricidae (Loricifera) from deep-sea sediments of volcanic origin in the Kilinailau Trench north of Papua New Guinea. *Helgol. Mar. Res.* **58**: 40–53.

Gad, G. (2004b): The Loricifera from the plateau of the Great Meteor Seamount. *Arch. Fish. Mar. Res.* **51**: 9–29.

Gad, G. (2005a): A parthenogenetic, simplified adult in the life cycle of *Pliciloricus pedicularis* sp.n. (Loricifera) from the deep sea of the Angola Basin. *Org. Div. Evol.* **5**: 77–103.

Gad, G. (2005b): Giant Higgins-larvae with paedogenetic reproduction from the deep sea of the Angola Basin – evidence for a new life cycle and for abyssal gigantism in Loricifera? *Org. Div. Evol.* **5**: 59–75.

Gagne, G. (1980): Ultrastructure of the sensory palps of *Tetranchyroderma papii* (Gastrotricha, Macrodasyida). *Zoomorphologie* **95**: 115–25.

Gaino, E., Burlando, B., Buffa, P. & Sará, M. (1986): Ultrastructural study of spermatogenesis in *Oscarella lobularis* (Porifera, Demospongiae). *Int. J. Invertebr. Reprod. Dev.* **10**: 297–305.

Garcia-Fernàndez, J. (2005): The genesis and evolution of homeobox gene clusters. *Nature Rev. Gen.* **6**: 881–92.

Garcia-Fernàndez, J. & Holland, P.W.H. (1994): Archaetypical organization of the amphioxus Hox gene cluster. *Nature* **370**: 563–66.

Garcia-Varela, M., Pérez-Ponce de León, G., de la Torre, P., Cummings, M.P., Sarma, S.S.S. & Laclette, J.P. (2000): Phylogenetic relationships of Acanthocephala based on analysis of 18S ribosomal RNA gene sequences. *J. Mol. Evol.* **50**: 532–40.

Gardiner, S.L. (1988): Respiratory and feeding appendages. In: The Ultrastructure of Polychaeta, Westheide, W. & Hermans, C.O. (eds.). *Microfauna Marina* **4**: 37–43.

Gardiner, S.L. (1992): Polychaeta: general organization, integument, musculature, coelom, and vascular system. In: *Microscopic Anatomy of Invertebrates*, Harrison, F.W. & Gardiner, S.L. (eds.), Vol. 7: Annelida. Wiley-Liss, New York: 1952.

Gardiner, S.L. & Jones, M.L. (1992): Vestimentifera. In: *Microscopic Anatomy of Invertebrates*, Harrison, F.W. & Rice, M.E. (eds.), Vol. 12: Onychophora, Chilopoda, and lesser Protostomata. Wiley-Liss, New York: 371–460.

Garey, J.R. (2001): Ecdysozoa: the relationship between Cycloneuralia and Panarthropoda. *Zool. Anz.* **240**: 321–30.

Garey, J.R., Near, T.J., Nonnemacher, M.R. & Nadler, S.A. (1996a): Molecular evidence for Acanthocephala as a subtaxon of Rotifera. *J. Mol. Evol.* **43**: 287–92.

Garey, J.R., Krotec, M., Nelson, D.R. & Brooks, J. (1996b): Molecular analysis supports a tardigrade-arthropod association. *Invertebr. Biol.* **115**: 79–88.

Garey, J.R. & Riggs, A.F. (1984): Structure and function of hemoglobin from *Urechis caupo*. *Arch. Biochem. Biophys.* **228**: 320–31.

Garey, J.R. & Schmidt-Rhaesa, A. (1998): The essential role of 'minor' phyla in molecular studies of animal evolution. *Am. Zool.* **38**: 907–17.

Garey, J.R., Schmidt-Rhaesa, A., Near, T.J. & Nadler, S.A. (1998): The evolutionary relationships of rotifers and acanthocephalans. *Hydrobiologia* **387/388**: 83–91.

Garlick, R.L. (1980): Structure of annelid high molecular weight hemoglobins (erythrocruorins). *Am. Zool.* **20**: 69–77.

Garlick, R.L., Williams, B.J. & Riggs, A.F. (1979): The hemoglobins of *Phoronopsis viridis*, of the primitive invertebrate phylum Phoronida: characterization and subunit structure. *Arch. Biochem. Biophys.* **194**: 13–23.

Garrone, R. (1978): Phylogenesis of connective tissue. Morphological aspects and biosynthesis of sponge intercellular matrix. *Front. Matrix Biol.* **5**. S. Karger, Basel.

Garrone, R. (1998): Evolution of metazoan collagens. In: Molecular Evolution: Towards the Origin of Metazoa, Müller, W.E.G. (ed.). *Progr. Mol. Subcell. Biol.* **21**. Springer, Berlin: 118–137.

Gauchat, D., Mazet, F., Berney, C., Schummer, M., Kreger, S., Pawlowski, J. & Galliot, B. (2000): Evolution of Antp-class genes and differential expression of *Hydra Hox/paraHox* genes in anterior patterning. *Proc. Natl. Acad. Sci. USA* **97**: 4493–98.

Gee, H. (1996): *Before the Backbone. Views on the Origin of the Vertebrates*. Chapman & Hall, London.

Gehring, W.J. (2001): The genetic control of eye development and its implication for the evolution of the various eye-types. *Zoology* **104**: 171–83.

Gehring, W.J. (2004): Historical perspective on the development and evolution of eyes and photoreceptors. *Int. J. Dev. Biol.* **48**: 707–717.

Gehring, W.J. (2005): New perspectives on eye development and the evolution of eyes and photoreceptors. *J. Heredity* **96**: 171–84.

Gehring, W.J. & Ikeo, K. (1999): *Pax 6* mastering eye morphogenesis and eye evolution. *Trends Genetics* **15**: 371–77.

Gerhart, J. (2000): Inversion of the chordate body axis: are there alternatives? *Proc. Natl. Acad. Sci. USA* **97**: 4445–48.

Germer, T. & Hündgen, M. (1978): The biology of colonial hydroids. II. The morphology and ultradtructure of the medusa of *Eirene viridula* (Thecata-Leptomedusa: Campanulinidae). *Mar. Biol.* **50**: 81–95.

Ghiselin, M.T. (1974): *The Economy of Nature and the Evolution of Sex*. University of California Press, Berkeley.

Ghiselin, M.T. (1988): The origin of molluscs in the light of molecular evidence. *Oxford Surv. Evol. Biol.* **5**: 66–95.

Ghysen, A. (2003): The origin and evolution of the nervous system. *Int. J. Dev. Biol.* **47**: 555–62.

Gibbons, L.M. (2002): General organization. In: *The Biology of Nematodes*, Lee, D.L. (ed.). Taylor & Francis, London: 31–59.

Giere, O., Rhode, B. & Dubilier, N. (1988): Structural peculiarities of the body wall of *Tubificoides benedii* (Oligochaeta) and possible relations to its life in sulphidic sediments. *Zoomorphology* **108**: 29–39.

Gilbert, J.J. (1983): Rotifera. In: *Reproductive Biology of Invertebrates*, Adiyodi, K.G. & Adiyodi, R.G. (eds.). Vol. 2: Spermatogenesis and sperm function. John Wiley & Sons, Chichester: 181–209.

Gilbert, J.J. (1989): Rotifera. In: *Reproductive Biology of Invertebrates*, Adiyodi, K.G. & Adiyodi, R.G. (eds.). Vol. 4A : Fertilization, development, and parental care. John Wiley & Sons, Chichester: 179–199.

Gilbert, S.F. (2000): *Developmental Biology*. 6th edition. Sinauer, Sunderland.

Giribet, G. (2002): Current advances in the phylogenetic reconstruction of metazoan evolution. A new paradigm for the Cambrian explosion? *Mol. Phyl. Evol.* **24**: 345–57.

Giribet, G. (2003): Molecules, development and fossils in the study of metazoan evolution; Articulata versus Ecdysozoa revisited. *Zoology* **106**: 303–26.

Giribet, G. & Ribera, C. (2000): A review of arthropod phylogeny: new data based on ribosomal DNA sequences and direct character optimization. *Cladistics* **16**: 204–31.

Giribet, G. & Wheeler, W.C. (1999): The position of arthropods in the animal kingdom: Ecdysozoa, islands, trees, and the 'parsimony ratchet'. *Mol. Phyl. Evol.* **13**: 619–23.

Giribet, G., Carranza, S., Baguñà, J., Riutort, M. & Ribera, C. (1996): First molecular evidence for the existence of a Tardigrada + Arthropoda clade. *Mol. Biol. Evol.* **13**: 76–84.

Giribet, G., Distel, D.L.D., Polz, M., Sterrer, W. & Wheeler, W.C. (2000): Triploblastic relationships with emphasis on the acoelomates and the position of Gnathostomulida, Cycliophora, Plathelminthes, and Chaetognatha: a cvombined approach of 18S rDNA sequences and morphology. *Syst. Biol.* **49**: 539–62.

Giribet, G., Edgecombe, G.D. & Wheeler, W.C. (2001): Arthropod phylogeny based on eight molecular loci and morphology. *Nature* **413**: 157–61.

Giribet, G., Okusu, A., Lindgren, A.R., Huff, S.W., Schrödl, M. & Nishigushi, M.K. (2006): Evidence for a clade composed of molluscs with serially repeated structures: monoplacophorans are related to chitons. *Proc. Nat. Acad. Sci. USA* **103**: 7723–28.

Giribet, G., Sørensen, M.V., Funch, P., Kristensen, R.M. & Sterrer, W. (2004): Investigations into the phylogenetic position of Micrognathozoa using four molecular loci. *Cladistics* **20**: 1–13.

Glätzer, K.H. (1971): Die Ei- und Frühentwicklung von *Corydendrium parasiticum* mit besonderer Berücksichtigung der Oocyten-Feinstruktur während der Vitellogenese. *Helgol. Wiss. Meeresunt.* **22**: 213–80.

Glenner, H., Hansen, A.J., Sørensen, M.V., Ronquist, F., Huelsenbeck, J.P. & Willerslev, E. (2004): Bayesian inference of the metazoan phylogeny: a combined molecular and morphological approach. *Curr. Biol.* **14**: 1644–49.

Glover, A.G., Källström, B., Smith, C.R. & Dahlgren, T.G. (2005): World-wide whale worms? A new species of

Osedax from the shallow north Atlantic. *Proc. R. Soc. London B* **272**: 2587–92.

Gnatzy, W. & Romer, F. (1984): Cuticle: formation, moulting and control. In: *Biology of the Integument*, Bereiter-Hahn, J., Matoltsy, A.G. & Sylvia Richards, K. (eds.). Vol. 1: Invertebrates. Springer, Berlin: 638–84.

Götting, K.-J. (1980): Argumente für die Deszendenz der Mollusken von metameren Antezendenten. *Zool. Jahrb. Anat.* **103**: 211–18.

Goldberg, W.M. (1976): Comparative study of the chemistry and structure of gorgonian and antipatharian coral skeletons. *Mar. Biol.* **35**: 253–67.

Goldberg, W.M. (1978): Chemical changes accompanying maturation of the connective tissue skeletons of gorgonian and antipatharian corals. *Mar. Biol.* **49**: 203–10.

Golding, D.W. (1992): Polychaeta: nervous system. In: *Microscopic Anatomy of Invertebrates*, Harrison, F.W. & Gardiner, S.L. (eds.), Vol. 7: Annelida. Wiley-Liss, New York: 153–79.

Goldsmith, T.H. (1990): Optimization, constraint, and history in the evolution of eyes. *Quart. Rev. Biol.* **65**: 281–322.

Goldstein, B. (2001): On the evolution of early development in the Nematoda. Philos. *Trans. R. Soc. London B* **356**: 1521–31.

Golz, R. & Thurm, U. (1994): The ciliated sensory cell of *Stauridiosarsia producta* (Cnidaria, Hydrozoa) – a nematocyst-free nematocyte? *Zoomorphology* **114**: 185–94.

Goodman, M., Pedwaydon, J., Czelusniak, J., Suzuki, T., Gotoh, T., Moens, L., Shishikura, F., Walz, D. & Vinogradov, S. (1988): An evolutionary tree for invertebrate globin sequences. *J. Mol. Evol.* **27**: 236–49.

Goodrich, E.S. (1895): On coelom, genital ducts, and nephridia. *Quart. J. Microsc. Sci.* **37**: 477–508.

Goodrich, E.S. (1945): The study of nephridia and genital ducts since 1895. *Quart. J. Microsc. Sci.* **86**: 113–392.

Gorman, A.L.F., McReynolds, J.S. & Barnes, S.N. (1971): Photoreceptors in primitive chordates: fine structure, hyperpolarizing receptor potentials, and evolution. *Science* **172**: 1052–54.

Goto, T. & Yoshida, M. (1984): Photoreception in Chaetognatha. In: *Photoreception and Vision in Invertebrates*, Ali, M.A. (ed.). Plenum Press, New York: 727–42.

Gould, S.J. (1977): *Ontogeny and Phylogeny*. Harvard University Press, Cambridge.

Govind, C.K. (1992): Nervous system. In: *Microscopic Anatomy of Invertebrates*, Harrison, F.W. & Humes, A.G. (eds.). Vol. 10: Decapod Crustacea. Wiley-Liss, New York: 395–438

Graebner, I. (1968): Erste Befunde über die Feinstruktur der Exkretionszellen der Gnathostomulidae (*Gnathostomula paradoxa*, Ax 1956 und *Austrognathia riedli*, Sterrer 1965). *Mikroskopie* **23**: 277–92.

Graebner, I. (1969): Vergleichende elektronenmikroskopische Untersuchung der Spermienmorphologie und Spermiogenese einiger Gnathostomula-Arten: *Gnathostomula paradoxa* (Ax 1956), *Gnathostomula axi* (Kirsteuer 1964), *Gnathostomula jenneri* (Riedl 1969). *Mikroskopie* **24**: 131–60.

Graebner, I. & Adam, H. (1970): Electron microscopical study of spermatogenesis and sperm morphology in gnathostomulids. In: *Comparative Spermatology*, Baccetti, B. (ed.). Academic Press, New York: 375–82.

Grave, C. & Riley, G. (1935): Development of the sense organs of the larva of *Botryllus schlosseri*. *J. Morphol.* **57**: 185–211.

Graveley, B.R. (2001): Alternative splicing: increasing diversity in the proteomic world. *Trends Gen.* **17**: 100–7.

Graveley, B.R. (2005): Mutually exclusive splicing of the insect *Dscam* Pre-mRNA directed by competing intronic RNA secondary structures. *Cell* **123**: 65–73.

Green, C.R. & Bergquist, P.L. (1982): Phylogenetic relationships within the invertebrata in relation to the structure of septate junctions and the development of 'occluding' junctional types. *J. Cell Sci.* **53**: 279–305.

Grell, K.G. (1971a): Über den Ursprung der Metazoen. *Mikrokosmos* **60**: 97–102.

Grell, K.G. (1971b): *Trichoplax adhaerens* F.E. Schulze und die Entstehung der Metazoen. *Naturwiss. Rundschau* **24**: 160–61.

Grell, K.G. & Benwitz, G. (1974): Elektronenmikroskopische Beobachtungen über das Wachstum der Eizelle und die Bildung der 'Befruchtungsmembran' von *Trichoplax adhaerens* F.E. Schulze (Placozoa). *Z. Morph. Tiere* **79**: 295–310.

Grell, K.G. & Benwitz, G. (1981): Ergänzende Untersuchungen zur Ultrastruktur von *Trichoplax adhaerens* F.E. Schulze (Placozoa). *Zoomorphology* **98**: 47–67.

Grell, K.G. & Ruthmann, A. (1991): Placozoa. In: *Microscopic Anatomy of Invertebrates*, Harrison, F.W. & Westfall, J.A. (eds.), Vol. 2: Placozoa, Porifera, Cnidaria, and Ctenophora. Wiley-Liss, New York: 13–27.

Greven, H. (1979): Notes on the structure of vasa Malpighii in the eutardigrade *Isohypsibius augusti* (Murray 1907). *Zesz. Nauk. Uniw. Jagiellonsk.* **25**: 87–95.

Greven, H. & Peters, W. (1986): Localization of chitin in the cuticle of Tardigrada using wheat germ agglutinin-gold conjugate as a specific electron-dense marker. *Tiss. Cell* **18**: 297–304.

Griffiths, G. (1996): On vesicles and membrane compartments. *Protoplasma* **195**: 37–58.

Grimmelikhuijzen, C.J.P. (1983): FMRFamide immunoreactivity is generally occurring in the nervous system of coelenterates. *Histochemistry* **78**: 361–381.

Grimmelikhuijzen, C.J.P. & Westfall, J.A. (1995): The nervous system of cnidarians. In: *The Nervous Systems of Invertebrates: An Evolutionary and Comparative Approach*, Breidbach, O. & Kutsch, W. (eds.). Birkhäuser Verlag, Basel: 7–24.

Grimmelikhuijzen, C.J.P., Graff, D., Koizumi, O., Westfall, J.A. & McFarlane, I.D. (1991): Neuropeptides in coelenterates: a review. *Hydrobiologia* **216/217**: 555–63.

Grimmer, J.C. & Holland, N.D. (1979): Haemal and coelomic circulatory systems in the arms and pinnules of *Flometra serratissima* (Echinodermata: Crinoidea). *Zoomorphologie* **94**: 93–109.

Gruhl, A., Grobe, P. & Bartolomaeus, T. (2005): Fine structure of the epistome in *Phoronis ovalis*: significance for the coelomic organization in Phoronida. *Invertebr. Biol.* **124**: 332–43.

Gruner, H.-E. (1984): Wirbellose Tiere. In: *Lehrbuch der Speziellen Zoologie*. Vol. 2: Cnidaria, Ctenophora, Mesozoa, Plathelminthes, Nemertini, Entoprocta, Nemathelminthes, Priapulida. Gustav Fischer Verlag, Stuttgart.

Gruner, H.-E., Moritz, M. & Dunger, W. (1993): Arthropoda (ohne Insecta). In: *Wirbellose Tiere*, Gruner, H.-E. (ed.). Vol. 4, 4th. Edn., Gustav Fischer Verlag, Stuttgart.

Gschwentner, R., Mueller, J., Ladurner, P., Rieger, R. & Tyler, S. (2003): Unique patterns of body-wall musculature in the Acoela (Plathelminthes): the ventral musculature of *Convolutriloba longifissura*. *Zoomorphology* **122**: 87–94.

Guidi, L., Ferraguti, M., Pierboni, L. & Balsamo, M. (2003a): Spermiogenesis and spermatozoa in *Acanthodasys aculeatus* (Gastrotricha, Macrodasyida): an ultrastructural study. *Acta Zool.* **84**: 77–85.

Guidi, L., Marotta, R., Pierboni, L., Ferraguti, M., Todaro, M.A. & Balsamo, M. (2003b): Comparative sperm ultrastructure of *Neodasys ciritus* and *Musellifer delamarei*, two species considered to be basal among Chaetonotida (Gastrotricha). *Zoomorphology* **122**: 135–43.

Guidi, L., Pierboni, L., Ferraguti, M., Todaro, A. & Balsamo, M. (2004): Spermatology of the genus *Lepidodasys* Remane, 1926 (Gastrotricha, Macrodasyida): towards a revision of the family Lepidodasyidae Remane, 1927. *Acta Zool.* **85**: 211–21.

Guraya, S.S. (1987): *Biology of Spermatogenesis and Spermatozoa in Mammals*. Springer, Berlin.

Gustavsson, L.M. (2001): Comparative study of the cuticle in some aquatic oligochaetes (Annelida: Clitellata). *J. Morphol.* **248**: 185–95.

Guthrie, D.M. & Banks, J.R. (1970): Observations on the function and physiological properties of a fast paramyosin muscle – the notochord of Amphioxus (*Branchiostoma lanceolatum*). *J. Evol. Biol.* **52**: 125–38.

Haag, E.S. (2005): Echinoderm rudiments, rudimentary bilaterians, and the origin of the chordate CNS. *Evol. Dev.* **7**: 280–81.

Haag, E.S. (2006): Reply to Nielsen. *Evol. Dev.* **8**: 3–5.

Haase, A., Stern, M., Wächtler, K. & Bicker, G. (2001): A tissue-specific marker of Ecdysozoa. *Dev. Genes Evol.* **211**: 428–33.

Hackman, R.H. (1984): Cuticle: biochemistry. In: *Biology of the Integument*, Bereiter-Hahn, J., Matoltsy, A.G. & Sylvia Richards, K. (eds.). Vol. 1: Invertebrates. Springer, Berlin: 583–610.

Hadfield, M.G. (1975): Hemichordata. In: *Reproduction of Marine Invertebrates*, Giese, A.C. & Pearse, J.S. (eds.). Vol. 2: Entoprocts and lesser coelomates. Academic Press, New York: 185–240.

Hadfield, M.G. & Switzer-Dunlap, M. (1984): Opisthobranchs. In: *The Mollusca*, Wilbur, K.M. (ed.), Vol. 7: Reproduction, Tompa, A.S., Verdonk, N.H. & van den Biggelaar, J.A.M. (volume eds.). Academic Press, Orlando: 209–350.

Hadzi, J. (1963): The *Evolution of the Metazoa*. Pergamon Press, New York.

Haeckel, E. (1874): Die Gastraea-Theorie, die phylogenetische Classification des Thierreiches und die Homologie der Keimblätter. *Jena Z. Naturwiss.* **8**: 1–55.

Hagner-Holler, S., Schoen, A., Erker, W., Marden, J.H., Rupprecht, R., Decker, H. & Burmester, T. (2004): A respiratory hemocyanin from an insect. *Proc. Nat. Acad. Sci. USA* **101**: 871–74.

Hahn-Keser, B. & Stockem, W. (1998): Intracellular pathways and degradation of endosomal contents in basal epithelial cells of freshwater sponges (Porifera, Spongillidae). *Zoomorphology* **117**: 223–36.

Hajduk, S.L. (1992): Ultrastructure of the tube-foot of an ophiuroid echinoderm, *Hemipholis elongata*. *Tiss. Cell* **24**: 111–20.

Halanych, K.M. (1993): Suspension feeding by the lophophore-like apparatus of the pterobranch hemichordate *Rhabdopleura normani*. *Biol. Bull.* **185**: 417–27.

Halanych, K.M. (1996a): Testing hypotheses of chaetognath origins: long branches revealed by 18S ribosomal DNA. *Syst. Biol.* **45**: 223–46.

Halanych, K.M. (1996b): Convergence in the feeding apparatuses of lophophorates and pterobranch hemichordates revealed by 18S rDNA: an interpretation. *Biol. Bull.* **190**: 1–5.

Halanych, K.M. (2004): The new view of animal phylogeny. *Ann. Rev. Ecol. Syst.* **35**: 229–56.

Halanych, K.M. (2005): Molecular phylogeny of siboglinid annelids (a.k.a. pogonophorans): a review. In: Morphology, Molecules, Evolution and Phylogeny in Polychaeta and Related Taxa, Bartolomaeus, T. & Purschke, G. (eds.). *Hydrobiologia* **535/536**: 297–307.

Halanych, K.M., Bacheller, J.D., Aguinaldo, A.M.A., Liva, S.M., Hillis, D.M. & Lake, J.A. (1995): Evidence from 18S ribosomal DNA that the lophophorates are protostome animals. *Science* **267**: 1641–43.

Halanych, K.M., Dahlgren, T.G. & McHugh, D. (2002): Unsegmented annelids? Possible origins of four lophotrochozoan worm taxa. *Integr. Comp. Biol.* **42**: 678–84.

Halanych, K.M., Feldmann, R.A. & Vrijenhoek, R.C. (2001): Molecular evidence that *Sclerolinum brattstormi* is closely related to vestimentiferans, not frenulate pogonophorans (Siboglinidae, Annelida).. *Biol Bull.* **201**: 65–75.

Halanych, K.M., Lutz, R.A. & Vrijenhoek, R.C. (1998): Evolutionary origins and age of vestimentiferan tubeworms. *Cah. Biol. Mar.* **39**: 355–58.

Halder, G., Callaerts, P. & Gehring, W.J. (1995): Induction of ectopic eyes by targeted expression of the *eyeless* gene in *Drosophila*. *Science* **267**: 1788–92.

Hall, B.K. (2005): *Bones and Cartilage: Developmental and Evolutionary Skeletal Biology.* Elsevier, Amsterdam.

Halton, D.W. & Gustafsson, M.K.S. (1996): Functional morphology of the platyhelminth nervous system. *Parasitology Suppl.* **113**: S47–S72.

Hamilton, P.V. (1991): Variation in sense organ design and associated sensory capabilities among closely related molluscs. *Am. Malacol. Bull.* **9**: 89–98.

Hamlett, E. (1993): Uterogestation and placentation in elasmobranchs. *J. Exp. Zool.* **266**: 347–69.

Han, Z. & Firtel, R.A. (1998): The homeobox-containing gene *Wariai* regulates anterior-posterior patterning and cell-type homeostasis in *Dictyostelium*. *Development* **125**: 313–25.

Hanelt, B., Thomas, F. & Schmidt-Rhaesa, A. (2005): Biology of the phylum Nematomorpha. *Adv. Parasitol.* **59**: 243–305.

Hanelt, B., van Schyndel, D., Adema, C.M., Lewis, L.A. & Loker, E.S. (1996): The phylogenetic position of *Rhopalura ophiocomae* (Orthonectida) based on 18S ribosomal DNA sequence analysis. *Mol. Biol. Evol.* **13**: 1187–91.

Hankeln, T., Ebner, B., Fuchs, C., Gerlach, F., Haberkamp, M., Laufs, T.L., Roesner, A., Schmidt, M., Weich, B., Wystub, S., Saaler-Reinhardt, S., Reuss, S., Bolognesi, M., de Sanctis, D., Marden, M.C., Kiger, L., Moens, L., Dewilde, S., Nevo, E., Avivi, A., Weber, R.E., Fago, A. & Burmester, T. (2005): Neuroglobin and cytoglobin in search of their role in the vertebrate globin family. *J. Inorganic Biochem.* **99**: 110–19.

Hanken, J. & Wake, D.B. (1993): Miniaturization of body size: organismal consequences and evolutionary significance. *Ann. Rev. Ecol. Syst.* **24**: 501–19.

Hansen, K. (1962): Elektronenmikroskopische Untersuchung der Hirudineen-Augen. *Zool. Beitr. N. F.* **7**: 83–128.

Hansen, U. (1995): New aspects of the possible sites of ultrafiltration in annelids (Oligochaeta). *Tiss. Cell* **27**: 73–78.

Hanson, E.D. (1963): Homologies and the ciliate origin of the Eumatazoa. In: *The Lower Metazoa. Comparative Biology and Phylogeny*, Dougherty, E.C. (ed.). University of California Press, Berkeley: 7–22.

Hanström, B. (1928): *Vergleichende Anatomie des Nervensystems der wirbellosen Tiere unter Berücksichtigung seiner Funktion*. Springer, Berlin.

Harbison, G.R. & Miller, R.L. (1986): Not all ctenophores are hermaphrodites. Studies on the systematics, distribution, sexuality and development of two species of *Ocyropsis*. *Mar. Biol.* **90**: 413–24.

Hardin, J. & Lockwood, C. (2004): Skin tight: cell adhesion in the epidermis of *Caenorhabditis elegans*. *Curr. Opin. Cell Biol.* **16**: 486–92.

Hardison, R.C. (1996): A brief history of hemoglobins: plant, animal, protist, and bacteria. *Proc. Natl. Acad. Sci. USA* **93**: 5675–79.

Hardisty, M.W. (1986): General introduction to lampreys. In: *The Freshwater Fishes of Europe*, Holčik, J. (ed.). Vol 1/1: Petromyzontiformes. Aula-Verlag, Wiesbaden: 19–83.

Harris, R.R. & Jaccarini, V. (1981): Structure and function of the anal sacs of *Bonellia viridis* (Echiura: Bonelliidae). *J. Mar. Biol. Assoc. UK* **61**: 413–30.

Harrison, F.W. & Chia, F.-S. (1994): *Microscopic Anatomy of Invertebrates*. Vol. 14: Echinodermata. Wiley-Liss, New York.

Harrison, F.W. & De Vos, L. (1991): Porifera. In: *Microscopic Anatomy of Invertebrates*, Harrison, F.W. & Westfall, J.A. (eds.), Vol. 2: Placozoa, Porifera, Cnidaria, and Ctenophora. Wiley-Liss, New York: 29–89.

Harrison, F.W. & Humes, A.G. (1992): *Microscopic Anatomy of Invertebrates*. Vol. 10: Decapod Crustacea. Wiley-Liss, New York.

Harrison, P.L. & Jamieson, B.G.M. (1999): Cnidaria and Ctenophora. In: *Reproductive Biology of Invertebrates*, Adiyodi, K.G., Adiyodi, R.G. & Jamieson, B.G.M. (eds.). Vol. 9, part A: Progress in male gamete ultrastructure and phylogeny. John Wiley & Sons, Chichester: 21–95.

Hart, M.W., Johnson, S.L., Addison, J.A. & Byrne, M. (2004): Strong character incongruence and character choice in phylogeny of sea stars of the Asterinidae. *Invertebr. Biol.* **123**: 343–56.

Hartman, H. & Fedorov, A. (2002): The origin of the eukaryotic cell: a genomic investigation. *Proc. Nat. Acad. Sci. USA* **99**: 1420–25.

Harzsch, S. (2006): Neurophylogeny: architecture of the nervous system and a fresh view on arthropod phylogeny. *Integr. Comp. Biol.* **46**: 162–94.

Harzsch, S. & Waloszek, D. (2001): Neurogenesis in the developing visual system of the branchiopod crustacean *Triops longicaudatus* (LeConte, 1846): corresponding patterns of compound-eye formation in Crustacea and Insecta? *Dev. Genes Evol.* **211**: 37–43.

Harzsch, S., Müller, C.H. G. & Wolf, H. (2005): From variable to constant cell numbers: cellular characteristics of the arthropod nervous system argue against a sister-ggroup relationship of Chelicerata and 'Myriapoda' but favour the Mandibulata concept. *Dev. Genes Evol.* **215**: 53–68.

Harzsch, S., Vilpoux, K., Blackburn, D.C., Platchetzki, D., Brown, N.L., Melzer, R., Kempler, K.E. & Battelle, B.A. (2006): Evolution of arthropod visual systems: development of the eyes and central visual pathways in the horseshoe crab *Limulus polyphemus* Linnaeus, 1758 (Chelicerata, Xiphosura). *Dev. Dynamics* **235**: 2641–55.

Hassanin, A. (2006): Phylogeny of Arthropoda inferred from mitochondrial sequences: strategies for limiting the misleading effects of multiple changes in pattern and rates of substitution. *Mol. Phyl. Evol.* **38**: 100–16.

Haszprunar, G. (1985a): The fine morphology of the osphradial sense organs of the Mollusca. I. Gastropoda, Prosobranchia. *Phil. Trans. R. Soc. London B* **307**: 457–96.

Haszprunar, G. (1985b): The fine morphology of the osphradial sense organs of the Mollusca. II. Allogastropoda (Architectonicidae, Pyramellidae). *Phil. Trans. R. Soc. London B* **307**: 497–505.

Haszprunar, G. (1986): Feinmorphologische Untersuchungen an Sinnesstrukturen ursprünglicher Solenogastres (Mollusca). *Zool. Anz.* **217**: 345–62.

Haszprunar, G. (1987a): The fine morphology of the osphradial sense organs of the Mollusca. III. Placophora and Bivalvia. *Phil. Trans. R. Soc. London B* **315**: 37–61.

Haszprunar, G. (1987b): The fine morphology of the osphradial sense organs of the Mollusca. IV. Caudofoveata and Solenogastres. *Phil. Trans. R. Soc. London B* **315**: 63–73.

Haszprunar, G. (1992): The first molluscs – small animals. *Boll. Zool.* **59**: 1–16.

Haszprunar, G. (1996a): Plathelminthes and Plathelminthomorpha – paraphyletic taxa. *J. Zool. Syst. Evolut. Res.* **34**: 41–48.

Haszprunar, G. (1996b): The Mollusca: coelomate turbellarians or mesenchymate annelids? In: *Origin and Evolutionary Radiation of the Mollusca*, Taylor, J. (ed.), Oxford University Press, Oxford: 1–28.

Haszprunar, G. (2000): Is the Aplacophora monophyletic? A cladisitc point of view. *Am. Malacol. Bull.* **15**: 115–30.

Haszprunar, G. & Schaefer, K. (1997): Monoplacophora. In: *Microscopic Anatomy of Invertebrates*, Harrison, F.W. & Kohn, A.J. (eds.), Vol. 6B: Mollusca II. Wiley-Liss, New York: 415–57.

Haszprunar, G., von Salvini-Plawen, L. & Rieger, R.M. (1995): Larval planctotrophy – a primitive trait in the Bilateria? *Acta Zool.* **76**: 141–54.

Hatschek, B. (1878): Studien über die Entwicklungsgeschichte der Anneliden. Ein Beitrag zur Morphologie der Bilateria. *Arb. Zool. Inst. Univ. Wien* **1**: 1–128.

Haupt, J. (1979): Phylogenetic aspects of recent studies on myriapod sense organs. In: *Myriapod Biology*, Camatini, M. (eds.). Academic Press, London: 391–406.

Hausdorf, B. (2000): Early evolution of the Bilateria. *Syst. Biol.* **49**: 130–42.

Hausen, H. (2005a): Comparative structure of the epidermis in polychaetes (Annelida). In: Morphology, Molecules, Evolution and Phylogeny in Polychaeta and Related Taxa, Bartolomaeus, T. & Purschke, G. (eds.). *Hydrobiologia* **535/536**: 25–35.

Hausen, H. (2005b): Chaetae and chaetogenesis in polychaetes (Annelida). In: Morphology, Molecules, Evolution and Phylogeny in Polychaeta and Related Taxa, Bartolomaeus, T. & Purschke, G. (eds.). *Hydrobiologia* **535/536**: 37–52.

Hausmann, K. (1981): Zur Struktur der Solenocyten (Cyrtocyten) von *Anaitides mucosa* (Annelida, Polychaeta). *Helgol. Meeresunters.* **34**: 485–89.

Hausmann, K., Hülsmann, N. & Radek, R. (2003): *Protistology*. 3rd Edn., Schweizerbart'sche Verlagsbuchhandlung, Berlin.

Havenhand, J.N. (1995): Evolutionary ecology of larval types. In: *Ecology of Marine Invertebrate Larvae*, McEdward, C. (ed.). CRC Press, Boca Raton: 79–122.

Hay-Schmidt, A. (1987): The ultrastructure of the protonephridium of the actinotroch larva (Phoronida). *Acta Zool.* **68**: 35–47.

Hay-Schmidt, A. (1989): The nervous system of the actinotroch larva of *Phoronis muelleri* (Phoronida). *Zoomorphology* **108**: 333–51.

Hay-Schmidt, A. (1990a): Catecholamine-containing, serotonin-like and neuropeptide FMRFamide-like immunoreactive cells and processes in the nervous system of the pilidium larva (Nemertini). *Zoomorphology* **109**: 231–44.

Hay-Schmidt, A. (1990b): Catecholamine-containing, serotonin-like, and FMRF-amide-like immunoreactive neurons and processes in the nervous system of the early actinotroch larva of *Phoronis vancouverensis* (Phoronida): distribution and development. *Can. J. Zool.* **68**: 1525–36.

Hay-Schmidt, A. (1990c): Distribution of catecholamine-containing, serotonin-like and neuropeptide FMRFamide-like immunoreactive neurons and processes in the nervous system of the actinotroch larva of *Phoronis muelleri* (Phoronida). *Cell Tiss. Res.* **259**: 105–18.

Hay-Schmidt, A. (1992): Ultrastructure and immunocytochemistry of the nervous system of the larvae of *Lingula anatina* and *Glottidia* sp. (Brachiopoda). *Zoomorphology* **112**: 189–205.

Hay-Schmidt, A. (2000): The evolution of the serotinergic nervous system. *Proc. R. Soc. London B* **267**: 1071–79.

Hayward, P.J. & Ryland, J.S. (1995): *Handbook of the Marine Fauna of North-West Europe.* Oxford University Press, Oxford.

Healy, J.M. (2000): Mollusca: relict taxa. In: *Reproductive Biology of Invertebrates*, Adiyodi, K.G., Adiyodi, R.G. & Jamieson, B.G.M. (eds.). Vol. 9, part B: Progress in male gamete ultrastructure and phylogeny. John Wiley & Sons, Chichester: 21–79.

Healy, J.M., Rowe, F.W.E. & Anderson, D.T. (1988): Spermatozoa and spermiogenesis in *Xyloplax* (class Concentricycloidea), new type of spermatozoon in the Echinodermata. *Zool. Scr.* **173**: 297–310.

Healy, J.M., Schaefer, K. & Haszprunar, G. (1995): Spermatozoa and spermatogenesis in a monoplacophoran mollusc, *Laevipilina antarctica*: ultrastructure and comparison with other Mollusca. *Mar. Biol.* **122**: 53–65.

Hegemann, P., Fuhrmann, M. & Kateriya, S. (2001): Algal sensory photoreceptors. *J. Phycol.* **37**: 668–76.

Heider, K. (1914): Phylogenie der Wirbellosen. In: *Die Kultur der Gegenwart*, Hinneberg, P. (ed.). 3rd Edn., Teubner, Berlin.

Heimler, W. (1983): Untersuchungen zur Larvalentwicklung von *Lanice conchilega* (Pallas) 1776 (Polychaeta, Terebellomorpha). Teil III: Bau und Struktur der Aulophora-Larve. *Zool. Jahrb. Anat.* **110**: 411–78.

Heimler, W. (1988): Larvae. In: The Ultrastructure of Polychaeta, Westheide, W. & Hermans, C.O. (eds.). *Meiofauna Marina* **4**: 353–71.

Heiner, I. (2004): *Armoloricus kristenseni* (Nanaloricidae, Loricifera), a new species from the Faroe Bank (North Atlantic). *Helgol. Mar. Res.* **58**: 192–205

Heiner, I. & Kristensen, R.M. (2005): Two new species of the genus *Pliciloricus* (Loricifera, Pliciloricidae) from the Faroe Bank, North Atlantic. *Zool. Anz.* **243**: 121–38.

Heinzeller, T. & Welsch, U. (1994): Crinoidea. In: *Microscopic Anatomy of Invertebrates*, Harrison, F.W. & Chia, F.-S. (eds.), Vol. 14: Echinodermata. Wiley-Liss, New York: 9–148.

Hejnol, A. & Schnabel, R. (2005): The eutardigrade *Thulinia stephaniae* has an indeterminate development and the potential to regulate early blastomere ablations. *Development* **132**: 1349–61.

Helfenbein, K., Brown, W.M. & Boore, J.L. (2001): The complete mitochondrial genome of the articulate brachiopod *Terebratulina transversa*. *Mol. Biol. Evol.* **18**: 1734–44.

Helfenbein, K.G., Fourcade, H.M., Vanjani, R.G. & Boore, J.L. (2004): The mitochondrial genome of *Paraspadella gotoi* is highly reduced and reveals that chaetognaths are a sister group to protostomes. *Proc. Natl. Acad. Sci. USA* **101**: 10639–43.

Heller, J. (1993): Hermaphroditism in molluscs. *Biol. J. Linn. Soc.* **48**: 19–42.

Hellingwerf, K.J., Hoff, W.D. & Crielaard, W. (1996): Photobiology of microorganisms: how photosensors catch photon to initialize signalling. *Mol. Microbiol.* **21**: 683–93.

Hendelberg, J. (1983): Platyhelminthes – Turbellaria. In: *Reproductive Biology of Invertebrates*, Adiyodi, K.G. & Adiyodi, R.G. (eds.). Vol. 2: Spermatogenesis and sperm function. John Wiley & Sons, Chichester: 75–104.

Hendelberg, J. & Åkesson, B. (1991): Studies on the budding process in *Convolutriloba retrogemma* (Acoela, Platyhelminthes). *Hydrobiologia* **227**: 11–17.

Hendelberg, J. & Hedlund, K.-O. (1974): On the morphology of the epidermal ciliary rootlet system of the acoelous turbellarian *Childia groenlandica*. *Zoon* **2**: 13–24.

Hendler, G. & Byrne, M. (1987): Fine structure of the dorsal arm plate of *Ophiocoma wendti*: evidence for a photoreceptor system (Echinodermata, Ophiuroidea). *Zoomorphology* **107**: 261–72.

Hennig, W. (1950): *Grundzüge einer Theorie der phylogenetischen Systematik*. Deutscher Zentralverlag, Berlin.

Hennig, W. (1966): *Phylogenetic Systematics*. University of Illinois Press, Urbana.

Henry, J.Q. & Martindale, M.Q. (1997): Nemertean, the ribbon worms. In: *Embryology – Constructing the Organism*, Gilbert, S.F. & Raunio, A.M. (eds.). Sinauer, Sunderland: 151–166.

Herlyn, H. (2002): The musculature of the praesoma in *Macracanthorhynchus hirudinaeus* (Acanthocephala, Archiacanthocephala): re-examination and phylogenetic significance. *Zoomorphology* **121**: 173–82.

Herlyn, H. & Ehlers, U. (1997): Ultrastructure and function of the pharynx of *Gnathostomula paradoxa* (Gnathostomulida). *Zoomorphology* **117**: 135–45.

Herlyn, H. & Ehlers, U. (2001): Organisation of the praesoma in *Acanthocephalus anguillae* (Acanthocephala, Palaeacanthocephala) with special reference to the muscular system. *Zoomorphology* **121**: 13–18.

Herlyn, H. Martini, N. & Ehlers, U. (2001): Organisation of the praesoma in *Paratenuisentis ambiguus* (Van Claeve, 1921) (Acanthocephala, Eoacanthocephala), with special reference to the lateral sense organs and musculature. *Syst. Parasitol.* **50**: 105–16.

Herlyn, H., Pisurek, O., Schmitz, J., Ehlers, U. & Zischler, H. (2003): The syndermatan phylogeny and the evolution of acanthocephalan endoparasitism as inferred from 18S rDNA sequences. *Mol. Phyl. Evol.* **26**: 155–64.

Hermans, C.O. & Eakin, R.M. (1969): Fine structure of the cerebral ocelli of a sipunculid, *Phascalosoma agassizii*. *Z. Zellforsch.* **100**: 325–39.

Hermans, C.O. & Eakin, R.M. (1975): Sipunculan ocelli: fine structure in *Phascolosoma agassizii*. In: *Proc. Int. Symp. Biol. Sipuncula and Echiura*, Rice, M.E. & Todorovic, M. (eds.), Naucno Delo, Belgrade: 229–37.

Hernandez-Nicaise, M.-L. (1973): Le système nerveux des cténaires. I. Structure et ultrastructure des réseaux epithéliaux. *Z. Zellforsch.* **137**: 223–50.

Hernandez-Nicaise, M.-L. (1984): Ctenophora. In: *Biology of the Integument*, Bereiter-Hahn, J., Matoltsy, A.G. & Sylvia Richards, K. (eds.). Vol. 1: Invertebrates. Springer, Berlin: 368–75.

Hernandez-Nicaise, M.-L. (1991): Ctenophora. In: *Microscopic Anatomy of Invertebrates*, Harrison, F.W. & Westfall, J.A. (eds.), Vol. 2: Placozoa, Porifera, Cnidaria, and Ctenophora. Wiley-Liss, New York: 359–18.

Hernandez-Nicaise, M.-L. & Amsellem, J. (1980): Ultrastructure of the giant smooth muscle fiber of the ctenophore *Beroe ovata*. *J. Ultrastruct. Res.* **72**: 151–68.

Herreid, C.F., LaRussa, V.F. & DeFesi, C.R. (1976): Blood vascular system of the sea cucumber, *Stichopus moebii*. *J. Morphol.* **150**: 423–52.

Herrel, A., Meyers, J.J., Timmermans, J.-P. & Nishikawa, K.C. (2002): Supercontracting muscle: producing tension over extreme muscle lengths. *J. Exp. Biol.* **205**: 2167–73.

Herrera-Alvarez, L., Fernández, I., Benito, J. & Pardos, F. (1996): Ultrastructure of the labrum and foregut of *Derocheilocaris remanei* (Crustacea, Mystacocarida). *J. Morphol.* **230**: 199–217.

Herrmann, K. (1997): Phoronida. In: *Microscopic Anatomy of Invertebrates*, Harrison, F.W. & Woollacott, R.M. (eds.), Vol. 13: Lophophorates, Entoprocta, and Cycliophora. Wiley-Liss, New York: 207–35.

Hertwig, O. & Hertwig, R. (1882): *Die Coelomtheorie*. G. Fischer, Jena.

Herzberg, A., Ruhberg, H. & Storch, V. (1980): Zur Ultrastruktur des weiblichen Genitaltraktes der Peripatopsidae (Onychophora). *Zool. Jahrb. Anat.* **104**: 266–79.

Heß, M., Melzer, R. & Smola, U. (1996): The eyes of a 'nobody', *Anoplodactylus petiolatus* (Pantopoda; Anoplodactylidae). *Helgol. Meeresunt.* **50**: 25–36.

Hessinger, D.A. & Lenhoff, H.M. (1988): *The Biology of Nematocysts*. Academic Press, San Diego.

Hessler, R.R. & Elofsson, R. (2001): The circulatory system and an enigmatic cell type of the cephalocarid crustacean *Hutchinsoniella macracantha*. *J. Crust. Biol.* **21**: 28–48.

Hessling, R. & Westheide, W. (1999): CLSM analysis of development and structure of the central nervous system of *Enchytraeus crypticus* ('Oligochaeta', Enchytraeidae). *Zoomorphology* **119**: 37–47.

Hessling, R. & Westheide, W. (2002): Are Echiura derived from a segmented ancestor? Immunohistochemical analysis of the nervous system in developmental stages of *Bonellia viridis*. *J. Morphol.* **252**: 100–13.

Hevert, F. (1984): Urine formation in the lamellibranchs: evidence for ultrafiltration and quantitative descritption. *J. Exp. Biol.* **111**: 1–12.

Hibberd, D.J. (1975): Observations on the ultrastructure of the choanoflagellate *Codosiga botrytis* (Ehr.) Saville-Kent with special reference to the flagellar apparatus. *J. Cell Sci.* **17**: 191–219.

Hickman, C.S. (1999): Larvae in invertebrate development and evolution. In: *The Origin and Evolution of Larval Forms*, Hall, B.K. & Wake, M.H. (eds.). Academic Press, San Diego: 21–59.

Higgins, R.P. & Storch, V. (1989): Ultrastructural observations of the larva of *Tubiluchus corallicola* (Priapulida). *Helgol. Wiss. Meeresunt.* **43**: 1–11.

Higgins, R.P. & Storch, V. (1991): Evidence for direct development in *Meiopriapulus fijiensis*. *Trans. Am. Microsc. Soc.* **110**: 37–46.

Hilario, A., Young, C.M. & Tyler, P.A. (2005): Sperm storage, internal fertilization, and embryonic dispersal in vent and seep tubeworms (Polychaeta: Siboglinidae: Vestimentifera). *Biol. Bull.* **208**: 20–28.

Hilken, G. (1998): Vergleich von Tracheensystemen unter phylogenetischem Aspekt. *Verh. Naturwiss. Ver. Hamburg* **37**: 5–94.

Hill, R.B., Sanger, J.W., Yantorno, R.E. & Deutsch, C. (1978): Contraction in a muscle with negligible sarcoplasmic reticulum: the longitudinal retractor of the sea cucumber *Isostichopus badionotus* (Selenka), Holothuria Aspidochirota. *J. Exp. Zool.* **206**: 137–50.

Hillis, D.M., Pollock, D.D., McGuire, J.A. & Zwickl, D.J. (2003): Is sparse taxon sampling a problem for phylogenetic inference? *Syst. Biol.* **52**: 124–26.

Himstedt, W. (1996): *Die Blindwühlen*. Vol. 630 in series: Die Neue Brehm-Bücherei. Westarp Wissenschaften, Magdeburg.

Hochberg, R. (2005): Musculature of the primitive gastrotrich *Neodasys* (Chaetonotida): functional adaptations to the interstitial environment and phylogenetic significance. *Mar. Biol.* **146**: 315–23.

Hochberg, R. (2006): On the serotinergic nervous system of two planktonic rotifers, *Conochilus coenobasis* and *C. dossuarius* (Monogononta, Flosculariacea, Conochiliidae). *Zool. Anz.* **245**: 53–62.

Hochberg, R. & Litvaitis, M.K. (2000): Phylogeny of Gastrotricha: a morphology-based framework of gastrotrich relationships. *Biol. Bull.* **198**: 299–305.

Hochberg, R. & Litvaitis, M.K. (2001a): The muscular system of *Dactylopodola baltica* and other macrodasyidan gastrotrichs in a functional and phylogenetic perspective. *Zool. Scr.* **30**: 325–36.

Hochberg, R. & Litvaitis, M.K. (2001b): Functional morphology of muscles in *Tetranchyroderma papii* (Gastrotricha). *Zoomorphology* **121**: 37–43.

Hochberg, R. & Litvaitis, M.K. (2001c): The musculature of *Draculiciteria tessalata* (Chaetonotoida, Paucitubulatina): implications for the evolution of dorsoventral muscles in Gastrotricha. *Hydrobiologia* **452**: 155–61.

Hochberg, R. & Litvaitis, M.K. (2003): Ultrastructural and immunocytochemical observations of the nervous system of three macrodasyidan gastrotrichs. *Acta Zool.* **84**: 171–78.

Hoffman, P.F., Kaufman, A.J., Halverson, G.P. & Schrag, D.P. (1998): A neoproterozoic snowball earth. *Science* **281**: 1342–46.

Hogg, N.A.S. & Wijdens, J. (1979): A study of gonadal organogenesis, and the factors influencing regeneration following surgical castration in *Deroceras reticulatum* (Pulmonata: Limacidae). *Cell Tiss. Res.* **198**: 295–307.

Holland, L.Z. (1988): Spermiogenesis in the salps *Thalia democratica* and *Cyclosalpa affinis* (Tunicata: Thaliacea): an electron microscopic study. *J. Morphol.* **198**: 189–204.

Holland, L.Z. (1990): Spermatogenesis in *Pyrosoma atlanticum* (Tunicata: Thaliacea: Pyrosomatida): implications for tunicate phylogeny. *Mar. Biol.* **105**: 451–70.

Holland, L.Z. (1991): The phylogenetic significance of tunicate sperm morphology. In: *Comparative Spermatology 20 Years After*, Baccetti, B. (eds.). Raven Press, New York: 961–65.

Holland, L.Z., Gorsky, G. & Fenaux, R. (1988): Fertilization in *Oikopleura dioica* (Tunicata, Appendicularia): acrosome reaction, cortical reaction and sperm-egg fusion. *Zoomorphology* **108**: 229–43.

Holland, N.D. (1984): Echinodermata: epidermal cells. In: *Biology of the Integument*, Bereiter-Hahn, J., Matoltsy, A.G. & Sylvia Richards, K. (eds.). Vol. 1: Invertebrates. Springer, Berlin: 756–74.

Holland, N.D. (1991): Echinodermata: Crinoidea. In: *Reproduction of Marine Invertebrates*, Giese, A.C., Pearse, J.S. & Pearse, V.B. (eds.). Vol. 6: Echinoderms and lophophorates. The Boxwood Press, Pacific Grove: 247–99.

Holland, N.D., Clague, D.A., Gordon, D.P., Gebruck, A., Pawson, D.L. & Vecchione, M. (2005): 'Lophoenteropneust' hypothesis refuted by collection and photos of new deep-sea hemichordates. *Nature* **434**: 374–76.

Holland, N.D. & Nealson, K.H. (1978): The fine structure of the echinoderm cuticle and the subcuticular bacteria of echinoderms. *Acta Zool.* **59**: 169–85.

Holland, P.W.H. (2001): Beyond the Hox: how widespread is homeobox gene clustering? *J. Anat.* **199**: 13–23.

Holland, P.W.H. & Garcia-Fernàndez, J. (1996): Hox genes and chordate evolution. *Dev. Biol.* **173**: 382–95.

Holland, P.W.H., Garcia-Fernàndez, J., Williams, N.A. & Sidow, A. (1994): Gene duplications and the origins of vertebrate development. *Development* **1994 Suppl.**: 125–33.

Holmberg, K. (1984): A transmission electron microscopic investigation of the sensory vesicle in the brain of *Oikopleura dioica* (Appendicularia). *Zoomorphology* **104**: 298–303.

Holstein, T. & Hausmann, K. (1988): The cnidocil apparatus of hydrozoans: a progenitor of higher metazoan mechanoreceptors? In: *The Biology of Nematocysts*, Hessinger, D.A. & Lenhoff, H.M. (eds.). Academic Press, San Diego: 53–73.

Holstein, T. & Tardent, P. (1984): An ultrahigh-speed analysis of exocytosis: nematocyst discharge. *Science* **223**: 830–33.

Holterman, M., van der Wurff, A., van der Elsen, S., van Megen, H., Bongers, T., Holovachov, O., Bakker, J. & Helder, J. (2006): Phylum-wide analysis of SSU rDNA reveals deep phylogenetic relationships among nematodes and accelerated evolution towards crown clades. *Mol. Biol. Evol.* **23**: 1792–1800.

Hooge, M.D. (2001): Evolution of body-wall musculature in the Platyhelminthes (Acoelomorpha, Catenulida, Rhabditophora). *J. Morphol.* **249**: 171–94.

Hooge, M.D., Haye, P.A., Tyler, S., Litvaitis, M.K. & Kornfield, I. (2002): Molecular systematics of the Acoela (Acoelomorpha, Platyhelminthes) and its concordance with morphology. *Mol. Phyl. Evol.* **24**: 333–42.

Hooper, S.L. & Thuma, J.B. (2005): Invertebrate muscles: muscle specific genes and proteins. *Physiol. Rev.* **85**: 1001–60.

Hope, W.D. (1969): Fine structure of the somatic muscles of the free-living marine nematode *Deontostoma californicum* Steiner and Albin, 1933 (Leptosomatidae). *Proc. Helminthol. Soc. Washington* **36**: 10–29.

Hope, W.D. & Gardiner, S.L. (1982): Fine structure of a proprioreceptor in the body wall of the marine nematode *Deontostoma californicum* Steiner and Albin, 1933 (Enoplida: Leptosomatidae). *Cell Tiss. Res.* **225**: 1–10.

Horn, E. (1985): Gravity. In: *Comprehensive Insect Physiology, Biochemistry and Pharmacology*, Kerkut, G.A. & Gilbert, L.I. (eds.). Pergamon Press, Oxford: 557–76.

Horridge, G.A. (1969): Statocysts of medusae and evolution of stereocilia. *Tiss. Cell* **1**: 341–53.

Horridge, G.A. & Mackay, B. (1962): Naked axons and symmetrical synapses in coelenterates. *Q. J. Microsc. Sci.* **103**: 531–41.

Hou, X. & Bergström, J. (1994): Palaeoscolecid worms may be nematomorphs rather than annelids. *Lethaia* **27**: 11–17.

Hou, X. & Bergström, J. (1995): Cambrian lobopodians – ancestors of extant onychophorans? *Zool. J. Linn. Soc.* **114**: 3–19.

Hou, X.-G., Bergström, J. & Ahlberg, P. (1995): *Anomalocaris* and other large animals in the lower Cambrian Chengjiang fauna of southwest China. *Geol. Foren. Stockholm Forh.* **117**: 163–83.

Houk, M.S. & Hinegardner, R.T. (1980): The formation and early differentiation of sea urchin gonads. *Biol. Bull.* **159**: 280–94.

Huebner, E. & Diehl-Jones, W. (1992): Developmental biology of insect ovaries: germ cells and nurse cell-oocyte polarity. In: *Microscopic Anatomy of Invertebrates*, Harrison, F.W. & Locke, M. (eds.). Vol. 11C: Insecta. Wiley-Liss, New York: 957–93.

Hughes, R.L. & Woollacott, R.M. (1978): Ultrastructure of potential photoreceptor organs in the larva of *Scrupocellaria bertholetti* (Bryozoa). *Zoomorphologie* **91**: 225–34.

Hughes, R.L. & Woollacott, R.M. (1980): Photoreceptors of bryozoan larvae (Cheilostomata, Cellularioidea). *Zool. Scr.* **9**: 129–38.

Hummon, M.R. & Hummon, W.D. (1983a): Gastrotricha. In: *Reproductive Biology of Invertebrates*, Adiyodi, K.G. & Adiyodi, R.G. (eds.). Vol. 2: Spermatogenesis and sperm function. John Wiley & Sons, Chichester: 195–205.

Hummon, M.R. & Hummon, W.D. (1992): Gastrotricha. In: *Reproductive Biology of Invertebrates*, Adiyodi, K.G. & Adiyodi, R.G. (eds.). Vol. 5: Sexual differentiation and behaviour. John Wiley & Sons, Chichester: 137–46.

Hummon, M.R. & Hummon, W.D. (1993): Gastrotricha. In: *Reproductive Biology of Invertebrates*, Adiyodi, K.G. & Adiyodi, R.G. (eds.). Vol. 6, part A: Asexual propagation and reproductive strategies. John Wiley & Sons, Chichester: 265–77.

Hummon, W.D. & Hummon, M.R. (1983b): Gastrotricha. In: *Reproductive Biology of Invertebrates*, Adiyodi, K.G. & Adiyodi, R.G. (eds.). Vol. 1: Oogenesis, oviposition, and oosorption. John Wiley & Sons, Chichester: 211–21.

Hummon, W.D. & Hummon, M.R. (1989): Gastrotricha. In: *Reproductive Biology of Invertebrates*, Adiyodi, K.G. & Adiyodi, R.G. (eds.). Vol. 4, part A: Fertilization, development, and parental care. John Wiley & Sons, Chichester: 201–6.

Huxley, H. & Hanson, J. (1954): Changes in the cross-striations of muscle during conttraction and stretch and their structural interpretation. *Nature* **173**: 973–76.

Hwang, U.W., Friedrich, M., Tautz, D., Park, C.J. & Kim, W. (2001): Mitochondrial protein phylogeny joins myriapods with chelicerates. *Nature* **413**: 154–57.

Hyman, L.H. (1940): *The Invertebrates, Vol. 1: Protozoa through Ctenophora*. McGraw-Hill, New York.

Hyman, L.H. (1951): *The Invertebrates, Vol. 2: Platyhelminthes and Rhynchocoela. The Acoelomate Bilatera*. McGraw-Hill, New York.

Hyman, L.H. (1959): *The Invertebrates. Vol. 5: Smaller Coelomate Groups*. McGraw-Hill, New York.

Ilan, E., David, M.M. & Daniel, E. (1981): Erythrocruorin from the crustacean *Caenestheria inopinata*. Quartery structure and arrangement of subunits. *Biochem.* **20**: 6190–94.

Immesberger, A. & Burmester, T. (2004): Putative phenoloxidases in the tunicate *Ciona intestinalis* and the origin of the arthropod hemocyanin superfamily. *J. Comp. Physiol. B* **174**: 169–80.

Imsiecke, G. (1993): Ingestion, digestion, and egestion in *Spongilla lacustris* (Porifera, Spongillidae) after pulse feeding with *Chlamydomonas reinhardtii* (Volvocales). *Zoomorphology* **113**: 233–44.

Inglis, W.G. (1964): The structure of the nematode cuticle. *Proc. Zool. Soc. London* **143**: 465–502.

Iomini, C., Raikova, O.I., Noury-Stairi, N. & Justine, J.-L. (1995): Immunocytochemistry of tubulin in spermatozoa of Platyhelminthes. In: Advances in Spermatozoal Phylogeny and Taxonomy, Jamieson, B.G.M., Ausio, J. & Justine, J.-L. (eds.). *Mém. Mus. Natn. Hist. Nat.* **166**: 97–104.

Ishii, A.I. & Sano, M. (1980): Isolation and identification of paramyosin from liver fluke muscle layer. *Comp. Biochem. Physiol. B* **65**: 537–41.

Israelsson, O. (1999): New light on the enigmatic *Xenoturbella* (phylum uncertain): ontogeny and phylogeny. *Proc. R. Soc. London B* **266**: 835–41.

Israelsson, O. & Budd, G.E. (2005): Eggs and embryos in *Xenoturbella* (phylum uncertain) are not ingested prey. *Dev. Genes Evol.* **215**: 358–63.

Ivanov, D.L. (1996): Origin of Aculifera and problems of monophyly of higher taxa in molluscs. In: *Origin and Evolutionary Radiation of the Mollusca*, Taylor, J. (ed.). Oxford University Press, Oxford: 59–65.

Jaeckle, W.B. & Rice, M.E. (2002): Phylum Sipuncula. In: *Atlas of Marine Invertebrate Larvae*, Young, C.M. (ed.). Academic Press, San Diego: 375–96.

Jägersten, G. (1936): Zur Kenntnis der Parapodialborsten bei *Myzostomum*. *Zool. Bidr. Uppsala* **16**: 283–99.

Jägersten, G. (1939): Zur Kenntnis der Larvenentwicklung bei *Myzostomum*. *Ark. Zool.* **31**: 1–21.

Jägersten, G. (1955): On the early phylogeny of the Metazoa. The bilaterogastrea theory. *Zool. Bidr. Uppsala* **30**: 321–54.

Jägersten, G. (1959): Further remarks on the early phylogeny of the Metazoa. *Zool. Bidr. Uppsala* **33**: 79–108.

Jägersten, G. (1972): *Evolution of the Metazoan Life Cycle*. Academic Press, London.

James, M.A. (1997): Brachiopoda. internal anatomy, embryology, and development. In: *Microscopic Anatomy of Invertebrates*, Harrison, F.W. & Woollacott, R.M. (eds.), Vol. 13: Lophophorates, Entoprocta, and Cycliophora. Wiley-Liss, New York: 297–407.

James, M.A., Ansell, A.D., Curry, G.B., Collins, M.J., Peck, L.S. & Rhodes, M.C. (1992): The biology of living brachiopods. *Adv. Mar. Biol.* **28**: 175–387.

Jamieson, B.G.M. (1992): Oligochaeta. In: *Microsocpic Anatomy of Invertebrates*, Harrison, F.W. & Gardiner, S.L. (eds.), Vol. 7: Annelida. Wiley-Liss, New York: 217–322.

Jamieson, B.G.M. (1995a): Evolution of tetrapod spermatozoa with particular reference to amniotes. In: Advances in Spermatozoal Phylogeny and Taxonomy, Jamieson, B.G.M., Ausio, J. & Justine, J.-L. (eds.). *Mém. Mus. Natn. Hist. Nat.* **166**: 343–58.

Jamieson, B.G.M. (1995b): The ultrastructure of spermatozoa of the Squamata (Reptilia) with phylogenetic considerations. In: Advances in Spermatozoal Phylogeny and Taxonomy, Jamieson, B.G.M., Ausio, J. & Justine, J.L. (eds.). *Mém. Mus. Natn. Hist. Nat.* **166**: 359–83.

Jamieson, B.G.M. (2000a): Lophophorata. In: *Reproductive Biology of Invertebrates*, Adiyodi, K.G., Adiyodi, R.G. & Jamieson, B.G.M. (eds.). Vol. 9, part C: Progress in male gamete ultrastructure and phylogeny. John Wiley & Sons, Chichester: 143–61.

Jamieson, B.G.M. (2000b): Echinodermata. In: *Reproductive Biology of Invertebrates*, Adiyodi, K.G., Adiyodi, R.G. & Jamieson, B.G.M. (eds.). Vol. 9, part C: Progress in male gamete ultrastructure and phylogeny. John Wiley & Sons, Chichester: 163–246.

Jamieson, B.G.M. (2000c): Cephalochordata. In: *Reproductive Biology of Invertebrates*, Adiyodi, K.G., Adiyodi, R.G. & Jamieson, B.G.M. (eds.). Vol. 9, part C: Progress in male gamete ultrastructure and phylogeny. John Wiley & Sons, Chichester: 299–314.

Jamieson, B.G.M. (2000d): Chordata – Hemichordata. In: *Reproductive Biology of Invertebrates*, Adiyodi, K.G., Adiyodi, R.G. & Jamieson, B.G.M. (eds.). Vol. 9, part C: Progress in male gamete ultrastructure and phylogeny. John Wiley & Sons, Chichester: 247–60.

Jamieson, B.G.M., Ausio, J. & Justine, J.-L. (1995): Advances in spermatozoal phylogeny and taxonomy. *Mém. Mus. Natn. Hist. Nat.* **166**: 1–565.

Jamieson, B.G.M. & Tudge, C.C. (2000): Crustacea-Decapoda. In: *Reproductive Biology of Invertebrates*, Adiyodi, K.G., Adiyodi, R.G. & Jamieson, B.G.M. (eds.). Vol. 9, part C: Progress in male gamete ultrastructure and phylogeny. John Wiley & Sons, Chichester: 1–95.

Jamieson, B.G.M., Hodgson, A. & Spottiswoode, C.N. (2006): Ultrastructure of the spermatozoon of *Myrmecocichla formicivora* (Vieillot, 1881) and *Philetairus socius* (Latham, 1790) (Aves; Passeriformes), with a new interpretation of the passerine acrosome. *Acta Zool.* **87**: 297–304.

Jamuar, M.P. (1966): Studies of spermiogenesis in a nematode, *Nippostrongylus brasiliensis*. *J. Cell Biol.* **31**: 381–96.

Jefferies, R.P.S. (1968): The subphylum Calcichordata (Jefferies, 1967) – primitive fossil chordates with echinoderm affinities. *Bull. Brit. Mus. Nat. Hist. Geol.* **16**: 243–339.

Jefferies, R.P.S. (1975): Fossil evidence concerning the origin of the chordates. *Symp. Zool. Soc. London* **36**: 253–318.

Jefferies, R.P.S. (1981): In defence of calcichordates. *Zool. J. Linn. Soc.* **73**: 351–96.

Jellies, J. & Kristan, W.B. (1988): Embryonic assembly of a complex muscle is directed by a single identified cell in the medical leech. *J. Neurosci.* **8**: 3317–26.

Jellies, J. & Kristan, W.B. (1991): The oblique muscle organizer in *Hirudo medicinalis*, an identified embryonic cell projecting multiple parallel growth cones in an ordinary array. *Dev. Biol.* **148**: 334–54.

Jenison, G. & Nolte, J. (1979): The fine structure of the parietal retinas of *Anolis carolinensis* and *Iguana iguana*. *Cell Tiss. Res.* **199**: 235–47.

Jenner, R.A. (2001): Bilaterian phylogeny and uncritical recycling of morphological data sets. *Syst. Biol.* **50**: 730–42.

Jenner, R.A. (2002): Boolean logic and character state identity: Pitfalls of character coding in metazoan cladistics. *Contrib. Zool.* **71**: 67–91.

Jenner, R.A. (2003): Unleashing the force of cladistics? Metazoan phylogenetics and hypothesis testing. *Integr. Comp. Biol.* **43**: 207–18.

Jenner, R.A. (2004a): Accepting partnership by submission? Morphological phylogenetics in a molecular millenium. *Syst. Biol.* **53**: 333–42.

Jenner, R.A. (2004b): Towards a phylogeny of the Metazoa: evaluating alternative phylogenetic positions of Platyhelminthes, Nemertea, and Gnathostomulida, with a critical rerappraisal of cladistic characters. *Contrib. Zool.* **73**: 3–163.

Jenner, R.A. (2004c): When molecules and morphology clash: reconciling conflicting phylogenies of the Metazoa by considering secondary character loss. *Evol. Dev.* **6**: 372–378.

Jenner, R.A. (2006): Challenging received wisdoms: some contributions of the new microscopy to the new animal phylogeny. *Integr. Comp. Biol.* **46**: 93–103.

Jenner, R.A. & Scholtz, G. (2005): Playing another round of metazoan phylogenetics: historical epistemology, sensitivity analysis, and the position of Arthropoda within the Metazoa on the basis of morphology. In: Crustacea and Arthropod Relationships, Koenemann, S. & Jenner, R.A. (eds.), *Crust. Issues* **16**: 355–85.

Jenner, R.A. & Schram, F.R. (1999): The grand game of metazoan phylogeny: rules and strategies. *Biol. Rev.* **74**: 121–42.

Jennings, J.B. (1957): Studies on feeding, digestion, and food storage in free-living flatworms (Platyhelminthes: Turbellaria). *Biol. Bull.* **112**: 63–80.

Jennings, J.B. (1968): Nutrition and digestion. In: *Chemical Zoology*, Florkin, M. & Scheer, B.T. (eds.). Academic Press, New York: 303–26.

Jennings, J.B. (1988): Nutrition and respiration in symbiotic Turbellaria. *Fortschr. Zool./Progr. Zool.* **36**: 3–13.

Jennings, J.B. (1989): Epidermal uptake of nutrients in an unusual turbellarian parasitic in the starfish *Coscinasterias calamaria* in Tasmanian waters. *Biol. Bull.* **176**: 327–36.

Jennings, J.B. & Cannon, L.R.G. (1985): Observations on the occurrence, nutritional physiology and respiratory pigment of three species of flatworms (Rhabdocoela: Pterastericolidae) endosymbiotic in starfish from temperate and tropical waters. *Ophelia* **24**: 199–215.

Jennings, J.B. & Cannon, L.R.G. (1987): The occurrence, spectral properties and probable role of haemoglobins in four species of entosymbiotic turbellarians (Rhabdocoela: Umagillidae). *Ophelia* **27**: 143–54.

Jespersen, A. & Lützen, J. (1987): Ultrastructure of the nephridio-circulatory connections in *Tubulanus annulatus* (Nemertini, Anopla). *Zoomorphology* **107**: 181–89.

Jeuniaux, C. (1975) : Principes de systématique biochimique et application a quelques problèmes particuliers concernant les Aschelminthes, les polychètes et les tardigrades. *Cah. Biol. Mar.* **16**: 597–12.

Jeuniaux, C. (1982a): La chitine dans la règne animal. *Bull. Soc. Zool. France* **107**: 363–86.

Jeuniaux, C. (1982b): Composition chimique comparée des formations squelettiques chez les lophophoriens et les endoproctes. *Bull. Soc. Zool. France* **107**: 233–49.

Jochmann, R. & Schmidt-Rhaesa, A. (2007): New ultrastructural data from the larva of *Paragordius varius* (Nematomorpha). *Acta Zool.* **88**: 137–44.

Jördens, J., Struck, T. & Purschke, G. (2004): Phylogenetic inference regarding Parergodrilidae and *Hrabeiella periglandulata* ('Polychaeta', Annelida) based on 18S rDNA, 28S rDNA and COI sequences. *J. Zool. Syst. Evolut. Res.* **42**: 270–80.

Johnson, A.D., Drum, M., Bachvarova, R.F., Masi, T., White, M.E. & Crother, B.I. (2003): Evolution of predetermined germ cells in vertebrate embryos: implications for macroevolution. *Evol. Dev.* **5**: 414–31.

Johnson, K.B. & Zimmer, R.L. (2002): Phylum Phoronida. In: *Atlas of Marine Invertebrate Larvae*, Young, C.M. (ed.). Academic Press, San Diego: 429–39.

Jondelius, U., Ruiz-Trillo, I., Baguñà, J. & Riutort, M. (2001): Nemertodermatida, a basal bilaterian group. *Belg. J. Zool.* **131** (Suppl. 1): 59.

Jondelius, U., Ruiz-Trillo, I., Baguñà, J. & Riutort, M. (2002): The Nemertodermatida are basal bilaterians and not members of the Platyhelminthes. *Zool. Scr.* **31**: 185–200.

Jones, C.M. & Smith, J.C. (1995): Revolving vertebrates. *Curr. Biol.* **5**: 574–76.

Jones, H.D. (1983): The circulatory system of gastropods and bivalves. In: *The Mollusca*, Wilbur, K.M. (ed.). Vol. 5: Physiology, part 2, Saleuddin, A.S.M. & Wilbur, K.M. (volume eds.). Academic Press, New York: 189–238.

Jones, H.L.J., Leadbeater, B.S.C. & Green, J.C. (1994): An ultrastructural study of *Marsupiomonas pellucida* gen. et sp. nov., a new member of the Pedinophyceae. *Eur. J. Phycol.* **29**: 171–81.

Jones, J. (2002): Nematode sense organs. In: *The Biology of Nematodes*, Lee, D.L. (ed.). Taylor & Francis, London: 353–68.

Jones, L.J.F., Carballido-Lopez, R. & Errington, J. (2001): Control of cell shape in bacteria: helical, actin-like filaments in *Bacillus subtilis*. *Cell* **104**: 913–22.

Josephson, R.K. & Young, D. (1987): Fiber ultrastructure and contraction kinetics in insect fast muscles. *Am. Zool.* **27**: 991–1000.

Joshi, J.G. & Sullivan, B. (1973): Isolation and preliminary characterization of hemerythrin from *Lingula unguis*. *Comp. Biochem. Physiol. B* **44**: 857–67.

Juberthie, C., Manier, J.F. & Boissin, L. (1976): Étude ultrastructurale de la double-spermatogenèse chez l'opilion cryphophtalme *Siro rubens* Latreille. *J. Microsc. Biol. Cell.* **25**: 137–148.

Julian, D., Passmann, W.E. & Arp, A.J. (1996): Water lung and body wall contributions to respiration in an echiuran worm. *Respir. Physiol.* **106**: 187–98.

Junqueira, L.C.U., Carneiro, J. & Gratzl, M. (2005): *Histologie*, 6th Edn., Springer, Berlin.

Justine, J.-L. (1991a): Phylogeny of parasitic Platyhelminthes: a critical study of synapomorphies proposed on the basis of the ultrastructure of spermiogenesis and spermatozoa. *Can. J. Zool.* **69**: 1421–40.

Justine, J.-L. (1991b): The spermatozoa of the schistosomes and the concept of progenetic spermiogenesis. In: *Comparative Spermatology 20 Years After*, Baccetti, B. (ed.). Serono Symp. Publ. Raven Press 75. Raven Press, New York: 977–79.

Justine, J.-L. (2001): Spermatozoa as phylogenetic characters for the Platyhelminthes. In: Interrelationships of the Platyhelminthes, Littlewood, D.T. & Bray, R.A. (eds.). *Syst. Assoc. Spec. Vol. Ser.* **60**. Taylor & Francis, London: 231–38.

Justine, J.-L. (2002): Male and female gametes and fertilization. In: The Biology of Nematodes, Lee, D.L. (ed.). Taylor & Francis, London: 73–119.

Justine, J.-L. & Jamieson, B.G.M. (2000): Nematoda. In: *Reproductive Biology of Invertebrates*, Adiyodi, K.G., Adiyodi, R.G. & Jamieson, B.G.M. (eds.). Vol. 9, part B: Progress in male gamete ultrastructure and phylogeny. John Wiley & Sons, Chichester: 183–66.

Kaiser, D. (2001): Building a multicellular organism. *Ann. Rev. Gen.* **35**: 103–23.

Kalk, M. (1970): The organization of a tunicate heart. *Tiss. Cell* **2**: 99–118.

Kanzawa, N., Takano-Ohmuro, H. & Maruyama, K. (1995): Isolation and characterization of sea sponge myosin. *Zool. Sci.* **12**: 765–69.

Kardong, K. (2005): *Vertebrates: Comparative Anatomy, Function, Evolution.* 4th Edn.. McGraw-Hill, Boston.

Karpov, S.A. & Leadbeater, B.S. (1998): Cytoskeleton structure and composition in choanoflagellates. *J. Eukar. Microbiol.* **45**: 361–67.

Karpova, E.V., Alekseeva, O.V., Kalebina, T.S. & Kulaev, I.S. (2003): Revealling of collagen-like sequences in the microorganisms of different taxonomic groups. *FEMS Congress* **1**: 174.

Karuppaswamy, S.A. (1977): Occurrence of β-chitin in the cuticle of a pentastomid *Raillietiella gowrii*. *Experientia* **33**: 735–36.

Katagiri, N., Katagiri, Y., Shimatani, Y. & Hashimoto, Y. (1995): Cell type and fine structure of the retina of *Onchidium* stalk-eye. *J. Electron Microsc.* **44**: 219–30.

Katayama, T., Wada, H., Furuya, H., Satoh, N. & Yamamoto, M. (1995): Phylogenetic position of the dicyemid Mesozoa inferred from 18S rDNA sequences. *Biol. Bull.* **189**: 81–90.

Kaufman, W.R. (1999): Chelicerate arthropods. In: *Encyclopedia of Reproduction*, Knobil, E. & Neill, J.D. (eds.). Vol. 1. Academic Press, San Diego: 564–71.

Kearn, G.C. & Baker, N.O. (1973): Ultrastructural and histochemical observations on the pigmented eyes of the oncomiracidium of *Entobdella solae*, a monogenean skin parasite of the common sole, *Solea solea*. *Z. Parasitenkd.* **41**: 239–54.

Keil, T.A. (1998): The structure of integumental mechanoreceptors. In: *Microscopic Anatomy of Invertebrates*, Harrison, F.W. & Locke, M. (eds.). Vol. 11B: Insecta. Wiley-Liss, New York: 385–404.

Kent, M.L., Andree, K.B., Bartholomew, J.L., El-Matbouli, M., Desser, S.S., Devlin, R.H., Feist, S.W., Hedrick, R.P., Hoffman, R.W., Khattra, J., Hallett, S.L., Lester, R.J.G., Longshaw, M., Palenzeula, O., Siddall, M.E. & Xiao, C. (2001): Recent advantages in our knowledge of the Myxozoa. *J. Eukar. Microbiol.* **48**: 395–13.

Keough, E.M. & Summers, R.G. (1976): An ultrastructural investigation of the striated subumbrellar musculature of the anthomedusan, *Pennaria tiarella*. *J. Morphol.* **149**: 507–26.

Kerkut, G.A. & Gilbert, L.I. (1985): Nervous system: sensory. In: *Comprehensive Insect Physiology, Biochemistry and Pharmacology*, Kerkut, G.A. & Gilbert, L.I. (eds.). Vol. 6. Pergamon Press, Oxford.

Kieneke, A., Ahlrichs, W.H., Martínez Arbizu, P. & Bartolomaeus, T. (2007): Ultrastructure of protonephridia in *Xenotrichula carolinensis syltensis* and *Chaetonotus maximus* (Chaetonotida, Gastrotricha) – comparative evaluation of the gastrotrich excretory organs. Zoomorphology (in press).

Kier, W.M. & Curtin, N.A. (2002): Fast muscle in squid (*Loligo pealei*): contractile properties of a specialized musclke fibre type. *J. Exp. Biol.* **205**: 1907–16.

Kim, J., Kim, W. & Cunningham, C.W. (1999): A new perspective on Lower metazoan relationships from 18S rDNA sequences. *Mol. Biol. Evol.* **16**: 423–27.

Kimmel, C.B. (1996): Was Urbilateria segmented? *Trends Gen.* **12**: 329–331.

King, G.M. (1999): Hemichordata. In: *Encyclopedia of Reproduction*, Knobil, E. & Neill, J.D. (eds.). Vol. 2. Academic Press, San Diego: 599–603.

King, K.L., Steward, M., Roberts, T.M. & Seavy, M. (1992): Structure and macromolecular assembly of two isoforms of the major sperm protein (MSP) from the amoeboid sperm of the nematode, *Ascaris suum*. *J. Cell Sci.* **101**: 847–57.

King, N. (2004): The unicellular ancestry of animal development. *Dev. Cell* **7**: 313–25.

King, N. & Carroll, S.B. (2001): A receptor tyrosine kinase from choanoflagellates: molecular insights into early animal evolution. *Proc. Nat. Acad. Sci. USA* **98**: 15032–37.

King, N., Hittinger, C.T. & Carroll, S.B. (2003): Evolution of key signaling and adhesion protein families predates animal origins. *Science* **301**: 361–63.

King, P.E. (1973): *Pycnogonids*. St. Martin's Press, New York.

Kingsolver, J.G. & Pfennig, D.W. (2004): Individual-level selection as a cause of Cope's rule of phyletic size increase. *Evolution* **58**: 1608–12.

Kirk, D.L. (1998): *Volvox. Molecular-Genetic Origins of Multicellularity and Cellular Differentiation*. Cambridge University Press, Cambridge.

Kirk, D.L. & Nishii, I. (2001): *Volvox carteri* as a model for studying the genetic and cytological control of morphogenesis. *Dev. Growth Differ.* **43**: 621–31.

Klein, D.C. (2004): The 2004 Aschoff/Pittsburgh lecture: theory of the origin of the pineal gland – a tale of conflict and resolution. *J. Biol. Rhythms* **19**: 264–79.

Klippenstein, G.L. (1980): Structural aspects of hemerythrin and myohemerythrin. *Am. Zool.* **20**: 39–51.

Kluge, B. & Fischer, A. (1990): The pronephros of the early ammocoete larva of lampreys (Cyclostomata, Petromyzontes): fine structure of the external glomus. *Cell Tiss. Res.* **260**: 249–59.

Kluge, B. & Fischer, A. (1991): The pronephros of the early ammocoete larva of lampreys (Cyclostomata, Petromyzontes): fine structure of the renal tubules. *Cell Tiss. Res.* **263**: 515–28.

Knauss, E.B. (1979): Indication of an anal pore in Gnathostomulida. *Zool. Scr.* **8**: 181–86.

Knauss, E.B. & Rieger, R.M. (1979): Fine structure of the male reproductive system in two species of *Haplognathia* Sterrer (Gnathostomulida, Filospermoidea). *Zoomorphologie* **94**: 33–48.

Knight-Jones, E.W. (1952): On the nervous system of *Saccoglossus cambrensis* (Enteropneusta). *Phil. Trans. R. Soc. London B* **236**: 315–54.

Knoll, A.H. (1991): End of the proterozoic eon. *Sci. Am.* **October 1991**: 42–49.

Knoll, A.H. (1992): The early evolution of eucaryotes: a geological perspective. *Science* **256**: 622–27.

Knust, E. & Bossinger, O. (2002): Composition and formation of intercellular junctions in epithelial cells. *Science* **298**: 1955–59.

Kobayashi, M., Furuya, H. & Holland, P.W.H. (1999): Dicyemids are higher animals. *Nature* **401**: 762.

Kobayashi, M., Wada, H. & Satoh, N. (1996): Early evolution of the Metazoa and phylogenetic status of diploblasts as inferred from amino-acid sequence of elongation factor-1α. *Mol. Phyl. Evol.* **5**: 414–22.

Koehler, L.D. (1995): Diversity of avian spermatozoa ultrastructure with emphasis on the members of the order Passeriformes. In: Advances in Spermatozoal Phylogeny and Taxonomy, Jamieson, B.G.M., Ausio, J. & Justine, J.-L. (eds.). *Mém. Mus. Natn. Hist. Nat.* **166**: 437–44.

Kojima, D., Terakita, A., Ishikawa, T., Tsukahara, Y., Maeda, A. & Shichida, Y. (1997): A novel G_0-mediated phototransduction cascade in scallop visual cells. *J. Biol. Chem.* **272**: 22979–82.

Kole, A.P. (1965): Flagella. In: *The Fungi. An Advanced Treatise*, Ainsworth, G.C. & Sussman, A.S. (eds.). Academic Press, New York: 77–93.

Kornegay, J.R., Schilling, J.W. & Wilson, A.C. (1994): Molecular adaptation of a leaf-eating bird: stomach lysozyme of the hoatzin. *Mol. Biol. Evol.* **11**: 921–28.

Koyama, H. & Kusunoki, T. (1993): Organization of the cerebral ganglion of the colonial ascidian *Polyandrocarpa misakiensis*. *J. Comp. Neurol.* **338**: 549–59.

Kozloff, E.N. (1969): Morphology of the orthonectid *Rhopalura ophiocomae*. *J. Parasitol.* **55**: 171–95.

Kraus, O. (2001): 'Myriapoda' and the ancestry of the Hexapoda. *Ann. Soc. Entomol. France* **37**: 105–27.

Kraus, O. & Kraus, M. (1994): Phylogenetic system of the Tracheata (Mandibulata): on 'Myriapoda' – Insecta interrelationships, phylogenetic age and primary ecological niches. *Verh. Naturwiss. Ver. Hamburg* **34**: 5–31.

Kraus, Y. & Technau, U. (2006): Gastrulation in the sea anemone *Nematostella vectensis* occurs by invagination and immigration: an ultrastructural study. *Dev. Genes Evol.* **216**: 119–32.

Kristensen, R.M. (1978): On the structure of *Batilipes noerrevangi* Kristensen 1978. 2. The muscle-attachments and the true cross-striated muscles. *Zool. Anz.* **200**: 173–84.

Kristensen, R.M. (1979): On the fine structure of *Batilipes noerrevangi* Kristensen, 1978 (Heterotardigrada). 3. Spermiogenesis. *Zesz. Nauk. Uniw. Jagiellonsk.* **79**: 99–105.

Kristensen, R.M. (1981): Sense organs of two marine arthrotardigrades (Heterotardigrada, Tardigrada). *Acta Zool.* **62**: 27–41.

Kristensen, R.M. (1982): The first record of cyclomorphosis in Tardigrada based on a new genus and species from Arctic meiobenthos. *Z. Zool. Syst. Evolutionsforsch.* **20**: 249–70.

Kristensen, R.M. (1983): Loricifera, a new phylum with Aschelminthes characters from the meiobenthos. *Z. Zool. Syst. Evolutionsforsch.* **21**: 163–80.

Kristensen, R.M. (1991): Loricifera. In: *Microscopic Anatomy of Invertebrates*, Harrison, F.W. & Ruppert, E.E. (eds.), Vol. 4: Aschelminthes. Wiley-Liss, New York: 351–75.

Kristensen, R.M. (1995): Are Aschelminthes pseudocoelomate or acoelomate? In: *Body Cavities: Function and Phylogeny*, Lanzavecchia, G., Valvassori, R. & Candia

Carnevali, M.D. (eds.). Selected Symposia and Monographs U.Z.I. 8. Mucchi, Modena: 41–43.

Kristensen, R.M. & Brooke, S. (2002): Phylum Loricifera. In: *Atlas of Marine Invertebrate Larvae*, Young, C.M. (ed.). Academic Press, San Diego: 179–87.

Kristensen, R.M. & Funch, P. (2000): Micrognathozoa: A new class with complicated jaws like those of Rotifera and Gnathostomulida. *J. Morphol.* **246**: 1–49.

Kristensen, R.M. & Gad, G. (2004): *Armoloricus*, a new genus of Loricifera (Nanaloricidae) from Trezen ar Skoden (Roscoff, France). *Cah. Biol. Mar.* **45**: 121–56.

Kristensen, R.M. & Hay-Schmidt, A. (1989): The protonephridia of the arctic kinorhynch *Echinoderes aquilonius* (Cyclorhagida, Echinoderidae). *Acta Zool.* **70**: 13–27.

Kristensen, R.M. & Higgins, R.P. (1991): Kinorhyncha. In: *Microscopic Anatomy of Invertebrates*, Harrison, F.W. & Ruppert, E.E. (eds.), Vol. 4: Aschelminthes. Wiley-Liss, New York: 377–404.

Kristensen, R.M. & Nørrevang, A. (1977): On the fine structure of *Rastrognathia macrostoma* gen. et sp. n. placed in Rastrognathiidae fam. n. (Gnathostomulida). *Zool. Scr.* **6**: 27–41.

Kroiher, M., Siefker, B. & Berking, S. (2000): Induction of segmentation in polyps of *Aurelia aurita* (Scyphozoa, Cnidaria) into medusae and formation of mirror-image medusa anlagen. *Int. J. Dev. Biol.* **44**: 485–90.

Krol, R.M., Hawkins, W.E. & Overstreet, R.M. (1992): Reproductive components. In: *Microscopic Anatomy of Invertebrates*, Harrison, F.W. & Humes, A.G. (eds.). Vol. 10: Decapod Crustacea. Wiley-Liss, New York: 295–343.

Kruse, M., Leys, S.P., Müller, I.M. & Müller, W.E.G. (1998): Phylogenetic position of the Hexactinellida within the phylum Porifera based on the amino acid sequence of the protein kinase C from *Rhabdocalyptus dawsoni*. *J. Mol. Evol.* **46**: 721–28.

Kümmel, G. (1964): Die Feinstruktur der Terminalzellen (Cyrtocyten) an den Protonephridien der Priapuliden. *Z. Zellforsch.* **62**: 468–84.

Kümmel, G. (1975): The physiology of protonephridia. *Fortschr. Zool./Progr. Zool.* **23**: 18–32.

Kumar, N.M. & Gilula, N.B. (1996): The gap junction communication channel. *Cell* **84**: 381–88.

Kusche, K. & Burmester, T. (2001): Diplopod hemocyanin sequence and the phylogenetic position of the Myriapoda. *Mol. Biol. Evol.* **18**: 1566–73.

Kusche, K., Ruhberg, H. & Burmester, T. (2002): A hemocyanin from the Onychophora and the emergence of respiratory proteins. *Proc. Nat. Acad. Sci. USA* **99**: 10545–48.

Kutsch, W. & Breidbach, O. (1994): Homologous structures in the nervous system of Arthropoda. *Adv. Insect. Physiol.* **24**: 1–113.

LaBarbera, M. (1986): The evolution and ecology of body size. In: *Patterns and Processes in the History of Life*, Raup, D.M. & Jablonski, D. (eds.). Dahlem-Konferenzen 1986. Springer-Verlag, Berlin: 69–98.

Labat-Robert, J., Robert, L., Auger, C., Lethias, C. & Garrone, R. (1981): Fibronectin-like protein in Porifera: its role in cell aggregation. *Proc. Natl. Acad. Sci. USA* **78**: 6261–65.

Lacalli, T.C. (1981): Structure and development of the apical organ in trochophores of *Spirobranchus polycerus*, *Phyllodoce maculata* and *Phyllodoce mucosa* (Polychaeta). *Proc. R. Soc. London B* **212**: 381–402.

Lacalli, T.C. (1982): The nervous system and ciliary band of Müller's larva. *Proc. R. Soc. London B* **217**: 37–58.

Lacalli, T.C. (1983): The brain and central nervous system of Müller's larva. *Can. J. Zool.* **61**: 39–51.

Lacalli, T.C. (1988): Ciliary band patterns and pattern rearrangements in the development of the doliolaria larva. In: *Echinoderm Biology*, Burke, R.D., Mladenov, P. V. Lambert, P. & Parsley, R.L. (eds.). Balkema, Rotterdam: 273–74.

Lacalli, T.C. (1994): Apical organs, epithelial domains, and the origin of the chordate central nervous system. *Am. Zool.* **34**: 533–41.

Lacalli, T.C. (1996): Dorsoventral axis inversion: a phylogenetic perspective. *BioEssays* **18**: 251–54.

Lacalli, T.C. (2002): Vetulicolians – are they deuterostomes? Chordates? *BioEssays* **24**: 208–11.

Lacalli, T.C. & Holland, L.Z. (1998): The developing dorsal ganglion of the salp *Thalia democratica*, and the nature of the ancestral chordate brain. *Phil. Trans. R. Soc. London B* **353**: 1943–67.

Lacalli, T.C. & West, J.E. (1985): The nervous system of a pilidium larva: evidence from electron microscope reconstructions. *Can. J. Zool.* **63**: 1909–16.

Ladurner, P. & Rieger, R. (2000): Embryonic muscle development of *Convoluta pulchra* (Turbellaria – Acoelomorpha, Platyhelminthes). *Dev. Biol.* **222**: 359–75.

Ladurner, P, Rieger, R. & Baguñà, J. (2000): Spatial distribution and differentiation potential of stem cells in hatchlings and adults in the marine platyhelminth *Macrostomum* sp.: a bromodeoxyuridine analysis. *Dev. Biol.* **226**: 231–41.

Lafay, B., Boury-Esnault, N., Vacelet, J. & Christen, R. (1992): An analysis of partial 28S ribosomal RNA sequences suggests early radiations of sponges. *BioSystems* **28**: 139–51.

Lai-Fook, J. & Beaton, C. (1998): Muscle insertions. In: *Microscopic Anatomy of Invertebrates*, Harrison, F.W. & Locke, M. (eds.), Vol. 11B: Insecta. Wiley-Liss, New York: 573–80.

Lammert, V. (1984): The fine structure of spiral ciliary receptors in Gnathostomulida. *Zoomorphology* **104**: 360–64.

Lammert, V. (1985): The fine structure of protonephridia in Gnathostomulida and their comparison within Bilateria. *Zoomorphology* **105**: 308–16.

Lammert, V. (1986): Ciliäre Rezeptoren der Gnathostomulida. *Verh. Dtsch. Zool. Ges.* **79**: 283–84.

Lammert, V. (1989): The fine structure of the epidermis in Gnathostomulida. *Zoomorphology* **109**: 131–44.

Lammert, V. (1991): Gnathostomulida. In: *Microscopic Anatomy of Invertebrates*, Harrison, F.W. & Ruppert, E.E. (eds.), Vol. 4: Aschelminthes. Wiley-Liss, New York: 19–39.

Land, M.F. (1984): Molluscs. In: *Photoreception and Vision in Invertebrates*, Ali, M.A. (ed.). Plenum Press, New York: 699–725.

Land, M.F. (1991): Optics of the eyes of the animal kingdom. In: *Vision and Visual Dysfunction, Vol. 2: Evolution of the Eye and Visual System*, Cronly-Dillon, J.R. & Gregory, R.L. (eds.). Macmillan Press, Houndmills: 118–35.

Land, M.F. & Nilsson, D.-E. (2004): *Animal Eyes*. Oxford University Press, Oxford.

Lanfranchi, A. & Bedini, C. (1986): Electron microscopic study of larval eye development in Turbellaria Polycladida. *Hydrobiologia* **132**: 121–26.

Lanfranchi, A. & Falleni, A. (1998): Ultrastructural observations on the male gametes in *Austrognathia* sp. (Gnathostomulida, Bursovaginoidea). *J. Morphol.* **237**: 165–76.

Lanfranchi, A., Bedini, C. & Ferrero, E. (1981): The ultrastructure of the eyes in larval and adult polyclads (Turbellaria). *Hydrobiologia* **84**: 267–75.

Lang, A. (1882): Der Bau von *Gunda segmentata* und die Verwandtschaft der Plathelminthen mit Coelenteraten und Hirudineen. *Mitt. Zool. Stat. Neapel* **3**: 187–251.

Lang, B.F., O'Kelly, C., Nerad, T., Gray, M.W. & Burger, G. (2002): The closest unicellular relatives of animals. *Curr. Biol.* **12**: 1773–78.

Langenbruch, P.-F. (1985): Die Aufnahme partikulärer Nahrung bei *Reniera* sp. (Porifera). *Helgol. Meeresunt.* **39**: 263–72.

Lans, D., Wedeen, C.J. & Weisblat, D.A. (1993): Cell lineage analysis of the expression of an *engrailed* homolog in leech embryos. *Development* **117**: 857–71.

Lanzavecchia, G. (1977): Morphological modulations in helical muscles (Aschelminthes and Annelida). *Int. Rev. Cytol.* **51**: 133–86.

Lanzavecchia, G. (1981): Morphofunctional and phylogenetic relations in helical muscles. *Boll. Zool.* **48**: 29–40.

Lanzavecchia, G. & Camatini, M. (1979): Phylogenetic problems and muscle cell ultrastructure in Onychophora. In: *Myriapod Biology*, Camatini, M. (ed.), Academic Press, London: 407–17.

Lanzavecchia, G., de Eguileor, M. & Scari, G. (1995): Body cavities of Nematomorpha. In: *Body Cavities: Function and Phylogeny*, Lanzavecchia, G., Valvassori, R. & Candia Carnevali, M.D. (eds.). Selected Symposia and Monographs U.Z.I. 8. Mucchi, Modena: 45–60.

Lanzavecchia, G., de Eguileor, M. & Valvassori, R. (1988): Muscles. In: The Ultrastructure of Polychaeta, Westheide, W. & Hermans, C.O. (eds.). *Microfauna Marina* **4**: 71–88.

Lanzavecchia, G., Valvassori, R., de Eguileor, M. & Lanzavecchia, P. (1979): Three-dimensional reconstruction of the contractile system of the Nematomorpha muscle fiber. *J. Ultrastruct. Res.* **66**: 201–23.

Larkman, A.U. (1983): An ultrastructural study of oocyte growth within the entoderm and entry into the mesogloea in *Actinia fragacea* (Cnidaria, Anthozoa). *J. Morphol.* **178**: 155–77.

Larkman, A.U. (1984): An ultrastructural study of the establishment of the testicular cysts during spermatogenesis in the sea anemone *Actinia fragacea* (Cnidaria: Anthozoa). *Gamete Res.* **9**: 303–27.

Laska-Mehnert, G. (1985): Cytomorphologische Veränderungen während der Metamorphose des Cubopolypen *Tripedalia cystophora* (Cubozoa, Carybdaeidae) in die Meduse. *Helgol. Meeresunters.* **39**: 129–64.

Laurie, G.W. & Leblond, C.P. (1985): Basement membrane nomenclature. *Nature* **313**: 372.

Lauterbach, K.-E. (1978): Gedanken zur Evolution der Euarthropoden-Extremität. *Zool. Jahrb. Anat.* **99**: 64–92.

Lauterbach, K.-E. (1983): Zum Problem der Monophylie der Crustacea. *Verh. Naturwiss. Ver. Hamburg* **26**: 293–320.

Lauterbach, K.-E. (1984): Das Phylogenetische System der Mollusca. *Mitt. Dtsch. Malak. Ges.* **37**: 66–81.

Lavrov, D.V., Forget, L., Kelly, M. & Lang, B.F. (2005): Mitochondrial genomes of two demosponges provide insight into an early stage of animal evolution. *Mol. Biol. Evol.* **22**: 1231–39.

Lawn, I.D., Mackie, G.O. & Silver, G. (1981): Conduction system in a sponge. *Science* **211**: 1169–71.

Leadbeater, B.S.C. (1983): Life-history and ultrastructure of a new marine species of Proterospongia (Choanoflagellida). *J. Mar. Biol. Assoc. UK* **63**: 135–60.

Leadbeater, B.S.C. (1994): Developmental studies on the loricate choanoflagellate *Stephanoeca diplocostata* Ellis. VIII. Nuclear division and cytokinesis. *Eur. J. Protistol.* **30**: 171–83.

Leasi, F., Rothe, B.H., Schmidt-Rhaesa, A. & Todaro, M.A. (2006): The musculature of three species of gastrotrichs surveyed with Confocal Laser Scanning Microscopy (CLSM). *Acta Zoologica* **87**: 171–80.

Lecointre, G., Philippe, H., Lê, H.L.V. & le Guyader, H. (1993): Species sampling has a major impact on phylogenetic inference. *Mol. Phyl. Evol.* **2**: 205–24.

Ledger, P.W. (1975): Septate junctions in the calcareous sponge *Sycon ciliatum*. *Tiss. Cell* **7**: 13–18.

Lee, D.L. & Smith, M.H. (1965): Hemoglobins of parasitic animals. *Exp. Parasitol.* **16**: 392–424.

Leitz, T. (2001): Endocrinology of the Cnidaria: state of the art. *Zoology* **103**: 202–21.

Lemburg, C. (1995a): Ultrastructure of sense organs and receptor cells of the neck and lorica of the *Halicryptus spinulosus* larva (Priapulida). *Microfauna Marina* **10**: 7–30.

Lemburg, C. (1995b): Ultrastructure of the introvert and associated structures of the larvae of *Halicryptus spinulosus* (Priapulida). *Zoomorphology* **115**: 11–29.

Lemburg, C. (1998): Electron microscopical localization of chitin in the cuticle of *Halicryptus spinulosus* and *Priapulus caudatus* (Priapulida) using gold-labelled wheat germ agglutinin: phylogenetic implications for the evolution of the cuticle within the Nemathelminthes. *Zoomorphology* **118**: 137–58.

Lemburg, C. (1999): *Ultrastrukturelle Untersuchungen an den Larven von Halicryptus spinulosus und Priapulus caudatus. Hypothesen zur Phylogenie der Priapulida und deren Bedeutung für die Evolution der Nemathelminthen.* Cuvillier Verlag, Göttingen.

Lemburg, C. & Schmidt-Rhaesa, A. (1999): Priapulida. In: *Encyclopedia of Reproduction*, Knobil, E. & Neill, J.D. (eds.). Vol. 1. Academic Press, San Diego: 1053–58.

Lemche, H., Hansen, B., Madsen, F.J., Tendal, O. & Wolff, T. (1976): Hadal life as analyzed from photographs. *Vidensk. Medd. Dansk. Naturh. Foren.* **139**: 263–336.

Lemche, H. & Wingstrand, K.G. (1959): The anatomy of *Neopilina galatheae* Lemche, 1957 (Mollusca, Tryblidiacea). *Galathea Rep.* **3**: 9–71.

Lentz, T.L. (1968): *Primitive Nervous Systems*. Yale University Press, New Haven.

Lesh-Laurie, G.E. & Suchy, P.E. (1991): Cnidaria: Scyphozoa and Cubozoa. In: *Microscopic Anatomy of Invertebrates*, Harrison, F.W. & Westfall, J.A. (eds.), Vol. 2: Placozoa, Porifera, Cnidaria, and Ctenophora. Wiley-Liss, New York: 185–266.

Lester, S.M. (1988): Ultrastructure of adult gonads and development and structure of the larva of *Rhabdopleura normani* (Hemichordata: Pterobranchia). *Acta Zool.* **69**: 95–109.

Levinton, J.S. (2001): *Genetics, Paleontology, and Macroevolution*. 2nd Edn. Cambridge University Press, Cambridge.

Leys, S.P. (2003): The significance of syncytial tissues for the position of the Hexactinellida in the Metazoa. *Integr. Comp. Biol.* **43**: 19–27.

Leys, S.P. & Degnan, B.M. (2001): Cytological basis of photosensitive behaviour in a sponge larva. *Biol. Bull.* **201**: 323–38.

Leys, S.P. & Degnan, B.M. (2002): Embryogenesis and metamorphosis in a haplosclerid demosponge: gastrulation and transdifferentiation of larval ciliated cells to choanocytes. *Invertebr. Biol.* **121**: 171–89.

Leys, S.P. & Mackie, G.O. (1997): Electrical recording from a glass sponge. *Nature* **387**: 29–30.

Leys, S.P., Cronin, T.W., Degnan, B.M. & Marshall, J.N. (2002): Spectral sensitivity in a sponge larva. *J. Comp. Physiol. A* **188**: 199–202.

Leys, S.P., Mackie, G.O. & Meech, R.W. (1999): Impulse conduction in a sponge. *J. Exp. Biol.* **202**: 1139–50.

Leys, S.P. & Reiswig, H.M. (1998): Transport pathways in the neotropical sponge *Aplysina*. *Biol. Bull.* **195**: 30–42.

Liem, K., Bemis, W., Walker, W.F. & Grande, L. (2000) *Functional Anatomy of the Vertebrates: an Evolutionary Perspective*. 3rd Edn. Thomson, Brooks & Cole, Belmont.

Liesenjohann, T., Neuhaus, B. & Schmidt-Rhaesa, A. (2006): Head sensory organs of *Dactylopodola baltica* (Macrodasyida, Gastrotricha): A combination of transmission electron microscopical and immunocytochemical techniques. *J. Morphol.* **267**: 897–908.

Ling, E.A. (1969): The structure and function of the cephalic organ of a nemertine *Lineus ruber*. *Tiss. Cell* **1**: 503–24.

Ling, E.A. (1970): Further investigations on the structure and function of cephalic organs of a nemertine *Lineus ruber*. *Tiss. Cell* **2**: 569–88.

Littlewood, D.T.J. & Bray, R.A. (2001): Interrelationships of the Platyhelminthes. *Syst. Assoc. Spec. Vol.* **60**, Taylor & Francis, London.

Littlewood, D.T.J. & Olson, P.D. (2001): Small subunit rDNA and the Platyhelminthes: signal, noise, conflict and compromise. In: Interrelationships of the Platyhelminthes, Littlewood, D.T.J. & Bray, R.A. (eds.). *Syst. Assoc. Spec. Vol.* **60**: 262–78.

Littlewood, D.T.J., Olson, P.D., Telford, M.J., Herniou, E.A. & Riutort, M. (2001): Elongation factor 1-α sequences alone do not assist in resolving the position of the Acoela within the Metazoa. *Mol. Biol. Evol.* **18**: 437–42.

Littlewood, D.T.J., Telford, M., Clough, K.A. & Rohde, K. (1998): Gnathostomulida – an enigmatic metazoan phylum from both morphological and molecular perspectives. *Mol. Phyl. Evol.* **9**: 72–79.

Locke, J.M. (2000): Ultrastructure of the statocyst of the marine enchytraeid *Grania americana* (Annelida: Clitellata). *Invertebr. Biol.* **119**: 83–93.

Locke, M. & Huie, P. (1977): Bismuth staining of Golgi complex is a characteristic arthropod feature lacking in *Peripatus*. *Nature* **270**: 341–43.

Lom, J. (1990): Phylum Myxozoa. In: *Handbook of Protoctista*, Margulis, L., Corliss, J.O., Melkonian, M. & Chapman, D.J. (eds.), Jones and Bartlett Publishers, Boston: 36–52.

Lom, J. & Dykova, I. (1997): Ultrastructural features of the actinosporean phase of *Myxosporea* (phylum Myxozoa): a comparative study. *Acta Protozool.* **36**: 83–103.

Lomolino, M.V. (1985): Body sizes of mammals on islands: the island rule reexamined. *Am. Nat.* **125**: 310–16.

Long, J.A. & Stricker, S.A. (1991): Brachiopoda. In: *Reproduction of Marine Invertebrates*, Giese, A.C., Pearse, J.S. & Pearse, V.B. (eds.). Vol. 6: Echinoderms and lophophorates. The Boxwood Press, Pacific Grove: 47–84.

Lopez-Bernad, F., del Cacho, E., Gallego, M., Quilez, J. & Sanchez-Acedo, C. (1996): Identification of a fibronectin-like molecule on *Eimeria tenella*. *Parasitology* **113**: 505–510.

Lora Lamia Donin, C. & Cotelli, F. (1977): The rod-shaped sperm of Gordioidea (Aschelminthes, Nematomorpha). *J. Ultrastruct. Res.* **61**: 193–200.

Lorenzen, S. (1978): Discovery of stretch receptor organs in nematodes – structure, arrangement and functional analysis. *Zool. Scr.* **7**: 175–78.

Lorenzen, S. (1985): Phylogenetic aspects of pseudocoelomate evolution. In: The origins and Relationships of Lower Invertebrates, Conway Morris, S., George, J.D., Gibson, D.I. & Platt, H.M. (eds.). *Syst. Assoc. Spec. Vol.* **28**. Clarendon Press, Oxford: 210–23.

Lorenzen, S. (1994): *The Phylogenetic Systematics of Freeliving Nematodes*. The Ray Society, London.

Lotmar, W. & Picken, L.E.R. (1950): A new crystallographic modification of chitin and its distribution. *Experientia* **6**: 58–59.

Lowe, C.J., Terasaki, M., Wu, M., Freeman, R.M., Runft, L., Kwan, K., Haigo, S., Aronowicz, J., Lander, E., Gruber, C., Smith, M., Kirschner, M. & Gerhart, J. (2006): Dorsoventral patterning in hemichordates: insights into early chordate evolution. *PloS Biology* **4**: 1603–19 (e291).

Lowe, C.J. & Wray, G.A. (1997): Radical alterations in the roles of homeobox containing genes during echinoderm evolution. *Nature* **389**: 718–21.

Lowe, C.J., Wu, M., Salic, A., Evans, L., Lander, E., Stange-Thormann, N., Gruber, C.E., Gerhart, J. & Kirschner, M. (2003): Anteroposterior patterning in hemichordates and the origin of the central nervous system. *Cell* **113**: 853–65.

Lüter, C. (1995): Ultrastructure of the metanephridia of *Terebratulina retusa* and *Crania anomala* (Brachiopoda). *Zoomorphology* **115**: 99–107.

Lüter, C. (1996): The median tentacle of the larva of *Lingula anatina* (Brachiopoda) from Queensland, Australia. *Aus. J. Zool.* **44**: 355–66.

Lüter, C. (1997): *Zur Ultrastruktur, Ontogenese und Phylogenie der Brachiopoda*. Cuvillier Verlag, Göttingen.

Lüter, C. (2000a): The origin of the coelom in Brachiopoda and its phylogenetic significance. *Zoomorphology* **120**: 15–28.

Lüter, C. (2000b): Ultrastructure of larval and adult setae of Brachiopoda. *Zool. Anz.* **239**: 75–90.

Lüter, C. (2001): Brachiopod larval setae – a key to the phylum's ancestral life cycle? In: Brachiopods Past and Present, Brunton, C.H.C., Cocks, L.R.M. & Long, S.L. (eds.). *Syst. Assoc. Spec. Vol.* **63**. Taylor & Francis, London: 46–55.

Lüter, C. (2004): Die Tentaculata im phylogenetischen System der Bilateria – gehören sie zu den Radialia oder den Lophotrochozoa? In: Kontroversen in der Phylogenetischen Systematik der Metazoa, Richter, S. & Sudhaus, W. (eds.). *Sitzungsber. Ges. Naturforsch. Freunde Berlin* **43**: 103–22.

Lüter, C. & Bartolomaeus, T. (1997): The phylogenetic position of Brachiopoda – a comparison of morphological and molecular data. *Zool. Scr.* **26**: 245–53.

Lützen, J. (1968): Unisexuality in the parasitic family Entoconchidae (Gastropoda: Prosobranchia). *Malacologia* **7**: 7–15.

Lumsden, R.D. & Foor, W.E. (1968): Electron microscopy of schistosome cercarial muscle. *J. Parasitol.* **54**: 780–94.

Lundin, K. (1997): Comparative ultrastructure of the epidermal ciliary rootlets and associated structures in species of the Nemertodermatida and Acoela (Plathelminthes). *Zoomorphology* **117**: 81–92.

Lundin, K. (1998): The epidermal ciliary rootlets of *Xenoturbella bocki* (Xenoturbellida) revisited: new support for a possible kinship with the Acoelomorpha. *Zool. Scr.* **27**: 263–270.

Lundin, K. (2001): Degenerating epidermal cells in *Xenoturbella bocki* (phylum uncertain), Nemertodermatida and Acoela (Platyhelminthes). *Belg. J. Zool.* **131** (Suppl. 1): 153–57.

Lundin, K. & Hendelberg, J. (1998): Is the sperm type of the Nemertodermatida close to that of the ancestral Platyhelminthes? *Hydrobiologia* **383**: 197–205.

Lundin, K. & Sterrer, W. (2001): The Nemertodermatida. In: Interrelationships of the Platyhelminthes, Littlewood, D.T.J. & Bray, R.A. (eds.). *Syst. Assoc. Spec. Vol.* **60**. Taylor & Francis, London: 24–27.

Lyons, K.M. (1972): Sense organs of monogeneans. In: Behavioural *Aspects of Parasite Transmission*, Canning, E.U. (ed.). Academic Press, London: 181–99.

Mackey, L.Y., Winnepenninckx, B., De Wachter, R., Backeljau, T., Emschermann, P. & Garey, J.R. (1996): 18S rRNA suggests that entoprocts are protostomes, unrelated to Ectoprocta. *J. Mol. Evol.* **42**: 552–59.

Mackie, G.O. (1973): Report on giant nerve fibres in *Nanomia*. *Publ. Seto Mar. Biol. Lab.* **20**: 745–56.

Mackie, G.O. (1978): Coordination in physonectid siphonophores. *Mar. Behav. Physiol.* **5**: 325–46.

Mackie, G.O. (1979): Is there a conduction system in sponges? In: Biologie des Spongiaires – Sponge Biology, Levi, C. & Boury-Esnault, N. (eds.). *Colloq. Int. C.N.R.S.* **291**: 145–51.

Mackie, G.O. (1990): The elementary nervous system revisited. *Am. Zool.* **30**: 907–20.

Mackie, G.O. (1995): On the 'visceral nervous system' of *Ciona*. *J. Mar. Biol. Assoc. UK* **75**: 141–51.

Mackie, G.O. & Burighel, P. (2005): The nervous system in adult tunicates: current research directions. *Can J. Zool.* **83**: 151–83.

Mackie, G.O. & Pasano, L.M. (1968): Epithelial conduction in hydromedusae. *J. Gen. Physiol.* **52**: 600–21.

Mackie, G.O. & Singla, C.L. (1983): Studies on hecxactinellid sponges. I. Histology of *Rhabdocalyptus dawsoni* (Lambe, 1873). *Philos. Trans. R. Soc. London B* **301**: 365–400.

Mackie, G.O., Mills, C.E. & Singla, C.L. (1988): Structure and function of the prehensile tentilla of *Euplokamis* (Ctenophora, Cydippida). *Zoomorphology* **107**: 319–37.

MacRae, E.K. (1967): The fine structure of sensory receptor processes in the auricular epithelium of the planarian, *Dugesia tigrina*. *Z. Zellforsch.* **82**: 479–94.

Maddrell, S. (1982): Insects: small size and osmoregulation. In: *A Companion to Animal Physiology*, Taylor, C.R., Johansen, K. & Bolis, L. (eds.). Cambridge University Press, Cambridge: 289–305.

Märkel, K. & Röser, U. (1991): Ultrastructure and organization of the epineural canal and the nerve cord in sea urchins (Echinodermata, Echinoida). *Zoomorphology* **110**: 267–79.

Maggenti, A.B. (1979): The role of cuticular strata nomenclature in the systematics of Nemata. *J. Nematology* **11**: 94–98.

Mainitz, M. (1979): The fine structure of gnathostomulid reproductive organs. I. New characters in the male copulatory organ of Scleroperalia. *Zoomorphologie* **92**: 241–72.

Mainitz, M. (1983): Gnathostomulida. In: *Reproductive Biology of Invertebrates*, Adiyodi, K.G. & Adiyodi, R.G. (eds.). Vol. 1: Oogenesis, oviposition, and oosorption. John Wiley & Sons, Chichester: 169–80.

Mainitz, M. (1989): Gnathostomulida. In: *Reproductive Biology of Invertebrates*, Adiyodi, K.G. & Adiyodi, R.G. (eds.). Vol. 4, part A: Fertilization, development, and parental care. John Wiley & Sons, Chichester: 167–77.

Malakhov, V.V. (1980): Cephalorhyncha, a new type of animal kingdom uniting Priapulida, Kinorhyncha, Gordiacea and a system of Aschelminthes worms. *Zool. Zh.* **59**: 485–99.

Malakhov, V.V. (1994): *Nematodes. Structure, Development, Classification, and Phylogeny*. Smithsonian Institution Press, Washington.

Malakhov, V.V. (1998): Embryological and histological peculiarities of the order Enoplida, a primitive group of nematodes. *Russ. J. Nematol.* **6**: 41–46.

Malakhov, V.V. & Adrianov, A.B. (1995): *Cephalorhyncha – A New Phylum of the Animal Kingdom*. KMK Scientific Press, Moscow.

Maldonaldo, M. (2004): Choanoflagellates, choanocytes, and animal multicellularity. *Invertebr. Biol.* **123**: 1–22.

Mallatt, J., Garey, J.R. & Shultz, J.W. (2004): Ecdysozoan phylogeny and Bayesian inference: first use of nearly complete 28S and 18S rRNA gene sequences to classify the arthropods and their kin. *Mol. Phyl. Evol.* **31**: 178–91.

Mallatt, J. & Giribet, G. (2006): Further use of nearly complete 28S and 18S rRNA genes to classify Ecdysozoa: 37 more arthropods and a kinorhynch. *Mol. Phyl. Evol.* **40**: 772–94.

Mallatt, J. & Winchell, C.J. (2002): Testing the new animal phylogeny: first use of combined large-subunit and small-subunit rRNA gene sequences to classify the protostomes. *Mol. Biol. Evol.* **19**: 289–301.

Mangum, C.P. (1983): Oxygen transport in the blood. In: *The Biology of Crustacea*, Bliss, D.E. (ed.). Vol. 5: Internal anatomy and physiological regulation, Mantel, L.H. (volume ed.). Academic Press, New York: 373–429.

Mangum, C.P. (1992): Physiological function of the hemerythrins. In: *Blood and Tissue Oxygen Carriers*, Mangum, C.P. (ed.). Springer, Berlin: 173–92.

Mann, T. (1984): *Spermatophores*. Springer, Berlin.

Manni, L., Caicci, F., Gasparini, F., Zaniolo, G. & Burighel, P. (2004): Hair cells in ascidians and the evolution of lateral line placodes. *Evol. Dev.* **6**: 379–81.

Manni, L., Lane, N.J., Sorrentino, M., Zaniolo, G. & Burighel, P. (1999): Mechanisms of neurogenesis during the embryonic development of a tunicate. *J. Comp. Neurol.* **412**: 527–41.

Manning, P.L. & Dunlop, J. (1995): The respiratory organs of eurypterids. *Palaeontology* **38**: 287–97.

Mansir, A. & Justine, J.-L. (1995): Centrioles with ten singlets in spermatozoa of the parasitic nematode *Heligmosoides polygyrus*. In: Advances in Spermatozoal Phylogeny and Taxonomy, Jamieson, B.G.M., Ausio, J. & Justine, J.-L. (eds.). *Mém. Mus. Natn. Hist. Nat.* **166**: 119–28.

Manuel, M. & Le Parco, Y. (2000): Homeobox diversification in the calcareous sponge, *Sycon raphus*. *Mol. Phyl. Evol.* **17**: 97–107.

Manuel, M., Borchiellini, C., Alivon, E., Le Parco, Y., Vacelet, J. & Boury-Esnault, N. (2003): Phylogeny and evolution of calcareous sponges: monophyly of Calcinea and Calcaronea, high level of morphological homoplasy, and the primitive nature of axial symmetry. *Syst. Biol.* **52**: 311–33.

Manuel, M., Kruse, M., Müller, W.E.G. & Le Parco, Y. (2000): The comparison of β-thymosin homologues among Metazoa supports an arthropod-nematode clade. *J. Mol. Evol.* **51**: 378–81.

Manwell, C. (1960): Histological specificity of respiratory pigments – II. Oxygen transfer systems involving hemerythrins in sipunculid worms of different ecologies. *Comp. Biochem. Physiol.* **1**: 277–85.

Manwell, C. (1966): Sea cucumber sibling species: polypeptide chain types and oxygen equilibrium of hemoglobin. *Science* **152**: 1393–96.

Manylov, O.G. (1995): Regeneration in Gastrotricha – I. Microscopical observations on the regeneration in *Turbanella* sp. *Acta Zool.* **76**: 1–6.

Manylov, O.G., Vladychenskaya, N.S., Milyutina, I.A., Kedrova, O.S., Korokhov, N.P., Dvoryanchikov, G.A., Aleshin, V.V. & Petrov, N.B. (2004): Analysis of 18S rRNA gene sequences suggests significant molecular differences between Macrodasyida and Chaetonotida (Gastrotricha). *Mol. Phyl. Evol.* **30**: 850–54.

Marcus, E. (1928): Zur Embryologie der Tardigraden. *Verh. Dtsch. Zool. Ges.* **32**: 134–46.

Margulis, L. (1981): *Symbiosis in Cell Evolution*. W.H. Freeman and Company, San Francisco.

Mariscal, R.N. (1975): Entoprocta. In: *Reproduction of Marine Invertebrates*, Giese, A.C. & Pearse, J.S. (eds.). Vol. 2: Entoprocts and lesser coelomates. Academic Press, New York: 1–41.

Mariscal, R.N. (1984): Cnidaria: cnidae. In: *Biology of the Integument*, Bereiter-Hahn, J., Matoltsy, A.G. & Sylvia Richards, K. (eds.). Vol. 1: Invertebrates. Springer, Berlin: 57–68.

Mark Welch, D.B. (2000): Evidence from a protein-coding gene that acanthocephalans are rotifers. *Invertebr. Biol.* **119**: 17–26.

Mark Welch, D.B. (2005): Bayesian and maximum likelihood analyses of rotifer-acanthocephalan relationships. *Hydrobiologia* **546**: 47–54.

Mark Welch, D. & Meselson, M. (2000): Evidence for the evolution of bdelloid rotifers without sexual reproduction or genetic exchange. *Science* **288**: 1211–14.

Marlétaz, F., Martin, E., Perez, X., Papillon, D., Caubit, X., Lowe, C.J., Freeman, B., Fasano, L., Dossat, C., Wincker, P., Weissenbach, J. & Le Parco, Y. (2006): Chaetognath phylogenomics: a protostome with deuterostome-like development. *Curr. Biol.* **16**: R577–R578.

Marotta, R. & Ruhberg, H. (2004): Sperm ultrastructure of an oviparous and an ovoviviparous onychophoran species (Peripatopsidae) with some phylogenetic considerations. *J. Zool. Syst. Evol. Res.* **42**: 313–22.

Marotta, R., Guidi, L., Pierboni, L., Ferraguti, M., Todaro, A. & Balsamo, M. (2005a): Sperm ultrastructure of *Macrodasys caudatus* (Gastrotricha: Macrodasyida) and a sperm-based phylogenetic analysis of Gastrotricha. *Meiofauna Marina* **14**: 9–21.

Marotta, R., Melone, G., Bright, M. & Ferraguti, M. (2005b): Spermatozoa and sperm aggregates in the vestimentiferan *Lamellibranchia luymesi* compared with *Riftia pachyptila* (Polychaeta: Siboglinidae: Vestimentifera). *Biol. Bull.* **209**: 215–26.

Marques, A.C. & Collins, A.G. (2004): Cladistic analysis of Medusozoa and cnidarian evolution. *Invertebr. Biol.* **123**: 23–42.

Marshall, D.J. & Hodgson, A.N. (1990): Structure of the cephalic tentacles of some species of prosobranch limpet (Patellidae and Fissurellidae). *J. Moll. Stud.* **56**: 415–24.

Martin, G.G. & Hose, J.E. (1992): Vascular elements and blood (hemolymph). In: *Microscopic Anatomy of Invertebrates*, Harrison, F.W. & Humes, A.G. (eds.). Vol. 10: Decapod Crustacea. Wiley-Liss, New York: 117–146.

Martin, S.M. & Spencer, A.N. (1983): Neurotransmitters in coelenterates. *Comp. Biochem. Physiol. C* **74**: 1–14.

Martin, V.J. (1997): Cnidarians, the jellyfish and hydras. In: *Embryology – Constructing the Organism*, Gilbert, S.F. & Raunio, A.M. (eds.). Sinauer, Sunderland: 57–86.

Martin, V.J. (2000): Reorganization of the nervous system during metamorphosis of a hydrozoan planula. *Invertebr. Biol.* **119**: 243–53.

Martin, V.J. (2002): Photoreceptors of cnidarians. *Can. J. Zool.* **80**: 1703–22.

Martin, V.J. & Koss, R. (2002): Phylum Cnidaria. In: *Atlas of Marine Invertebrate Larvae*, Young, C.M. (ed.). Academic Press, San Diego: 51–108.

Martindale, M.Q. (2005): The evolution of metazoan axial properties. *Nature Rev. Gen.* **6**: 917–27.

Martindale, M.Q. & Henry, J. (1997): Ctenophorans, the comb jellies. In: *Embryology – Constructing the Organism*, Gilbert, S.F. & Raunio, A.M. (eds.). Sinauer, Sunderland: 87–111.

Martindale, M.Q., Finnerty, J.R. & Henry, J.Q. (2002): The Radiata and the evolutionary origins of the bilaterian body plan. *Mol. Phyl. Evol.* **24**: 358–65.

Martindale, M.Q., Pang, K. & Finnerty, J.R. (2004): Investigating the origins of triploblasty: 'mesodermal' gene expression in a diploblastic animal, the sea anemone *Nematostella vectensis* (phylum, Cnidaria; class, Anthozoa). *Development* **131**: 2463–74.

Mashanov, V.S., Zueva, O.R., Heinzeller, T. & Dolmatov, I.Y. (2006): Ultrastructure of the circumoral nerve ring and the radial nerve cords in holothurians (Echinodermata). *Zoomorphology* **125**: 27–38.

Maslakova, S.A., Martindale, M.Q. & Norenburg, J. (2004a): Vestigial prototroch in a basal nemertean, *Carinoma tremaphoros* (Nemertea; Palaeonemertea). *Evol. Dev.* **6**: 219–26.

Maslakova, S.A., Martindale, M.Q. & Norenburg, J. (2004b): Fundamental properties of the spiralian developmental program are displayed by the basal nemertean *Carinoma tremaphoros* (Palaeonemertea, Nemertea). *Dev. Biol.* **267**: 342–60.

Maslin, T.P. (1971): Conclusive evidence of parthenogenesis in three species of *Cnemidophorus* (Teiidae). *Copeia* **1**: 156–58.

Matricon, I. (1973): Quelques données ultrastructurales sur un myoépithélium: le pharynx d'un bryozoaire. *Z. Zellforsch.* **136** : 569–78.

Matsumoto, G.I. (1999): Ctenophora. In: *Encyclopedia of Reproduction*, Knobil, E. & Neill, J.D. (eds.). Vol. 1. Academic Press, San Diego: 792–800.

Mattei, X. (1970): Spermiogenèse comparée des poissons. In: *Comparative Spermatology*, Baccetti, B. (ed.). Academic Press, New York: 57–69.

Mattei, X. (1991a): Spermatozoon ultrastructure and its systematic implications in fishes. *Can. J. Zool.* **69**: 3038–55.

Mattei, X. (1991b): Spermatozoa ultrastructure and taxonomy in fishes. In: *Comparative Spermatology 20 Years After*, Baccetti, B. (ed.). Serono Symp. Publ. Raven Press 75. Raven Press, New York: 985–90.

Mattei, X. & Marchand, B. (1988): La spermiogenèse de *Myzostomum* sp. (Procoelomata, Myzostomida). *J. Ultrastruct. Mol. Struct. Res.* **100**: 75–85.

Mattei, X., Mattei, C., Reizer, C. & Chevalier, J.-L. (1972): Ultrastructure des spermatozoïdes aflagellés des mormyres (poissons téléostéens). *J. Microscopie* **15** : 57–78.

Matus, D.Q., Copley, R.R., Dunn, C.W., Hejnol, A., Eccleston, H., Halanych, K.M., Martindale, M.Q. & Telford, M.J. (2006): Broad taxon sampling and gene sampling indicate that chaetognaths are protostomes. *Curr. Biol.* **16**: R575–R576.

Maxmen, A., Browne, W.E., Martindale, M.Q. & Giribet, G. (2005): Neuroanatomy of sea spiders implies an appendicular origin of the protocerebral segment. *Nature* **437**: 1144–48.

Mayer, G. (2006a): Structure and development of onychophoran eyes: what is the ancestral visual organ in arthropods? *Arthropod Struct. Dev.* **35**: 231–45.

Mayer, G. (2006b): Origin and differentiation of nephridia in the Onychophora provide no support for the Articulata. *Zoomorphology* **125**: 1–12.

Mayer, G. & Bartolomaeus, T. (2003): Ultrastructure of the stomochord and the heart-glomerulus complex in *Rhabdopleura compacta* (Pterobranchia): phylogenetic implications. *Zoomorphology* **122**: 125–33.

Mayer, G., Ruhberg, H. & Bartolomaeus, T. (2004): When an epithelium ceases to exist – an ultrastructural study on the fate of the embryonic coelom in *Epiperipatus biolleyi* (Onychophora, Peripatidae). *Acta Zool.* **85**: 163–70.

Maynard-Smith, J. (1978): *The Evolution of Sex*. Cambridge University Press, Cambridge.

Mayne, R. (1984): The different types of collagen and collageneous peptides. In: *The Role of Extracellular Matrix in Development*, Trelstad, R.L. (ed.). Alan R. Liss, Inc., New York: 33–42.

.Mazet, F. & Shimeld, S.M. (2002): The evolution of chordate neural segmentation. *Dev. Biol.* **251**: 258–70.

Mazet, F., Hutt, J.A., Milloz, J., Millard, J., Graham, A. & Shimeld, S.M. (2005): Molecular evidence from *Ciona intestinalis* for the evolutionary origin of vertebrate sensory placodes. *Dev. Biol.* **282**: 494–508

Mazzini, M., Carcupino, M. & Fausto, A.M. (2000): Myriapoda. In: *Reproductive Biology of Invertebrates*, Adiyodi, K.G., Adiyodi, R.G. & Jamieson, B.G.M. (eds.). Vol. 9, part C: Progress in male gamete ultrastructure and phylogeny. John Wiley & Sons, Chichester: 117–41.

McFaddien-Carter, M. (1979): Scaphopoda. In: *Reproduction of Marine Invertebrates*, Giese, A.C. & Pearse, J.S. (eds.). Vol. 5: Molluscs: pelycopods and lesser classes. Academic Press, New York: 95–111.

McGinnis, W. & Krumlauf, R. (1992): Homeobox genes and axial patterning. *Cell* **68**: 283–302.

McHugh, D. (1997): Molecular evidence that echiurans and pogonophorans are derived annelids. *Proc. Natl. Acad. Sci. USA* **94**: 8006–9.

McHugh, D. (1999): Annelida. In: Encyclopedia of Reproduction, Knobil, E. & Neill, J.D. (eds.). Vol. 1. Academic Press, San Diego: 219–23.

McKenzie, J.D. & Hughes, D.J. (1999): Integument of *Maxmuelleria lankesteri* (Echiura), with notes on bacterial symbionts and possible evidence of viral activity. *Invertebr. Biol.* **118**: 296–309.

McLaren, D.J. (1976): Sense organs and their secretions. In: *The Organization of Nematodes*, Croll, N.A. (ed.). Academic Press, London: 139–61.

McLaughlin, P.A. (1980): *Comparative Morphology of Recent Crustacea*. W.H. Freeman and Company, San Francisco.

McShea, D.W. (2002): A complexity drain on cells in the evolution of multicellularity. *Evolution* **56**: 441–52.

Medina, M., Collins, A.G., Silberman, J.D. & Sogin, M.L. (2001): Evaluating hypotheses of basal animal phylogeny using complete sequences of large and small subunit rRNA. *Proc. Natl. Acad. Sci. USA* **98**: 9707–12.

Mehl, D. & Reiswig, H.M. (1991): The presence of flagellar vanes in choanomeres of Porifera and their possible phylogenetic implications. *Z. Zool. Syst. Evolutionsforsch.* **29**: 312–19.

Meinertzhagen, I.A. & Okamura, Y. (2001): The larval ascidian nervous system: the chordate brain from its small beginnings. *Trends Neurosci.* **24**: 401–10.

Meinhardt, H. (1986): Models of segmentation. In: *Somites in Developing Embryos*, Bellairs, R., Ede, D.A. & Lash, J.W. (eds.). Plenum Press, New York: 179–89.

Meisenheimer, J. (1899): Zur Morphologie der Urniere der Pulmonaten. *Z. Wiss. Zool.* **65**: 709–24.

Melone, G. & Ferraguti, M. (1994): The spermatozoa of *Brachionus plicatilis* (Rotifera, Monogononta) with some notes on sperm ultrastructure in Rotifera. *Acta Zool.* **75**: 81–88.

Melone, G., Ricci, C. & Wallace, R.L. (1998): Phylogenetic relationships of Acanthocephala and Rotifera: morphological vs. molecular approaches. *Mem. Mus. Civ. Stor. Nat. Verona Ser.* **2**: 57–62.

Mendoza, L., Taylor, J.W. & Ajello, L. (2002): The class Mesomycetozoea: a heterogeneous group of microorganisms at the animal-fungal boundary. *Ann. Rev. Microbiol.* **56**: 315–44.

Messenger, J.B. (1991): Photoreception and vision in molluscs. In: *Vision and Visual Dysfunction*, Cronly-Dillon, J.R. & Gregory, R.L. (eds.). Vol. 2: Evolution of the eye and visual system. Macmillan Press, Houndmills: 364–97.

Mickoleit, G. (2004): *Phylogenetische Systematik der Wirbeltiere*. Pfeil-Verlag, München.

Miller, D.M. & Dunagan, T.T. (1976): Body wall organization of the acanthocephalan *Macracanthorhynchus hirudinaceus*: a reexamination of the lacunar system. *Proc. Helminthol. Soc. Washington* **43**: 99–106.

Miller, D.M. & Dunagan, T.T. (1977): The lacunar system and tubular muscles in Acanthocephala. *Proc. Helminthol. Soc. Washington* **44**: 201–5.

Miller, D.M. & Dunagan, T.T. (1985): Functional morphology. In: *Biology of the Acanthocephala*, Crompton, D.W.T. & Nickol, B.B. (eds.). Cambridge University Press, Cambridge: 73–123.

Millot, J. (1968): Ordre des Amblypyges. I : Traité de Zoologie, Grassé, P. (ed.). Vol. 6 : *Onychophores, Tardigrades, Arthropodes, Trilobitomorphes, Chélicérates*. Masson et Cie Éditeurs, Paris : 563–88.

Millott, N. (1968): The dermal light sense. *Symp. Zool. Soc. London* **23**: 1–13.

Min, G.-S., Kim, S.-H. & Kim, W. (1998): Molecular phylogeny of arthropods and their relatives: polyphyletic origin or arthropodization. *Mol. Cells* **8**: 75–83.

Minelli, A. (1993): Chilopoda. In: *Microscopic Anatomy of Invertebrates*, Harrison, F.W. & Rice, M.E. (eds.). Vol. 12: Onychophora, Chilopoda, and lesser Protostomata. Wiley-Liss, New York: 57–114.

Minelli, A. (2003a): *The Development of Animal Form. Ontogeny, Morphology, and Evolution*. Cambridge University Press, Cambridge.

Minelli, A. (2003b): The origin and evolution of appendages. *Int. J. Dev. Biol.* **47**: 573–81.

Minning, D.M., Gow, A.J., Bonaventura, J., Braun, R., Dewhirst, M., Goldberg, D.E. & Stamler, J.S. (1999): *Ascaris* haemoglobin is a nitric oxide-activated 'deoxygenase'. *Nature* **401**: 497–502.

Miyazaki, K. (2002): On the shape of foregut lumen in sea spiders (Arthropoda: Pycnogonida). *J. Mar. Biol. Assoc. UK* **82**: 1037–038.

Moens, L., Vanfleteren, J., van de Peer, Y., Peeters, K., Kapp, O., Czelusniak, J., Goodman, M., Blaxter, M. & Vinogradov, S. (1996): Globins in noninvertebrate species: dispersal by horizontal gene transfer and evolution of the structure-function relationships. *Mol. Biol. Evol.* **13**: 324–33.

Monks, S. (2001): Phylogeny of the Acanthocephala based on morphological characters. *Syst. Parasitol.* **48**: 81–116.

Monteiro, A.S., Okamura, B. & Holland, P.W.H. (2002): Orphan worm finds a home: *Buddenbrockia* is a myxozoan. *Mol. Biol. Evol.* **19**: 968–71.

Monteiro, A.S., Schierwater, B., Dellaporta, S.L. & Holland, P.W.H. (2006): A low diversity of ANTP class homeobox genes in Placozoa. *Evol. Dev.* **8**: 174–82.

Montgomery, T.H. (1903): The adult organization of *Paragordius varius*. *Zool. Jahrb. Anat. Ontog.* **18**: 387–474.

Montgomery, T.H. (1904): The development and structure of the larva of *Paragordius*. *Proc. Acad. Nat. Sci. Philadelphia* **56**: 738–55.

Monticelli, F.S. (1897): Adelotacta Zoologica. *Mitt. Zool. Stat. Napoli* **12**: 432–62.

Moon, S.Y. & Kim, W. (1996): Phylogenetic position of the Tardigrada based on the 18S ribosomal RNA gene sequences. *Zool. J. Linn. Soc.* **116**: 61–69.

Moor, B. (1983): Organogenesis. In: *The Mollusca*, Wilbur, K.M. (ed.). Vol. 3: Development, Verdonk, N.H., van den Biggelaar, J.A.M. & Tompa, A.S. (volume eds.). Academic Press, New York: 123–77.

Morisawa, S. (2005): Spermiogenesis in the hagfish *Eptatretus burgeri* (Agnatha). *Biol. Bull.* **209**: 204–14.

Moritz, K. & Storch, V. (1971): Elektronenmikroskopische Untersuchung eines Mechanorezeptors von Evertebraten (Priapuliden, Oligochaeten). *Z. Zellforsch.* **117**: 226–34.

Morrow, E.H. (2004): How the sperm cell lost its tail: the evolution of aflagellate sperm. *Biol. Rev.* **79**: 795–814.

Morse, M.P. & Reynolds, J.P. (1996): Ultrastructure of the heart-kidney complex in smaller classes supports symplesiomorphy of molluscan coelomic characters.

In: *Origin and Evolutionary Radiation of the Mollusca*, Taylor, J. (ed.). Oxford University Press, Oxford: 89–97.

Morse, M.P. & Zardus, J.D. (1997): Bivalvia. In: *Microscopic Anatomy of Invertebrates*, Harrison, F.W. & Kohn, A.J. (eds.), Vol. 6A: Mollusca II. Wiley-Liss, New York: 7–118.

Moser, W.E. & Desser, S.S. (1995): Morphological, histochemical, and ultrastructural characterization of the salivary glands and proboscises of three species of glossiphoniid leeches (Hirudinea: Rhynchobdellida). *J. Morphol.* **225**: 1–18.

Moshel, S.M., Levine, M. & Collier, J.R. (1998): Shell differentiation and *engrailed* expression in the *Ilyanassa* embryo. *Dev. Genes Evol.* **208**: 135–41.

Moura, G. & Christoffersen, M.L. (1996): The system of the mandibulate arthropods: Tracheata and Remipedia as sister groups, 'Crustacea' non-monophyletic. *J. Comp. Biol.* **1**: 95–113.

Müller, C.H.G., Rosenberg, J., Richter, S. & Meyer-Rochow, V.B. (2003): The compound eye of *Scutigera coleoptrata* (Linnaeus, 1758) (Chilopoda: Notostigmophora): an ultrastructural reinvestigation that adds support to the Mandibulata concept. *Zoomorphology* **122**: 191–209.

Müller, K.J. & Hinz-Schallreuter, I. (1993): Palaeoscolecid worms from the middle Cambrian of Australia. *Palaeontology* **36**: 549–92.

Müller, K.J. & Walossek, D. (1991): 'Orsten' arthropods – small size but of great impact on biological and phylogenetic interpretations. *Geol. Foren. Stockholm Forh.* **113**: 88–90.

Müller, M.C.M. (2006): Polychaete nervous systems: ground pattern and variations – cLS microscopy and the importance of novel characteristics in phylogenetic analyses. *Integr. Comp. Biol.* **46**: 125–33.

Müller, M.C.M. & Schmidt-Rhaesa, A. (2003): Reconstruction of the muscle system in *Antygomonas* sp. (Kinorhyncha, Cyclorhagida) by means of phalloidin labelling and cLSM analysis. *J. Morphol.* **256**: 103–10.

Müller, M.C.M. & Sterrer, W. (2004): Musculature and nervous system of *Gnathostomula peregrina* (Gnathostomulida) shown by phalloidin labelling, immunocytochemistry, and cLSM, and their phylogenetic significance. *Zoomorphology* **123**: 169–77.

Müller, M.C. & Westheide, W. (2000): Structure of the nervous system of *Myzostoma cirriferum* (Annelida) as revealed by immunohistochemistry and cLSM analysis. *J. Morphol.* **245**: 87–98.

Müller, M.C.M., Jochmann, R. & Schmidt-Rhaesa, A. (2004): The musculature of horsehair worm larvae (*Gordius aquaticus*, *Paragordius varius*, Nematomorpha): F-actin staining and reconstruction by cLSM and TEM. *Zoomorphology* **123**: 45–54.

Müller, W.E.G. & Müller, I.M. (1999): Origin of the Metazoa: a review of molecular biological studies with sponges. *Mem. Queensl. Mus.* **44**: 381–97.

Mukai, H., Terakado, K. & Reed, C.G. (1997): Bryozoa. In: *Microscopic Anatomy of Invertebrates*, Harrison, F.W. & Woollacott, R.M. (eds.), Vol. 13: Lophophorates, Entoprocta, and Cycliophora. Wiley-Liss, New York: 45–206.

Mushegian, A.R., Garey, J.R., Martin, J. & Liu, L.X. (1998): Large-scale taxonomic profiling of eucaryotic model organisms: a comparison of orthologous proteins encoded by human, fly, nematode, and yeast genomes. *Genome Res.* **8**: 590–98.

Musio, C., Santillo, S., Taddei-Ferretti, C., Robies, L.J., Vismara, R., Barsanti, L. & Gualtieri, P. (2001): First identification and localization of a visual pigment in *Hydra* (Cnidaria, Hydrozoa). *J. Comp. Physiol. A* **187**: 79–81.

Nakano, H., Hibino, T., Oji, T., Hara, Y. & Amemiya, S. (2003): Larval stages of a living sea lily (stalked crinoid echinoderm). *Nature* **421**: 158–60.

Nardi, F., Spinsanti, G., Boore, J.L., Carapelli, A., Dallai, R. & Frati, F. (2003): Hexapod origins: monophyletic or paraphyletic? *Science* **299**: 1887–89.

Natesan, A., Geetha, L. & Zatz, M. (2002): Rhythm and soul in the avian pineal. *Cell Tiss. Res.* **309**: 35–45.

Near, T.J., Garey, J.R. & Nadler, S.A. (1998): Phylogenetic relationships of the Acanthocephala inferred from 18S ribosomal DNA sequences. *Mol. Phyl. Evol.* **10**: 287–98.

Nebelsick, M. (1993): Introvert, mouth cone, and nervous system of *Echinoderes capitatus* (Kinorhyncha, Cyclorhagida) and implications for the phylogenetic relationships of Kinorhyncha. *Zoomorphology* **113**: 211–32.

Nebelsick, M., Blumer, M., Novak, R. & Ott, J. (1992): A new glandular sensory organ in *Catanema* sp. (Nematoda, Stilbonematinae). *Zoomorphology* **112**: 17–26.

Neuhaus, B. (1987): Ultrastructure of the protonephridia in *Dactylopodola baltica* and *Mesodasys laticaudatus* (Macrodasyida): implications for the ground pattern of the Gastrotricha. *Microfauna Marina* **3**: 419–38.

Neuhaus, B. (1988): Ultrastructure of the protonephridia in *Pycnophyes kielensis* (Kinorhyncha, Homalorhagida). *Zoomorphology* **108**: 245–53.

Neuhaus, B. (1994): Ultrastructure of alimentary canal and body cavity, ground pattern, and phylogenetic relationships of Kinorhyncha. *Microfauna Marina* **9**: 61–156.

Neuhaus, B. (1997): Ultrastructure of the cephalic sensory organs of adult *Pycnophyes dentatus* and of the first juvenile stage of *P. kielensis* (Kinorhyncha, Homalorhagida). *Zoomorphology* **117**: 33–40.

Neuhaus, B. (1999): Kinorhyncha. In: *Encyclopedia of Reproduction*, Knobil, E. & Neill, J.D. (eds.). Vol. 3. Academic Press, San Diego: 933–37.

Neuhaus, B. & Higgins, R.P. (2002): Ultrastructure, biology, and phylogenetic relationships of Kinorhyncha. *Integr. Comp. Biol.* **42**: 619–32.

Neuhaus, B., Bresciani, J. & Peters, W. (1997a): Ultrastructure of the pharyngeal cuticle and lectin labelling with wheat germ agglutinin-gold conjugate indicating chitin in the pharyngeal cuticle of *Oesophagostomum dentatum* (Strongylidea, Nematoda). *Acta Zool.* **78**: 205–13.

Neuhaus, B., Kristensen, R.M. & Lemburg, C. (1996): Ultrastructure of the cuticle of the Nemathelminthes and electron microscopical localization of chitin. *Verh. Dtsch. Zool. Ges.* **89.1**: 221.

Neuhaus, B., Kristensen, R.M. & Peters, W. (1997b): Ultrastructure of the cuticle of Loricifera and demonstration of chitin using gold-labelled wheat germ agglutinin. *Acta Zool.* **78**: 215–25.

Neumeister, H. & Budelmann, B.U. (1997): Structure and function of the *Nautilus* statocyst. *Phil. Trans. R. Soc. London B* **352**: 1565–588.

Neville, A.C. (1975): *Biology of the Arthropod Cuticle*. Springer, Berlin.

Neville, A.C. (1984): Cuticle: organization. In: *Biology of the Integument*, Bereiter-Hahn, J., Matoltsy, A.G. & Sylvia Richards, K. (eds.). Vol. 1: Invertebrates. Springer, Berlin: 611–25.

Newberry, A.T. (1999): Tunicata. In: *Encyclopedia of Reproduction*, Knobil, E. & Neill, J.D. (eds.). Vol. 4. Academic Press, San Diego: 872–79.

Newell, R.C. & Courtney, W.A.M. (1965): Respiratory movements in *Holothuria forskali* delle Chaje. *J. Exp. Biol.* **42**: 45–57.

Nezlin, L.P. (2000): Tornaria of hemichordates and other dipleurula-type larvae: a comparison. *J. Zool. Syst. Evolut. Res.* **38**: 149–56.

Nezlin, L.P. & Yushin, V.V. (2004): Structure of the nervous system in the tornaria larva of *Balanoglossus proterogonius* (Hemichordata: Enteropneusta) and its phylogenetic implications. *Zoomorphology* **123**: 1–13.

Nicaise, G. & Amsellem, J. (1983): Cytology of muscle and neuromuscular junction. In: *The Mollusca*, Wilbur, K.M. (ed.). Vol. 4, Physiology, Part 1, Saleuddin, A.S.M. & Wilbur, K.M. (volume eds.), Academic Press, New York: 1–33.

Nicander, L. (1970): Comparative studies on the fine structure of vertebrate spermatozoa. In: *Comparative Spermatology*, Baccetti, B. (ed.). Academic Press, New York: 47–54.

Nicholas, W.L. (1991): Interstitial meiofauna. In: *Metazoan Life Without Oxygen*, Byrant, C. (ed.). Chapman and Hall, London: 129–45.

Nickel, M. (2004): Kinetics and rhythm of body contractions in the sponge *Tethya wilhelma* (Porifera: Demospongiae). *J. Exp. Biol.* **207**: 4515–24.

Nicol, D. & Meinertzhagen, I.A. (1988): Development of the central nervous system of the larva of the ascidian, *Ciona intestinalis* L. *Dev. Biol.* **130**: 721–36.

Nielsen, C. (1971): Entoproct life-cycles and the entoproct-ectoproct relationship. *Ophelia* **9**: 209–341.

Nielsen, C. (1985): Animal phylogeny in the light of the trochaea theory. *Biol. J. Linn. Soc.* **25**: 243–99.

Nielsen, C. (1987): Structure and function of metazoan ciliary bands and their phylogenetic significance. *Acta Zool.* **68**: 205–62.

Nielsen, C. (1990a): Bryozoa Entoprocta. In: *Reproductive Biology of Invertebrates*, Adiyodi, K.G. & Adiyodi, R.G. (eds.). Vol. 4, part B: fertilization, development, and parental care. John Wiley & Sons, Chichester: 201–9.

Nielsen, C. (1990b): Bryozoa Ectoprocta. In: *Reproductive Biology of Invertebrates*, Adiyodi, K.G. & Adiyodi, R.G. (eds.). Vol. 4, part B: fertilization, development, and parental care. John Wiley & Sons, Chichester: 185–200.

Nielsen, C. (1991): The development of the brachiopod *Crania* (*Neocrania*) *anomala* (O.F. Müller) and its phylogenetic significance. *Acta Zool.* **72**: 7–28.

Nielsen, C. (1994): Larval and adult characters in animal phylogeny. *Am. Zool.* **34**: 492–501.

Nielsen, C. (1995): *Animal Evolution. Interrelationships of the Living Phyla*. 1st Edn.. Oxford University Press, Oxford.

Nielsen, C. (1997): The phylogenetic position of the Arthropoda. In: Arthropod Relationships, Fortey, R.A. & Thomas, R.H. (eds.). *Syst. Assoc. Spec. Vol.* **55**, Chapman & Hall, London: 11–21.

Nielsen, C. (1998): Origin and evolution of animal life cycles. *Biol. Rev.* **73**: 124–55.

Nielsen, C. (1999): Origin of the chordate central nervous system – and the origin of chordates. *Dev. Genes Evol.* **209**: 198–205.

Nielsen, C. (2001): *Animal Evolution. Interrelationships of the Living Phyla*. 2nd Edn.. Oxford University Press, Oxford.

Nielsen, C. (2002a): The phylogenetic position of Entoprocta, Ectoprocta, Phoronida, and Brachiopoda. *Integr. Comp. Biol.* **42**: 685–91.

Nielsen, C. (2002b): Ciliary filter-feeding structures in adult and larval gymnolaemate bryozoans. *Invertebr. Biol.* **121**: 255–61.

Nielsen, C. (2002c): Phylum Entoprocta. In: *Atlas of Marine Invertebrate Larvae*, Young, C.M. (ed.). Academic Press, San Diego: 397–409.

Nielsen, C. (2003): Proposing a solution to the Articulata-Ecdysozoa controversy. *Zool. Scr.* **32**: 475–482.

Nielsen, C. (2006): Homology of echinoderm radial nerve cords and the chordate neural tube??? *Evol. Dev.* **8**: 1–2.

Nielsen, C. & Jespersen, Å. (1997): Entoprocta. In: *Microscopic Anatomy of Invertebrates*, Harrison, F.W. & Woollacott, R.M. (eds.), Vol. 13: Lophophorates, Entoprocta, and Cycliophora. Wiley-Liss, New York: 13–43.

Nielsen, C. & Nørrevang, A. (1985): The trochaea theory: an example of life cycle phylogeny. In: The Origins and Relationships of Lower Invertebrates, Conway Morris, S., George, J.D., Gibson, R. & Platt, H.M. (eds.). *Syst. Assoc. Spec. Vol.* **28**: 28–41.

Nielsen, C. & Riisgård, H.U. (1998): Tentacle structure and filter-feeding in *Crisia eburnea* and other cyclostomatous bryozoans; with a review of upstream-collecting mechanisms. *Mar. Ecol. Progr. Ser.* **168**: 163–86.

Nielsen, C. & Rostgaard, J. (1976): Structure and function of an entoproct tentacle with a discussion of ciliary feeding types. *Ophelia* **15**: 115–40.

Nielsen, C., Scharff, N. & Eibye-Jacobsen, D. (1996): Cladistic analysis of the animal kingdom. *Biol. J. Linn. Soc.* **57**: 385–410.

Nieuwenhuys, R. (2002): Deuterostome brain: synopsis and commentary. *Brain Res. Bull.* **57**: 257–70.

Nieuwkoop, P.D. & Sutasurya, L.A. (1979): *Primordial Germ Cells in the Chordates: Embryogenesis and Phylogenesis*. Cambridge University Press, Cambridge.

Nieuwkoop, P.D. & Sutasurya, L.A. (1981): *Primordial Germ Cells in the Invertebrates: From Epigenesis to Preformation*. Cambridge University Press, Cambridge.

Nilsson, D.-E. (1994): Eyes as optical alarm systems in fan worms and ark clams. *Phil. Trans. R. Soc. London B* **346**: 195–212.

Nilsson, D.-E. & Pelger, S. (1994): A pessimistic estimate of time required for an eye to evolve. *Proc. R. Soc. London B* **256**: 53–58.

Nilsson, D.-E., Gislen, L., Coates, M.M., Skogh, C. & Garm, A. (2005): Advanced optics in a jellyfish eye. *Nature* **435**: 201–5.

Nishii, I. & Ogihara, S. (1999): Actomyosin contraction of the posterior hemisphere is required for inversion of the *Volvox* embryo. *Development* **126**: 2117–27.

Noda, A.O., Ikeo, K. & Gojobori, T. (2006): Comparative genome analysis of nervous system-specific genes. *Gene* **365**: 130–36.

Nogales, E., Downing, K.H., Amos, L.A. & Löwe, J. (1998): Tubulin and FtsZ form a distinct family of GTPases. *Nature Struct. Biol.* **5**: 451–58.

Noguchi, Y., Endo, K., Tajima, F. & Ueshima, R. (2000): The mitochondrial genome of the brachiopod *Laqueus rubellus*. *Genetics* **155**: 245–59.

Norén, M. & Jondelius, U. (1997): *Xenoturbella's* molluscan relatives . . . *Nature* **390**: 31–32.

Norenburg, J.L. & Roe, P. (1998): Reproductive biology of several species of recently collected pelagic nemerteans. *Hydrobiologia* **365**: 73–91.

Normark, B.B. (2003): The evolution of alternative genetic systems in insects. *Ann. Rev. Entomol.* **48**: 397–423.

Nørrevang, A. (1963): Fine structure of the solenocyte tree in *Priapulus caudatus* Lamarck. *Nature* **198**: 700–1.

Nørrevang, A. (1965): Oogenesis in *Priapulus caudatus* Lamarck. *Vidensk. Medd. Dansk. Naturh. Foren.* **128**: 1–75.

Nørrevang, A. (1974): Photoreceptors of the phaosome (hirudinean) type in a pogonophore. *Zool. Anz.* **193**: 297–304.

Nübler-Jung, K. & Arendt, D. (1994): Is ventral in insects dorsal in vertebrates? *Roux's Arch. Dev. Biol.* **203**: 357–66.

Nübler-Jung, K. & Arendt, D. (1999): Dorsoventral axis inversion: enteropneust anatomy links invertebrates to chordates turned upside down. *J. Zool. Syst. Evolut. Res.* **37**: 93–100.

Nunzi, M.G., Burighel, P. & Schiaffing, S. (1979): Muscle cell differentiation in the ascidian heart. *Dev. Biol.* **68**: 371–80.

Nuttman, C.J. (1974): The fine structure and organization of the tail musculature of the cercaria of *Schistosoma mansoni*. *Parasitology* **68**: 147–54.

Nyholm, K.-G. (1977): Receptaculum seminis and morphological hermaphroditism in Homalorhaga Kinorhyncha. *Zoon* **5**: 7–10.

Nyholm, K.-G. & Nyholm, P.-G. (1982): Spermatozoa and spermatogenesis in Homalorhagha Kinorhyncha. *J. Ultrastruct. Res.* **78**: 1–12.

Nylund, A., Ruhberg, H., Tjonneland, A., & Meidell, B. (1988): Heart ultrastructure in four species of Onychophora (Peripatopsidae and Peripatidae) and phylogenetic implications. *Zool. Beitr.* **32**: 17–30.

Obst, M., Funch, P. & Giribet, G. (2005): Hidden diversity and host specificity in cycliophorans: a phylogeographic analysis along the North Atlantic and Mediterranean Sea. *Mol. Ecol.* **14**: 4427–40.

Obst, M., Funch, P. & Kristensen, R.M. (2006): A new species of Cyclophora from the mouthparts of the American lobster, *Homarus americanus* (Nephropidae, Decapoda). *Org. Div. Evol.* **6**: 83–97.

Oeschger, R., Peper, H., Graf, G. & Theede, H. (1992): Metabolic responses of *Halicryptus spinulosus* (Priapulida) to reduced oxygen levels and anoxia. *J. Exp. Mar. Biol. Ecol.* **162**: 229–41.

Ohtsuki, H. (1990): Statocyte and ocellar pigment cell in embryos and larvae of the ascidian, *Stylea plicata* (Lesueur). *Dev. Growth Differ.* **32**: 85–90.

Okafor, N. (1965): Isolation of chitin from the shell of the cuttlefish, *Sepia officinalis* L. *Biochim. Biophys. Acta* **101**: 193–200.

Okamura, B. & Canning, E.U. (2003): Orphan worms and homeless parasites enhance bilaterian diversity. *Trends Ecol. Evol.* **18**: 633–39.

Okamura, B., Curry, A., Wood, T.S. & Canning, E.U. (2002): Ultrastructure of *Buddenbrockia* identifies it as a myxozoan and verifies the bilaterian origin of the Myxozoa. *Parasitology* **124**: 215–23.

Økland, S. (1980): The heart ultrastructure of *Lepidochiton asellus* (Spengler) and *Tonicella marmorea* (Fabricius) (Mollusca: Polyplacophora). *Zoomorphology* **96**: 1–19.

Økland, S. (1981): Ultrastructure of the pericardium in chitons (Mollusca: Polyplacophora), in relation to filtration and contraction mechanisms. *Zoomorphology* **97**: 193–203.

Oliphant, L.W. & Cloney, R.A. (1972): The ascidian myocardium: sarcoplasmic reticulum and excitation-contraction coupling. *Z. Zellforsch.* **129**: 395–412.

Olive, P.J.W. (1975): Reproductive biology of *Eulalia viridis* (Müller) (Polychaeta: Phyllodocidae) in the North Eastern U.K. *J. Mar. Biol. Assoc. UK* **55**: 313–26.

Olson, P.D., Littlewood, D.T.J., Bray, R.A. & Mariaux, J. (2001): Interrelationships and evolution of the tapeworms (Platyhelminthes: Cestoda). *Mol. Phyl. Evol.* **19**: 443–67.

OOta, S. & Saitou, N. (1999): Phylogenetic relationship of muscle tissues deduced from superimposition of gene trees. *Mol. Biol. Evol.* **16**: 856–67.

O'Rand, M.G. & Miller, R.L. (1974): Spermatozoan vesicle loss during penetration of the female gonangium in the hydroid *Campanularia flexuosa*. *J. Exp. Zool.* **188**: 179–94.

Orrhage, L. (1962): Über das Vorkommen von Muskelzellen vom 'Nematoden-Typus' bei Polychaeten als phylogenetisch-systematisches Merkmal. *Zool. Bidr. Uppsala* **35**: 321–27.

Orrhage, L. (1971): Light and electron microscope studies of some annelid setae. *Acta Zool.* **52**: 157–69.

Orrhage, L. & Müller, M.C.M. (2005): Morphology of the nervous system of Polychaeta (Annelida). In: Morphology, Molecules, Evolution and Phylogeny in Polychaeta and Related Taxa, Bartolomaeus, T. & Purschke, G. (eds.). *Hydrobiologia* **535/536**: 79–111.

Osborne, M.P. (1967): Supercontraction in the muscles of the blowfly larva: an ultrastructural study. *J. Insect Physiol.* **13**: 1471–82.

Palmberg, I., Reuter, M. & Wikgren, M. (1980): Ultrastructure of epidermal eyespots of *Microstomum lineare* (Turbellaria, Macrostomida). *Cell Tiss. Res.* **210**: 21–32.

Pancer, Z., Kruse, M., Müller, I.M. & Müller, W.E.G. (1997): On the origin of metazoan adhesion receptors: cloning of integrin α subunit from the sponge *Geodia cydonium*. *Mol. Biol. Evol.* **14**: 391–98.

Panganiban, G., Irvine, S.M., Lowe, C., Roehl, H., Corley, L.S., Sherbon, B., Grenier, J.K., Fallon, J.F., Kimble, J., Walker, M., Wray, G.A., Swalla, B.J., Martindale, M.Q. & Carroll, S.B. (1997): The origin and evolution of animal appendages. *Proc. Natl. Acad. Sci. USA* **94**: 5162–66.

Paniagua, R., Royuela, M., Garcia-Anchuelo.R.M. & Fraile, B. (1996): Ultrastructure of invertebrate muscle cell types. *Histol. Histopathol.* **11**: 181–201.

Pante, N. (1994): Paramyosin polarity in the thick filament of molluscan smooth muscle. *J. Struct. Biol.* **113**: 148–63.

Papillon, D., Perez, Y., Caubit, X. & Le Parco, Y. (2004): Identification of chaetognaths as protostomes is supported by the analysis of their mitochondrial genome. *Mol. Biol. Evol.* **21**: 2122–29.

Papillon, D., Perez, Y., Fasano, L., Le Parco, Y. & Caubit, X. (2003): *Hox* survey in the chaetognath *Spadella cephaloptera*: evolutionary implications. *Dev. Genes Evol.* **213**: 142–48.

Pardos, F., Roldán, C., Benito, J., Aguirre, A. & Fernández, I. (1993): Ultrastructure of the lophophoral tentacles in the genus *Phoronis* (Phoronida, Lophophorata). *Can. J. Zool.* **71**: 1861–68.

Pardos, F., Roldán, C., Benito, J. & Emig, C. (1991): Fine structure of the tentacles of *Phoronis australis* Haswell (Phoronida, Lophophorata). *Acta Zool.* **72**: 81–90.

Parker, G.H. (1910): The reactions of sponges, with a consideration of the origin of the nervous system. *J. Exp. Zool.* **8**: 765–805.

Pasquinelli, A.E., McCoy, A., Jiménez, E., Saló, E., Ruvkun, G., Martindale, M.Q. & Baguñà, J. (2003): Expression of the 22 nucleotide *let-7* heterochronic RNA throughout the Metazoa: a role in life history evolution? *Evol. Dev.* **5**: 372–78.

Pass, G. (1991): Antennal circulatory organs in Onychophora, Myriapoda and Hexapoda: functional morphology and evolutionary implications. *Zoomorphology* **110**: 145–64.

Passamaneck, Y.J. & Halanych, K.M. (2004): Evidence from Hox genes that bryozoans are lophotrochozoans. *Evol. Dev.* **6**: 275–81.

Patel, N.H., Martin-Blanco, E., Coleman, K.G., Poole, S.J., Ellis, M.C., Kornberg, T.B. & Goodman, C.S. (1989): Expression of *engrailed* proteins in arthropods, annelids, and chordates. *Cell* **58**: 955–68.

Paulus, H.F. (2000): Phylogeny of the Myriapoda – Crustacea – Insecta: a new attempt using photoreceptor structure. *J. Zool. Syst. Evolut. Res.* **38**: 189–208.

Paulus, T. & Müller, M.C.M. (2006): Cell proliferation dynamics and morphological differentiation during regeneration in *Dorvillea bermudensis* (Polychaeta, Dorvilleidae). *J. Morphol.* **267**: 393–403.

Paulus, W. & Weissenfels, N. (1986): The spermatogenesis of *Ephydatia fluviatilis* (Porifera). *Zoomorphology* **106**: 155–62.

Pavans de Ceccatty, M. (1986): Cytoskeletal organization and tissue patterns of epithelia in the sponge *Ephydatia mülleri. J. Morphol.* **189**: 45–65.

Pawlowski, J., Montoya-Burgos, J.-I., Fahrni, J.F., Wüest, J. & Zaninetti, L. (1996): Origin of the Mesozoa inferred from 18S rRNA gene sequences. *Mol. Biol. Evol.* **13**: 1128–32.

Paxton, H. (2004): Jaw growth and replacement in *Ophryotrocha labronica* (Polychaeta, Dorvilleidae). *Zoomorphology* **123**: 147–54.

Paxton, H. (2005): Molting polychaete jaws – ecdysozoans are not the only molting animals. *Evol. Dev.* **7**: 337–340.

Pearse, J.S. (1979): Polyplacophora. In: *Reproduction of Marine Invertebrates*, Giese, A.C. & Pearse, J.S. (eds.). Vol. 5: Molluscs: pelycopods and lesser classes. Academic Press, New York: 27–85.

Pearse, J.S. & Cameron, R.A. (1991): Echinodermata: Echinoidea. In: *Reproduction of Marine Invertebrates*, Giese, A.C., Pearse, J.S. & Pearse, V.B. (eds.). Vol. 6: Echinoderms and lophophorates. The Boxwood Press, Pacific Grove: 513–662.

Pechenik, J.A. (2005): *Biology of the Invertebrates*. 5th Edn., McGraw-Hill, New York.

Pedersen, K.J. (1991): Invited review: structure and composition of basement membranes and other basal matrix systems in selected invertebrates. *Acta Zool.* **72**: 181–201.

Pedersen, K.J. & Pedersen, L.R. (1986): Fine structural observations on the extracellular matrix (ECM) of *Xenoturbella bocki* Westblad, 1949. *Acta Zool.* **67**: 103–13.

Pedersen, K.J. & Pedersen, L.R. (1988): Ultrastructural observations on the epidermis of *Xenoturbella bocki* Westblad, 1949; with a discussion of epidermal cytoplasmic filament systems of invertebrates. *Acta Zool.* **69**: 231–46.

Penn, P.E. & Alexander, C.G. (1980): Fine structure of the optic cushion in the astroid *Nepanthia belcheri. Mar. Biol.* **58**: 251–256.

Pennington, J.T. & Stricker, S.A. (2002): Phylum Brachiopoda. In: *Atlas of Marine Invertebrate Larvae*, Young, C.M. (ed.). Academic Press, San Diego: 441–61.

Pennisi, E. (2003): Modernizing the tree of life. *Science* **300**: 1692–97.

Pentreath, R.J. (1971): Respiratory surfaces and respiration in three New Zealand intertidal ophiuroids. *J. Zool. London* **163**: 397–412.

Pentreath, V.W. & Cobb, J.L.S. (1982): Echinodermata. In: *Electrical Conduction and Behaviour in 'Simple' Invertebrates*, Sheldon, G.A.B. (ed.). Clarendon Press, Oxford: 440–72.

Pentreath, V.W. & Cottrell, G.A. (1970): The blood supply to the central nervous system of *Helix pomatia. Z. Zellforsch.* **111**: 160–78.

Perkins, F.O., Ramsey, R.W. & Street, S.F. (1971): The ultrastructure of fishing tentacle muscles in the jellyfish *Chrysaora quinquecirrha*: a comparison of contracted and relaxed states. *J. Ultrastruct. Res.* **35**: 431–50.

Pernet, B. (2003): Persistent ancestral feeding structures in nonfeeding annelid larvae. *Biol. Bull.* **205**: 295–307.

Pesce, A., Bolognesi, M., Bocedi, A., Ascenzi, P., Dewilde, S., Moens, L., Hankeln, T. & Burmester, T. (2002b): Neuroglobin and cytoglobin. Fresh blood for the vertebrate globin family. *EMBO Rep.* **3**: 1146–51.

Pesce, A., Nardini, M., Dewilde, S., Geuens, E., Yamauchi, K., Ascenzi, P., Riggs, A.F., Moens, L. & Bolognesi, M. (2002a): The 109 residue nerve tissue minihemoglobin from *Cerebratulus lacteus* highlights striking structural plasticity of the α-helical globin fold. *Structure* **10**: 725–35.

Peters, W. & Latka, I. (1986): Electron microscopic localization of chitin using colloidal gold labelled with wheat germ agglutinin. *Histochemistry* **84**: 155–60.

Peterson, K.J. (2004): Isolation of *Hox* and *Parahox* genes in the hemichordate *Ptychodera flava* and the evolution of deuterostome Hox genes. *Mol. Phyl. Evol.* **31**: 1208–15.

Peterson, K.J. & Butterfield, N.J. (2005): Origin of the Eumetazoa: testing ecological predictions of molecular clocks against the Proterozoic fossil record. *Proc. Nat. Acad. Sci. USA* **102**: 9547–52.

Peterson, K.J. & Davidson, E.H. (2000): Regulatory evolution and the origin of the bilaterians. *Proc. Natl. Acad. Sci. USA* **97**: 4430–33.

Peterson, K.J. & Eernisse, D.J. (2001): Animal phylogeny and the ancestry of bilaterians: inferences from morphology and 18S rDNA sequences. *Evol. Dev.* **3**: 170–205.

Peterson, K.J., McPeek, M.A. & Evans, D.A.D. (2005): Tempo and mode of early animal evolution: inferences from rocks, Hox, and molecular clocks. *Paleobiology* Suppl. **31**: 36–55.

Peteya, D.J. (1973): A possible proprioreceptor in *Ceriantheopsis americanus* (Cnidaria, Ceriantharia). *Z. Zellforsch.* **144**: 1–10.

Petrov, A.M., Hooge, M.D. & Tyler, S. (2004): Ultrastructure of sperms in Acoela (Acoelomorpha) and its concordance with molecular systematics. *Invertebr. Biol.* **123**: 183–197.

Pflüger, H.-J. & Stevenson, P.A. (2005): Evolutionary aspects of octopaminergic systems with emphasis on arthropods. *Arthropod Struct. Dev.* **34**: 379–96.

Pflugfelder, O. (1948): Entwicklung von *Paraperipatus amboinensis* n.sp. *Zool. Jahrb. Anat.* **69**: 443–92.

Philip, G.K., Creevey, C.J. & McInerney, J.O. (2005): The Opisthokonta and the Ecdysozoa may not be clades: stronger support for the grouping of plant and animal than for animal and fungi and stronger support for Coelomata than Ecdysozoa. *Mol. Biol. Evol.* **22**: 1175–84.

Philip, G.M. (1979): Carpoids – echinoderms or chordates? *Biol. Rev.* **54**: 439–471.

Philippe, H. & Telford, M. (2006): Large-scale sequencing and the new animal phylogeny. *Trends Ecol. Evol.* **21**: 614–20.

Philippe, H., Chenuil, A. & Adoutte, A. (1994): Can the cambrian explosion be inferred through molecular phylogeny? *Development* **1994** Suppl.: 15–25.

Philippe, H., Lartillot, N. & Brinkmann, H. (2005): Multigene analysis of bilaterian animals corroborate the monophyly of Ecdysozoa, Lophotrochozoa, and Protostomia. *Mol. Biol. Evol.* **22**: 1246–53.

Phillips, J.I. (1978): The occurrence and distribution of haemoglobin in the endosymbiotic rhabdocoel *Paravortex scrobiculariae* (Graff) (Platyhelminthes: Turbellaria). *Comp. Biochem. Physiol. A* **61**: 679–83.

Phillips, J. (1981): Comparative physiology of insect renal function. *Am. J. Physiol.* **241**: R241–R257.

Piccinini, M., Kleinschmidt, T., Jurgens, K.D. & Braunitzer, G. (1990): Primary structure and oxygen-binding properties of the hemoglobin from guanaco (*Lama guanacoe*, Tylopoda). *Biol. Chem. Hoppe Seyler* **371**: 641–48.

Pietsch, A. & Westheide, W. (1987): Protonephridial organs in *Myzostoma cirriferum* (Myzostomida). *Acta Zool.* **68**: 195–203.

Pike, A.W. & Wink, R. (1986): Aspects of photoreceptor structure and phototactic behavior in Platyhelminthes, with particular reference to the symbiotic turbellarian *Paravortex*. *Hydrobiologia* **132**: 101–4.

Pilato, G., Binda, M.G., Biondi, O., D'Urso, V., Lisi, O., Marletta, A., Maugeri, S., Nobile, V., Rappazzo, G., Sabella, G., Sammartano, F., Turrisi, G. & Viglianisi, F. (2005): The clade Ecdysozoa, perplexities and questions. *Zool. Anz.* **43**: 43–50.

Pilger, J.F. (1993): Echiura. In: *Microscopic Anatomy of Invertebrates*, Harrison, F.W. & Rice, M.E. (eds.), Vol. 12: Onychophora, Chilopoda, and lesser Protostomata. Wiley-Liss, New York: 185–236.

Pilger, J.F. (2002): Phylum Echiura. In: *Atlas of Marine Invertebrate Larvae*, Young, C.M. (ed.). Academic Press, San Diego: 471–73.

Pilger, J.F. & Rice, M.E. (1987): Ultrastructural evidence for the contractile vessel of sipunculans as a possible ultrafiltration site. *Am. Zool.* **27**: 152A.

Pilgrim, D. (1999): *Caenorhabditis elegans*. In: *Encyclopedia of Reproduction*, Knobil, E. & Neill, J.D. (eds.). Vol. 1. Academic Press, San Diego: 449–57.

Pochon-Masson, J. (1983): Arthropoda – Crustacea. In: *Reproductive Biology of Invertebrates*, Adiyodi, K.G. & Adiyodi, R.G. (eds.). Vol. 2: Spermatogenesis and sperm function. John Wiley & Sons, Chichester: 407–49.

Podar, M., Haddock, S.H.D., Sogin, M.L. & Harbison, G.R. (2001): A molecular phylogenetic framework for the phylum Ctenophora using 18S rRNA genes. *Mol. Phyl. Evol.* **21**: 218–30.

Poinar, G.O. & Hansen, E. (1983): Sex and reproductive modifications in nematodes. *Helminthol. Abstr., Ser. B (Plant Nematology)* **52**: 145–63.

Poinar, G.O. & Hess-Poinar, R.T. (1993): The fine structure of *Gastromermis* sp. (Nematoda: Mermithidae) sperm. *J. Submicrosc. Cytol. Pathol.* **25**: 417–31.

Pollmanns, D. & Hündgen, M. (1981): Licht- und elektronenmikroskopische Untersuchung der Rhopalien von *Aurelia aurita* (Scyphozoa, Semaeostomae). *Zool. Jahrb. Anat.* **105**: 508–25.

Polyak, S. (1968): *The Vertebrate Visual System*: its origin, structure, and function and its manifestations in disease with an analysis of its role in the life of animals and in the origin of man. University of Chicago Press, London.

Popova, N.V. & Mamkaev, Y.V. (1985): Ultrastructure and primitive features of the eyes of *Convoluta convoluta* (Turbellaria, Acoela). *Dokl. Akad. Nauk. SSSR* **283**: 756–59. (in russian).

Por, F.D. & Masry, D. (1968): Survival of a nematode and an oligochaete species in the anaerobic benthal of Lake Tiberias. *Oikos* **19**: 388–91.

Potts, W.T.W. (1975): Excretion in gastropods. *Fortschr. Zool./Progr. Zool.* **23**: 76–88.

Proctor, H.C. (1998): Indirect sperm transfer in arthropods: behavioral and evolutionary trends. *Ann. Rev. Entomol.* **43**: 153–74.

Protasoni, M., de Eguileor, M., Congiu, T., Grimaldi, A. & Reguzzoni, M. (2003): The extracellular matrix of the cuticle of *Gordius panighettensis* (Gordioiidae, Nematomorpha): observations by TEM, SEM and AFM. *Tiss. Cell* **35**: 306–11.

Prud'Homme, B., de Rosa, R., Arendt, D., Julien, J.-F., Pajaziti, R., Dorresteijn, A.W.C., Adoutte, A., Wittbrodt, J. & Balavoine, G. (2003): Arthropod-like expression patterns of *engrailed* and *wingless* in the annelid *Platynereis dumerilii* suggest a role in segment formation. *Curr. Biol.* **13**: 1876–81.

Prud'Homme, B., Gompel, N., Rokas, A., Kassner, V.A., Williams, T.M., Yeh, S.-D., True, J.R. & Carroll, S.B. (2006): Repeated morphological evolution through *cis*-regulatory changes in a pleiotrophic gene. *Nature* **440**: 1050–53.

Purschke, G. (2002a): On the ground pattern of Annelida. *Org. Div. Evol.* **2**: 181–196.

Purschke, G. (2002b): Male genital organs, spermatogenesis and spermatozoa in the enigmatic terrestrial polychaete *Parergodrilus heideri* (Annelida, Parergodirlidae). *Zoomorphology* **121**: 125–38.

Purschke, G. (2003): Ultrastructure of phaosomous photoreceptors in *Stylaria lacustris* (Naididae, 'Oligochaeta', Clitellata) and their importance for the position of Clitellata in the phylogenetic system of Annelida. *J. Zool. Syst. Evol. Res.* **41**: 100–8.

Purschke, G. (2005): Sense organs in polychaetes (Annelida). In: Morphology, Molecules, Evolution and Phylogeny in Polychaeta and Related Taxa, Bartolomaeus, T. & Purschke, G. (eds.). *Hydrobiologia* **535/536**: 53–78.

Purschke, G. & Hausen, H. (2007): Lateral organs in sedentary polychaetes (Annelida) – ultrastructure and phylogenetic significance of an insufficiently known sense organ. *Acta Zool.* **88**: 23–39.

Purschke, G. & Müller, M.C.M. (2007): Evolution of body wall musculature. *Integr. Comp. Biol.* **46**: 497–507.

Purschke, G. & Tzetlin, A.B. (1996): Dorsolateral ciliary folds in the polychaete foregut: structure, prevalence and phylogenetic significance. *Acta Zool.* **77**: 33–49.

Purschke, G., Wolfrath, F. & Westheide, W. (1997): Ultrastructure of the nuchal organ and cerebral organ in *Onchnesoma squamatum* (Sipuncula, Phascolionidae). *Zoomorphology* **117**: 23–31.

Purschke, G., Arendt, D., Hausen, H. & Müller, M.C.M. (2006): Photoreceptor cells and eyes in Annelida. *Arthropod Struct. Dev.* **35**: 211–30.

Quast, B. & Bartolomaeus, T. (2001): Ultrastructure and significance of the transitory nephridia in *Erpobdella octoculata* (Hirudinea, Annelida). *Zoomorphology* **120**: 205–13.

Quiring, R., Walldorf, U., Kloter, U. & Gehring, W.J. (1994): Homology of the *eyeless* gene of *Drosophila* to the *small eye* gene in mice and *Aniridia* in humans. *Science* **265**: 785–89.

Rähr, H. (1979): The circulatory system of Amphioxus (*Branchiostoma lanceolatum* (Pallas)). *Acta Zool.* **60**: 1–18.

Raible, F., Tessmar-Raible, K., Arboleda, E., Kaller, T., Bork, P., Arendt, D. & Arnone, M.I. (2006) : Opsins and clusters of sensory G-protein-coupled receptors in the sea urchin genome. *Dev. Biol.* **300**: 461–75.

Raible, F., Tessmar-Raible, K., Osoegawa, K., Wincker, P., Jubin, C., Balavoine, G., Ferrier, D., Benes, V., de Jong, P., Weissenbach, J., Bork, P. & Arendt, D. (2005): Vertebrate-type intron-rich genes in the marine annelid *Platynereis dumerilii*. *Science* **310**: 1325–26.

Raikova, E.V. (1995): Occurrence and ultrastructure of collar cells in the stomach gastrodermis of *Polypodium hydriforme* Ussov (Cnidaria). *Acta Zool.* **76**: 11–18.

Raikova, O.I., Flyatchinskaya, L.P. & Justine, J.-L. (1998): Acoel spermatozoa: ultrastructure and immunocytochemistry of tubulin. *Hydrobiologia* **383**: 207–14.

Raikova, O.I., Reuter, M., Gustafsson, M.K.S., Maule, A.G., Halton, D.W. & Jondelius, U. (2004): Basiepidermal nervous system in *Nemertoderma westbladi* (Nemertodermatida): GYIRFamide immunireactivity. *Zoology* **107**: 75–86.

Raikova, O.I., Reuter, M., Jondelius, U. & Gustafsson, M.K.S. (2000a): An immunocytochemical and ultrastructural study of the nervous and muscular systems of *Xenoturbella westbladi* (Bilateria inc. sed.). *Zoomorphology* **120**: 107–18.

Raikova, O.I., Reuter, M., Jondelius, U., Gustafsson, M.K.S. (2000b): The brain of the Nemertodermatida (Platyhelminthes) as revealed by anti-5HT and anti-FMRFamide immunostainings. *Tiss. Cell* **32**: 358–365.

Raikova, O.I., Reuter, M., Kotikova, E.A. & Gustafsson, M.K.S. (1998): A commissural brain! The pattern of 5-HT immunoreactivity in Acoela (Plathelminthes). *Zoomorphology* **118**: 69–77.

Rebecchi, L. (1997): Ultrastructural study of spermiogenesis and the testicular and spermathecal spermatozoon of the gonochoristic tardigrade *Xerobiotus pseudohufelandi* (Eutardigrada, Macrobiotidae). *J. Morphol.* **234**: 11–24.

Rebecchi, L. & Bertolani, R. (1999): Spermatozoon morphology of three species of Hypsibiidae (Tardigrada, Eutardigrada) and phylogenetic evaluation. *Zool. Anz.* **238**: 319–28.

Rebecchi, L. & Guidi, A. (1995): Spermatozoon ultrastructure in two species of *Amphibolus* (Eutardigrada, Eohypsibiidae). *Acta Zool.* **76**: 171–176.

Rebecchi, L., Guidi, A. & Bertolani, R. (2000): Tardigrada. In: *Reproductive Biology of Invertebrates*, Adiyodi, K.G., Adiyodi, R.G. & Jamieson, B.G.M. (eds.). Vol. 9, part B: Progress in male gamete ultrastructure and phylogeny. John Wiley & Sons, Chichester: 267–91.

Rebecchi, L., Guidi, A. & Bertolani, R. (2003): The spermatozoon of the Echiniscidae (Tardigrada, Heterotardigrada) ant its phylogenetic significance. *Zoomorphology* **122**: 3–9.

Reed, C.G. (1991): Bryozoa. In: Reproduction of Marine *Invertebrates*, Giese, A.C., Pearse, J.S. & Pearse, V.B. (eds.). Vol. 6: Echinoderms and lophophorates. The Boxwood Press, Pacific Grove: 85–245.

Reed, C.G. & Cloney, R.A. (1977): Brachiopod tentacles: ultrastructure and functional significance of the connective tissue and myoepithelial cells in *Terebratulina*. *Cell Tiss. Res.* **185**: 17–42.

Rees, G. (1971): Locomotion of the cercaria of *Parorchis acanthus*, Nicoll and the ultrastructure of the tail. *Parasitology* **62**: 489–503.

Reger, J.F. (1969): Studies on the fine structure of muscle fibres and contained crystalloids in basal socket muscle of the entoproct, *Barentsia gracilis*. *J. Cell Sci.* **4**: 305–25.

Reger, J.F. (1971): A fine structure study on spermiogenesis in the entoproct, *Bugula* sp. *J. Submicrosc. Cytol.* **3**: 193–200.

Regier, J.C. & Shultz, J.W. (1997): Molecular phylogeny of the major arthropod groups indicates polyphyly of crustaceans and a new hypothesis for the origin of hexapods. *Mol. Biol. Evol.* **14**: 902–13.

Regier, J.C. & Shultz, J.W. (2001): Elongation factor-2: a useful gene for arthropod phylogenetics. *Mol. Phyl. Evol.* **20**: 136–48.

Regier, J.C., Shultz, J.W. & Kambric, R.E. (2005): Pancrustacean phylogeny: hexapods are terrestrial crustaceans and maxillopods are not monophyletic. *Proc. R. Soc. B* **272**: 395–401.

Rehkämper, G. (1968): *Nervensysteme im Tierreich. Bau, Funktion und Entwicklung.* Quelle & Meyer Verlag, Heidelberg.

Rehkämper, G., Storch, V., Alberti, G. & Welsch, U. (1989): On the fine structure of the nervous system of *Tubiluchus philippinensis* (Tubiluchidae, Priapulida). *Acta Zool.* **70**: 111–20.

Reichert, H. & Simeone, A. (2001): Developmental genetic evidence for a monophyletic origin of the bilaterian brain. *Philos. Trans. R. Soc. London B* **356**: 1533–44.

Reise, K. (1985): Tidal flat ecology. An experimental approach to species interactions. *Ecological Studies* **54**. Springer-Verlag, Berlin.

Reise, K. & Ax, P. (1979): A meiofaunal 'thiobios' limited to the anaerobic sulfide system of marine sand does not exist. *Mar. Biol.* **54**: 225–37.

Reisinger, E. (1960): Was ist *Xenoturbella*? *Z. Wiss. Zool.* **164**: 188–98.

Reisinger, E. (1961): Allgemeine Morphologie der Metazoen. Morphologie der Coelenteraten, acoelomaten und pseudocoelomaten Würmer. *Fortschr. Zool.* **13**: 1–82.

Reisinger, E. (1972): Die Evolution des Orthogons der Spiralier und das Archicölomatenproblem. *Z. Zool. Syst. Evolutionsforsch.* **10**: 1–43.

Reiswig, H.M. & Miller, T.L. (1998): Freshwater sponge gemmules survive months of anoxia. *Invertebr. Biol.* **117**: 1–8.

Reiter, D., Boyer, B., Ladurner, P., Mair, G., Salvenmoser, W. & Rieger, R. (1996): Differentiation of the body wall musculature in *Macrostomum hystricinum marinum* and *Hoploplana inquilina* (Plathelminthes), as models for muscle development in lower Spiralia. *Roux's Arch. Dev. Biol.* **205**: 410–23.

Reitner, J. & Mehl, D. (1996): Monophyly of the Porifera. *Verh. Naturwiss. Ver. Hamburg* **36**: 5–32.

Remane, A. (1950): Die Entstehung der Metamerie der Wirbellosen. *Verh. Dtsch. Zool. Ges.* **1949** (Zool. Anz. Suppl. 14): 16–23.

Remane, A. (1951): Die Bursa-Darmverbindung und das Problem des Enddarms bei Turbellarien. *Zool. Anz.* **146**: 275–91.

Remane, A. (1958): Zur Verwandtschaft und Ableitung der niederen Metazoen. *Verh. Dtsch. Zool. Ges.* **1957** (Zool. Anz. Suppl. **21**): 179–96.

Remane, A. (1963): The enterocelic origin of the celom. In: *The Lower Metazoa*, Dougherty, E.C. (ed.). University of California Press, Berkeley: 78–90.

Retzius, G. (1904): Zur Kenntnis der Spermien der Evertebraten I. *Biologische Untersuchungen von Gustav Retzius, Neuw Folge* **11**: 1–32.

Retzius, G. (1905): Zur Kenntnis der Spermien der Evertebraten II. *Biologische Untersuchungen von Gustav Retzius, Neuw Folge* **11**: 79–102.

Reuter, M. & Gustafsson, M.K.S. (1995): The flatworm nervous system: pattern and phylogeny. In: *The Nervous Systems of Invertebrates: An Evolutionary and Comparative Approach*, Breidbach, O. & Kutsch, W. (eds.). Birkhäuser Verlag, Basel: 25–59.

Reuter, M., Mäntylä, K. & Gustafsson, M.K.S. (1998): Organization of the orthogon – main and minor nerve cords. *Hydrobiologia* **383**: 175–82.

Reuter, M., Raikova, O.I. & Gustafsson, M.K.S. (2001b): Patterns in the nervous and muscle systems in lower flatworms. *Belg. J. Zool.* **131** (Supplement 1): 47–53.

Reuter, M., Raikova, O.I., Jondelius, U., Gustafsson, M.K.S., Maule, A.G. & Halton, D.W. (2001a): Organisation of the nervous system in the Acoela: an immunocytochemical study. *Tiss. Cell* **33**: 119–28.

Revel, J.-P. (1988): The oldest multicellular animal and its junctions. In: Gap Junctions, Hertzberg, E.L. & Johnson, R.G. (eds.), *Modern Cell Biol.* **7**: 135–49.

Reynolds, P.D. (1990): Fine structure of the kidney and characterization of secretory products in *Dentalium rectius* (Mollusca, Scaphopoda). *Zoomorphology* **110**: 53–62.

Reynolds, P.D. (1992): Distribution and ultrastructure of ciliated sensory receptors in the posterior mantle epithelium of *Dentalis rectius* (Mollusca, Scaphopoda). *Acta Zool.* **73**: 263–70.

Reynolds, P.D., Morse, M.P. & Norenburg, J. (1993): Ultrastructure of the heart and pericardium of an aplacophoran mollusc (Neomeniomorpha): evidence for ultrafiltration of blood. *Proc. R. Soc. London B* **254**: 147–52.

Reznick, D.N., Mateos, M. & Springer, M.S. (2002): Independent origins and rapid evolution of the placenta in the fish genus *Poeciliopsis*. *Science* **298**: 1018–20.

Rhode, B. (1992): Development and differentiation of the eye of *Platynereis dumerilii* (Annelida, Polychaeta). *J. Morphol.* **212**: 71–85.

Rhodes, J.D., Thain, J.F. & Wildon, D.C. (1996): The pathway for systemic electrical signal conduction in the wounded tomato plant. *Planta* **200**: 50–57.

Ribeiro, P., El-Shehabi, F. & Patocka, N. (2005): Classical transmitters and their receptors in flatworms. *Parasitology* **131**: S19–S40.

Ricci, C. (1992): Rotifera: parthenogenesis and heterogony. In: *Sex Origin and Evolution*, Dallai, R. (ed.). Selected Symposia and Monographs U.Z.I. 6. Mucchi, Modena: 329–41.

Ricci, C., Melone, G. & Sotgia, C. (1993): Old and new data on Seisonidea (Rotifera). *Hydrobiologia* **255/56**: 495–511.

Rice, M.E. (1989): Sipuncula. In: *Reproductive Biology of Invertebrates*, Adiyodi, K.G. & Adiyodi, R.G. (eds.). Vol. 4, part A: fertilization, development, and parental care. John Wiley & Sons, Chichester: 263–80.

Rice, M.E. (1993): Sipunculida. In: *Microscopic Anatomy of Invertebrates*, Harrison, F.W. & Rice, M.E. (eds.), Vol. 12: Onychophora, Chilopoda, and lesser Protostomata. Wiley-Liss, New York: 237–325.

Rice, M.E. (1999): Sipuncula. In: *Encyclopedia of Reproduction*, Knobil, E. & Neill, J.D. (eds.). Vol. 4. Academic Press, San Diego: 492–97.

Rice, S.A. (1992): Polychaeta: spermatogenesis and spermiogenesis. In: *Microscopic Anatomy of Invertebrates*, Harrison, F.W. & Gardiner, S.L. (eds.), Vol. 7: Annelida. Wiley-Liss, New York: 129–51.

Richter, S. (2002): The Tetraconata concept: hexapod-crustacean relationships and the phylogeny of Crustacea. *Org. Div. Evol.* **2**: 217–37.

Richter, S. & Wirkner, C. (2004): Kontroversen in der Phylogenetischen Systematik der Euarthropoda. In: Kontroversen in der Phylogenetischen Systematik der Metazoa, Richter, S. & Sudhaus, W. (eds.). *Sitzungsber. Ges. Naturforsch. Freunde Berlin* N.F. **43**: 73–102.

Ridley, R.K. (1969): Electron microscopic studies on dicyemid Mesozoa. II. Infusorigen and infusoriform stages. *J. Parasitol.* **55**: 779–93.

Riedl, R.J. (1969): Gnathostomulida from America. *Science* **163**: 445–52.

Riegel, J.A. & Cook, M.A. (1975): Recent studies of excretion in Crustacea. *Fortschr. Zool./Progr. Zool.* **23**: 48–75.

Rieger, G.E. & Rieger, R.M. (1977): Comparative fine structure study of the gastrotrich cuticle and aspects of cuticle evolution within the Aschelminthes. *Z. Zool. Syst. Evolutionsforsch.* **15**: 8–124.

Rieger, R.M. (1976): Monociliated epidermal cells in Gastrotricha: significance for concepts of early metazoan evolution. *Z. Zool. Syst. Evolutionsforsch.* **14**: 198–226.

Rieger, R.M. (1978): Multiple ciliary structures in developing spermatozoa of marine Catenulida (Turbellaria). *Zoomorphologie* **89**: 229–36.

Rieger, R.M. (1981): Fine structure of the body wall, nervous system, and digestive tract in the Lobatocerbridae Rieger and the organization of the glioitestinal system in Annelida. *J. Morphol.* **167**: 139–65.

Rieger, R.M. (1984): Evolution of the cuticle in the lower Eumetazoa. In: *Biology of the Integument*, Bereiter-Hahn, J., Matoltsy, A.G. & Sylvia Richards, K. (eds.). Vol. 1: Invertebrates. Springer, Berlin: 389–99.

Rieger, R.M. (1986): Über den Ursprung der Bilateria: die Bedeutung der Ultrastrukturforschung für ein neues Verstehen der Metazoenevolution. *Verh. Dtsch. Zool. Ges.* **79**: 31–50.

Rieger, R.M. (1994a): The biphasic life cycle – a central theme of metazoan evolution. *Am. Zool.* **34**: 484–91.

Rieger, R.M. (1994b): Evolution of the 'lower' Metazoa. In: *Early Life on Earth*, Bengtson, S. (ed.). Nobel Symp. 84. Columbia University Press, New York: 475–88.

Rieger, R.M. (2003): The phenotypic transition from uni- to multicellular animals. In: *The New Panorama of Animal Evolution*, Legakis, A., Sfenthourakis, S., Polymeni, R. & Thessalou-Legakis, M. (eds.). Proc. 18th Int. Congr. Zool., Pensoft Publishers, Sofia: 247–58.

Rieger, R. & Ladurner, P. (2001): Searching for the stem species of the Bilateria. *Belg. J. Zool.* (Suppl. 1) **131**: 27–34.

Rieger, R.M. & Lombardi, J. (1987): Ultrastructure of coelomic lining in echinoderm podia: significance for concepts in the evolution of muscle and peritoneal cells. *Zoomorphology* **107**: 191–208.

Rieger, R.M. & Mainitz, M. (1977): Comparative fine structure of the body wall in Gnathostomulida and their phylogenetic position between Platyhelminthes and Aschelminthes. *Z. Zool. Syst. Evolutionsforsch.* **15**: 9–35.

Rieger, R.M. & Tyler, S. (1995): Sister-group relationship of Gnathostomulida and Rotifera-Acanthocephala. *Invertebr. Biol.* **114**: 186–88.

Rieger, R.M. & Weyrer, S. (1998): The evolution of the lower Metazoa: evidence from the phenotype. In: Molecular Evolution: Towards the Origin of Metazoa, Müller, W.E.G. (ed.), *Progr. Mol. Subcell. Biol.* **21**: 20–43.

Rieger, R.M., Haszprunar, G. & Schuchert, P. (1991a): On the origin of the Bilateria: traditional views and recent alternative concepts. In: *The Early Evolution of Metazoa and the Significance of Problematic Taxa*,

Simonetta, A.M. & Conway Morris, S. (eds.). Cambridge University Press, Cambridge: 107–12.

Rieger, R.M., Tyler, S., Smith, J.P.S. & Rieger, G.E. (1991b): Platyhelminthes: Turbellaria. In: *Microscopic Anatomy of Invertebrates*, Harrison, F.W. & Bogitsh, B.J. (eds.), Vol. 3: Platyhelminthes and Nemertinea. Wiley-Liss, New York: 7–140.

Riggs, A.F. (1991): Aspects of the origin and evolution of non-vertebrate hemoglobins. *Am. Zool.* **31**: 535–45.

Riisgård, H.U. (2002): Methods of ciliary filter feeding in adult *Phoronis muelleri* (phylum Phoronida) and in its free-swimming actinotroch larva. *Mar. Biol.* **141**: 75–87.

Riisgård, H.U. & Manriquez, P. (1997): Filter-feeding in fifteen marine ectoprocts (Bryozoa): particle capture and water pumping. *Mar. Ecol. Progr. Ser.* **154**: 223–39.

Riisgård, H.U., Nielsen, K.K., Fuchs, J., Rasmussen, B.F., Obst, M. & Funch, P. (2004): Ciliary feeding structures and particle capture mechanism in the freshwater bryozoan *Plumatella repens* (Phylactolaemata). *Invertebr. Biol.* **123**: 156–67.

Roberts, A. & Mackie, G.O. (1980): The giant axon escape system of a hydrozoan medusa, *Aglantha digitale*. *J. Exp. Biol.* **84**: 303–18.

Rodrigo, A.G., Bergquist, P.R., Bergquist, P.L. & Reeves, R.A. (1994): Are sponges animals? An investigation into the vagaries of phylogenetic inference. In: *Sponges in Time and Space*, van Soest, R.W.M., van Kempen, T.M.G. & Braekman, J.-C. (eds.). Proc. 4th Int. Porifera Congr. Amsterdam. Balkema, Rotterdam: 47–54.

Röhlich, P. & Török, L.J. (1964): Elektronenmikroskopische Beobachtungen an den Sehzellen des Blutegels, *Hirudo medicinalis* L. *Z. Zellforsch.* **63**: 618–35.

Rohde, K. & Watson, N.A. (1995): Sensory receptors and epidermal structures of a meiofaunal turbellarian (Proseriata: Monocelididae: Minoninae). *Aust. J. Zool.* **43**: 69–81.

Rohde, K., Johnson, A.M., Baverstock, P.R. & Watson, N.A. (1995): Aspects of the phylogeny of Platyhelminthes based on 18S ribosomal DNA and protonephridial ultrastructure. *Hydrobiologia* **305**: 27–35.

Rohde, K., Watson, N. & Cannon, L.R.G. (1988): Ultrastructure of epidermal cilia of *Pseudactinoposthia* sp. (Platyhelminthes, Acoela); implications for the phylogenetic status of Xenoturbellida and Acoelomorpha. *J. Submicrosc. Cytol. Pathol.* **20**: 759–67.

Rohde, K., Watson, N.A. & Faubel, A. (1993): Ultrastructure of the statocyst in an undescribed species of Luridae (Platyhelminthes: Rhabdocoela: Luridae). *Aust. J. Zool.* **41**: 215–24.

Rokas, A. & Carroll, S.B. (2005): More genes or more taxa? The relative contribution of gene number and taxon number to phylogenetic accuracy. *Mol. Biol. Evol.* **22**: 1337–44.

Rokas, A., King, N., Finnerty, J.R. & Carroll, S.B. (2003): Conflicting phylogenetic signals at the base of the metazoan tree. *Evol. Dev.* **5**: 346–59.

Rokas, A., Krüger, D. & Carroll, S.B. (2005): Animal evolution and the molecular signature of radiations compressed in time. *Science* **310**: 1933–38.

Rokas, A, Williams, B.L., King, N. & Carroll, S.B. (2003): Genome-scale approaches to resolving incongruence in molecular phylogenies. *Nature* **425**: 798–804.

Roldan, E.R.S., Vitullo, A.D. & Gomendio, M. (1991): Sperm shape and size: evolutionary processes in mammals. In: *Comparative Spermatology 20 years After*, Baccetti, B. (ed.). Serono Symp. Publ. Raven Press 75. Raven Press, New York: 1001–10.

Romer, A.S. (1977): *The Vertebrate Body*. 5th ed., Saunders, Philadelphia.

Rose, R.D. & Stokes, D.R. (1981): A crustacean statocyst with only three hairs: light and scanning electron microscopy. *J. Morphol.* **169**: 21–28.

Rosen, M.D., Stasek, C.R. & Hermans, C.O. (1978): The ultrastructure and evolutionary significance of the cerebral ocelli of *Mytilus edulis*, the bay mussel. *Veliger* **21**: 10–18.

Rosen, M.D., Stasek, C.R. & Hermans, C.O. (1979): The ultrastructure and evolutionary significance of the ocelli in the larva of *Katharina tunicata* (Mollusca: Polyplacophora). *Veliger* **22**: 173–78.

Rosenberg, M.S. & Kumar, S. (2001): Incomplete taxon sampling is not a problem for phylogenetic inference. *Proc. Natl. Acad. Sci. USA* **98**: 10751–56.

Rosenbluth, J. (1972): Obliquely struated muscle. In: *The Structure and Function of Muscle*, Bourne, G.H. (ed.). Academic Press, New York: 389–420

Rothe, B.H. & Schmidt-Rhaesa, A. (2004): Probable development from continuous to segmental longitudinal musculature in *Pycnophyes kielensis* (Kinorhyncha). *Meiofauna Marina* **13**: 21–28.

Rothe, B.H., Schmidt-Rhaesa, A. & Todaro, M.A. (2006): The general muscular architecture in *Tubiluchus troglodytes* (Priapulida). *Meiofauna Marina* **15**: 79–86.

Rouse, G.W. (1999a): Trochophore concepts: ciliary bands and the evolution of larvae in spiralian Metazoa. *Biol. J. Linn. Soc.* **66**: 411–64.

Rouse, G.W. (1999b): Polychaete sperm: phylogenetic and functional considerations. *Hydrobiologia* **402**: 215–24.

Rouse, G.W. (2000): Polychaeta, including Pogonophora and Myzostomida. In: *Reproductive Biology of Invertebrates*, Adiyodi, K.G., Adiyodi, R.G. & Jamieson, B.G.M. (eds.). Vol. 9, part B: Progress in male gamete ultrastructure and phylogeny. John Wiley & Sons, Chichester: 81–124.

Rouse, G.W. (2005): Annelid sperm and fertilization biology. In: Morphology, Molecules, Evolution and

Phylogeny in Polychaeta and Related Taxa, Bartolomaeus, T. & Purschke, G. (eds.). *Hydrobiologia* **535/536**: 167–78.

Rouse, G.W. & Fauchald, K. (1997): Cladistics and polychaetes. *Zool. Scr.* **26**: 139–204.

Rouse, G.W. & Fauchald, K. (1998): Recent views on the status, delineation and classification of the Annelida. *Am. Zool.* **38**: 953–64.

Rouse, G. & Fitzhugh, K. (1994): Broadcasting fables: is external fertilization really primitive? Sex, size, and larvae in sabellid polychaetes. *Zool. Scr.* **23**: 271–312.

Rouse, G.W., Goffredi, S.K. & Vrijenhoek, R.C. (2004): *Osedax*: bone-eating marine worms with dwarf males. *Science* **305**: 668–71.

Rouse, G.W. & Jamieson, B.G.M. (1987): An ultrastructural study of the spermatozoa of the polychaete *Eurythoe complanata* (Amphinomidae), *Clymenella* sp. and *Micromaldane* sp. (Maldanidae), with definition of sperm types in relation to reproductive biology. *J. Submicrosc. Cytol.* **19**: 573–84.

Rousset, V., Rouse, G.W., Siddall, M.E., Tillier, A. & Pleijel, F. (2004): The phylogenetic position of Siboglinidae (Annelida) inferred from 18S rRNA, 28S rRNA and morphological data. *Cladistics* **20**: 518–33.

Rowe, F.W.E., Healy, J.M. & Anderson, D.T. (1994): Concentricycloidea. In: *Microscopic Anatomy of Invertebrates*, Harrison, F.W. & Chia, F.-S. (eds.), Vol. 14: Echinodermata. Wiley-Liss, New York: 149–68.

Royuela, M., Astier, C., Grandier-Vazeille, X., Benyamin, Y., Fraile, B., Paniagua, R. & Duvert, M. (2003): Immunocytochemistry of chaetognath body wall muscles. *Invertebr. Biol.* **122**: 74–82.

Royuela, M., Fraile, B., Arenas, I. & Paniagua, R. (2000b): Characterization of several invertebrate muscle cell types: a comparison with vertebrate muscles. *Microsc. Res. Tech.* **48**: 107–15.

Royuela, M., Meyer-Rochow, V.B., Fraile, B. & Paniagua, R. (2000a): Muscle cells in the tiny antarctic mite *Halacarellus thomasi*: an ultrastructural and immunocytochemical study. *Polar Biol.* **23**: 759–65.

Ruby, E.G. & Fox, D.L. (1976): Anaerobic respiration in the polychaete *Euzonus* (*Thoracophelia*) *mucronata*. *Mar. Biol.* **35**: 149–53.

Ruhberg, H. (1990): Onychophora. In: *Reproductive Biology of Invertebrates*, Adiyodi, K.G. & Adiyodi, R.G. (eds.). Vol. 4, part B: Fertilization, development, and parental care. John Wiley & Sons, Chichester: 61–76.

Ruiz-Trillo, I., Paps, J., Loukota, M., Ribera, C., Jondelius, U., Baguñà, J. & Riutort, M. (2002): A phylogenetic analysis of myosin heavy chain type II sequences corroborates that Acoela and Nemertodermatida are basal bilaterians. *Proc. Natl. Acad. Sci. USA* **99**: 11246–51.

Ruiz-Trillo, I., Riutort, M., Littlewood, D.T.J., Herniou, E.A. & Baguñà, J. (1999): Acoel flatworms: earliest extant bilaterian metazoans, not members of Platyhelminthes. *Science* **283**: 1919–23.

Ruppert, E.E. (1978a): A review of metamorphosis of turbellarian larvae. In: *Settlement and Metamorphosis of Marine Invertebrate Larvae*, Chia, F.-S. & Rice, M.E. (eds.). Elsevier, New York: 65–81.

Ruppert, E.E. (1978b): The reproductive system of gastrotrichs. II. Insemination in *Macrodasys*: a unique mode of sperm transfer in Metazoa. *Zoomorphologie* **89**: 207–28.

Ruppert, E.E. (1982): Comparative ultrastructure of the gastrotrich pharynx and the evolution of myoepithelial foreguts in Aschelminthes. *Zoomorphology* **99**: 181–220.

Ruppert, E.E. (1991a): Introduction to the aschelminth phyla: a consideration of mesoderm, body cavities, and cuticle. In: *Microscopic Anatomy of Invertebrates*, Harrison, F.W. & Ruppert, E.E. (eds.), Vol. 4: Aschelminthes. Wiley-Liss, New York: 1–17.

Ruppert, E.E. (1991b): Gastrotricha. In: *Microscopic Anatomy of Invertebrates*, Harrison, F.W. & Ruppert, E.E. (eds.), Vol. 4: Aschelminthes. Wiley-Liss, New York: 41–109.

Ruppert, E.E. (1994): Evolutionary origin of the vertebrate nephron. *Am. Zool.* **34**: 542–53.

Ruppert, E.E. (1997): Cephalochordata (Acrania). In: *Microscopic Anatomy of Invertebrates*, Harrison, F.W. & Ruppert, E.E. (eds.), Vol. 15: Hemichordata, Chaetognatha, and the invertebrate chordates. Wiley-Liss, New York: 349–504.

Ruppert, E.E. (2005): Key characters uniting hemichordates and chordates: homologies or homoplasies? *Can. J. Zool.* **83**: 8–23.

Ruppert, E.E. & Balser, E.J. (1986): Nephridia in the larvae of hemichordates and echinoderms. *Biol. Bull.* **171**: 188–96.

Ruppert, E.E. & Barnes, R.D. (1994): *Invertebrate Zoology*, 6th Edn., Saunders College Publ., Fort Worth.

Ruppert, E.E. & Carle, K.J. (1983): Morphology of metazoan circulatory systems. *Zoomorphology* **103**: 193–208

Ruppert, E.E. & Rice, M.E. (1995): Functional organization of dermal coelomic canals in *Sipunculus nudus* (Sipuncula) with a discussion of respiratory designs in sipunculans. *Invertebr. Biol.* **114**: 51–63.

Ruppert, E.E. & Schreiner, S.P. (1980): Ultrastructure and potential significance of cerebral light-refracting bodies of *Stenostomum virginianum* (Turbellaria, Catenulida). *Zoomorphology* **96**: 21–31.

Ruppert, E.E. & Smith, P.R. (1988): The functional organization of filtration nephridia. *Biol. Rev.* **63**: 231–58

Ruppert, E.E., Cameron, C.B. & Frick, J.E. (1999): Endostyle-like features of the dorsal epibranchial ridge of an enteropneust and the hypothesis of dorso-ventral axis inversion in chordates. *Invertebr. Biol.* **118**: 202–12.

.Ruppert, E.E., Fox, R.S. & Barnes, R.D. (2004): *Invertebrate Zoology*, 7th Edn., Brooks/Cole, Belmont.

.Ruthmann, A., Behrendt, G. & Wahl, R. (1986): The ventral epithelium of *Trichoplax adhaerens* (Placozoa): cytoskeletal structures, cell contacts and endocytosis. *Zoomorphology* **106**: 115–22.

Ruud, J.T. (1954): Vertebrates without erythrocytes and blood pigment. *Nature* **173**: 848–50.

Ryland, J.S. (1970): *Bryozoans*. Hutchinson & Co Publishers, London.

Saito, M., Kojima, S. & Endo, K. (2000): Mitochondrial COI sequences of brachiopods: genetic code shared with protostomes and limits of utility for phylogenetic reconstruction. *Mol. Phyl. Evol.* **15**: 331–44.

Saleuddin, A.S.M. (1999): Mollusca. In: *Encyclopedia of Reproduction*, Knobil, E. & Neill, J.D. (eds.). Vol. 3. Academic Press, San Diego: 276–83.

Sandeman, D.C. (1982): Organization of the central nervous system. In: *The Biology of Crustacea*, Atwood, H.L. & Sandeman, D.C. (eds.), Vol. 3: Neurobiology: structure and function. Academic Press, New York: 1–61.

Sanderson, M.J. (1984): Cilia. In: *Biology of the Integument*, Bereiter-Hahn, J., Matoltsy, A.G. & Sylvia Richards, K. (eds.). Vol. 1: Invertebrates. Springer, Berlin: 18–42.

Santagata, S. (2002): Structure and metamorphic remodelling of the larval nervous system and musculature of *Phoronis pallida* (Phoronida). *Evol. Dev.* **4**: 28–42.

Santagata, S. & Zimmer, R.L. (2002): Comparison of the neuromuscular systems among actinotroch larvae: systematic and evolutionary implications. *Evol. Dev.* **4**: 43–54.

Sará, M. (1992): Sex in Porifera. In: *Sex Origin and Evolution*, Dallai, R. (ed.). Selected Symposia and Monographs U.Z.I. 6. Mucchi, Modena: 45–57.

Saranak, J. & Foster, K.W. (1997): Rhodopsin guides fungal phototaxis. *Nature* **387**: 465–66.

Sarbu, S.M., Kane, T.C. & Kinkle, B.K. (1996): A chemoautotrophically based cave ecosystem. *Science* **272**: 1953–55.

Sarras, M.P., Madden, M.E., Zhang, X., Gunwar, S., Huff, J.K. & Hudson, B.G. (1991): Extracellular matrix (mesogloea) of *Hydra vulgaris*. I. Isolation and characterization. *Dev. Biol.* **148**: 481–94.

Sauber, F., Reuland, M., Berchtold, J.-P., Hertu, C., Tsoupras, G., Luu, B., Moritz, M.-E. & Hoffmann, J.A. (1983): Cycle de mue et ecdystéroïdes chez une sangsue, *Hirudo medicinalis*. *C. R. Acad. Sci. Paris* **296**: 413–18.

Sauer, K.P. & Kullmann, H. (2005): Analyse der historischökologischen Ursachen der Evolution der gastroneuralen Metazoa – Testen einer phylogenetischen Hypothese. *Bonner Zool. Beitr.* **53**: 149–63.

Savagner, P. (2005): *Rise and Fall of Epithelial Phenotype: Concepts of Epithelial-Mesenchymal Transition*. Kluwer Academic/Plenum Press, New York.

Sawada, N. (1980): An electron microscopical study on spermatogenesis in *Golfingia ikedai*. *Acta Zool.* **61**: 127–32.

Sawyer, R.T. (1986): *Leech Biology and Behaviour*. Vol. 2: Feeding biology, ecology, and systematics. Clarendon Press, Oxford.

Scanabissi, F., Eder, E. & Cesari, M. (2005): Male occurrence in Austrian populations of *Triops cancriformis* (Branchiopoda, Notostraca) and ultrastructural observations of the male gonad. *Invertebr. Biol.* **124**: 57–65.

Schäfer, W. (1983): Vergleichende Untersuchungen über Struktur und Genese der Anthozoenoocyten. Teil 1: Frühentwicklung und innere Differenzierung der Oocyten. *Zool. Jahrb. Anat.* **109**: 407–48.

Schedl, T. (1997): Developmental genetics of the germ line. In: *C. elegans* II, Riddle, D.L., Blumenthal, T., Meyer, B.J. & Priess, J.R. (eds.). Cold Spring Harbor Laboratory Press, Cold Spring Harbor: 241–69.

Scheltema, A.H. (1993): Aplacophora as progenetic aculiferans and the coelomate origin of molluscs as the sister taxon of Sipuncula. *Biol. Bull.* **184**: 57–78.

Scheltema, A.H. (1996): Phylogenetic position of Sipunculida, Mollusca and the progenetic Aplacophora. In: *Origin and Evolutionary Radiation of the Mollusca*, Taylor, J. (ed.). Oxford University Press, Oxford: 53–58.

Scheltema, A.H. & Ivanov, D.L. (2002): An aplacophoran postlarva with iterated dorsal groups of spicules and skeletal similarities to Paleozoic fossils. *Invertebr. Biol.* **121**: 1–10.

Scheltema, A.H., Tscherkassky, M. & Kuzirian, A.M. (1994): Aplacophora. In: *Microscopic Anatomy of Invertebrates*, Harrison, F.W. & Kohn, A.J. (eds.), Vol. 5: Mollusca I. Wiley-Liss, New York: 13–54.

Scheltinga, D.M. & Jamieson, B.G.M. (2003): Spermatogenesis and the mature spermatozoon: form, function and phylogenetic implications. In: *Reproductive Biology and Phylogeny of Anura*, Jamieson, B.G.M. (ed.). Vol. 2 of series: Reproductive biology and phylogeny, Science Publishers, Enfield: 119–251.

Schembri, P.J. & Jaccarini, V. (1977): Locomotory and other movements of the trunk of *Bonellia viridis* (Echiura, Bonelliidae). *J. Zool.* **182**: 477–94.

Schiemer, F. (1973): Respiration rates of two species of gnathostomulids. *Oecologia* **13**: 403–6.

Schierwater, B. (2005): My favourite animal, *Trichoplax adhaerens*. *BioEssays* **27**: 1294–1302.

Schierwater, B. & DeSalle, R. (2001): Current problems with the zootype and the early evolution of Hox genes. *J. Exp. Zool.* **291**: 169–74.

Schierwater, B. & Kuhn, K. (1998): Homology of Hox genes and the zootype concept in early metazoan evolution. *Mol. Phyl. Evol.* **9**: 375–81.

Schierwater, B., Dellaporta, S. & DeSalle, R. (2002): Is the evolution of *Cnox*-2 Hox/ParaHox genes 'multicolored' and 'polygenealogical'? *Mol. Phyl. Evol.* **24**: 374–78.

Schipp, R. & Hevert, F. (1981): Ultrafiltration in the branchial heart appendage of dibranchiate cephalopods: a comparative ultrastructural and physiological study. *J. Exp. Biol.* **92**: 23–35.

Schlegel, M., Lom, J., Stechmann, A., Bernhard, D., Leipe, D., Dykova, I. & Sogin, M.L. (1996): Phylogenetic analysis of complete small subunit ribosomal RNA coding region of *Myxidium lieberkuehni*: evidence that Myxozoa are Metazoa and related to the Bilateria. *Arch. Protistenkd.* **147**: 1–9.

Schlupp, I. (2005): The evolutionary ecology of gynogenesis. *Ann. Rev. Ecol. Syst.* **36**: 399–417.

Schmid, V. & Alder, H. (1984): Isolated, mononucleated, striated muscle can undergo pluripotent transdifferentiation and form a complex regenerate. *Cell* **38**: 801–9.

Schmid, V. & Alder, H. (1986): The potential for transdifferentiation of differentiated medusa tissues in vitro. *Curr. Topics Dev. Biol.* **20**: 117–35.

Schmid, V., Bally, A., Beck, K., Haller, M., Schlage, W.K. & Weber, C. (1991): The extracellular matrix (mesogloea) of hydrozoan jellyfish and its ability to support cell adhesion and spreading. *Hydrobiologia* **216/217**: 3–10.

Schmidt, H. & Moraw, B. (1982): Die Cnidogenese der Octocorallia (Anthozoa, Cnidaria): II. Reifung, Wanderung und Zerfall von Cnidoblast und Nesselkapsel. *Helgol. Meeresunt.* **35**: 97–118.

Schmidt, H. & Zissler, D. (1979): Die Spermien der Anthozoen und ihre phylogenetische Bedeutung. *Zoologica* **44**: 1–97.

Schmidt, M., Giessl, A., Laufs, T., Hankeln, T., Wolfrum, U. & Burmester, T. (2003): How does the eye breathe? Evidence for neuroglobulin-mediated oxygen supply in the mammalian retina. *J. Biol. Chem.* **278**: 1932–935.

Schmidt-Nielsen, K. (1997): *Animal Physiology: Adaptation and Environment*. Cambridge University Press, Cambridge.

Schmidt-Rhaesa, A. (1993): Ultrastructure and development of the spermatozoa of *Multipeniata* (Prolecitophora, Plathelminthes). *Microfauna Marina* **8**: 131–38.

Schmidt-Rhaesa, A. (1996a): Ultrastructure of the anterior end in three ontogenetic stages of *Nectonema munidae* (Nematomorpha). *Acta Zool.* **77**: 267–78.

Schmidt-Rhaesa, A. (1996b): The nervous system of *Nectonema munidae* and *Gordius aquaticus*, with implications on the ground pattern of the Nematomorpha. *Zoomorphology* **116**: 133–42.

Schmidt-Rhaesa, A. (1997a): Phylogenetic relationships of the Nematomorpha – a discussion of current hypotheses. *Zool. Anz.* **236**: 203–16.

Schmidt-Rhaesa, A. (1997b): Ultrastructural features of the female reproductive system and female gametes of *Nectonema munidae* Brinkmann 1930 (Nematomorpha). *Parasitol. Res.* **83**: 77–81.

Schmidt-Rhaesa, A. (1997c): Ultrastructural observations of the male reproductive system and spermatozoa of *Gordius aquaticus* L. 1758 (Nematomorpha). *Invertebr. Reprod. Dev.* **32**: 31–40.

Schmidt-Rhaesa, A. (1998): Muscular ultrastructure in *Nectonema munidae* and *Gordius aquaticus* (Nematomorpha). *Invertebr. Biol.* **117**: 38–45.

Schmidt-Rhaesa, A. (1999): Nematomorpha. In: *Encyclopedia of Reproduction*, Knobil, E. & Neill, J.D. (eds.), Vol. 3. Academic Press, San Diego: 13–21.

Schmidt-Rhaesa, A. (2001): Tardigrades – are they really miniaturized dwarfs? *Zool. Anz.* **240**: 549–55.

Schmidt-Rhaesa, A. (2002): Two dimensions of biodiversity research exemplified by Nematomorpha and Gastrotricha. *Integr. Comp. Biol.* **42**: 633–40.

Schmidt-Rhaesa, A. (2003): Old trees, new trees – is there any progress? *Zoology* **106**: 291–301.

Schmidt-Rhaesa, A. (2004): Ultrastructure of an integumental organ with probable sensory function in *Paragordius varius* (Nematomorpha). *Acta Zool.* **85**: 15–19.

Schmidt-Rhaesa, A. (2005): Morphogenesis of *Paragordius varius* (Nematomorpha) during the parasitic phase. *Zoomorphology* **124**: 33–46.

Schmidt-Rhaesa, A. (2006a): Perplexities concerning the Ecdysozoa: a reply to Pilato et al. *Zool. Anz.* **244**: 205–8.

Schmidt-Rhaesa, A. (2006b): Das Kopfproblem der Arthropoda. *Naturwiss. Rundschau* **59**: 217–18.

Schmidt-Rhaesa, A. & Gerke, S. (2006): Cuticular ultrastructure of *Chordodes nobilii* Camerano, 1901, with a comparison of cuticular ultrastructures in horsehair worms (Nematomorpha). *Zool. Anz.* **245**: 269–76.

Schmidt-Rhaesa, A. & Kulessa, J. (2007): Muscular architecture of *Milnesium tardigradum* (Tardigrada) and other eutardigrades. *Zoomorphology* (in press).

Schmidt-Rhaesa, A., Bartolomaeus, T., Lemburg, C., Ehlers, U. & Garey, J.R. (1998): The position of the Arthropoda in the phylogenetic system. *J. Morphol.* **238**: 263–85.

Schmidt-Rhaesa, A. & Rothe, B.H. (2006): Postembryonic development of longitudinal musculature in *Pycnophyes kielensis* (Kinorhyncha, Homalorhagida). *Integr. Comp. Biol.* **46**: 144–50.

Schminke, H.K. (1973): Evolution, System und Verbreitungsgeschichte der Familie Parabathynellidae

(Bathynellacea, Malacostraca). *Mikrofauna Meeresboden* **24**: 1–192.

Schminke, H.K. (1981): Adaptation of Bathynellaceae (Crustacea, Syncarida) to life in the interstitial ('zoëa theory'). *Int. Rev. Ges. Hydrobiol.* **66**: 578–637.

Schneider, A. (1860): Ueber die Muskeln und Nerven der Nematoden. *Arch. Anat. Physiol. Wiss. Med.* **1860**: 224–42.

Schöttler, U. & Bennet, E.M. (1991): Anelids. In: *Metazoan Life Without Oxygen*, Bryant, C. (ed.). Chapman and Hall, London: 165–85.

Scholtz, G. (2002): The Articulata hypotesis – or what is a segment? *Org. Div. Evol.* **2**: 197–215.

Scholtz, G. (2004): Coelentherata versus Acrosomata – zur Position der Rippenquallen (Ctenophora) im phylogenetischen System der Metazoa. In: Kontroversen in der Phylogenetischen Systematik der Metazoa, Richter, S. & Sudhaus, W. (eds.). *Sitzungsber. Ges. Naturforsch. Freunde Berlin* **43**: 15–33.

Scholtz, G. & Edgecombe, G.D. (2005): Heads, Hox and the phylogenetic position of trilobites. In: Crustacea and Arthropod Relationships, Koenemann, S. & Jenner, R.A. (eds.), *Crust. Issues* **16**: 139–65

Scholtz, G. & Kamenz, C. (2006): The book lungs of Scorpiones and Tetrapulmonata (Chelicerata, Arachnida): evidence for homology and a single terrestrialization event of a common arachnid ancestor. *Zoology* **109**: 2–13.

Schram, F. R. & Hof, C.J. (1998): Fossils and the interrelationship of major crustacean groups. In: *Arthropod Fossils and Phylogeny*, Edgecombe, G.D. (ed.). Columbia University Press, New York: 233–302.

Schreiber, A., Storch, V., Powilleit, M. & Higgins, R.P. (1991): The blood of *Halicryptus spinulosus* (Priapulida). *Can. J. Zool.* **69**: 201–7.

Schröder, O. (1910): *Buddenbrockia plumatellae*, eine neue Mesozoenart aus *Plumatella repens* L. und *Pl. fungosa* Pall. *Z. Wiss. Zool.* **96**: 525–37.

Schubert, M., Escriva, H., Xavier-Neto, J. & Laudet, V. (2006): Amphioxus and tunicates as evolutionary model systems. *Trends Ecol. Evol.* **21**: 269–77.

Schuchert, P. (1993): Phylogenetic analysis of the Cnidaria. *Z. Zool. Syst. Evolutionsforsch.* **31**: 161–73.

Schuchert, P. & Rieger, R.M. (1990a): Ultrastructural observations on the dwarf male of *Bonellia viridis* (Echiura). *Acta Zool.* **71**: 5–16.

Schuchert, P. & Rieger, R.M. (1990b): Ultrastructural examination of spermatogenesis in *Retronectes atypica* (Catenulida, Platyhelminthes). *J. Submicrosc. Cytol. Pathol.* **22**: 379–87.

Schürmann, F.W. (1987): Histology and ultrastructure of the onychophoran brain. In: *Arthropod Brain, its Evolution, Development, Structure, and Functions*, Gupta, A.P. (ed.). John Wiley & Sons, New York: 159–80.

Schürmann, F.W. (1995): Common and special features of the nervous system of Onychophora: A comparison with Arthropoda, Annelida and some other invertebrates. In: *The Nervous System of Invertebrates: An Evolutionary and Comparative Approach*, Breidbach, O. & Kutsch, W. (eds.). Birkhäuser Verlag, Basel: 139–58.

Scotland, R.W., Olmstead, R.G. & Bennett, J.R. (2003): Phylogeny reconstruction: the role of morphology. *Syst. Biol.* **52**: 539–48

Scott, M.P. (1994): Intimations of a creatures. *Cell* **79**: 1121–124.

Seaver, E.C. (2003): Segmentation: mono- or polyphyletic? *Int. J. Dev. Biol.* **47**: 583–95.

Sedgwick, A. (1884): On the origin of metameric segmentation and some other morphological questions. *Quart. J. Micr. Sci.* **24**: 43–82.

Seifert, G. & Rosenberg, J. (1976): Feinstruktur des 'Sacculus' der Nephridien von *Peripatoides leuckardi* (Onychophora: Peripatopsidae). *Ent. Germ.* **3**: 202–11.

Seipel, K., Eberhardt, M., Müller, P., Pescia, E., Yanze, N. & Schmid, V. (2004): Homologs of vascular endothelial growth factor and receptor, VEGF and VEGFR, in the jellyfish *Podocoryne carnea*. *Dev. Dynamics* **231**: 303–12.

Seipel, K. & Schmid, V. (2005): Evolution of striated muscle: jellyfish and the origin of triploblasty. *Dev. Biol.* **282**: 14–26.

Seitz, K.A. (1986): Excretory organs. In: *Ecophysiology of Spiders*, Nentwig, W. (ed.). Springer, New York: 239–48.

Sekiguchi, H. & Terazawa, T. (1997): Statocyst of *Jasus edwardsii* (Crustacea, Palinuridae), with a review of crustacean statocysts. *Mar. Freshw. Res.* **48**: 715–19.

Semon, R. (1888): Die Entwicklung der *Synapta digitata* und ihre Bedeutung für die Phylogenie der Echinodermen. *Jenaische Z. Naturwiss.* **22**: 175–309 (with plates VI–XII).

Senz, W. & Tröstl, R.A. (1997): Überlegungen zur Struktur des Gehirns und Orthogons der Nemertinen. *Sitzungsber. Österr. Akad. Wiss., Math.-Nat. Kl.* (Abt. 1) **204**: 63–78.

Shankland, M. (2003): Evolution of body axis segmentation in the bilaterian radiation. In: *The New Panorama of Animal Evolution*, Legakis, A., Sfenthourakis, S., Polymeni, R. & Thessalou-Legakis, M. (eds.). Proc. 18th Int. Congr. Zool.: 187–95.

Shapeero, W.L. (1961): Phylogeny of Priapulida. *Science* **133**: 879–80.

Shapeero, W.L. (1962): The epidermis and cuticle of *Priapulus caudatus*. Lamarck. *Trans. Am. Microsc. Soc.* **81**: 352–55.

Shaw, C. (1996): Neuropeptides and their evolution. *Parasitology Suppl.* **113**: S35–S45.

Shaw, S.R. & Stowe, S. (1982): Photoreception. In: *The Biology of Crustacea*, Atwood, H.L. & Sandeman, D.C. (eds.). Vol. 3: Neurobiology: structure and function. Academic Press, New York: 291–367.

Sheetz, M.P. & Spudich, J.A. (1983): Movement of myosin-coated fluorescent beads on actin cables in vitro. *Nature* **303**: 31–35.

Shick, J.M. (1983): Respiratory gas exchange in echinoderms. *Echinoderm Studies* **1**: 67–110.

Shimek, R.L. & Steiner, G. (1997): Scaphopoda. In: *Microscopic Anatomy of Invertebrates*, Harrison, F.W. & Kohn, A.J. (eds.), Vol. 6B: Mollusca II. Wiley-Liss, New York: 719–81.

Shimeld, S.M. & Holland, N.D. (2005): Amphioxus molecular biology: insights into vertebrate evolution and developmental mechanisms. *Can. J. Zool.* **83**: 90–100.

Shimeld, S.M., Purkiss, A.G., Dirks, R.P.H., Bateman, O.A., Slingsby, C. & Lubsen, N.H. (2005): Urochordate βγ-crystallin and the evolutionary origin of the vertebrate lens. *Curr. Biol.* **15**: 1684–89.

Shimotori, T. & Goto, T. (2001): Developmental fates of the first four blastomeres of the chaetognath *Paraspadella gotoi*: relationship to protostomes. *Dev. Growth Differ.* **43**: 371–82.

Shinn, G.L. (1997): Chaetognatha. In: *Microscopic Anatomy of Invertebrates*, Harrison, F.W. & Ruppert, E.E. (eds.), Vol. 15: Hemichordata, Chaetognatha, and the invertebrate chordates. Wiley-Liss, New York: 103–220.

Shinn, G.L. (1999): Chaetognatha. In: *Encyclopedia of Reproduction*, Knobil, E. & Neill, J.D. (eds.). Vol. 1. Academic Press, San Diego: 559–64.

Shirley, T.C. & Storch, V. (1999): *Halicryptus higginsi* n.sp. (Priapulida) – a giant new species from Barrow, Alaska. *Invertebr. Biol.* **118**: 404–13.

Shu, D.-G., Conway Morris, S., Han, J., Chen, L., Zhang, X.-L., Zhang, Z.-F., Liu, H.-Q., Li, Y. & Liu, J.-N. (2001): Primitive deuterostomes from the Chengjiang Lagerstätte (Lower Cambrian, China). *Nature* **414**: 419–24.

Shu, D.-G., Conway Morris, S., Han, J., Zhang, Z.-F. & Liu, J.-N. (2002): Ancestral echinoderms from the Chengjiang deposits of China. *Nature* **430**: 422–28.

Siddall, M.E., Martin, D.S., Bridge, D., Desser, S.S. & Cone, D.K. (1995): The demise of a phylum of protists: phylogeny of Myxozoa and other parasitic Cnidaria. *J. Parasitol.* **81**: 961–67.

Siddall, M.E. & Whiting, M.F. (1999): Long-branch abstractions. *Cladistics* **15**: 9–24.

Siddiqui, I.A. & Viglierchio, D.R. (1970): Fine structure of photoreceptors in *Deontostoma californicum*. *J. Nematol.* **2**: 274–76.

Siebert, A.E. (1974): A description of the embryology, larval development, and feeding of the sea anemones *Anthopleura elegantissima* and *A. xanthogrammica*. *Can. J. Zool.* **52**: 1383–88.

Siewing, R. (1976): Probleme und neuere Erkenntnisse in der Großsystematik der Wirbellosen. *Verh. Dtsch. Zool. Ges.* **69**: 59–83.

Siewing, R. (1980): Das Archicoelomatenkonzept. *Zool. Jahrb. Anat.* **103**: 439–82.

Siewing, R. (1985): Cladus: Echiuroida. In: *Lehrbuch der Zoologie*, Siewing, R. (ed.). Vol. 2: Systematik. 3rd Edn., Gustav Fischer, Stuttgart: 574–76.

Silén, L. (1950): On the nervous system of *Glossobalanus marginatus* Meek (Enteropneusta). *Acta Zool.* **31**: 149–75.

Simpson, T.L. (1984): *The Cell Biology of Sponges*. Springer, New York.

Sinakevitch, I., Douglass, J.K., Scholtz, G., Loesel, R. & Strausfeld, N.J. (2003): Conserved and convergent organization in the optic lobes of insects and isopods, with reference to other crustacean taxa. *J. Comp. Neurol.* **467**: 150–72.

Sineshchekov, O.A., Govorunova, E.G., Jung, K.-H., Zauner, S., Maier, U.-G. & Spudich, J.L. (2005): Rhodopsin-mediated photoreception in cryptophyte flagellates. *Biophys. J.* **89**: 4310–19.

Sineshchekov, O.A., Jung, K.-H. & Spudich, J.L. (2002): Two rhodopsins mediate phototaxis to low- and high-intensity light in *Chlamydomonas reinhardtii*. *Proc. Nat. Acad. Sci. USA* **99**: 8689–94.

Singla, C.L. (1974): Ocelli in hydromedusae. *Cell Tiss. Res.* **149**: 413–29.

Singla, C.L. (1975): Statocysts of hydromedusae. *Cell Tiss. Res.* **158**: 391–407.

Singla, C.L. (1978a): Locomotion and neuromuscular system of *Aglantha digitale*. *Cell Tiss. Res.* **188**: 317–27.

Singla, C.L. (1978b): Fine structure of the neuromuscular system of *Polyorchis penicillatus* (Hydromedusae, Cnidaria). *Cell Tiss. Res.* **193**: 163–74.

Skorokhod, A., Gamulin, V., Gundacker, D., Kavsan, V., Müller, I.M. & Müller, W.E.G. (1999): Origin of insulin receptor-like tyrosine kinases in marine sponges. *Biol. Bull.* **197**: 198–206.

Skovsted, C.B. & Holmer, L.E. (2003): The early Cambrian (Botomian) stem group brachiopod *Mickwitzia* from Northeast Greenland. *Acta Palaeontol. Pol.* **48**: 1–20.

Slyusarev, G.S. (1994): Fine structure of the female *Intoshia variabili* (Alexandrov & Sljusarev) (Mesozoa: Orthonectida). *Acta Zool.* **75**: 311–21.

Slyusarev, G.S. (2000): Fine structure and development of the cuticle of *Intoshia variabili* (Orthonectida). *Acta Zool.* **81**: 1–8.

Slyusarev, G.S. & Ferraguti, M. (2002): Sperm structure of *Rhopalura littoralis* (Orthonectida). *Invertebr. Biol.* **121**: 91–94.

Smiley, S. (1994): Holothuroidea. In: *Microscopic Anatomy of Invertebrates*, Harrison, F.W. & Chia, F.-S. (eds.), Vol. 14: Echinodermata. Wiley-Liss, New York: 401–71.

Smiley, S., McEuen, F.S., Chaffee, C. & Krishnan, S. (1991): Echinodermata: Holothuroidea. In: *Reproduction of Marine Invertebrates*, Giese, A.C., Pearse, J.S. & Pearse, V.B. (eds.). Vol. 6: Echinoderms and lophophorates. The Boxwood Press, Pacific Grove: 663–750.

Smith, C.U.M. (2000): *Biology of Sensory Systems*. John Wiley & Sons, Chichester.

Smith, D.S. (1966): The organization and function of the sarcoplasmic reticulum and t-systems of muscle cells. *Prog. Biophys. Mol. Biol.* **16**: 107–42.

Smith, J. & Tyler, S. (1985): The acoel turbellarians: kingpins of metazoan evolutions or a specialized offshot? In: The Origins and Relationships of Lower Invertebrates, Conway Morris, S., George, J.D., Gibson, R. & Platt, H.M. (eds.). *Syst. Assoc. Spec. Vol.* **28**: 123–42.

Smith, J.P.S.I. & Tyler, S. (1986): Frontal organs in the Acoelomorpha (Turbellaria): ultrastructure and phylogenetic significance. *Hydrobiologia* **132**: 71–78.

Smith, J.P.S., Tyler, S. & Rieger, R.M. (1986): Is the Turbellaria polyphyletic? *Hydrobiologia* **132**: 13–21.

Smith, P.R. (1992): Polychaeta: excretory system. In: *Microscopic Anatomy of Invertebrates*, Harrison, F.W. & Gardiner, S.L. (eds.), Vol. 7: Annelida. Wiley-Liss, New York: 71–108.

Smith, P.R. & Ruppert, E.E. (1988): Nephridia. In: The Ultrastructure of Polychaeta, Westheide, W. & Hermans, C.O (eds.). *Microfauna Marina* **4**: 231–62.

Smith, P.R., Lombardi, J. & Rieger, R.M. (1986): Ultrastructure of the body cavity lining in a secondary acoelomate, *Microphthalmus* cf. *listensis* Westheide (Polychaeta: Hesionidae). *J. Morphol.* **188**: 257–71.

Smith, P.R., Ruppert, E.E. & Gardiner, S.L. (1987): A deuterostome-like nephridium in the mitraria larva of *Owenia fusiformis* (Polychaeta, Annelida). *Biol. Bull.* **172**: 315–23.

Smothers, J.F., von Dohlen, C.D., Smith, L.H. & Spall, R.D. (1994): Molecular evidence that the myxozoan protists are metazoans. *Science* **265**: 1719–21.

Snell, E.A., Furlong, R.F. & Holland, P.W.H. (2001): Hsp70 sequences indicate that choanoflagellates are closely related to animals. *Curr. Biol.* **11**: 967–70.

Snodgrass, R.E. (1938): Evolution of the Annelida, Onychophora, and Arthropoda. *Smithson. Misc. Collect.* **97**: 1–159.

Snyder, G.K. & Sheafor, B.A. (1999): Red blood cells: centerpiece in the evolution of the vertebrate circulatory system. *Am. Zool.* **39**: 189–98.

Sopott-Ehlers, B. (1982): Ultrastruktur potentiell photoreceptorischer Zellen unterschiedlicher Organisation bei einem Proseriat (Plathelminthes). *Zoomorphology* **101**: 165–75.

Sopott-Ehlers, B. (1984): Epidermale Collar-Rezeptoren der Nematoplanidae und Polystyliphoridae (Plathelminthes, Unguiphora). *Zoomorphology* **104**: 226–30.

Sørensen, M.V. (2000): An SEM study of the jaws of *Haplognathia rosea* and *Rastrognathia macrostoma* (Gnathostomulida), with a preliminary comparison with the rotiferan trophi. *Acta Zool.* **81**: 9–16.

Sørensen, M.V. (2002): On the evolution and morphology of the rotiferan trophi, with a cladistic analysis of Rotifera. *J. Zool. Syst. Evol. Res.* **40**: 129–54.

Sørensen, M.V. 2003. Further structures in the jaw apparatus of *Limnognathia maerski* (Micrognathozoa), with notes on the phylogeny of the Gnathifera. *J. Morphol.* **255**: 131–45.

Sørensen, M.V. & Giribet, G. (2006): A modern approach to rotiferan phylogeny: combining morphological and molecular data. *Mol. Phyl. Evol.* **40**: 585–608.

Sørensen, M.V. & Sterrer, W. (2002): New characters in the gnathostomulid mouth parts revealed by scanning electron microscopy. *J. Morphol.* **253**: 310–34.

Sørensen, M.V., Funch, P., Willerslev, E., Hansen, A.J. & Olesen, J. (2000): On the phylogeny of the Metazoa in the light of Cycliophora and Micrognathozoa. *Zool. Anz.* **239**: 297–318.

Sorrentino, M., Manni, L., Lane, N.J. & Burighel, P. (2000): Evolution of cerebral vesicles and their sensory organs in an ascidian larva. *Acta Zool.* **81**: 243–58.

Sosinsky, G. (2000): Gap junction structure: new structures and new insights. *Curr. Topics Membranes* **49**: 1–22.

Southward, E.C. (1984): Pogonophora. In: *Biology of the Integument*, Bereiter-Hahn, J., Matoltsy, A.G. & Sylvia Richards, K. (eds.). Vol. 1: Invertebrates. Springer, Berlin: 376–87.

Southward, E.C. (1993): Pogonophora. In: *Microscopic anatomy of invertebrates*, Harrison, F.W. & Rice, M.E. (eds). **12**: Onychophora, Chilopoda, and lesser Protostomata. Wiley-Liss, New York: 327–69.

Southward, E.C., Schulze, A. & Gardiner, S.L. (2005): Pogonophora (Annelida): form and function. In: Morphology, Molecules, Evolution and Phylogeny in Polychaeta and Related Taxa, Bartolomaeus, T. & Purschke, G. (eds.). *Hydrobiologia* **535/536**: 227–51.

Spangenberg, D.B. (1968): Statolith differentiation in *Aurelia aurita. J. Exp. Zool.* **169**: 487–500.

Spencer, A.N. (1979): Neurobiology of *Polyorchis*. II. Structure of effector systems. *J. Neurobiol.* **10**: 95–117.

Spencer, A.N. (1981): The parameters and properties of a group of electrically coupled neurons in the central nervous system of a hydrozoan jellyfish. *J. Exp. Biol.* **93**: 33–50.

Spring, J., Yanze, N., Jösch, C., Middel, A.M., Winninger, B. & Schmid, V. (2002): Conservation of *brachyury*, *Mef2*, and *snail* in the myogenic lineage of jellyfish: a connection to the mesoderm of Bilateria. *Dev. Biol.* **244**: 372–84.

Spudich, J.L., Yang, C.-S., Jung, K.-H. & Spudich, E.N. (2000): Retinylidene proteins: structures and functions from Archaea to humans. *Ann. Rev. Cell Dev. Biol.* **16**: 365–92.

Squire, L.R., Bloom, F.E., McConnell, S.K., Roberts, J.L., Spitzer, N.C. & Zigmond, M.J. (2003): *Fundamental Neuroscience*. 2nd Edn., Academic Press, Amsterdam.

Stach, T., Dupond, S., Israelsson, O., Fauville, G., Nakano, H., Kånneby, T. & Thorndyke, M. (2005): Nerve cells of *Xenoturbella bocki* (phylum uncertain) and *Harrimania kupfferi* (Enteropneusta) are positively immunoreactive to antibodies raised against echinoderm neuropeptides. *J. Mar. Biol. Assoc. UK* **85**: 1519–524.

Stach, T. & Eisler, K. (1998): The ontogeny of the nephridial system of the larval amphioxus (*Branchiostoma lanceolatum*). *Acta Zool.* **79**: 113–18.

Stach, T. & Turbeville, J.M. (2002): Phylogeny of Tunicata inferred from molecular and morphological characters. *Mol. Phyl. Evol.* **25**: 408–28.

Stanley, H.P. (1967): The fine structure of spermatozoa in the lamprey *Lampetra planeri*. *J. Ultrastruct. Res.* **19**: 84–99.

Starck, D. (1982): *Vergleichende Anatomie der Wirbeltiere*. Vol. 3. Springer, Berlin.

Stearns, S.C. (1987): *The Evolution of Sex and its Consequences*. Birkhäuser Verlag, Basel.

Stebbing, A.R.D. (2006): Genetic parsimony: a factor in the evolution of complexity, order and emergence. *Biol. J. Linn. Soc.* **88**: 295–308.

Stecher, H.-J. (1968): Zur Organisation und Fortpflanzung von *Pisione remota* (Southern) (Polychaeta, Pisionidae). *Z. Morphol. Tiere* **61**: 347–410.

Steele, V.J. (1984): Morphology and ultrastructure of the organ of Bellonci in the marine amphipod *Gammarus setosus*. *J. Morphol.* **181**: 97–131.

Steinböck, O. (1924): Untersuchungen über die Geschlechtstrakt-Darmverbindung bei Turbellarien nebst einem Beitrag zur Morphologie des Trikladendarmes. *Z. Morph. Ökol. Tiere* **2**: 461–504.

Steinböck, O. (1963a): Origin and affinities of the lower Metazoa: the 'acoeloid' ancestry of Eumetazoa. In: *The Lower Metazoa. Comparative Biology and Phylogeny*, Dougherty, E.C. (ed.). University of California Press, Berkeley: 40–54.

Steinböck, O. (1963b): Regeneration experiments and phylogeny. In: *The Lower Metazoa*, Dougherty, E.C. (ed.). University of California Press, Berkeley: 108–12.

Steinbrecht, R.A. (1984): Chemo-, hygro-, and thermoreceptors. In: *Biology of the Integument*, Bereiter-Hahn, J., Matoltsy, A.G. & Sylvia Richards, K. (eds.). Vol. 1: Invertebrates. Springer, Berlin: 523–53.

Steinbrecht, R.A. (1998): Bimodal thermo- and hygrosensitive sensilla. In: *Microscopic Anatomy of Invertebrates*, Harrison, F.W. & Locke, M. (eds.). Vol. 11B: Insecta. Wiley-Liss, New York: 405–22.

Steiner, G. (1993): Spawning behaviour of *Pulsellum lofotensis* (M. Sars) and *Cadulus subfusiformis* (M. Sars) (Scaphopoda, Mollusca). *Sarsia* **78**: 31–33.

Steiner, G. (2004): Neuralgische Punkte in der Phylogenie der Mollusca. In: Kontroversen in der Phylogenetischen Systematik der Metazoa, Richter, S. & Sudhaus, W. (eds.). *Sitzungsber. Ges. Naturforsch. Freunde Berlin* **43**: 51–71.

Steiner, S.C.C. (1993): Comparative ultrastructural studies on scleractinian spermatozoa (Cnidaria, Anthozoa). *Zoomorphology* **113**: 129–36.

Sterrer, W. (1969): Beiträge zur Kenntnis der Gnathostomulida. I. Anatomie und Morphologie des Genus *Pterognathia* Sterrer. *Ark. Zool.* **22**: 1–125.

Sterrer, W. (1971): On the biology of Gnathostomulida. *Vie Milieu Suppl.* **22**: 493–507.

Sterrer, W. (1974): Gnathostomulida. In: *Reproduction of Marine Invertebrates*, Giese, A.C. & Pearse, J.S. (eds.). Vol. 1: Acoelomate and pseudocoelomate metazoans. Academic Press, New York: 345–57.

Sterrer, W. (1998): New and known Nemertodermatida (Platyhelminthes – Acoelomorpha) – a revision. *Belg. J. Zool.* **128**: 55–92.

Sterrer, W. (1999): Gnathostomulida. In: *Encyclopedia of Reproduction*, Knobil, E. & Neill, J.D. (eds.). Vol. 2. Academic Press, San Diego: 461–63.

Sterrer, W. & Rieger, R. (1974): Retronectidae – a new cosmopolitan marine family of Catenulida (Turbellaria). In: *Biology of the Turbellaria*, Riser, N.W. & Morse, M.P. (eds.). McGraw-Hill Book Company, New York: 63–92.

Steward, C.-B. & Wilson, A.C. (1987): Sequence convergence and functional adaptation of stomach lysozymes from foregut fermenters. *Cold Spring Harbor Symp. Quant. Biol.* **52**: 891–99.

Sterrer, W., Mainitz, M. & Rieger, R.M. (1985): Gnathostomulida: enigmatic as ever. In: The Origins and Relationships of Lower Invertebrates, Conway Morris, S., George, J.D., Gibson, D.I. & Platt, H.M. (eds.). *Syst. Assoc. Spec. Vol.* **28**. Clarendon Press, Oxford: 181–99.

Stokes, M.D. (1999): Cephalochordata. In: *Encyclopedia of Reproduction*, Knobil, E. & Neill, J.D. (eds.). Vol. 1. Academic Press, San Diego: 529–35.

Stollewerk, A. & Chipman, A.D. (2006): Neurogenesis in myriapods and chelicerates and its importance for understanding arthropod relationships. *Integr. Comp. Biol.* **46**: 195–206.

Storch, V. (1984): Echiura and Sipuncula. In: *Biology of the Integument*, Bereiter-Hahn, J., Matoltsy, A.G. & Sylvia Richards, K. (eds.). Vol. 1: Invertebrates. Springer, Berlin: 368–75.

Storch, V. (1991): Priapulida. In: *Microscopic Anatomy of Invertebrates*, Harrison, F.W. & Ruppert, E.E. (eds.), Vol. 4: Aschelminthes. Wiley-Liss, New York: 333–50.

Storch, V. & Abraham, R. (1972): Elektronenmikroskopische Untersuchungen über die Sinneskante des terricolen Turbellars *Bipalium kewense* Moseley (Tricladida). *Z. Zellforsch.* **133**: 267–75.

Storch, V. & Alberti, G. (1978): Ultrastructural observations on the gills of polychaetes. *Helgol. Wiss. Meeresunters.* **31**: 169–79.

Storch, V. & Herrmann, K. (1978): Podocytes in the blood vessel linings of *Phoronis muelleri* (Phoronida, Tentaculata). *Cell Tiss. Res.* **190**: 553–56.

Storch, V. & Higgins, R.P. (1989): Ultrastructure of developing and mature spermatozoa of *Tubiluchus corallicola* (Priapulida). *Trans. Am. Microsc. Soc.* **108**: 45–50.

Storch, V. & Jamieson, B.G.M. (1992): Further spermatological evidence for including the Pentastomida (tongue worms) in the *Crustacea. Int. J. Parasitol.* **22**: 95–108.

Storch, V. & Moritz, K. (1971): Zur Feinstruktur der Sinnesorgane von *Lineus ruber* O.F. Müller (Nemertini, Heteronemertini). *Z. Zellforsch.* **117**: 212–25.

Storch, V. & Ruhberg, H. (1977): Fine structure of the sensilla of *Peripatopsis moseleyi* (Onychophora). *Cell Tiss. Res.* **177**: 539–53.

Storch, V. & Ruhberg, H. (1993): Onychophora. In: *Microscopic Anatomy of Invertebrates*, Harrison, F.W. & Rice, M.E. (eds.), Vol. 12: Onychophora, Chilopoda, and lesser Protostomata. Wiley-Liss, New York: 11–56.

Storch, V. & Schlötzer-Schrehardt, U. (1988): Sensory structures. In: The Ultrastructure of Polychaeta, Westheide, W. & Hermans, C.O. (eds.). *Microfauna Marina* **4**: 121–33.

Storch, V. & Welsch, U. (1997): *Systematische Zoologie*. 5th Edn. Gustav Fischer Verlag, Stuttgart.

Storch, V., Alberti, G., Rosito, R.M. & Sotto, F.B. (1985): Some ultrastructural observations on *Tubiluchus philippinensis* (Priapulida), a new faunal element of Philippine coastal waters. *Philippine Sci.* **22**: 144–56.

Storch, V., Higgins, R.P. & Morse, M.P. (1989): Internal anatomy of *Meiopriapulus fijiensis* (Priapulida). *Trans. Am. Microsc. Soc.* **108**: 245–61.

Storch, V., Kempendorf, C., Higgins, R.P., Shirley, T.C. & Jamieson, B.G.M. (2000a): Priapulida. In: *Reproductive Biology of Invertebrates*, Adiyodi, K.G., Adiyodi, R.G. & Jamieson, B.G.M. (eds.). Vol. 9, part B: Progress in male gamete ultrastructure and phylogeny. John Wiley & Sons, Chichester: 1–19.

Storch, V., Ruhberg, H. & Alberti, G. (1978): Zur Ultrastruktur der Segmentalorgane der Peripatopsidae (Onychophora). *Zool. Jahrb. Anat.* **100**: 47–63.

Storch, V., Ruhberg, H., Alberti, G. & Jamieson, B.G.M. (2000b): Onychophora. In: *Reproductive Biology of Invertebrates*, Adiyodi, K.G., Adiyodi, R.G. & Jamieson, B.G.M. (eds.). Vol. 9, part B: Progress in male gamete ultrastructure and phylogeny. John Wiley & Sons, Chichester: 293–310.

Strathmann, R.R. (1973): Function of lateral cilia in suspension feeding of lophophorates (Brachiopoda, Phoronida, Ectoprocta). *Mar. Biol.* **23**: 129–36.

Strathmann, R.R. (1978): The evolution and loss of feeding larval stages of marine invertebrates. *Evolution* **32**: 894–906.

Strathmann, R.R. (1985): Feeding and nonfeeding larval development and life-history evolution in marine invertebrates. *Ann. Rev. Ecol. Syst.* **16**: 339–61.

Strathmann, R.R. (1987): Larval feeding. In: *Reproduction of Marine Invertebrates*, Giese, A.C., Pearse, J.S. & Pearse, V.B. (eds.). Vol. 9: General aspects: seeking unity in diversity. Blackwell Scientific Publications, Palo Alto and Boxwood Press, Pacific Grove: 465–550.

Strathmann, R.R. (1989): Existence and functions of a gel filled primary body cavity in development of echinoderms and hemichordates. *Biol. Bull.* **176**: 25–31.

Strathmann, R.R. (2005): Ciliary sieving and active ciliary response in capture of particles by suspension-feeding brachiopod larvae. *Acta Zool.* **86**: 41–54.

Strathmann, R.R. (2006): Versatile behaviour in capture of particles by the bryozoan cyphonautes larva. *Acta Zool.* **87**: 83–89.

Strathmann, R.R. & Grünbaum, D. (2006): Good eaters, poor swimmers: compromises in larval form. *Integr. Comp. Biol.* **46**: 312–22.

Strausfeld, N.J., Strausfeld, C.M., Loesel, R., Rowell, D. & Stowe, S. (2006b): Arthropod phylogeny: onychophoran brain organization suggests an archaic relationship with a chelicerate stem lineage. *Proc. R. Soc. B* **273**: 1857–66.

Strausfeld, N.J., Strausfeld, C.M., Stowe, S., Rowell, D. & Loesel, R. (2006a): The organization and evolutionary implications of neuropils and their neurons in the brain of the onychophoran *Euperipatus rowelli*. *Arthr. Struct. Dev.* **35**: 169–96.

Strayer, D.L. & Hummon, W.D. (1991): Gastrotricha. In: *Ecology and Classification of North American Freshwater Invertebrates*, Thorp, J.H. & Covich, A.P. (eds.). Academic Press, San Diego: 173–85.

Stretton, O.W. (1976): Anatomy and development of the somatic musculature of the nematode *Ascaris. J. Exp. Biol.* **64**: 773–88.

Stricker, S.A. (1986): An ultrastructural study of oogenesis, fertilization, and egg laying in a nemertean ectosymbiont of crabs, *Carcinonemertes epialti* (Nemertea, Hoplonemertea). *Can. J. Zool.* **64**: 1256–69.

Stricker, S.A. (1999): Brachiopoda. In: *Encyclopedia of Reproduction*, Knobil, E. & Neill, J.D. (eds.). Vol. 1. Academic Press, San Diego: 382–88.

Stricker, S.A. & Folsom, M.W. (1998): A comparative ultrastructural analysis of spermiogenesis in nemertean worms. *Hydrobiologia* **365**: 55–72.

Striedter, G.F. (2005): *Principles of Brain Evolution*. Sinauer, Sunderland.

Struck, T., Hessling, R. & Purschke, G. (2002): The phylogenetic position of the Aeolosomatidae and Parergodrilidae, two enigmatic oligochaete-like taxa of the 'Polychaeta', based on molecular data from 18S rDNA sequences. *J. Zool. Syst. Evol. Res.* **40**: 155–63.

Struck, T., Purschke, G. & Halanych, K.M. (2006): Phylogeny of Eunicida (Annelida) and exploring data congruence using a partition addition bootstrap alteration (PBBA) approach. *Syst. Biol.* **55**: 1–20.

Sun, Y., Jin, K., Mao, X.O., Zhu, Y. & Greenberg, D.A. (2001): Neuroglobin is up-regulated by and protects neurons from hypoxic-ischemic injury. *Proc. Nat. Acad. Sci. USA* **98**: 15306–11.

Svane, I. & Young, C.M. (1991): Sensory structures in tadpole larvae of the ascidians *Microcosmus exasperatus* Heller and *Herdmania momus* (Savigny). *Acta Zool.* **72**: 129–35.

Swanson, C. J. (1971): Occurrence of paramyosin among the Nematomorpha. *Nature New Biol.* **232**: 122–23.

Swedmark, B. & Teissier, G. (1966): The Actinulida and their evolutionary significance. *Symp. Zool. Soc. London* **16**: 119–33.

Syed, T. & Schierwater, B. (2002a): *Trichoplax adhaerens*: discovered as a missing link, forgotten as a hydrozoan, re-discovered as a key to metazoan evolution. *Vie Milieu* **52**: 177–87.

Syed, T. & Schierwater, B. (2002b): The evolution of the Placozoa: a new morphological model. *Senckenbergiana Lethaea* **82**: 315–24.

Sylvia Richards, K. (1984): Annelida: cuticle. In: *Biology of the Integument*, Bereiter-Hahn, J., Matoltsy, A.G. & Sylvia Richards, K. (eds.). Vol. 1: Invertebrates. Springer, Berlin: 310–22.

Taddei-Ferretti, C. & Musio, C. (2000): Photobehaviour of *Hydra* (Cnidaria, Hydrozoa) and correlated mechanisms: a case of extraocular photosensitivity. *J. Photochem. Photobiol. B (Biol.)* **55**: 88–101.

Tamm, S.L. (1982): Ctenophora. In: *Electrical Conduction and Behaviour in 'Simple' Invertebrates*, Shelton, G.A.B. (ed.). Clarendon Press, Oxford: 266–359.

Tamm, S. & Tamm, S. (1991): Actin pegs and ultrastructure of presumed sensory receptors of *Beroë* (Ctenophora). *Cell Tiss. Res.* **264**: 151–59.

Tamm, S.L. & Tamm, S. (1993): Diversity of macrociliary size, tooth patterns, and distribution in *Beroe* (Ctenophora). *Zoomorphology* **113**: 79–89.

Tanaka, M., Wechsler, S.B., Lee, I.W., Yamasaki, N., Lawitts, J.A. & Izumo, S. (1999): Complex modular *cis*-acting elements regulate expression of the cardiac specifying homeobox gene *Csx/Nkx2.5*. *Development* **126**: 1439–50.

Tardent, P. (1978): Coelentherata, Cnidaria. In: *Morphogenese der Tiere*, Seidel, F. (ed.). First series. Gustav Fischer Verlag, Jena: 71–391.

Tardent, P. (1995): The cnidarian cnidocyte, a high-tech cellular weaponry. *BioEssays* **17**: 351–62.

Tardent, P. & Schmid, V. (1972): Ultrastructure of mechanoreceptors of the polyp *Coryne pintneri* (Hydrozoa, Athecata). *Exp. Cell Res.* **72**: 265–75.

Tavernarakis, N. & Driscoll, M. (1997): Molecular modeling of mechanotransduction in the nematode *Caenorhabditis elegans*. *Ann. Rev. Physiol.* **59**: 659–89.

Taylor, J.R. & Kier, W.M. (2003): Switching skeletons: hydrostatic support in molting crabs. *Science* **301**: 209–10.

Technau, U. (2001): *Brachyury*, the blastopore and the evolution of the mesoderm. *BioEssays* **23**: 788–94.

Technau, U. & Scholz, C.B. (2003): Origin and evolution of endoderm and mesoderm. *Int. J. Dev. Biol.* **47**: 531–39.

Technau, U., Rudd, S., Maxwell, P., Gordon, P.M.K., Saina, M., Grasso, L.C., Hayward, D.C., Sensen, C.W., Saint, R., Holstein, T.W., Ball, E.E. & Miller, D.J. (2005): Maintenance of ancestral complexity and non-metazoan genes in two basal cnidarians. *Trends Gen.* **21**: 633–39.

Tekle, Y.I., Raikova, O.I., Justine, J.-L., Hendelberg, J. & Jondelius, U. (2007): Ultrastructural and immunocytochemical investigation of acoel sperms with 9 + 1 axoneme structure: new sperm characters for unraveling phylogeny in Acoela. *Zoomorphology* **126**: 1–16.

Telfer, W.H. & Kunkel, J.G. (1991): The function and evolution of insect storage hexamers. *Ann. Rev. Entomol.* **36**: 205–28.

Telford, M.J. (2000): Turning Hox 'signatures' into synapomorphies. *Evol. Dev.* **2**: 360–64.

Telford, M.J. (2004): The multimeric β-thymosin found in nematodes and arthropods is not a synapomorphy of the Ecdysozoa. *Evol. Dev.* **6**: 90–94.

Telford, M.J. & Holland, P.W.H. (1993): The phylogenetic affinities of the chaetognaths: a molecular analysis. *Mol. Biol. Evol.* **10**: 660–76.

Telford, M.J. & Holland, P.W.H. (1997): Evolution of 28S ribosomal DNA in chaetognaths: duplicate genes and molecular phylogeny. *J. Mol. Evol.* **44**: 135–44.

Telford, M. & Thomas, R. H. (1998): Expression of homeobox genes shows chelicerate arthropods retain their deutocerebral segment. *Proc. Natl. Acad. Sci. USA* **95**: 10671–75.

Telford, M., Herniou, E.A., Russell, R.B. & Littlewood, D.T.J. (2000): Changes in mitochondrial genetic codes as phylogenetic characters: two examples from the flatworms. *Proc. Natl. Acad. Sci. USA* **97**: 11359–64.

Temkin, M.H. (1994): Gamete spawning and fertilization in the gymnolaemate bryozoan *Membranipora membranacea. Biol. Bull.* **187**: 143–55.

Temkin, M.H. & Zimmer, R.L. (2002): Phylum Bryozoa. In: *Atlas of Marine Invertebrate Larvae*, Young, C.M. (ed.). Academic Press, San Diego: 411–27.

Tepass, U., Tannentzapf, G., Ward, R. & Fehon, R. (2001): Epithelial cell polarity and cell junctions in *Drosophila. Ann. Rev. Gen.* **35**: 747–84.

Terakita, A. (2005): The opsins. *Genome Biol.* **6**: 213.1–213.9.

Terwilliger, R.C. (1980): Structures of invertebrate hemoglobins. *Am. Zool.* **20**: 53–67.

Terwilliger, R.C. & Read, K.R.H. (1972): The hemoglobin of the holothurian echinoderm, *Mopalia oölitica* Pourtales. *Comp. Biochem. Physiol. B* **42**: 65–72.

Teuchert, G. (1967): Zum Protonephridialsystem mariner Gastrotrichen der Ordnung Macrodasyidea. *Mar. Biol.* **1**: 110–12.

Teuchert, G. (1968): Zur Fortpflanzung und Entwicklung der Macrodasyoidea (Gastrotricha). *Z. Morphol. Tiere* **63**: 343–418.

Teuchert, G. (1973): Die Feinstruktur des Protonephridialsystems von *Turbanella cornuta* Remane, einem marinen Gastrotrich der Ordnung Macrodasyoidea. *Z. Zellforsch.* **136**: 277–89.

Teuchert, G. (1975): Differenzierung von Spermien bei dem marinen Gastrotrich *Turbanella cornuta* Remane (Ordnung Macrodasyoidea). *Verh. Anat. Ges.* **69**: 743–48.

Teuchert, G. (1976a): Sinneseinrichtungen bei *Turbanella cornuta* Remane (Gastrotricha). *Zoomorphologie* **83**: 193–207.

Teuchert, G. (1976b): Elektronenmikroskopische Untersuchung über die Spermatogenese und Spermatohistogenese von *Turbanella cornuta* Remane (Gastrotricha). *J. Ultrastruct. Res.* **56**: 1–14.

Teuchert, G. (1977): The ultrastructure of the marine gastrotrich *Turbanella cornuta* Remane (Macrodasyoidea) and its functional and phylogenetic importance. *Zoomorphologie* **88**: 189–246.

Teuchert, G. (1978): Strukturanalyse von Bewegungsformen bei Gastrotrichen. *Zool. Jahrb. Anat.* **99**: 12–22.

Thane, A. (1974): Rotifera. In: *Reproduction of Marine Invertebrates*, Giese, A.C. & Pearse, J.S. (eds.). Vol. 1: Acoelomate and pseudocoelomate metazoans. Academic Press, New York: 471–84.

Thiemann, M. & Ruthmann, A. (1989): Microfilaments and microtubules in isolated fiber cells of *Trichoplax adhaerens* (Placozoa). *Zoomorphology* **109**: 89–96.

Thomas, M.B. & Edwards, N.C. (1991): Cnidaria: Hydrozoa. In: *Microscopic Anatomy of Invertebrates*, Harrison, F.W. & Westfall, J.A. (eds.), Vol. 2: Placozoa, Porifera, Cnidaria, and Ctenophora. Wiley-Liss, New York: 91–183.

Thompson, R.F. & Langford, G.M. (2002): Myosin superfamily evolutionary history. *Anat. Rec.* **268**: 276–89.

Thurst, R. (1968): Submikroskopische Untersuchungen über die Morphogenese des Integumentes von *Ascaris lumbricoides* L. 1758. *Z. Wiss. Zool.* **178**: 1–39.

Tiemann, H., Sötje, I., Becker, A., Jarms, G. & Epple, M. (2006): Calcium sulfate hemihydrate (bassanite) statoliths in the cubozoan *Carybdea* sp. *Zool. Anz.* **245**: 13–17.

Timm, R.W. (1953): Observations on the morphology and histological anatomy of a marine nematode, *Leptosomatum acephalatum* Chitwood, 1936, new combination (Enoplidae: Leptosomatinae). *Am. Midl. Nat.* **49**: 229–48.

Todaro, A., Littlewood, D.T.J., Balsamo, M., Herniou, E.A., Cassanelli, S., Manicardi, G., Wirz, A. & Tongiorgi, P. (2003): The interrelationships of the Gastrotricha using small rRNA subunit sequence data, with an interpretation based on morphology. *Zool. Anz.* **242**: 145–56.

Todt, C. & Tyler, S. (2006): Morphology and ultrastructure of the pharynx in Solenofilomorphidae (Acoela). *J. Morphol.* **267**: 776–92.

Todt, C. & Tyler, A. (2007): Ciliary receptors associated with the mouth and pharynx of Acoela (Acoelomorpha): a comparative ultrastructural study. *Acta Zool.* **88**: 41–58.

Tomassetti, P., Voigt, O., Collins, A.G., Porrello, S., Pearse, V.B. & Schierwater, B. (2005): Placozoans (*Trichoplax adhaerens* Schulze, 1883) in the Mediterranean Sea. *Meiofauna Marina* **14**: 5–7.

Tompa, A.S. (1984): Land snails (Stylommatophora). In: *The Mollusca*, Wilbur, K.M. (ed.), Vol. 7: Reproduction, Tompa, A.S., Verdonk, N.H. & van den Biggelaar, J.A.M. (volume eds.). Academic Press, Orlando: 47–140.

Tonosaki, A. (1967): Fine structure of the retina in *Haliotis discus. Z. Zellforsch.* **79**: 469–80.

Tops, S., Curry, A. & Okamura, B. (2005): Diversity and systematics of the Malacosporea (Myxozoa). *Invertebr. Biol.* **124**: 285–95.

Torrence, S.A. (1986): Sensory endings of the ascidian static organ (Chordata, Ascidiacea). *Zoomorphology* **106**: 61–66.

Tranter, P.R.G., Nicholson, D.N. & Kinchington, D. (1982): A description of spawning and post-gastrula development of the cool temperate coral, *Caryophyllia smithi*. *J. Mar. Biol. Assoc. UK* **62**: 845–54.

Trent, J.T. & Hargrove, M.S. (2002): A ubiquitously expressed human hexacoordinate hemoglobin. *J. Biol. Chem.* **277**: 19538–45.

Troyer, D. & Schwager, P. (1979): Ultrastructure and evolution of a sperm: phylogenetic implications of altered motile machinery in *Ophryotrocha puerilis* spermatozoon. *Eur. J. Cell Biol.* **20**: 174–76.

Tsunekawa, N., Naito, M., Sakai, Y., Nishida, T. & Noce, T. (2000): Isolation of chicken *vasa* homolog gene and tracing the origin of primordial germ cells. *Development* **127**: 2741–50.

Tuchschmid, P.E., Kunz, P.A. & Wilson, K.J. (1978): Isolation and characterization of the hemoglobin from the lanceolate fluke *Dicrocoelium dendriticum*. *Eur. J. Biochem.* **88**: 387–94.

Turbeville, J.M. (1986): An ultrastructural analysis of coelomogenesis in the hoplonemertine *Prosorhochmus americanus* and the polychaete *Magelona* sp. *J. Morphol.* **187**: 51–60.

Turbeville, J.M. (1991): Nemertinea. In: *Microscopic Anatomy of Invertebrates*, Harrison, F.W. & Bogitsh, B.J. (eds.), Vol. 3: Platyhelminthes and Nemertinea. Wiley-Liss, New York: 285–328.

Turbeville, J.M. (2002): Progress in nemertean biology: development and phylogeny. *Integr. Comp. Biol.* **42**: 692–703.

Turbeville, J.M. & Ruppert, E.E. (1983): Epidermal muscles and peristaltic burrowing in *Carinoma tremaphoros* (Nemertini): correllates of effective burrowing without segmentation. *Zoomorphology* **103**: 103–20.

Turbeville, J.M. & Ruppert, E.E. (1985): Comparative ultrastructure and the evolution of Nemertines. *Am. Zool.* **25**: 53–71.

Turbeville, J.M., Pfeifer, D.M., Field, K.G. & Raff, R.A. (1991): The phylogenetic status of arthropods, as inferred from 18S rRNA sequences. *Mol. Biol. Evol.* **8**: 669–86.

Turlejski, K. (1996): Evolutionary ancient roles of serotonin: long-lasting regulation of activity and development. *Acta Neurobiol. Exp.* **56**: 619–36.

Turon, X., Lopez-Legentil, S. & Banaigs, B. (2005): Cell types, microsymbionts, and pyridoacridine distribution in the tunic of three color morphs of the genus *Cystodytes* (Ascidiacea, Polycitoridae). *Invertebr. Biol.* **124**: 355–69.

Tuzet, O., Garrone, R. & Pavans de Ceccatty, M. (1970): Observations ultrastructurales sur la spermatogenèse chez la démosponge *Aplysilla rosea* Schulze (Dendroceratidae): une métaplasie exemplaire. *Ann. Sci. Nat. Zool.*, 12e Ser. **12**: 27–50.

Tyler, S. (1979): Distinctive features of cilia in metazoans and their significance for systematics. *Tiss. Cell* **11**: 385–400.

Tyler, S. (1999): Platyhelminthes. In: *Encyclopedia of Reproduction*, Knobil, E. & Neill, J.D. (eds.). Vol. 3. Academic Press, San Diego: 901–8.

Tyler, S. (2001): The early worm: origins and relationships of the lower flatworms. In: Interrelationships of the Platyhelminthes, Littlewood, D.T.J. & Bray, R.A. (eds.). *Syst. Assoc. Spec. Vol. Ser.* **60**: 3–12.

Tyler, S. (2003): Epithelium – the primary building block for metazoan complexity. *Integr. Comp. Biol.* **43**: 55–63.

Tyler, S. & Rieger, R.M. (1975): Uniflagellate spermatozoa in *Nemertoderma* (Turbellaria) and their phylogenetic significance. *Science* **188**: 730–31.

Tyler, S. & Rieger, R.M. (1977): Ultrastructural evidence for the systematic position of the Nemertodermatida (Turbellaria). *Acta Zool. Fenn.* **154**: 193–207.

Tyler, S. & Rieger, R.M. (1999): Functional morphology of musculature in the acoelomate worm, *Convoluta pulchra* (Plathelminthes). *Zoomorphology* **119**: 127–41.

Tzetlin, A.B. & Filippova, A.V. (2005): Muscular system in polychaetes (Annelida). In: Morphology, Molecules, Evolution and Phylogeny in Polychaeta and Related Taxa, Bartolomaeus, T. & Purschke, G. (eds.). *Hydrobiologia* **535/536**: 113–26.

Tzetlin, A. & Purschke, G. (2005): Pharynx and intestine. In: Morphology, Molecules, Evolution and Phylogeny in Polychaeta and Related Taxa, Bartolomaeus, T. & Purschke, G. (eds.). *Hydrobiologia* **535/536**: 199–225.

Tzetlin, A. & Purschke, G. (2006): Fine structure of the pharyngeal apparatus of the pelagosphera larva in *Phascolosoma agassizii* (Sipuncula) and its phylogenetic significance. *Zoomorphology* **125**: 109–17.

Tzetlin, A.B., Zhadan, A., Ivanov, I., Müller, M.C.M. & Purschke, G. (2002): On the absence of circular muscle elements in the body wall of *Dysponetus pygmaeus* (Chrysopetalidae, 'Polychaeta', Annelida). *Acta Zool.* **83**: 81–85.

Ulrich, W. (1950): Über die systematische Stellung einer neuen Tierklasse (Pogonofora K.E. Johansson), den Begriff der Archicoelomaten und die Einteilung der Bilaterien. *Sitzungsber. Akad. Wiss. Berlin, Math.-Naturwiss. Kl.* **1949**: 1–25.

Ulrich, W. (1973): Archicoelomaten W. Ulrich 1949 (1950–1970). *Aufs. Reden Senckenberg. Naturforsch. Ges.* **22**: 7–50.

Umesono, Y., Watanabe, K. & Agata, K. (1999): Distinct structural domains in the planarian brain defined by the expression of evolutionarily conserved homeobox genes. *Dev. Genes Evol.* **209**: 31–39.

Ung, C.Y. & Molteno, A.C. (2004): An enigmatic eye: the histology of the tuatara pineal complex. *Clin. Exp. Ophthalmol.* **32**: 614–18.

Unwin, P.N.T. (1987): Gap junction structure and the control of cell-to-cell communication. In: *Junctional Complexes of Epithelial Cells*, Bock, G. & Clark, S. (eds.). John Wiley & Sons, Chichester: 78–91.

Ursprung, H. & Schabtach, E. (1965): Fertilization in tunicates: loss of the paternal mitochondrion prior to sperm entry. *J. Exp. Zool.* **159**: 379–84.

Vacelet, J. (2006): New carnivorous sponges (Porifera, Poecilosclerida) collected from manned submersibles in the deep Pacific. *Zool. J. Linn. Soc.* **148**: 553–84.

Vacelet, J. & Boury-Esnault, N. (1995): Carnivorous sponges. *Nature* **373**: 333–35.

Valembois, P. & Boiledieu, D. (1980): Fine structure and functions of haemerythrocytes and leucocytes of *Sipunculus nudus*. *J. Morphol.* **163**: 69–77.

Valentine, J.W. (2004): On the origin of phyla. University of Chicago Press, Chicago.

Valvassori, R., de Eguileor, M., Grimaldi, A. & Lanzavecchia, G. (1999): Nematomorpha. In: *Reproductive Biology of Invertebrates*, Adiyodi, K.G., Adiyodi, R.G. & Jamieson, B.G.M. (eds.). Vol. 9, part A: Progress in male gamete ultrastructure and phylogeny. John Wiley & Sons, Chichester: 213–27.

Van de Velde, M.C. & Coomans, A. (1988): Ultrastructure of the photoreceptor of *Diplolaimella* sp. (Nematoda). *Tiss. Cell* **20**: 421–29.

Van den Biggelaar, J.A.M., Dictus, W.J.A.G. & van Loon, A.E. (1997): Cleavage patterns, cell-lineages and cell specification are clues to phyletic lineages in Spiralia. *Cell Dev. Biol.* **8**: 367–78.

Van den Biggelaar, J.A.M., Edsinger-Gonzales, E. & Schram, F.R. (2002): The improbability of dorso-ventral axis inversion during animal evolution, as presumed by Geoffroy Saint Hillaire. *Contrib. Zool.* **71**: 29–36.

Van den Ent, F., Amos, L.A. & Löwe, J. (2001): Prokaryotic origin of the actin cytoskeleton. *Nature* **413**: 39–44.

Van der Horst, G., Wilson, B. & Channing, A. (1995): Amphibian sperm: phylogeny and fertilization environment. In: Advances in Spermatozoal Phylogeny and Taxonomy, Jamieson, B.G.M., Ausio, J. & Justine, J.-L. (eds.). *Mém. Mus. Natn. Hist. Nat.* **166**: 333–42.

Van der Land, J. & Nørrevang, A. (1985): Affinities and intraphyletic relationships of the Priapulida. In: The Origins and Relationships of Lower Invertebrates, Conway Morris, S., George, J.D., Gibson, D.I. & Platt, H.M. (eds.). *Syst. Assoc. Spec. Vol.* **28**: 261–73.

Van Deurs, B. (1972): On the ultrastructure of the mature spermatozoon of a chaetognath, *Spadella cephaloptera*. *Acta Zool.* **53**: 93–104.

Van Deurs, B. (1973): Axonemal 12+0 pattern in the flagellum of the motile spermatozoon of *Nymphon leptocheles*. *J. Ultrastruct. Res.* **42**: 594–98.

Van Deurs, B. (1974a): Spermatology of some Pycnogonida (Arthropoda), with special reference to a microtubule-nuclear envelope complex. *Acta Zool.* **55**: 151–62.

Van Deurs, B. (1974b): Pycnogonid sperm. An example of inter- and intraspecific axonemal variation. *Cell Tiss. Res.* **149**: 105–11.

Van Holde, K.E. (1997): Respiratory proteins of invertebrates: structure, function and evolution. *Zoology* **100**: 287–97.

Van Holde, K.E., Miller, K.I. & Decker, H. (2001): Hemocyanins and invertebrate evolution. *J. Biol. Chem.* **276**: 15563–66.

Van Valkenburgh, B., Wang, X. & Damuth, J. (2004): Cope's rule, hypercarnivory, and extinction in North American canids. *Science* **306**: 101–4.

Van Voorhies, W.A. & Ward, S. (2000): Broad oxygen tolerance in the nematode *Caenorhabditis elegans*. *J. Exp. Biol.* **203**: 2467–78.

Vandergon, T.L., Riggs, C.K., Gorr, T.A., Colacino, J.M. & Riggs, A.F. (1998): The mini-hemoglobins in neural and body wall tissue of the nemertean worm, *Cerebratulus lacteus*. *J. Biol. Chem.* **273**: 16998–17011.

Vandermeulen, J.H. (1974): Studies on reef corals. II. Fine structure of planctonic planula larva of *Pocillopora damicornis*, with emphasis on the aboral epidermis. *Mar. Biol.* **27**: 239–49.

Vanfleteren, J.R. (1982): A monophyletic line of evolution? Ciliary induced photoreceptor membranes. In: *Visual Cells in Evolution*, Westfall, J.A. (ed.). Raven Press, New York: 107–36.

Vanfleteren, J.R. & Coomans, A. (1976): Photoreceptor evolution and phylogeny. *Z. Zool. Sys. Evolutionsforsch.* **14**: 157–69.

Vaupel-von Harnack, M. (1963): Über den Feinbau des Nervensystems des Seesternes (*Asterias rubens* L.). III. Mitteilung: die Struktur der Augenpolster. *Z. Zellforsch.* **60**: 432–51.

Verger-Bocquet, M. (1992): Polychaeta: sensory structures. In: *Microscopic Anatomy of Invertebrates*, Harrison, F.W. & Gardiner, S.L. (eds.), Vol. 7: Annelida. Wiley-Liss, New York: 181–96.

Vernberg, V.B. (1968): Platyhelminthes: respiratory metabolism. In: *Chemical Zoology*, Florkin, M. & Scheer, B.T. (eds.). Academic Press, New York: 359–3.

Vernet, G. (1970): Ultrastructure des photorécepteurs de *Lineus ruber* (O.F. Müller) (Hétéronémertes Lineïdae). I. Ultrastructure de l'œil normal. *Z. Zellforsch.* **104**: 494–506.

Vernet, G. (1974): Étude ultrastructurale de cellules présumées photoréceptrices dans les ganglions cérébroïdes des Lineidae (Hètèronemertes). *Ann. Sci. Nat. Zool.* 12ᵉ, Ser. **16**: 27–36.

Vickaryous, M.K. & Hall, B.K. (2006): Human cell type diversity, evolution, development, and classification with special reference to cells derived from the neural crest. *Biol. Rev.* **81**: 425–55.

Vigoreaux, J.O. (1994): The muscle Z band: lessons in stress management. *J. Musc. Res. Cell Motil.* **15**: 237–55.

Vinnikov, Y.A. (1982): *Evolution of Receptor Cells. Cytological, Membranous and Molecular Levels*. Springer, Berlin.

Vinogradov, S.N. (1985): The structure of invertebrate extracellular hemoglobins (erythrocruorins and chlorocruorins). *Comp. Biochem. Physiol. B* **82**: 1–15.

Vinson, C.R. & Bonaventura, J. (1987): Structure and oxygen equilibrium of the three coelomic cell hemoglobins of the echiuran worm *Thalassema mellitu* (Conn). *Comp. Biochem. Physiol. B* **87**: 361–66.

Voigt, O., Collins, A.G., Pearse, V.B., Pearse, J.S., Ender, A., Hadrys, H. & Schierwater, B. (2004): Placozoa – no longer a phylum of one. *Curr. Biol.* **14**: R944–R945.

Voltzow, J. (1994): Gastropoda: Prosobranchia. In: *Microscopic Anatomy of Invertebrates*, Harrison, F.W. & Kohn, A.J. (eds.), Vol. 5: Mollusca I. Wiley-Liss, New York: 111–252.

Von Döhren, J. & Bartolomaeus, T. (2006): Ultrastructure of sperm and male reproductive system in *Lineus viridis* (Heteronemertea, Nemertea). *Zoomorphology* **125**: 175–85.

Von Salvini-Plawen, L. (1969): Solenogastres und Caudofoveata (Mollusca, Aculifera): Organisation und phylogenetische Bedeutung. *Malacologia* **9**: 191–216.

Von Salvini-Plawen, L. (1971): *Schild- und Furchenfüsser*. Vol. 441 in series: Die Neue Brehm-Bücherei. A. Ziemsen Verlag, Wittenberg.

Von Salvini-Plawen, L. (1972): Zur Morphologie und Phylogenie der Mollusken: Beziehungen der Caudofoveata und der Solenogastres als Aculifera, als Mollusca und als Spiralia. *Z. Wiss. Zool.* **184**: 205–394.

Von Salvini-Plawen, L. (1978): On the origin and evolution of the lower Metazoa. *Z. Zool. Syst. Evolutionsforsch.* **16**: 40–88.

Von Salvini-Plawen, L. (1980a): A reconsideration of systematics in Mollusca (phylogeny and higher classification). *Malacologia* **19**: 247–78.

Von Salvini-Plawen, L. (1980b): Was ist eine Trochophora? Eine Analyse der Larvenformen mariner Protostomier. *Zool. Jahrb. Anat.* **103**: 389–423.

Von Salvini-Plawen, L. (1982): On the polyphyletic origin of photoreceptors. In: *Visual Cells in Evolution*, Westfall, J.A. (ed.). Raven Press, New York: 137–54.

Von Salvini-Plawen, L. (1985): Early evolution and the primitive groups. In: *The Mollusca*, Trueman, E.R. & Clarke, M.R. (eds.). Academic Press, Orlando: 59–150.

Von Salvini-Plawen, L. (1988): The structure and function of molluscan digestive systems. In: *The Mollusca*, Wilbur, K.M. (ed.). Vol. 11: Form and function, Trueman, E.R. & Clarke, M.R. (volume eds.). Academic Press, San Diego: 301–79.

Von Salvini-Plawen, L. (1989): Mesoderm heterochrony and metamery in Chordata. *Fortschr. Zool.* **35**: 207–63.

Von Salvini-Plawen, L. (2003): On the phylogenetic significance of the aplacophoran Mollusca. *Iberus* **21**: 67–97.

Von Salvini-Plawen, L. & Bartolomaeus, T. (1995): Mollusca: Mesenchymata with a 'coelom'. In: *Body Cavities: Function and Phylogeny*, Lanzavecchia, G., Valvassori, R. & Candia Carnevali, M.D. (eds.). Selected Symposia and Monographs U.Z.I. 8. Mucchi, Modena: 75–92.

Von Salvini-Plawen, L. & Mayr, E. (1977): On the evolution of photoreceptors and eyes. *Evol. Biol.* **10**: 207–63.

Von Salvini-Plawen, L. & Nopp, H. (1974): Chitin bei Caudofoveata (Mollusca) und die Ableitung ihres Radulaapparates. *Z. Morphol. Tiere* **77**: 77–86.

Von Salvini-Plawen, L. & Splechtna, H. (1979): Zur Homologie der Keimblätter. *Z. Zool. Syst. Evolutionsforsch.* **17**: 10–30.

Von Salvini-Plawen, L. & Steiner, G. (1996): Synapomorphies and plesiomorphies in higher classification of molluscs. In: *Origin and Evolutionary Radiation of the Mollusca*, Taylor, J. (ed.). Oxford University Press, Oxford: 29–51.

Von Ubisch, L. (1913): Die Entwicklung von *Strongylocentrus lividus* (*Echinus microtuberculatus, Arbacia pustulosa.*). *Z. Wiss. Zool.* **106**: 409–48.

Voronezhskaya, E.E., Tyurin, S.A. & Nezlin, L.P. (2002): Neuronal development in larval chiton *Ischnochiton hakodadensis* (Mollusca: Polyplacophora). *J. Comp. Neurol.* **444**: 25–38.

Voronov, D.A. & Panchin, Y.V. (1998): Cell lineage in marine nematode *Enoplus brevis*. *Development* **125**: 143–50.

Voronov, D.A., Panchin, Y.V. & Spiridonov, S.E. (1998): Nematode phylogeny and embryology. *Nature* **395**: 28.

Wada, H., Saiga, H., Satoh, N. & Holland, P.W.H. (1998): Tripartite organization of the ancestral chordate brain and the antiquity of placodes: insights from ascidian *Pax-2/5/8*, *Hox* and *Otx* genes. *Development* **125**: 1113–22.

Wägele, J.-W. (1993): Rejection of the 'Uniramia' hypothesis and implications of the Mandibulata concept. *Zool. Jahrb. Syst.* **120**: 253–88.

Wägele, J.-W. & Misof, B. (2001): On the quality of evidence in phylogeny reconstruction: a reply to Zrzavý's defence of the 'Ecdysozoa' hypothesis. *J. Zool. Syst. Evol. Res.* **39**: 165–76.

Wägele, J.-W. & Rödding, F. (1998): Origin and phylogeny of metazoans as reconstructed with rDNA sequences. In: Molecular Evolution: Towards the Origin of Metazoa, Müller, W.E.G. (ed.). *Progr. Mol. Subcell. Biol.* **21**: 44–70.

Wägele, J.-W., Erikson, T., Lockhart, P. & Misof, B. (1999): The Ecdysozoa: artifact or monophylum? *J. Zool. Syst. Evolut. Res.* **37**: 211–23.

Waggoner, B.M. (1996): Phylogenetic hypotheses of the relationships of arthropods to Precambrian and Cambrian problematic fossil taxa. *Syst. Biol.* **45**: 190–222.

Wainwright, P.O., Hinkle, G., Sogin, M.L. & Stickel, S.K. (1993): Monophyletic origin of the Metazoa: an evolutionary link with fungi. *Science* **260**: 340–42.

Walker, M. & Campiglia, S. (1988): Some aspects of segment formation and post-placental development in *Peripatus acacioi* Marcus and Marcus (Onychophora). *J. Morphol.* **195**: 123–140.

Walker, R.J., Brooks, H.L. & Holden-Dye, L. (1996): Evolution and overview of classical transmitter molecules and their receptors. *Parasitology Suppl.* **113**: S3–S33.

Wallace, R.L. (1999): Rotifera. In: *Encyclopedia of Reproduction*, Knobil, E. & Neill, J.D. (eds.). Vol. 4. Academic Press, San Diego: 290–301.

Wallace, R.L. (2002): Rotifers: exquisite metazoans. *Integr. Comp. Biol.* **42**: 660–67.

Wallace, R.L. & Ricci, C. (2002): Rotifera. In: *Freshwater Meiofauna: Biology and Ecology*, Rundle, S.D., Robertson, A.L. & Schmidt-Araya, J.M. (eds.). Backhuys Publishers, Leiden: 15–44.

Wallace, R.L. & Snell, T.W. (1991): Rotifera. In: *Ecology and Classification of North American Freshwater Invertebrates*, Thorp, J.H. & Covich, A.P. (eds.). Academic Press, San Diego: 187–248.

Wallace, R.L., Ricci, C. & Melone, G. (1996): A cladistic analysis of pseudocoelomate (aschelminth) morphology. *Invertebr. Biol.* **115**: 104–12.

Wallberg, A., Thollesson, M., Farris, J.S. & Jondelius, U. (2004): The phylogenetic position of the comb jellies (Ctenophora) and the importance of taxonomic sampling. *Cladistics* **20**: 558–78.

Walls, G.L. (1963): *The Vertebrate Eye and its Adaptive Radiation*. Hafner Publishing Company, New York (reprint from the 1942 edition).

Walne, P.L. & Arnott, H.J. (1967): The comparative ultrastructure and possible function of eyespots: *Euglena granulata* and *Chlamydomonas eugametos*. *Planta* **77**: 325–53.

Waloszek, D. (1999): On the Cambrian diversity of Crustacea. In: *Crustaceans and the Biodiversity Crisis*, Schram, F.R. & von Vaupel Klein, J.C. (eds.). Proc. 4th Int. Crust. Congr. Amsterdam: Brill, Leiden. 3–27.

Waloszek, D. & Dunlop, J. (2002): A larval sea spider (Arthropoda: Pycnogonida) from the upper Cambrian 'Orsten' of Sweden, and the phylogenetic position of pycnogonuids. *Palaeontology* **45**: 421–46.

Walossek, D. & Müller, K.J. (1997): Cambrian 'Orsten'-type arthropods and the phylogeny of Crustacea. In: Arthropod Relationships, Fortey, R.A. & Thomas, R.H. (eds.). *Syst. Assoc. Spec. Vol. Ser.* **55**. Chapman & Hall, London: 139–53.

Walter, U. & Wägele, J.W. (1990): Ultrastructure of the maxillary gland of *Asellus aquaticus* (Crustacea, Isopoda). *J. Morphol.* **204**: 281–93.

Walthall, W.W. (1995): Repeating patterns of motoneurons in nematodes: the origin of segmentation? In: *The Nervous System of Invertebrates: An Evolutionary and Comparative Approach*, Breidbach, O. & Kutsch, W. (eds.). Birkhäuser Verlag, Basel: 61–75.

Walz, B. (1974): The fine structure of somatic muscles of Tardigrada. *Cell Tiss. Res.* **149**: 81–89.

Walz, B. (1975a): Ultrastructure of muscle cells in *Macrobiotus hufelandi*. *Mem. Inst. Ital. Idrobiol.* **32** (Suppl.): 425–43.

Walz, B. (1975b): Modified ciliary structures in receptor cells of *Macrobiotus hufelandi* (Tardigrada). *Cytobiologie* **11**: 181–85.

Walz, B. (1978): Electron microscopic investigation of cephalic sense organs of the tardigrade *Macrobiotus hufelandi* C.A.S. Schultze. *Zoomorphologie* **89**: 1–19.

Wang, D.Y.C., Kumar, S. & Hedges, S.B. (1999): Divergence time estimates for the early history of animal phyla and the origin of plants, animals and fungi. *Proc. R. Soc. London B* **266**: 163–71.

Wanninger, A. (2004): Myo-anatomy of juvenile and adult loxosomatid Entoprocta and the use of muscular body plans for phylogenetic inferences. *J. Morphol.* **261**: 249–57.

Wanninger, A. (2005): Immunocytochemistry of the nervous system and the musculature of the chordoid larva of *Symbion pandora* (Cycliophora). *J. Morphol.* **265**: 237–43.

Wanninger, A. & Haszprunar, G. (2001): The expression of an *engrailed* protein during embryonic shell formation of the tusk-shell, *Antalis entalis* (Mollusca, Scaphopoda). *Evol. Dev.* **3**: 312–21.

Wanninger, A. & Haszprunar, G. (2002): Chiton myogenesis: perspectives for the development and evolution of larval and adult muscle systems in molluscs. *J. Morphol.* **251**: 103–13.

Wanninger, A., Koop, D., Bromham, L., Noonan, E. & Degnan, B.M. (2005a): Nervous and muscle system development in *Phascolion strombus* (Sipuncula). *Dev. Genes Evol.* **215**: 509–18.

Wanninger, A., Koop, D. & Degnan, B.M. (2005b): Immunocytochemistry and metamorphic fate of the larval nervous system of *Triphyllozoon mucronatum* (Entoprocta: Gymnolaemata: Cheilostomata). *Zoomorphology* **124**: 161–70.

Warbrick, E.V., Barker, G.C., Rees, H.H. & Howells, R.E. (1993): The effect of invertebrate hormones and potential hormone inhibitors on the third larval moult of the filarial nematode, *Dirofilaria immitis*, in vitro. *Parasitology* **107**: 459–63.

Warner, F.D. (1969): The fine structure of the protonephridia in the rotifer *Asplanchna*. *J. Ultrastruct. Res.* **29**: 499–524.

Warrant, E.J. & Nilsson, D.-E. (2006): *Invertebrate Vision*. Cambridge University Press, Cambridge.

Waschuk, S.A., Bezerra, A.G., Shi, L. & Brown, L.S. (2005): *Leptosphaeria* rhodopsin: bacteriorhodopsin-like proton pump from an eukaryote. *Proc. Nat. Acad. Sci. USA* **102**: 6879–83.

Wasserthal, L.T. (1998): The open hemolymph system of Holometabola and its relation to the tracheal space. In: *Microscopic Anatomy of Invertebrates*, Harrison, F.W. & Locke, M. (eds.), Vol. 11B: Insecta. Wiley-Liss, New York: 583–620.

Wasson, K. (1999a): Asexual reproduction. In: *Encyclopedia of Reproduction*, Knobil, E. & Neill, J.D. (eds.). Vol. 1. Academic Press, San Diego: 311–19.

Wasson, K. (1999b): Kamptozoa (Entoprocta). In: *Encyclopedia of Reproduction*, Knobil, E. & Neill, J.D. (eds.). Vol. 2. Academic Press, San Diego: 924–32.

Watabe, N. (1984): Shell. In: *Biology of the Integument*, Bereiter-Hahn, J., Matoltsy, A.G. & Sylvia Richards, K. (eds.). Vol. 1: Invertebrates. Springer, Berlin: 448–85.

Watabe, N. (1988): Shell structure. In: *The Mollusca*, Wilbur, K.M. (ed.). Vol. 11: Form and function, Trueman, E.R. & Clarke, M.R. (volume eds.). Academic Press, San Diego: 69–104.

Watson, N.A. (1999): Platyhelminthes. In: *Reproductive Biology of Invertebrates*, Adiyodi, K.G., Adiyodi, R.G. & Jamieson, B.G.M. (eds.). Vol. 9, part A: Progress in male gamete ultrastructure and phylogeny. John Wiley & Sons, Chichester: 97–142.

Watson, N.A. (2001): Insights from comparative spermatology in the 'turbellarian' Rhabdocoela. In: Interrelationships of the Platyhelminthes, Littlewood, D.T. & Bray, R.A. (eds.). *Syst. Assoc. Spec. Vol. Ser.* **60**. Taylor & Francis, London: 217–230.

Watson, N.A. & Rohde, K. (1995): Sperm and spermiogenesis of the 'Turbellaria' and implications for the phylogeny of the phylum Platyhelminthes. In: Advances in Spermatozoal Phylogeny and Taxonomy, Jamieson, B.G.M., Ausio, J. & Justine, J.-L. (eds.). *Mém. Mus. Natn. Hist. Nat.* **166**: 37–54.

Watson, N.A. & Schockaert, E.R. (1996): Spermiogenesis and sperm ultrastructure in *Thylacorhynchus ambronensis* (Schizorhynchia, Kalyptorhynchia, Platyhelminthes). *Invertebr. Biol.* **115**: 263–72.

Watts, J.A., Koch, R.A., Greenberg, M.J. & Pierce, S.K. (1981): Ultrastructure of the heart of the marine mussel, *Geukensia demissa*. *J. Morphol.* **170**: 301–19.

Watts, P.C., Buley, K.R., Sanderson, S., Boardman, W., Ciofi, C. & Gibson, R. (2006): Parthenogenesis in Komodo dragons. *Nature* **444**: 1021–22.

Weaving, J.N. & Cullen, M.J. (1978): Unusual high volume of sarcoplasmic reticulum in a wasp leg muscle. *Experientia* **34**: 796–97.

Weber, C., Singla, C.L. & Kerfoot, P.A.H. (1982): Microanatomy of the subumbrellar motor innervation in *Aglantha digitale* (Hydromedusae: Trachylina). *Cell Tiss. Res.* **223**: 305–12.

Weber, R.E. (1971): Oxygenational properties of vascular and coelomic haemoglobins from *Nephtys hombergii* (Polychaeta) and their functional significance. *Netherlands J. Sea Res.* **5**: 240–51.

Weber, R.E. (1978): Respiratory pigments. In: *Physiology of Annelids*, Mill, P. (ed.). Academic Press, London: 393–446.

Weber, R.E., Fänge, R. & Rasmussen, K.K. (1979): Respiratory significance of priapulid hemerythrin. *Mar. Biol. Lett.* **1**: 87–97.

Weglarska, B. (1980): Light and electron microscopic studies on the excretory system of *Macrobiotus richtersi* Murray, 1911 (Eutardigrada). *Cell Tiss. Res.* **207**: 171–82.

Wehner, R & Gehring, W. (1995): *Zoologie*. 23 Edn., Georg Thieme Verlag, Stuttgart.

Weiss, M.J. (2001): Widespread hermaphroditism in freshwater gastrotrichs. *Invertebr. Biol.* **120**: 308–41.

Weiss, M.J. & Levy, D.P. (1979): Sperm in 'parthenogenetic' freshwater gastrotrichs. *Science* **205**: 302–3.

Weissenfels, N. (1976): Bau und Funktion des Süßwasserschwammes *Ephydatia fluviatilis* L. (Porifera). III. Nahrungsaufnahme, Verdauung und Defäkation. *Zoomorphologie* **85**: 73–88.

Weissenfels, N. (1992): The filtration apparatus for food collection in freshwater sponges (Porifera, Spongillidae). *Zoomorphology* **112**: 51–55.

Wells, M.J. (1983): Circulation in cephalopods. In: *The Mollusca*, Wilbur, K.M. (ed.). Vol. 5: Physiology, part 2, Saleuddin, A.S.M. & Wilbur, K.M. (volume eds.). Academic Press, New York: 239–290.

Wells, R.M.G. & Dales, R.P. (1974): Oxygenational properties of haemerythrin in the blood of *Magelona papillicornis* Müller (Polychaeta: Magelonidae). *Comp. Biochem. Physiol. A* **49**: 57–64.

Welsch, U. (1995): Evolution of the body cavities in Deuterostomia. In: *Body Cavities: Function and*

Phylogeny, Lanzavecchia, G., Valvassori, R. & Candia Carnevali, M.D. (eds.). Selected Symposia and Monographs U.Z.I. 8. Mucchi, Modena: 111–34.

Welsch, U. & Rehkämper, G. (1987): Podocytes in the axial organ of echinoderms. *J. Zool. London* 213: 45–50.

Werner, B. (1984): Stamm Cnidaria. In: *Lehrbuch der Speziellen Zoologie*, Gruner, H.-E. (ed.), Vol. I.2. Gustav Fischer Verlag, Stuttgart: 11–305.

Werner, B., Chapman, D.M. & Cutress, C.E. (1976): Muscular and nervous systems of the cubopolyp (Cnidaria). *Experientia* 32: 1047–49.

Wessing, A. & Eichelberg, D. (1975): Ultrastructural aspects of transport and accumulation of substances in the malpighian tubules. *Fortschr. Zool./Progr. Zool.* 23: 148–172.

West, D.L. (1978): The epitheliomuscular cell of *Hydra*: its fine structure, three-dimensional architecture and relation to morphogenesis. *Tiss. Cell* 10: 629–46.

West, J.B. (2006): Human responses to high altitudes. *Integr. Comp. Biol.* 46: 25–34.

Westblad, E. (1949): Studien über skandinavische Turbellaria Acoela. *V. Ark. Zool.* 41: 1–82.

Westblad, E. (1950): *Xenoturbella bocki* n.g., n.sp. a peculiar, primitive turbellarian type. *Ark. Zool.* (N.S.) 1: 11–29.

Westfall, J.A. (1970): The nematocyte complex in a hydromedusan, *Gonionemus vertens*. *Z. Zellforsch.* 110: 457–70.

Westfall, J.A., Elliott, S.R., Kumar, P.S.M. & Carlin, R.W. (2000): Immunocytochemical evidence for biogenic amines and immunogold labelling of serotinergic synapses in tentacles of *Aiptasia pallida* (Cnidaria, Anthozoa). *Invertebr. Biol.* 119: 370–78.

Westfall, J.A., Sayyar, K.L. & Elliott, C.F. (1998): Cellular origins of kinocilia, stereocilia, and microvilli on tentacles of sea anemones of the genus *Calliactis* (Cnidaria: Anthozoa). *Invertebr. Biol.* 117: 186–93.

Westheide, W. (1965): *Parapodrilus psammophilus* nov. gen. nov. spec., eine neue Polychaeten-Gattung aus dem Mesopsammal der Nordsee. *Helgol. Wiss. Meeresunt.* 12: 207–13.

Westheide, W. (1982): *Microphthalmus hamosus* sp.n. (Polychaeta, Hesionidae) – an example of evolution leading from the interstitial fauna to a macrofaunal interspecific relationship. *Zool. Scr.* 11: 189–93.

Westheide, W. (1984): The concept of reproduction in polychaetes with small body size: adaptations in interstitial species. *Fortschr. Zool./Progr. Zool.* 29: 265–87.

Westheide, W. (1985a): The systematic position of the Dinophilidae and the archiannelid problem. In: The Origins and Relationships of Lower Invertebrates, Conway Morris, S., George, J.D., Gibson, D.I. & Platt, H.M. (eds.). *Syst. Assoc. Spec. Vol.* 28. Clarendon Press, Oxford: 310–26.

Westheide, W. (1985b): Ultrastructure of the protonephridia in the dorvilleid polychaete *Apodotrocha progenerans* (Annelida). *Zool. Scr.* 14: 273–78.

Westheide, W. (1986): The nephridia of the interstitial polychaete *Hesionides arenaria* and their phylogenetic significance (Polychaeta, Hesionidae). *Zoomorphology* 106: 35–43.

Westheide, W. (1987): Progenesis as a principle in meiofauna evolution. *J. Nat. Hist.* 21: 843–54.

Westheide, W. (1990): Polychaetes: interstitial families. In: *Synopsis of the British Fauna*, Kermack, D.M. & Barnes, R.S.K. (eds.). Vol. 44. Universal Book Services/W. Backhuys, Oegstgeest.

Westheide, W. (1997): The direction of evolution within the Polychaeta. *J. Nat. Hist.* 31: 1–15.

Westheide, W. & Rieger, R.M. (1978): Cuticle ultrastructure of hesionid polychaetes (Annelida). *Zoomorphologie* 91: 1–18.

Westheide, W. & Rieger, R. (2004): *Spezielle Zoologie*. Part 2: Wirbel- oder Schädeltiere. Spektrum Akademischer Verlag, Heidelberg.

Westheide, W. & Rieger, R. (2007): *Spezielle Zoologie*. Part 1: Einzeller und wirbellose Tiere. 2nd. edition. Elsevier – Spektrum Akademischer Verlag, Heidelberg.

Westheide, W. & Riser, N.W. (1983): Morphology and phylogenetic relationships of the neotenic interstitial polychaete *Apodotrocha progenerans* n. gen., n. sp. (Annelida). *Zoomorphology* 103: 67–87.

Westheide, W., McHugh, D., Purschke, G. & Rouse, G. (1999): Systematization of the Annelida: different approaches. *Hydrobiologia* 402: 291–307.

Wetzel, M.A., Jensen, P. & Giere, O. (1995): Oxygen/sulfide regime and nematode fauna associated with *Arenicola marina* burrows: new insights in the thiobios case. *Mar. Biol.* 124: 301–12.

Weygoldt, P. (1958): Die Embryonalentwicklung des Amphipoden *Gammarus pulex* (L). *Zool. Jahrb. Anat.* 77: 51–110.

Weygoldt, P. (1986): Arthropod interrelationships – the phylogenetic-systematic approach. *Z. Zool. Syst. Evolutionsforsch.* 24: 19–35.

Weyrer, S., Rützler, K. & Rieger, R. (1999): Serotonin in Porifera? Evidence from developiong *Tedania ignis*, the caribbean fire sponge (Demospongiae). *Mem. Queensl. Mus.* 44: 659–65.

Wharton, D.A. (1986): *A Functional Biology of Nematodes*. Croom Helm, London.

Wheeler, W.C. (1997): Sampling, groundplans, total evidence and the systematics of arthropods. In: Arthropod Relationships, Fortey, R.A. & Thomas, R.H. (eds.). *Syst. Assoc. Spec. Vol.* 55, Chapman & Hall, London: 87–96.

Wheeler, W.C., Cartwright, P. & Hayashi, C.Y. (1993): Arthropod phylogeny: a combined approach. *Cladistics* 9: 1–39.

White, J.G., Southgate, E., Thomson, J.N. & Brenner, S. (1986): The structure of the nervous system of the nematode *Caenorhabditis elegans*. *Philos. Trans. R. Soc. London B* **314**: 1–340.

Whitfield, J.B. (1971): Spermiogenesis and spermatozoan ultrastructure in *Polymorphus minutus* (Acanthocephala). *Parasitology* **62**: 415–30.

Whiting, M.F., Bradler, S. & Maxwell, T. (2003): Loss and recovery of wings in stick insects. *Nature* **421**: 264–67.

Whittington, H.B. & Almond, J.E. (1987): Appendages and habits of the Upper Ordovician trilobite *Triarthrus eatoni*. *Phil. Trans. R. Soc. London B* **317**: 1–46.

Widersten, B. (1965): Genital organs and fertilization in some Scyphozoa. *Zool. Bidr. Uppsala* **37**: 45–58.

Wiedermann, A. (1995): Zur Ultrastruktur des Nervensystems bei *Cephalodasys maximus* (Macrodasyida, Gastrotricha). *Microfauna Marina* **10**: 173–233.

Wildon, D.C., Thain, J.F., Minchin, P.E.H., Gubb, I.R., Reilly, A.J., Skipper, Y.D., Doherty, H.M., O' Donnell, P.J.O. & Bowles, D.J. (1992): Electrical signalling and systemic proteinase inhibitor induction in the wounded plant. *Nature* **360**: 62–65.

Wiley, C.A. & Ellisman, M.H. (1980): Rows of dimeric-particles within the axolemma and juxtaposed particles within glia, incorporated into a new model for the paranodal glial-axonal junction at the node of Ranvier. *J. Cell Biol.* **84**: 261–80.

Wilke, U. (1972a): Der Eicheldarm der Enteropneusten als Stützorgan für Glomerulus und Perikardvesikel. *Verh. Dtsch. Zool. Ges.* **66**: 93–96.

Wilke, U. (1972b): Die Feinstruktur des Glomerulus von *Glossobalanus minutus* Kowalewsky (Enteropneusta). *Cytobiol.* **5**: 439–47.

Wilkens, J.L. (1999): Evolution of the cardiovascular system in Crustacea. *Am. Zool.* **39**: 199–214.

Willenz, P. & van de Vyver, G. (1984): Ultrastructural localization of lysosomal digestion in the fresh water sponge *Ephydatia fluviatilis*. *J. Ultrastruct. Res.* **87**: 13–22.

Williams, A. (1984): Lophophorates. In: *Biology of the Integument*, Bereiter-Hahn, J., Matoltsy, A.G. & Sylvia Richards, K. (eds.). Vol. 1: Invertebrates. Springer, Berlin: 728–45.

Williams, A. (1997): Brachiopoda: introduction and integumentary system. In: *Microscopic Anatomy of Invertebrates*, Harrison, F.W. & Woollacott, R.M. (eds.), Vol. 13: Lophophorates, Entoprocta, and Cycliophora. Wiley-Liss, New York: 237–96.

Williams, A. & Holmer, L.E. (2002): Shell structure and inferred growth, functions and affinities of the sclerites of the problematic *Micrina*. *Palaeontology* **45**: 845–73.

Williams, A., Cusack, M. & Mackay, S. (1994): Collageneous chitinophosphatic shell of the brachiopod *Lingula*. *Phil. Trans. R. Soc. London B* **346**: 223–66

Wills, M.A. (1998): Cambrian and recent disparity: the picture from priapulids. *Paleobiology* **24**: 177–199.

Wills, M.A., Briggs, D.E.G., Fortey, R.A. & Wilkinson, M. (1995): The significance of fossils in understanding arthropod evolution. *Verh. Dtsch. Zool. Ges.* **88.2**: 203–215.

Wills, M.A., Briggs, D.E.G., Fortey, R.A., Wilkinson, M. & Sneath, P.H.A. (1998): An arthropod phylogeny based on fossil and recent taxa. In: *Arthropod Fossils and Phylogeny*, Edgecombe, G.D. (ed.). Columbia University Press, New York: 33–105.

Wilmer, P. (1990): *Invertebrate Relationships. Patterns in Animal Evolution*. Cambridge University Press, Cambridge.

Wilson, R.A. & Webster, L.A. (1974): Protonephridia. *Biol. Rev.* **49**: 127–60.

Winchell, C.J., Sullivan, J., Cameron, C.B., Swalla, B.J. & Mallatt, J. (2002): Evaluating hypotheses of deuterostome phylogeny and chordate evolution with new LSA and SSU ribosomal DNA data. *Mol. Biol. Evol.* **19**: 762–76.

Wingstrand, K.G. (1972): Comparative spermatology of a pentastomid, *Raillietiella hemidactyli*, and a branchiuran crustacean, *Argulus foliaceus*, with a discussion of pentastomid relationships. *Kong. Dansk. Vidensk. Selsk. Biol. Skr.* **19**: 1–72.

Winnepenninckx, B., Backeljau, T. & De Wachter, R. (1994): Small ribosomal subunit RNA and the phylogeny of Mollusca. *Nautilus Suppl.* **2**: 98–110.

Winnepenninckx, B., Backeljau, T., Mackey, L.M., Brooks, J.M., De Wachter, R., Kumar, S. & Garey, J.R. (1995): 18S rRNA data indicate that Aschelminthes are polyphyletic in origin and consist of at least three distinct clades. *Mol. Biol. Evol.* **12**: 1132–37.

Winnepenninckx, B., Backeljau, T. & De Wachter, R. (1996): Investigation of molluscan phylogeny on the basis of 18S rRNA sequences. *Mol. Biol. Evol.* **13**: 1306–17.

Winnepenninckx, B., Van de Peer, Y. & Backeljau, T. (1998a): Metazoan relationships on the basis of 18S rRNA sequences: a few years later ... *Am. Zool.* **38**: 888–906.

Winnepenninckx, B., Backeljau, T. & Kristensen, R. M. (1998b): Relations of the new phylum Cycliophora. *Nature* **393**: 636–38.

Wirth, U. (1984): Die Struktur der Metazoen-Spermien und ihre Bedeutung für die Phylogenetik. *Verh. Nat.-Wiss. Ver. Hamburg* **27**: 295–362.

Wirz, A., Pucciarelli, S., Miceli, C., Tongiorgi, P. & Balsamo, M. (1999): Novelty in phylogeny of Gastrotricha: evidence from 18 rRNA gene. *Mol. Phyl. Evol.* **13**: 314–18.

Wittenberg, B.A., Briehl, R.W. & Wittenberg, J.B. (1965): Haemoglobins of invertebrate tissues. Nerve haemobins of *Aphrodite*, *Aplysia* and *Halosydna*. *Biochem. J.* **96**: 363–71.

Wittmann, K.J., Schlacher, T.A. & Ariani, A.P. (1993): Structure of recent and fossil mysid statoliths (Crustacea, Mysidacea). *J. Morphol.* **215**: 31–49.

Wolf, Y.I., Rogozin, I.B. & Koonin, E.V. (2004): Coelomata and not Ecdysozoa: evidence from genome-wide phylogenetic analysis. *Genome Res.* **14**: 29–36.

Wolken, J.J. (1971): *Invertebrate Photoreceptors. A Comparative Analysis.* Academic Press, New York.

Wolken, J.J. (1995): *Light Detectors, Photoreceptors, and Imaging Systems in Nature.* Oxford University Press, New York.

Womersley, C.Z., Wharton, D.A. & Higa, L.M. (1998): Survival biology. In: *The Physiology and Biochemistry of Free-living and Plant-parasitic Nematodes*, Perry, R.N. & Wright, D.J. (eds.). CABI Publishing: 271–302.

Wood, R.L. (1985): The use of *Hydra* for studies of cellular ultrastructure and cell junctions. *Arch. Sci. Genéve* **38**: 371–83.

Woollacott, R.M. (1999): Bryozoa (Ectoprocta). In: *Encyclopedia of Reproduction*, Knobil, E. & Neill, J.D. (eds.). Vol. 1. Academic Press, San Diego: 439–48.

Woollacott, R.M. & Eakin, R.M. (1973): Ultrastructure of a potential photoreceptoral organ in the larva of an entoproct. *J. Ultrastruct. Res.* **43**: 412–25.

Woollacott, R.M. & Pinto, R.L. (1995): Flagellar basal apparatus and its utility in phylogenetic analysis of the Porifera. *J. Morphol.* **226**: 247–65.

Woollacott, R.M. & Zimmer, R.L. (1972): Fine structure of a potential photoreceptor organ in the larva of *Bugula neritina* (Bryozoa). *Z. Zellforsch.* **123**: 458–69.

Worsaae, K. & Kristensen, R.M. (2005): Evolution of interstitial Polychaeta (Annelida). In: Morphology, Molecules, Evolution and Phylogeny in Polychaeta and Related Taxa, Bartolomaeus, T. & Purschke, G. (eds.). Springer, Berlin. *Hydrobiologia* **535/536**: 319–40.

Wourms, J.P. (1981): Viviparity: The maternal-fetal relationship in fishes. *Am. Zool.* **21**: 473–515.

Wright, D.J. (1998): Respiratory physiology, nitrogen excretion and osmotic and ionic regulation. In: *The Physiology and Biochemistry of Free-living and Plant-parasitic Nematodes*, Perry, R.N. & Wright, D.J. (eds.). CABI Publishing: 103–31.

Wright, D.J. (1999): Nematodes and related phyla. In: *Encyclopedia of Reproduction*, Knobil, E. & Neill, J.D. (eds.). Vol. 3. Academic Press, San Diego: 326–33.

Wright, J.C. & Luke, B.M. (1989): Ultrastructural and histochemical investigations of *Peripatus* integument. *Tiss. Cell* **21**: 605–25.

Wright, K.A. (1976): Somatic centrioles in the parasitic nematode, *Capillaria hepatica* Bancroft, 1893. *J. Nematology* **8**: 92–93.

Wright, K.A. (1980): Nematode sense organs. In: *Nematodes as Biological Models*, Zuckerman, B.M. (eds.). Academic Press, New York: 237–95.

Wright, K.A. (1983): Nematode chemosensilla: form and function. *J. Nematol.* **15**: 151–58.

Wright, K.A. (1991): Nematoda. In: *Microscopic Anatomy of Invertebrates*, Harrison, F.W. & Ruppert, E.E. (eds.), Vol. 4: Aschelminthes. Wiley-Liss, New York: 111–95.

Xiao, S. & Knoll, A.H. (2000): Phosphatized animal embryos from the Neoproterozoic Doushantou formation at Weng'an, Guizhou, South China. *J. Paleontol.* **74**: 767–88.

Xiao, S., Zhang, Y. & Knoll, A.H. (1998): Three-dimensional preservation of algae and animal embryos in a neoproterozoic phosphorite. *Nature* **391**: 553–58.

Xiong, J., Kurtz, D.M., Ai, J. & Sanders-Loehr, J. (2000): A hemerythrin-like domain in a bacterial chemotaxis protein. *Biochem.* **39**: 5117–25.

Xylander, W. & Bartolomaeus, T. (1995): Protonephridien: neue Erkenntnisse über Funktion und Evolution. *Biologie in unserer Zeit* **25**: 107–114.

Yamamoto, M. & Yoshida, M. (1978): Fine structure of the ocelli of a synaptid holothurian, *Opheodesoma spectabilis*, and the effects of light and darkness. *Zoomorphologie* **90**: 1–17.

Yamasu, T. & Yoshida, M. (1973): Electron microscopy of the photoreceptors of an anthomedusa and a scyphomedusa. *Publ. Seto Mar. Biol. Lab.* **20**: 757–78.

Yamasu, T. & Yoshida, M. (1976): Fine structure of complex ocelli of a cubomedusan, *Tamoya bursaria* Haeckel. *Cell Tiss. Res.* **170**: 325–39.

Yanase, T. & Sakamoto, S. (1965): Fine structure of the visual cells of the dorsal eye in the mollusc *Onchidium verruculatum*. *Zool. Mag.* **74**: 238–42 (in japanese with english summary).

Yoshida, M. (1979): Extraocular photoreception. In: *Handbook of Sensory Physiology*, Autum, H., Jung, R., Loewenstein, W.R., MacKay, D.M. & Teuber, H.-L. (eds.). Vol. VII/6A: Comparative physiology and evolution of vision in invertebrates, Autrum, H. (volume ed.). Springer, Berlin: 581–640.

Yoshida, M., Takasu, N. & Tamotsu, S. (1984): Photoreception in echinoderms. In: *Photoreception and Vision in Invertebrates*, Ali, M.A. (ed.). Plenum Press, New York: 743–71.

Young, C.M. (1994): The biology of external fertilization in deep-sea echinoderms. In: *Reproduction, Larval Biology, and Recruitment of the Deep-sea Benthos*, Young, C.M. & Eckelbarger, K.J. (eds.). Columbia University Press, New York: 179–200.

Young, C.M. (2002): *Atlas of Marine Invertebrate Larvae.* Academic Press, San Diego.

Young, J.Z. (1989): The Bayliss-Starling lecture: Some special senses in the sea. *J. Physiol.* **411**: 1–25.

Yushin, V.V. & Coomans, A. (2000): Ultrastructure of sperm development in the free-living marine nematodes of the family Chromadoridae (Chromadorida: Chromadorina). *Nematology* **2**: 285–96.

Yushin, V.V. & Coomans, A. (2005): Ultrastructure of sperm development in the free-living marine nematode *Metachromadora itoi* (Chromadoria, Desmodorida). *Acta Zool.* **86**: 255–65.

Yushin, V.V., Coomans, A. & Malakhov, V.V. (2002): Ultrastructure of spermatogenesis in the free-living marine nematode *Pontonema vulgare* (Enoplida, Oncholaimidae). *Can. J. Zool.* **80**: 1371–82.

Yushin, V.V. & Malakhov, V.V. (1994): Ultrastructure of sperm cells in the female gonoduct of free-living marine nematodes from genus *Enoplus* (Nematoda: Enoplida). *Fundam. Appl. Nematol.* **17**: 513–19.

Yushin, V.V. & Malakhov, V.V. (1998): Ultrastructure of sperm development in the free-living marine nematode *Enoplus anisospiculus* (Enoplida: Enoplidae). *Fundam. Appl. Nematol.* **21**: 213–25.

Zal, F., Lallier, F.H., Wall, J.S., Vinogradov, S.N. & Toulmond, A. (1996): The multi-hemoglobin system of the hydrothermal vent tube worm *Riftia pachyptila*. I. Reexamination of the number and masses of its constituents. *J. Biol. Chem.* **271**: 8869–74.

Zantke, J., Wolff, C. & Scholtz, G. (2007): Three dimensional reconstruction of the central nervous system of *Macrobiotus hufelandi* (Eutardigrada, Parachela): implications for the phylogenetic position of Tardigrada. *Zoomorphology* (in press).

Zapotosky, J.E. (1974): Fine structure of the larval stage of *Paragordius varius* (Leidy, 1851) (Gordioidea: Paragordidae). I. The preseptum. *Proc. Helminthol. Soc. Washington* **41**: 209–21.

Zapotosky, J.E. (1975): Fine structure of the larval stage of *Paragordius varius* (Leidy, 1851) (Gordioidea: Paragordidae). II. The postseptum. *Proc. Helminthol. Soc. Washington* **42**: 103–11.

Zebe, E. (1991): Arthropods. In: *Metazoan Life Without Oxygen*, Bryant, C. (ed.). Chapman and Hall, London: 219–37.

Zerbst-Boroffka, I. (1975): Function and ultrastructure of the nephridium in *Hirudo medicinalis* L. III. Mechanisms of the formation of primary and final urine. *J. Comp. Physiol.* **100**: 307–15.

Zerbst-Boroffka, I. & Haupt, J. (1975): Morphology and function of the metanephridia in annelids. *Fortschr. Zool./Progr. Zool.* **23**: 33–47.

Zhang, X.-G. & Pratt, B.R. (1994): Middle Cambrian arthropod embryos with blastomeres. *Science* **266**: 637–39.

Zhao, B. & Liu, B. (1992): Ultrastructure of the spermatid and spermatozoon of *Macracanthorhynchus hirudinaceus*. *J. Helminthol.* **66**: 267–72.

Zhukov, V.V., Borissenko, S.L., Ziegler, M.V., Vakoliuk, I.A. & Meyer-Rochow, V.B. (2006): The eye of the freshwater prosobranch gastropod *Viviparus viviparus*: ultrastructure, electrophysiology and behaviour. *Acta Zool.* **87**: 13–24.

Zimmer, R.L. (1978): The comparative structure of the preoral hood coelom in Phoronida and the fate of this cavity during and after metamorphosis. In: *Settlement and Metamorphosis of Marine Invertebrate Larvae*, Chia, F.-S. & Rice, M.E. (eds.). Elsevier, New York: 23–40.

Zimmer, R.L. (1991): Phoronida. In: *Reproduction of Marine Invertebrates*, Giese, A.C., Pearse, J.S. & Pearse, V.B. (eds.). Vol. 6: Echinoderms and lophophorates. The Boxwood Press, Pacific Grove: 1–45.

Zrzavý, J. (2001): The interrelationships of metazoan parasites: a review of phylum- and higher-level hypotheses from recent morphological and molecular phylogenetic analyses. *Folia Parasitologica* **48**: 81–103.

Zrzavý, J. (2003): Gastrotricha and metazoan phylogeny. *Zool. Scr.* **32**: 61–81.

Zrzavý, J. & Hypša, V. (2003): Myxozoa, *Polypodium*, and the origin of Bilateria: the phylogenetic position of 'Endocnidozoa' in light of the rediscovery of *Buddenbrockia*. *Cladistics* **19**: 164–69.

Zrzavý, J., Hypša, V. & Tietz, D. F. (2001): Myzostomida are not annelids: molecular and morphological support for a clade of animals with anterior sperm flagella. *Cladistics* **17**: 170–98.

Zrzavý, J., Hypša, V. & Vlášková, M. (1997): Arthropod phylogeny: taxonomic congruence, total evidence and conditional combination approaches to morphological and molecular data sets. In: Arthropod Relationships, Fortey, R.A. & Thomas, R.H. (eds.). *Syst. Assoc. Spec. Vol.* **55**, Chapman & Hall, London: 97–107.

Zrzavý, J., Mihulka, S., Kepka, P., Bezdek, A. & Tietz, D. (1998): Phylogeny of the Metazoa based on morphological and 18S ribosomal DNA evidence. *Cladistics* **14**: 249–85.

Index

5-hydroxytryptamine 112
18S rDNA gene
  Caenorhabditis elegans 6
  Choanoflagellata and Metazoa 7
  'Mesozoa' 11
  resolution 5
  Symbion 27
  Trochozoa 25, 26
  Xenoturbella 12, 70

Acanthamoeba 74
Acanthocephala
  body cavities 156, 160, 164, 165
  circulatory systems 192, 200
  epidermis 65, 72, 73
  excretory systems 173, 176–7, 189
  gametes 270, 271, 285, 286, 289, 292
  general body organization 53
  intestinal systems 218, 224, 227–8, 232, 236
  musculature 88, 89, 90, 92, 94
  nervous system 101, 114
  phylogenetic frame 16, 17, 27
  reproductive organs 244, 247, 254, 257
  respiratory systems 213, 217
  sensory organs 121
  size 37
Acanthocephalus anguillae 89
Acanthochiton fascicularis 138
Acari
  intestinal systems 226, 227
  reproductive organs 245, 250, 251
  respiratory systems 207
accessory cells (in protonephridia) 183–4
accessory centrioles 262, 268, 273, 275, 280
acetylcholine 112, 117
acetylcholinesterase 117
aciliary spermatozoa 262, 264, 267, 269, 274, 277
Acipenser 211
Acoela
  epidermis 57–8, 68, 69–70, 71
  excretory systems 173
  gametes 265, 268, 287
  intestinal systems 220, 221, 225, 236
  musculature 83, 87
  nervous system 100–1, 110
  phylogenetic frame 22–4
  reproductive organs 241, 249, 256
  sensory organs 128, 129, 134
  skeletons 51
acoelomate 2, 148, 149, 151, 154–5, 163, 167, 170
Acoelomorpha
  body cavities 154, 160, 164, 165
  epidermis 70, 71, 73
  excretory systems 169, 173, 188, 189
  gametes 265, 286–7
  intestinal systems 220, 221, 223, 232, 236
  musculature 81, 83, 87, 90, 91
  nervous system 100–1, 110–11, 113
  phylogenetic frame 12, 13, 21, 22, 23, 24
  reproductive organs 244, 246, 247, 249, 252, 256
  respiratory systems 213
  sensory organs 120, 128, 134–5
Acrania
  anteroposterior axis 40
  body cavities 152, 160, 162, 163, 164, 165
  circulatory systems 198–9, 200
  dorsoventral axis 42, 43
  epidermis 59, 73
  excretory systems 185–6, 189, 190
  gametes 280, 281, 286, 291
  intestinal systems 225, 228, 232, 237
  musculature 90, 94
  nervous system 108–9, 113, 116
  phylogenetic frame 30, 32, 33
  reproductive organs 245, 247, 248, 260
  respiratory systems 209, 211, 213, 215
  segmentation 50
  sensory organs 123, 143–4, 146
Acropora millepora 34
Acrosomata 9, 10, 292
acrosomes 262–4, 266, 268, 270–5, 277, 279–81, 292
actin
  cell–cell junctions 55
  epidermal cell 66, 67
  musculature 74–5, 77, 80, 82, 87, 90, 94
actinarians 39
Actinistia 199, 211, 280
actin–myosin system 74–5, 94
actinopterygian fishes 280
actinotrocha
  body cavities 159
  excretory systems 183, 184
  intestinal systems 233
  larva 29
  musculature 93
  sensory organs 122
action potentials 95
active transport 170, 172, 174, 178
Acylus fluviatilis 178
adhaerens junctions 55, 56, 57, 73
adrenaline 112, 117
Aegina 127
Aeolidia papillosa 178
Afrocimex constrictus 137
agametic-asexual reproduction 240, 241, 243
Aglantha digitale 71, 97
agnathans 248, 260
Alciopidae 179
alpaca 214
α-actinin 84, 85
Amaroucium constellatum 142
Amaurobius 137
Amblypygi 207, 226, 227
Ambulacralia 32
Ambystoma 243
Amia calva 210, 211
amino acids 112, 169
ammonia 169, 170, 174
amniotes 199, 248, 281

amoebocytes 155, 157
amorphin 84, 85
Ampharetidae 227
Amphibia 144, 199, 210, 245, 260
Amphibolus 275
A. volubilis 276
Amphipoda 130, 195
Amphiporus 204
amphystomy 43, 222
Anaitides 188
A. mucosa 179, 181
Anaperus 100
anchoring junctions 56
Annelida
  body cavities 150, 151, 152, 154, 157–8, 160, 161, 163, 164, 165, 166, 168
  circulatory systems 191, 194–5, 200, 201
  complexity 35
  epidermis 62, 64, 65, 68, 73
  excretory systems 172, 173, 179–82, 187, 188, 189, 190
  gametes 273–5, 286, 289–90
  intestinal systems 222, 224, 225, 226, 227, 230, 232, 234, 235, 237
  locomotory appendages 52
  musculature 79, 83, 84, 85, 86, 89, 90, 92
  nervous system 102–3, 106, 115
  phylogenetic frame 18, 22, 24, 25–6, 27, 28, 29, 30
  reproductive organs 241, 242, 244, 247, 251, 258
  respiratory systems 206–7, 212, 213, 214, 215
  segmentation 47, 48–50
  sensory organs 122, 129, 130, 139–40, 145
  size 37
Annulonemertes minusculus 48
Anomalocaris 20, 106
Anopla 224
antagonistic systems 85–6
Antedon 31
antennapedia-like genes 220
anteroposterior (AP) axis 39–42, 50
anthomedusae 46
Anthopleura elegantissima 96
anthox1 gene 42
anthox6 gene 42
Anthozoa
  anteroposterior axis 42
  complexity 34
  epidermis 59, 63, 68, 71
  gametes 263, 264
  germ layers 47
  intestinal systems 223, 235
  musculature 75, 76
  nervous system 96
  phylogenetic frame 10
  reproductive organs 241, 246, 256
  sensory organs 120
  symmetry 38
Antipatharia 60, 61
Anura 144
anus 43, 221
Apharyngostrigea cornu 101
Aphidina 242
aphids 245
Aphrodite 207
  A. aculeata 212, 214
Aphroditidae 207
apical organ 111–12, 120
Apicomplexa 58
Aplacophora
  circulatory systems 193
  gametes 271, 272
  reproductive organs 258
  respiratory systems 204
  sensory organs 122
Aplysia californica 212
Aplysilla rosea 263
Aplysina cavernicola 59
Apoda 209
Apodida 207, 237
Apodotrocha 36, 37
Aporrhais 138
Appendicularia
  gametes 280, 281
  musculature 94
  nervous system 108
  phylogenetic frame 31
  reproductive organs 245
  sensory organs 130, 142
apterygote insects 245
Arabidopsis thaliana 74
Arachnida 124, 137, 182, 207, 208, 224
Araneae 137
Araneus diadematus 183
Arapaima 210
Arca 145
archaebacteria 131
archaeocytes 218, 246
Archaster typicus 279
archenteron 44, 218, 219, 222, 230
Archiacanthocephala 89, 156, 176–7
archicoelomates 166
archimery 28, 32
archinephros 186
Arenicola 129
  A. marina 52, 130, 158, 216
Areniocolidae 49
Argyrotheca 245
Arhynchobdellida 227
arkshells 138
Armadillidium vulgare 80
Arthropoda 2, 52
  body cavities 167
  epidermis 55, 61, 63, 64, 67–8
  excretory systems 190
  gametes 285
  intestinal systems 233
  locomotory appendages 52
  musculature 88
  nervous system 104–6, 110
  phylogenetic frame 13, 17–18, 19–21, 25
  respiratory systems 211, 215, 216
  segmentation 47, 48–9, 50
  sensory organs 125, 131
  size 37
  skeletons 51
Arthrotardigrada 93
Articulata 17–18, 51, 106, 234, 278
Asajirella 93
  A. gelatinosa 84
Asbestopluma 219
Ascaris 79, 217
  A. lumbricoides 37
  A. suum 38, 203, 214, 253
Aschelminthes 13, 16
Ascidia
  dorsoventral axis 43
  excretory systems 185
  musculature 94
  nervous system 108, 109, 112
  sensory organs 129, 130, 141, 142
  size changes 36
  gametes 280, 281
Ascidia 281
ascidiids 185
asexual reproduction 240–5, 247, 261
Aspidochirotida 198, 209
Asplanchna 92, 176, 236, 243
  A. brightwelli 136, 176, 270
Asterias 242
  A. forbesi 185
Asterina 235
Asteroida
  circulatory systems 198
  epidermis 59, 62
  excretory systems 185
  gametes 279
  intestinal systems 224, 237
  nervous system 107
  photoreceptors 142, 146
  reproductive organs 242, 245
  respiratory systems 208, 209
Astropecten 39

asymmetrical synapses 96
Atubaria 31, 211
Aurelia 91
  A. aurita
    medusa 8
    musculature 82
    reproduction 241
    sensory organs 125, 126, 134
auricularia 230, 231
Austrognathia 269
autapomorphies
  Arthropoda 21
  body cavities 150, 166
  diploblastic animals 8–9, 10
  entocodon 47
  epidermis 56, 57, 58, 68, 72
  intestinal systems 220, 233
  nervous system 109, 110, 111, 117
  Platyhelminthes 23
  Radialia 28
  reproductive organs 240
  sensory organs 122, 140, 142
  syncytial epidermis 72
Autolytus purpureomaculatus 242
automated sequencing of DNA 3
Avagina incola 100, 101
Aves
  circulatory systems 199–200
  gametes 281–2
  photoreceptors 144
  reproductive organs 248
  respiratory systems 210–11
axons 95

bacteria 120
bacteriorhodopsin 131
Balanoglossus 237
Barbatia 145
Barentsia
  B. gracilis 92
  B. laxa 273
Barlasia 204
basal body 262
basal lamina/basement membrane 57, 72, 73
Bathynellacea 36
Batillipes
  B. mirus 19
  B. noerrevangi 275
Bdelloida
  intestinal systems 236
  photoreceptors 136
  phylogenetic frame 16, 17
  reproductive organs 242, 244, 257
Bellonci, organ of 130
belt desmosomes 55, 56, 57, 73
Beroe 67, 68, 69, 70, 97

B. ovata 263
Bilateria
  anteroposterior axis 39, 40, 41, 42
  body cavities 148, 154–66, 167
  dorsoventral axis 42
  epidermis 57, 60, 70, 71
  excretory systems 169–70, 172–87, 188
  gametes 283, 292
  germ layers 43, 44, 46, 47
  intestinal systems 218, 220, 222, 225, 233, 235, 239
  musculature 81, 82, 86–94
  nervous system 96, 99, 110–11, 113–16, 117
  phylogenetic frame 5, 9, 10, 11, 12–13, 23, 24, 30
  reproductive organs 240–5, 246, 249, 251–2, 255, 261
  respiratory systems 212
  segmentation 51
  symmetry 38, 39
bilaterian ancestor
  anteroposterior axis 40, 41
  circulatory systems 201
  complexity 35
  epidermis 63, 72
  excretory system 188
  fossil record 36
  gametes 283, 284
  intestinal systems 230
  musculature 86, 90
  nervous system 99, 110–11, 117
  photoreceptors 147
  reproductive organs 240
  respiratory system 214
  segmentation 50, 51
  size 37–8
  symmetry 38–9, 40
bilaterogastraea-hypothesis 38, 40, 220
biogenic amines 112–17
biphasic life cycle 38
bipinnaria 230
Bittium reticulatum 138
Bivalvia
  body cavities 157
  circulatory systems 193
  excretory systems 178
  gametes 271, 272
  intestinal systems 234, 235, 236–7
  musculature 89
  reproductive organs 251
  respiratory systems 204, 205, 215
  sensory organs 122, 129, 137, 138, 145
blastea 230
blastocoel 148, 150, 154, 167

excretory systems 184–5
germ layers 44, 47
blastomere 46
blastopore 43, 44, 47, 221–2
blastula 44
blood 191–2, 201
blood vascular systems (BVS) 191–8
body cavities 2, 148, 154–66
  absent 149
  aceolomate – pseudocoelomate – coelomate succession? 167
  enterocoel hypothesis and coelom ancestry 166–7
  functions 154
  gonocoel hypothesis 167–8
  mixocoel 152–3
  primary 149–51
  secondary 151–2
  types and terms 148–9
body organization 34
  anteroposterior axis 39–42
  complexity 34–5
  dorsoventral axis 42–3
  germ layers 43–7
  locomotory appendages 52
  parasitism 52–3
  segmentation 47–51
  size 35–8
  skeletons 51–2
  symmetry 38–9, 40
Boltenia 280
  B. echinata 142
Bonellia 173, 234, 258, 273
  B. viridis
    circulatory system 194
    complexity 34
    epidermis 67
    excretory system 178–9
    gametes 272, 273
    nervous system 103
    phylogenetic frame 26
bone morphogenetic protein-4 (BMP-4) gene 42
book lungs 207
Botryllus 129, 143
  B. schlosseri 31, 130, 142
Bougainvillia 134
  B. principis 134
Bowman's capsule 186
Brachionus 92, 236
  B. plicatilis 16, 270
Brachiopoda
  body cavities 158–9, 160, 161, 164, 165, 166
  circulatory systems 196, 200
  epidermis 59, 62, 63, 65, 71, 73
  excretory systems 184, 189

Brachiopoda (cont.)
  gametes 277, 278, 286, 290
  intestinal systems 224, 232, 233–5, 237, 239
  misidentification problems 5
  musculature 83, 84, 85, 90, 93
  nervous system 106–7, 110, 111, 112, 115
  phylogenetic frame 12, 13, 27, 28, 29, 30
  reproductive organs 245, 247, 251, 259
  respiratory systems 208, 213, 214, 215
  sensory organs 123, 130
  skeletons 51
Brachyury gene 47
brain 99, 100–2, 105–6, 109, 110, 111, 112, 117
branchiobdellids 273
Branchiopoda 208, 214
Branchiostoma 40, 42, 185
  B. lanceolatum 32, 85, 112
  B. moretonensis 281
brood protection 255
Bryozoa
  body cavities 158–9, 160, 161, 164, 165, 166
  circulatory systems 196, 200
  epidermis 59, 62, 63, 71, 73
  excretory systems 184, 189
  gametes 277, 278, 286, 290
  intestinal systems 224, 226, 227, 232, 233, 234, 237, 238, 239
  musculature 83, 84, 90, 93
  nervous system 106–7, 110, 111, 112, 115
  phylogenetic frame 12, 13, 27, 28–9, 30
  reproductive organs 244, 247, 251, 259
  respiratory systems 213
  sensory organs 123, 141–2, 146
  skeletons 51
Buddenbrockia plumatellae 11
budding 241, 242
Bugula 277, 278
  B. flabellata 28
  B. neritina 142
Bursovaginoida 91, 113, 255, 261
Busycon 51

cadherins 55
Caenogastropoda 205
Caenorhabditis 56
  C. elegans
    anteroposterior axis 40

bilaterian ancestor 12
body cavities 166–7
complexity 34, 35
nervous system 104, 105
phylogenetic frame 5, 6, 15
reproduction 245, 246–7
respiratory system 217
sensory organs 119, 124
Calcarea 8, 9, 70
calcichordate hypothesis 229
caldesmon 85
Callinectes sapidus 52
Calloria inconspicua 184
calponin 85
camels 214
Campanularia flexuosa 264
canal system 171–2, 174–5, 176, 177–8, 179–81, 182, 183, 188
canaliculus cells 182
Cannichthyes 211
Capillaria hepatica 70
Capitella capitata 274
Capitellidae 274
CapZ 84
Cardium 138
Carinoma 22
  C. tremaphoros 156, 161, 231
Carinonemertes epialti 271
cartilage 51
Carybdea marsupialis 133, 134, 263, 264
Caryophyllidea 48
Caspr 56
Catenulida
  excretory systems 173, 175
  gametes 268, 269
  intestinal systems 225
  musculature 88
  nervous system 101
  phylogenetic frame 22–3, 24
  reproductive organs 241, 249, 252
  sensory organs 128, 129, 136
  skeletons 51
caudal organ 255, 261
Caudofoveata
  epidermis 61
  gametes 271, 272
  intestinal systems 236
  musculature 89
  phylogenetic frame 25
  reproductive organs 244
  respiratory systems 204, 205
  sensory organs 122
cell–cell junctions 55–7, 72–3
cement glands 254
central nervous system (CNS) 99, 117
centrioles 70, 262
cephalic sensory organs 120, 135

cephalocarids 195
Cephalochordata *see* Acrania
Cephalodasys maximus 105
Cephalodella hyalina 228
Cephalodiscida 31, 211, 225, 228, 229
Cephalodiscus
  C. gracilis 184
  circulatory system 197
  gametes 279
  musculature 94
  phylogenetic frame 31
  respiratory systems 208, 211
Cephalopoda
  circulatory systems 191, 193, 194, 200
  excretory systems 178
  gametes 271, 272
  intestinal systems 236
  musculature 89, 92
  reproductive organs 244
  respiratory systems 204, 205
  sensory organs 122, 129, 137, 138, 145, 147
  skeletons 51
Cephalothrix rufifrons 271
Cerastoderma 145
  C. edule 138
cercaria 80, 83, 91
cerebral eyes 139, 145, 147
Cerebratulus lacteus 204, 271
Ceriantheopsis americanus 120
Cestoda 48, 53, 218, 242
chaetae 62, 64, 65
chaetoblasts 65
Chaetodermomorpha 272
Chaetognatha
  body cavities 159, 160, 161, 163
  circulatory systems 196, 200
  epidermis 62, 65
  excretory systems 183
  gametes 277, 278
  intestinal systems 224, 237
  musculature 78, 79, 80, 93
  nervous system 115
  phylogenetic frame 5, 30
  reproductive organs 244, 259
  sensory organs 122, 141, 146
Chaetonotida
  excretory systems 174
  gametes 266
  intestinal systems 223, 227
  musculature 83
  reproductive organs 242, 244
Chaetonotus maximus 61
chameleons 80
chaonocytes 246

character coding, phylogenetic analysis 4
Cheilostomata 278
Chelicerata
  circulatory systems 195
  excretory systems 183
  gametes 276, 277, 284
  nervous system 106
  photoreceptors 140
  phylogenetic frame 20, 21
  reproductive organs 248, 250, 251, 259
  respiratory systems 215, 216
  skeleton 51
cheliceres 20
Chelidura acanthopygia 15
chelifers 20–1, 106
chelipeds 52
Chelonia 199
chemical synapses 96
chemoreceptors 119–23
Childia groenlandica 100, 268
Chilopoda
  excretory systems 183
  gametes 277
  photoreceptors 140
  phylogenetic frame 20
  reproductive organs 245
  respiratory systems 207
chitin 63–4
  cuticle 60, 61, 63, 72, 73
chitons 122, 137, 284
Chlamydomonas 131
chlorocruorin 207, 214
choanocytes
  autapomorphies 8–9
  and collar complex, resemblance between 7, 9
  epidermis 71
  gametes 263
  intestinal systems 218
  organization 8–9
Choanoflagellata 7, 9, 58, 71
choline esters 112
Chondrichthyes 199
Chondrostei 211
Chordata
  dorsoventral axis 42–3
  intestinal systems 228, 229
  nervous system 107, 108–9
  photoreceptors 147
  phylogenetic frame 31, 32–3
  respiratory systems 211, 212
  see also Myomerata
chordin (*chd*) gene 42
Chordodes nobilii 121
chordoid larva 51

Chromadorida 135, 267
chromophores 131
chrondrichthyes 280
Chrysaora quinquecirrha 76
cicadas 81
cilia 66–71, 72
  collar complex of choanoflagellates 7
  excretory systems 171–2, 173, 174, 175–90
  gametes 262, 264–6, 268–73, 276–7, 279–81, 285
  intestinal systems 229, 231–8
  sensory organs 118, 120–5, 134, 135, 137–46, 147
  striated rootlets 9, 70–1, 72
ciliated urns 157
Ciona intestinalis 142, 143, 147, 209, 215
circomyarian muscle cells 78, 79
circulatory systems 191–201
cirratulids 206
cirri 68, 253
Cirripedia 34, 38
Cladistia 210, 211
Cladocera 242
cladorhizids 9
claudins 55, 56
Clitellata
  body cavities 158
  circulatory systems 195
  excretory systems 182
  gametes 273, 274, 286, 289
  musculature 86, 89
  nervous system 103
  reproductive organs 244
  segmentation 50
  sensory organs 130, 139, 145
clitellum 65
closed circulatory system 191–201
Clymenura clypeata 181
Cnemidophorus 242
Cnidaria
  anteroposterior axis 41, 42
  biphasic life cycle 38
  body cavities 166
  circulatory systems 192
  complexity 34
  epidermis 56, 57, 58, 59, 60, 63, 71, 73
  excretion 169
  gametes 263, 264, 282, 286, 287, 292
  germ layers 43, 44, 45, 46, 47
  intestinal systems 219, 223, 225, 232, 234, 235, 236
  musculature 75–6, 81, 82, 84, 85, 87, 90, 91

  nervous system 96–9, 107, 112
  phylogenetic frame 7, 8, 9, 10, 11
  reproductive structures 240, 241, 244, 246, 247, 249, 250, 256
  sensory structures 118, 120, 125, 126–7, 131, 132, 133–4, 145, 146
  size changes 36
  symmetry 38, 39
cnidocil 120
cnidocytes 235
Codosiga botrytis 7
Coelenterata 9, 10
coeloblastula 44, 45
coelom 2, 148, 149, 151–2, 154–68
  excretory systems 171
  reproductive organs 251, 252
Coelomata 148, 166–7
coelomic circulatory systems 191, 199
coelomocytes 154, 155, 184, 212
coelomyarian muscle cells 78, 79
coelothel 151, 152, 157, 159, 163
Coleoida 138, 145, 205
collagen 57, 58, 72, 73
collar 108, 109
collar complex 7, 9
collar receptors 67, 120, 123, 124
collemboles 277
colloblasts 235
colonies 241
commisures 102, 104, 105, 106, 110
comparative morphology 3, 293
compartmenting lamella 68
compensation sac 194
complexity 34–5
compound eyes 140, 145
Concentricycloidea 278–9, 284, 291
Conchifera 61, 224
conchin 61
connectives 104
connexons 56–7
Conophoralia 255, 261, 269, 270, 288
contamination problems, phylogenetic analysis 5
Convoluta 100
  C. convoluta 134, 236
  C. pulchra 51, 83, 87, 91
  C. retrogemma 241
Convolutriloba longifissura 241
Copepoda 53
Cope's rule 35
copulation 253, 254, 261
  see also reproductive organs
Corallochytrea 7
corals 60
corellids 185
Corymorpha palma 126

Corynesoma 156
Cothurnocystis 31, 212, 229
Crania anomala 30, 277, 278
Craniacea 234
Craniota
    anteroposterior axis 40
    body cavities 160, 162, 163, 164, 165
    circulatory systems 191, 199–200, 201
    complexity 34, 35
    dorsoventral axis 42
    epidermis 55, 56, 73
    excretory systems 186, 189, 190
    gametes 280–2, 286, 291
    intestinal systems 225, 228, 232, 237
    locomotory appendages 52
    musculature 80, 81, 85, 90, 94
    nervous system 109, 110, 116, 117
    phylogenetic frame 30, 31, 32, 33
    reproductive organs 243, 245, 247, 248, 260
    respiratory systems 209–11, 212, 213–14, 215
    segmentation 50
    sensory organs 123, 130–1, 138, 144–5, 146, 147
    size 35
    skeletons 51
Crepidula fornicata 204
Crinoida
    circulatory systems 198
    epidermis 59, 62
    excretory system 185
    gametes 278, 279
    intestinal systems 224, 230, 231, 237
    musculature 93
    nervous system 108
    respiratory systems 208
Crisia 93
Cristidiscoidea 7
Crocodylia 199
cross-striated muscle 77, 78, 80, 81, 82–4, 90, 91–4
    molecular components 84, 85
crown cells (in protonephridia) 178
Crustacea
    circulatory systems 195, 196
    excretory systems 183
    gametes 276, 277, 292
    general body organization 53
    musculature 80
    photoreceptors 132, 140
    phylogenetic frame 20, 21
    reproductive organs 242, 245, 248, 259
    respiratory systems 207, 208, 215, 216
    symmetry 39

cryptobiosis 217
Cryptochelides loweni 269
Cryptochiton stelleri 271, 284
Cryptomonas 236
Cryptosula 93
crystallin 96
ctenidia 204–5
Cteno-Hox1 gene 41
Ctenophora
    anteroposterior axis 41–2
    epidermis 59, 67, 68–9, 70, 71, 73
    excretion 169
    gametes 263, 286, 287, 292
    germ layers 43, 44, 46
    intestinal systems 219, 221, 223, 232, 235, 236
    musculature 75, 81, 82, 90, 91
    nervous system 96–9
    phylogenetic frame 7, 8, 9, 10
    reproductive structures 244, 247, 249, 250, 256
    sensory structures 120, 127, 128, 134, 145
    symmetry 38
cubopolyps 82, 91
Cubozoa
    gametes 263, 264
    musculature 75
    nervous system 97
    phylogenetic frame 10
    sensory structures 126, 127, 133, 134, 145
Cucumaria
    C. lubrica 279
    C. pseudocurata 279
cuticle 51, 52, 54, 58–63, 64, 67–8, 72, 73
    intestinal systems 227
    special structure 64–5
cuticular ciliary receptors 121, 122, 124–5
Cyanea capillata 97
Cyanobacteria 58
Cycliophora
    excretory systems 177
    nervous system 114
    phylogenetic frame 26, 27
    reproductive organs 244
    sensory organs 121
    skeletons 51
Cycloneuralia
    body cavities 163, 167
    epidermis 60–1, 63, 64, 67–8
    gametes 284
    intestinal systems 221, 227, 233
    musculature 87
    nervous system 104–6, 110, 111
    phylogenetic frame 13–15, 18

reproductive organs 245
sensory organs 125
size 37
skeletons 51
Cyclorhagida 223
Cyclostomata 93, 186, 210, 278, 280
cydippids 97, 98
cyrtopodocytes 185, 186
Cyrtotreta 31
Cystiplex axi 51
cytochemical advances 3
cytoglobin 211, 213–14, 215
cytoplasmic rods 174
cytoskeleton 74

Dactylopodola 83, 261
    D. baltica
        excretory system 174
        gametes 265–6
        intestinal system 226
        musculature 83
        photoreceptors 135
        reproduction 261
    D. typhle 261
Dalyellioida 72, 128
decapentaplegic (*dpp*) gene 42
Decapoda 39, 130
Demospongia
    epidermis 59, 70, 71
    gametes 263
    germ layers 44
    metabotropic glutamate receptor 96
    photoreception 132
    phylogenetic frame 8, 9
dendrites 95
Dendrochirotida 209
Dendronereides heteropoda 206, 207
Dendrostomum 206
dense bodies 271
Deontostoma californicum 79
Dermaptera 15
dermonephridia 173
Deroceras reticulatum 246
Derocheilocaris 227
Desmarella moniliformis 7
desmosomes 55, 56, 57, 73
Desulphovibrio vulgaris 215
Deuterostomia
    body cavities 166
    circulatory systems 201
    dorsoventral axis 42, 43
    epidermis 63
    excretory systems 188, 190
    intestinal systems 221, 228–9, 230, 233–5, 239
    musculature 89

INDEX 371

nervous system 107–9, 110, 111, 112
photoreceptors 146
phylogenetic frame 12, 13, 22, 28, 29, 30–3
reproductive organs 248
respiratory systems 215
size 38
deuterostomy 221–2
developmental biology 3
Dibranchiata 205
Dictyostelium 40
  D. discoideum 74
Dicyemida 10, 11, 53, 264
Digenea 23
  circulatory systems 192
  intestinal systems 222
  photoreceptors 136
  reproductive organs 242, 243
Dinophilidae 36
Dinophilus 36, 37
Diopatra aciculata 227
Dipetalonema viteae 124
dipleurula 230, 231, 232, 233
diploblasts
  body cavities 154
  excretion 172
  gametes 283
  germ layers 43, 44, 46
  musculature 75–7, 81
  nervous system 109, 117
  phylogenetic frame 7–10
  reproductive structures 240, 246
  respiration 217
Diplocirrus glaucus 158
Diplolaimella 135
diplopods 80, 215, 216
Diplura 277
Dipnoi 199, 210, 211, 280
Discinacea 234
Discinisca 62
Discs large 56
distal centrioles 262
distal-less gene 52
distance methods, molecular analysis 3
Distaplia occidentalis 142
DNA
  automated sequencing 3
  long-branch problems 5
  whole genome sequencing 5
doliolaria 230, 231
doliolids 94, 130
Dollo's rule 4
dopamine 112, 117
dorsal strand 108
dorsal strand plexus 108
dorsal vessel 194–5

dorsoventral axis 42–3
Dorvillea bermudensis 243
Dorvilleida 36, 37
dpp gene 42
Draculiciteria 82
Drepanophorida 102
Drepanophorus 104
Drosophila
  bilaterian ancestor 12
  body cavities 167
  circulatory systems 201
  complexity 34, 35
  dorsoventral axis 42
  epidermis 56
  eyeless gene 146
  regaining of wingspots 4
  reproduction 246, 247
Dugesia polychroa 225
dyneins 74, 75

earthworms 86
Ecdysozoa
  body cavities 166
  epidermis 61, 63, 64, 68
  intestinal systems 221, 227
  nervous system 106, 110
  phylogenetic frame 13, 17–18, 19
  segmentation 51
Echiniscidae 275
Echiniscoides 275
Echinococcus 242
Echinoderes aquilonius 104, 174
Echinodermata
  body cavities 149, 159–60, 162, 164, 165, 166, 168
  circulatory system 198, 200
  dorsoventral axis 42, 43
  epidermis 59, 62, 63, 70, 71, 73
  excretory systems 185, 189, 190
  gametes 277–9, 282, 284, 286, 291, 292
  intestinal systems 224, 228, 229, 230, 231, 232, 233, 234, 235, 237
  locomotory appendages 52
  musculature 85, 89–90, 93
  nervous system 107–8, 109, 110, 112, 113, 115
  phylogenetic frame 12, 28, 30, 31, 32, 33
  reproductive organs 250, 251
  respiratory systems 208–9, 212, 213, 214
  sensory organs 123, 130, 132, 142, 146
  size 38
  skeletons 51
  symmetry 38

Echinoida
  circulatory systems 198
  epidermis 59, 62
  gametes 278, 279
  intestinal systems 224, 230, 237
  nervous system 107
  photoreceptors 142
  respiratory systems 208
Echinus 112
Echiurida
  body cavities 157, 160, 161, 164, 165, 166
  circulatory systems 194, 200
  complexity 34
  epidermis 61, 64, 65, 67, 73
  excretory systems 173, 178, 189, 190
  gametes 272–3, 286, 289
  intestinal systems 224, 230, 232, 234, 237
  musculature 86, 89, 90, 92
  nervous system 103, 114
  phylogenetic frame 24, 25–6, 27
  reproductive organs 244, 247, 258
  respiratory systems 206, 213, 214
  sensory organs 122
Echiurus 179
  E. abyssalis 103, 179
  E. echiurus 103, 273
ect-aquasperm 282
ectoderm 43, 44, 46, 47
  intestinal systems 218, 223, 225, 227
  reproductive organs 246
ectomesoderm 44–5, 46
ectoneural nervous system 107–8
Ectoprocta *see* Bryozoa
Edwardsia 263
Eimeria 58
Eirene viridula 126
Eisenia foetida 85, 139
Elasipoda 209
Electra 93
electrical signalling 95
electrical synapses 96
electroreceptors 118, 119
elongation-factor-1α gene (EF1α-gene) 24, 26
Encentrum marinum 176
enchytraeids 103
Enchytraeus crypticus 49, 103
endocytosis 218, 239
endoderm 43, 44, 46
  intestinal systems 218, 223
  reproductive organs 246
endomesoderm 44–5, 46
endoskeleton 51
endosymbiosis 74
engrailed gene 50

Enopla 224, 231, 234
Enoplida 135, 154, 267
Enoplus
  E. brevis 203
  E. communis 203
  E. demani 267
Ensis directus 157
ent-aquasperm 282
enterocoely 44, 151, 163, 164, 166
Enteropneusta
  body cavities 149, 159, 160, 162, 164, 165
  circulatory systems 196–8
  dorsoventral axis 42–3
  epidermis 59, 65, 66, 68, 69–70, 71, 73
  excretory systems 184–5, 189
  gametes 279, 280, 286, 291
  intestinal systems 225, 228, 230, 231, 232, 234, 237
  musculature 89–90
  musculature 90, 93
  nervous system 108, 109, 116
  phylogenetic frame 28, 30–2, 33
  reproductive organs 245, 247, 259
  respiratory systems 208, 211, 213
  sensory organs 123, 142, 146
  size 38
Entobdella solae 136
entocodon 46, 47
entocodon cavity 46, 47
Entocolax 3
Entoconchidae 53
entoneural nervous system 108
Entoprocta see Kamptozoa
Eoacanthocephala 89, 156
Ephydatia muelleri 75, 216
epicuticle 60, 61, 68
epidermal cells 54, 65–71
epidermis 54–5, 72–3
  cell–cell junctions 55–7
  chitin 63–4
  epidermal cell 65–71
  extracellular matrix 57–8
  glycocalyx and cuticle 58–63
  moulting 64
  special cuticular structures 64–5
  syncytial 71–2
epigenesis 246, 248, 261
Epimenia australis 271, 283
Epiperipatus
  E. biolleyi 153
  E. imthurni 245
Epiphanes senta 270
epithelial–mesenchymal transition (EMT) 47
epithelio-muscle cells (EMCs) 75–7, 87, 90, 94

Epithelioza
  epidermis 57, 70, 71, 73
  phylogenetic frame 9, 10
epithelium 54–5
  body cavities 148, 149, 151, 154
  intestinal systems 220
  reproductive organs 249–52
Erpobdella octoculata 182, 274
erythrocytes 155, 157, 212
Euacoela 220, 221, 223
Euarthropoda
  body cavities 152, 153, 155–6, 160, 161, 164, 165, 166
  circulatory systems 195–6, 200, 201
  epidermis 64, 73
  excretory systems 172, 183, 189
  gametes 275–7, 286, 290
  intestinal systems 226, 227, 232, 237
  locomotory appendages 52
  musculature 78, 80, 81, 83, 84, 85, 88, 90, 93
  nervous system 106, 115
  phylogenetic frame 18, 19, 20–1
  reproductive organs 245, 247, 248, 259
  respiratory systems 207–8, 213, 214, 216
  sensory organs 122, 124, 129, 130, 140–1, 145
  skeletons 51, 52
eubacteria 131
Eubilateria 220
Euborlasia 204
Eucestoda 48, 288
Euclymene oerstedii 56
eudrilids 273
euglenids 132
eukaryotes 59, 67, 74
Eukrohnia 141
Eulalia 180, 188
  E. viridis 158, 181
Eumetazoa
  epidermis 55, 57, 73
  intestinal systems 219, 220, 230
  nervous system 95, 96, 104, 117
  phylogenetic frame 9–10
  sensory organs 120, 132–46
Eumollusca 178, 224, 258
Eunicida 139, 227
Euplectella aspergillum 7
Euplokamis 82, 91
Eurotatoria 16, 17, 36, 270, 288
Eurotifera
  body cavities 156, 160, 164, 165
  circulatory systems 192
  complexity 34
  epidermis 73
  excretory systems 173, 176, 189

  gametes 286
  intestinal systems 224, 227, 228, 231, 232, 236
  musculature 83, 90, 92
  nervous system 101, 114
  phylogenetic frame 27
  reproductive organs 242, 244, 247, 255, 257
  respiratory systems 213
  sensory organs 121, 136, 145
  size 37
eurypterids 207
eusperm 273
Euspiralia 22, 221
Eutardigrada
  excretory systems 182
  gametes 275, 276, 286
  intestinal systems 224
  musculature 93
  photoreceptors 140
  reproductive organs 258
Eutrochozoa 22
Euzonus mucronata 217
even-skipped gene 50
eversion of introverts 88, 154
everse eyes 132, 133, 134, 139, 140, 144, 145–6, 147
excretory systems 169–72, 188–90
  Bilateria 172–87
  nephridia evolution models 187–8
Exogone 137
exopinacocytes 55
exoskeleton 51, 52
expressed sequence tags (EST) approaches 5, 293
expression patterns of genes 3, 293
extracellular matrix (ECM) 54, 55, 57–8, 72, 73
  body cavities 148, 149–53, 154, 163, 165, 167
  circulatory systems 191, 192, 194, 195, 198
  excretory systems 170, 171–2, 177, 179, 186–7, 188
  musculature 86, 94
  nervous system 101
  reproductive organs 246, 249
  skeletons 51
eyeless gene 146
eyes 132
  see also photoreceptors

Fabricia sabella 152
Faerlea glomerata 100, 101
feeding
  in adults 235–9
  in larvae 229–35

fertilization 253–5, 261, 262–92
fibre muscle cells 75–7, 90, 94
fibronectin 57, 58, 72, 73
filaments 55
filiform spermatozoa 262, 265–6, 270–7, 280, 283, 286, 292
Filina longiseta 136
Filospermoida
   gametes 269, 270, 286, 288
   musculature 91
   nervous system 113
   reproductive organs 255, 261
filter feeding 67, 235–9
Fissurellidae 204
Flabelligeridae 139, 207
flagella 262
flatworms
   epidermis 65, 67, 69–70, 71
   musculature 80, 83, 85, 87, 88
   nervous system 101, 110–11
   phylogenetic frame 22–4
   segmentation 48, 49, 50
   skeletons 51
   see also Platyhelminthes
Flometra serratissima 279
flosculi 121
Flustrellidra 93
   F. hispida 28, 239
focal adhaerens junctions 57, 72–3
focal adhesions 55
follicular epithelium 249
food groves 204
Forkhead transcription factor 47
Fortiforceps 20, 106
fossils
   Arthropoda 20, 21
   Cycloneuralia 15
   deuterostomes 31
   incomplete fossil record 6
   locomotory appendages 52
   phylogenetic frame 3, 15, 20, 21
   respiratory systems 217
   size bias 35–6
   tentaculate taxa 29, 30
free nerve endings 121
Frenulata 25
frogs 280–1
frontal organ 255
FtsZ 74
fungi 58, 63, 71
funiculus 196

gametes 262–82, 287–92
   acrosome evolution 292
   correlation between sperm type and fertilization mode 282–5

function and phylogeny of derived spermatozoa 285–6
   origin of 246–9, 256–60
   transfer and receiving 253–61
gametogenic areas 249
γ-aminobutyric acid (GABA) 112, 117
Gammarus pulex 195
ganglia 99, 101, 102, 103, 105, 106, 109
Ganglioneura 89, 102
ganthiferans 29
gap junctions 56–7, 73
gastraea 230
Gastromermis 267
Gastroneuralia *see* Protostomia
Gastropoda
   body cavities 157
   circulatory systems 193
   complexity 34
   excretory systems 178
   gametes 272
   general body organization 53
   intestinal systems 234, 235, 236
   musculature 89, 92
   reproductive organs 251
   respiratory systems 204, 205–6, 212
   sensory organs 122, 126, 129, 133, 137, 138, 145
   skeletons 51
Gastrotricha
   body cavities 154, 160, 164, 165
   epidermis 61, 67, 68, 71, 73
   excretory systems 173, 174, 188, 189
   gametes 262, 265–6, 285, 286, 287
   intestinal systems 223, 226, 227, 231, 232, 233, 236
   musculature 82, 83, 87, 90, 91
   nervous system 104–6, 110, 111, 112, 113
   phylogenetic frame 13, 15–16, 22
   reproductive organs 242, 244, 245, 247, 252, 253, 255–61
   respiratory systems 202, 213, 214
   sensory organs 120, 135, 145
   size 36, 38
   skeletons 51
gastrula 221, 222
gastrulation 44
general body organization *see* body organization
genes
   evolution 293
   expression patterns 3, 293
   multigene analyses 5–6
   resolution, phylogenetic analysis 5
   targeted search for 3

genetic complexity 34–5
genetic parsimony 34
Genitoconia atriolonga 102
genomic data 3
Geodia cydonium 96
germarium 254
germ cells 251–2, 261
germinal epithelium 249–50
germ layers 43–7
Gigantorhynchus echinodiscus 177
gills 202, 204–5, 206, 207–8, 209–12
gill slits in deuterostomes 228–9
Ginglymodi 210, 211
Glockenkern 46
Glomeris margina 80
glomerulus 186
Glossobalanus sarniensis 280
Glottidia 62
   G. pyramidata 29, 233
glutamate receptors 96
Glycera tridactyla 68
glycerids 195
glycocalyx 54, 58–63, 72, 73
Gnathifera
   body cavities 166, 167
   epidermis 72
   intestinal systems 227–8
   musculature 88
   nervous system 101
   phylogenetic frame 13, 17, 21–2, 27
   reproductive organs 243
Gnathostomata 186, 225, 237
Gnathostomula paradoxa 16, 176, 269
Gnathostomulida
   body cavities 154, 160, 164, 165
   epidermis 59, 71, 73
   excretory systems 173, 176, 188, 189
   gametes 269, 270, 285, 286, 288
   intestinal systems 220–1, 223, 227, 231, 232, 236
   musculature 78, 81, 88, 90, 91
   nervous system 101, 113
   phylogenetic frame 5, 16, 17, 22, 23
   reproductive organs 244, 246, 247, 252, 253, 255, 257, 261
   respiratory systems 202, 213, 216
   sensory organs 121
   size 36, 38
Golfingia
   G. gouldi 273
   G. ikedai 272, 273
   G. minuta 157, 244
gonadal stem cells 246
gonads 149, 167–8
   evolution of 249–53, 256–60
Goniadides falcigera 225
gonochorism 244–6, 247, 261

gonocoel hypothesis 149, 167–8, 187, 252
Gordiida 87, 104, 256, 267, 268
Gordius 155
   G. aquaticus 267
Gorgonacea 60
gorgonin 60
Grania americana 130
green algae 58
ground pattern reconstruction, phylogenetic analysis 4
Gryllus firmus 156
guanaco 214
guanine 169, 183
guanocytes 183
Gunda segmentata 49
Gymnolaemata
   gametes 277, 290
   intestinal systems 233, 239
   phylogenetic frame 28, 29
Gymnophione 210, 281
gynogenesis 243

hagfishes 280
hairy gene 50
Halammohydra 36, 126
   H. schulzei 264
Halecomorphi 210, 211
Halicryptus 268
   H. higginsi 14
   H. spinulosus 175, 203
Haliotis 138, 204
Halisarcida 44
Halkieriids 30
Hallucigenia 20
Halobacterium salinarum 131
Halobiotus crispae 140, 155
Halycryptus spinulosis 216, 217
Hamingia 273
   H. arctica 273
Haplognathia 221
   H. rosea 176, 202, 269
Haplogonaria 51
Harmothoe 180, 188
   H. sarsi 181
Harrimaniidae 31
Hautmuskelschlauch (HMS) 86–94
head kidneys 179
heart 193, 195–201
heat shock protein 7, 70
helicoidal muscle *see* obliquely striated muscle
Heligmosoides polygyrus 267
Helisoma trivolvis 206
Helix 133, 200
   H. aspersa 85
   H. pomatia 138, 193

Helobdella triserialis 50
hemal system 196, 200
hemerythrins 203, 206, 207, 208, 212, 213, 214, 215
hemerythrocytes 206
Hemichordata
   body cavities 159–60, 166, 168
   circulatory systems 196–8, 200, 201
   dorsoventral axis 43
   excretory systems 184, 185, 190
   intestinal systems 228–9, 233, 235
   nervous system 107, 108, 110, 112
   phylogenetic frame 12, 31–2, 33
   reproductive organs 241
   respiratory systems 208
hemidesmosomes 55
Hemipholis elongata 209
Hemirotifera 17
hemocyanins 205–6, 207, 208, 209, 212, 213, 214, 215–16
hemoglobins 202–8, 209, 211, 212–15, 217
hemolymph 191–201
Hennig, Willi 2, 3
hermaphroditism 244–6, 247, 250, 251, 261
Hesionides 51, 68
Hesse ocelli 144, 146
Heterocyemida 10
heterogony 242, 243
Heteronemertini 231, 234, 271
Heteroptera 137
heterosarcomeric muscle 79
Heterotardigrada 122, 140, 258, 275, 276
Heteroxenotrichula squamosa 266
Hexacorallia 263
Hexactinellida 7
   epidermis 70, 72
   nervous system 95
   phylogenetic frame 8, 9
Hirudinea 79, 80–1, 227, 258, 274
Hirudo medicinalis 64, 182
histamine 112
Holochordata 32
   see also Myomerata
Holothuroida
   circulatory systems 198
   epidermis 62, 63
   excretory system 185
   gametes 278, 279
   intestinal systems 224, 230, 231, 235, 237
   locomotory appendages 52
   musculature 81, 86, 89–90, 93

nervous system 107
photoreceptors 146
reproductive organs 245, 248, 259
respiratory systems 208–9
sensory organs 130, 142
Homalorhagida 223, 227, 267, 268
Homalozoa 31, 212
Homarus gammarus 83
Homeobox genes 39–40, 41, 101
Homoscleromorpha 263, 264
hooks 65
hoplonemerteans 271
Hormiphora 97
Hox cluster 40, 41
Hox gene
   Acoela 24
   anteroposterior axis 39–42
   Arthropoda 20
   Bilateria 12–13
   Chaetognatha 30
   deuterostomes 32
   'Mesozoa' 11
   Myxozoa 11
humans 74, 186, 215
Hutchinsoniella macracantha 195
hydatids 242
Hydractinia echinata 97
Hydra 56, 58
   H. echinata 44
   H. vulgaris 57
hydromedusae 82, 84, 91, 131
hydropolyps 71, 126
hydropores 184
hydroskeleton 51–2, 154, 164, 166
Hydrozoa
   epidermis 57, 58
   gametes 263, 264
   germ layers 44, 46, 47
   musculature 75, 82
   nervous system 97, 111
   reproductive organs 246, 256
   sensory organs 126, 127, 133, 134, 145
Hymenoptera 245
hypodermal injection 253
hyponeural nervous system 107–8
HZO-1 56

Ikeda 89
   I. taenioides 179
Ikedosoma 273
immunocytochemical advances 3
Inarticulata 278
incomplete fossil record 6
inner microvilli 171

Insecta
  body cavities 156
  complexity 35
  excretory systems 183
  gametes 277, 285
  intestinal systems 224
  photoreceptors 137, 140
  phylogenetic frame 19, 21
  reproductive organs 242, 246, 250, 252
  respiratory systems 207, 208, 214, 215, 216
  size 35
integrins 57, 58
internal ciliary receptors 120
interstitia 150–1
interstitial cells (I-cells) 246
intestinal systems 218, 239
  endocytosis 218
  feeding in adults 235–9
  feeding in larvae 229–35
  gill slits in deuterostomes 228–9
  origin of intestinal tract 218–20
  pharyngeal hard structures 227–8
  pharynges 225–7
  from sac-shaped intestine to one-way gut 220–5
introns 35
introsperm 282
introvert 88
inverse eyes 132, 133, 134, 139, 144, 145–6, 147
Isopoda 130, 292
Isostichopus badionotus 81
Ixodes ricinus 21

japygids 277
Joseph cells 144, 146

Kalyptorhynchia 72, 269
Kamptozoa
  body cavities 157, 160, 164, 165
  circulatory systems 194, 200
  epidermis 59, 61, 63, 64, 68, 73
  excretory systems 173, 178, 189
  gametes 271–2, 273, 286, 289
  intestinal systems 224, 230, 232, 234, 236, 238
  musculature 89, 90, 92
  nervous system 114
  phylogenetic frame 24, 25, 26, 27, 28, 29
  reproductive organs 244, 247, 258
  respiratory systems 213
  sensory organs 122, 137, 145
Katherina tunicata 29
Keratella 236

Kerygmachela kierkegaardi 207
kettin 85
kidney 178, 186
Kiemenplatten 207
kinesins 74, 75
Kinorhyncha
  body cavities 155, 160, 163, 164, 165
  epidermis 60, 64, 73
  excretory systems 173, 174–5, 189
  gametes 267, 268, 286, 288
  intestinal systems 223, 226, 227, 231, 232, 236
  musculature 78, 80, 88, 90, 91
  nervous system 104, 105, 113
  phylogenetic frame 13, 15
  reproductive organs 244, 247, 257
  respiratory systems 213
  segmentation 48, 49
  sensory organs 121, 135, 145
  size 36, 38
  skeletons 51
Kinorhynchus phyllotropis 268
Kölliker's pit 108
Komodo dragon 242

Lacertilia 143
Lacuna divaricata 138
Lacunifera 63
Laevipilina antarctica 271, 272
lamellar cells 144
lamellibranch bivalves 84, 92, 178
lamellibranch gills 204
lamina densa 57
  see also basal lamina
lamina fibroreticularis 57, 149, 150, 192
lamina lucida 57
laminin 57, 72, 73
Lampetra 143
  L. fluviatilis 210, 281
lampreys 280
Lanice conchilega 130
Laomedea geniculata 39
larvae
  feeding in 229–35
  nervous system 111–12
Lecithoepitheliata 269
lecithotrophy 229, 233–5, 239
leeches
  body cavities 158
  excretory systems 182
  gametes 273
  intestinal systems 226
  musculature 87
  photoreceptors 139
Lepidochitona cinereus 178, 193
Lepidodasys 83
Lepidosauria 199

Lepisosteus 210, 211
Leptomedusae 126, 127
Leptosomatum acephalatum 79
Leptosynapta clarki 279
Lernaeocera branchialis 53
Lethal giant larvae 56
Leuckartiara octona 134
likelihood-based methods 3
Limax flavus 178
Limnognathia (*L. maersui*)
  body cavities 154, 160, 164, 165
  epidermis 59, 72, 73
  excretory system 173, 176, 189
  gametes 286
  intestinal system 221, 223, 227, 231, 232, 236
  musculature 88, 90, 92
  nervous system 101, 113
  photoreceptors 136
  phylogenetic frame 17
  reproduction 247
  reproductive organs 244, 257
  respiratory systems 213
  sensory organs 121
Limulus 51, 183
Lineus 133, 244
  L. ruber 66, 67, 136
  L. viridis 157, 177, 178
Lingulacea 234
Lingula 62
  L. anatina 123, 130, 184
lingulids 84, 93
'Linnean' hierarchical levels 6
Listeriolobus 273
  L. pelodes 273
lithocytes 125, 128, 130
Littorina 138
lama 214
lobatocerebrids 68
lobopodia 52
locomotory appendages 52
Loligo
  L. forbesii 39
  L. vulgaris 80
long-branch problems 5
Lopadorhynchus 103
'lophoenteropneusts' 32
lophophor 239
Lophotrochozoa *see* Spiralia
lorica 51
Loricifera
  body cavities 155, 160, 163, 164, 165
  epidermis 60, 64, 73
  excretory systems 173, 174, 189
  gametes 268, 286, 288
  intestinal systems 223, 226, 227, 231, 232, 236

Loricifera (*cont.*)
  musculature 88, 90, 91
  nervous system 104–5, 113
  phylogenetic frame 13, 14–15
  reproductive organs 244, 247, 257
  respiratory systems 213
  sensory organs 121
  size 36, 38
  skeletons 51
low oxygen content, adaptations to 214–15, 216–17
Lox-5 11
Loxosomella
  L. atkinsae 26, 61
  L. fauveli 178
  L. harmeri 137
Lumbricus 195
  L. terrestris 158
lungfishes 280
lungs 202, 205, 210–11
Lurus 128
Lymnaea 138
  L. stagnalis 178
lymphatic system 192
lysosomes 218
Lytechinus variegatus 279

Macracanthorhynchus 192
  M. hirudinaeus 89, 94, 177
Macrobiotus hufelandi 275
macrocilia 68–9
Macrodasyida
  excretory systems 174
  gametes 265, 266, 286, 287
  intestinal systems 223, 227
  musculature 83
  reproductive organs 255
Macrodasys 51, 261, 266
macrofauna 35
Macrostomida 136, 176, 225, 241, 269
madreporite 185
Magelona 151, 214, 215
  M. mirabilis 71, 179, 188
  M. papillicornis 207, 274
magnetoreceptors 119–20
major-sperm protein (MSP) 267, 285
Malacoceros fuliginosus 206, 207
Malacostraca 21, 224
malpighian tubules 182, 183
Mammalia 199–200, 248, 281
Mandibulata 21
manubrium 46
Marirhynchus longaseta 72
Markstrang 99, 106
Markuelia 15, 35
Marsupiomonas pelliculata 7
Marthasterias 242

Maxmuelleria lankesteri 122
Mbl 74
Meara 87
M. stichopi 100, 265
mechanoreceptors 118, 119, 120–3, 124
median eyes 140, 145–6
medusae
  circulatory systems 192
  germ layers 46, 47
  musculature 82
  photoreceptors 133
  reproduction 240, 241, 249
Medusozoa *see* Tesserazoa
megafauna 35
meiofauna 35
Meiopriapulus
  body cavities 155
  gametes 268, 284, 288
  M. fijiensis
    body cavities 155, 160, 161, 165
    excretory system 175
    gametes 268, 282
    phylogenetic frame 14
Membranipora 93
merism 48
  see also segmentation
Mermis nigrescens 203
Mermithida 203, 267
mesenchymal mode 45
mesentoblast 46
mesocoel 166
Mesodasys 266
  M. laticaudatus 174, 266
mesoderm 43, 44–6, 47
mesodermal stripe 167
mesogloea 246, 249
Mesomycetozoea 7
Mesozoa 10, 264
Metabranchia 204, 205
metacoel 166
metagenesis 240, 241
metamery 48
  see also segmentation
metanephridia 166
  excretory systems 172, 178, 179, 181, 182, 184, 187, 188, 190
metanephridial system 170, 172, 173, 178–81, 183, 184, 186, 189, 190
metatroch 230
Metazoa
  body cavities 150
  circulatory systems 192–200
  complexity 34
  epidermis 54, 56, 57, 58, 59, 61, 62–3, 64, 65, 67
  gametes 269, 283
  germ layers 46

intestinal systems 218, 220, 235, 239
locomotory appendages 52
musculature 74, 75, 94
nervous system 117
reproductive organs 249, 253
respiratory systems 202–11
segmentation 47
sensory organs 120–5, 131–2, 147
size 37–8
symmetry 38
metazoan ancestor
  epidermis 57, 59, 63, 71, 72, 73
  intestinal systems 230
  musculature 74, 75
  reproductive organs 240, 249
Methridium senile 76
Mickwitzia 30
Micrina 30
microfauna 35
microfilaments 55
Micrognathozoa 17
Microphthalmus 68, 226, 227
  M. hamosus 36
Micropilina arntzi 272
microtubule-organizing centrioles (MTOC) 262
microtubules
  epidermis 67, 68–70
  gametes 262, 264, 265, 268, 269, 271, 276, 281, 282
microvilli
  epidermis 59, 61–2, 63, 65, 66, 67
  excretory systems 171, 173, 174, 175–90
  sensory organs 118, 135, 137–40, 142–6, 147
Micrura
  M. fasciolata 271
  M. leidyi 161
Milnesium tardigradum 88, 140
miniaturization 36
Ministeriida 7
Minona trigonopora 99
misidentification problems, phylogenetic analysis 5
mitochondrial genome 7, 9
mixocoel 148, 152–3, 155–6, 166, 182
Mnemiopsis leidyi 8
molecular biology
  and morphology, bridging the gaps between 293
  phylogenetic frame 3, 4–5
molecular components 293
molecular data, phylogenetic analysis 4–5
molgulids 185

Mollusca
  body cavities 150, 157, 160, 161, 164, 165, 166, 168
  circulatory systems 193–4, 200, 201
  epidermis 61, 63, 64, 65, 68, 73
  excretory systems 173, 178, 189, 190
  gametes 271, 272, 282, 283, 286, 289
  intestinal systems 224, 227, 230, 232, 234, 235, 236–7
  musculature 83, 84, 89, 90, 92
  nervous system 102, 114
  phylogenetic frame 24, 25, 26, 27, 28, 29–30
  reproductive organs 244, 247, 258
  respiratory systems 204–6, 211, 213, 214, 215, 216
  segmentation 48, 50
  sensory organs 122, 128–30, 132, 137–8, 145
  size changes 36
Molpadida 209, 237
Monhysterida 135
Monilifera 25
Moniliformis moniliformis 17
monogeneans 136
Monogononta
  gametes 270, 288
  intestinal systems 236
  photoreceptors 136
  phylogenetic frame 16, 17
  reproductive organs 242, 243, 244, 257
monophyly of Metazoa 6–7
Monoplacophora
  circulatory systems 193
  excretory systems 178
  gametes 272
  musculature 89
  nervous system 102
  phylogenetic frame 25
  reproductive organs 244
  sensory organs 128
Monosiga 7
Mopalia 24
Mormyridae 280
morphological data, phylogenetic analysis 4–5
morphology
  and molecular biology, bridging the gaps between 293
  phylogenetic frame 3, 4–5
morula 45
motor-proteins 74
moults 18, 52, 64, 73
mouse 12, 146
MreB 74

Muggiaea 287
  M. kochi 263, 264, 292
multigene analyses 5–6
Multipeniata 285
Multitubulatina 266
musculature 74, 94, 293
  actin and myosin as 'old' molecules 74–5
  antagonistic systems 85–6
  bilaterians 86–94
  diploblasts 75–7
  molecular components other than actin and myosin 84–5
  myofilament arrangement patterns 77–80
  striation patterns 80–4
  T-tubules and sarcoplasmatic reticulum 81
Musellifer 2
  M. delamerei 266
myocytes 75, 165
myoepithelium
  body cavities 152, 158, 159, 166
  circulatory system 194, 201
myofibrils 94
myofilament arrangement patterns 77–80
myoglobin 211, 213, 214
Myomerata
  body cavities 166
  circulatory systems 201
  excretory systems 190
  musculature 78
  phylogenetic frame 32, 33
  segmentation 50
myosin 74–5, 90, 94
  myofilament arrangement patterns 77, 80
  striation patterns 80, 81
myotomes 90–4
myriapods 19, 21, 259
Mysidacea 129, 130
Mystacocarida 226, 276, 277
Mytilus 145
  M. edulis 25
    epidermis 67
    photoreceptors 138
Myxini 209
Myxinoida 31
Myxobacteria 58
Myxozoa 11, 53
Myzostoma cirriferum 181
Myzostomida
  excretory system 173, 181
  gametes 275, 290
  general body organization 53
  phylogenetic frame 25, 26, 27

Naididae 139
Nanaloricus 14, 155, 268
  N. mysticus 174, 268
Nanomia 98
  N. bijuga 97
  N. cara 98
Narcomedusae 4, 11, 127
Nausithoë 264
  N. punctatum 126, 127
Nautiloidea 204, 205
Nautilus 122, 129, 138, 145
nebulin 84, 85
Nectonema
  gametes 267, 268
  musculature 87
  N. munidae 60, 79
  nervous system 104, 113
  reproductive organs 256
  sensory organs 121, 135
Nemathelminthes
  epidermis 63
  gametes 284
  intestinal systems 221, 227
  phylogenetic frame 13
  reproductive organs 243
Nematoda
  anteroposterior axis 40
  body cavities 148, 150, 151, 154, 160, 163, 164, 165, 166
  circulatory systems 192
  complexity 34, 35
  epidermis 57, 60, 64, 65, 70, 72, 73
  excretory systems 174, 189
  gametes 267, 284, 285, 286, 287, 292
  intestinal systems 218, 222, 223, 226, 227, 231, 232, 236
  musculature 79, 85, 87, 90, 91
  nervous system 104–5, 112, 113
  phylogenetic frame 5, 6, 13, 14, 15
  reproductive organs 244, 245, 246–7, 250, 252, 253, 256
  respiratory systems 202–3, 212, 213, 214, 215, 216, 217
  segmentation 48, 50
  sensory organs 119, 121, 123, 124, 125, 135, 145
  size 36, 37, 38
Nematomorpha 15
  body cavities 154–5, 160, 163, 164, 165
  circulatory systems 192
  epidermis 59, 60, 64, 65, 73
  excretory systems 189
  gametes 267–8, 284, 286, 287, 292
  general body organization 53
  intestinal systems 218, 223, 231, 232, 236

Nematomorpha (*cont.*)
  musculature 79, 84, 85, 87, 90, 91
  nervous system 104, 105, 113
  phylogenetic frame 13, 15
  reproductive organs 244, 247, 251, 256
  respiratory systems 213
  segmentation 48
  sensory organs 121, 135
Nematostella vectensis 34, 42, 47
Nemertini
  body cavities 156–7, 160, 161, 164, 165, 168
  circulatory systems 192–3, 200
  epidermis 59, 66, 67, 73
  excretory systems 173, 177–8, 189
  gametes 271, 282, 286, 289
  germ layers 44
  intestinal systems 224, 231, 232, 233, 234, 235, 236
  musculature 81, 84, 86, 89, 90, 92
  nervous system 101–2, 112, 114
  phylogenetic frame 21–2
  reproductive organs 241, 244, 247, 257
  respiratory systems 204, 213, 214
  sensory organs 121, 128, 129, 133, 136–7, 145, 147
Nemertinoides elongatus 5
Nemertoderma 100, 265, 287
Nemertodermatida
  epidermis 69–70, 71
  excretory systems 173
  gametes 265, 287
  intestinal systems 220, 221, 223, 236
  misidentification problems 5
  nervous system 100
  phylogenetic frame 22, 24
  reproductive organs 249, 256
  sensory organs 127, 128
nemoglobins 203
neoblasts 246
Neocrania anomala 159
Neocribellata 207
Neodasys 82, 202, 214, 266
  N. ciritus 266
Neodermata 37, 72, 176, 203–4, 217
neodermis 72
Neomeniomorpha 272
Neomysis 129
Neoophora 255
Neopilina 48
Neopilinida 204, 205, 236, 271
Neotrigonia bednalli 271
nephridia 170, 178, 179, 181, 182, 186, 187–8

nephridiopore cells 174
nephromixia 168, 187
nephrons 186
nephropore cell 176
nephropores 178, 179, 188
Nephrozoa 13, 23, 169
nephtyids 179
Nephtys 274
  N. hombergi 212
Nereidae 179
Nereis
  N. diversicolor 179
  N. pelagica 179
Nerilla antennata 103
nerillids 36
nervous system 95, 111–12, 117
  Bilateria 99, 110–11, 113–16
  cnidarians and ctenophores 96–9
  deuterostomes 107–9
  gastrotrichs, cycloneuralians and arthropods 104–6
  neurotransmitters 112–17
  pre-metazoan nervous components 95
  spiralians 99–104
  sponges 95–6
  tentaculates 106–7
  unusual innervation patterns 112
neural gland 185
neural tubes 108–9
Neurexin IV 56
neuroglobin 211, 214, 215
neuropeptides 97, 112
neurotransmitters 112–17
Nippostrongylus brasiliensis 267
Nitella 74
non-bilaterian animals *see* diploblasts
noradrenaline 112, 117
norephidrine 117
Notaulax nudicollis 274
Notochordata 32
  see also Myomerata
Notommata 236
  N. copeus 176
Notostigmophora 140
Nucella lapillus 25
Nucula 236
Nymphon 275, 276

Obelia 82
obliquely striated muscle 78, 80–1, 82–5, 90, 91–4
Obturata *see* Vestimentifera
occludin 55
ocelli 132
  see also photoreceptors
octopamine 112, 117

Octopus 137
Ocypode quadrata 39
Ocyropsis 244
Osedax 25
Oesophagostomum 60
Oikopleura dioica 130, 245, 280, 281
Olfactores 33
Oligacanthorhynchidae 176–7
Oligacanthorhynchus 192
  O. taenioides 177
Oligobrachia gracilis 140, 182
Oligochaeta
  epidermis 62, 64
  excretory system 182
  gametes 273
  musculature 79, 86
  photoreceptors 139
  reproductive organs 258
  respiratory systems 216
Onchidium verruculatum 138
Onchnesoma 178
one-way gut 220–5
Onychophora
  body cavities 152, 153, 155–6, 160, 161, 164, 165, 166
  circulatory systems 195, 200
  dorsoventral axis 43
  epidermis 64, 73
  excretory systems 172, 182, 183, 189
  gametes 275, 276, 286, 290
  intestinal systems 222, 224, 226, 227, 232, 237
  locomotory appendages 52
  musculature 84, 86, 88, 90, 93
  nervous system 106, 115
  phylogenetic frame 17, 19, 20
  reproductive organs 245, 247, 250, 255, 258
  respiratory systems 207, 213, 214, 215, 216
  segmentation 50
  sensory organs 122, 140, 145
oocytes 240, 248, 249–51, 252–61, 263–4, 274, 277, 279, 280
oogenesis 240, 251
Ooperipatellus insignis 276
Opabinia 20
  O. regalis 207, 208
open circulatory system 191–201
Ophactis virens 209
opheliids 139
Opheodesoma 146
  Opheodesoma spectabilis 142
Ophiactis rubropoda 209
Ophidiaster granifer 245
Ophiocoma wendtii 142

Ophiuroida
  circulatory systems 198
  epidermis 62, 63
  intestinal systems 224, 230, 235, 237
  musculature 93
  nervous system 107
  photoreceptors 142
  reproductive organs 245
  respiratory systems 208, 209
Ophryotrocha 36, 37, 274, 228
  O. labronica 227
Opiliones 207, 284
Opisthobranchia 178, 204, 205, 272
opsins 131, 142, 147
orbiniids 26
Ordovician 35
orthogon 99–106, 110, 117
Orthonectida
  epidermis 68
  gametes 264, 265
  general body organization 53
  phylogenetic frame 10, 11
Oscarella 263
  O. lobularis 263, 264, 292
osculum 75
osmosis 169
Osphradia 122
Osteognathostomata 199, 210, 211
Ototyphlonemertes 86, 128, 129
oviduct 249
ovipary 255
Owenia fusiformis 71, 179

Palaeacanthocephala 89, 156
Palaeonemertini
  gametes 271
  intestinal systems 231, 234
  musculature 89, 92
  phylogenetic frame 22
palaeontology 3
  see also fossils
palaeoscolecids 15
Panarthropoda 19
paraacrosomal bodies 266
Parachordodes gemmatus 15
Paragordius varius 59, 60, 121, 135
ParaHox cluster 41
Paramecium 120
paramyosin 80, 85
paranodal glial-axonal junction 56
Parapeytoida yunnanensis 207
parapodia 52
Parapodrilus 37
Pararotatoria 17
parasperm 273
Paratenuisentis ambiguus 89, 94

Paratomella
  excretory systems 173
  intestinal system 220, 221, 223
  phylogenetic frame 24
  sensory organs 128
paratomy 241, 242
Paraturbanella 266
parenchymal cells 249
Parergodrilus heideri 274, 282
Parotoplana 126
parsimony methods, morphological analysis 3
parthenogenesis 240, 243–5, 247, 261, 266
passerine birds 281–2
patellids 138
Patellogastropoda 204, 205
Patiriella 235
Paucitubulatina 174, 266
Pax6 genes 146
PCR technique 3
Pecten 129, 138, 145
Pectinaria korenia 179
Pedicellina 26
  P. cernua 25
penes 253
Pentastomida 53, 64, 292
pericard (pericardium) 157, 163, 164, 166
periostracum 61, 62
Peripatidae 255, 258
Peripatopsidae 250, 258, 276
Peripatopsis moseleyi 172
peripheral microvilli 171
peripheral nervous system (PNS) 99, 101, 102–3, 117
peritoneal cells 158
peritoneum 152, 158, 159
Perophora 280
Petromyzontida
  gametes 280–1
  intestinal systems 225, 237
  phylogenetic frame 31
  respiratory systems 209, 210
  photoreceptors 143, 144
Phaenocora 203
phaosomes 139–40
pharyngeal hard structures 227–8
pharynges, evolution of 222–3, 225–7
Pharyngotremata 31, 212
Phascolion 178
  P. strombus 87
Phascolosoma 157
  P. agassizii 138
Phasmida 4
phenoloxidases 215–16

phenotypic complexity 34–5
Philinoglossa 51
  P. helgolandica 157
Philodina 92
  P. roseola 136
Pholoe 180, 188
  P. inornata 181
Phoronida
  body cavities 149, 158–9, 160, 161, 164, 165, 166
  circulatory systems 196, 197, 200
  epidermis 59, 63, 68, 71, 73
  excretory systems 173, 183–4, 188, 189, 190
  gametes 277, 278, 286, 290, 292
  intestinal systems 224, 232, 233–5, 237, 239
  misidentification problems 5
  musculature 79, 86, 89–90, 93
  nervous system 106–7, 110, 111, 112, 115
  phylogenetic frame 12, 13, 27, 28, 29, 30
  reproductive organs 241, 245, 247, 251, 259
  respiratory systems 208, 213, 214
  sensory organs 122
Phoronis
  P. muelleri 158–9, 183
  P. ovalis 59, 93, 197, 245
  P. pallida 278
phosphatized fossils 35
photoreceptors 119, 120–3, 124, 131–2
Phragmatopoma lapidosa 274
Phronopsis harmeri 245
Phylactolaemata 28–9, 159, 277
Phyllodoce 180
Phyllodocidae 179, 227
phylogenetic frame 1–2, 3–6
  Acoels and other flatworms 22–4
  Arthropoda 19–21
  Bilateria 12–13
  Chaetognatha 30
  Cycloneuralia/Nemathelminthes/Aschelminthes/Gnathifera 13–17
  deuterostomes 30–3
  diploblastic animals 7–10
  Ecdysozoa or Articulata? 17–18
  'Mesozoa' 10–11
  monophyly of Metazoa 6–7
  Myxozoa 11
  Spiralia 21–2
  Symbion 27
  tentaculates 27–30
  Trochozoa 24–7
  Xenoturbella 11–12

Physa 178
pilidium 231
pinacocytes 34, 71, 75, 218
pineal organ 143, 144
Pisione 188
  P. remota 181
Pisionidae 179
placenta 255
Placentalia 255
planctotrophy 229, 233–5, 239
Planorbarius corneus 178, 206
Plathelminthomorpha 17, 22, 220
platyctenids 264
Platyhelminthes
  body cavities 154, 160, 164, 165, 166, 168
  circulatory systems 192
  epidermis 59, 73
  excretory systems 173, 175–6, 189
  gametes 268–70, 285, 286, 288
  intestinal systems 220–1, 222, 223, 224, 225, 226, 231, 232, 233, 236
  musculature 81, 84, 86, 90, 91
  nervous system 99, 101, 104, 110, 113
  phylogenetic frame 16, 17, 21–2, 23, 24, 26
  reproductive organs 241, 244, 246, 247, 249, 252, 254, 254, 255, 257
  respiratory systems 203–4, 212, 213, 214
  segmentation 50
  sensory organs 121, 126, 128, 129, 131, 135–6, 145
  size 36, 38
platymyarian muscle cells 78–9
Platynereis dumerilii 35, 50, 139, 147
Platyzoa 16, 22
plesiomorphies
  Arthropoda 19
  body cavities 158, 163, 166
  Cycloneuralia 16
  deuterostomes 32
  excretory systems 188
  gametes 269
  glycocalyx 58, 59, 63, 72
  hemoglobin gene 212, 213
  intestinal systems 221
  locomotory appendages 52
  monociliarity 71
  musculature 82, 87, 94
  nervous system 106, 111
  size 37
  tentaculates 27
Pleurobrachia 97
Pleurotomaria 204
Pliciloricidae 155

Pliciloricus 288
  P. enigmaticus 268
pluteus 230
Podocoryne carnea 57, 84, 201
podocytes
  body cavities 152, 157, 158, 159, 163, 166
  circulatory system 196
  excretory systems 172, 173, 178, 179, 181, 182, 183, 184–6, 190
Poeciliidae 255
Poeciliopsis 243
Poecilochaetus serpens 62
Pogonophora
  body cavities 166
  epidermis 61
  excretory systems 181–2
  gametes 275, 286, 290
  phylogenetic frame 25–6, 27
  photoreceptors 140
  reproductive organs 244
Polia sanguiruba 204
Polyarthra 236
Polychaeta
  body cavities 151, 157–8, 167
  circulatory systems 195
  complexity 35
  dorsoventral axis 43
  epidermis 62, 64, 71, 73
  excretory systems 173, 179–81, 187, 188
  gametes 273–5, 282, 284, 289
  intestinal systems 225, 226, 227, 228, 230, 233, 235
  locomotory appendages 52
  musculature 79, 89, 92
  nervous system 103, 104, 115
  progenesis 36
  reproductive organs 242, 243, 244, 258
  respiratory systems 206, 212, 215, 217
  segmentation 49, 50
  sensory organs 129, 130, 137, 139, 145, 147
  skeletons 51
Polycladida
  body cavities 149
  circulatory systems 192, 193
  gametes 269
  intestinal systems 222, 231–3, 234
  nervous system 112
  photoreceptors 136
Polygordius 103, 167
  P. lacteus 274
  P. minutus 94, 270
Polynoidae 181, 207

Polyorchis penicillatus 97
Polyplacophora
  body cavities 157
  circulatory systems 193
  epidermis 61
  excretory systems 178
  gametes 271, 272, 282
  intestinal systems 234, 236
  musculature 89
  nervous system 102
  phylogenetic frame 24, 25
  reproductive organs 244
  respiratory systems 204, 205
  segmentation 48
  sensory organs 122, 145
polyps
  circulatory systems 192
  germ layers 46, 47
  musculature 75, 87
  nervous system 97
  reproduction 249
  segmentation 48
  symmetry 39
Polypterus 210, 211, 280
Polystoma integerrimum 136
Polystyliphora filum 49
Pomatoceros triqueter 64
Porifera (sponges)
  biphasic life cycle 38
  and Choanoflagellata, relationship between 7, 9
  complexity 34
  epidermis 55, 56, 57, 58, 59, 60, 63, 70, 71, 73
  excretion 169
  gametes 263–4, 286, 287, 292
  germ layers 43, 44, 46
  intestinal systems 218, 232, 234, 235, 236
  larvae, similarities with other metazoans 9
  myocytes 75
  nervous system, signalling 95–6
  neurotransmitters 112
  photoreceptors 132
  phylogenetic frame 7, 8–9, 10
  reproduction 240, 244, 245–6, 247, 249, 256
  symmetry 38
postsynaptic membrane 96, 112
preformation 246, 248, 261
presynaptic membrane 96
Priapulida
  body cavities 149, 150, 155, 160, 161, 163, 164, 165
  circulatory systems 192
  epidermis 60, 64, 73

excretory system 173, 174, 175, 189
gametes 267, 268, 282, 284, 286, 288
intestinal systems 223, 227, 231, 232, 236
musculature 84, 86, 87–8, 90, 91
nervous system 104–5, 113
phylogenetic frame 13, 14, 15
reproductive organs 244, 247, 257
respiratory systems 203, 213, 214, 215, 216
sensory organs 121
Priapulopsis 203, 268
Priapulus 203, 268
  P. caudatus 175, 203, 267
primary body cavities 148, 149–51, 154, 160, 163
  excretory systems 170
primary sensory cells 118–19
primary urine 170, 182, 184, 186
primordial germ cells (PGCs) 246, 248–9, 251–2, 261
procaryotes 131
progenesis 36
progenetic spermiogenesis 284
proglottids 48
Progoneata
  excretory systems 183
  gametes 277
  phylogenetic frame 20
  reproductive organs 245
  respiratory systems 207
prokaryotes 74
Prolecithophora 136, 269, 285
Proseriata
  excretory systems 176
  segmentation 49
  sensory organs 126, 128, 129, 136
Prostomatella arenicola 22, 177
protein kinase C 8
Proterospongia choanojuncta 7
Protobranchia 205
protocerebrum 106
protocoel 184
Protodrilus 26, 62
proto-Hox cluster 41
proto-Hox gene 41
protonephridia
  body cavities 168
  epidermis 67
  excretory systems 170–2, 173–4, 175–84, 186, 187–9
Protopterus annectens 281
protostome ancestor
  intestinal systems 233
  nervous system 104
  respiratory system 215

protostome-deuterostome ancestor (PDA) 12
Protostomia (Gastroneuralia)
  phylogenetic frame 13, 28, 29, 30
  complexity 35
  dorsoventral axis 43
  epidermis 56, 63
  excretory systems 188
  gametes 292
  intestinal systems 222, 230, 231, 235, 239
  nervous system 107, 110–11, 112, 117
  photoreceptors 146, 147
  size 38
protostomy 221–2
prototroch 230–1, 232
Provortex tubiferus 72
proximal centrioles 262
  see also accessory centrioles
Pseudechiniscus 275
  P. juanitae 276
Pseudoacanthocephalus bufonis 254
Pseudobiotus megalonyx 275, 276
Pseudoceros canadensis 136
Pseudochordodes bedriagae 155
pseudocoel 149, 150
  see also primary body cavities
pseudocoelom 2
Pseudocoelomata
  body cavities 148, 163, 167
  gametes 284
  phylogenetic frame 13
  reproductive organs 245
pseudocopulation 271, 279
pseudohemocyanins 215, 216
Pseudoscorpiones 207
Pseudostomella
  P. arenicola 137
  P. etrusca 266
pseudostratified epithelia 66
Pterobranchia
  body cavities 160, 162, 164, 165
  circulatory systems 196–8
  epidermis 59, 70, 71, 73
  excretory systems 184, 189
  gametes 278, 279–80, 286, 291, 292
  intestinal systems 225, 228, 232, 233, 234, 237, 239
  musculature 84, 90, 94
  nervous system 108, 109, 116
  phylogenetic frame 28, 30, 31–2, 33
  reproductive organs 245, 247, 259
  respiratory systems 208, 211, 213
  sensory organs 123
Pterognathia rosea 202
Pterotrachea 129
Ptychodera flava 142

Ptychoderidae 31, 108
Pulmonata 204–5, 246, 251, 272
Pycnogonida
  gametes 275, 285, 290
  intestinal systems 226, 227
  photoreceptors 140, 141
  phylogenetic frame 20–1
  reproductive organs 259
Pycnogonum littorale 275–6
Pycnophyes 35
  P. greenlandicus 174, 175
  P. kielensis 14
    excretory system 174
    gametes 267
    musculature 80
    nervous system 104
    segmentation 49
pyloric gland 185
Pyrosoma atlanticum 280, 281

radial canals 46, 47
Radialia
  epidermis 65
  excretory systems 188
  nervous system 110, 111
  phylogenetic frame 13, 28
Raillietiella gowrii 64
Rana temporaria 144
Rastrognathia 113
  R. macrostoma 176
Rathkea octopunctata 46, 47
rays 255
receptor cells 67
receptor tyrosine kinase 7, 8
Rectronectes cf. *sterreri* 269
regaining of characters, phylogenetic analysis 4
Remipedia 21, 277
renal sac 185
Reneira 132
renette cells 174
reproductive organs 240, 261
  asexual versus sexual reproduction 240–5
  gametes, origin of 246–9
  gametes, transfer and receiving 253–61
  gonads, origin of 249–53
  hermaphroditism versus gonochorism 245–6
  see also gametes
respiratory systems 202
  low oxygen content, adaptation to 216–17
  respiratory organs 202–12
  respiratory pigments 202–16
respiratory trees 209

retinal 131
retraction 88
retractors 75, 76, 89
Retronectes atypica 269
Rhabditophora
  excretory systems 173, 176
  misidentification problems 5
  musculature 88
  phylogenetic frame 23, 24
Rhabdocoela 128, 136, 269
rhabdome 138, 140
rhabdomeres 138
rhabdomeric photoreceptors 146, 147
Rhabdopleura
  circulatory system 197
  intestinal system 228, 229
  phylogenetic frame 31
  respiratory systems 208, 211
  R. compacta 184, 197
  R. normani 279
  sensory organs 123
rhachis 251
Rhinoglena 92
  R. frontalis 136
rhodopsins 131, 147
Rhombozoa 10, 11
rhopalia 126, 133
Rhopaliophora 126
Rhopalomenia aglaopheniae 102
Rhopalonema velatum 126
Rhopalura littoralis 264, 265
Rhynchobdellida 227
Rhynchobdelliformes 182
Rhynchocephalia 143, 144
rhynchocoel 156, 166
Rhynchonellidae 184
Riftia pachyptila 215
Rossia 271
Rotifera
  body cavities 148, 149, 156
  epidermis 72
  musculature 88–9
  phylogenetic frame 16, 17, 27
round-headed spermatozoa 262–4, 268, 272–5, 277, 279–86, 292

Sabellaria cementarium 181
Sabellariidae 181
Sabellida
  gametes 274, 275, 284
  intestinal systems 235
  phylogenetic frame 26
  photoreceptors 139
  respiratory systems 207
Saccharomyces cerevisiae 74
Saccocirrus papillocerus 102–3, 104

Saccoglossus 237
  S. cambrensis 108
  S. kovalevskii 93
sacculi
  body cavities 152–3, 155, 156
  excretory systems 182, 183
Sacculina carcini 38
sac-shaped intestine 220–5
Sagitta setosa 30
sagittids 141
Salmo salar 281
Salpida 143
Salpingoeca frequentissima 7
salps 94, 108, 143
sarcomeres 77–8, 79–80, 81, 82, 84, 94
sarcoplasmic reticulum (SR) 81, 91–4
sauropsids 144
Saxipedium coronatum 279, 280
Scalibregma inflatum 206, 207
Scalidophora
  body cavities 163
  fossil record 35
  gametes 284
  phylogenetic frame 15
  sensory organs 121
  symmetry 38
scallops 92
Scaphopoda
  excretory systems 178
  gametes 272
  intestinal systems 234, 237
  musculature 89
  reproductive organs 244
  respiratory systems 204, 205
  sensory organs 122, 128
Schistosoma mansoni 23, 244, 245
schizocoely 47, 151, 152, 164–5, 251
schyphopolyps 82
Scleractinia 264
Sclerolinum 25
Scleroperalia 255, 261, 269, 270, 288
Scoloplos armiger 103
Scorpiones 207
Scribble 56
Scrupocellaria bertholetti 142
Scutigera 140
Scyphozoa
  gametes 264
  intestinal systems 223
  musculature 75
  nervous system 97
  phylogenetic frame 10
  reproductive organs 241, 246, 249, 256
  segmentation 48
  sensory organs 125, 126, 127, 133, 134, 145

sea urchins 235
Secernentea 121, 174
secondary body cavities 148, 151–2, 154, 160
  see also coelom
secondary loss of characters, phylogenetic analysis 3
secondary muscle 79
secondary sensory cells 118–19
secondary urine 170
segmentation 18, 47–51
Seisonidea
  body cavities 156, 160, 164, 165
  circulatory systems 192
  epidermis 73
  excretory system 173, 176, 189
  gametes 270–1, 286, 289, 292
  intestinal systems 224, 227, 231, 232, 236
  musculature 90, 92
  nervous system 101, 114
  phylogenetic frame 16–17
  reproductive organs 244, 247, 249, 257
  respiratory systems 213
  sensory organs 121
Seison 16, 17, 72, 88–9, 270
  S. annulatus 176, 236
  S. nebaliae 236, 270
seminal receptacle 254, 261
seminal vesicle 255
sensory organs 118–19
  basic principles 119–20
  photoreceptors 131–47
  static 125–31
  summary for metazoans 120–5
sensory receptor molecules 118
septate junctions 56, 57, 73
sequence data 3
serial repetition 48
  see also segmentation
Seriata 136, 269
serotonin 112, 117
Serpula vermicularis 230
Serpulidae 139, 207
sexual reproduction 240–5, 247, 261
sharks 255
shearing partners hypothesis 81
shedding 64
short gastrulation (*sog*) gene 42
Siboglinidea *see* Pogonophora
Siboglinum
  S. ekmani 182
  S. fiordicum 140
Sigalionida 181
Sigmadocia caerulea 71
signalling proteins 7

signal transduction cascade 118
siliceous sponges 8
 see also Demospongia; Hexactinellida
Siphonobranchia lauensis 182
Siphonophora 97, 263, 264, 292
Siphonosoma 206
Sipunculida
 body cavities 157, 160, 161, 164, 165, 166
 circulatory systems 194, 200
 epidermis 61, 73
 excretory systems 178, 189, 190
 gametes 272, 273, 286, 289
 intestinal systems 224, 226, 230, 232, 234, 237
 musculature 86, 87, 88, 89, 90, 92
 nervous system 103, 114
 phylogenetic frame 24, 26, 27
 reproductive organs 244, 247, 251, 258
 respiratory systems 206, 213, 214, 215
 sensory organs 122, 138–9, 145
Sipunculus 157
 S. nudus 27, 238
Siro rubens 284
size 35–8
 germ layers 43–4
 skeletons 51–2
sliding filament model 74
small eye gene 146
smooth muscle 77, 78, 80, 82–4, 90, 91–4
 molecular components 84–5
Snail transcription factor 47
solenocytes 171
solenocyte tree 175
Solenofilomorphidae 68
Solenogastres
 epidermis 61
 excretory systems 178
 gametes 272
 intestinal systems 234, 236
 musculature 89
 nervous system 102
 phylogenetic frame 25
 respiratory systems 204, 205
 segmentation 48
 sensory organs 122
Solifugae 207
Spadella 62
 S. cephaloptera 278
spadellids 141
Spatangoida 224
Spathebothriidea 48
spermatophores 254, 261, 265–6, 268, 270, 274, 275, 277, 282

spermatozeugmata 274
spermatozoa 1, 249, 252, 253–61, 262–82, 287–92
 acrosome evolution 292
 correlation between sperm type and fertilization mode 282–5
 epidermis 67
 function and phylogeny of derived spermatozoa 285–6
sperm cyst 249
spermiogenesis 240, 249
Sphenodon 144
Sphyrion lumpi 53
Spiralia
 body cavities 154
 germ layers 46
 intestinal systems 220, 221, 233
 musculature 88
 nervous system 99–104, 105, 110, 111
 phylogenetic frame 11, 13, 15, 17, 21–2, 28, 29
 reproductive organs 246
 respiratory systems 215
Spirorbis spirorbis 132, 179
sponges *see* Porifera
spongin 57
sporocysts 242, 243
Squamata 143, 245
static sense organs 125–31
statoconia 125
statocysts 120–3, 124, 125–31, 142–3
statoliths 125, 126–7, 129
Stauromedusa 10, 133
stenolaemates 28, 29, 233, 277
stereom hypothesis 229
sternites 51
sterroblastula 44, 45
Stichopus
 S. californicus 279
 S. moebii 198
stick insects 4
Stolonica socialis 142
stomatogastric nervous system 99
Stomochordata 31
striated rootlets 9, 70–1, 72
striation patterns 77–80, 90, 91–4
 evolution 81–4
 functions 80–1
strobilation 48
Strongylida 267
Strongylocentrotus purpuratus 142
Strongyloidea 60
Styela 142
Styelidae 143
Stylea plicata 142
Stylochus mediterraneus 136
Stylocoronella riedli 133

Suberites massa 263
sulphide-biome 217
supercontraction 80
swim-bladder 210
syllids 139, 241, 242
Symbion 27, 92, 173
 S. americanus 27
 S. pandora 27, 51
symmetrical synapses 96
symmetry 38–9, 40
Symsagittifera schultzei 265
synapomorphies 6
 Annelida and Arthropoda 18
 body cavities 157
 diploblastic animals 10
 epidermis 63, 64
 gametes 292
 Nemertini and Trochozoa 22
 nervous system 106
 tentaculates 29
Synchaeta 236
syncytial tissue 71–2, 73, 220
Syndermata
 body cavities 156, 166
 epidermis 72
 gametes 292
 intestinal systems 221
 phylogenetic frame 16, 17, 26
 reproductive organs 245

tadpole larvae 108
tapetum 140
Tardigrada
 body cavities 155, 160, 164, 165
 epidermis 64, 73
 excretory systems 182, 189
 gametes 275, 276, 285, 286, 290, 292
 intestinal systems 224, 226, 227, 232, 237
 locomotory appendages 52
 musculature 84, 88, 90, 93
 nervous system 106, 115
 phylogenetic frame 17, 19, 20
 reproductive organs 242, 245, 247, 258
 respiratory systems 213
 sensory organs 122, 140, 145
 size 37
targeted search for genes 3
taxon sampling, phylogenetic analysis 4
Tedania ignis 75, 96
Teleostei
 gametes 281, 291
 reproductive organs 245, 248, 255, 260
 respiratory systems 210, 211

Tentaculata
  body cavities 166
  excretory systems 190
  intestinal systems 233, 235, 239
  musculature 89
  nervous system 106–7, 110–11, 112
  phylogenetic frame 13, 27–30
  see also Brachiopoda; Bryozoa; Phoronida
Terbellida 179
Terebratulina 51
  T. retusa 29
  T. caputserpentis 277, 278
  T. transversa 130
tergites 51
terminal cells 171, 173, 174, 175–6, 177–81, 183–4, 188, 190
terminal complexes 174, 183–4
terminal organs 173, 174, 175, 178
terminal regions 174, 176
Tesserazoa 10, 120, 240, 244
Tetraconata-hypothesis 140
Tetraplatia volitans 126
Tetrapoda 94, 210, 211, 280–1
Thalassema mellita 206
Thaliacea 94, 143, 280, 281
Thalia democratica 280
Themiste lageniformis 244
thermoreceptors 119
Thulinia stephaniae 155
Thylacorhynchus ambronensis 269
Thyonicola 53
Thysanozoon brocchii 136
Tiaropsis multicirrata 134
tight junctions 55, 56, 57, 73
'tinsel-type' cilium 71
tissue evolution 293
titin 84, 85
Tobrilus gracilis 123
Tomopteridae 49
Tomopteris 180, 188
  T. helgolandica 158, 181
tornaria 142, 184, 230, 231
total evidence method 3
totipotent cells 246
tracheae 207–8
Tracheata 21, 195, 207
trachymedusae 126, 127
transcriptomes 293
trematodes 246, 257
Trepaxonemata 270
Treptoplax reptans 7
Trichocerca 92
  T. rattus 136
Trichoplax (*T. adhaerens*)
  anteroposterior axis 41
  epidermis 55, 56, 57, 58, 59, 70, 71, 73

excretory systems 169, 189
gametes 264, 286
germ layers 46
intestinal system 219–20, 232, 235, 236
musculature 75, 90
phylogenetic frame 7, 9, 10
reproduction 244, 247, 256
respiratory systems 213
size 38
Tricladida 49, 192, 193, 222, 223
Trididemnum 280, 281
trilobites 21, 140, 207
Trilobodrilus 36, 37
Tripedalia cystophora 126, 127
triploblasts 170
triploblasty 43, 46
trochaea 230
trochophora
  epidermis 71
  excretory systems 179
  intestinal systems 230, 231, 232, 233, 234
  nervous system 103
  phylogenetic frame 27
Trochozoa
  circulatory systems 201
  epidermis 61, 63, 64, 68, 71
  intestinal systems 230, 231, 232, 233, 235
  nervous system 112
  phylogenetic frame 12, 21–2, 24–7
tropomodulin 84, 85
tropomyosin 80
troponin 85
Trox-2 gene 41, 220
tryptophan 112
t-tubules 81, 91–4
tuatara 144
tubificids 273
Tubificoides benedii 64
Tubiluchus
  body cavities 155
  gametes 268, 286, 288
  intestinal systems 223
  musculature 91
  respiratory system 203
  T. corallicola 84, 267, 284
  T. philippinensis 175
  T. troglodytes 87
Tubulanus 193
  T. annulatus 177
Tubularia crocea 47
tubulin 74, 75
Tubulipora littacea 278
tunic 62, 73

Tunicata
  anteroposterior axis 41
  body cavities 160, 162, 164, 165
  circulatory systems 198, 200, 201
  epidermis 62, 63, 73
  excretory systems 185, 189
  gametes 280, 281, 282, 286, 291
  intestinal systems 225, 228, 232, 237
  locomotory appendages 52
  musculature 83, 90, 94
  nervous system 108, 109, 116
  phylogenetic frame 30, 31, 33
  reproductive organs 245, 247, 260
  respiratory systems 209, 211, 213, 215
  sensory organs 123, 129, 130, 141, 142–3, 146, 147
  size 38
  skeletons 51
tunicin 62
Turbanella 83, 266
  T. cornuta 16, 105, 112, 174, 261
  T. ocellata 202
tyrosine metabolism 112

U12648 sequence 5
U70083 sequence 5
ultrafiltration 170, 172, 174–82, 184–7, 188
unciliated receptors 121
Uniramia 19
Urbilateria 12
urea 169, 174
Urechis 273
  U. caupo 194, 206, 237
uric acid 169, 183
Urnatella 26
  U. gracilis 64, 178
Urochordata *see* Tunicata
urodeles 281
Uropygi 207

vacuolated tissue 51
vanes 7, 9, 71
vasa gene 246
vascular endothelial growth factors (VEGF) 201
ventral vessel 194–5
Vernanimalcula guizhouena 167
Vertebrata
  body cavities 167
  epidermis 57, 65
  phylogenetic frame 30, 32–3
  segmentation 48
  see also Myomerata

Vestimentifera 25, 182, 215
vetellarium 254, 255
Vetigastropoda 205
vetulocolids 229
vetulocystids 229
vitellocytes 254, 255
Viviparus viviparus 138
vivipary 255
Volvox 58, 75, 131
   V. carteri 58

wariai gene 40
Wasmannia auropunctata 243
wasps 81
water lungs 209
weirs 171, 174, 176
wheel organ eyes 136, 145
whole genome sequencing 5
wingless gene 50

Xenodasys 82, 83
Xenotrichula intermedia 266
Xenotrichulidae 266, 287
Xenoturbella
   body cavities 154, 160
   epidermis 70
   excretory system 169, 173, 188
   gametes 264–5
   intestinal systems 223, 225, 236
   musculature 81, 87, 91
   nervous system 110, 111, 113
   phylogenetic frame 11–12, 13
   reproduction 244, 249, 256
   sensory organs 120, 127, 128
   X. bocki 5, 11, 69–70
   X. westbladi 11, 12
Xerobiotus pseudohufelandi
   275, 276
Xiphosura 140, 224, 259, 275, 290

Xyloplax turnerae 279, 284

Yoldia eightsi 215

Z-bodies 78, 81, 82, 84
Z-discs 78, 80, 84, 90, 93, 94
   molecular components 84, 85
Z-dots 77
Z-elements 77, 80, 81, 83–4, 90, 94
   molecular components 84, 85
zeugmatin 84, 85
Z-lines 78
ZO-1 56
Zoantharia 59
zonites 48
   see also segmentation
Z-rods 78
Zygeupolia rubens 161
zygotes 240